Managing Soil Drought

Global drylands, covering over 40% of Earth's land surface, are important among worldwide ecoregions and support large human and livestock populations. However, these ecologically sensitive ecoregions are undergoing a rapid transformation resulting from climate change, socioeconomic and political factors, increases in population, and ever-growing demands for goods and services.

Managing Soil Drought addresses basic processes and provides specific case studies throughout covering the protection, restoration, and sustainable management goals of global drylands under changing and harsh climatic conditions, including fragile and vulnerable ecosystems. The book is written by numerous researchers, academicians, practitioners, advocates, land managers, and policymakers involved in bringing about transformation in these regions important to human and nature. It includes information on basic strategies for sustainable management of global drylands aimed at improving water use efficiency through choosing appropriate species, developing new varieties, using organic and inorganic amendments, and scaling up innovative farming systems.

This volume in the *Advances in Soil Sciences* series is an essential read for development organizations and policymakers involved in improving crop productivity and sustainability in drought-prone regions; students, researchers, and academicians interested in sustainable management of water resources; and those involved in emerging concepts of regenerative agriculture, agroecology, and conservation agriculture.

Advances in Soil Science

Series Editors:
Rattan Lal and B.A. Stewart

Soil Erosion and Carbon Dynamics
E.J. Roose, R. Lal, C. Feller, B. Barthès, and B. A. Stewart

Soil Quality and Biofuel Production
R. Lal and B. A. Stewart

World Soil Resources and Food Security
R. Lal and B. A. Stewart

Soil Water and Agronomic Productivity
R. Lal and B. A. Stewart

Principles of Sustainable Soil Management in Agroecosystems
R. Lal and B. A. Stewart

Soil Management of Smallholder Agriculture
R. Lal and B. A. Stewart

Soil-Specific Farming: Precision Agriculture
R. Lal and B. A. Stewart

Soil Phosphorus
R. Lal and B. A. Stewart

Urban Soils
R. Lal and B. A. Stewart

Soil Nitrogen Uses and Environmental Impacts
R. Lal and B. A. Stewart

Soil and Climate
Rattan Lal and B.A. Stewart

Food Security and Soil Quality
R. Lal and B. A. Stewart

Soil Degradation and Restoration in Africa
Rattan Lal and B.A. Stewart

Soil and Fertilizers: Managing the Environmental Footprint
Rattan Lal

The Soil-Human Health-Nexus
Rattan Lal

Soil Organic Matter and Feeding the Future: Environmental and Agronomic Impacts
Rattan Lal

Soil Organic Carbon and Feeding the Future: Basic Soil Processes
Rattan Lal

Soil and Drought: Basic Processes
Rattan Lal

Managing Soil Drought
Rattan Lal

For more information about this series, please visit: https://www.crcpress.com/Advances-in-Soil-Science/book-series/CRCADVSOILSCI

Managing Soil Drought

Edited by
Rattan Lal

CRC Press
Taylor & Francis Group
Boca Raton London New York

CRC Press is an imprint of the
Taylor & Francis Group, an **informa** business

Designed cover image: Shutterstock

First edition published 2024
by CRC Press
2385 NW Executive Center Drive, Suite 320, Boca Raton FL 33431

and by CRC Press
4 Park Square, Milton Park, Abingdon, Oxon, OX14 4RN

CRC Press is an imprint of Taylor & Francis Group, LLC

© 2024 Taylor & Francis Group, LLC

Reasonable efforts have been made to publish reliable data and information, but the author and publisher cannot assume responsibility for the validity of all materials or the consequences of their use. The authors and publishers have attempted to trace the copyright holders of all material reproduced in this publication and apologize to copyright holders if permission to publish in this form has not been obtained. If any copyright material has not been acknowledged please write and let us know so we may rectify in any future reprint.

Except as permitted under U.S. Copyright Law, no part of this book may be reprinted, reproduced, transmitted, or utilized in any form by any electronic, mechanical, or other means, now known or hereafter invented, including photocopying, microfilming, and recording, or in any information storage or retrieval system, without written permission from the publishers.

For permission to photocopy or use material electronically from this work, access www.copyright.com or contact the Copyright Clearance Center, Inc. (CCC), 222 Rosewood Drive, Danvers, MA 01923, 978-750-8400. For works that are not available on CCC please contact mpkbookspermissions@tandf.co.uk

Trademark notice: Product or corporate names may be trademarks or registered trademarks and are used only for identification and explanation without intent to infringe.

Library of Congress Cataloging-in-Publication Data
Names: Lal, R., editor. Title: Managing soil drought / Rattan Lal. Other titles: Advances in soil science (Boca Raton, Fla.) ; 21. Description: First edition. | Boca Raton, FL : CRC Press, 2024 | Series: Advances in soil sciences ; 21 | Includes bibliographical references and index. | Summary: "Global drylands, covering over 40% of Earth's land surface, are important among worldwide ecoregions, and support large populations of human and livestock. However, these ecologically sensitive ecoregions are undergoing a rapid transformation resulting from climate change, socioeconomic and political factors, increases in population, and ever-growing demands for goods and services. Managing Soil Drought addresses basic processes and provides specific case studies throughout covering protection, restoration, and sustainable management goals of global drylands under changing and harsh climatic conditions, including fragile and vulnerable ecosystems. The book is written by numerous researchers, academicians, practitioners, advocates, land managers, and policy makers involved in bringing about transformation in these regions important to human and nature. It includes information on basic strategies of sustainable management of global drylands aimed at improving water use efficiency through choosing appropriate species, developing new varieties, using organic and inorganic amendments, and scaling up innovative farming systems. This volume in the Advances in Soil Sciences series is an essential read for development organizations and policy makers involved in improving crop productivity and sustainability in drought-prone regions; students, researchers, and academicians interested in sustainable management of water resources; and those involved in emerging concepts of regenerative agriculture, agroecology, and conservation agriculture"– Provided by publisher. Identifiers: LCCN 2023038636 (print) | LCCN 2023038637 (ebook) | ISBN 9781032352404 (hardback) | ISBN 9781032352411 (paperback) | ISBN 9781003326007 (ebook) Subjects: LCSH: Drought management. | Arid regions. | Soil moisture conservation. Classification: LCC QC929.24 .M36 2024 (print) | LCC QC929.24 (ebook) | DDC 632/.12–dc23/eng/20231025 LC record available at https://lccn.loc.gov/2023038636LC ebook record available at https://lccn.loc.gov/2023038637

ISBN: 9781032352404 (hbk)
ISBN: 9781032352411 (pbk)
ISBN: 9781003326007 (ebk)

DOI: 10.1201/b23132

Typeset in Times
by codeMantra

Contents

Preface vii
Editor ix
Contributors x

Chapter 1 Enhancing Resilience to Pedological and Agronomic Droughts in
Dryland Farming 1

Rattan Lal

Chapter 2 Drought Management through Genetic Improvement in Dryland Cereals and
Grain Legumes 14

O.P. Yadav, P.H. Zaidi, R. Madhusudhana, Manoj Prasad, and Abhishek Bohra

Chapter 3 Economic and Policy Issues of Drought Management 41

S.K. Chaudhari, D.V. Singh, O.P. Yadav, S. Bhaskar, C.A. Rama Rao, and B.M.K. Raju

Chapter 4 Soil, Water, and Nutrient Management in Drylands 85

S.K. Chaudhari, Priyabrata Santra, Deepesh Machiwal, Mahesh Kumar, V.K. Singh, K. Sammi Reddy, and S. Kundu

Chapter 5 Water Storage in the Rest Zone by Enhancing Soil Organic
Matter Content in the Rest Zone 133

Ch. Srinivasarao, S. Rakesh, G. Ranjith Kumar, M. Jagadesh, K.C. Nataraja, R. Manasa, S. Kundu, S. Malleswari, K.V. Rao, JVNS Prasad, R.S. Meena, G. Venkatesh, P.C. Abhilash, J. Somasundaram, and R. Lal

Chapter 6 Drought Management in Soils of the Semi-Arid Tropics 161

Kaushal K. Garg, K.H. Anantha, M.L. Jat, Shalander Kumar, Gajanan Sawargaonkar, Ajay Singh, Venkataradha Akuraju, Ramesh Singh, Md. Irshad Ahmed, Ch Srinivas Rao, R.S. Meena, Martin M. Moyo, Bouba Traore, Gizaw Desta, Rebbie Harawa, Bruno Gerard, Y.S. Saharawat, Alison Laing, and Mahesh K. Gathala

Chapter 7 Managing Drought in Semi-Arid Regions through Improved Varieties and
Choice of Species 212

Faisal Nadeem, Abdul Rehman, Aman Ullah, Muhammad Farooq, and Kadambot H.M. Siddique

Chapter 8 Engineering Abiotic Stresses in Crops by Using Biotechnological Approaches 235

Aladdin Hamwieh, Naglaa A. Abdallah, Nourhan Fouad, Khaled Radwan, Tawffiq Istanbuli, Sawsan Tawkaz, and Michael Baum

Chapter 9 Groundwater's Geochemical Status in Agricultural and Sustainable Use (Western Mitidja, Algeria) 255

Ahcène Semar, Hakim Bachir, and Rattan Lal

Chapter 10 Saharan Agriculture: Strategic Choice, Environmental Issues in a Perspective of Sustainable and Resilient Agriculture. Case of Northeastern Sahara of Algeria 277

Hakim Bachir, Ahcène Semar, Nour el houda Abed, Wassima Lakhdari, Dalila Smadhi, and Rattan Lal

Chapter 11 Soil and Water Management by Climate Smart and Precision Agriculture Systems under Arid Climate 310

Y.G.M. Galal and S.M. Soliman

Chapter 12 Managing Soil Drought in Agro-Ecosystems of North China Plains 362

Zheng-Rong Kan, Xing Wang, Jian-Ying Qi, Ling-Tao Zhong, Rattan Lal, and Hai-Lin Zhang

Chapter 13 Soil Drought and Human Health 386

Marium Husain

Chapter 14 Drought and Soil Structure 400

Naba R. Amgain and Rattan Lal

Index 417

Preface

Global drylands, covering over 40% of Earth's land surface, are important among global eco-regions and support large populations of human and livestock. These ecologically sensitive ecoregions are also undergoing a rapid transformation because of climate change, socio-economic and political factors, increase in population and of their ever-growing demands for goods and services. Ecologically, global drylands are also the largest sources of inter-annual variability in the global carbon (C) sink because of uncertainties associated with changes in vegetation, soil organic carbon (SOC) stocks and effects on the formation of secondary carbonates through both biotic and abiotic drivers. Furthermore, drylands may expand over the next century primarily into formally productive ecosystems. Thus, gross primary productivity of global drylands may decrease and lead to a net reduction in ecosystem services because of the increase in frequency and severity of extreme climatic events. However, strategies for mitigation of and adaptation to these extreme events may vary because of regional differences due to biophysical and socio-economic, political, and cultural factors.

Basic strategies for sustainable management of global drylands include those aimed at improving water use efficiency (WUE) through the choice of appropriate species, the development of new varieties, the use of organic and inorganic amendments, and the identification/upscaling of innovative farming systems. Furthermore, sustainable agriculture must also be based on conservation agriculture, mulch farming, and innovative options that conserve soil, water, nutrients, biodiversity, and other critical resources.

This is the second volume in the series aimed at sustainable management of global drylands. This timely publication addresses the basic processes and provides examples of specific case studies for protection, restoration, and sustainable management of global drylands under changing and harsh climatic conditions and fragile and vulnerable ecosystems. The information collated and synthesized in this book is based on collective action of numerous researchers, academicians, practitioners, advocates, land managers, and policymakers involved in bringing about a transformation change in these regions important to human and nature. This volume is indicative of dedication, commitment, and professional experience of the authors in conducting research, analyzing and synthesizing a vast amount of data from field and laboratory studies and farm survey. Several authors have provided examples of case studies under site-specific conditions, representing diverse biophysical and socio-economic conditions.

Authors from around the world have confirmed that sustainable management of global drylands is critical to strengthening critical ecosystem services and eliminating disservices. As complementary to the first volume on Dryland Farming dedicated to Dr. B.A. Stewart, this book is also a major contribution to processes governing the ecosystem functions of global drylands and outlines the principles and practices of their sustainable management. Indeed, these two volumes are important reference material for researchers, students, practitioners (e.g., farmers, ranchers, and foresters) and policymakers. The material presented is also of interest to researchers and students in soil science, agronomy, forestry, animal husbandry, ecology, and management of natural resources. These two volumes present useful information on Global Drylands and their management with specific focus on food and nutritional security, soil quality, carbon sequestration, water resources and their management. The information presented herein will stimulate discussions and resolve toward advancing sustainable development goals (SDGs) of the United Nations. The information presented herein is also important to accomplishing the mission of transformation of the World Food Systems as outlined in the U.N. Food System Summit. The importance of the information presented in this volume is also relevant to achieving zero net land degradation as promoted by the United Nations Convention to Combat Desertification (U.N.C.C.D.).

I thank the authors for sharing their knowledge and wisdom and for their timely submission and revision of their chapter. I also thank the staff of the CFAES Rattan Lal Center for Carbon

Management and Sequestration (Lal Carbon Center) and of the School of Environment and Natural Resources for their support. Thanks are also due to Ms. Regina Loayza for her help in formatting the book chapters and the front material. I also thank the staff of Taylor and Francis (Ms. Randy Brehm and Tom Connelly) for corresponding with authors, managing the flow of manuscript and their support in timely publication of the book.

Rattan Lal
March 2023
Columbus, OH

Editor

Rattan Lal, Ph.D., is a Distinguished University Professor of Soil Science and Director of the CFAES Rattan Lal Center for Carbon Management and Sequestration at The Ohio State University. He received B.Sc. from PAU, Ludhiana (1963); M.Sc. from IARI, New Delhi (1965); and Ph.D. from OSU, Columbus, Ohio (1968). He was Sr. Research Fellow at the University of Sydney (1968–69), soil physicist at IITA, Ibadan, Nigeria (1970–87), and Professor of Soil Science at OSU (1987 to date). He has authored/co-authored about 1,100 refereed journal articles and 575 book chapters; written and edited/co-edited about 110 books; received an Honoris Causa degree from nine universities throughout Europe, USA, South America, and Asia; the Medal of Honor from UIMP, Santander, Spain (2018); the Distinguished Service Medal of the IUSS (2018); and is a fellow of five professional societies. Dr. Lal has mentored about 120 graduate students, 185 visiting scholars, and 75 postdoctoral scholars and research scientists. His total citations are about 150,000. He was President of the WASWC (1987–90), ISTRO (1988–91), SSSA (2006–08), and the IUSS (2017–18). He holds the Chair in Soil Science and is the Goodwill Ambassador for Sustainability Issues for IICA, member of the 2021 U.N. Food System Summit Science Committee and Action Tracks 1 & 3, and member of the U.N. Food System Summit Coordination Hub (2023). Dr. Lal is laureate of the GCHERA World Agriculture Prize of Nanjing, China (2018); Glinka World Soil Prize of FAO, Rome (2018); Japan Prize of Tokyo, Japan (2019); U.S. Awasthi IFFCO Prize of New Delhi, India (2020); Arrell Global Food Innovation Award of Guelph, Canada (2020); World Food Prize of Des Moines, Iowa (2020); India's Padma Shri Award from the President of India (2021). He is an honorary member of the Moldova Academy of Sciences (2022) and a member of the Academia Europaea (2022). He received the Presidential Award of SSSA (2022) and the IPCC-Nobel Peace Prize Certificate (2007). The PAU, Ludhiana, named its Soil Science and Agronomy as Rattan Lal Laboratories in 2020.

Contributors

Naglaa A. Abdallah
Agricultural Genetic Engineering Research Institute & Department of Genetics
Faculty of Agriculture
Cairo University
Giza, Egypt

Nour el houda Abed
National Institute of Agronomic Research
Oued Smar, Algeria

P.C. Abhilash
Banaras Hindu University
Varanasi, India

Venkataradha Akuraju
International Crops Research Institute for the Semi-Arid Tropics (ICRISAT)
Hyderabad, India

Md. Irshad Ahmed
International Crops Research Institute for the Semi-Arid Tropics (ICRISAT)
Hyderabad, India

Naba Raj Amgain
College of Food, Agriculture & Environment Science
Ohio State University
Columbus

K.H. Anantha
International Crops Research Institute for the Semi-Arid Tropics (ICRISAT)
Hyderabad, India

Hakim Bachir
National Institute of Agronomic Research
Algeria

Michael Baum
International Centre for Agricultural Research in the Dry Areas (ICARDA)
Maadi, Egypt

S. Bhaskar
Division of Natural Resource Management, Indian Council of Agricultural Research
New Delhi, India

Abhishek Bohra
ICAR-Indian Institute of Pulses Research (IIPR)
Kanpur, India

S.K. Chaudhari
Division of Natural Resource Management
Indian Council of Agricultural Research
New Delhi, India

Gizaw Desta
International Crops Research Institute for the Semi-Arid Tropics (ICRISAT)
Addis Ababa, Ethiopia

Muhammad Farooq
Sultan Qaboos University
Al-Khoud, Oman

Nourhan Fouad
International Centre for Agricultural Research in the Dry Areas (ICARDA)
Maadi, Egypt

Y.G.M. Galal
Atomic Energy Authority, Nuclear Research Center
Department of Soil and Water Research
Abou-Zaabl, Egypt

Mahesh K. Gathala
International Maize and Wheat Improvement Centre (CIMMYT)
Dhaka, Bangladesh

Kaushal K. Garg
International Crops Research Institute for the Semi-Arid Tropics (ICRISAT)
Hyderabad, India

Contributors

Bruno Gerard
UM6P
Beguirier, Morocco

Aladdin Hamwieh
International Centre for Agricultural Research in the Dry Areas (ICARDA)
Maadi, Egypt

Rebbie Harawa
International Crops Research Institute for the Semi-Arid Tropics (ICRISAT)
Nairobi, Kenya

Marium Husain
Ohio State University Comprehensive Cancer Center
Columbus, Ohio

Tawffiq Istanbuli
International Centre for Agricultural Research in the Dry Areas (ICARDA)
Maadi, Egypt

M. Jagadesh
Tamil Nadu Agricultural University
Coimbatore, India

M.L. Jat
International Crops Research Institute for the Semi-Arid Tropics (ICRISAT)
Hyderabad, India

Zheng-Rong Kan
Nanjing Agricultural University
Nanjing, PR China

G. Ranjith Kumar
ICAR-National Academy of Agricultural Research Management
Hyderabad, India

Mahesh Kumar
ICAR-Central Arid Zone Research Institute (CAZRI)
Jodhpur, India

Shalander Kumar
International Crops Research Institute for the Semi-Arid Tropics (ICRISAT)
Hyderabad, India

S. Kundu
ICAR-Central Research Institute in Dryland Agriculture
Hyderabad, India

Wassima Lakhdari
Biopesticide Laboratory, Research Division for Plant Protection
National Institute of Agronomic Research of Algeria (INRAA)
Algeria

Alison Laing
CSIRO
Brisbane, Australia

Rattan Lal
Ohio State University
Columbus, Ohio

D. Machiwal
ICAR-Central Arid Zone Research Institute (CAZRI)
Jodhpur, India

R. Madhusudhana
ICAR-Indian Institute of Millets Research (IIMR)
Hyderabad, India

S. Malleswari
Acharya N. G. Ranga Agricultural University
Guntur, India

R. Manasa
ICAR-National Academy of Agricultural Research Management
Hyderabad, India

R.S. Meena
Banaras Hindu University
Varanasi, India

Martin M. Moyo
International Crops Research Institute for the Semi-Arid Tropics (ICRISAT)
Bulawayo, Zimbabwe

Faisal Nadeem
University of Agriculture
Dera Ismail Khan, Pakistan

K.C. Nataraja
Acharya N. G. Ranga Agricultural University
Guntur, India

JVNS Prasad
ICAR-Central Research Institute in Dryland Agriculture
Hyderabad, India

Manoj Prasad
National Institute of Plant Genome Research (NIPGR)
New Delhi, India

Jian-Ying Qi
China Agricultural University
Beijing, PR China

Khaled Radwan
Agricultural Genetic Engineering Research Institute
Giza, Egypt

B.M.K. Raju
ICAR-Central Research Institute in Dryland Agriculture
Hyderabad, India

S. Rakesh
ICAR-National Academy of Agricultural Research Management
Hyderabad, India

C.A. Rama Rao
ICAR-Central Research Institute in Dryland Agriculture
Hyderabad, India

K. V. Rao
ICAR-Central Research Institute in Dryland Agriculture
Hyderabad, India

K. Sammi Reddy
ICAR-Central Research Institute in Dryland Agriculture
Hyderabad, India

Abdul Rehman
The Islamia University of Bahawalpur
Bahawalpur, Pakistan

Y.S. Saharawat
International Fertilizer Development Center
Muscle Shoals, Alabama

Priyabrata Santra
ICAR-Central Arid Zone Research Institute (CAZRI)
Jodhpur, India

Gajanan Sawargaonkar
International Crops Research Institute for the Semi-Arid Tropics (ICRISAT)
Hyderabad, India

Ahcène Semar
Ecole Nationale Supérieure Agronomique
Algiers, Algeria

Kadambot H.M. Siddique
The University of Western Australia
Perth, Australia

Ajay Singh
International Crops Research Institute for the Semi-Arid Tropics (ICRISAT)
Hyderabad, India

D.V. Singh
ICAR-Central Arid Zone Research Institute (CAZRI)
Jodhpur, India

Contributors

Ramesh Singh
International Crops Research Institute for the Semi-Arid Tropics (ICRISAT)
Hyderabad, India

V.K. Singh
ICAR-Central Research Institute in Dryland Agriculture
Hyderabad, India

Dalila Smadhi
Research Division of Bioclimatology and Agricultural Hydraulics
National Institute of Agronomic Research of Algeria (INRAA)
Algeria

S.M. Soliman
Atomic Energy Authority, Nuclear Research Center
Department of Soil and Water Research
Abou-Zaabl, Egypt

J. Somasundaram
Indian Institute of Soil Science
Nabibagh, India

Ch. Srinivasarao
ICAR-National Academy of Agricultural Research Management
Hyderabad, India

Sawsan Tawkaz
International Centre for Agricultural Research in the Dry Areas (ICARDA)
Maadi, Egypt

Bouba Traore
International Crops Research Institute for the Semi-Arid Tropics (ICRISAT)
Niger

Aman Ullah
Center for Agriculture and Bioscience International (CABI) Central and West Asia
Rawalpindi, Pakistan

G. Venkatesh
ICAR-Central Research Institute in Dryland Agriculture
Hyderabad, India

Xing Wang
China Agricultural University
Beijing, PR China

O.P. Yadav
ICAR-Central Arid Zone Research Institute (CAZRI)
Jodhpur, India

P.H. Zaidi
Global Maize Program
International Maize and Wheat Improvement Centre (CIMMYT)
Hyderabad, India

Hai-Lin Zhang
China Agricultural University
Beijing, PR China

Ling-Tao Zhong
Ningxiang Agricultural Technology
Clangsha, PR China

1 Enhancing Resilience to Pedological and Agronomic Droughts in Dryland Farming

Rattan Lal

1.1 INTRODUCTION

The term "dryland farming" means growing crops on soil water storage without any supplemental irrigation. The term is also synonymous with "dryland agriculture," rain-fed agriculture or "dry farming." Rain-fed farming is practiced during the rainy season, and dry farming involves growing crops during the dry season by using the residual moisture in the soil. Thus, dry farming may be limited to eco-regions that receive at least 500 mm/year of rainfall. By nature, dryland farming produces lower agronomic yield than that with supplemental irrigation. Nonetheless, the adoption of innovative and sustainable practices of soil, crop (species and varieties of cultivators), and nutrient management by adopting regenerative practices and innovation can minimize losses of soil water by reducing evaporation and surface runoff and optimizing water use efficiency (WUE). The overall strategy is to adopt management practices (soil, crop, plant nutrients, species, cultivators, rotation, etc.), which ensure minimal productivity even in the worst year (below average precipitation) than during the best season characterized by optimal precipitations. Cultivation without supplemental irrigation and growing crops during the dry season makes dry farming a more challenging system and demands high technical skills and the use of innovative strategies to enhance and curtail soil health and restore its plant-available water capacity (PAWC).

The objective of this chapter is to deliberate technological options for soil, water, crop, and nutrient management that enhance WUE and sustain agronomic productivity. The specific objective is to explain the significance of site-specific management practices to conserve water in the root zone, conserve and manage soil water judiciously, and sustain agronomic productivity by adopting practices that restore and sustain soil health by increasing green water supply in the root zone.

1.2 SOIL MANAGEMENT

The strategy of soil management is to protect, restore, and manage soil structure, minimize risks of water and wind erosion, and improve structural stability with a specific focus on retention pores. Soil management is also aimed at improving water infiltrability, minimizing risks of crust formation, and reducing losses of soil water by evaporation and uptake by weeds. Thus, the soil surface must be protected by mulch of one type or another to moderate temperature and decrease evaporation. Mulching is also useful to improve water infiltrability.

1.2.1 CROP RESIDUES AS MULCH

Rather than removing or in-field burning, crop residues must be returned to the soil as mulch, as a component of conservation agriculture. In addition to providing protection against wind and water erosion, mulch also conserves water, recycles nutrients, moderates soil temperature, increases soil biodiversity, and sequesters carbon as soil organic matter (SOM) and secondary bicarbonates. There are numerous examples of the beneficial impacts of residue mulch on soil health and productivity in diverse soils and ecoregions around the world. In China, Yang et al. (2022) reported the moisture-conserving effects of a mulch-based no-till (NT) system on the proportion of fertile spikes and grain yield increase of 20% in wheat for environments with rainfall of less than 200 mm. This increase was attributed to an increased first tiller emergence rate resulting from increased N uptake, leaf N content and N remobilization from tillers to their grain. Furthermore, second and third tillers, with additional photosynthesis contributed to the tiller survival rate because of more leaf numbers. In semi-arid East Africa, Tuure et al. (2021) observed that the use of crop residue mulching increased the efficient utilization of seasonal precipitation and even reduced the risk of complete crop failure. Tuure and colleagues concluded that maize residue mulching is an accessible and feasible method for conserving soil moisture in the effective root zone in dryland small holder systems in East Africa. Mulching patterns can all be beneficial for rainwater harvesting by prolonging the growing season (Ren et al., 2016).

1.2.2 Plastic Mulch

Rather than using crop residues, plastic film has been widely used for sustainable dryland agriculture because of its multiple benefits in conserving water, moderating soil temperature and controlling weeds. In the Loess Plateau of China, with a cool and semi-arid to arid climate prone to pedologic drought, Li et al. (2020) observed positive effects of ridge-furrow plastic film mulching (RFM) for dryland agriculture. Li and colleagues observed that crop productivity in the RFM system was double or more than that for the un-mulched control. Furthermore, the RFM system promoted the coordinated development of grain, forage and livestock and more profit. Li and colleagues claimed that the RFM system alleviated poverty and helped develop a moderately prosperous society in these harsh environments. In another study, Zhang et al. (2020) also observed increased soil water content and improved yield and WUE of potatoes compared with those of control. Zhang and colleagues concluded that plastic film mulching is a promising method to address seasonal drought stress and increase potato production in semi-arid rainfed areas of Loessial soils. In another study, Zhang et al. (2022) observed the soil carbon sequestration effects of plastic mulching in the Loess Plateau region of China. The use of plastic mulching along with rotation cropping maintained better soil conditions, sustained crop development, and increased soil C sequestration in semi-arid rainfed agriculture. In an enrichment experiment, Zhang et al. (2019b) repeated the positive effects of plastic film cover in rainfed agriculture in the semi-arid Khorchin area in northeast China for rainfed maize. Zhang Z and colleagues observed that autumn mulching with plastic film advanced crop development, increased crop yield and WUE, and reduced climate risks.

Similar positive effects were reported by Zhang et al. (2019a) who considered plastic mulching of a ridge-furrow system as a superior technique for overcoming simultaneous drought and cold stresses in northwestern China, Ren et al. (2016) reported that plastic mulching led to an increase in the soil water at 0–20 cm depth for wheat and 0–80 cm depth for potatoes. This increased the available water to guarantee the crops' water demand at the dry seedling stage for maize and the revival stage for wheat against the threat of seasonal drought and ensured high crop yield.

1.2.3 Gravel Mulch

In regions with low rainfall (< 250 mm/annum) and where crop residue mulches are scarce, gravel mulch is a big industry and extremely useful to grow a range of crops and vegetables. In the low rainfall regions of China, the production and use of gravel mulch is a profitable enterprise in the development of dry, arid areas. Zhao et al. (2013) reported that gravel mulch is a unique mode of conservation tillage, but the ecological effect gradually decreases over time. Zhao WJ suggested paying attention to increasing the replenishment fertilization to the gravel-mulch field, improving the planting patterns, selecting new varieties of drought-resistant crops, establishing modern water-saving supporting systems, etc. for sustainable development of gravel-mulched field agriculture (Figure 1.1).

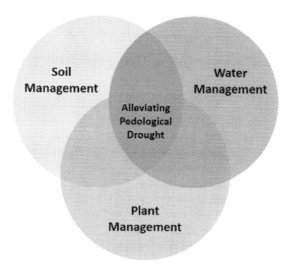

FIGURE 1.1 Strategies of sustainable dryland farming through adaptation and mitigation of drought.

1.3 CONSERVATION TILLAGE METHODS

1.3.1 Conservation Agriculture

These mulch farming techniques are often used in conjunction with no-till on ridge-furrow methods of seedbed preparation to enhance their effectiveness in conserving water in the root zone and ensuring agronomic productivity. A system-based conservation agriculture (CA), practiced on some 200 Mha globally, is effective in conserving soil and water and sequestering atmospheric CO_2 in the soil as humus (Lal, 2015). Schillinger et al. (2022) explored the impact of biosolids vs synthetic fertilizers in Washington State and observed that the application of biosolids combined with low disturbance is an agronomically and environmentally sound practice for dryland wheat production. The presence of residue mulch is critical to the effectiveness of CA. Papendick and Parr (1997) observed overwhelming evidence that mechanical tillage is destroying soil resource base and causing adverse environmental impacts. They recommended that continuing no-till is the most effective and practical approach for restoring and improving soil quality, increasing SOC content, enhancing soil structure, controlling soil erosion, and improving water relations and nutrient availability.

In China's Loess Plateau, Liu et al. (2018) observed that straw-mulched furrows greatly enhanced soil water storage in 0–60 cm depth but decreased from 61–100 cm depth. Mulching also decreased soil temperature in the cold season by 2–2.2°C but increased WUE by 44.04%. Liu and colleagues concluded that ridge-furrow planting with plastic film-mulched ridge and straw-mulched furrow has a good potential for raising wheat production on the Loess Plateau. Also in China, Ren et al. (2016) observed that using a plastic-covered ridge for rainwater harvesting and a furrow as a planting zone increases water availability for crops and stabilizes crop production in northwest China. However, the rainwater-harvesting effect increased with the ridge width increasing to 60 cm. In the sub-humid region of China, Xiaoli et al. (2012) observed that WUE increased with the plastic-covered ridge and furrow rainwater harvesting. This system, used with biodegradable film and straw mulches, is a viable option with high potential to increase agronomic productivity in dryland farming systems without irrigation capability (Xiaoli et al., 2012). Also, in the Loess Plateau of China, Liu et al. (2011) reported that film mulching and straw mulching had different trends in soil temperature. The seasonally averaged soil temperature was the highest under film mulching and the lowest under straw mulching treatment. Film mulching also improved the crop grain yield and yield components.

In northern China, Wang et al. (2011) observed that maize grain yields were greatly influenced by the soil water contents at sowing. Further, grain yields under no-till were generally higher (+19%) in dry years but lower (−7%) in wet years. The no-till treatment has 8%–12% more water in the soil profiles and improved WUE than the conventional and reduced tillage system. Thus, the no-till system has the potential for drought mitigation and economic use of fertilizers in drought-prone rainfed conditions in northern China. Similar observations on the use of plastic mulch ridges were made in northwest China by Zhang et al. (2019) and Wei et al. (2018).

1.3.2 Traditional Tillage

Based on a study conducted in northeastern Tanzania, Enfors et al. (2011) observed crop yield benefits of CA in dryland farming and concluded that the CA system can boost productivity during already good seasons rather than stabilizing harvests during poor rainfall seasons by improving water availability in the crop root zone. Traditional tillage systems, based on local knowledge on ecosystem management, have also been found relevant in Tanzania. A study conducted in dryland areas of Mpwapwa District, Central Tanzania, showed that the use of a no-till system by small landholder contributes to low soil fertility, low soil moisture retention, and poor crop yield. Thus, the choice of site-specific tillage systems to improve soil water retention and enhance nutrient availability is essential to achieving agronomic sustainability under resource-poor small landholder conditions. In the North Wollo zone of the Ethiopian high lands, McHugh et al. (2007) observed that during a season with moderate intensity rainfall open and tied ridges increased sorghum yield by 67%–73% over that of the control (730 kg/ha) while no-till decreased yield by 25%. On the contrary, during a season when high rainfall intensity damaged the ridges, sub-soiling had the best sorghum yield with a 42% increase over the control (1,430 kg/ha). McHugh and colleagues concluded that on slopes below 8% gradient, oxen-drawn ridge-tillage and sub-soiling, to a lesser degree, can mitigate the adverse impacts of short dry spells, especially during seasons with less intense rainfall events (Figure 1.2).

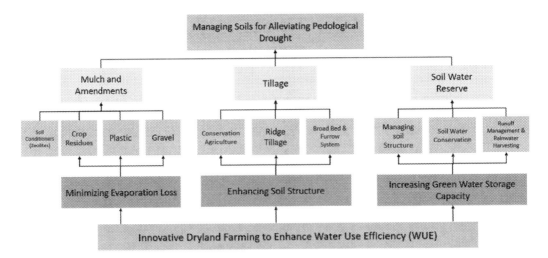

FIGURE 1.2 Managing pedological drought through innovative soil management options for a successful dryland farming.

1.4 GREEN WATER STORAGE

Green water is the amount of PAWC in the rootzone soil, and on it depend crop growth and agronomic yield in dryland farming. PAWC is the difference between field water capacity and the permanent wilting point, expressed on a volumetric basis and summed for all soil depths on the root zone. Green water supply, affected by soil structure and factors affecting it such as SOC and clay contents, can be sustained by the management of soil, crops, and cropping/farming systems. Therefore, increasing retention pores (Greenland, 1979) would enhance field water capacity and thus increase green water supply. In general, increasing SOC content would enhance moisture retention at the field water capacity (Lal, 2020). Bagnall et al. (2022) developed carbon-sensitive pedotransfer functions and showed substantial effects of soil calcareousness and SOC on PAWC. Bagnall and colleagues saw an increase in SOC of 10 g/kg (1%) in calcareous soil. The average increase in SOC-related increase in PAWC is about double the previous estimates. In other words, 1–2 mm per 100 mm soil is associated with a 10 g/kg increase in SOC across all soil classes. This model provides a quantitative measure of the benefits of soil management practices that increase SOC content for drought resilience. Similar to SOC, soil amendments are also used to enhance soil structure and improve PAWC. Ma et al. (2020) observed that multiple years of annual application of polyacrylamide (PAM) significantly increased soil profile water storage while also reducing soil bulk density. Ma B. and colleagues concluded that repeated annual PAM application for 2–3 years would be an effective strategy to combat drought and land degradation and foster sustainable crop production in dryland agriculture.

1.5 CROP MANAGEMENT

In addition to the choice of drought-resilience species and varieties, improved crop management also involves other practices such as management of soil fertility, root system characteristics, and canopy attributes, aimed at enhancing WUE and crop yield. For example, Yan et al. (2023) conducted experiments to assess the effects of root pruning and observed that it significantly decreased root: shoot ratio and increased grain yield of maize by 12.9%.

In Southeast Africa, Ndoye et al. (2022) observed that breeding for specific root traits could improve crop resilience. Ndoye and colleagues observed that basal root whorl number and longer and dense root hairs increase P acquisition efficiency and yield of common bean. With regard to water-saving strategies, root hair density and deep root growth could improve sorghum and pearl millet yield in West Africa. Similarly, denser root systems and mycorrhizal fungi could benefit rice growth.

Sun et al. (2020) studied the root traits of eight cultivars of winter wheat adapted to dryland conditions in Shaanxi Province of China. They observed that the overall root size of dryland wheat cultivars in Shaanxi Province changed with the planting decade. For example, modern cultivars developed after the 2000s had larger root surface areas than older cultivars under drought conditions, especially at 0–40 cm depth. Consequently, there was an improvement in WUE of about 47.0% from the earliest to the most recent cultivars. Furthermore, water stress promoted larger root sizes than those found in the irrigation treatment.

Yan and Zhang (2017) recommended the introduction of dwarfing genes to achieve genetic advantages in grain yields and WUE under rainfed conditions. These genes reduce plant height and affect root and coleoptile length and enhance yield and WUE. Drought stress can inhibit physiological traits (plant uptake) (Yan et al., 2016). Thus, genetic improvement can enhance resilience.

Some effects of root pruning at the stem elongation stage were also reported on drought tolerance and WUE of winter wheat by Ma et al. (2013). Whereas root pruning had no effect on grain yield in well-watered and medium drought soil, but it significantly decreased grain yield under severe drought conditions. Thus, Ma, S. and colleagues suggested possible direction toward drought-resistance breeding.

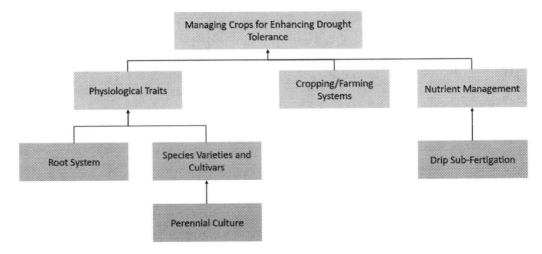

FIGURE 1.3 Adapting crops through choice of species and cultivars, cropping/farming systems, and plant nutrient management.

The benefits of crop rotations have also been observed in enhancing SOC sequestration and erosion control (Schillinger, 2016). Van der Pol et al. (2022) observed that incorporating legumes into a continuous rotation influences the form and amount of soil organic matter (SOM) as well as productivity in farms of the Central Great Plains region of the U.S. Van der Pol and colleagues reported that intensifying rotations with continuous grains led to 1.5-fold increase in aggregate size but did not change SOC stocks. In comparison, incorporating a legume into the continuous grain rotation resulted in 1 mgC/ha more SOC on average in surface soil compared to wheat-fallow rotations, but no significant changes in SOC content were observed at depths. Van der Pol et al. hypothesized that longer-term adoption of legume-based rotations could allow for 10% greater SOC gains over time compared to wheat-fallow systems.

Deng et al. (2021) studied the effects of extreme drought on SOC and N cycles and observed that the effects of drought were regulated by the ecosystem type, and drought duration and intensity. Deng and colleagues reported that drought reduced SOC content mainly because of reduced plant litter input. Drought increased mineral N contents but reduced N mineralization rate and nitrification rate, and this left total N unchanged. However, there is a lack of understanding of the effects of long-term drought on ecosystem C and N dynamics (Figure 1.3).

1.6 DROUGHT-TOLERANT SPECIES

Adoption of drought-tolerant crops, forages, and trees can have strong development potential in dryland. Emam et al. (2012) reported that common bean cultivars with a determinate growth habit appeared to have a potential as a dryland rotation crop for farming in arid regions. In semi-arid areas of China, Huang et al. (2020) compared soil water consumption of sweet sorghum, sudan grass, and forage maize under natural rainfall conditions. They observed that the yield of sweet sorghum was significantly higher than that of sudan grass and forage maize. Soil water consumption mostly happened in 0–150 cm layer in the forage maize, and in 0–100 cm layer for sweet sorghum and sudan grass. Furthermore, the average daily evapotranspiration of forage maize was about 10% and 15% higher than that of sweet sorghum and sudan grass, respectively. They recommended sweet sorghum for forage production because it presented the highest yield, less soil water consumption, and similar nutritional quality to that of forage maize.

Growing mixed species plantations can also alleviate drought stress and create many economic benefits. In the Loess Plateau of China, Gong et al. (2020) conducted a meta-analysis based on 457 field observations to assess the effects of different planting patterns on the soil moisture regime to 5 m depth. They observed that compared with monoculture plantations, mixed species plantations were better able to maintain the soil moisture at 0–4 m depth. Gong et al. concluded that mixed-species plantations (arbors with shrubs) were conducive to enhancing drought resistance in arid and semi-arid regions.

Perennial wheat and cereals are recommended for saving labor and tillage imports (Glover et al., 2010). In Australia, Bell et al. (2010) suggested perennial wheat for rectifying several ecological issues including hydrological imbalance, nutrient losses, soil erosion, depleting SOC content and degrading soil health. Perennial wheat may also have direct production benefits from lower external inputs, providing extra grazing for livestock in mixed farming systems and whole farm benefits that may offset lower grain yield (Bell et al., 2010). Perennial wheat can also diversify current cropping systems.

Similar to perennial cereals, there are also perennial pastures. Hayes et al. (2010) argued that perennial-based pasture swards provide land managers control in temperate cropping zone environments to satisfy the dual role of fostering increased agricultural productivity and reduced deep drainage in NSW, Australia. Breeding and adaptation of perennial pasture species under site-specific conditions could diversify farming systems under harsh arid environments. There are also new pasture plant species to achieve sustainable systems which require strengthening of screening and breeding program (Dear and Ewing, 2008). Perennial legumes are also important in the Mediterranean region and in environment ranging from mountains to deserts (Cocks, 2003). In the Mediterranean environment, and elsewhere in dry regions, genetic improvements including changing the phenological development to better match the rainfall, increases early vision, deeper rooting, osmotic adjustment, increased transpiration, efficiency and improved assimilage storage and remobilization (Turner, 2004). Breeding of new varieties for high WUE of wheat under limited water availability is critical (Deng et al., 2003).

History of 125 years of dryland wheat farming in the Inland Pacific Northwest of the U.S. indicates that the yield of wheat has increased from <1.0 to 3.4 mg/ha by innovative management (Schillinger and Papendick, 2008). Therefore, a substantial yield improvement in dryland farming is possible through the adoption of innovation in the management of soil, water, crops, species, and farming systems.

In accordance with the concept of adopting the approach of integrated agro-ecosystem approach to the management of drought (Solh and Van Ginkel, 2014), and along with K-fertilization (Zhang et al., 2014), application of N fertilization may in some cases ameliorate negative effects of long-term drought. Zhang et al. (2012) reported that moderate N application also plays a physiological role in the alleviation of drought stress effects on plant growth by improving water status on N metabolism, especially for drought-sensitive cultivars.

Rotations are also an integral component of the integrated approach (see Section 1.5). Data from a wheat-based rotation under drought-stress conditions in Northern Syria's medium rainfall zone obtained by Christiansen et al. (2011) recommended that barley rather than wheat is the desired cereal in rotation with legume in regions with rainfall of 350 mm. Because of the importance of sheet in the region's farming systems of Syria, incorporating vetch in the rotation cycle is of critical importance (Figure 1.4).

1.7 DROUGHT RESILIENT RAINFED TECHNOLOGIES

Key drought-resilient technologies include the choice of appropriate crops (species and varieties) and cropping systems and integrated farming systems. Soil management options include CA, ridge-tillage, mulch farming. Crop management includes rotations and inter-cropping (Rao and Gopinath, 2016). Reducing runoff losses and achieving effective erosion control, by mulching and the use of organic amendments and deep-rooted crops (pigeon peas), can improve rainwater consumption and alleviate drought. In Cabo Verde, Baptista et al. (2015) observed that for sloping land, mulch with pigeon peas and organic amendments reduced runoff and soil erosion.

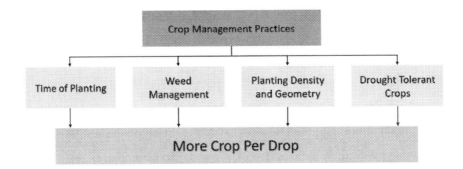

FIGURE 1.4 Innovative crop management for sustainable dry farming.

Rainfed agriculture is practiced on 58% of cropland and is home to 40% of human and 60% of livestock population in India. Recommended interventions for drought management are those related to natural resource management; crop, livestock and fisheries production systems; and capacity building (Prasad et al., 2015).

It is all about water, and the strategy is to improve water productivity. For the Australian Grain industry, a nationally coordinated approach was adapted for the 300–700 mm annual rainfall zone. Kirkegaard et al. (2014) outlined The Water Use Efficiency Initiative which challenged growers and researchers to increase WUE by 10%. As stated above, the judicious use of K fertilizer can also mitigate the adverse effects of drought. Zhang et al. (2014) provided direct evidence of the beneficial physiological formation of K fertilization in mitigating the adverse effects of drought stress in maize by increased nitrate assimilation and osmotic regulation, but not due to its nutritive role. Hao et al. (2005) reported that long-term P application enhanced yield in normal years but not so in the drought years. In most cases, P uptake efficiency rarely exceeds 15%–20% in the first year, with progressively smaller percentages in subsequent years (Ryan, 2003). The focus, therefore, should be in developing innovative options of managing soil fertility which increase the use efficiency of fertilizer but decrease the rate of fertilizer inputs.

In addition to using PAM, green water storage can also be increased by using amendments such as zeolite (He and Huang, 2001, 2004). Zeolite can absorb water because of its high CEC, free structural water storage and surface absorption. The data by He H. and Huang Z. showed that input of zeolite increased water infiltration by 7%–30% on gentle slope and >50% on steep land. Furthermore, zeolite-treated soil increases moisture by 0.4%–1.3% under extreme drought and 5%–15% in mild conditions. It also reduces runoff and minimizes the risks of soil erosion.

1.8 POLICY AND INSTITUTIONAL SUPPORT

Dryland regions contribute 41% of the world's land surface (Solh and Van Ginkel, 2014). Drought preparedness by farmers, especially the resource-poor and small landholders, would also benefit from structured institutional support along with favorable government policies, and cooperation with the private sector. The dryland regions also have a high poverty, and thus coping with drought and water scarcity are critical to advancing SDGs of the United Nations. Policy interventions are needed to promote the adoption of an integrated approach to addressing the challenges of dryland agro-ecosystems.

Policy focus must involve integrated management of soils, crops, livestock, rangeland, and trees based on system thinking. It must also involve all stakeholders in the value chain along a research-to-impact pathway for enhanced food security and improved livelihood in dry areas (Solh and Van Ginkel, 2014).

Policy interventions are needed for the adoption of CA to enhance soil water conservation which can lead to dry spell mitigation and erosion control in drought-prone regions such as the North

Wedlo zone of the Ethiopian highlands (McHugh et al., 2007) and drought-prone regions of China (Wang et al., 2011), India and elsewhere. The overall strategy is to improve WUE (Hsiao et al., 2007) leading to more crop per drop.

Policy interventions are also needed in the adaptation of crop management to drought. Adaptive management includes (Debaeke and Aboudrare, 2004): drought escape, avoidance or tolerance, and crop rotation. These strategies can be grouped under the following: (i) increasing green water storage, (ii) increasing soil water uptake, (iii) reducing evaporation (iv) optimizing the water use patterns between the pre- and post-antithesis, and (v) tolerating drought stress and enhancing chances of recovery. An additional option of water harvesting by using plastic mulch on ridge top and seeding crop in the furrow (see section 1.2.1 to 1.2.3 on mulching).

Marginal lands are increasingly being used for crop production in dryland and are less available for livestock grazing (Hamadeh et al., 1999). Policy interventions are needed to promote farming systems which can address the issue of shrinking rangeland and decreasing feed availability. These emerging and existing situations necessitate a comprehensive sustainable development of dryland agriculture. Such farming systems must focus on a form of sustainable agriculture and comprise a technical system for increased WUE and also for diverse crop products and by-products. For the dryland of Northwest China, Song et al. (1997) suggested that key features of the prototype system would include: (i) drought-resistant crops with low water requirements and high yield potential, (ii) intercropping systems with high yield and high benefits, (iii) nutrient management which can enhance WUE, CA, polythene or biodecomposable film, vetch or a perennial forage, etc. Pro-farmer policy must be in place to promote the adoption of appropriate practices and farming systems.

The Green Revolution of 1960 in India was centered on land equipped for irrigation, but drylands and drought-prone regions were not included in this endeavor (Ninan and Chandrashekar, 1993). Thus, there is a strong need to develop technologies, and policies, of eco-intensification that enhance productivity and restore the environment quality of drylands, under diverse environments and constraints.

It is precisely in this context that the concept of "climate-resilient villages" was implemented in India (Rao et al., 2016). Indian agriculture faces the daunting task of feeding 17.5% of the world population (1.4 B out of 8 B) on 24% of land and 4% of water resources with 60% of cropland under rainfed conditions. Sub-Saharan Africa is faced with an event bigger challenge.

1.9 CONCLUSIONS

Drylands of the world, predicted to be expanding from 40% in the 2020s to 50% of the Earth's surface by 2050, are also vulnerable to global warming and extreme events of drought and heat waves. Yet, dryland farming has a vast potential to increase productivity and sustainability. While additional research is needed to develop innovative technologies and farming systems which produce more crops per drop by conserving water in the root zone and enhancing the use efficiency of water through sustainable management of finite and fragile soil and water resources, there is a strong need to translate science into action. In addition to cooperation between researchers and farmers, involvement of the private sector is also critical to promote the adoption of innovative technologies. The private sector can facilitate access to essential inputs (seed, fertilizers, amendments, machinery, etc.), facilitate payments to farmers for ecosystem services and also provide support for additional research and upscaling of innovative technologies. Four-way cooperation between land managers (farmers, ranchers, and foresters), researchers, policymakers and private sector would be the best option.

Dryland farming has a bright future. More changes will happen between 2020 and 2050 than all the innovations before. In addition to innovation in science, policy intervention and involvement of private sector will also play a critical role in the transformation of dryland farming.

REFERENCES

Bagnall, D. K., Morgan, C. L.S., Cope, M., Bean, G. M., Cappellazzi, S., Greub, K., Liptzin, D., Norris, C. L., Rieke, E., Tracy, P., Aberle, E., Ashworth, A., Bañuelos Tavarez, O., Bary, A., Baumhardt, R. L., Borbón Gracia, A., Brainard, D., Brennan, J., Briones Reyes, D., … & Honeycutt, C. W. (2022). Carbon-sensitive pedotransfer functions for plant available water. *Soil Science Society of America Journal*, **86**(3):612–629. doi:10.1002/saj2.20395.

Baptista, I., Ritsema, C., Querido, A., Ferreira, A., & Geissen, V. (2015). Improving rainwater- use in Cabo Verde drylands by reducing runoff and erosion. *Geoderma*, **237**:283–297. doi:10.1016/j.geoderma.2014.09.015.

Bell, L., Wade, L., & Ewing, M. (2010). Perennial wheat: A review of environmental and agronomic prospects for development in Australia. *Crop & Pasture Science*, **61**(9):679–690. doi:10.1071/CP10064.

Christiansen, S., Ryan, J., & Singh, M. (2011). Forage and food legumes in a multi-year, wheat-based rotation under drought-stressed conditions in northern Syria's medium rainfall zone. *Journal of Agronomy and Crop Science*, **197**(2):146–154. doi:10.1111/j.1439-037X.2010.00447.x.

Cocks, P. (2003). The adaptation of perennial legumes to Mediterranean conditions. In: Bennett, S., (Ed.). New Perennial Legumes for Sustainable Agriculture,. Universtiy of Western Australia,Perth, p. 35–54 ...

Dear, B. & Ewing, M. (2008). The search for new pasture plants to achieve more sustainable production systems in southern Australia. *Australian Journal of Experimental Agriculture*, **48**(4):387–396. doi:10.1071/EA07105.

Debaeke, P. & Aboudrare, A. (2004). Adaptation of crop management to water-limited environments. *European Journal of Agronomy*, **21**(4):433–446. doi:10.1016/j.eja.2004.07.006.

Deng, L., Peng, C., Kim, D., Li, J., Liu, Y., Hai, X., Liu, Q., Huang, C., Shangguan, Z., & Kuzyakov, Y. (2021). Drought effects on soil carbon and nitrogen dynamics in global natural ecosystems. Earth-Science Reviews, 214.2021 103501doi:10.1016/j.earscirev.2020.103501.

Deng, X., Shan, L., Inanaga, S., & Ali, M. (2003). Highly efficient use of limited water in wheat production of semiarid area. *Progress in Natural Science-Materials International*, **13**(12):881–888. doi:10.1080/10020070312331344590.

Emam, Y., Shekoofa, A., Salehi, F., Jalali, A., & Pessarakli, M. (2012). Drought stress effects on two common bean cultivars with contrasting growth habits. *Archives of Agronomy and Soil Science*, **58**(5):527–534. doi:10.1080/03650340.2010.530256.

Enfors, E., Barron, J., Makurira, H., Rockstrom, J., & Tumbo, S. (2011). Yield and soil system changes from conservation tillage in dryland farming: A case study from North Eastern Tanzania. *Agricultural Water Management*, **98**(11):1687–1695. doi:10.1016/j.agwat.2010.02.013.

Glover, J. D., Reganold, J. P., Bell, L. W., Borevitz, J., Brummer, E. C., Buckler, E. S., … & Xu, Y. (2010). Increased food and ecosystem security via perennial grains. *Science*, **328**(5986):1638–1639.

Gong, C., Tan, Q., Xu, M., & Liu, G. (2020). Mixed-species plantations can alleviate water stress on the Loess Plateau. Forest Ecology and Management **458**(1):117767 doi:10.1016/j.foreco.2019.117767.

Greenland, D.J. (1979). Structural organization of soils and crop production. In: Lal, R., & Greenland, D. J. (Eds.) Soil Physical Properties and Crop Production in the Tropics, **2.1**:47–56. I.Wiley and Sons,Chichester,U.K. ISBN: 0471997579.

Hamadeh, S., Zurayk, R., El-Awar, F., Talhouk, S., Ghanem, D., & Abi-Said, M. (1999). Farming system analysis of drylands agriculture in Lebanon: An analysis of sustainability. *Journal of Sustainable Agriculture*, **15**(2–3):33–43. doi:10.1300/J064v15n02_05.

Hao, M., Fan, J., Wei, X., Pen, L., & Lu, L. (2005). Effect of fertilization on soil fertility and wheat yield of dryland in the Loess Plateau. *Pedosphere*, **15**(2):189–195.

Hayes, R., Dear, B., Li, G., Virgona, J.M., Conyers, M.K., Hackney, B.F., & Tidd, J. (2010). Perennial pastures for recharge control in temperate drought-prone environments. Part 1: Productivity, persistence and herbage quality of key species. *New Zealand Journal of Agricultural Research*, **53**(4):283–302. doi:10.1080/00288233.2010.515937.

Hayes, R., Li, G., Dear, B., Conyers, M., Virgona, J., & Tidd, J. (2010). Perennial pastures for recharge control in temperate drought-prone environments. Part 2: Soil drying capacity of key species. *New Zealand Journal of Agricultural Research*, **53**(4):327–345. doi:10.1080/00288233.2010.525784.

He, X., & Huang, Z. (2001). Zeolite application for enhancing water infiltration and retention in loess soil. *Resources Conservation and Recycling*, **34**(1):45–52.

He, X., & Huang, Z., & CSTP. (2004). Efficient analysis on soil water conditioner of zeolite in loess soil. Resouce Conservation and Recycling 34: 223–227.

Hsiao, T., Steduto, P., & Fereres, E. (2007). A systematic and quantitative approach to improve water use efficiency in agriculture. *Irrigation Science*, **25**(3):209–231. doi:10.1007/s00271-007-0063-2.

Huang, Z., Dunkerley, D., Lopez-Vicente, M., & Wu, G. (2020). Trade-offs of dryland forage production and soil water consumption in a semi-arid area. *Agricultural Water Management*, 241. 1 November 2020, 106349:doi:10.1016/j.agwat.2020.106349.

Kirkegaard, J., Hunt, J., McBeath, T.M., Lilley, J.M., Moore, A., Verburg, K., Robertson, M., Oliver, Y., Ward, P.R., Milroy, S., & Whitbread, A.M. (2014). Improving water productivity in the Australian grains industry-a nationally coordinated approach. *Crop & Pasture Science*, **65**(7):583–601. doi:10.1071/CP14019.

Lal, R. (2015). A system approach to conservation agriculture. *Journal of Soil and Water Conservation*, **70**(4):82A–88A.

Lal, R. (2020). Soil organic matter and water retention. *Agronomy Journal*, **112**(5):3265–3277.

Li, G.X., Xu, B., Yin, L., Wang, S.W., Zhang, S.Q., Shan, L., Kwak, S.S., Ke, Q., & Deng, X.P. (2020). Dryland agricultural environment and sustainable productivity. *Plant Biotechnology Reports*, **14**(2):169–176. doi:10.1007/s11816-020-00613-w.

Liu, G.Y., Zuo, Y., Zhang, Q., Yang, L., Zhao, E., Liang, L., & Tong, Y. (2018). Ridge-furrow with plastic film and straw mulch increases water availability and wheat production on the Loess Plateau. *Scientific Reports*, 8.Article Number 6503(2018) doi:10.1038/s41598-018-24864-4.

Liu, Y., Shen, Y., Yang, S., Li, S., & Chen, F. (2011). Effect of mulch and irrigation practices on soil water, soil temperature and the grain yield of maize (Zea mays L) in Loess Plateau, China. *African Journal of Agricultural Research*, **6**(10):2175–2182.

Ma, B., Ma, B., McLaughlin, N., Mi, J., Yang, Y., & Liu, J. (2020). Exploring soil amendment strategies with polyacrylamide to improve soil health and oat productivity in a dryland farming ecosystem: One-time versus repeated annual application. *Land Degradation & Development*, **31**(9):1176–1192. doi:10.1002/ldr.3482.

Ma, S., Li, F., Yang, S., Li, C., Xu, B., & Zhang, X. (2013). Effects of root pruning on non- hydraulic root-sourced signal, drought tolerance and water use efficiency of winter wheat. *Journal of Integrative Agriculture*, **12**(6):989–998. doi:10.1016/S2095-3119(13)60476-1.

McHugh, O., Steenhuis, T., Abebe, B., & Fernandes, E. (2007). Performance of in situ rainwater conservation tillage techniques on dry spell mitigation and erosion control in the drought-prone North Wello zone of the Ethiopian highlands. *Soil & Tillage Research*, **97**(1):19–36. doi:10.1016/j.still.2007.08.002.

Ndoye, M., Burridge, J., Bhosale, R., Grondin, A., & Laplaze, L. (2022). Root traits for low input agroecosystems in Africa: Lessons from three case studies. *Plant Cell and Environment*, **45**(3):637–649. doi:10.1111/pce.14256.

Ninan, K. & Chandrashekar, H. (1993). Green-revolution, dryland agriculture and sustainability – Insights from India. *Economic and Political Weekly*, **28**(12–13):A2–A7.

Papendick, R. & Parr, J. (1997). No-till farming: The way of the future for a sustainable dryland agriculture. *Annals of Arid Zone*, **36**(3):193–208.

Prasad, Y., Srinivasarao, C., & Dixit, S., (2015). Evidences from farmer participatory technology demonstrations to combat increasing climate uncertainty in rainfed agriculture in India. Procedia Environmental Sciences 29(2015)291-292:doi:10.1016/j.proenv.2015.07.221.

Rao, C. & Gopinath, K. (2016). Resilient rainfed technologies for drought mitigation and sustainable food security. *Mausam*, **6767**(1540):169–182.

Rao, C., Gopinath, K., Prasad, J., Prasannakumar, & Singh, A. (2016). Climate resilient villages for sustainable food security in tropical India: Concept, process, technologies, institutions, and impacts. In: Sparks, D. (Ed.) *Advances in Agronomy*, **140**:101–214. doi:10.1016/bs.agron.2016.06.003.

Ren, X., Cai, T., Chen, X., Zhang, P., & Jia, Z. (2016). Effect of rainfall concentration with different ridge widths on winter wheat production under semiarid climate. *European Journal of Agronomy*, **77**:20–27. Science Direct.com, Elsevier, Netherlands, doi:10.1016/j.eja.2016.03.008.

Ren, X.L., Zhang, P., Liu, X., Ali, S., Chen, X., & Jia, Z. (2016). Impacts of different mulching patterns in rainfall-harvesting planting on soil water and spring corn growth development in semihumid regions of China. *Soil Research*, **55**(3):285–295. doi:10.1071/SR16127.

Ryan, J. (2003). Phosphorus fertilizer use in dryland agriculture: The perspective from Syria. P500-503. In"Innovative Soil Plant Systems for Sustainable Agricultural Practices " (Ed). J.M.Lynch, J.S.Scheppers, Nd I.Unver. OECD, Paris, France, and Tubitak, Ankra, Turkdey.

Schillinger, W. (2016). Seven rainfed wheat rotation systems in a drought-prone Mediterranean climate. *Field Crops Research*, **191**:123–130. doi:10.1016/j.fcr.2016.02.023.

Schillinger, W., Cogger, C., Bary, A. (2022). Biosolids and conservation tillage for rainfed wheat farming in dry Mediterranean climates. *Soil & Tillage Research*, **223**, p.105478. doi:10.1016/j.still.2022.105478.

Schillinger W, & Papendick, R. (2008). Then and now: 125 years of dryland wheat farming in the Inland Pacific Northwest. *Agronomy Journal*, **100**(3):S166–S182. doi:10.2134/agronj2007.0027c.

Solh, M. & van Ginkel, M. (2014). Drought preparedness and drought mitigation in the developing world's drylands. *Weather and Climate Extremes*, **3**:62–66. doi:10.1016/j.wace.2014.03.003.

Song, S.Y., Fan, T., & Wang, Y. (1997). Comprehensive sustainable development of dryland agriculture in Northwest China. *Journal of Sustainable Agriculture*, **9**(4):67–84.

Sun, Y.Y., Zhang, S., & Chen, W. (2020). Root traits of dryland winter wheat (*Triticum aestivum* L.) from the 1940s to the 2010s in Shaanxi Province, China. Scientific Reports, *10*(1), p.5328.doi:10.1038/s41598-020-62170-0.

Turner, N. (2004). Sustainable production of crops and pastures under drought in a Mediterranean environment. *Annals of Applied Biology*, **144**(2):139–147. doi:10.1111/j.1744-7348.2004.tb00327.x.

Tuure, J., Rasanen, M., Hautala, M., Pellikka, P., Makela, P., & Alakukku, L. (2021). Plant residue mulch increases measured and modelled soil moisture content in the effective root zone of maize in semi-arid Kenya. Soil & Tillage Research, 209, p. 104945doi:10.1016/j.still.2021.104945.

van der Pol, L., Robertson, A., Schipanski, M., Calderon, F., Wallenstein, M., & Cotrufo, M. (2022). Addressing the soil carbon dilemma: Legumes in intensified rotations regenerate soil carbon while maintaining yields in semi-arid dryland wheat farms. *Agriculture Ecosystems & Environment*, 330, p. 107906.doi:10.1016/j.agee.2022.107906.

Wang, X.B., Dai, K., Zhang, D., Zhang, X., Wang, Y., Zhao, Q., Cai, D., Hoogmoed, W.B., & Oenema, O. (2011). Dryland maize yields and water use efficiency in response to tillage/crop stubble and nutrient management practices in China. *Field Crops Research*, **120**(1):47–57. doi:10.1016/j.fcr.2010.08.010.

Wei, T., Dong, Z., Zhang, C., Ali, S., Chen, X., Han, Q., Zhang, F., Jia, Z., Zhang, P., & Ren, X. (2018). Effects of rainwater harvesting planting combined with deficiency irrigation on soil water use efficiency and winter wheat (*Triticum aestivum* L.) yield in a semiarid area. *Field Crops Research*, **218**:231–242. doi:10.1016/j.fcr.2017.12.019.

Xiaoli, C., Pute, W., Xining, Z., Xiaolong, R., & Zhikuan, J. (2012). Rainfall harvesting and mulches combination for corn production in the subhumid areas prone to drought of China. *Journal of Agronomy and Crop Science*, **198**(4):304–313. doi:10.1111/j.1439-037X.2012.00508.x.

Yan, J. & Zhang, S. (2017). Effects of dwarfing genes on water use efficiency of bread wheat. *Frontiers of Agricultural Science and Engineering*, **4**(2):126–134. doi:10.15302/J-FASE-2017134.

Yan, M.F., Zhang, C., Li, H., Zhang, L., Ren, Y., Chen, Y., Cai, H., & Zhang, S. (2023). Root pruning improves maize water-use efficiency by root water absorption. Frontiers in Plant Science, 13. doi:10.3389/fpls.2022.1023088.

Yan, W.M., Zhong, Y., & Shangguan, Z. (2016). Evaluation of physiological traits of summer maize under drought stress. *Acta Agriculturae Scandinavica Section B-Soil and Plant Science*, **66**(2):133–140. doi:10.1080/09064710.2015.1083610.

Yang, H., Xiao, Y., He, P., Ai, D., Zou, Q., Hu, J., Liu, Q., Huang, X., Zheng, T., & Fan, G. (2022). Straw mulch-based no-tillage improves tillering capability of dryland wheat by reducing asymmetric competition between main stem and tillers. *Crop Journal*, **10**(3):864–878. doi:10.1016/j.cj.2021.09.011.

Zhang, L.X., Gao, M., Li, S., Alva, A., & Ashraf, M. (2014). Potassium fertilization mitigates the adverse effects of drought on selected Zea mays cultivars. *Turkish Journal of Botany*, **38**(4):713–723. doi:10.3906/bot-1308-47.

Zhang, L.X., Li, S., Li, S., & Liang, Z. (2012). How does nitrogen application ameliorate negative effects of long-term drought in two maize cultivars in relation to plant growth, water status, and nitrogen metabolism? *Communications in Soil Science and Plant Analysis*, **43**(12):1632–1646. doi:10.1080/00103624.2012.681735.

Zhang, P., Wei, T., Li, Y., & Zhang, Y. (2019). Effects of deficit irrigation combined with rainwater harvesting planting system on the water use efficiency and maize (*Zea mays* L.) yield in a semiarid area. *Irrigation Science*, **37**(5):611–625. doi:10.1007/s00271-019-00628-4.

Zhang, X.C., Guo, J., Ma, Y., Yu, X.F. (2020). Effects of vertical rotary subsoiling with plastic mulching on soil water availability and potato yield on a semiarid Loess plateau, China. *Soil & Tillage Research*, *199*, p. 104591.doi:10.1016/j.still.2020.104591.

Zhang, X.C., Hou, H., Yin, J., Fang, Y., Yu, X., Wang, H., Ma, Y., & Lei, K. (2022). Crop rotation with plastic mulching increased soil organic carbon and water sustainability: A field trial on the Loess Plateau. *Soil Use and Management*. **39**(2), pp. 717–728. doi:10.1111/sum.12873.

Zhang, X.D., Zhao, J., Yang, L., Kamran, M., Xue, X., Dong, Z., Jia, Z., & Han, Q. (2019a). Ridge-furrow mulching system regulates diurnal temperature amplitude and wetting-drying alternation behavior in soil to promote maize growth and water use in a semiarid region. *Field Crops Research*, **233 issue 1**:121–130. doi:10.1016/j.fcr.2019.01.009.

Zhang, Z., Zhang, Y., Sun, Z., Zheng, J., Liu, E., Feng, L., Feng, C., Si, P., Bai, W., Cai, Q., Yang, N., van der Werf, W., & Zhang, L. (2019b). Plastic film cover during the fallow season preceding sowing increases yield and water use efficiency of rain-fed spring maize in a semi-arid climate. *Agricultural Water Management*, **212**:203–210. doi:10.1016/j.agwat.2018.09.001.

Zhao, W.J., Wang, L., Zhan, G., Li, N., & Wang, F. (2013). The ecological effects and improving measures of gravel-mulched field in agricultural development of the arid areas. Zhao, J., Iranpour, R., Li, X., Jin, B. (Eds.) 726–731:3780. *726*, pp. 3780–3786. doi:10.4028/www.scientific.net/AMR.726-731.3780.

2 Drought Management through Genetic Improvement in Dryland Cereals and Grain Legumes

O.P. Yadav, P.H. Zaidi, R. Madhusudhana, Manoj Prasad, and Abhishek Bohra

2.1 INTRODUCTION

Drought is one of the most important constraints in crop production in drylands in different parts of the world, adversely impacting not only crop productivity but also food security, livelihood, and economic growth (Bodner et al., 2015). Climate change is likely to make drought more severe in future, particularly in semi-arid and arid tropics of the drylands, in the form of its larger spread, greater intensity, longer duration and higher frequency (Cook et al., 2018; Rama Rao et al., 2019).

Drought hampers plant growth, development, and yield by changing the inherent agro-physiological and biochemical processes and pathways (Afzal et al., 2017). In addition, the temporal and spatial variation in drought across different environments has made it a complex problem to deal with (Zhou et al., 2021). Furthermore, crop production in drylands is likely to become more challenging due to predicted intense drought stress, increased temperature and incidences of diseases and insect-pests (Sultan et al., 2013; Rama Rao et al., 2019). Therefore, drought management remains the key intervention to make the dryland production system more resilient and less vulnerable to climatic vagaries through technological, institutional and policy options (Shiferaw et al., 2014).

From a technology point of view, both agronomic and genetic improvement approaches have a great role to play in drought management. Agronomic approaches such as mulching, tillage, intercropping, nutrient management, water conservation, early sowing and micro-irrigation are technically feasible and economically viable options to overcome the drought problem (Tyagi et al., 2020) which require additional resources and physical interventions.

Genetic improvement of field crops is an attractive option to develop and deploy crop cultivars that are inherently more tolerant to drought (Tuteja and Gill, 2013). Therefore, the development of crops with better adaptation to drought is critical to have sustainable food production in drylands. This article reviews the research efforts for understanding the adaptation to drought and breeding for drought tolerance in major dryland cereals and legumes that are grown largely under rainfed ecology, and which are naturally subjected to different degrees of water deficit during their growth period.

2.2 CHOICE OF FIELD CROP SPECIES FOR DROUGHT ECOLOGY

2.2.1 Drought-Tolerant Field Crops and Their Distribution

Several cereals and legumes are important components of dryland farming systems. The different cereal-legume combinations have multiple benefits like maintenance of soil fertility, better use

Drought Management through Genetic Improvement

TABLE 2.1
Area, Production and Productivity of Major Dryland Cereals and Legumes in World

Crop	Area (million ha)	Production (million mg)	Productivity (mg/ha)	Top 5 Grower Countries	Major Production Constraints
Maize	202.00	1162.4	5.8	China, USA, Brazil, India, and Argentina	Drought, heat, excessive moisture, nutrient imbalance, diseases, and insect pests
Sorghum	46.00	57.9	1.4	USA, Nigeria, Ethiopia, India, and Mexico	Drought, heat, diseases, and insect-pests
Millets	32.10	30.5	2.3	India, Niger, China, Nigeria, and Mali	Drought, diseases, and weeds
Chickpea	14.84	15.1	1016.3	India, Turkey, Pakistan, Myanmar, and Ethiopia	Drought, heat, Fusarium wilt, Ascochyta blight, and pod borer
Pigeon pea	6.09	5.0	822.2	India, Malawi, Myanmar, Tanzania, and Kenya	Drought, water logging, diseases, and insect pests

of resources and nutrients, and management of the ecosystem. The choice of crops in drylands is determined by the crop duration, the length of season, and the productivity and ability of the crop to meet the food and fodder requirements of the household crop-livestock farming system. The main dryland cereals are maize (*Zea mays* L.), sorghum (*Sorghum bicolor* L. Moench), pearl millet (*Pennisetum glaucum* (L.). R. Br.), and other millets; while the major legumes in dryland include chickpea (*Cicer arietinum* L.) and pigeon pea (*Cajanus cajan* L. Millsp.).

Maize is the cereal with the largest global production and area (Table 2.1) and plays a critical role in the sustenance and livelihoods of millions of resource-poor smallholders in drylands, especially in tropical regions of Asia, Latin America, and Sub-Saharan Africa (SSA). Sorghum is adapted to dryland agro-ecosystems of the arid and semi-arid tropics of the world due to its higher inbuilt genetic resilience to drought and changing climatic conditions. Across the globe, sorghum is produced in >100 countries in Africa, Asia, Oceania, and the Americas. Pearl millet is an important crop grown in the semiarid and arid regions of South Asia (SA) and SSA that are characteristically challenged by low and erratic rainfall and high mean temperature and simultaneously have soils with low organic carbon content and poor water-holding capacity. Small millets like foxtail millet (*Setaria italica*), finger millet (*Eleusine coracana*), barnyard millet (*Echinochloa crusgalli*), proso millet (*Panicum miliaceum*), kodo millet (*Paspalum scrobiculatum*), and little millet (*Panicum sumatrense*) possess excellent potential to grow under water deficit conditions and provide an alternative choice in drought-prone areas. Once small millets were the regular part of farming in drylands and human diet but have been significantly marginalized in the post-Green Revolution period in arid and semi-arid regions across the globe. In addition to their in-built intolerance to abiotic stress, these crops are nutritionally superior to rice and wheat as they are rich in protein, fibre, vitamins, and antioxidant compounds (Singh et al., 2022). Chickpea, pigeon pea, and mung bean are important for food security and livelihood generation to resource-poor people in the semi-arid and subtropical world. As a source of affordable proteins, these crops are key ingredients of vegetarian diets in developing world. The inherent traits of legume crops, including biological nitrogen fixation (BNF) and a deep root system, make them crucial for the sustainability of farming systems in these regions.

All dryland cereals and legumes are known for their drought tolerance with built-in adaptive traits to produce yield under adverse conditions, yet drought stress adversely affects their production and productivity. In the context of climate change, their inherent resilience to drought needs to be further improved in the dryland regions in view of existing variation within a crop species.

2.2.2 Crop Response to Drought

Drought stress occurs in different patterns and intensities at different crop growth stages. Crops have been reported to exhibit a differential response to drought depending on the growth stage at which drought occurs. Much work has been done to understand the nature of drought in target dryland environments and the response of crops to moisture stress that occurs in different stages of growth to understand their adaptation to types of drought environments.

The probability of drought in dryland environments is highest at the start and in the latter part of the rainy season, and therefore, crops are highly prone to face water deficiency during the establishment and flowering/early grain-filling stages (Bänziger et al., 2000a). In dryland areas, seedling death is particularly high under the combined effects of drought and heat stress (Ndlovu et al., 2021). The basic requirement to obtain good yield in drylands is to have a sufficient plant stand. At the time of germination, emergence and early vegetative stages, moisture availability is a critical factor for proper growth and development of maize, sorghum, pearl millet and minor millets (Gregory, 1983; Carberry et al., 1985; Bänziger et al., 1997). Root, and shoot length and root/shoot ratio, leaf formation and secondary root development are strongly affected by drought stress and there are reported genetic differences for these traits. If drought stress severely reduces stand at the beginning of the season, farmers have a choice, though at additional cost, to replant fields with a shorter duration cultivar or a different species. Agronomic management plays an important role in reducing seedling mortality due to early-season moisture stress.

However, drought during the vegetative stage of growth may have a more pronounced effect on drought-sensitive than drought-tolerant cultivars (Fadoul et al., 2018). On the other hand, little adverse effect on productivity is observed by drought during the vegetative stage in pearl millet as there is a significant increase in the number of panicles (Bidinger et al., 1987a), which has been established as a compensation mechanism for a damaged main shoot (van Oosterom et al., 2003, 2006). Water stress during the vegetative phase also results in delayed flowering in pearl millet and sorghum (Mahalakshmi et al., 1987). Such developmental plasticity increases the chances for escape from the most sensitive stage of growth (Henson and Mahalakshmi, 1985).

Crops are relatively more sensitive to water deficit during the reproductive stage, especially around flowering, compared with other growth stages (Shaw, 1977; Grant et al., 1989). For example, during tassel emergence, anthesis and silking, maize is highly sensitive to drought that results in cob barrenness (Banziger et al., 2000a). Similarly, terminal drought stress at the reproductive stage, when drought occurs at flowering and grain-filling stages of pearl millet, has a strong adverse impact on grain yield (Mahalakshmi et al., 1987; Kholová and Vadez, 2013) due to a decrease in the number of fertile florets and grain size (Bidinger et al., 1987a; Fussell et al., 1991). Sorghum production is affected by drought stress during both panicle development and the post-flowering stages (Adugna and Tirfessa, 2014). A study on sorghum (Kapanigowda et al., 2013) showed that both pre- and post-flowering drought stress significantly reduces grain production due to a reduction in the number of grains per panicle (Manjarrez-Sandoval et al., 1989), a trait that directly contributes to grain yield.

Leguminous crops (pulses) often experience drought stress because of their cultivation on marginal lands under rainfed conditions. Drought stress influences various aspects of growth and development in legumes, including poor germination, marked decline in stomatal conductance, chlorophyll content, and photosynthesis, reduced number of pods, impaired root nodule development, poor nutrient uptake, increased leaf senescence and enhanced reactive oxygen species (ROS) activity (Khatun et al., 2021). All these reflect finally into a significant compromise in yielding capacities of leguminous crops under drought stress conditions. As a result, drought stress is reported to cause substantial yield losses in legume crops. For example, up to 50% of chickpea yield is reported to be lost to drought stress (Jha et al., 2019). In legumes, flowering and reproductive stages such as anthesis, pollen germination, and pollen fertility are highly vulnerable to drought stress.

Drought Management through Genetic Improvement

2.2.3 Understanding Drought-Coping Mechanism of Dryland Crops

Grain yield is a complex trait influenced by several component traits at the bottom of the structural organization of the plant and is also the consequence of an interaction between the environment and the genotype.

A range of morpho-physiological, biochemical, and molecular mechanisms operate in crops to impart adaptation to diverse environmental stresses including drought. Drought-coping mechanisms that allow plants to mitigate the negative effects of drought can be classified into three broad categories as escape, avoidance and tolerance (Levitt, 1980). Escape involves the completion of life cycle prior to the onset of drought stress, while avoidance is based on maintaining hydration despite water deficit through some specific morpho-physiological features such as deep rooting, stomatal closure etc., and finally, tolerance involves features that allow the plant to maintain, at least partially, proper functionality in a dehydrated state (Levitt, 1980).

2.2.3.1 Drought Escape

Early phenology (early flowering and maturity) has been reported to be the most important mechanism to escape terminal drought stress in cereal and leguminous crops (Bidinger et al., 1987b; Fussell et al., 1991; Banziger et al., 2000; Araus et al., 2002; Gaur et al., 2015). Early maturing genotypes with higher yields are preferred because of their ability to escape drought by completing their life cycle before stress is intensified. However, early maturing genotypes have relatively less total evapo-transpiration and leaf area index (LAI) with the result that there is a trade-off with yield potential (van Oosterom et al., 1995; Banziger et al., 2000).

2.2.3.2 Drought Avoidance

The maintenance of a proper water balance in plants is essential for adequate growth and development. In water stress situations, plants tend to increase water uptake and decrease water loss through coordinated regulation of root development (Blum, 2009; Zaidi et al., 2022) and stomatal conductance (Hepworth et al., 2016). One of the strategies for improved yields in drought-prone dryland system is to develop a deeper and more intense rooting system to access water from the deep soil profile (Vadez et al., 2011, 2015; Zaidi et al., 2016). Sorghum roots can grow to depths of 1–2 m by the booting stage and can efficiently extract water at a lateral distance of 1.6 m from the plant (Routley et al., 2003). Genotypes that have a large number of seminal roots and a large diameter vessel in both seminal and nodal roots show a better survival rate under drought stress conditions (Bawazir and Idle, 1989). The thick and deep root system in legumes such as chickpea, pigeon pea, mung bean, and common bean is conducive to the extraction of more water from soil, and length, density and biomass of roots are the main determinants of drought avoidance (Turner et al., 2001; Kavar et al., 2008). Research in chickpea and mung bean has demonstrated significant variations in root traits and indicate that prolific roots with maximum root depth with higher root length to weight ratio are the characteristics that ensure greater water uptake under prevailing drought stress conditions (Ramamoorthy et al., 2017; Bangar et al., 2019).

Stomatal conductance reduces transpiration and plays an essential role in regulating plant water balance in field crops experiencing drought (Hadi et al., 2016). Stomatal closure also reduces cell expansion and growth rate leading to a significant reduction in photosynthesis. There is genetic variation within species of dryland crops in terms of their drought avoidance (Nemeskeri et al., 2015; Rauf et al., 2015).

2.2.3.3 Drought Tolerance

Leaf-rolling and survival rate are two common physiological indexes that are used to measure drought tolerance at the seedling stage. Leaf rolling helps plants to temporarily reduce water loss and avoid stress injuries. At the cellular level, drought signals promote stomatal closure to save water, stimulate the production of stress-protectant metabolites, up-regulate the antioxidant system, and deploy peroxidase enzymes to prevent acute cellular damage and loss of membrane integrity (Gupta et al., 2020).

In the genetic improvement programme, the final target trait is the grain yield. Understanding drought tolerance in terms of physiology, phenology, and morphology of the crop has led to enhanced knowledge of the yield formation process under drought (van Oosterom et al., 2003; Yadav, 2011). This scientific progress has helped breeders to identify and target specific traits in different drought environments. In maize, the anthesis-silking interval (ASI) is the trait used to assess the degree of tolerance to drought. This simple and easy-to-measure trait at a large scale in field is an indirect measure of complex physiological functions such as – rate of current photosynthesis under drought stress and sink strength of developing kernels. Ears per plant (that is measurement of barrenness under stress), with high heritability is also a suitable target trait for improving maize drought tolerance (Monneveux et al., 2008; Xue et al., 2013; Jia et al., 2020), which is an indirect measure of another complex physiological trait, i.e. - assimilate remobilization efficiency towards kernel development. Low ASI (<5.0 days) under stress has been found to be significantly correlated with grain yield under drought conditions and other abiotic stresses as well (Bruce et al., 2002; Zaidi et al., 2004). Stay-green trait, i.e. reduced leaf senescence especially at the early grain-filling stage, when developing kernel are highly dependent on current supply of photo-assimilates helps in reducing kernel abortion after fertilization (Zaidi et al., 2003). Stay green, high chlorophyll content and chlorophyll fluorescence and cooler canopy temperature coupled with high transpiration efficiency are key physiological traits for drought tolerance in sorghum (Harris et al., 2007; Kapanigowda et al., 2013). In pearl millet, morphological traits such as high tillering, small grain size, and shorter grain filling periods that can be easily measured have been successfully manipulated in breeding programmes that target improved drought tolerance (Yadav et al., 2012).

As explained above, legumes cope with drought-challenged scenarios through a variety of mechanisms that include escape, avoidance, and tolerance. Completion of life cycle before the onset of dry conditions forms a key adaptation mechanism to escape drought in leguminous crops. In this context, early flowing and short maturity duration have been identified in several chickpea varieties and lines such as ICC 96029, ILC 1799, ILC 3832, KAK 2 that demonstrate the escape mechanism concerning drought stress. Drought avoidance helps curtail water loss while maximizing water use under water-limiting conditions. Drought escape and drought avoidance mechanisms are successful where crops are grown in stored soil moisture and high-water holding capacities. However, soils with low water retention capacities require plants with intrinsic tolerance mechanisms to withstand drought stress. Morphological features, such as root system architecture (RSA), also play an important role in imparting tolerance against dry conditions. A variety of physiological traits, such as photosynthetic efficiency, relative water content (RWC), and water use efficiency (WUE), are reported to have great relevance with respect to mitigating drought stress in legume crops. In mung bean, increase in activities of catalase, ascorbate peroxidase, superoxide dismutase and peroxidase has been associated with drought tolerance (Ali et al., 2017).

2.3 GENETIC IMPROVEMENT FOR DROUGHT TOLERANCE

Like any other trait, progress in drought tolerance is determined by the availability of germplasm sources with drought tolerance, variation in traits determining performance under drought and efficiency of selection to enhance drought tolerance in crops.

2.3.1 Genetic Resources of Dryland Crops

The genetic resources of dryland crops include local landraces, improved elite material, local cultivars, genetic stocks, and wild relatives. Systematic efforts at the global level led to the availability of germplasm of dryland crops. For instance, global germplasm collections of maize consist of 327,932 accessions. CIMMYT works as the global repository for maize germplasm collection and maintains

28,193 accessions from 64 countries. Apart from the germplasm collection, there is one primary maize bank, especially for genes, the Maize Genetic Stock Centre. This centre has conserved and annotated nearly 80,000 maize mutant stocks and are available to maize geneticists.

Global sorghum germplasm collections consist of 235,688 accessions. The largest global collection of sorghum from 93 countries is collected and conserved at ICRISAT, Patancheru (Upadhyaya et al., 2017). ICRISAT has a total of 41,023 accessions in the gene bank which include 35,632 landraces or traditional cultivars, 4,841 breeding material, 461 wild relatives and 89 improved cultivars (GENESYS-PGR, 2019).

At the global level, pearl millet germplasm collection consists of 65,447 accessions in more than 1,750 gene banks of 46 countries. Six large ex-situ holders are ICRISAT, India (33%); CNPMS, Brazil (11%); NBPGR, India (9%); ORSTOM, France (6%) and ICRISAT, Nairobi (4%). ICRISAT holds 23,841 germplasm accessions that include 20,628 traditional cultivar/landraces, 2,268 breeding materials, 816 wild relatives, and 129 advanced or improved cultivars from 50 countries (ICRISAT, 2019). Indigenous collections of ICAR-NBPGR are from 17 states and union territories (Yadav et al., 2017).

Dwivedi et al. (2012) summarized the collection of cultivated and wild relatives of different small millets across the continents in national and international gene banks. The major collections of germplasm accessions are stored in gene banks, viz., foxtail millet in China, India, France and Japan; finger millet in India and African countries; proso millet in the Russian Federation, China, Ukraine, and India; barnyard millet in Japan and India; kodo millet in India and USA; and little millet in India (Upadhyaya et al., 2016; Vetriventhan et al., 2016). ICRISAT is holding the global germplasm of small millets. Indigenous collection in ICAR-NBPGR for foxtail millet and finger millet is from 26 states, little millet from 20 states, and kodo millet from 13 states.

Globally over 86,533 cultivated and 1,032 wild germplasm accessions of chickpea are conserved in world gene banks. ICRISAT (20,267), International Centre for Agricultural Research in Dry Area (13,362) and NBPGR (15,131) have the major holdings of chickpea collections. Worldwide, a total of 43,027 mungbean accessions are held ex situ (Nair et al., 2012). ICRISAT holds 13,783 accessions of pigeon pea while more than 10,000 accessions are being maintained by the All India Coordinated Pigeon pea Improvement centres. The selection of suitable germplasm from the large collections is truly a herculean task. Hence tailor-made smaller sets like core, mini core, reference, and composite collections with minimum repetitiveness and maximum diversity, have been made available to researchers and breeders as workable germplasm subsets.

Research conducted so far has indicated that the genetic resources from drylands hold a unique advantage as they have evolved over centuries by natural and human selection under drought, high temperature or saline conditions. They are better adapted to the local conditions and would contribute to enhancing resilience at the farm level. These resources could be of immense importance, especially as sources of native genes conditioning resistance to various biotic and abiotic stresses and make unique study material to understand the mechanism of adaptation to abiotic stresses (Yadav et al., 2020).

Pearl millet landraces that evolved in dry areas because of natural and man-made selection over thousands of years demonstrate better adaptability to water stress (Yadav et al., 2000; Yadav, 2010, 2014). Efforts were made to utilize these landraces in pearl millet breeding practices in a regular approach. Cycles of mass selection in genetically heterogeneous landraces were found to increase yield considerably (Bidinger et al., 1995; Yadav and Bidinger, 2007) and have also been revealed as a valuable germplasm source to breed drought-tolerant lines (Yadav, 2004) and developing inbred pollinator lines for hybrid breeding (Yadav et al., 2009, 2012). There are also reports of successful introgression of drought tolerance in the agronomically desirable background from elite genetic resources (Presterl and Weltzien 2003; Yadav and Rai, 2011). Crosses between adapted landraces and elite genetic materials showed enhanced adaptation to drought combined with higher productivity up to 20%–30% (Yadav and Weltzien, 2000; Yadav, 2010). Dwivedi et al. (2016) have also proposed a systematic landrace evaluation to facilitate the identification of alleles for enhancing abiotic stress adaptation and yield to raise productivity and stability in vulnerable environments.

Sorghum landraces that are collected from arid/semi-arid environment showed greater osmotic adjustment than the landraces from humid environment (Blum and Sullivan, 1986), and registered 24% higher yield than genotypes with low osmotic adjustment (Ludlow and Muchow 1990). Landraces from Maharashtra and Andhra Pradesh states of India are drought tolerant (Elangovan et al., 2009). The Ethiopian *durras* are an excellent source for the stay-green (non-senescence) trait related to post-flowering drought-tolerance (Dahlberg et al., 2020). *Caudatum* sorghums are adapted to drought-stressed conditions. Drought-tolerant accessions have been widely identified (Reddy et al., 2004; Kumar et al., 2011; Upadhyaya et al., 2014; Venkateswaran et al., 2014).

Several potential donors have been reported across legume crops that carry specific traits that confer tolerance against water stress conditions. Crop wild relatives (CWRs) have a large potential for improving drought tolerance traits in different crops. These donors have applications in introgression breeding and in the development of experimental populations to understand the complex genetic architecture of drought tolerance. For example, wild *Cicer* species are the reservoir of the many beneficial genes for broadening the genetic base of the cultivars to survive in challenging environments. Vernalization treatment i.e., exposure to low temperatures will induce early flowering in the wild species therefore interspecific crosses help to escape drought in chickpea. A major QTL from an interspecific RIL population [ICC 4958 (*C. arietinum*)×PI 489777 (*C. reticulatum*)] has been identified on CaLG03 that explains 55% of phenotypic variation for vernalization response (Samineni et al., 2015). Early phenology such as early flowering, early podding and early maturity has been reported to be the crucial mechanism to escape drought stress across legumes and cereals. The early maturing genotypes have been used to identify the genomic regions or QTL for earliness trait, such as in chickpea (Gaur et al., 2015) and pigeon pea (Kumawat et al., 2012).

2.3.2 Phenotypic Traits Associated with Drought Tolerance

Although yield is a trait of primary interest, partitioning it into its component traits that are significantly associated with yield under stress gives a better understanding of the targeted trait and helps to keep track with stress intensity for mid-term correction, if needed. Also partitioning complex traits such as grain yield under drought into components adds to the genomic region discovery efforts. Secondary traits can also be used as preliminary selection criteria when the turn-around time between seasons is short. A secondary trait could give greater gains for the primary trait (grain yield) than selection for yield alone when $hGY < rG \times hST$, where hGY and hST are the square roots of heritability of grain yield and the secondary trait, and rG is the genetic correlation between grain yield and the secondary trait (Falconer and McKay, 1996). This condition is rarely met except when yield is low, and the secondary traits are expressed best under stress. However, in most cases, secondary traits are added to a selection index along with the primary trait in the belief that the heritability of the index will exceed that of the primary trait and yield.

A range of secondary traits have been proposed for different types of abiotic stresses, including drought stress; all putatively related to improved survival or tolerance. However, for a secondary trait to be useful in a breeding programmeme, it must comply with some key requirements (Araus et al., 2002, 2008; Lafitte et al., 2003), such as:

- a high genetic correlation with grain yield under the environmental conditions of the target environment, i.e., the relationship with yield must be causal not casual,
- a lower effect of environment than grain yield is i.e., having a higher heritability than the yield itself, and so less genotype×environment interaction effects,
- a high genetic variability for the trait must exist within the species,
- a lesser association with poor yields in unstressed environments in case of traits being addressed in breeding for stress-prone environments,

- amenable to measuring the trait rapidly and more economically than yield itself, and in a reliable way, and
- enable to be assessed in individual plants or in very small plots, preferably by non-destructive means.

Studies have shown that key secondary traits for maize under drought are reduced barrenness (increased ears per plant under stress), anthesis-silking interval, stay green (reduced lower leaf senescence), leaf erectness and to a lesser extent, leaf rolling under drought (Banziger et al., 2000b). In index selection for drought tolerance weightage is assigned based on the correlation of traits with grain yield and heritability under drought stress (Bolanos and Edmeades, 1996). In addition, plant height and days to 50% anthesis are also used in the selection index to avoid extremes in selection for plant height and maintain maturity group, respectively. Other traits such as root growth are only useful when they have been field-tested and have met the criteria prescribed for an ideal secondary trait. Of course, roots have a very important role in water acquisition and a significant component of tolerance to water-deficit stress (Barker and Varughese, 1992; McCully, 1999), however, due to complications in the observation of root traits especially on a large number of genotypes in field conditions it is logically not possible to use them in routine selection process, except in strategic research such as selection of trait donors for new breeding start etc.

Tillering ability is the most important trait in pearl millet that has been strategically manipulated in mid-season drought stress breeding. There is a large variation in tillering capacity of pearl millet, which has been reported to be a moderately heritable trait (Appa Rao et al., 1986; Rai et al., 1997; Yadav et al., 2017). The greater tillering provides elasticity to the growth and development of pearl millet and is part of its mechanism for adaptation to severe drought conditions. Several drought tolerance studies conducted in the Sahel have indicated that pearl millet is tolerant to water deficit until early grain filling, predominantly because the main shoot can be compensated by basal tillering (Winkel et al., 1997).

Earliness and short and rapid grain filling periods have been manipulated to improve tolerance to terminal drought. Early flowering essentially determines grain productivity under water stress (Bidinger et al., 1987b; Fussell et al., 1991; van Oosterom et al., 1995). Genetic variation with respect to earliness is widely available in the germplasm (Rai et al., 1997; Yadav et al., 2017) and phenotype-based selection has been accomplished (Rattunde et al., 1989). The frequently exploited basis of earliness is the *Iniadi*-type landraces collected from western Africa (Andrews and Kumar, 1996). Promising lines with the early flowering trait have been developed from *Iniadi* landraces and adopted in Indian and African agroecosystems.

The proportion of the panicle threshing denoting seed setting potential under low soil moisture contents and integrating the effects of assimilation and translocation of photosynthates in drought environments is a measure of drought tolerance (Bidinger et al., 1987b). It exhibits a large variation in grain yield (Fussell et al., 1991; Bidinger and Mahalakshmi, 1993), is highly heritable (Yadav, 1994), and selection is effective (Bidinger and Mahalakshmi, 1993). Some mathematical models have also been used to recognize lines that are performing well in adverse conditions by comparing grain yields between stress and non-stress (optimum) conditions (Bidinger et al., 1987b; Yadav and Bhatnagar, 2001). Accordingly, multi-environmental data from a diverse range of growing conditions is used to identify the drought-tolerant genotypes.

Stay-green is an adaptive mechanism in sorghum and is an effective strategy for increasing grain and fodder production, particularly under water-limited conditions (Borrell et al., 2014). It also efficiently remobilizes and assimilates during the grain-filling stage, to maintain normal grain weight, quality, and nutrients. Root architecture is a key factor in understanding the interplay of drought stress, and there is significant variability for the root architectural traits. With higher root traits, CRS67, Phule Suchitra and STG44 were potential genotypes for use in breeding for drought tolerance in sorghum (Kiran et al., 2022). IS13540 was found to be a drought-tolerant line, and its tolerance was related to a deep prolific root system and reduced transpiration rate

(Gowsiga et al., 2021). Sorghum has a dense and deep root system and has the ability to reduce metabolic processes, transpiration through leaf rolling, and stomatal closure under drought. While tolerance to drought is mainly routed through osmotic adjustments, protective solutes, high proline, desiccation-tolerant enzymes and high stomatal conductance, the escape mechanism primarily includes early flowering, early maturity, high leaf nitrogen level, high photosynthetic capacity, and remobilization of assimilates.

Phenotypic traits that can serve as signature to identify stress situations in legumes include leaf rolling, stomatal conductance, root characteristics, osmotic adjustment, dehydration tolerance, transpiration efficiency, solute accumulation and stay green mechanism. Research has demonstrated stomatal conductance and leaf rolling as one of the most reliable physiological indicators of drought tolerance. Studies indicate that leaf rolling is caused by the reduction in leaf water potential, which can vary from species to species (Kadioglu et al., 2012). In legumes, drought induces a reduction in the leaf area and causes early leaf senescence. As has been observed in pigeon pea and cowpea, abscission, and senescence are promoted in leaves at the time of flower blooming and pod-filling stage. Drought stress is also reported to affect nitrogen uptake, leaf senescence, and chlorophyll efficiency in legumes (Khatun et al., 2021). Because dryness and monocarpy cause comparable patterns of acropetal leaf senescence in cowpea, their combined action appears to increase senescence under drought (Khatun et al., 2021). Many germplasm accessions among legumes such as chickpea, ICCs 8261, 4958, 16374B, 15510, 9586, 867, 14778 and ICCV 10 impart drought tolerance by controlling root traits root length density, dry weight, and deep rooting system (Jha et al., 2020). Changes in photosynthesis, osmotic regulatory substances, drought-induced proteins, and antioxidant enzymes represent the varying levels of influence under drought stress in legumes. Photosynthetic and transpiration rates decrease with the decrease in the relative water content of the soil. Studies have revealed that the rate of photosynthesis is reduced in response to drought which could be stomatal under drought stress and can be non-stomatal under severe drought stress. Under water-limiting conditions, it leads to a decrease in photosynthesis due to reduced CO_2 availability, resulting in diffusion limitations of the stomata and the mesophyll. Stress also reduces nodule formation, as is evident from the study in faba bean, which revealed a lesser number of root hairs under stress conditions. The impact on chlorophyll level is also revealed by Mafakheri et al. (2010) who compared a sensitive and the resistance chickpea under different stages of water deficiency.

2.3.3 Breeding for Enhancing Drought Tolerance

2.3.3.1 Characterizing Target Environment

The interaction of genotypes with the environment restricts the genetic gain in developing insights into drought adaptation. Therefore, it is important to characterize the environment in which the crop is grown. A clear understanding of the target population of environment (TPE) is essential for identifying the best selection environment where the phenotyping site should be established. The phenotyping site does not necessarily have to be in the target environment but should have a relevant relationship that represents the key constraints, such as the timing and intensity of drought stress in TPE. Therefore, a minimum amount of information about the TPE required includes the following:

- Daily weather data, preferably from the past 5 years, including maximum temperature (Tmax), minimum temperature (Tmin), relative humidity (RH), and rainfall with its distribution pattern, for understanding and defining the most relevant type of drought stress in TPE.
- Soil type, cropping season and cropping system, especially the planting window for maize.
- Other relevant information, such as major biotic stresses and socio-economic constraints.

Analysing these data helps to understand the requirements for establishing a phenotyping site that is significantly related to the TPE.

A crop modelling approach would help in the detailed characterization of the growing area and identify production constraints (drought stress patterns) to enable the breeder to understand the need for breed-specific cultivars for each target agro-ecoregion using a suitable breeding strategy. Crop modelling is highly useful for designing ideal plant ideotypes based on the evaluation of past genetic improvements for the selected environment. The efficiency of crop improvement for constantly changing environments can be improved if the physiological and morphological traits associated with drought adaptation are identified and integrated into breeding programmes. Singh et al. (2017) used a modified CSM-CERES-Pearl millet model to study the effect of altered traits determining the maturity of the crop, its yield and adaptation to heat and drought prevailing in semi-arid regions of India and Africa. It was found that decreasing crop maturity duration had a negative impact on yield although increasing the maturity duration benefitted yield in a few locations in current and future climatic conditions. In addition, increasing radiation use efficiency (RUE) resulted in higher yields under climate change conditions. Also improving the length and depth of roots are recommended as important mechanisms for drought adaptation and achieving better yield (Vadez et al., 2012). The interaction of genotypes with the environment usually results in hampering the progress of crop breeding programmes. Therefore, it is essential to understand the underlying physiological process behind the interactions (Basford and Cooper, 1998).

Most crop improvement programmes have divided all crop-growing regions of India into different mega zones based on the geographic boundaries, rainfall pattern and local adaptation of the crop (Gupta et al., 2013). The differences in the water use response and growth clearly showed that breeding for various agro-ecological zones also resulted in the breeding of specific plant strategies associated with traits like plant water use (Medina et al., 2017). A detailed characterization of crop growing area through a modelling approach and identification of crop production constraints (drought stress patterns) will enable the breeder to understand each target agro ecology for breeding specific cultivars.

2.3.3.2 Selection Environment

Choosing an appropriate selection environment to improve productivity under drought has been the subject of a major debate in plant breeding, and several theoretical and empirical studies have been reported.

The difficulty in choosing the appropriate selection environment has often restricted the progress of breeding for tolerance to drought in highly variable TPEs. Even though there is extensive evidence that selection under targeted stress may accelerate breeding gains for TPE (Bänziger et al., 1997), the difficulty of choosing appropriate environments, given a highly variable target environment, may limit the identification of superior genotypes. While breeding programmemes in high-income countries can access real-time geographic information system (GIS) data for adequately weighting results from multi-environments trials (Podlich et al., 1999), those opportunities rarely exist in low-income countries because there is a lack of both real-time GIS information and resources for conducting a large number of multi-environment trials. Therefore, based on a systematic analysis of TPE, a suitable field phenotyping location can be selected to establish a dedicated phenotyping site for managed drought stress phenotyping. Location for managed stress phenotyping needs to be chosen carefully so that the targeted crop stages (e.g., flowering, and early grain-filling stage) coincide with a rain-free period to avoid early relief from the indented stress. This is done based on long-term weather data (at least the last 5 years), including Tmax, Tmin, relative humidity, and rainfall, which could be used in identifying suitable planting window. *For example* – at Hyderabad location in India (17.3850°N, 78.4867°E, 545 masl), November to February is usually the dry season, i.e. most part is almost rain-free. Also, Tmax is <35°C and Tmin is >8°C in most part of this period. Such a site is suitable for managed drought stress phenotyping, where planting can be taken up during the last week of November and a field trial with medium maturity group of entries reaches the flowering stage sometime in the first week of February, and most critical stages of the reproductive stage complete within February, which is usually a dry period.

a. Stress timing is managed in such a way that the targeted growth stage(s) are exposed to the desired level of drought stress.
b. Stress intensity is severe enough so that traits that are important for yield under stress become distinct from those which affect yield under non-stressed conditions, e.g., mean ASI, increased senescence, etc.
c. Stress uniformity occurs over space and time for the expression of genotypic variation within a trial.

Some researchers believe that cultivars targeted for drought conditions can be identified under non-drought conditions (indirect approach) while others think that selection for drought environments should be undertaken under drought stress (direct approach). The indirect approach involves selection for high yielding potential under non-stress conditions with the assumption that genotypes selected under optimum conditions would also perform well under drought. In this approach, drought resistance is an unidentified component of performance over different environments and more emphasis is laid on yield potential. The main advantage of this approach is that yield potential, and its components have higher heritability under optimum conditions than that under stress conditions (Ceccarelli, 1994). Since yield potential has been reported to be a significant factor in determining the yield under moisture stress (Fussell et al., 1991), improvement in yield potential may have some spill over effects under water stress conditions.

The direct approach recommends that varieties for drought-prone areas must be selected, developed, and tested under the target drought environments. Theoretical analyses also indicate that selection for stress environments should be done in stress environments (Rosielle and Hamblin, 1981; Simmonds, 1991). In this approach, improvement of yield under moisture stress requires dissociation from yield potential under optimum conditions as a major selection criterion (Ceccarelli and Grando, 1991; Ceccarelli et al., 1992) and the emphasis is placed on drought adaptation and yield under drought.

Many studies have compared relative gains in performance under drought conditions through selection in drought versus non-drought environments. Low correlations are often reported between yields measured in stress and optimum conditions which indicate that yield performance under drought and non-drought conditions are separate genetic entities, and direct selection for yield performance in the target drought environments would be required to make greater gains in productivity. This is further substantiated by the existence of significant cross-over genotype×environment interactions observed across optimum and stress environments (Virk and Mangat, 1991). Using evaluation data from drought-stressed and non-stressed environments, many studies showed that drought tolerance and escape were major determinants of performance in drought environments (Virk et al., 1991; van Oosterom et al., 1995). On the other hand, high yield potential accounted for 10%–15% variation towards performance under drought. This has highlighted the importance of evaluation and selection in drought-prone locations and early maturity and suggested *in-situ* breeding for drought environments.

An osmotic solution with polyethylene glycol (PEG) is often used for inducing drought conditions and also for maintaining constant water potential during the entire experimental process (Lu and Neumann, 1998). It has been observed that the percentage of germination and plant growth are affected by drought (Zhang et al., 2015). This simple and cost-effective *in vitro* method is useful in the screening of large germplasm materials.

Screening for drought tolerance using pots is simple and cost-effective but is difficult to evaluate large populations with sufficient replication for traits like leaf area as it involves a destructive method and assessing transpiration is also laborious. Therefore, high throughput and automated phenotyping platform, LeasyScan are considered as more effective to screen various plant materials in a non-destructive manner during the vegetative stage using an optical system (PlantEye®; www.phenospex.com). This can be used for precise measurements of plant canopy traits such as digital biomass, 3D-leaf area, plant height, leaf area index, leaf angle, leaf inclination and light penetration depth (Vadez et al., 2015). Screening by using a lysimeter is also similar to the field environment in which an additional benefit of water use could also be followed throughout the crop cycle (Vadez et al., 2013).

Screening under field conditions is done by evaluating the test material through multi-locational trials conducted in drought-prone regions or by growing crops in a rain-free season under adequate water supply but withholding irrigations to expose the test material to drought at the desired stage. Field screening is still the most preferred method to assess the drought response of breeding materials and experimental test cultivars in large breeding experiments.

At CIMMYT in El Batan, Mexico, an intensive effort for developing improved maize germplasm with tolerance to drought stress was launched during the mid- to late-1970s (Fischer et al., 1983; Bänziger et al., 1997; Edmeades et al., 1997, 2017). The targeted breeding for drought tolerance in maize was started in 1975 with recurrent selection in a tropical white dent population, namely - *Tuxpeño Sequia*. A total of eight cycles of full-sib recurrent selection under managed drought stress were completed at the CIMMYT station in Tlaltizapán, Morelos, Mexico, where the rain-free period between November and April allowed precise timing and intensity of stress levels. Later during the 1980s, S_1 recurrent selection was implemented in other populations, including the Drought Tolerant Population (DTP), La Posta Sequia, and Pool 26 Sequia. The recurrent selection programmes produced improved populations, notably Tuxpeño Sequia c6, La Posta Sequia c7, and the DTPY c9 and DTPW c9, that served as source germplasm for deriving drought tolerant lines and moving towards hybrid breeding for drought tolerance. Drought-tolerant lines extracted from these populations have been used as donors in tropical maize breeding programme of SSA, Asia and Latin America, and some were elite enough to be released as CIMMYT Maize Lines (CMLs). Later a series of bi- and multi-parent populations were developed, followed by several cycles of recurrent selections and improvement for tolerance to drought, in SSA, Asia and Latin America (Edmeades et al., 2017). Using base population developed at CIMMYT-Mexico, breeding for drought stress tolerance for the mid-altitudes of eastern Africa, was initiated in 1998 in Kenya (Banziger and Diallo, 2004). Over time, more lines from La Posta Sequía c7 and DTPW c9 were used to increase the frequency of alleles for drought tolerance in new breeding starts. Thus, using various selection approaches across diverse testing environments, many inbred lines with good combining ability for drought tolerance and other adaptive traits were identified, and several elite CMLs and improved maize hybrids/synthetics were released (Prasanna et al., 2021).

Breeding and selection under carefully managed high-priority abiotic stresses, such as drought stress, have significantly increased maize yields in highly variable drought-prone environments and particularly in severely stressed environment conditions with lower average yields (Figure 2.1). Similar results were also reported from a recent study on gains with trait-based breeding under managed stress environment conducted in CIMMYT's Asia regional maize programme (Zaidi et al., 2020).

Many of the new drought-tolerant maize lines were recycled through conventional pedigree or doubled haploid (DH) to develop better drought-tolerant donor lines with higher productivity. These new donor lines have been used to develop multiple stress-tolerant hybrids and deployed across SSA (Cairns and Prasanna, 2018). In the CIMMYT Asia maize programme, breeding for drought tolerance was initiated in 2008 with the introgression of drought-tolerant yellow and white donors from CIMMYT-Mexico and white donor lines from CIMMYT-Zimbabwe and Kenya. These donors were crossed to elite Asia-adapted CIMMYT lines primarily bred for yield potential and local adaptation, especially for resistance to diseases like downy mildews and *Turcicum* leaf blight (TLB). Three sets of new drought-tolerant Asia-adapted CMLs, including CML-562 to 565, CML-578 to 582, and CML-615 to 618 were released in the past 5 years and made available to public and private sector maize breeding programme in the region.

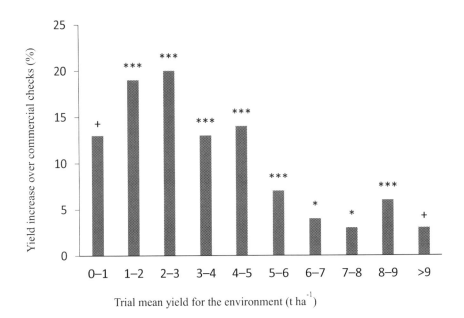

FIGURE 2.1 Selection differentials in trials grown across eastern and southern Africa of 42 CIMMYT hybrids selected for drought (and nitrogen stress) tolerance in Zimbabwe as compared to 41 private company check hybrids. *** and * indicate statistically significant (P<0.001) gains; + indicates non-significant gains. (Banziger et al., 2006)

2.3.3.3 Molecular Techniques

Conventional breeding methods like pedigree and backcross for improved grain production resulted in the development of improved cultivars with a significant increase in the productivity of crops. The progress in genetic improvement for drought tolerance using traditional plant breeding practices has been slow due to the complex interaction between genotype and environment, stress levels, and stages at which drought occurs. In view of the difficulties in unravelling genetic mechanisms controlling drought tolerance, genomics-assisted breeding offers a greater opportunity to develop drought-tolerant crops faster.

2.3.3.3.1 Trait-QTL Association

During the past two decades or more, different types of molecular (DNA) marker systems have been devised and used for various genetic applications like assessing genetic variability, population structure, detecting genomic regions associated with agriculturally important traits etc.

A series of studies have been conducted on QTL mapping of drought tolerance in maize (Liu and Qin, 2021). By analysing a RIL population derived from a cross between CML444 (drought tolerant) and SC-Malawi (drought sensitive), 81, 57, 51, and 34 QTLs were uncovered for six target traits (male flowering, ASI, grain yield, kernel number, 100-kernel fresh weight, and plant height) (Messmer et al., 2009). Nine QTLs related to leaf temperature have been reported on chromosomes 1, 2, 9, and 10 that were identified based on 248 SSR markers using 187 recombinant inbred lines (RILs) (Liu et al., 2011). Five QTLs related to grain yield have been reported on chromosomes 1, 3, 5, 6, and 8, which explain 50% of the phenotypic variance (Agrama and Moussa 1996). Ribaut et al. (1996) reported QTLs associated with flowering time and ASI in maize under well-watered conditions and two water-stressed regimes (Ribaut et al., 1996). In further study, the group identified several small to moderate effects QTL associated with grain yield under different levels of drought stress. However, these QTLs, in general, were not stable across different drought environments (Ribaut et al., 1997). Using a larger population and higher marker density, Messmer et al.

(2009) identified six QTLs associated with grain yield under optimal and drought environments, with limited overlap of genomic regions identified across environments. Almeida et al. (2013) identified 83 QTL associated with yield under drought stress, each QTL explained 2.6 to 17.8% of the phenotypic variation. Seven meta-QTL (mQTL) were identified across three populations, with six mQTL expressed under drought and optimal conditions. A meta-analysis of 18 bi-parental populations evaluated under a range of drought and optimal environments revealed 15 mQTL associated with grain yield (Semagn et al., 2013). However, mQTL were not stable across environments and genetic backgrounds. Genome wide association mapping studies (GWAS) on grain yield under drought, heat and optimal conditions identified several single nucleotide polymorphisms (SNPs) and candidate genes across locations; however, no overlapping SNPs were observed across treatments (Yuan et al., 2019).

The high genetic variability among sorghum genotypes and the relatively small size of its genome (730 Mb) (Paterson et al., 2009) has helped in the identification of several QTLs in sorghum related to drought tolerance. CO_2 assimilation, transpiration rate (Kapanigowda et al., 2014), stomatal conductance and density (Lopez et al., 2017), epicuticular wax (Uttam et al., 2017), crown root angle at seedling (Mace et al., 2012) and maturity (Lopez et al., 2017), nodes with brace roots (Li et al., 2014), seedling root dry weight (Mace et al., 2012), root length, roots/plant, root: shoot ratio, root volume and weight (Fakrudin et al., 2013), pre-flowering drought (Kebede et al., 2001) and post-flowering drought tolerance (Hayes et al., 2016). Furthermore, *Stg2*, *Stg3* and *StgB* were identified as the three key QTLs for MAS to improve terminal drought tolerance (Reddy et al., 2014). Genes associated with delayed senescence have been reported (Kiranmayee et al., 2020; Johnson et al., 2015; Abebe et al., 2021; Aquib and Nafis, 2022). These *stg* loci were also found to reduce the canopy size during flowering, reduce tillering and promote the overall root growth (Harris-Shultz et al., 2019).

In pearl millet also, DNA markers have been used to dissect QTL to investigate the molecular and biochemical basis of tolerance to abiotic stress and to devise an efficient approach for crop improvement for yields under drought stresses. Subsequently, the dissection of quantitative trait loci pertaining to drought tolerance (DT-QTLs) and high grain yield was identified under independent studies using different pearl millet mapping populations (Yadav et al., 2002, 2004; Bidinger et al., 2007). The milestone breakthrough to detect major DT-QTL on LG2 related to grain yield, explaining 32% phenotypic variance was carried out (Yadav et al., 2002) using bi-parental populations of individuals of different crosses. Afterward, a minor QTL on LG5 linked with drought tolerance, explaining 14.8% phenotypic variance was also detected (Yadav et al., 2004). The putative drought-tolerant QTL on LG2 was evaluated using near-isogenic lines derived from H 77/833-2 into which DT-QTL has been introgressed from PRLT 2/89-33 (Serraj et al., 2005). In the same direction, three major QTLs (positioned on LG2, LG3, and LG4) pertaining to grain yield with limited QTL×environment interactions were analysed as key components of MAB for improved grain yield under variable post-flowering water stresses in pearl millet (Serraj et al., 2005; Bidinger et al., 2007). The major QTLs mapped on LG2 and LG3 accounted for a wide (13%–25%) range of phenotypic variations for grain yield traits under drought stress conditions. At the same time, minor QTLs were also co-mapped for harvest index under drought stress, and QTLs for both grain number and individual grain mass in water deficits were identified (Bidinger et al., 2007).

These QTL mapping findings have been validated in follow-up research that mapped QTLs linked to high grain yield and its related traits under terminal drought stresses in pearl millet (Yadav et al., 2011). In this study, one major QTL associated with grain yield and drought tolerance of grain yield in water stress conditions was detected on LG2 which explains about 32% of the phenotypic variance in testcross progenies (Yadav et al., 2011). This major QTL explaining 32% of the variance under drought stress was confirmed in different marker-aided backcross programmes, where the 30% increase in the general combining ability (GCA) of grain yield anticipated of DT-QTL in water deficits was recovered in the QTL introgression lines (Yadav et al., 2011). The QTL associated with low transpiration rates that contribute to water stress tolerance by lodging soil water contents to be used at the grain filling stage by limiting moisture loss at the vegetative phase has been co-mapped with that of terminal DT-QTL (Kholová et al., 2012). A low rate of transpiration is maintained by physiological and morphological interactions (Kholová et al., 2012). The DT-QTLs associated with terminal drought tolerance have been introgressed into elite pearl millet lines to improve terminal drought tolerance (Jangra et al., 2019). Marker-aided foreground selection has been performed with robust SSR markers mapped on LG2 and LG5 to select the plants harbouring alleles that render resistance to bi-parental progenies (BC_4F_2) along with rigorous phenotypic selection to recover the genome of the recurrent parent in pearl millet (Jangra et al., 2019).

Genetic dissection of drought tolerance in grain legumes has been primarily driven by the analysis of biparental populations. Several studies have detected QTL controlling traits that are important in tolerance against drought in grain legumes. For more details, the reader may refer to other reviews (i.e., Jha et al., 2019). Notable among these examples is a study by Varshney et al. (2014). The authors reported a QTL cluster on LG 4, referred to as "*QTL-hotspot*" that harboured QTLs for 12 traits and explained about 58.20 % phenotypic variation. This QTL region was identified by the analysis of two RIL populations viz. ICC 4958 × ICC 1882 and ICC 283 × ICC 8261, which segregated for a variety of drought tolerance-related traits. Similarly, analysis of a RIL population (VC2917 × ZL) under irrigation and drought conditions in mung bean led authors to identify 58 QTLs for several drought tolerance-related traits including maximum leaf area, relative water content and yield (Liu et al., 2017). Consistent QTL explained up to 20.1 of the phenotypic variation observed in the population. The genome-wide association studies (GWAS) have also emerged as a popular technique to dissect the genetic architecture of drought tolerance traits in grain legumes. Li et al. (2018) measured yield and yield-related traits of 132 Australian chickpea varieties under drought stress and implemented GWAS and genomic selection (GS). The study examined drought response of these 132 lines by analysing the phenotypic and whole genome resequencing (WGRS) data. Advances in DNA sequencing technologies have facilitated the identification of candidate genomic regions/causative loci that can be exploited in legume breeding programmes to improve genotypes for their tolerance level against drought stress. For instance, the functional genomics approach in pigeon pea uncovered a set of drought-responsive candidate genes including *CcHyPRP*, *CcCDR*, *CcCYP*, *CcMT1*, *DLP*, *APB*, and *LTP1* under drought response (Deeplanaik et al., 2013). Among the various expressed sequence tags (ESTs) identified, three of the selected stress-responsive genes, viz. CcHyPRP, CcCDR, and CcCYP showed remarkable tolerance against multiple abiotic stresses in transgenic Arabidopsis (Mir et al., 2014). Analysis of whole genome re-sequencing (WGRS) data of 292 pigeon pea genotypes superior haplotypes for 10 drought-responsive genes (Sinha et al., 2020). The study led to the identification of a total of 83, 132 and 60 haplotypes specific to breeding lines, landraces, and wild species, respectively. Candidate gene-based association analysis using these ten genes in a set of 137 accessions revealed significant associations of five genes with seven drought-tolerance-related features. Furthermore, haplo-pheno analysis for the strongly associated genes resulted in the identification of most promising haplotypes for three genes regulating five component drought traits. The haplotypes such as C. cajan_23080-H2, C. cajan_30211-H6, C. cajan_26230-H11, C. cajan_26230-H5 were identified as superior haplotypes which could be targeted for assembly in future pigeon pea cultivars for improved response to drought stress.

Drought tolerance is a complex quantitative trait regulated by coordinated effects of many genes or several quantitative trait loci (QTLs). Millets being a drought-tolerant species, QTL mapping in millets might provide some novel genomic locus or alleles controlling drought response and could be employed in future crop improvement programmemes. An interspecific hybridization between *S. italica* cv. Yugu 1 and its wild relative *S. viridis* cv. W53 leads to the generation of a mapping population which is further employed in the detection of 18 QTLs associated with drought and dehydration stress (Qie et al., 2014). Similarly, recombinant inbred lines (RILs) developed following a cross between 863B and ICMB 841 were utilized as mapping population to harness water use-related QTLs in pearl millet (Aparna et al., 2015). The study identified four major QTLs associated with water use in which a QTL mapped on linkage group 6 was found to control plant growth, transpiration, and drought responses. Six more drought-associated QTLs were identified in a fine mapping population of pearl millet which also related to tiller contribution and plant biomass under low water conditions (Tharanya et al., 2018). A functional synonymous SNP (A/G transition) of *SiDREB2* at 558th position was found to be linked with dehydration tolerance in foxtail millet. This SNP serves as a potential marker to distinguish drought-responsive genotypes and was validated in 170 foxtail millet accessions (Lata et al., 2011; Lata and Prasad, 2013). Likewise, many SNPs were identified through genotyping-by-sequencing (GBS) strategies in diverse accessions of pearl- and finger millet (Hu et al., 2015; Gimode et al., 2016; Kumar et al., 2016). These genetic variations can be employed to determine population diversity and offspring selection during marker assisted breeding for the development of drought-tolerant verities.

2.3.3.3.2 Marker-Assisted Selection

Several studies have reported molecular markers-based analysis of drought stress tolerance in maize, including various secondary traits associated with grain yield under drought-stressed environments in the tropics (Prasanna et al., 2021). Though, the genetic dissection of drought tolerance in maize has provided good insight on this challenging trait, so far a few applications have emerged in practical maize breeding programmes. The key factors behind this include the complex genetic basis of the traits, crop stage for drought stress in field, significant effect of genetic background, cost of fine mapping of QTLs, and QTL×environment effects (Tuberosa et al., 2002). Most of the QTLs are often genetic background-specific and applying MAS for several small effects QTLs may be more expensive than conventional breeding methods for improving drought tolerance (Xu et al., 2009). MAS for any trait, including drought tolerance, needs major QTLs with large effects, and stable across genetic backgrounds and environments. Unfortunately, no QTLs with sufficiently large effects are found to be effectively used in MAS programmes for drought tolerance in maize. Therefore, the lack of consistent and major phenotypic effects of the QTL in diverse recipient genetic backgrounds suggests that QTL-based marker-assisted selection is unlikely to play a major role in breeding for drought tolerance in maize.

Of the stay-green genotypes (B35, SC56, and E36-1) studied, B35 (BTx642) is a useful source of stay-green for research and development of sorghum hybrids (Jordan et al., 2012). Stay-green QTL individually reduced leaf senescence in introgression lines and contributed significantly towards breeding drought tolerance (Harris et al., 2007; Kassahun et al., 2010). More recently, the potential use of Stay-green QTL in improving transpiration efficiency and water extraction capacity in sorghum for terminal drought tolerance (Vadez et al., 2011) and grain yield particularly under low yield environments has been demonstrated (Jordan et al., 2012). Marker-assisted breeding is a better approach to enhance post-drought tolerance in sorghum (Kassahun et al., 2010). Therefore, efforts have also been initiated to transfer this trait through marker-assisted backcrossing (MAB) into elite cultivars and study their expression in different backgrounds (Ngugi et al., 2013; Kassahun et al., 2010; Isaac et al., 2019). Current studies at ICAR-IIMR, Hyderabad on marker-assisted introgression of stay-green QTL, Stg3a and Stg3b from B35 to Indian post-rainy sorghum lines, M35-1, CSv-29R, CSV-26, CRS4 and RSLG262 have shown promise in imparting terminal drought tolerance. The introgression lines had higher green leaf area retention at maturity, and improved stover yield and seed size along with grain yield under both stress and no-stress conditions.

Three major QTLs related to grain yield with low quantitative trait loci and environmental (QTL×E) interactions were detected across different post-flowering water stresses in pearl millet (Bidinger et al., 2007). One of the major QTLs explained ~32% of the phenotypic variation for grain yield under water deficit conditions. The impacts of these dominant QTLs have been validated in the marker-assisted back cross (MABC) programmes wherein 30% enhancement of the general combining ability for grain yield is anticipated from this QTL under terminal drought stress. It was recovered in introgression lines using informative data generated with markers flanking the QTL (Yadav et al., 2011). This QTL has been fine-mapped using the LG2 QTL NIL-derived F2 mapping population with ddRAD SNPs (Srivastava et al., 2017). Out of 52,028 SNPs that were identified between the NILs, a total of ten SNPs were anchored to the QTL interval and are being used in the forward breeding programmes using the HTPG platform.

Many potential marker-assisted backcrossing (MABC) methods have been employed in QTLs introgression from a donor to an elite recurrent parent in pearl millet. Several validated QTLs have been introgressed into elite hybrid parental lines (A-/B-, R-) resulting in the improved version of the hybrids or (essentially derived varieties (EDVs).

One of the notable examples of genomics-assisted breeding for drought tolerance in legumes includes the transfer of 'QTL hotspot' region in chickpea. This has resulted in the development and release of *Pusa Chickpea* 10216 in India following genomics-guided transfer of genomic regions from ICC 4958 that control several component traits associated with drought tolerance (Bharadwaj et al., 2021).

2.3.3.3.3 Genomic Selection

Genomic selection (GS) is another marker-based strategy that incorporates all the available marker information simultaneously into a model to predict the genetic value of progenies for selection (Meuwissen et al., 2001; Lorenz, 2013). Each marker is considered a putative QTL, reducing the risk of missing small-effect QTLs (Guo et al., 2012). De los Campos et al. (2009) and Crossa et al. (2010) examined several statistical models for genomic selection in diverse panels of maize germplasm from the CIMMYT using a random cross-validation scheme that mimics the prediction of unobserved phenotypes based on markers and pedigrees. Beyene et al. (2015) implemented GS for drought tolerance on eight bi-parental maize populations and demonstrated the efficiency of GS in maize, with an average gain per cycle of 0.086 mg/ha under drought stress without significant changes in maturity and plant height. The study showed that overall gain in average grain yield using GS was two- to four folds higher than the previously reported gain in average GY under drought stress using conventional phenotypic selection. Vivek et al. (2017) used GS to enhance drought tolerance in Asian maize germplasm and suggested that a positive selection response can be obtained with the use of markers for GY under drought. Hence, the statistical model used to determine the effects of the markers works in practice and thus is validated. The use of GEBV allowed the selection of superior plant phenotypes, in the absence of the target stress, resulting in rapid genetic gains for DT in maize. Das et al. (2020, 2021) compared genetic gains with GS and conventional phenotypic selection for multiple stress tolerance, including drought and waterlogging stress and found that the gains with GS were relatively higher under drought, whereas PS showed *at par* or better response under waterlogging stress. The study concluded that the careful constitution of multiparent population involving trait donors for targeted stresses along with elite high-yielding parents and its improvement using GS is an effective breeding approach for building multiple stress tolerance without compromising on yields under optimal conditions.

2.4 FUTURE PERSPECTIVES

Modern agriculture involving digital tools for automated high-throughput and reliable phenotypic technologies is increasingly used to accelerate genetic gain in various crop breeding programmemes. Phenomics platforms employ simple and fast quantitative and qualitative methods to evaluate plant growth and development. This facilitates the detailed observation and measurement of the different traits resulting from the expression of genetic characteristics of plants, both physical and environmental factors. Of the several field-based plant phenomics approaches, aerial drones are highly promising for measuring traits like panicle emergence and its traits, plant height, biomass, biotic and abiotic stress incidence, canopy cover etc. Aerial drones can cover large areas quickly, allowing all genotypes in a study to be measured simultaneously, and are not impeded by plant height, which allows them to capture data throughout the entire growth season (Liebisch et al., 2015).

Although genotyping has become considerably cheaper and more precise in the recent past, precision phenotyping has been a major challenge, especially for drought. The full advantage of genomic resources can only be taken when quick, accurate, and cost-effective phenotypic data including root systems are available for genetic dissection of drought tolerance and selection of drought-resilient genotypes (Tuberosa, 2012). The use of automated and high-throughput phenotyping platforms such as LeasyScan has been demonstrated in the selection of a large number of genotypes for tolerance to drought (Vadez et al., 2015). There exists a much greater need to enhance the capacity for drought tolerance breeding programmes to generate quick and accurate data using drones, near-infrared imaging, and remote sensing.

The wild relatives of the crop are plant species that are closely related to a domesticated crop and have been used to produce new lines, improving traits such as resistance to disease and pests, nutritional value, yield, and tolerance to abiotic stresses. Genetic diversity in wild species is much higher than the cultivated ones.

CRISPR/Cas9 is considered as one of the most important technologies for precise genome editing for genetic improvement in crop plants. In sorghum so far, five published reports on gene editing are available. Jiang et al. (2013) provided the first successful demonstration of gene editing in sorghum. In a report by Li et al. (2018), *k1C* genes were successfully edited through CRISPR/Cas9 to create variants with reduced *kafirin* levels and improved protein quality and digestibility in sorghum. The genomic regions enriched with heat shock protein (HSP), expansin, and aquaporin genes responsive to drought stress could be used as powerful targets for improving drought tolerance in sorghum and other cereals (Zhang et al., 2019). Brant et al. (2021) targeted the Liguleless1 (LG1) gene to edit, and the edited events would display rapidly scorable phenotype traits such as altered leaf inclination angle, ligule, and auricle size. The LG1 edited lines have the potential to increase canopy-level photosynthesis and yield by enabling increased field planting density, as has been demonstrated in maize (Tian et al., 2019).

The genetic gains of any breeding programme are significantly dependent on the number of crop breeding cycles a programme can undertake in a year. Under this current scenario, breeding a new crop cultivar takes about a decade or more, with 6 or 7 years spent in seasonal generational advancements to arrive at elite materials that go for testing and release. Now, new environmentally controlled facilities, known as "RapidGen," have been developed, which will shorten the 6–7-year window significantly. When used with the full suite of breeding acceleration techniques, RapidGen can make it possible to take four crop cycles in a year (https://www.icrisat.org/first-public-research-facilityto-put-agriculture-on-fast-forward-launched-at-icrisat/). With such new facilities in place, we are now moving toward the new era of speed breeding in crops.

A sustained increase in crop productivity requires the integration of suitable cultural practices in diverse production environments for disease-resistant and improved cultivars to achieve greater genetic gain. On-farm demonstrations of improved cultivars and production technologies have established that crop yields at farm levels can easily be increased by 20%–25% by adopting suitable agrotechniques. Integration of legumes and cereals to maintain soil fertility and microdosing of nutrients are very important to further enhance productivity gains in drought-prone regions. Machine-harvestable plant type and lodging resistance are the need of the hour to reduce cultivation costs and enhance profitability. Crops in drylands are mainly grown on soils that are inherently low in their nitrogen (N) and phosphorous (P) contents. Their adaptation to low nutrients is seldom addressed assuming that this issue can be easily addressed through the external use of fertilizers. Limited studies conducted on this aspect (Gemenet et al., 2015) have indicated the possibility of breeding nutrient-use-efficient (NUE) crop cultivars. Looking to soil degradation and water contamination due to the leaching of N in subsurface or groundwater, there is a need for a systematic study to understand the variability for NUE using core breeding materials including mini core collections available at GenBank.

2.5 CONCLUSIONS

Drought is predicted to be more intense due to climate change. In addition to the agronomic interventions to reduce the adverse impact of drought, genetic improvement of field crops is an effective mechanism to develop crop cultivars that are inherently more tolerant to drought. The purpose of this review is to assess the research efforts that have gone into understanding the adaptation to drought and breeding for drought tolerance in major dryland cereals and legumes. Choice of crops and cultivars are important interventions in drought management. Various dryland crops have different inherent capacity to tolerate drought, at species level, which can be further improved by developing cultivars with greater drought tolerance. Studies on the response of crops to drought at different growth stages have led to identification of the most sensitive stage to drought. Significant progress has been made in understanding drought-coping mechanism of dryland crops and several morpho-physiological, biochemical, and molecular mechanisms operating in crop plants have been identified that impart adaptation to drought. Early maturity has been established to be the most important mechanism to escape terminal drought stress in cereal and legume crops. The genetic resources of dryland crops that include local landraces, traditional cultivars, genetic stocks, and wild relatives make a very suitable base genetic material to contribute drought tolerance in developing cultivars suited for drylands. Characterizing the target drought environment is the most important step in breeding for enhancing drought tolerance to minimize genotype×environment interaction. New tools like precision phenotyping, molecular breeding, genomic selection, genome editing, and speed breeding would play a greater role in developing more resilient crops for drylands.

REFERENCES

Abebe, T., B. Gurja, T. Taye, and K. Gemechu. 2021. Stay-green genes contributed for drought adaptation and performance under post-flowering moisture stress on sorghum (*Sorghum bicolor* L. Moench). *J Plant Breed Crop Sci.* 13: 190–202.

Adugna, A., and A. Tirfessa. 2014. Response of stay-green quantitative trait locus QTL introgression sorghum lines to post-anthesis drought stress. *Afr J Biotechnol.* 14: 14157.

Afzal, F., B. Reddy, A. Gul, M. Khalid, A. Subhani, K. Shazadi, U.M. Quraishi, A.M. Ibrahim, and A. Rasheed. 2017. Physiological, biochemical and agronomic traits associated with drought tolerance in a synthetic-derived wheat diversity panel. *Crop Pasture Sci.* 68: 213–224.

Andrews, D.J., and K. Anand Kumar. 1996. Use of the West African pearl millet landrace *Iniadi* in cultivar development. *Plant Gen Resour Newsl.* 105: 15–22.

Aparna, K., T. Nepolean, R.K. Srivastsava, J. Kholová, V. Rajaram, S. Kumar, B. Rekha, S. Senthilvel, C.T. Hash, and V. Vadez. 2015. Quantitative trait loci associated with constitutive traits control water use in pearl millet [*Pennisetum glaucum* (L.). R. Br]. *Plant Biol (Stuttg).* 17: 1073–1084.

Appa Rao, S., M.H. Mengesha, K.L. Vyas, and C. Rajagopal Reddy. 1986. Evaluation of pearl millet germplasm from Rajasthan. *Indian J Agric Sci.* 56: 4–9.

Aquib, A., and S. Nafis. 2022. A meta-analysis of quantitative trait loci associated with stay-green in sorghum. *J Trop Crop Sci.* 9: 39.

Araus, J.L., G.A. Slafer, M.P. Reynolds, and C. Royo. 2002. Plant breeding and water stress in C3 cereals: What to breed for? *Ann Bot.* 89: 925–940.

Araus, J.L., G.A. Slafer, C. Royo, and M.D. Serret. 2008. Breeding for yield potential and stress adaptation in cereals. *Critic Rev Plant Sci.* 27: 1–36.

Bangar, P., A. Chaudhury., B. Tiwari., S. Kumar., R. Kumari, and K.V. Bhat. 2019. Morphophysiological and biochemical response of mungbean [*Vigna radiata* (L.) Wilczek] varieties at different developmental stages under drought stress. *Turk J Biol.* 43: 58–69.

Bänziger, M., G.O. Edmeades, and S. Quarrie. 1997. Drought stress at seedling stage - are there genetic solutions? In: *Developing Drought and Low-N Tolerant Maize*, Edmeades, G. O., Bänziger, M., Mickelson, H. R., and Peña-Valdivia, C. B. (eds.) CIMMYT, Mexico DF, pp. 348–354.

Banziger, M., S. Mugo, and G.O. Edmeades. 2000a. Breeding for drought tolerance in tropical maize -conventional approaches and challenges to molecular approaches. In: *Molecular Approaches for the Genetic Improvement of Cereals for Stable Production in Water Limited Environments. Proceedings of a Symposium, June 21–25, 1999*, Ribaut, J. M. and Poland, D. (eds.) CIMMYT, El Batan, Mexico, pp. 69–72.

Banziger, M., K.V. Pixley, B. Vivek, B.T. Zambezi. 2000b. Characterization of elite maize germplasm grown in eastern and southern Africa: Results of the 1999 regional trials conducted by CIMMYT and the Maize and Wheat Improvement Research Network for SADC (MWIRNET). CIMMYT, Harare, Zimbabwe, pp. 1–44. https://repository.cimmyt.org/bitstream/handle/10883/769/72610.pdf?sequence=1&isAllowed=y

Banziger, M., P.S. Setimela, D. Hodson, and B.S. Vivek. 2006. Breeding for improved abiotic stress tolerance in maize adapted to southern Africa. *Agric Water Manag.* 80: 212–224.

Barker, T.C., and G. Varughese. 1992. Combining ability and heterosis among eight complete spring hexaploid triticale lines. *Crop Sci.* 32: 340–344.

Basford, K.E., and M. Cooper. 1998. Genotype x environment interactions and some considerations of their implications for wheat breeding in Australia. *Aust J Agric Res.* 49: 153–174.

Bawazir, A.A.A., and D.B. Idle. 1989. Drought resistance and root morphology in sorghum. *Plant Soil.* 119: 217–221.

Beyene, Y., K. Semagn, S. Mugo, A. Tarekegne, R. Babu, B. Meisel, et al. 2015. Genetic gains in grain yield through genomic selection in eight biparental maize populations under drought stress. *Crop Sci.* 55: 154–163.

Bidinger, F.R., and V. Mahalakshmi. 1993. Selection for drought tolerance. In: *Cereals Programme, ICRISAT 1993. Annual Report 1992.* ICRISAT, Patancheru, India, pp. 57–59.

Bidinger, F.R, V. Mahalakshmi, B.S. Talukdar, and R.K. Sharma. 1995. Improvement of landrace cultivars of pearl millet for arid and semi-arid environments. *Annals Arid Zone.* 34: 105–110.

Bidinger, F.R, V. Mahalakshmi, and G.D.P. Rao. 1987a. Assessment of drought resistance in pearl millet [*Pennisetum amencanum* (L.) Leeke]: I. Factors affecting yield under stress. *Aust. J Agric Res.* 38: 37–48.

Bidinger, F.R., V. Mahalakshmi, and G.D.P. Rao. 1987b. Assessment of drought resistance in pearl millet [*Pennisetum americanum* (L.) Leeke]. II. Estimation of genotype response to stress. *Aust J Agric Res.* 38: 49–59.

Bidinger, F.R., T. Nepolean, C.T. Hash, R.S. Yadav, and C.J. Howarth. 2007. Identification of QTL for grain yield of pearl millet [*Pennisetum glaucum* (L.) R. Br.] in environments with variable moisture during grain filling. *Crop Sci.* 47: 969–980.

Blum, A. 2009. Effective use of water EUW and not water-use efficiency WUE is the target of crop yield improvement under drought stress. *Field Crop Res.* 112: 119–123.

Blum, A. 2011. Drought resistance and its improvement. In: *Plant Breeding for Water-Limited Environments.* Springer, New York, NY. pp. 53–152. https://doi.org/10.1007/978-1-4419-7491-4_3.

Blum, A., and C.Y. Sullivan. 1986. The comparative drought resistance of landraces of sorghum and millet from dry and humid regions. *Ann Bot.* 57: 835–846.

Bodner, G., A. Nakhforoosh, and H.P. Kaul. 2015. Management of crop water under drought: A review. *Agron Sustain Dev.* 35: 401–442.

Bolanos, J., and G.O. Edmeades. 1996. The importance of the anthesis-silking interval in breeding for drought tolerance in tropical maize. *Field Crops Res.* 48: 65–80.

Borrell, A.K., G.L. Hammer, C.L. Andrew, and A.C.L. Douglas. 2000. Does maintaining green leaf area in sorghum improve yield under drought? I. Leaf growth and senescence. *Crop Sci.* 40: 1026–1037.

Borrell, A.K., J.E. Mullet, B. George-Jaeggli, E.J. van Oosterom, G.L. Hammer, P.E. Klein, and D.R. Jordan. 2014. Drought adaptation of stay-green sorghum is associated with canopy development, leaf anatomy, root growth, and water uptake. *J Exp Bot.* 65: 6251–6263.

Brant, E.J., C.B. Mehmet, A. Parikh, and F. Altpeter. 2021. CRISPR/Cas9 mediated targeted mutagenesis of LIGULELESS-1 in sorghum provides a rapidly scorable phenotype by altering leaf inclination angle. *Biotech J.* 16: 2100237.

Bruce, W.B., G.O. Edmeades, T.C. Barker. 2002. Molecular and physiological approaches to maize improvement for drought tolerance. *J Exp Bot.* 53: 13–25.

Cairns, J.E., and B.M. Prasanna. 2018. Developing and deploying climate-resilient maize varieties in the developing world. *Curr Opin Plant Biol.* 45: 226–230.

Carberry, P.S., L.E. Cambell, and F.R. Bidinger. 1985. The growth and development of pearl millet as affected by plant population. *Field Crops Res.* 11: 193–220.

Ceccarelli, S. 1994. Specific adaptation and breeding for marginal conditions. *Euphytica.* 77: 205–219.

Ceccarelli, S., and S. Grando. 1991. Environment of selection and type of germplasm in barley breeding for low-yielding conditions. *Euphytica.* 57: 207–219.

Ceccarelli, S., S. Grando, and J. Hamblin. 1992. Relationship between barley grain yield measured in low and high yielding environments. *Euphytica.* 64: 49–58.

Cook, B.I., J.S. Mankin, and K.J. Anchukaitis. 2018. Climate change and drought: From past to future. *Curr Climate Change Rep.* 4: 164–179.

Crossa, J., G.L. Campos, P. Perez, D. Gianola, J. Burgueno, J.L. Araus, et al. 2010. Prediction of genetic values of quantitative traits in plant breeding using pedigree and molecular markers. *Genetics*. 186: 713–724.

Dahlberg, J., M. Harrison, H.D. Upadhyaya, M. Elangovan, S. Pandey, and H.S. Talwar. 2020. Global status of sorghum genetic resources conservation. In: *Sorghum in the 21st Century: Food-Fodder-Feed-Fuel for a Rapidly Changing World,* Tonapi, V. A., Talwar, H. S., Are, A. K., Bhat, V. B., Reddy, Ch. R., and Dalton, T. J. (eds.) Springer, Singapore. pp. 43–64.

Das, R.R., M.T. Vinayan, M.B. Patel, R.K. Phagna, S.B. Singh, J.P. Shahi, A. Sarma, N.S. Barua, R. Babu, K. Seetharam, J.A. Burgueño, and P.H. Zaidi. 2020. Genetic gains with rapid-cycle genomic selection for combined drought and waterlogging tolerance in tropical maize (*Zea mays* L.). *Plant Gen.* 13: 1–15.

Das, R.R., M.T. Vinayan, M.B. Patel, R.K. Phagna, S.B. Singh, J.P. Shahi, A. Sarma, N.S. Barua, R. Babu, and P.H. Zaidi. 2021. Genetic gains with genomic versus phenotypic selection for drought and waterlogging tolerance in tropical maize *Zea mays* L. *Crop J.* 9(6): 1438–1448. DOI: 10.1016/j.cj.2021.03.012.

De los Campos, G., H. Naya, D. Gianola, J. Crossa, A. Legarra, E. Manfredi, K. Weigel, and M., Cotes. 2009. Predicting quantitative traits with regression models for dense molecular markers and pedigree. *Genetics*. 182: 375–385.

Dwivedi, S., H. Upadhayaya, S. Senthivel, C. Hash, K. Fukunga, X. Diao, D. Santra, and D. Baltensperger. 2012. Millets genetic and genomic resources. In: *Plant Breeding Reviews*, vol 25, 1st ed. Janic, J. (ed.) Wiley-Blackwell, Hoboken, New Jersey, pp. 247–375. ISBN 9781118100509John Wiley & Sons Inc.

Dwivedi, S.L., S. Ceccarelli, M.W., Blair, H.D. Upadhyaya, A.K. Are, and R. Ortiz. 2016. Landrace germplasm for improving yield and abiotic stress adaptation. *Trends Plant Sci.* 21(1): 31–42.

Edmeades, G.O., W.L. Trevision, B.M. Prasanna, and H. Campos. 2017. Tropical maize (*Zea mays* L.). In: *Genetic Improvement of Tropical Crops*, H. Campos and P, D. S. Caligari (eds), Springer Cham. pp. 57–109. DOI: 10.1007/978-3-319-59819-2_3.

Elangovan, M., V.A. Tonapi, and D.C.R. Reddy. 2009. Collection and characterization of Indian sorghum landraces. *Indian J Plant Genet Resour.* 22: 173–181.

Fadoul, H.E., M.A.E. Siddig, A.W.H. Abdalla, and A.A.E. Hussein. 2018. Physiological and proteomic analysis of two contrasting Sorghum bicolor genotypes in response to drought stress. *Aust J Crop Sci.* 12: 1543–1551.

Fakrudin, B., S.P. Kavil, Y. Girma, S.S. Arun, D. Dadakhalandar, B.H. Gurusiddesh, A.M. Patil, M. Thudi, S.B. Bhairappanavar, and Y.D. Narayana. 2013. Molecular mapping of genomic regions harbouring QTLs for root and yield traits in sorghum (*Sorghum bicolor* L. Moench). *Physiol Mol Biol Plants*. 19: 409–419.

Falconer, D.S., and T.F.C. MacKay. 1996. *Introduction to Quantitative Genetics*, 4th ed. Prentice Hall, London.

Fischer, K.S., E.C. Jonson, and G.O. Edmeades. 1983. *Breeding and Selection for Drought Resistance in Tropical Maize*. CIMMYT, Mexico DF.

Fischer, R.A., D. Byerlee, and G.O. Edmeades. 2014. *Crop Yields and Global Food Security: Will Yield Increase Continue to Feed the World? ACIAR Monograph No. 158*. Australian Centre for International Agricultural Research, Canberra, 634 pp.

Fussell, L.K., F.R. Bidinger, and P. Bieler. 1991. Crop physiology and breeding for drought tolerance. Research and development. *Field Crops Res.* 27: 183–199.

Gaur, P.M., S. Samineni, S. Tripathi, R.K. Varshney, and C.L.L. Gowda. 2015. Allelic relationships of flowering time genes in chickpea. *Euphytica*. 203: 295–308.

GENESYS-PGR. 2019. https://www.genesys-pgr.org/explore/overview. Surveyed on 6.5.2019.

Gimode, D., D.A. Odeny, E.P. de Villiers, et al. 2016. Identification of SNP and SSR markers in finger millet using next generation sequencing technologies. *PLoS One*. 11: e0159437.

Gowsiga, S., M. Djanaguiraman, N. Thavaprakaash, P. Jeyakumar, and M. Govindaraj. 2021. Sorghum drought tolerance is associated with deeper root system and decreased transpiration rate. *Preprints.* 2021110144 https://doi.org/10.20944/preprints202111.0144.v1

Grant, R.F., B.S. Jackson, J.R. Kiniry, and G.F. Arkin. 1989. Water deficit timing effects on yield components in maize. *Agron J.* 81: 61–65.

Gregory, P.J. 1983. Response to temperature in a stand of pearl millet (*Pennisetum typhoides* S. & H.): III. Root development. *Exp Bot.* 34: 744–756.

Gupta, A., A. Rico-Medina, and A.I. Cano-Delgado. 2020. The physiology of plant responses to drought. *Science*. 368: 266–269.

Gupta, S.K., A. Rathore, O.P. Yadav, K.N. Rai, I.S. Khairwal, B.S. Rajpurohit, and R.R. Das. 2013. Identifying mega-environments and essential test locations for pearl millet cultivar selection in India. Crop Sci. 53: 2444–2453.

Hadi, P., S. Armin, H.P. Moucheshi, and P. Mohammad. 2016. Stomatal responses to drought stress. In: *Water Stress and Crop Plants: A Sustainable Approach*, vol 1, 1st ed. Ahmad, P. (ed.) John Wiley & Sons, Ltd. DOI:10.1002/9781119054450

Harris, K., P.K. Subudhi, A. Borrell, D. Jordan, D. Rosenow, H. Nguyen, P. Klein, R. Klein, and J. Mullet. 2007. Sorghum stay-green QTL individually reduce post-flowering drought-induced leaf senescence. *J Exp Bot.* 58: 327–338.

Harris-Shultz, R.K., C.M. Hayes, and J.E. Knoll. 2019. Mapping QTLs and identification of genes associated with drought resistance in sorghum. *Methods Mol Biol.* 1931: 11–40.

Henson, I.E. and V. Mahalakshmi. 1985. Evidence for panicle control of stomatal behaviour in water stressed plants of pearl millet. *Field Crops Res.* 11: 281–290.

Hu, Z., B. Mbacké, R. Perumal, et al. 2015. Population genomics of pearl millet (*Pennisetum glaucum* (L.) R. Br.): Comparative analysis of global accessions and Senegalese landraces. *BMC Genom.* 16: 1048.

Isaac, K.G., P.G. Allan, C.T. Hash, F.R. Bidinger, and C.J. Howarth. 2019. A comparative assessment of the performance of a stay-green sorghum (*Sorghum bicolor* (L) Moench) introgression line developed by marker-assisted selection and its parental lines. *African J Biotech.* 18: 548–563.

Jangra, S., A. Rani, R.C. Yadav, N.R. Yadav, and D. Yadav. 2019. Introgression of terminal drought stress tolerance in advance lines of popular pearl millet hybrid through molecular breeding. *Plant Physiol Rep.* 24: 359–369.

Jha, U.C., Bohra, A., and Nayyar, H. 2020. Advances in "omics" approaches to tackle drought stress in grain legumes. *Plant Breed.* 139: 1–27.

Jia, H., M. Li, W. Li, L. Liu, Y. Jian, Z. Yang, X. Shen, Q. Ning, Y. Du, R. Zhao, D. Jackson, X. Yang, and Z. Zhang. 2020. A serine/threonine protein kinase encoding gene kernel number per row regulates maize grain yield. *Nat Commun.* 11(1): 988.

Jiang, W., H. Zhou, H. Bi, M. Fromm, B. Yang, and D.P. Weeks. 2013. Demonstration of CRISPR/Cas9/sgRNA-mediated targeted gene modification in Arabidopsis, tobacco, sorghum and rice. *Nucleic Acids Res.* 41(20): e188–e188.

Johnson, S.M., I. Cummins, F.L. Lim, A.R. Slabas, and M.R. Knight. 2015. Transcriptomic analysis comparing stay-green and senescent Sorghum bicolor lines identifies a role for proline biosynthesis in the stay-green trait. *J Exp Bot.* 66: 7061–7073.

Jordan, D.R., C.H. Hunt, A.W. Cruickshank, A.K. Borrell, and R.G. Henzell. 2012. The relationship between the stay-green trait and grain yield in elite sorghum hybrids grown in a range of environments. *Crop Sci.* 52: 1153–1161.

Kadioglu, A., R. Terzi, N. Saruhan, and A. Saglam. 2012. Current advances in the investigation of leaf rolling caused by biotic and abiotic stress factors. *Plant Sci.* 182: 42–48.

Kapanigowda, M.H., W.A. Payne, W.L. Rooney, J.E. Mullet, and M. Balota. 2014. Quantitative trait locus mapping of the transpiration ratio related to preflowering drought tolerance in sorghum (*Sorghum bicolor*). *Funct Plant Biol.* 41: 1049–1065.

Kapanigowda, M.H., R. Perumal, M. Djanaguiraman, R.M. Aiken, T. Tesso, P.V.V. Prasad, and C.R. Little. 2013. Genotypic variation in sorghum [*Sorghum bicolor* L. Moench] exotic germplasm collections for drought and disease tolerance. *Springerplus.* 2(1): 1–13 DOI: 10.1186/2193-1801-2-650.

Kashiwagi, J., L. Krishnamurthy, P.M. Gaur, S. Chandra, and H.D. Upadhyaya. 2008. Estimation of gene effects of the drought avoidance root characteristics in chickpea (*C. arietinum* L.). *Field Crops Res.* 105: 64–69.

Kavar, T., M. Maras, M. Kidrič, J. Šuštar-Vozlič, and V. Meglič. 2008. Identification of genes involved in the response of leaves of *Phaseolus vulgaris* to drought stress. *Mol Breed.* 21: 159–172.

Kebede, H., P.K. Subudhi, D.T. Rosenow, and H.T. Nguyen 2001. Quantitative trait loci influencing drought tolerance in grain sorghum (*Sorghum bicolor* L. Moench). *Theor Appl Genet.* 103: 266–276.

Khatun, M., S. Sarkar, F.M. Era, A.M. Islam, M.P. Anwar, S. Fahad, R. Datta, and A.K.M. Islam. 2021. Drought stress in grain legumes: Effects, tolerance mechanisms and management. *Agronomy.* 11: 2374.

Kholová, J., and V. Vadez. 2013. Water extraction under terminal drought explains the genotypic differences in yield, not the anti-oxidant changes in leaves of pearl millet (*Pennisetum glaucum*). *Funct Plant Biol.* 40: 44–53.

Kholová, J., T. Nepolean, C.T. Hash, A. Supriya, V. Rajaram, S. Senthilvel, A. Kakkera, R. Yadav, and V. Vadez. 2012. Water saving traits co-map with a major terminal drought tolerance quantitative trait locus in pearl millet [*Pennisetum glaucum* (L.) R. Br.]. *Mol Breed.* 30: 1337–1353.

Kiran, B.O., S.S. Karabhantanal, S.B. Patil, V.H. Ashwathama, R.B. Jolli, G.M. Sajjnar, and V.A. Tonapi. 2022. Phenotyping sorghum for drought-adaptive physiological and root architectural traits under water-limited environments. *Cereal Res Commun.* 50: 885–893. DOI: 10.1007/s42976-021-00228-z.

Kiranmayee, K.N.S., Usha, C.T. Hash, S. Sivasubramani, P. Ramu, B.P. Amindala, A. Rathore, P.B.K. Kishor, R. Gupta, and S.P. Deshpande. 2020. Fine-mapping of sorghum stay-green QTL on chromosome 10 revealed genes associated with delayed senescence. *Genes.* 11: 1026.

Kumar, A., D. Sharma, A. Tiwari, J.P. Jaiswal, N.K. Singh, and S. Sood. 2016. Genotyping-by-sequencing analysis for determining population structure of finger millet germplasm of diverse origins. *Plant Genome*. 9(10): 3835.

Kumar, R.R., K. Karajol, and G.R. Naik. 2011. Effect of polyethylene glycol induced water stress on physiological and biochemical responses in pigeonpea (*Cajanus cajan* L. Millsp.). *Recent Res Sci Tech*. 3(1): 148–152.

Kumawat, G., R.S. Raje, S. Bhutani, J.K. Pal, A.S. Mithra, K. Gaikwad, and N.K. Singh. 2012. Molecular mapping of QTLs for plant type and earliness traits in pigeonpea (*Cajanus cajan* L. Millsp.). *BMC Genet*. 13(1). 1–11.

Lafitte, R.H., A. Blum, and G. Atlin. 2003. Breeding rice for drought-prone environments. In: *Breeding Rice for Drought-Prone Environments*, Fisher, K.S., Lafitte, R.H., Fukai, S., Atlin, G, and Hardy, B. (eds.) IRRI, Los Baños, The Philippines, pp. 14–22.

Lata, C., S. Bhutty, R.P. Bahadur, M. Majee, and M. Prasad. 2011. Association of an SNP in a novel DREB2-like gene SiDREB2 with stress tolerance in foxtail millet [*Setaria italica* (L.)]. *J Exp Bot*. 62: 3387–3401.

Lata, C., and M. Prasad. 2013. Validation of an allele-specific marker associated with dehydration stress tolerance in a core set of foxtail millet accessions. *Plant Breed*. 132: 496–499.

Levitt, J. 1980. Responses of plants to environmental stresses. In: *Water, Radiation, Salt, and Other Stresses*, vol II. Academic Press, London.

Li, A., S. Jia, A. Yobi, Z. Ge, S.J. Sato, C. Zhang, R. Angelovici, T.E. Clemente, and D.R. Holding. 2018. Editing of an alpha-kafirin gene family increases, digestibility and protein quality in sorghum. *Plant Physiol*. 177: 1425–1438.

Li, R., Y. Han, P. Lv, R. Du, and G. Liu. 2014. Molecular mapping of the brace root traits in sorghum (*Sorghum bicolor* L. Moench). *Breed Sci*. 64(2): 193–198.

Liebisch, F., N. Kirchgessner, D. Schneider, A. Walter, and A. Hund. 2015. Remote, aerial phenotyping of maize traits with a mobile multi-sensor approach. *Plant Methods*. 11: 1–20.

Liu, Y., C. Subhash, J. Yan, C. Song, J. Zhao, and J. Li. 2011. Maize leaf temperature responses to drought: Thermal imaging and quantitative trait loci (QTL) mapping. *Environ Exp Bot*. 71: 158–165. DOI: 10.1016/j.envexpbot.2010.11.010.

Liu, C., Wu, J., Wang, L., Fan, B., Cao, Z., Su, Q., and S. Wang. 2017. Quantitative trait locus mapping under irrigated and drought treatments based on a novel genetic linkage map in mungbean (*Vigna radiata* L.). *Theor Appl Genet*. 130: 2375–2393.

Liu, S., and F. Qin. 2021. Genetic dissection of maize drought tolerance for trait improvement. *Mol Breed*. 41: 8.

Lopez, J.R., J.E. Erickson, P. Munoz, A. Saballos, T.J. Felderhoff, and W. Vermerris. 2017. QTLs associated with crown root angle, stomatal conductance, and maturity in Sorghum. *Plant Genome*. 10(2). doi: 10.3835/plantgenome2016.04.0038. PMID: 28724080.

Lorenz, A.J. 2013. Resource allocation for maximizing prediction accuracy and genetic gain of genomic selection in plant breeding: A simulation experiment. *G3: Genes|Genomes|Genetics*. 3: 481–491.

Lu, Z., and P.M. Neumann. 1998. Water-stressed maize, barley and rice seedlings show species diversity in mechanisms of leaf growth inhibition. *J Exp Bot*. 49: 945–1952.

Ludlow, M.M. 1989. A strategies of response to water stress. In: Kreeeb, K. H., Richter, H., and Hinckley, T. M., (eds.) *Structural and Functional Responses to Environmental Stresses*, SPB Academic Publishing, The Hague.

Ludlow, M.M., and R.C. Muchow. 1990. A critical evaluation of traits for improving crop yields in water-limited environments. *Adv Agron*. 43: 107–153.

Mace, E.S., V. Singh, E.J. Van Oosterom, G.L. Hammer, C.H. Hunt, and D.R. Jordan. 2012. QTL for nodal root angle in sorghum (*Sorghum bicolor* L. Moench) co-locate with QTL for traits associated with drought adaptation. *Theor Appl Genet*. 124: 97–109.

Mafakheri, A., A.F. Siosemardeh, B. Bahramnejad, P.C. Struik, and Y. Sohrabi. 2010. Effect of drought stress on yield, proline and chlorophyll contents in three chickpea cultivars. *Austr J Crop Sci*. 4(8): 580–585.

Mahalakshmi, V., F.R. Bidinger, and D.S. Raju. 1987. Effect of timing of water deficit on pearl millet (*Pennisetum americanum*). *Field Crops Res*. 15: 327–339.

Manjarrez-Sandoval, P., V.A. González-Hernández, L.E. Mendoza-Onofre, and E.M. Engleman. 1989. Drought stress effects on the grain yield and panicle development of sorghum. *Can J Plant Sci*. 69: 631–641.

McCully, M.E. 1999. Root in soil: Unearthing the complexity of root and their rhizosphere. *Ann Rev Plant Physiol Plant Mol Biol*. 50: 695–718.

Medina, S., S.K. Gupta, and V. Vadez. 2017. Transpiration response and growth in pearl millet parental lines and hybrids bred for contrasting rainfall environments. *Front Plant Sci*. 8, 1846.

Messmer, R., Y. Fracheboud, M. Banziger, M. Vargas, P. Stamp, and J.M. Ribaut 2009. Drought stress and tropical maize: QTL by- environment interactions and stability of QTLs across environments for yield components and secondary traits. *Theor Appl Genet.* 119: 913–930.

Meuwissen, T.H.E., B.J. Hayes, and M.E. Goddard. 2001. Prediction of total genetic value using genome-wide dense marker maps. *Genetics.* 157: 1819–1829.

Monneveux, P., C. Sanchez, A. Tiessen. 2008. Future progress in drought tolerance in maize needs new secondary traits and cross combinations. *J Agric Sci.* 146(3): 287–300.

Nair, R.M., R. Schafleitner, L. Kenyon, R. Srinivasan, W. Easdown, A.W. Ebert, and P. Hanson 2012. Genetic improvement of mungbean. *SABRAO J Breed Genet.* 44: 177–190.

Ndlovu, E., J. van Staden, M. Maphosa. 2021. Morpho-physiological effects of moisture, heat and combined stresses on *Sorghum bicolor* [Moench (L.)] and its acclimation mechanisms. *Plant Stress.* 2, 2021, 100018, ISSN 2667-064X, https://doi.org/10.1016/j.stress.2021.100018. DOI: 10.1016/j.stress.2021.100018.

Nemesker, E., K. Molnar, R. Vigh, J. Nagy, and A. Dobos. 2015. Relationships between stomatal behaviour, spectral traits and water use and productivity of green peas (*Pisum sativum* L.) in dry seasons. *Acta Physiol Plant.* 37: 1–16.

Ngugi, K., W. Kimani, D. Kiambi, and E.W. Mutitu. 2013. Improving drought tolerance in *Sorghum bicolor* L. Moench: Marker-assisted transfer of the stay-green quantitative trait loci (QTL) from a characterized donor source into a local farmer variety. *Int J Sci Res Knowl.* 1(6): 154–162.

Paterson, A.H., J.E. Bowers, R. Bruggmann, I. Dubchak, J. Grimwood, H. Gundlach, G. Haberer, U. Hellsten, T. Mitros, and A. Poliakov. 2009. The Sorghum bicolor genome and the diversification of grasses. *Nature.* 457(7229): 551–556.

Podlich, D.W., M. Cooper, and K.E. Basford. 1999. Computer simulation of a selection strategy to accommodate genotype x environment interactions in a wheat recurrent selection programmeme. *Plant Breed.* 118: 17–28.

Prasanna, B.M. et al. 2021. Beat the stress: Breeding for climate resilience in maize for the tropical rainfed environments. *Theor Appl Genet.* 134, 1729–1752.

Presterl, T., and E. Weltzien. 2003. Exploiting heterosis in pearl millet population breeding in arid environments. *Crop Sci.* 43: 767–776.

Qi, X., S. Xie, Y. Liu, F. Yi, and J. Yu. 2013. Genome-wide annotation of genes and noncoding RNAs of foxtail millet in response to simulated drought stress by deep sequencing. *Plant Mol Biol.* 83: 459–473.

Rai, K.N., K. Anand Kumar, D.J. Andrews, S.C. Gupta, and B. Ouendeba. 1997. Breeding pearl millet for grain yield and stability. In: Proceedings of International Conference on Genetic Improvement of Sorghum and *Pearl Millet, Lubbock, Texas, 23–27 September 1996.* INTSORMIL, Lincoln, NE. pp. 71–83.

Rama Rao, C.A., B.M.K. Raju, A.V.M.S. Rao, D.Y. Reddy, Y.L. Meghana, N. Swapna et al. 2019. Yield variability of sorghum and pearl millet to climate change in India. *Indian J Agril Econ.* 74: 350–362.

Ramamoorthy, P., K. Lakshmanan, H.D. Upadhyaya, V. Vadez, and R.K. Varshney. 2017. Root traits confer grain yield advantages under terminal drought in chickpea *Cicer arietinum* L. *Field Crops Res.* 201: 146–161.

Rattunde, H.F., P. Singh, and J.R. Witcombe. 1989. Feasibility of mass selection in pearl millet. *Crop Sci.* 29: 1423–1427.

Rauf, S., J.M. Al-Khayri, M. Zaharieva, P. Monneveux, and F. Khalil. 2015. Breeding strategies to enhance drought tolerance in crops. In: *Advances in Plant Breeding Strategies; Agronomic, Abiotic and Biotic Stress Traits,* Al-Khayri, J.M., et al. (eds.) Springer, Berlin, pp. 1–70.

Reddy, B.M., A.A., Johnson, N.J. Kumar, B. Venkatesh, N. Jayamma, M. Pandurangaiah and C. Sudhakar. 2022. De novo transcriptome analysis of drought-adapted cluster bean cultivar RGC-1025 reveals the wax regulatory genes involved in drought resistance. *Front Plant Sci.* 13:868142. doi: 10.3389/fpls.2022.868142.

Ribaut, J.M, C. Jiang, D. Gonzalez-de-Leon., G.O. Edmeades, and D.A. Hoisington. 1997. Identification of quantitative trait loci under drought conditions in tropical maize. 2. Yield components and marker-assisted selection strategies. *Theor Appl Genet.* 94: 887–896.

Ribaut, J.M., D.A. Hoisington, J.A. Deutsch, C. Jiang, and D. Gonzalez De Leon. 1996. Identification of quantitative trait loci under drought conditions in tropical maize. 1. Flowering parameters and the anthesis-silking interval. *Theor Appl Genet.* 92: 905–914.

Rockstrom, J., L. Karlberg, S.P. Wani, J. Barron, N. Hatibu, T. Oweis, et al. 2010. Managing water in rainfed agriculture-The need for a paradigm shift. *Agric Water Manag.* 97: 543–550.

Rosielle, A.A. and J. Hamblin. 1981. Theoretical aspect of selection for yield in stress and non-stress environments. *Crop Sci.* 21: 943–948.

Routley, R., I.J. Broad, G. McLean, J. Whish, and G. Hammer. 2003. The effect of row configuration on yield reliability in grain sorghum: I. Yield, water use efficiency and soil water extraction. In: *Solutions for a Better Environment: Proceedings of the 11th Australian Agronomy Conference, 2–6 February 2003, Horsham, Victoria, Australia, Australian Society of Agronomy* http://www.regional.org.au/au/asa/2003.

Semagn, K., Y. Beyene, M. Warburton, et al. 2013. Meta-analyses of QTL for grain yield and anthesis silking interval in 18 maize populations evaluated under water-stressed and well-watered environments. *BMC Genom.* 14: 313. DOI: 10.1186/1471-2164-14-313.

Serba, D.D., Yadav, R.S., Varshney, R.K., Gupta, S.K., Mahalingam, G., Srivastava, R.K., et al. 2020. Genomic designing of pearl millet: A resilient crop for arid and semi-arid environments. In: *Genomic Designing of Climate-Smart Cereal Crops*, Kole, C. (ed.) Springer, Cham, pp. 221–286. DOI: 10.1007/978-3-319-93381-8_6.

Serraj, R., C.T. Hash, S.M.H. Rizvi, A. Sharma, R.S. Yadavand, and F.R. Bidinger. 2005. Recent advances in marker-assisted selection for drought tolerance in pearl millet. *Plant Prod Sci.* 8(3): 334–337.

Shaw, R.H. 1977. Water use and requirements of maize – A review. In: *Agro-Meteorology of the Maize (Corn) Crop*. World Meteorological Organization No. 481, pp. 198–208.

Shiferaw, B., K. Tesfaye, M. Berresaw, T. Abate, B.M. Prasanna, and A. Menkir 2014. Managing vulnerability to drought and enhancing livelihood resilience in Sub-Saharan Africa: Technological, institutional and policy options. *Weather Clim Extremes.* 3: 67–79.

Simmonds, N.W. 1991. Selection for local adaptation in plant breeding programme. *Theor Appl Genet.* 82: 363–367.

Singh, R., A. Dhaka, M. Muthamilarasan, and M. Prasad. 2022. Nutritional improvement of cereal crops to combat hidden hunger during COVID-19 pandemic: Progress and prospects. *Adv Food Secur Sustain.* 7: 61–82. DOI: 10.1016/bs.af2s.2022.02.001.

Sinha, P., V.K. Singh, R.K. Saxena, A.W. Khan, R. Abbai, A. Chitikineni, and R.K. Varshney 2020. Superior haplotypes for haplotype-based breeding for drought tolerance in pigeonpea (*Cajanus cajan* L.). *Plant Biotech J.* 18(12): 2482–2490.

Srivastava, R.K., V. Vadez, J. Kholova, V.B.R. Lachagari, M.D. Mahendrakar, P. Katiya, S.P. Lekkala, M. Praveen, R. Gupta, and E. Blumwald. 2017. Fine mapping of the linkage group 2 drought tolerance QTL in pearl millet [*Pennisetum glaucum* (L.) R. Br.]. In: *InterDrought-V, February 21–25, 2017, Hyderabad, India*.

Sultan, B., P. Roudier, P. Quirion, A. Alhassane, B. Muller, M. Dingkuhn, et al. 2013. Assessing climate change impacts on sorghum and millet yields in the Sudanian and Sahelian savannas of West Africa. *Environ Res Lett.* 8: 014040. DOI: 10.1088/1748-9326/8/1/014040.

Tian, J., C. Wang, J. Xia, L. Wu, G. Xu, W. Wu, D. Li, W. Qin, X. Han, and Q. Chen. 2019. Teosinte ligule allele narrows plant architecture and enhances high-density maize yields. *Science.* 365(6454): 658–664.

Tuberosa, R., S. Salvi, M.C. Sanguineti, P. Landi, M. Maccaferri, and S. Conti. 2002. Mapping QTLs regulating morpho-physiological traits and yield: Case studies, shortcomings and perspectives in drought-stressed maize. *Ann Bot.* 89 Spec No(7): 941–963.

Turner, N.C., G.C. Wright, and K.H.M. Siddique. 2001. Adaptation of grain legumes (pulses) to water-limited environments. *Adv Agron.* 71: 193–231.

Tuteja, N. and S.S. Gill. 2013. *Crop Improvement under Adverse Conditions.* Springer New York Heidelberg Dordrecht London. ISBN-13: 978-1461446323.

Tyagi, V., M. Nagargade, and R.K. Singh. 2020. Agronomic interventions for drought management in crops. In: *New Frontiers in Stress Management for Durable Agriculture*, Rakshit, A., Singh, H. S., Singh, A. K., Singh, U. S., and Fraceto, L. (eds.) Springer Nature Singapore Pte Ltd., Springer Singapore.

Ulemale, C.S., S.N. Mate, and D.V. Deshmukh. 2013. Physiological indices for drought tolerance in chickpea (*Cicer arietinum* L.). *World J Agric Sci.* 9(2): 123–131.

Upadhyaya, H.D., M. Vetriventhan, S.L. Dwivedi, S.K. Pattanashetti, and S.K. Singh. 2016. Porso, barnyard, little and kodo millets. In: *Genetic and Genomic Resources for Grain Cereals Improvement*. Singh, M. and Upadhyaya, H.D. (eds.) Elsevier Inc. DOI: 10.1016/B978-0-12-802000-5.00008-3.

Upadhyaya, H.D., K.N. Reddy, M. Vetriventhan, M.I. Ahmed, G.M. Krishna, M.T. Reddy, S.K. Singh. 2017. Sorghum germplasm from West and Central Africa maintained in the ICRISAT genebank: Status, gaps, and diversity. *Crop J.* 5: 518–532.

Upadhyaya, H.D., S. Sharma, S.L. Dwivedi, and S.K. Singh. 2014. Sorghum genetic resources: Conservation and diversity assessment for enhanced utilization in sorghum improvement. In: *Genetics, Genomics and Breeding of Sorghum. Series on Genetics, Genomics and Breeding of Crop Plants*, Y. Wang, H. D. Upadhyaya, and C. Kole (eds). CRC Press (Taylor & Francis), Boca Raton, pp. 28–55. ISBN 9781482210088.

Vadez, V., S.P. Deshpande, J. Kholova, G.L. Hammer, A.K. Borrell, H.S. Talwar, and C.T. Hash. 2011a. Stay-green quantitative trait loci's effects on water extraction, transpiration efficiency and seed yield depend on recipient parent background. *Funct Plant Biol*. 38(7): 553–566.

Vadez, V., T. Hash, and J. Kholova. 2012. II. 1.5 Phenotyping pearl millet for adaptation to drought. *Front Physiol*. 3: 386.

Vadez, V., J. Kholová, R.S. Yadav, and C.T. Hash. 2013. Small temporal differences in water uptake among varieties of pearl millet (*Pennisetum glaucum* (L.) R Br) are critical for grain yield under terminal drought. *Plant Soil*. 371: 447–462.

Vadez, V., L. Krishnamurthy, C.T. Hash, H.D. Upadhyaya, and A.K. Borrell. 2011b. Yield, transpiration efficiency, and water-use variations and their interrelationships in the sorghum reference collection. *Crop Past Sci*. 62: 645–655.

van Oosterom, E.J., F.R. Bidinger, and E.R. Weltzien. 2003. A yield architecture framework to explain adaptation of pearl millet to environmental stress. *Field Crops Res*. 80, 33–56.

van Oosterom, E.J., V. Mahalakshmi, G.K. Arya, H.R. Dave, B.D. Gothwal, A.K. Joshi, P. Joshi, R.L. Kapoor, P. Sagar, M.B.L. Saxena, D.L. Singhania, and K.L. Vyas. 1995. Effect of yield potential, drought escape and drought tolerance on yield of pearl millet (*Pennisetum glaucum*) in different stress environments. *Indian J Agric Sci*. 65: 629–635.

van Oosterom, E.J., E. Weltzien, O.P. Yadav, and F.R. Bidinger. 2006. Grain yield components of pearl millet under optimum conditions can be used to identify germplasm with adaptation to arid zones. *Field Crops Res*. 96: 407–421.

Varshney, R.K., M. Thudi, S.N. Nayak, P.M. Gaur, J. Kashiwagi, L. Krishnamurthy, ..., K.P. Viswanatha. 2014. Genetic dissection of drought tolerance in chickpea (*Cicer arietinum* L.). *Theor Appl Genet*. 127(2): 445–462.

Venkateswaran, K., M. Muraya, S.L. Dwivedi, and H.D. Upadhyaya. 2014. Wild sorghums-their potential use in crop improvement. In: Y. Wang, H. D. Upadhyaya, and C. Kole (eds), *Genetics, Genomics and Breeding of Sorghum. Series on Genetics, Genomics and Breeding of Crop Plants*. CRC Press (Taylor & Francis), Boca Raton, pp. 56–89. ISBN 9781482210088.

Vetriventhan, M., H.D. Upadhyaya, S.L. Dwivedi, S.K. Pattanashetti, and S.K. Singh 2016. Finger and foxtail millets. In: Singh, M. and Upadhyaya, H.D. (eds.) *Genetic and Genomic Resources for Grain Cereals Improvement*. Elsevier Inc. DOI: 10.1016/B978-0-12-802000-5.00007-1.

Virk, D.S. and B.K. Mangat. 1991. Detection of cross over genotype×environment interactions in pearl millet. *Euphytica*. 52: 193–199.

Vivek, B.S., G.K. Krishna, V. Vengadessan, et al. 2017. Use of genomic estimated breeding values results in rapid genetic gains for drought tolerance in maize. *Plant Genome*. DOI: 10.3835/plantgenome2016.07.0070. PMID: 28464061

Winkel, T., J.F. Renno, and W.A. Payne. 1997. Effect of the timing of water deficit on growth, phenology and yield of pearl millet (*Pennisetum glaucum* (L) R Br) grown in Sahelian conditions. *J Exp Bot*. 48: 1001–1009.

Xin, Z., R. Aiken, and J. Burke. 2009. Genetic diversity of transpiration efficiency in sorghum. *Field Crops Res*. 111: 74–80.

Xin, Z., C. Franks, P. Payton, and J.J. Burke. 2008. A simple method to determine transpiration efficiency in sorghum. *Field Crops Res*. 107: 180–183.

Xu, Y., D.J. Skinner, H. Wu, N. Palacios-Rojas, J.L. Araus, J. Yan, S. Gao, M.L. Warburton, and J.H. Crouch. 2009. Advances in maize genomics and their value for enhancing genetic gains from breeding. *Int J Plant Genom*. 957602. DOI: 10.1155/2009/957602. PMID: 19688107

Xu, W., P.K. Subudhi, O.R. Crasta, D.T. Rosenow, J.E. Mullet, and H.T. Nguyen. 2000. Molecular mapping of QTLs conferring stay-green in grain sorghum (*Sorghum bicolor* L. Moench). *Genome*. 43: 461–469.

Xue, Y., M.L. Warburton, M. Sawkins, X. Zhang, T. Setter, Y. Xu, P. Grudloyma, J. Gethi, J.M. Ribaut, W. Li, Y. Zheng, and J Yan. 2013. Genome-wide association analysis for nine agronomic traits in maize under well-watered and water-stressed conditions. *Theor Appl Genet*. 126: 2587–2596.

Yadav, O.P. 1994. Inheritance of threshing percentage in pearl millet. *Euphytica*. 78, 77–80.

Yadav, O.P. 2004. CZP 9802- A new drought-tolerant cultivar of pearl millet. *Indian Farm*. 54, 15–17.

Yadav, O.P. 2010. Drought response of pearl millet landrace-based populations and their crosses with elite composites. *Field Crops Res*. 118, 51–57.

Yadav, O.P. 2011. Breeding dual-purpose pearl millet (*Pennisetum glaucum*) for northwestern India: Understanding association of biomass and phenotypic traits. *Indian J Agric Sci*. 81, 816–820.

Yadav, O.P. 2014. Developing drought-resilient crops for improving productivity of drought-prone ecologies. *Indian J Genet Plant Breed*. 74, 548–552.

Yadav, O.P. and S.K. Bhatnagar. 2001. Evaluation of indices for identification of pearl millet cultivars adapted to stress and non-stress conditions. *Field Crops Res.* 70: 201–208.

Yadav, O.P. and F.R. Bidinger. 2007. Utilization, diversification and improvement of landraces for enhancing pearl millet productivity in arid environments. *Ann Arid Zone.* 46: 49–57.

Yadav, O.P., F.R. Bidinger, and D.V. Singh. 2009. Utility of pearl millet landraces in breeding dual-purpose hybrids for arid zone environments of India. *Euphytica.* 166: 239–247.

Yadav, O.P., and K.N. Rai. 2011. Hybridization of Indian landraces and African elite composites of pearl millet results in biomass and stover yield improvement under arid zone conditions. *Crop Sci.* 51, 1980–1987.

Yadav, O.P., K.N. Rai, and S.K. Gupta. 2012. Pearl Millet: Genetic improvement for tolerance to abiotic stresses. In: *Improving Crop Resistance to Abiotic Stress.* Tuteja, N., Gil, S. S., and Tuteja, R. (eds.) Wlley·VCH Verlag GmbH & Co. KGaA. Springer New York Heidelberg Dordrecht London. pp. 261–288.

Yadav, O.P., B.S. Rajpurohit, G.R. Kherwa, and A. Kumar. 2012. Prospects of enhancing pearl millet (*Pennisetum glaucum*) productivity under drought environments of north-western India through hybrids. *Indian J Genet Plant Breed.* 72: 25–30.

Yadav, O.P., J.P. Singh, R.K. Kakani, H.R. Mahla, M.P. Rajora, A. Singh, P.R. Meghwal, and A. Verma. 2020. Managing agrobiodiversity of Indian drylands for climate-change adaptation. *Indian J Plant Genetic Resour.* 33: 3–16.

Yadav, O.P., H.D. Upadhyaya, K.N. Reddy, A.K. Jukanti, S. Pandey, and R.K. Tyagi. 2017. Genetic resources of pearl millet: Status and utilization. *Indian J Plant Genetic Resour.* 30: 31–47.

Yadav, O.P. and Weltzien, R.E. 2000. Differential response of pearl millet landrace-based populations and high yielding varieties in contrasting environments. *Annals Arid Zone.* 39: 39–45.

Yadav, O.P., E. Weltzien-Rattunde, F.R. Bidinger, and V. Mahalakshmi. 2000. Heterosis in landrace-based top-cross hybrids of pearl millet across arid environments. *Euphytica.* 112: 285–295.

Yadav, R.S., C.T. Hash, F.R. Bidinger, G.P. Cavan, and C.J. Howarth. 2002. Quantitative trait loci associated with traits determining grain and stover yield in pearl millet under terminal drought stress conditions. *Theor Appl Genet.* 104: 67–83.

Yadav, R.S., C.T. Hash, F.R. Bidinger, K.M. Devos, and C.J. Howarth. 2004. Genomic regions associated with grain yield and aspects of post-flowering drought tolerance in pearl millet across stress environments and tester background. *Euphytica.* 136: 265–277.

Yadav, R.S., D. Sehgal, and V Vadez. 2011. Using genetic mapping and genomics approaches in understanding and improving drought tolerance in pearl millet. *J Exp Bot.* 62: 397–408.

Zaidi, P.H., K. Seetharam, G. Krishna, L. Krishnamurthy, S. Gajanan, R. Babu, et al. 2016. Genomic regions associated with root traits under drought stress in tropical maize (*Zea mays* L.). *PLoS One.* 11(10): e0164340.

Zaidi, P.H., T. Nguyen, D.N. Ha, S. Thaitad, S. Ahmed, M. Arshad, K.B. Koirala, T.R. Rijal, P.H. Kuchanur, A.M. Patil, S.S. Mandal, R. Kumar, S.B. Singh, B. Kumar, J.P. Shahi, M.B. Patel, M.K. Gumma, K. Pandey, R. Chaurasia, A. Haque, K. Seetharam, R.R. Das, M.T. Vinayan, Z. Rashid, S.K. Nair, and B.S. Vivek. 2020. Stress-resilient maize for climate-vulnerable ecologies in the Asian tropics. *Aust J Crop Sci.* 14(08): 1264–1274.

Zaidi, P.H., K. Seetharam, M.T. Vinayan, Z. Rashid, L. Krishnamurthi, and B.S. Vivek. 2022. Contribution of root system architecture and function in the performance of tropical maize (*Zea may* L.) genotypes under different moisture regimes. *Aust J Crop Sci.* 16: 809–818.

Zare, M., B. Dehghani, O. Alizadeh, and A. Azarpanah. 2013. The evaluation of various agronomic traits of mungbean *Vigna radiata* L. genotypes under drought stress and non-stress conditions. *Intern J Farm Allied Sci.* 2: 764–770.

Zhan, A., H. Schneider, and J.P. Lynch. 2015. Reduced lateral root branching density improves drought tolerance in maize. *Plant Physiol.* 168: 1603–1615.

Zhang, F., J. Yu, C.R. Johnston, Y. Wang, K. Zhu, F. Lu, Z. Zhang, and J. Zou. 2015. Seed Priming with polyethylene glycol induces physiological changes in sorghum (*Sorghum bicolor* L. Moench) seedlings under suboptimal soil moisture environments. *PLoS One.* 15(10): e0140620.

Zhou, K., Y. Wang, J. Chang, S. Zhou, and A. Guo. 2021. Spatial and temporal evolution of drought characteristics across the Yellow River basin. *Ecolog Indicat.* 131: 108207.

3 Economic and Policy Issues of Drought Management

S.K. Chaudhari, D.V. Singh, O.P. Yadav, S. Bhaskar, C.A. Rama Rao, and B.M.K. Raju

3.1 INTRODUCTION

Between 1998 and 2017, climate-related and geophysical disasters killed 1.3 million people and left 4.4 billion injured, homeless, displaced, or in need of emergency assistance (UNISDR and CRED, 2018). The cost to the global economy from disasters every year is estimated to be about US $520 billion (Hallegatte et al., 2017). The costs could be 50% higher if indirect and small-scale losses are also included (Ray et al., 2021). The losses may occur at a larger scale in future due to the accumulation of socio-economic value and changing weather conditions in the coming years (Bevere, 2021).

In South Asia, floods and droughts are more frequent than other natural disasters (i.e., cyclones, earthquakes, landslides, and tsunamis) (Mirza, 2010). Naveendrakumar et al. (2019) also concluded that droughts have become more notable in South Asia during the recent decades.

Indian agriculture is described as a gamble with monsoon. Although the country has achieved remarkable progress in agriculture since independence, the occurrence of natural disasters still continues to pose threat to the sustainability of agriculture and livelihoods. More than half of its net cultivated area is still dependent on rainfall, 75% of which is received during the 4 months (June–September) of the southwest monsoon season (MoAFW, 2017). So, if the monsoon fails, so does the main growing season. India is among the most drought-prone regions in the world (UNDP, 2002), and climate change is expected to increase the frequency and intensity of droughts in future (Aadhar and Mishra, 2021).

Drought monitoring and its management are important areas of research, development planning and policy formulation in the country, and it's relevant to overview the comprehensive drought management policy of India. The objectives of this chapter are to examine the vulnerability of Indian agriculture to drought and the impacts of major droughts; to investigate the evolution of drought management policies over time; to examine existing drought management policies and institutions in the country; and to analyze the mechanism of drought management finances in the country.

3.1.1 DROUGHT AND ITS TYPES

Drought concept, its definition and characteristics have always been a challenging task. Drought is a natural phenomenon which represents reduced water availability; but the onset and duration of each drought event, the area and population affected, and the extent of such effects varies from event to event (Sinha Ray, 2006). 'Drought is an insidious natural hazard characterized by lower-than-normal precipitation that, when extended over a season or longer period of time, is insufficient to meet the demands of human activities and the environment' (WMO, 2006). Unlike aridity, which is a permanent characteristic of a climate, drought is a temporary climatic feature (Wilhite, 1992) and unlike other natural hazards, it is slow in onset and is a creeping phenomenon. The frequency and intensity of extreme weather events like droughts, floods, etc. have increased in the recent past due to global warming (NRAA, 2009), and climate change is projected to increase the frequency and intensity of extreme weather events across regions (IPCC, 2021).

The National Commission on Agriculture (1976) in India classified droughts into three categories viz., meteorological, agricultural, and hydrological. Meteorological drought is a situation when there is a significant decrease in precipitation from long-period average (i.e. >25%) over an area. Hydrological drought results from prolonged meteorological drought resulting in the depletion of surface and sub-surface water resources for normal needs. Agricultural drought is a situation when soil moisture and precipitation are inadequate to meet the evapotranspiration needs of crops. Agricultural drought may be early-, mid- and late-season drought depending on the time of onset. Agricultural drought can occur even when rainfall is normal due to its uneven temporal distribution.

In addition to these, recent studies have proposed socio-economic drought which defines the imbalances in supply and demand of drought-dependent socio-economic commodities (Wilhite and Glantz, 1985; Roy and Hirway, 2007) and ecological drought, which reflects an extended and widespread shortage of water availability in ecosystems, leading to ecological stresses (Smakhtin and Schipper, 2008; Crausbay et al., 2017). When agricultural drought is widespread, it results in socio-economic drought adversely affecting people directly dependent on agriculture as well as those outside agriculture as agricultural and non-agricultural sectors are highly interlinked.

Deficiency of the southwest monsoon rainfall is one of the main drivers of meteorological drought in India, which, if prolonged, can transform into more impactful agricultural and hydrological droughts (Mishra and Singh, 2010; Mishra et al., 2010) which can have lasting impacts on food production and water availability (Van Loon, 2015; Samaniego et al., 2018; Mishra, 2020).

3.1.2 Rainfall Seasons and Distribution in India

India experiences two monsoon seasons: the Indian summer monsoon (ISM) or southwest monsoon season and the northeast monsoon (NEM) or winter monsoon season (Gadgil and Gadgil, 2006; Rajeevan et al., 2012). The ISM, occurring from June to September, is the major source of precipitation in India (Table 3.1). The NEM occurs from October to December and is important for parts of southern India (Kripalani and Kumar, 2004; Zubair and Ropelewski, 2006; Yadav, 2012).

The ISM is a synoptic-scale weather system (Laskar and Bohra, 2021) which is a manifestation of the seasonal migration of the Inter-tropical Convergence Zone (Webster et al., 1998; Gadgil, 2018). These seasonal to multi-decadal climate anomalies have been reported to be associated with El-Nino Southern Oscillation (ENSO), Indian Ocean Dipole (IOD), North Atlantic Oscillation, Pacific Decadal Oscillation, Inter Decadal Pacific Oscillation, and Indo-Pacific Warm Pool (Laskar and Bohra, 2021). However, the effects of most of these forces on ISM are non-stationary (Kumar et al., 2006; Laskar and Bohra, 2021). The ENSO and IOD phenomena are well-known drivers of ISM rainfall (Ashok et al., 2001; Kumar et al., 2007) and the year-to-year variability of the ISM precipitation is linked with these large-scale climate factors (Kumar et al., 1999; Kumar, 2003; Mishra et al., 2012; Roxy et al., 2015). It is well known that the El Nino (warmer) phase of ENSO has an adverse impact on ISM resulting in droughts, while La Nina (cooler) phase usually results in normal or enhanced rainfall (Kripalani and Kulkarni, 1997; Barlow et al., 2002; Aggarwal, 2003; Sigdel

TABLE 3.1
Rainfall Distribution in India in Different Seasons

S. No.	Season	Period	Share of Rainfall (%)
1.	Winter	January–February	2.9
2.	Pre-monsoon	March–May	10.4
3.	ISM (South-west monsoon)	June–September	73.4
4.	NEM (Post-monsoon)	October–December	13.3

Source: Adapted from GoI (2016) and Deshpande (2022).

Economic and Policy Issues of Drought Management

and Ikeda, 2010; Kumar et al., 2013; Roxy et al., 2015). Relatively fewer studies have focused on the causes of NEM rainfall variability (Dimri et al., 2016) and Mishra et al. (2021) found the negative IOD and La Nina conditions as facilitators for NEM rainfall deficits, which inhibit moisture transport from the Bay of Bengal into peninsular India. Both monsoon seasons have experienced major changes over the past few decades (Rajeevan et al., 2012; Roxy et al., 2015; Singh et al., 2019; Mishra, 2020) and the rainfall pattern may undergo changes in future with ongoing increases in greenhouse gases (Mishra and Singh, 2010; Dai, 2011; Diffenbaugh et al., 2015; Cai et al., 2018; Mukherjee et al., 2018; Timmermann et al., 2018).

India receives an average annual rainfall of around 1,200 mm. Northeastern states (6,000–8,000 mm) and slopes of Western Ghats (5,500 mm) receive largest quantity of annual rainfall in India (Figure 3.1), while the lowest quantity (200–350 mm) of rainfall is received in Western Rajasthan and adjacent parts of other states (NAAS, 2011). About 33% and 35% area of India receives low and medium annual rainfall (Table 3.2, Khanna and Khanna, 2011; DAFW, 2020a).

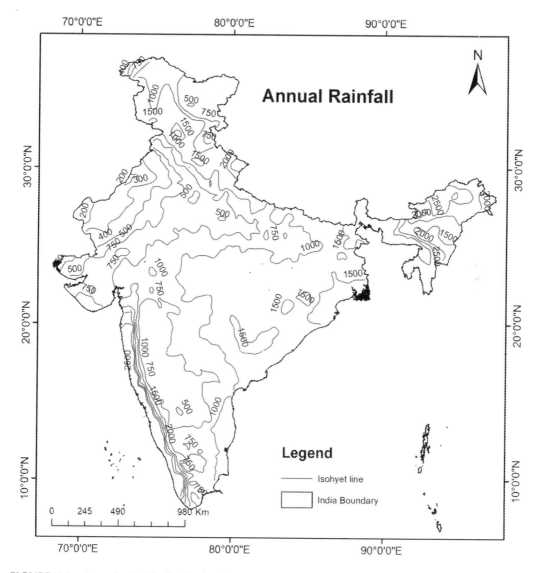

FIGURE 3.1 Annual rainfall distribution in India. Source: IMD gridded (0.25°×0.25°) rainfall data of 1991–2020; https://cdsp.imdpune.gov.in/home_gridded_data.php.

TABLE 3.2
Normal Annual Rainfall Distribution in Different Regions of India

Region	Average Annual Rainfall (mm)	Area (%)
Low rainfall region	<750	33
Medium rainfall region	751–1,125	35
High rainfall region	1,126–2,000	24
Very high rainfall region	>2,000	8

Source: Adapted from Khanna and Khanna (2011) and DAFW (2020a).

3.1.3 Drought Impacts

Droughts are the most detrimental of all the natural disasters (Cook et al., 2007; Mishra and Singh, 2010). The decline of Bronze-Age civilizations in Egypt, Greece, Mesopotamia, and Indus Valley has been attributed to long-term droughts that began around 2000 BC (Dixit, et al., 2014; Marris, 2014). Globally, about one-fifth of the damage caused by natural hazards can be attributed to droughts (Wilhite, 2000). Conway (2008) estimated that between 1993 and 2003, drought-induced famines affected 11 million people in Africa, while World Meteorological Organization (WMO) indicated that droughts may have caused 280,000 human deaths between 1991 and 2000 globally (Logar and van den Bergh, 2011).

It is well known that no two droughts are same, and their impacts depend on duration, intensity, and location. The impacts are more severe in water-scarce basins (Schmidt et al., 2012), regions lacking water storage infrastructures (Estrela and Sancho, 2016), or in the absence of contingency plans (Martin-Carrasco et al., 2013). These are strongly modulated by the socio-economic characteristics of affected areas, such as their vulnerability and resilience to drought, as well as their level of drought preparedness (WMO and GWP, 2014).

Drought impacts cut across many sectors and across normal divisions of responsibility of local, state, and federal agencies (Wilhite and Vanyarkho, 2000). The impacts on farming, rural community and economy are profound. It adversely impacts the growth of agricultural sector and often slows down the growth of other sectors of the economy (Pelling et al., 2002; Loayza et al., 2012; Fomby et al., 2013). Though drought is primarily associated with the loss or reduction of agricultural production, it has significant impacts on energy, transportation, health, recreation/tourism, malnutrition of human being and livestock, fodder availability, land degradation, loss of other economic activities, spread of diseases, and migration of people and livestock (NAAS, 2011; WMO and GWP, 2014).

Broadly, drought impacts may be referred to as direct or primary (e.g., reduced crop productivity or increased fire hazard) or indirect or secondary (e.g., reduced income of farmers, migration). Presently, relatively little knowledge is available on the costs of indirect and longer-term drought impacts. The impacts can also be classified as economic, social, and environmental impacts. The timely and accurate estimation of drought-induced losses is a must to initiate appropriate post-drought relief works, but it is a challenging task (Fankhauser et al., 2014; Hirsch et al., 2017). Only a few studies in India have evaluated the impact of drought on crop production, income, food security, malnutrition, on-farm employment, financial status, etc. (UNFCCC, 2013; Udmale et al., 2015; Zhang et al., 2017; Bahinipati, 2020).

Economic impacts include costs to and losses in agricultural and livestock production (NAAS, 2011); food security and famine (IFRC, 2006); excessive consumption of electricity and fossil fuels to pump groundwater for irrigation (NRAA, 2013) resulting in increased cost of cultivation (Gautam, 2012; NRAA, 2013); shortage of quality seeds in the subsequent season; distress sales of household assets (Cain, 1981; Kinsey et al., 1998); poverty (Pandey et al., 2007); conflict for irrigation water; general economic effects, such as decreased land prices; losses to the recreation and tourism industries; reduced hydroelectricity generation (Shadman et al., 2016); losses to water suppliers; losses in the transportation industry; and increased food prices due to declining food production (NAAS, 2011).

Social impacts include adverse effects on short- and long-term health effects (Ebi and Bowen, 2015; Lohmann and Lechtenfeld, 2015; Algur et al., 2021), including mental and physical stress; reductions in consumption and diet quality (Kazianga and Udry, 2006; Carpena, 2019); stunting and wasting in children (Stanke et al., 2013) leading to reduction in human capital (Victora et al., 2008); chronic energy deficiency in adults (Stanke et al., 2013); increased conflicts over water; conflict and civil unrest (Linke et al., 2015), migration (Gautam, 2012; Gray and Mueller, 2012; Boustan et al., 2020), gender disparities (Fisher and Carr, 2015), reduced quality of life; disruption of cultural belief systems and re-evaluation of social values; and institutional restraints on water use. Female children are usually more vulnerable than other family members during drought (Algur et al., 2021).

Environmental impacts include damage to animal species; hydrological effects (Panda et al., 2007), such as lower water levels in reservoirs, lakes, and ponds; increase in concentrations of salts and toxic elements such as arsenic, fluoride, and nitrate in groundwater (NAAS, 2011; Gautam, 2012); damage to plant communities; depletion of forage resources, overgrazing and indiscriminate cutting of vegetation leading to land degradation (NAAS, 2011); increased number and severity of fires; and increased dust and pollutants. Land abuse during periods of good rains and its continuation during periods of deficient rainfall is the combination that contributes to desertification (UN, 1990a). The use of groundwater is associated with additional pumping costs (Howitt et al., 2015), but the future costs of such groundwater substitution are unknown.

Therefore, economic, social, and environmental consequences of water scarcity can be substantial and depend on the drought magnitude and resilience of the affected area i.e., adaptation and mitigation strategies in place (Rossi and Cancelliere, 2013).

India has a long history of major droughts that have had lasting impacts on water resources, agriculture, gross domestic product, and rural livelihood (Bhalme and Mooley, 1980; Mooley and Pant, 1981; Mooley and Parthasarathy, 1983; Mishra and Singh, 2010). The World Bank has reported that the frequency of droughts has been increasing in India, and the magnitude of drought costs also seems to be increasing over time (World Bank, 2003), mainly due to the increasing value of drought-vulnerable assets. The changing climate is also likely to expand the geographical extent of drought-prone areas (Mishra and Singh, 2009; IPCC, 2014).

A few examples of recent droughts in India may give some insight into the losses accrued. India faced one of the severest droughts in recorded history in 2002, which affected 56% of the geographical area and the livelihoods of 300 million people in 18 states (Samra, 2004). About 150 million cattle were affected, and food grain production declined to 174 million tons from 212 million tons in the previous year (Murthy and Sesha Sai, 2010). Agricultural GDP was reduced by 3.2% (Murthy and Sesha Sai, 2010) and total GDP by 1% (Chandrasekara et al., 2021) due to agricultural income loss of Indian Rupees (INR) 390 billion (US$ 8.7 billion). The central government allocated a financial relief of INR 200 billion (US$4.4 billion) (Samra, 2004). About 1.5 billion L of drinking water was transported every day by tankers and railways during the drought period. The groundwater table in drought-affected areas declined by 2–4 m below the normal levels. The drought was declared in 32 districts of Rajasthan solely based on the deficit of July rainfall.

As a result of drought in 2009, the food grain production was estimated to be less by 7% compared to 2008–09 (Venkateswarlu, 2010; NAAS, 2011). The crop holiday observed by farmers in the resource-rich Krishna delta area of Andhra Pradesh during the 2009 drought was triggered by the empty reservoirs. Cropped area in Karnataka declined to a third of what is normally sown due to the poor onset of monsoon rains in 2012 and when the rainfall deficit extended, it resulted in a hydrological drought with most of the reservoirs going empty affecting irrigated agriculture as well (Lokesh and Poddar, 2018).

The recent drought of 2015–16 affected crop production and water availability in the Indo-Gangetic Plain and Maharashtra region (Mishra et al., 2016; Prakash, 2018) and about 330 million people in ten states were affected (UNICEF, 2016). It also resulted in lower reservoir storage and hydropower production. Furthermore, it caused a significant depletion in groundwater in the Indo-Gangetic Plain and southern states of India (UNICEF, 2016).

Mostly economic losses are reported which resulted from droughts, in addition to a few others, if any. It is easier to estimate economic losses which cover goods and services that can be sold in the market, but there is no market for goods and services belonging to non-economic losses (loss of cultural heritage, biodiversity, human mobility, soil erosion, etc.) and therefore, their monitory value cannot be evaluated directly (Hallegatte, 2014). The economic losses are usually high in developed countries, but the non-economic losses are expected to be significant in developing countries (Serdeczny et al., 2018) which are rarely estimated to the full extent.

3.2 POLICY FRAMEWORK FOR DROUGHT MANAGEMENT

Drought is a naturally occurring phenomenon due to climate variability and little can be done to reduce the recurrence of drought events in a region. Therefore, drought cannot be effectively managed as a temporary crisis. Its management involves long-term water resources management planning with an emphasis on monitoring and managing emerging stress conditions and other hazards associated with climate variability in the focus region. However, the drought scenario has been handled as a temporary crisis since time immemorial.

3.2.1 Evolution of Drought Management Policies

3.2.1.1 Pre-Independence Era

Though certain elements of response to drought can be traced back to the dawn of civilization in ancient India as Verse 2.3.1 of the *Atharvaveda* instructs for proper management of various water bodies such as brooks, wells, pools and an efficient use of their water resources for reducing the drought intensity and water scarcity (Sharma and Shruthi, 2017). One of the first treatises on governance (*Arthshastra*), written more than 2,000 years ago by Kautilya, pronounces that when famine threatens, a good king should 'institute the building of forts or waterworks with the grant of food, or share provisions with the people, or entrust the country to another king' (Chetty and Ratha, 1987). In medieval times, the major famine relief measures adopted by Indian rulers included free distribution of food grains, opening of free kitchens, remission of land revenue, remission of other taxes, payment of advances or loans, construction of public works, wells, canals, and embankments, encouragement of migration, increase in the pay of soldiers, etc. (Srivastava, 1968). However, there were no fixed norms for rulers and the type and magnitude of relief works varied widely from ruler to ruler. Curley (1977) concluded that the evidence is inadequate to judge the real efficacy of relief efforts before the 19th century, and it is not implausible that they were often far from systematic and comprehensive.

The first planned drought response started with the report of the first Famine Commission in 1880 established by the Colonial British Government in response to some major famines and large causalities. On its recommendations, the detailed Indian Famine Code (FC) of 1883 was adapted as a national model for different regions of British India. The backbone of the famine relief strategy embodied in the Famine Code was to provide employment at subsistence wages and at a reasonable distance from their homes to all those who wanted it, and public employment was directed to the creation of public assets such as roads and canals; and gratuitous relief for those who could not work (Dreze, 1991; Khera, 2006). It also included suspension of land revenue and grant of loans, relaxation of forest laws and provision for protection of cattle. The FC defined three levels of food insecurity: near-scarcity, scarcity, and famine. 'Near scarcity' was a condition of a mild drought; 'scarcity' was defined as three successive years of crop failure, crop yields of one-third or one-half normal, and large populations in distress; while 'famine' included a rise in food prices above 140% of normal, the movement of people in search of food, and widespread mortality (Deshpande, 2022). The emphasis was given on development of rail and road network as well as the creation of irrigation facilities to mitigate the adverse effect of drought in a region. The FC further provided for the declaration of famine by the local authorities based on unusual migration of people and cattle, drying of wells, non-availability of fodder, withering away of crops, higher density of beggars and so on (Brennan, 1984; Murton, 1984). It also emphasized the need to understand the causes of famine and to predict it. Thus, the FC was one of the first attempts to predict famine or drought.

The first FC or Scarcity Manual, as it is sometimes called, was prepared in 1883, which was followed by other manuals by some provincial governments (Hirway, 2001). Recommendations of subsequent Famine Commissions modified codes, for example the Famine Commissions of 1945 recommended that the relief could be given in any of the following ways: (i) gratuitous relief in the form of gruel, uncooked food grains and cash, (ii) wages for relief in kind or cash, (iii) agricultural loans for (a) maintenance - in kind or cash, (b) purchase of cattle - in cash, (c) agricultural operations in kind or cash, (iv) sale of food grains at low rates to the poor. In 1928, the Royal Commission on Agriculture recommended the promotion of dryland farming to enhance the resilience of agriculture in drought-affected regions.

3.2.1.2 Post-Independence Era

In recent times, different nations have developed codes, manuals, procedures, and policies for monitoring and management of drought learning from their past experiences and procedures followed by others. India too has learned its lesions over the years and has strived to develop a fairly elaborate system of institutionalized drought monitoring, declaration and mitigation at different levels (Samra, 2004). India's response to the need for enhanced drought management has contributed to the overall development of the country (Gupta et al., 2011).

The drought management policies continued to evolve over time after independence with the new insights gained after each major drought episode. For example, the drought of 1965–67 encouraged the Green Revolution, after the 1972 drought employment generation programs were developed for the rural poor; after the 1987–88 drought relief effort focused on preserving the quality of life (UN, 1990b).

As Samra (2004) succinctly put it, drought management strategies in India have evolved out of the experience of 17 major droughts, five severe droughts and technological advances since 1871. Initially, the emphasis was given on minimizing deaths from starvation. The 1965–66 drought induced restructuring of the Indian strategy through a public distribution system to ensure the physical availability of food. A certain amount of food grain was provided to each household at highly subsidized rates. The 1972 drought prioritized economic access to food through employment generation for the drought-affected population. 'Food for Work' program was launched in 1977–78. Under this program, the government provided work to poor people within a reasonable distance from their villages and gave food grains instead of wages. This program was later restructured as the National Food for Work Programme in 2001. Since 2006, this program has been absorbed into the Mahatma Gandhi National Rural Employment Guarantee Scheme (MGNREGS). The experience of monitoring the 1979 drought underpinned robustness, resilience and stability of farming systems and livelihood opportunities. Several studies that followed the severe drought of 1987 looked into various dimensions of drought such as incidence, impacts, monitoring, mitigation and coping responses. It paved the way for watershed management approach for mitigating the adverse impacts of droughts on livelihood, livestock and vegetation. Isolated droughts in 2000 in vulnerable areas like Rajasthan consolidated the institutional and capacity-building infrastructure. The 2002 drought was unique since rains failed at the very beginning of the rainy season, crops could not be sown, and the existing formalities of monitoring and declaration were relaxed, revised or waived off. The severity of the 2002 drought also called for the setting up of special task forces, additional monitoring systems and responses. The 2002 drought and other disasters that struck the country led to the promulgation of Disaster Management (DM) Act in 2005.

The Second Irrigation Commission in 1972 and the Report of the Agricultural Commission in 1976 helped in the identification of drought-prone areas of the country. Many programs such as Drought Prone Areas Programme (DPAP), Desert Development Programme (DDP), and National Watershed Development Programme for Rainfed Areas were implemented with varying degree of success. Identifying and popularizing technological interventions were at the core of such programs. Considering the importance of forecasting and monitoring drought, institutional arrangements were put in place in the form Crop Weather Watch Group (CWWG) that monitors drought in the country. Establishment of National Rainfed Area Authority (NRAA) was also a significant development for managing drought in rainfed areas. Thus, the emphasis gradually shifted from post-drought response to monitoring and enhanced preparedness to deal with drought.

3.2.2 Institutional Framework for Disaster Management

Broad disaster management mechanism in India is same for every sort of disaster, and hence a brief description is given here. The DM Act was enacted in 2005 and Section 11 of the Act mandates that there shall be a National Disaster Management Plan (NDMP) for the country. Therefore, a National Policy on Disaster Management (NPDM) was prepared in 2009. Subsequently, the Government of India (GoI) adopted three landmark international agreements in 2015, having bearing on disaster management, viz., (i) Sendai Framework for Disaster Risk Reduction; (ii) Sustainable Development Goals (SDGs) (2015–30); and (iii) Paris Agreement on Climate Change at the 21st Conference of Parties, under the United Nations Framework Convention on Climate Change (UNFCCC). Keeping this in mind, the revised NPDM was announced by the prime minister in November 2016 in New Delhi (NDMA, 2019). The GoI established improved institutional arrangements and Disaster Risk Reduction (DRR) mechanisms to deal with any disaster as outlined in NDMP 2019 (NDMA, 2019). This was designed to help all concerned stakeholders in central government as well as in states and Union Territories (UTs) to achieve the national goals.

The DM Act 2005 has laid all the institutional, legal, financial and coordination mechanisms at various levels i.e., national, state and district levels. These institutions are supposed to work in close harmony for coordination, decision making and communication for disaster management and do not imply any chain of command. This institutional framework is expected to shift the pattern of disaster management from a relief centric to proactive regime with an emphasis on preparedness, prevention and mitigation.

At the national level, the NDMP complies with the NPDM and conforms to the provisions of the DM Act making it mandatory for the various central ministries and departments to have adequate DM plans. The overall coordination of disaster management lies with the MHA, while the Cabinet Committee on Security (CCS) and the National Crisis Management Committee (NCMC) are core committees involved in the top-level decision-making for disaster management. The National Disaster Management Authority (NDMA) is responsible for the approval of the NDMP and facilitating its implementation at the national level. CCS is responsible for evaluating an incident from a national security perspective. NCMC oversees the command, control and coordination of the disaster response and gives direction for specific actions to face crisis. NDMA is headed by the prime minister and has the responsibility of preparing the policies, plans and guidelines for disaster management, coordinating their enforcement and effective implementation within the stipulated time. Guidelines prepared by NDMA assist various central ministries, departments and states to formulate their respective disaster management plans. It approves the NDMP and the DM plans for various kinds of disasters. NDMA is assisted by the National Executive Committee (NEC) in the discharge of its functions. It includes secretaries of various GoI ministries/departments. It ensures compliance with the directions issued by the central government and performing such other functions as may be required by the NDMA. National Disaster Response Force assists the relevant state government/district administration in case of a hazard event or in its aftermath. The National Institute of Disaster Management is responsible for human resource development and capacity building for disaster management within the broad policies and guidelines laid down by the NDMA.

At the state level, the DM Act 2005 mandates each state/Union Territory to have its own institutional framework for disaster management. The Act requires the setting of a State Disaster Management Authority (SDMA) and the preparation of its own State Disaster Management Plan. The SDMA approves and coordinates the disaster management plans prepared by various state departments, following the guidelines laid down by the NDMA, recommends the provision of funds for mitigation and preparedness measures, reviews the developmental plans to ensure the integration of prevention, preparedness and mitigation measures (DAFW, 2020b). The State Executive Committee (SEC) assists the SDMA in the performance of its functions, coordinates and monitors the implementation of the national policy, the national plan, and the state plan. At the district level, there is a District Disaster Management Authority (DDMA) and the District Collector, District Magistrate or the Deputy Commissioner, as applicable, is responsible for the preparation of DM plans for the district and overall coordination of the disaster management planning at the district level. The DDMA ensures that the guidelines for prevention, mitigation, preparedness, and response measures laid down by the NDMA and the SDMA are followed by all the district-level offices of the state Government. The relief operations are executed at district and sub-district levels under the DDMA authorities.

3.2.2.1 Institutional Arrangements for Drought Management

While the NDMP broadly covers guidelines for all major types of disasters at the national level, the hazard-specific nodal ministries and departments notified by the GoI, prepare detailed DM plans for the assigned disaster. Ministry of Agriculture and Farmers Welfare (MoAFW) is the nodal ministry for preparing detailed DM plans for drought, cold-wave, frost, hailstorm and pest attack, and getting them approved by the NDMA. National drought management guidelines were prepared by MoAFW in 2010, approved by NDMA, to provide direction to the central ministries/departments, and state governments for preparing detailed action plans to handle drought (MoAFW, 2017). Department of Agriculture and Farmers Welfare (DAFW, erstwhile Department of Agriculture, Cooperation and Farmers Welfare) under the MoAFW prepared a Manual for Drought Management in 2009 (DAC, 2009), which was updated in 2016 (DACFW, 2016) and in 2020 (DAFW, 2020).

Additional Secretary in the DAFW serves as the Central Drought Relief Commissioner (CDRC), assisted by a Drought Management Cell in the department. CDRC in DAFW is responsible for collecting information from different sources, monitoring drought conditions, issuing advisories, as well as monitoring and coordinating the central government response to drought. The Crop Weather Watch Group for Drought Management (CWWGDM), set up in the DAFW, acts as an inter-ministerial mechanism and meets at least once a week during the ISM period to monitor drought situation in different parts of the country (MoAFW, 2017). The CWWGDM evaluates information and data from different scientific and technical bodies to determine the likely impact of meteorological and other environmental parameters on agriculture. A Crisis Management Group functions under the Chairmanship of the CDRC with representatives of associated ministries and organizations, and the group meets from time to time to review the drought situation in the country and progress of relief measures. The Crisis Management Plan (CMP) for Drought prepared by the DAFW, is an actionable plan in the event of a drought, to minimize damage to life and property. It is updated every year and delineates the roles and responsibilities of central and state governments and their agencies in managing the drought situation effectively (MoAFW, 2017).

In a nutshell, the DAFW frames and monitors the drought policy. It sets up the institutional mechanism at the three levels of government i.e. central, state and district. The policy aims on drought vulnerability assessment of different districts, drought monitoring, drought preparedness, mitigation and prevention measures at the pre-drought stage, coordination of relief measures during the post-drought stage, capacity building and creating awareness.

At the state level, Department of Disaster Management and Relief (name varies from state to state), headed by a Secretary or Relief Commissioner is responsible for directing drought operations in the state. The relief commissioner/secretary monitors the drought situation and regulates the release of all financial assistance to the district administration. NDMA has recommended the establishment of a separate Drought Monitoring Cell at the state level, under SDMA, with adequate staff. Drought Monitoring Cells are responsible for the preparation of vulnerability maps for their respective states and operate a drought management planning and monitoring system involving various stakeholders for drought risk reduction.

At the district level, District Collector implements all decisions related to drought management through a number of line departments and field agencies. District collector heads a district drought/disaster management committee consisting of public representatives and line departments. At the sub-district level, *Panchayati Raj* (rural government system) institutions are involved in the implementation of drought management programs.

3.2.3 Drought Hazard, Risk and Vulnerability Assessment

The risk of climate hazards is the product of hazard, exposure, and vulnerability. Without clear information about hazard, exposure, and the vulnerability of people and natural resources at risk, we face several limitations, ranging from weaknesses in the prediction of impacts to the evaluation of disaster mitigation strategies (Enenkel et al., 2020). In drought planning, making the transition from crisis to risk management is difficult because of the limited understanding of the risks associated with drought. Areas of high risk should be identified which depends on the exposure of a location to drought hazard and the vulnerability of the location to droughts (Blaikie et al., 1994). It is important to know the level of exposure (i.e., frequency of droughts of various intensities and durations) to the drought hazard; vulnerability (i.e., social factors such as land use patterns, water use, population, economic development, diversity of economic base, government policies, social behavior, etc.) of the target region.

A drought vulnerability assessment in Rajasthan (Rathore, 2004) found that both low- and middle-income households were vulnerable to droughts. Marginal and small farmers were normally most vulnerable, while in the desert region, even large farmers were vulnerable due to poor quality of soils with no irrigation facilities and dependence on erratic rainfall. Landless households were better off as they were less dependent on agriculture and livestock for their livelihood and were more mobile.

3.2.3.1 Major Droughts in the Past

Droughts and floods occur almost every year in different parts of the country. However, mild droughts have been observed at a higher frequency than moderate and severe droughts (Mundetia and Sharma, 2015). The entire country is prone to drought except the eastern regions, while the western arid regions experience frequent droughts (Venkateswarlu, 2010). Moderate droughts were observed over many regions of the country, while intensified droughts were predominant in the north, northwest and central India (Pai et al., 2017).

There has been some variation in reported droughts at the national level, probably due to different criteria used. For example, from 1871 to 2002, Samra (2004) reported 22 major droughts in 1875, 1877, 1899, 1901, 1904, 1905, 1911, 1918, 1920, 1941, 1951, 1965, 1966, 1968, 1972, 1974, 1979, 1982, 1985, 1986, 1987, and 2002; while NRAA (2009, 2013) reported 22 large scale droughts from 1891 to 2002 in 1891, 1896, 1899, 1905, 1911, 1915, 1918, 1920, 1941, 1951, 1965, 1966, 1972, 1974, 1979, 1982, 1986, 1987, 1988, 1999, 2000 and 2002. Deshpande (2022) reported that India faced 54 droughts or famines during the 19th and 20th centuries and four in the current century. The list in Table 3.3 may not include all droughts, as some of which were local, unreported or short-lived.

IMD analyzed the drought in India and found that droughts have become more frequent and severe since 1965 (Shewale and Kumar, 2005). Since 1965, droughts of 1972, 1987, 2002 and 2009

TABLE 3.3
Historical Occurrence of Droughts in India (1800–2021)

Period	Drought Years	Numbers
1801–50	1801, 1803, 1804, 1806, 1812, 1818, 1822, 1825, 1832, 1833, 1837, 1839, 1845	13
1851–1900	1862, 1866, 1867, 1868, 1875, 1877, 1878, 1883, 1891, 1896, 1898, 1899, 1900	13
1901–50	1901, 1904, 1905, 1907, 1911, 1912, 1913, 1918, 1920, 1939, 1941	11
1951–2000	1951, 1965, 1966, 1968, 1972, 1974, 1979, 1982, 1985, 1986, 1987, 1992, 2000	13
2001–20	2002, 2009, 2014, 2015, 2018	5

Source: Adapted from Samra (2004), Deshpande (2022), and IMD annual reports.

were classified as severe, while those observed in 1965, 1966, 1974, 1979, 1982, 1985, 2000 and 2012 were classified as moderate droughts. Mishra et al. (2019) have reported seven major drought-prone periods in India during 1870–2016, covering 1876–82, 1895–1900, 1908–24, 1937–45, 1982–90, 1997–2004 and 2011–15 periods. Based on Standardized Precipitation Index (SPI) of all India rainfall from 1871 to 2010, Sarkar (2011) reported five extreme droughts (SPI:<−2.0); seven severe droughts (SPI: −1.50 to −1.99); and nine moderate droughts (SPI: −1.0 to −1.49). SPI values of area weighted rainfall data indicate that the country faced three extremely severe droughts (1965, 1972 and 2009), three severe and 16 moderate droughts during 1901–2021 (Figure 3.2). Mishra (2020) identified 18 meteorological and 16 hydrological droughts during the 1870–2018 period. The five deadly meteorological droughts occurred in 1899, 1876, 2000, 1918 and 1965 while the five major hydrological droughts occurred in 1899, 2000, 1876, 1965, and 1918. He concluded that the drought of 1899 was the most severe during the entire period.

Despite a substantial expansion in the irrigation infrastructure, droughts occur almost every year in one part or other of the country. This is evident from the number of districts that have been declared drought affected by different state governments since 2000 (Figure 3.3). During 2000–17 period, maximum number of districts that were declared drought affected were in Karnataka (48.9%), followed by Rajasthan (47.3%), Maharashtra (35.6%), Jharkhand (34.5%), Kerala (26.6%),

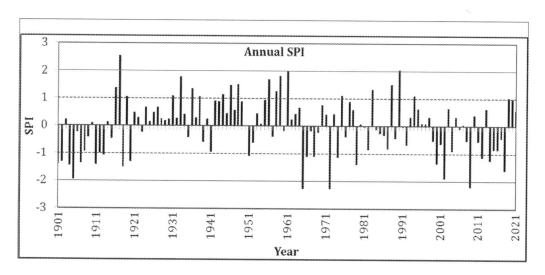

FIGURE 3.2 Annual SPI of all India area weighted rainfall from 1901 to 2021. Source: 1901–2014 data: https://data.gov.in/resource/all-india-area-weighted-monthly-seasonal-and-annual-rainfall-mm; 1915–2021 data: https://hydro.imd.gov.in/hydrometweb/ (from annual rainfall statistics of India).

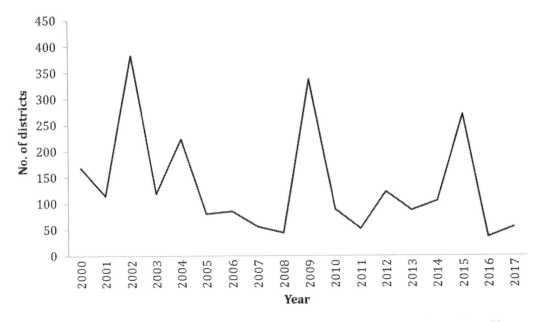

FIGURE 3.3 Number of districts declared drought affected by state governments. Source: https://farmer.gov.in/Drought/Droughtreport.aspx.

Madhya Pradesh (26.3%), Uttarakhand (26.1%), Bihar (22.4%), Himachal Pradesh (22.2%), Odisha (21.5%) and Uttar Pradesh (21.1%).

3.2.3.2 Drought to Famine

Low and erratic rainfall, particularly the ISM, causes droughts and crop failure. Such droughts, sometimes in consecutive years lead to famines. Famines in India resulted in more than 30 million deaths (Mohapatra et al., 2022) over the 18th, 19th and early 20th centuries (Table 3.4).

India has a long history of famines that led to the starvation of millions of people (Passmore, 1951). Droughts occurred in all the periods, but there was an alarming increase in the frequency of

TABLE 3.4
Major Famines in the Pre-Independence India

Name of the Famine	Year	No. of Victims (million)
Bengal Famine	1769–70	2–10
Chalisa Famine	1783–84	11
Doji Bara Famine	1791–92	11
Agra Famine	1837–38	0.8
Upper Doab Famine	1860–61	2
Orrisa Famine	1865–67	4–5
Rajasthan Famine	1868–70	1.5
Bihar Famine	1873–74	2.5
Southern India Famine	1876–78	6–10
Indian Famine	1896–97	12–16
Indian Famine	1899–1900	3–10
Great Bengal Famine	1942–43	1.5-3

Source: Adapted from Samra (2004) and Mohapatra et al. (2022).

droughts resulting in famines during the British period (Bhatia, 1985). Severe famines frequently took place even when crop failures were localized and short-lived, and food grain was far from wanting in the country as a whole (Dreze, 1991). There were a few exceptions like the last two famines of the 19th century (1896–97 and 1899–1900) when famine occurred due to massive crop failures, virtually throughout the country, at times when stocks were already diminished (Dreze, 1991). During the British rule in India (1765–1947), 12 major famines occurred (in 1769–70, 1783–84, 1791–92, 1837–38, 1860–61, 1865–67, 1868–70, 1873–74, 1876–78, 1896–97, 1899–1900 and 1943–44) which led to the deaths of millions of people (Maharatna, 1996). Many of these famines were caused by the failure of the ISM, which led to droughts and crop failures (Cook et al., 2010), and in almost all cases, major famine followed massive crop failure resulting from drought (Bhatia, 1967). Crop failures not only resulted in less food availability in the affected region but also, and more importantly, disrupted the rural economy (Bhatia, 1967). Landless agricultural laborers found little employment, while general impoverishment simultaneously enlarged the supply of casual labor. People resorted to selling their meager resources to meet their food needs, which became costly due to less availability, while trade was often slow to move food to the affected area from other regions. This recurring scenario was aptly summarized by Baird Smith's well-known statement, after the 1860–61 famine, to the effect that famines in India were 'rather famines of work than of food' (Srivastava, 1968). The successive Famine Enquiry Commissions of the 18th and 19th centuries arrived almost at the same conclusion which has also been endorsed by most independent analysts.

Per capita food production declined or remained stagnant in India till decades after its independence in 1947 due to substantial population growth (Dreze, 1991). India has faced several droughts since independence (as in 1966–67, 1972–73, 1979–80, 1985–87 and 2002–03), but famine deaths have been almost eliminated in modern India. This achievement is attributed to better food distribution and buffer food stocks, rural employment generation, transportation, and groundwater-based irrigation (Dreze, 1991; Aiyar, 2012). Sen (2000) has pointed out the importance of democracy in linking the incentives of agents (e.g., the disaster-affected) and political incentives through electoral accountability.

3.2.3.3 Drought Prone Areas in India

Drought occurs in regions with high as well as low rainfall. Different agencies and researchers have used different criteria for the classification of drought-prone areas in the country owing to a lack of universally agreed criteria. Based on annual rainfall amount, 35% of the area that receives rainfall between 750 and 1125 mm are considered drought-prone, and 33% area that receives <750 mm rainfall is considered chronically drought-prone area (Khanna and Khanna, 2011; Rawat et al., 2022).

IMD monitors both meteorological and agricultural droughts (Sarkar, 2011). It defined meteorological drought as a situation when the ISM rainfall over an area is less than 75% of its long-term average. Drought is classified as moderate if the rainfall deficit is 26%–50% and severe if the deficit exceeds 50% of the average rainfall. Based on these criteria, droughts in different regions over a period of 135 years (1875–2009) were identified by IMD, and West Rajasthan (34 cases) and Saurashtra and Kutch (31 cases) had the highest occurrences of drought followed by adjoining Gujarat region (28 cases) (Sarkar, 2011). Other areas with large incidences of meteorological drought were Haryana, Delhi, Punjab, Himachal Pradesh and East Rajasthan in northwest India, and Rayalaseema in the southern peninsula. The sub-humid and humid areas of east and northeast India had the lowest occurrences of drought. Based on the drought probability, IMD (2005) classified different regions into chronically drought-prone areas (>20% probability of drought occurrence), frequently drought-prone areas (10%–20% probability of drought occurrence), and least drought-prone areas (<10% probability of drought occurrence). West Rajasthan and Gujarat state came under chronically drought-prone areas; while the frequently drought-prone areas covered East Uttar Pradesh, Uttarakhand, Haryana, Punjab, Himachal Pradesh, East Rajasthan, West Madhya Pradesh, Marathwada, Vidarbha, Telangana, Coastal Andhra Pradesh and Rayalaseema (IMD, 2005).

Rainfed areas are more vulnerable to agricultural drought. In rainfed areas, drylands are more prone to drought, which are found from north to south and in the western part of the country (Gautam and Bana, 2014). NRAA (2009) estimated that the possibility of occurrence of agricultural drought varies from once in 2 years in Western Rajasthan to once in 15 years in northeast India. Samra (2004) reported that about 28% of agricultural land in India is drought prone and as such suffers from critical water shortages, while NDMA (2019) estimates indicate that 68% of the cultivable area in India is vulnerable to drought, which includes parts of Rajasthan (chronically drought affected), Gujarat, Maharashtra, Madhya Pradesh, Uttar Pradesh, Chhatisgarh, Jharkhand and Andhra Pradesh.

When the Government of India wanted to give special attention to drought-prone areas, rainfall and percent area under irrigation were used as the indicators to select districts for implementing the DPAP and DDP schemes in the 1970s. Later in the 1990s, a committee constituted under the chairmanship of C.H. Hanumantha Rao suggested Moisture Index (MI) and level of irrigation as the new measures for selecting the districts under the aforesaid schemes. The moisture index [$MI = \{(P-PET)/PET\} \, 100$] was calculated from precipitation (P) and potential evapotranspiration (PET) data and the regions were classified as arid ($MI < -66.7$), semi-arid (MI: -66.6 to -33.3) and dry sub-humid (MI: -33.3 to 0.0). Arid, semi-arid and dry sub-humid regions are characterized by acute, moderate and little shortage of soil moisture, respectively (Chary et al., 2014). Therefore, MoRD (1994) suggested the scientific criteria (Table 3.5) on the basis of moisture index-based ecosystem and level of irrigation for making a district eligible for implementation of DPAP or DDP.

Later, Raju et al. (2013a,b) observed climatic shifts in about 27% of the geographical area in the country using the same criteria. The arid region was observed to increase in Gujarat and decrease in Haryana state. Semi-arid region increased in Madhya Pradesh, Tamil Nadu and Uttar Pradesh due to climate shift from dry sub-humid to semi-arid (Figure 3.4), while the moist sub-humid pockets in Chhattisgarh, Orissa, Jharkhand, Madhya Pradesh and Maharashtra states turned dry sub-humid to a larger extent. Similarly, Venkateswarlu et al. (2014) revisited the eligibility of a district for DPAP and DDP. The number of eligible districts for DDP and DPAP was 22 and 121, respectively (Figure 3.5). Three DPAP districts of Gujarat state viz., Porbander, Amreli and Bhavanagar became eligible for DDP, while 27 districts, mainly from the eastern region (Orissa, Chhattisgarh and Jharkhand) became eligible for DPAP. The number of existing DPAP and DDP districts that did not qualify was 72 and 16, respectively.

Rama Rao et al. (2019a) assessed drought proneness of different parts of the country and observed several districts with >25% drought incidence (Figure 3.6).

Ministry of Home Affairs (MHA), in an effort to compute composite drought index, observed that drought hazard zonation of districts has not been carried out by any agency (MHA, 2019), therefore, it used (i) moisture index, (ii) cropped area not irrigated, (iii) drought-prone area, and (iv) number of drought years to compute drought index for different districts. Finance Commission (2020) found Andhra Pradesh, Gujarat, Karnataka, Maharashtra, Odisha, Rajasthan and Telangana to be chronically drought-prone states based on such index.

TABLE 3.5
Criteria for Eligibility of Districts for DPAP and DDP

Ecosystem	Moisture Index	Irrigated Area as % of Net Sown Area	Program Permissible
Arid	< −66.7	Up to 50	DDP
Semi-arid	−66.6 to −33.3	Up to 40	DPAP
Dry sub-humid	−33.3 to 0.0	Up to 30	DPAP

Source: Adapted from MoRD (1994) and NRAA (2020).

Economic and Policy Issues of Drought Management 55

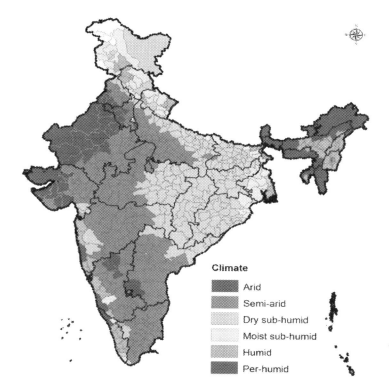

FIGURE 3.4 Climatic classification at district level (1971–2005). Source: Raju et al. (2013b).

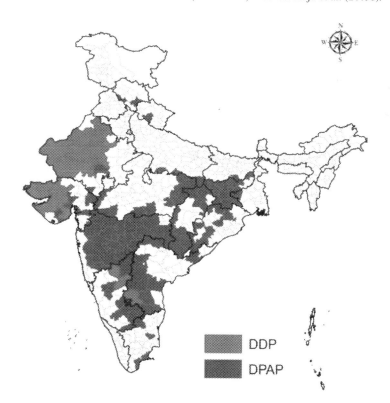

FIGURE 3.5 Eligibility of districts to DPAP and DDP revisited. Source: Venkateswarlu et al. (2014).

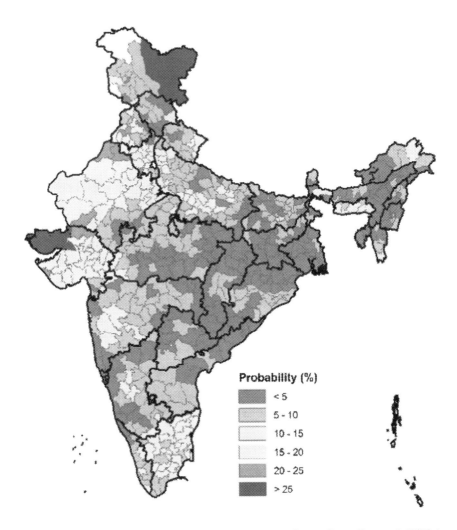

FIGURE 3.6 Map showing the probability of drought incidence. Source Rama Rao et al. (2019a).

3.2.4 DROUGHT PREPAREDNESS

The MoAFW released a Drought Management Plan (DMP) in November 2017 (MoAFW, 2017), which delineates the roles and responsibilities of different states and GoI departments/agencies involved in drought management for drought prevention and mitigation, preparedness, response and relief measures. The major focus of DMP is on preparation and timely communication among stakeholders to reduce the time taken in mobilizing resources for an effective response and enable a harmonious relationship among stakeholders. The MoAFW has also prepared a national Crisis Management Plan (CMP) for Drought (DAFW, 2022) and District Agriculture Contingency Plans (DACPs) to provide necessary guidelines for drought management to implementation authorities. The CMP is an actionable program document for central and state government ministries/departments, includes major critical steps that need to be taken for drought preparedness, and delineates the roles and responsibilities of various stakeholders, including central and state governments and

Economic and Policy Issues of Drought Management

their agencies in managing the drought effectively in a structured and planned manner with the most efficient and optimum utilization of time, effort and resources. It identifies trigger-points for various response actions and integrates such actions in an appropriate response matrix. The CMP is part of the overall spectrum of drought management, but is focused on management interventions required during the time of crisis i.e. the plan is pressed into action in the event of a drought. Crisis management framework given in CMP is aimed at the identification of basic aspects of crisis situation namely, phases of crisis (normal, alert/watch, warning, emergency, acute or potential disaster, recovery or post-disaster) and vulnerability magnitude, impact, trigger mechanism and strategic response actions for each phase of crisis. It is updated every year before *kharif* (rainy) season including the contact details of all the relevant central government officials. The list of nodal officers is also maintained in the Drought Monitoring Centre (DMC) of the DAFW, GoI and is updated every month. A Crisis Management Group (CMG) for Drought Management is identified and listed to manage various phases of drought. DMCs and CMGs are also supposed to work at state and district levels on the same lines.

Contingency crop planning is an essential component of risk management in agriculture. The MoAFW has prepared exhaustive, district-level contingency plans for most of the districts in the country, comprising of several measures that the farmers can take during situations such as the late onset of monsoon, early withdrawal, long dry spells during cropping season, untimely excess rains, high or low-temperature stress, etc. DACPs include suggestion for field and horticulture crops to cope with drought in rainfed and irrigated farming situations in the district, on account of delayed onset of monsoon by 2, 4, 6 or 8 weeks. The suggestions given in DACPs include sowing date adjustments, plant population adjustments, changes in recommended fertilizer, irrigation or pest management practices, choices of alternate crops and varieties, etc. DACPs also include suggestions for the establishment of seed bank, fodder bank and nutrient centers, at strategically advantageous locations for providing relief to farmers during distress period. The DACPs for various districts of India are available at the website of the Department of Agriculture and Farmers Welfare (http://agricoop.gov.in/en/DocAgriContPlan#gsc.tab=0).

3.2.5 Drought Monitoring and Early Warning

Development of hydro-meteorological capacities and early warning systems in developing countries to levels similar to those in developed countries can yield annual benefits of between US$ 4 and 36 billion, with benefit-cost ratios between 4 and 35 (Hallegatte, 2012).

A number of indicators and indices have been proposed for drought monitoring (Rossi and Cancelliere, 2013; WMO and GWP, 2016; Tsakiris, 2017). Indicators are variables or parameters such as precipitation, temperature, stream flow, groundwater and reservoir levels, soil moisture, etc., used to describe drought conditions (WMO and GWP, 2016) and snowpack. Indices are computed values using climatic or hydro-meteorological inputs including the indicators. Indices are used to provide a quantitative assessment of the severity, location, timing and duration of drought events. The drought may be quantified based on (i) an individual index, (ii) multiple indices, and (iii) a composite index (Chandrasekara et al., 2021). A large number of drought indices like Palmer Drought Severity Index (Palmer, 1965), SPI (McKee et al., 1993, 1995), Standardized Precipitation Evapotranspiration Index (Vicente-Serrano et al., 2010), Standardized Runoff Index (Shukla and Wood, 2008), Standardized Drought Index (Nalbantis and Tsakiris, 2009), are based on land surface states and fluxes, while Normalized Difference Vegetation Index (NDVI) (Rouse et al., 1973), Vegetation Drought Response Index (Brown et al., 2008) are based on remote sensing techniques.

In India, Jain et al. (2015) compared various drought indices like SPI, Effective Drought Index, China Z-Index, Rainfall Departure, and Rainfall Decile based Drought Index and suggested that the Effective Drought Index better correlates with other indices for all time scales. Patel et al. (2012) identified that Vegetation Temperature Condition Index precisely identified the intensity and spatial extent of drought stress in 2000, 2002, and 2004 in India. Bhuiyan et al. (2006) compared the SPI, Standardized Water-Level Index, Vegetation Condition Index, Temperature Condition Index, and Vegetation Health Index using NDVI data. They observed that on some occasions, the negative SPI anomalies were observed even in the absence of drought, while Mahajan et al. (2016) found that the SPI could perform drought monitoring at multiple timescales more precisely than the Percent of Normal Precipitation.

Earlier the drought monitoring was limited mainly to simple variables like rainfall, crop loss, reservoir levels, fodder availability, etc., but the list of indicators/indices has been updated since 2009. Checklist for drought monitoring (DAFW, 2020a) now includes meteorological data like daily, weekly and monthly rainfall, temperature, snow fall; hydrological data such as water storage in reservoirs, river flow, groundwater level, yield and draft from aquifers; agricultural data like soil moisture, crop sown area and type of crop, status of crop growth, crop yield; remote sensing data like vegetation monitoring, surface wetness and temperature; socio-economic data such as availability and prices of food grains, availability of fodder, migration of livestock and human. The data is analyzed to monitor, predict, or declare drought.

At the national level, IMD issues short (1–3 days), medium (5–7 days), long range (2–4 weeks), monthly and seasonal rainfall and temperature forecasts, which are useful for drought prediction. The National Agricultural Drought Assessment and Monitoring System (NADAMS) became operational in 1989 (Das et al., 2007). National Remote Sensing Centre, Department of Space, has been assessing and monitoring agricultural drought since 1989 under the NADAMS. Since 2012, Mahalanobis National Crop Forecast Centre has been carrying out drought assessment and monitoring under the NADAMS after the technology transfer. District/sub-district level monthly drought assessments are carried out for 14 major drought-prone agricultural states, using satellite data for multiple vegetation indices (Gautam and Bana, 2014), rainfall deficiency (or SPI) using meteorological data, soil moisture index (from agro-meteorological modeling and satellite data), irrigation statistics and sown area data (MoAFW, 2017). Reservoirs storage level is measured by Central Water Commission. At the state level, agriculture, irrigation, economics and statistics departments monitor drought through rainfall data, crop condition, rural employment situation, migration, fodder and drinking water availability, etc.

3.2.6 Drought Declaration

During the pre-DM Act era, the directives provided in the FC were the basis for drought declaration (Rathore, 2005). Designated government officials from village to tehsil (sub-district), district, region, and state level, used to undertake monitoring and reporting on a regular basis. The *Patwari* (village revenue record keeper) was responsible for documenting changes in crop conditions, fodder and drinking water availability. He was responsible for providing detailed records on crop production (including losses) at the end of the cropping season, which were further compiled at tehsil, district and Relief Commissioner level. These reports would then become the basis for drought declaration, following stepwise actions, authority responsible for taking action and schedule of responses (Rathore, 2004). Because of very high variability in low rainfall areas of Rajasthan, a village was the unit of drought declaration.

A few states like Chhattisgarh solely relied on rainfall deficiency to declare a drought (Sharma, 2019). Other States also considered crop losses while declaring the drought. For crop loss estimation, they followed the traditional *annawari*, *paisewari* or *girdawari* system (Sharma, 2019). These systems were based on crop-cutting experiments to estimate yield reduction compared to a normal year and were highly subjective due to different benchmarks and methods employed for yield loss assessment. The threshold for drought declaration varied from state to state ranging from 25% to 75% yield loss compared to the previous normal year and the state governments had the right to declare or not to declare drought if yield losses were moderate (Samra, 2004). In addition to rainfall deficiency and crop loss estimation, the availability of drinking water, irrigation water, food availability and fodder availability for animals at the local level were also considered before declaring the drought (DAC, 2009). It is clear that following multiple methods and parameters of crop losses estimation, drought declaration and post-drought relief policy of states also varied significantly, and it was highly subjective and flexible. The FC guidelines were often violated for various political, financial and administrative reasons. Earlier when mainly the states were responsible for incurring drought relief expenditure, they used to delay or avoid drought declaration (Rathore, 2005). Under-reporting of drought-affected areas and limited relief work were also common.

The situation somewhat changed after 1995, with the provision of Calamity Relief Fund. The share of relief expenditure by state and central governments became 25% and 75%, respectively. This prompted the state governments to declare drought much earlier and at a bigger scale i.e. at the district level as compared to a village. An early declaration and start of relief work in 2002–03 in Rajasthan is a good example, very much welcomed by the drought-affected population (Rathore, 2005). However, because of the opacity of procedures, the process of drought declaration effectively became a routine heavily politicized, exercise of bureaucracy, based on other considerations (Rathore, 2005).

The MoAFW published a Manual for Drought Management in 2009 and suggested certain criteria for drought declaration. However, the guidelines of 2009 were suggestive and not binding in nature. As a result, until 2015, the drought assessment method and declaration process varied from state to state, and it gave a wide scope to states to declare or not to declare drought. For example, during the 2015–16 drought, Karnataka declared drought in August 2015, Gujarat in April 2016, while Haryana and Bihar were even unable to decide for a long time whether drought was there or not (Sharma, 2019). The Supreme Court of India, while hearing several petitions filed before it in 2016, directed the GoI to frame a national drought policy including a new system for drought declaration (Sen and Bera, 2016). As a result, the MoAFW updated the 2009 drought manual in 2016. The Manual for Drought Management was subsequently updated in 2020 as well. The new manual prescribes standardized, transparent and mandatory rules for drought declaration within a reasonable time frame. The traditional system of crop loss estimations (*annawari, paisawari, girdawari*, etc.) was replaced with more advanced field and remote sensing techniques. It recommended that the *kharif* (rainy, ISM) season drought should be declared by 30 October and the *rabi* (winter) season drought by 31 March.

Considering the highly complex nature of drought, sometimes compounded by poor availability of reliable data, the manual recommended five categories of indices viz., rainfall, vegetation, water, crop and others for drought monitoring and declaration. Rainfall is considered to be the most important indicator and other indices are to be evaluated in conjunction with the rainfall-related data to assess the impact of rainfall deficiency.

Drought can be declared in any given administrative unit based on rainfall data (rainfall deviation, SPI, dry spell); remote sensing data based vegetation indices such as NDVI, Normalized Difference Wetness Index (NDWI), Vegetation Condition Index (VCI); crop situation related indices such as sown area; soil moisture based indices like Percent Available Soil Moisture (PASM), Moisture Adequacy Index (MAI); hydrological indices such as Reservoir Storage Index (RSI), Groundwater Drought Index (GWDI), Stream-Flow Drought Index (SFDI); other factors like fodder availability, food grain availability and prices, drinking water availability, migration of rural labor and increase in job demand; followed by ground truthing in at least 10% drought affected area (DAFW, 2020a). Rainfall deviation or SPI or dry spell will be the primary criteria to assess if the first drought trigger is set off. In case the first drought trigger is set off, the impact indicators will be examined as per Trigger 2 (Table 3.6).

Trigger 2 becomes the base for drought severity declaration. The states may choose any three of the four Trigger 2 (one from each) for the assessment of drought and its severity. Drought will be considered severe if at least two of the selected three impact indicators are in 'severe' and one in 'moderate' category. Drought will be moderate if at least two of the selected three impact indicators are in 'moderate' category.

TABLE 3.6
Rainfall and Impact Indicators for Drought Declaration

Level	Category	Index/Parameter
Trigger1 (cause)	Rainfall	Rainfall deviation/SPI/Dry spell
Trigger 2 (effect)	Remote sensing	NDVI/NDWI/VCI
	Crop situation	Crop area sown
	Soil moisture	PASM/ MAI
	Hydrological	SFDI/RSI/GWDI
Verification	Field data	Ground truthing in at least 10% of affected villages

Source: Adapted from *Manual for Drought Management* 2020 (DAFW, 2020a).

The state governments then can declare moderate or severe drought through a notification clearly specifying the geographical extent and administrative units such as villages, tehsils (sub-districts) or districts by 31 October. However, deficit rainfall in June and July with prolonged dry spells leading to a significant reduction in crop sown area, can trigger the declaration of early drought (DAFW, 2020a).

For drought declaration during *rabi* (winter) season, different criteria are used for the four cropping situation namely (i) Rainfed crops depending on residual moisture, (ii) Surface irrigated command areas, (iii) Ground water irrigated areas, and (iv) North-East monsoon dependent areas. In addition to crop sown area, one or more indices based on soil moisture (PASM or MAI), remote sensing (NDVI or NDWI or VCI), hydrology (RSI or GWDI) and rainfall (ISM or NEM) are used for four different cropping situations. *Rabi* season drought can be declared by 31 March and the process of drought declaration is similar to the earlier one.

3.2.7 Drought Relief

After independence, the state governments adapted and amended the FC at different times, which remained the basis of famine prevention/drought response until about the 1970s. Agriculture being the State subject under the constitution of India, the state governments had varied drought monitoring and management approaches. For example, Revenue Department was responsible for every sort of relief work in Rajasthan state up to 1951, when Relief Department, manned by Relief Commissioner, was created (DM & Relief Department Rajasthan, 2022a). In 1962, the state government prepared its own FC for drought relief work. Till then, the relief work continued on ad-hoc basis broadly on British era FC. Based on the directions of the MHA (GoI), issued to all the states, the state government renamed Relief Department as Disaster Management and Relief Department in 2003. Disaster Management and Relief Department is Nodal department for drought, hailstorm, heat and cold wave, thunder and lightning, cyclone disasters in Rajasthan. Relief work and activities are carried out in coordination with different departments/organizations like Public Works Department, forest department, soil conservation department, Public Health Engineering Department, *Panchayati Raj* (self-government of villages or village council) Department, Revenue Department, local bodies, etc.

In Rajasthan, the FC enacted in 1962, was intended to replace the ad hoc nature of the relief operations that were undertaken from time to time and to codify the procedures, duties and responsibilities of the state government. The Code required the Relief Commissioner to: (i) arrange for the provision of funds to undertake relief measures, (ii) to formulate proposals to set-up an organization to deal with the scarcity or famine conditions, and (iii) to coordinate activities of different departments and local bodies to provide effective relief. The following types of relief measures were supposed to be taken during scarcity/drought conditions:

I. Relief works: (Mainly related to providing employment to the needy)
II. Public Works Department: (i) Roads (ii) Other works;
III. Irrigation Agency: Tanks, canals, etc.;
IV. Revenue Agency: (i) Construction of tanks, (ii) Other works, (iii) Construction of roads.
V. Relief to the people employed otherwise than on relief works.
VI. Gratuitous relief.
VII. Miscellaneous: (i) Water supply arrangements, (ii) Cattle conservation and fodder arrangements, and (iii) Transportation charges on movement of food grains in scarcity areas.

As in the British era, governments relied mainly on giving employment through public works for ensuring adequate food availability to households. The Code states that: 'The program of public works shall provide for 20% of the population of the areas for a period of six months after taking into consideration the likelihood of scarcity of famine in each area'. It is evident from Table 3.7 that, during the pre-DM Act era, except a few years (1983–84, 1988–89, 1990–91, 1994–95, 2003–04) from 1981–82 to 2009–10, several districts of Rajasthan faced drought-related scarcity and the government had to provide employment (DM & Relief Department Rajasthan, 2022b). It is also evident from a meager number of labor engaged per drought-affected village that relief employment was most probably not provided in all the drought-affected villages.

TABLE 3.7
Labor Employed in Rajasthan for Drought Relief from 1981–82 to 2009–10

Year	No. of Districts Affected	No. of Villages Affected	Labor Ceiling ('000)	Labor Engaged ('000)	Labor Engaged/ Affected Village
1981–82	26	23,246	739	662	32
1982–83	26	22,606	692	625	31
1984–85	21	10,276	294	193	29
1988-86	26	26,859	1,100	1,044	41
1986–87	27	31,936	1,735	1,597	54
1987–88	27	36,252	2,043	1,992	56
1989–90	25	14,024	440	414	31
1991–92	30	30,041	790	540	26
1992–93	12	4,376	115	71	26
1993–94	25	22,586	885	792	39
1995–96	29	25,478	807	728	32
1996–97	21	5,905	55	37	9
1997–98	24	4,633	133	83	29
1998–99	20	20,069	639	612	32
1999-00	26	23,406	1,597	1,516	68
2000–01	31	30,583	1,740	1,699	57
2001–02	18	7,964	167	161	21
2002–03	32	40,990	3,735	3,796	91
2003–04	3	649	67	67	103
2004–05	31	19,814	1,800	1,674	91
2005–06	22	15,778	1,859	1,801	118
2006–07	28	11,944	518	408	43
2007–08	18	6,790	150	97	22
2008–09	12	7,402	150	124	20
2009–10	27	33,464	283	178	8

Source: DM & Relief Department Rajasthan (2022b).

As per the revised guidelines effective from 2016 (DAFW, 2020a), drought relief and response measures such as rural employment, provision of drinking water, cattle camp and fodder banks, remission of land revenue, other taxes and dues, deferment and restricting of crop loans, agricultural input subsidy etc., are initiated following the declaration of drought by the state government. The objective of the drought relief policy is to provide financial and material assistance for livelihood and survival. The state governments are supposed to undertake drought relief measures using funds available in State Disaster Response Fund (SDRF), if drought is moderate. In case of severe drought or large-scale moderate drought, central assistance may be sought, following established guidelines. Drought relief requires coordination among departments of finance, agriculture, horticulture, animal husbandry, water resources, irrigation, power, drinking water, rural development, social welfare, public distribution, school education, public health, etc. It is the responsibility of the state (relief commissioner) and district (collector) authorities to coordinate activities among different departments. All the relief activities in a district are performed under the supervision of the district collector.

The most important post-drought relief component is still the employment generation. Due to drought, people look for alternative employment. As soon as drought is declared, it is necessary to start relief employment programs immediately and provide employment to the needy in the near vicinity.

Mahatma Gandhi National Rural Employment Guarantee Act (MGNREGA), effective from 2006, provides the legal framework for at least 100 days of guaranteed wage employment in a financial year in rural areas to every household whose adult members volunteer to do unskilled manual work. In addition to this, it is necessary for the state government to converge all schemes and programs of both center and state governments, which have the potential for employment generation such as *Pradhan Mantri Krishi Sinchai Yojana* (PMKSY, Prime Minister Agricultural Irrigation Scheme), Local Area Development Schemes for the Members of Parliament and Members of State Legislature, etc. It has been suggested to give priority to water conservation, water harvesting works (such as check dams, gabion structures, and percolation tanks), and minor irrigation works (such as tanks and farm ponds, canal excavation, community wells, afforestation).

Water resource management is another critical task of relief operations. Assured supply of drinking water for human and cattle populations, even through emergency measures like water-trains, tankers, bullock carts, etc. is one of the most important responsibilities of the government. It requires efficient utilization and management of water resources in both urban and rural areas, augmentation of water supply, rationing of water use, etc.

State government has to ensure the availability of sufficient food grains at reasonable price. The National Food Security Act 2013 covers about 75% of the rural population and up to 50% of the urban population, for providing subsidized food grains under Targeted Public Distribution System. State government can request the GoI for additional allocation of food grains, over and above allocation under the National Food Security Act, to deal with special situations arising due to drought. The nutritional aspects of food security should be addressed through Integrated Child Development Services and the mid-day meal scheme. The Integrated Child Development Service is implemented for pre-school children, while mid-day meal has been introduced for school children. The district collector can start community kitchens during drought situation for old, disabled and extremely distressed people. There is provision of gratuitous assistance in the form of cash or food to old, disabled and destitute persons in drought-affected areas as per National Disaster Response Fund (NDRF) or SDRF norms.

Providing necessary assistance for fodder, feed, and cattle health services is another responsibility of the state governments. Establishment of fodder depots in the drought-affected areas for selling fodder, cattle feed and concentrate at a fixed price; procurement of fodder from neighboring districts/states, forest department, etc. are done to ensure sufficient feed and fodder availability. Cattle camps or feeding centers are set up either through district administration or NGOs for feeding the stray or disowned cattle. Surveillance of water and vector-borne diseases, provision of safe drinking water, health camps in drought-affected areas for common ailments, etc. are done by the health department. State government may take decisions on remission of land revenue, waiver; deferment of water, irrigation, electricity and other farm-related charges; restructuring of agricultural loans, concessions, etc., taking into account the fiscal situation of the state and severity of the drought.

Finally, states need to conduct drought impact assessment and evaluation after each drought event through sample-based household surveys to capture the impact of drought and evaluate the response system. The information available through various drought indices needs to be considered in generating a random sample for inventory.

3.2.8 Risk Prevention and Mitigation

Prevention and mitigation are two important aspects of disaster risk reduction. Risk prevention means preventing the creation of new risks of disasters from the beginning, risk mitigation implies that existing risks of disasters are reduced to minimize or lessen the adverse impacts of disasters on lives, livelihoods and assets of individuals, communities and countries (MHA, 2019). Risks are created when vulnerable conditions are exposed to the hazards, they can be prevented by three approaches, either in isolation or in combination (MHA, 2019). First, risks can be prevented if hazards are eliminated, but elimination of natural hazards like drought is nearly impossible. Second, exposure of vulnerabilities to hazards can be reduced through policies and regulations. Third, vulnerabilities can be reduced through a combination of various structural and non-structural measures, so that even if exposed to hazards, people can withstand the adverse impacts. Preventative measures mainly belong to the water planning domain for reinforcing the structural system to increase its response capacity towards droughts.

When vulnerable conditions are exposed to hazards, the risks can be mitigated by reducing vulnerabilities and strengthening capacities. Vulnerability to drought is the susceptibility to be negatively affected by drought (Adger, 2006), its opposite being resilience, i.e. the ability to cope successfully with drought and overcome its impacts. Both of these reflect the adaptive capacity of a community (Engle, 2013), which is decided by the actions taken to mitigate drought impacts and increase drought preparedness (Wilhite et al., 2014). Wilhite et al. (2005) classified actions for drought preparedness in a ten-step process, which was further refined for national drought management policies by WMO and GWP (2014). High-level meeting on national drought policy recommended considering both long- and medium-term measures for drought preparedness and mitigation (Sivakumar et al., 2014).

3.2.8.1 Soil, Water and Land Resources Management

Long-term investment in the improvement of soil, water and land resources are some of the measures for drought mitigation as they can support agricultural activities better and longer during rainfall deficit periods. The MoAFW has suggested some structural measures (DAFW, 2020b) for drought mitigation which include (i) creation of storage facilities such as drinking water storage and distribution facilities, fodder storage facilities, and rainwater harvesting systems, (ii) water conservation structures such as water harvesting and storage structures, check dams and reservoirs, groundwater recharge augmentation systems, and (iii) social housing schemes for rainwater harvesting and storage especially in drought-prone areas. Major non-structural measures include (i) mitigation measures like promotion and better implementation of watershed development projects, convergence of lessons learnt by research institutions, promote private participation in drought management, risk management for dryland/rainfed farmers through agricultural extension and financial institutions, (ii) promote soil and water conservation, and efficient water use such as judicious use of surface and groundwater, promote water efficient irrigation systems (sprinklers, drip, etc.), and protected agriculture, provide advice to farmers for crop management under drought conditions, training in water and soil moisture conservation, promote afforestation and other options using economically useful vegetation, (iii) agricultural inputs, credit, finance, marketing and crop insurance such as ensuring availability of quality agricultural inputs, provide need-based credit and financing products in drought-prone areas, promote agricultural insurance programs, marketing support, (iv) implementation of risk transfer arrangements including multi-hazard insurance for life and property, (v) reducing climate change impact such as initiate and promote measures for reducing the impact of climate change on drought, implement various water and soil conservation programs taking into account climate change impacts.

3.2.8.2 Need for Soil Health Act in India

Land is finite and one of the most important natural resources. The National Commission on Agriculture (1976) gave emphasis on scientific land use planning for achieving food security, self-reliance and enhanced livelihood security. The GoI also felt the need for a policy framework to ensure optimal utilization of land resources through appropriate land use planning and management (Department of Land Resources, 2013). However, the emphasis of 2013 Land Utilization Policy was on integrated and comprehensive development planning to meet land use requirements of different sectors, and it also gave emphasis on the protection of prime farmlands, such as the command areas of irrigation projects, double cropped land etc. from land use conversions.

In agricultural lands, crop productivity has increased substantially during the last five to six decades due to the availability of high yielding varieties, fertilizers, irrigation facilities, plant protection chemicals, etc. However, yields are stagnating and factor productivity is declining due to inherently low nutrient content in soils, nutrient depletions and emerging nutrient deficiencies because of crop removal and imbalanced fertilizer use (Das et al., 2022). The majority of agricultural holdings in India are low in soil organic matter and nitrogen contents. The farmers' field data collected from 12 states ($n = 178,037$) showed that 55% of soil samples were low in soil organic carbon content with an average value of 0.54% (Das et al., 2022), and the majority of soils were low in major and micronutrients as well. Over 70% of soils also suffered either from soil acidity or alkalinity.

The Sustainable Agriculture Mission is one of the eight missions outlined under India's National Action Plan on Climate Change. The National Mission for Sustainable Agriculture has been formulated for enhancing agricultural productivity, especially in rainfed areas, focusing on integrated farming, water use efficiency, soil health management and resource conservation (DAC, 2014). Soil Health Management (SHM) is one of the four major programs of this Mission. The SHM aims at promoting location and crop-specific sustainable soil health management, including appropriate land use based on land capability classes, minimizing soil erosion/land degradation, residue management, organic farming, judicious use of fertilizers, etc. It supports soil and land resource surveys for creating a comprehensive database; promotes integrated nutrient management through judicious use of chemical fertilizers, organic manures and bio-fertilizers; augmentation and strengthening of soil and fertilizer testing facilities; quality control of fertilizers, bio-fertilizers and organic fertilizers; reclamation of problem soils through soil amendments and land development; capacity building of stakeholders, etc.

Globally, a need is felt for the management of soil health through judicious and prudent governance (Lal, 2020) and appropriate legal framework is being developed to serve this purpose. A bill introduced in the United States Senate modifies the Environmental Quality Incentives Program to provide permanent incentive payments to producers to adopt practices designed to improve soil health through increasing carbon levels in soils (Congress.gov., 2020). It emphasizes studying changes in soil health and soil organic carbon levels due to adopted practices; development of criteria that prioritize practices which score highest in soil carbon sequestration; create a soil organic carbon conservation activity plan to measure and monitor sequestration and mitigation improvement levels from the practices etc. Similarly, the European Commission proposed the Soil Health Law (European Commission, 2022) to specify the conditions for healthy soil, determine options for monitoring soil and lay out rules conducive to sustainable soil use and restoration.

Prevention of soil erosion and land degradation as well as maintenance of favorable soil physical, chemical and biological properties are essential for good soil health. The soil organic matter plays a very crucial role in maintaining soil health. Its quantity, quality and turnover moderate soil quality and functionality comprising of physical, chemical, biological and ecological properties and processes (Lal, 2020). But the long-term experiments conducted in different agro-climatic regions of India have shown that improving the organic carbon status of soils through external inputs, even at a fairly high rate of 5.0–10.0 t/ha per year, is difficult (Das et al., 2022). It may be primarily due to the tropical and sub-tropical climate of the country where temperatures remain high during most of the year. Multi-location trials conducted for 20 years at 28 benchmark sites covering arid, semi-arid and moist humid tropical locations in India revealed that the Vertisols had higher carbon sequestering potential than the Alfisols, the legume-based systems sequestered more carbon than the cereals, and horticultural crops and grasslands sequestered more carbon than the field crops (Bhattacharyya et al., 2007). In the Indian context, an assessment of soil health should be done keeping in mind the management goals with associated soil functions for different soil types and cropping systems followed by the screening of relevant soil physical, chemical and biological indicators along with their optimum and threshold value (Das et al., 2022). Therefore, legislation is needed in India to restore, maintain and/or improve soil health; to encourage management practices that contribute to this goal, while strongly discouraging the practices which have adverse impacts on soil health. Such act will ensure sustainable land use and increased crop productivity in normal rainfall years, as well as more drought-resilient production systems.

3.2.8.3 Major Government Programs

Some of the significant national programs that may have a decisive bearing on drought mitigation are PMKSY, Rainfed Area Development Programme, MGNREGS, etc. (DAFW, 2020a). Many of these programs can contribute to drought mitigation strategy at the state level due to the flexibility which has been in-built into the centrally sponsored schemes for the purposes of mitigation of calamities like drought.

PMKSY was initiated in the financial year 2015–16 with the vision of extending irrigation coverage ('*har khet ko pani*', water to every farm) and improving water use efficiency ('more crop per drop'). The major objectives of PMKSY include the convergence of investments in major and medium irrigation projects with command area development, improving water use efficiency, enhancing adoption of precision-irrigation and other water-saving technologies, scientific development of watersheds, enhancing recharge of aquifers and introduction of water conservation practices and greater private investment in precision irrigation system. District Irrigation Plans are the backbone for planning and implementation of PMKSY. These plans include a holistic irrigation development perspective including watershed development, medium- and long-term irrigation plans integrating three components viz., water sources, distribution network and water use for all the users.

The objectives of the Rainfed Area Development Programme are to increase agricultural productivity in rainfed areas in a sustainable manner by adopting appropriate farming system-based approaches, minimize adverse impact on crops due to drought and other calamities through diversified and integrated farming systems, etc.

Some states have done excellent work on water harvesting, soil and water conservation, etc., during the last few years using MGNREGS funds. Based on such success stories, there is a paradigm shift from relief works approach to an integrated natural resource management approach in the implementation of MGNREGS. Planned and systematic development of land and harnessing rainwater following watershed principles has now become the central focus of MGNREGS works.

National Mission on Sustainable Agriculture emphasizes the development of drought and heat-tolerant varieties, crop diversification, soil health management, soil and water conservation, micro-irrigation, integrated faring systems, etc. Numerous other central and state government programs are contributing towards making agriculture drought resilient. Several institutional arrangements such as the establishment of seed banks, fodder banks, community nurseries, custom hiring centers, etc. play a vital role in dealing with drought risk (Srinivas et al., 2017). Another institutional arrangement that brings together public and private players, knowledge and technologies is the delivery of weather-based agro-advisories. The decisions taken by farmers following such advisories were found to be useful in reducing crop losses and increasing returns (Dupdal et al., 2020).

3.2.8.4 Financial Products

Financial products such as credit and insurance cover are useful in dealing with income risk emanating from drought incidence. Farmers' access to institutional credit has increased tremendously over the years. However, the penetration of these products is rather limited in India due to the supply side constraints to institutional credit such as linking credit to collateral security, income risk, high transaction costs, and the demand side constraints such as low financial literacy, procedural difficulties, low and variable profitability, smaller farm size, cultural and habitual rigidities. Agricultural insurance has evolved considerably since the drought of 1965–66, learning from the experiences over the years. For combating the adverse financial impacts of drought and other calamities, the system of crop insurance has been in place since 1972, it was not successful in risk mitigation owing to lack of transparency, high premiums, delay in estimating crop losses, and non-payment or delayed payment of claims to farmers. Subsequently, the National Agricultural Insurance Scheme was introduced in 1999 and Weather Based Crop Insurance Scheme in 2007 (Rathore et al., 2014). In 2016, *Pradhan Mantri Fasal Bima Yojana* (Prime Minister Crop Insurance Scheme), an area-based scheme and Restructured Weather Based Crop Insurance Scheme, was introduced (Gulati et al., 2018). Though *Pradhan Mantri Fasal Bima Yojana* has covered some shortcomings of previous schemes, it has fallen short of meeting its own targets. During the implementation year of 2016–17, an increase in coverage of cultivated land from 23% to 29% was observed, but the coverage remains much lower than its own target of 50% (Alexander, 2019). Direct payments in the form of cash transfers, relief payments, input subsidies are also playing a prominent role in managing drought and other risks to agriculture.

3.2.8.5 Development and Spread of Drought Mitigation Technologies

A large number of natural resource and crop management technologies have been developed, promoted and spread by agricultural research institutes and universities and state line departments for minimizing the adverse impacts of drought on crop production and farm incomes. The National Innovations in Climate Resilient Agriculture project has also tested and spread such technologies in several parts of the country.

Natural resource management technologies that are found suitable for drought-prone areas include in-situ moisture conservation structures such as contour bunds, ridges and furrows, broad bed furrow, conservation furrow, vegetative barriers and mulches (Gudade et al., 2016; Mishra et al., 2020); rainwater harvesting structures like excavated or dug out ponds, surface ponds, spring or creek fed ponds and off stream storage ponds (Rama Rao et al., 2019b); conservation agriculture (Chatterjee et al., 2020) including no tillage, ridge tillage, zone tillage, mulch tillage; artificial groundwater recharge, to name a few. Similarly, crop management technologies include the development of short duration, drought tolerant and early maturing crop varieties (Prasad et al., 2014); crop diversification; agro-climatic zone-specific integrated farming systems including suitable intercropping systems (Rama Rao et al., 2003), integrated water, nutrient and pest management technologies; direct seeded rice (Kaur and Singh, 2017); drum seeding of rice (Tayade, 2017), etc. Such technologies have great potential to minimize drought-induced yield losses in agriculture (Rama Rao et al., 2018). It was further observed that the adoption of resource conservation technologies in tandem with the tolerant varieties would be more effective in securing crop yields in case of drought incidence. Some of such technological measures can be adopted *ex ante* or *ex post*.

3.3 ECONOMIC POLICY OF DROUGHT MANAGEMENT

Millions of people suffered during the Bengal famine of 1943, a consequence of severe drought and drought management policies. India had faced food shortage before and after independence. 'Grow-more-food' campaign was launched in 1948, as a program of crop production, field demonstration and contact with the farmers to introduce the improved techniques. Central and state governments started several schemes to increase food grain production such as creation of irrigation facilities, soil and water conservation schemes, development and popularization of improved varieties and crop management practices, etc. Numerous such schemes and programs continued and are still continuing, which have enhanced productivity and resilience of Indian agriculture. Several such policies and programs have addressed drought monitoring and forecasting, vulnerability reduction and drought mitigation, rural employment generation, etc.

WMO and GWP (2017) classified the costs of action against droughts into three categories: (i) preparedness costs; (ii) drought risk mitigation costs; and (iii) drought relief costs. A large number of government programs and schemes cover drought preparedness and proactive mitigation costs, which is not possible to cover here. Only the economic policies directly related to drought relief costs or disaster management costs are discussed in this section. Financial allocations in India and guidelines for their utilization, are made to cover all the specified disasters and disaster-specific allocations are rarely made.

3.3.1 Financial Mechanism and Provision for Disaster Relief

As per the federal structure envisaged in the Indian constitution, taxation is mostly controlled by the central government, while spending is largely done by state governments. Hence, the seamless transfer of resources from the center to the states is very crucial. Finance Commission (FC) of India, under article 280 of the constitution, has a mandate to assess the financial needs of the states and allocates relief funds to states for coming five financial years (April–March) towards natural disaster relief including drought, and therefore, the post-independence history of financing relief expenditure is the history of awards of the FCs (Rathore, 2005). The FC also lays down rules by which the center provides grants-in-aid to the states from the Consolidated Fund of India. There has been an understanding that the primary responsibility of dealing with natural calamities rests with the states, providing financial support for this purpose substantially depends on the center.

The first FC (FC-I, 1952–57) made provision of central assistance equivalent to 50% of requirements for relief employment works. This was provided in the form of loans and a grant (not exceeding INR 20 million per state per annum). An option of further assistance was also kept in the form of advances to states which need to tackle severe natural calamities. The FC-II (1957–62) felt a need to make some recurring funds available to states to meet immediate requirements of relief in less severe calamities (Finance Commission, 2020). This was known as a 'margin money' scheme and each state was sanctioned specified margin money based on its relief expenditure during the past decade. Total allocation gradually increased from INR 61.5 million per annum (FC-II, 1957) to 2407.5 million per annum in 1984–89 (FC-VIII). If expenditure exceeded the margin money with the state, central assistance in the form of loan (50%) and grant (25%) was made available. The FC-IV introduced a system of visit of calamity affected areas by central teams, if the state demanded additional funds. Amounts in excess of margin money were to be determined by central teams. Additional central assistance was provided only when the relief requirements of a severe calamity could not be met from margin money.

The FC-IX (1989–95) extended the concept of margin money and recommended the establishment of a Calamity Relief Fund (CRF) for each state, and its size was decided on the basis of the average of the actual ceiling of expenditure approved for the state during the decade ending 1988–89. Contributions of central and state governments in The CRF were 75% and 25%, respectively. The contributions of the center and states were to be credited twice a year in May and November. The CRF concept was not very different from the margin money scheme, but it increased the contribution of the GoI. The FC-IX also introduced the 'normative approach' to relief expenditure, which envisaged expenditure from CRF on predetermined items (norms) at predetermined scales. The FC-X (1995–2000) continued the CRF with certain modifications in operational arrangements and also recommended the setting up of a National Fund for Calamity Relief (NFCR) to assist a state affected by a calamity of rare severity. The size of this fund was INR 7,000 million, which was to be built over 5 years, with an initial corpus of INR 2,000 million. The FC-XI (2000–05) continued

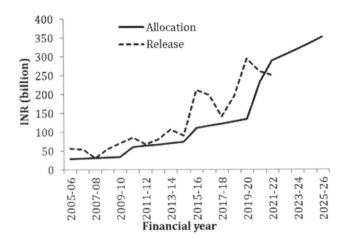

FIGURE 3.7 Allocation (2005–06 to 2025–26) of total funds for disaster management and release (2005–06 to 2021–22) from central pool for all the states. Source: For 2005–06 to 2020–21: DMD, 2022; for 2021–22 to 2025–26: Finance Commission (2020).

CRF, but found that the entire corpus of the NFCR was exhausted in 3 years and the fund failed to make available adequate amount for meeting the requirements of severe calamities. It dissolved the NFCR and recommended the setting up of a National Calamity Contingency Fund (NCCF) with an initial corpus of INR 5,000 million which was to be recouped through the levy of a special surcharge on central taxes. The FC-XII (2005–10) recommended the continuation of the CRF and NCCF in their existing forms.

When the FC-XIII (2010–15) started deliberations on disaster relief and management, the DM Act 2005 had already come into effect, with provisions for Response and Mitigation Funds. Based on these statutory provisions in the Act, it recommended the merger of CRFs into the SDRFs of individual states and that of NCCF into the NDRF. The contribution to the SDRFs was shared between the center and states in the ratio of 75:25 for general category states and 90:10 for special category states (eight Northeast states and two hill states - Himachal Pradesh and Uttrakhand). The total size of the SDRFs was kept at INR 335,810 million. It also recommended an additional grant of INR 5,250 million for capacity building to the states. The FC-XIV (2015–20) arrived at an aggregate corpus of INR 612,190 million for SDRFs. It also recommended that up to 10% of the funds available under the SDRF could be used by states for local natural disasters, not included in the notified list of disasters of the MHA. Total fund allocation and released from the central pool (SDRF and NDRF) are given in Figure 3.7.

3.3.2 Financial Provision for Mitigation

The FC-XV has given its recommendations for the 2020–25 period (Finance Commission, 2020). It has recommended the establishment of the National Disaster Mitigation Fund (NDMF) and State Disaster Mitigation Fund (SDMF), which were not created earlier despite statutory provisions in the DM Act 2005. It recommended that the mitigation funds should typically provide small grants for community-based local initiatives to mitigate hazards through soft measures, rather than through hard measures; however, large-scale mitigation interventions should be pursued through regular development schemes.

Overall allocation of INR 1,601,530 million to State Disaster Risk Management Fund has been split in 80:20 ratio to SDRF (INR 1,281,220 million) and SDMF (INR 320,310 million). Similarly, total allocation of INR 684,630 million to the National Disaster Risk Management Fund has been divided between NDRF (INR 547,700 million) and NDMF (INR 136,930 million) in 80:20 ratio.

Relief and mitigation funds are not interchangeable. Contribution by center and state to SDMF will be in the ratio of 90:10 for NE and Hill states and in 75:25 ratio for other states. It divided the NDRF and SDRF into three sub-windows (i) response and relief (40%), (ii) recovery and reconstruction (30%), and (iii) preparedness and capacity building (10%). However, there is some flexibility for re-allocation within three sub-windows.

Additional financial assistance from NDMF will be done on a graded cost-sharing basis i.e. states to contribute 10% for assistance up to INR 2,500 million; 20% for assistance up to INR 5,000 million; and 25% for assistance exceeding INR 5,000 million.

NDMF can be used for projects involving two or more states or which require national agencies to collaborate; important state projects, where SDMF is insufficient; projects of national and strategic significance, eco-systems, and natural resource base for the notified disasters; and research and studies related to disaster mitigation. Minimum finance of INR 100 million will be done for a mitigation project from NDMF for structural measures, but the funds cannot be a source of funding for existing programs for the maintenance of any structure. SDMF can be used for projects for mitigation measures against notified disasters, which promote practices to reduce risks and their impacts; important state-level projects, protecting assets, eco-systems, and settlements; creating safe conditions of living for people from weaker sections; building community resilience through information and knowledge; research and studies related to disaster mitigation. At least 10% of the NDMF/SDMF should be earmarked for non-structural measures and not more than 50% of funds may be utilized for measures to mitigate risks from a single hazard. Up to 5% of the NDMF and SDMF may be earmarked for small grants window for research/studies.

3.3.3 Funds for Drought Mitigation

The FC-XV has earmarked allocation of INR 12,000 million from NDMF as catalytic assistance to 12 most drought-prone states (INR 1,000 million for each), namely Andhra Pradesh, Bihar, Gujarat, Jharkhand, Karnataka, Madhya Pradesh, Maharashtra, Odisha, Rajasthan, Tamil Nadu, Telangana, and Uttar Pradesh.

3.3.4 Framework for Drought Management Fund Utilization

State government can start utilizing the SDRF immediately after declaring drought, its coverage and intensity, as per the SDRF guidelines. A memorandum for assistance from NDRF has to be submitted to DAFW, a department of MoAFW, within a week of the declaration of drought, if the drought is severe. However, if SDRF is not sufficient to meet drought relief, even if the drought is of moderate category, the state may submit the memorandum for assistance under NDRF. Only those items may be included in the memorandum, which are admissible as per extant guidelines of the MHA. The DAFW constitutes and dispatches an inter-ministerial central team to the drought-affected area within a week of receipt of the memorandum and the team submits its report to the central government within 7 days after collecting all relevant information from the state government, following its visit to the drought-affected areas. The central government takes a final decision within a month of the receipt of the central team's report. The state government has to start fund utilization within 1 month from the date of receipt of assistance from the NDRF. The process was almost similar for CRF and NCCF.

States demand from NDRF is mostly for drought relief when it comes to managing agricultural calamities. The funds released from NDRF for agricultural calamities i.e. drought, hailstorm, pest attack and cold wave/frost; were mostly for drought relief during 2010–11 to 2021–22 (MoAFW, 2022). During this period, there was damage to crops due to hailstorms in several states in 2013–14 and 2014–15, and consequent request for assistance from the DAFW (MoAFW), otherwise drought was mostly the only cause for the release of funds from central government (Figure 3.8). Out of the total release of INR 559.5 billion by MoAFW during this period, INR 470.6 billion were released from NDRF for drought management only.

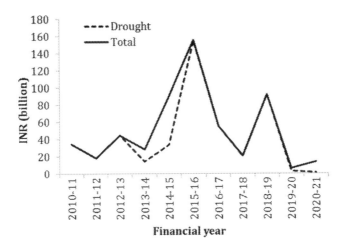

FIGURE 3.8 Release (INR in billion) from NDRF for all agricultural calamities (total) and drought alone (drought). Source: MoAFW (2022).

Among the 12 drought-prone states which received most of the drought-related assistance from NDRF, Maharashtra tops the list with a total receipt of INR 134.5 billion during 2010–11 to 2021–22 (Table 3.8), followed by Karnataka (INR 82.5 billion), Rajasthan (INR 41.8 billion) and Andhra Pradesh (INR 33.1 billion). About 55% of the total NDRF release for drought management went to the three worst drought-affected states (Maharashtra, Karnataka and Rajasthan). During these 11 years, Andhra Pradesh needed NDRF funds for 8 years, Karnataka and Rajasthan for seven times and Maharashtra for 6 years.

Funds allocated and utilized for drought relief by a very drought-prone state of Rajasthan during pre-DM Act and post-DM Act era are given in Figures 3.9 and 3.10. The requirement of this drought-prone state is very high, while a major contribution comes from the central government to the drought relief fund. The central government plays a dominant role in the whole process of decision-making as it has been a major contributor in the state funds (CRF/SDRF) and national funds (NCCF/NDRF). Relations between central and state governments, and the political parties in power, play a major role in sanction/approval from central government (Rathore, 2005). The situation becomes more complicated when a state also demands foodgrain as the stocks of foodgrain are owned by the central government. Quantum and timely release of foodgrain had become a big political issue in the past.

The objective of the new drought management policy was to remove the subjectivity and do a scientific and accurate assessment of the losses and provide reasonable relief. Now, if a moderate drought occurs, the state government is responsible for managing it with the SDRF or employing its own resources. In case of severe drought, the states may approach the central government for additional funds. States' dependence on the central government for relief finances is very high (Prabhakar and Shaw, 2018). As per the contemporary drought management framework, the central government has limited its financial obligation towards states. It almost withdrew itself it the drought is moderate and may help the states if severe drought occurs. With these changes, the central government's commitment to drought management has become limited. It now formulates the guidelines and systems for monitoring, early warning, declaration and relief process instead of directly intervening in the drought declaration and relief process. Even in case of severe drought, the central government acts only on state's demand rather than on its own for the additional allocation of relief funds.

Often the disbursement from NDRF may be lower than what has been demanded by states. It has also been pointed out that fund allocation in SDRF may not be in line with the vulnerability of the state to natural disasters, as Kamepalli (2019) pointed out that the FC-XIV allocated more

TABLE 3.8
Funds Released (INR in Billion) from NDRF for Drought Relief to 12 Most Drought Prone States during 2010–11 to 2020–21

State	2010–11	2011–12	2012–13	2013–14	2014–15	2015–16	2016–17	2017–18	2018–19	2019–20	2020–21	Total
Maharashtra	-	5.7	18.2	-	19.6	43.2	-	0.6	47.1	-	-	134.5
Karnataka	-	4.7	5.3	2.3	2.0	22.6	25.8	-	19.9	-	-	82.5
Rajasthan	-	-	3.2	-	-	11.9	5.9	5.3	12.1	2.3	1.1	41.8
Andhra Pradesh	-	7.1	1.4	2.5	2.4	4.3	5.2	1.1	9.0	-	-	33.1
Madhya Pradesh	-	-	-	-	-	20.3	-	8.4	-	-	-	28.7
Uttar Pradesh	-	-	-	-	7.8	19.3	-	1.6	-	-	-	28.6
Bihar	14.6	-	-	9.3	-	-	-	-	-	-	-	23.9
Tamil Nadu	-	-	6.2	-	-	-	17.5	-	-	-	-	23.7
Chhattisgarh	-	-	-	-	-	12.8	-	4.0	-	-	-	16.7
Jharkhand	8.6	-	-	-	-	3.4	-	-	2.7	-	-	14.6
Odisha	3.8	-	-	-	-	8.2	-	-	-	-	-	11.9
Gujarat	-	-	8.6	-	-	-	-	-	1.3	-	-	9.9
Total	34.2	17.5	44.6	14.1	33.5	146.7	55.6	20.9	92.1	2.3	1.1	462.7
No of States	4	3	7	3	5	10	6	6	6	1	1	

Source: MoAFW (2022).

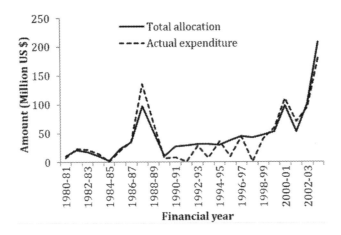

FIGURE 3.9 Source of drought relief funds in Rajasthan (million US $) from 1980–81 to 2003–04. Source: Adapted from Rathore (2005).

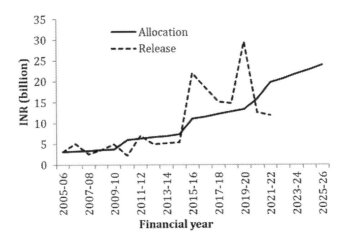

FIGURE 3.10 Allocation (2005–06 to 2025–26) of total funds for disaster management and release (2005–06 to 2021–22) from central pool to Rajasthan state. Source: For 2005–06 to 2020–21: DMD, 2022; for 2021–22 to 2025–26: Finance Commission (2020).

SDRF funds (INR 21,540 million) to Punjab compared to Karnataka (INR 15,270 million), though the vulnerability of Karnataka from all natural disasters, especially drought, is significantly higher.

3.4 CONCLUSION

India has very diverse agro-climatic conditions; some regions are very drought prone while others rarely face a drought situation. Northwestern parts of the country, in and around the Thar Desert, receive 150–500 mm annual rainfall, with very wide inter- and intra-seasonal variations, and are most drought-prone areas in the country. Rain-shadow areas of the Western Ghat Mountain Range, located in the peninsular India, also receive low and erratic rainfall, making it a drought-prone region.

    The country had faced very devastating famines, mostly a consequence of droughts, during the pre-independence period. Food security was one of the highest priorities of both central and state governments after independence in 1947, which took almost two to three decades to achieve. Several programs/schemes were launched to increase irrigated area, for soil and water conservation, water and nutrient management, development of location-specific crop varieties and agro-techniques, to name a few. These efforts not only increased foodgrain production, but made agriculture more resilient to weather vagaries. However, when a state or part of it faced drought, the old ad-hoc approach of FC continued for decades to provide relief and save human and livestock lives, with some improvement, here and there, after every major drought incidence. The enactment of DM Act in 2005, after numerous national and international consultations and India's commitment to some international agreements (Sendai Framework for DRR, SDGs, climate change) is considered a watershed moment in disaster management policy. Despite that, the drought management policies are far from perfect, but it is a good sign that they are still evolving.

Strengthening weather monitoring, forecasting and early warning system is the first important step to minimize adverse impacts of weather aberrations. A dense network for near real-time collection of hydrological and meteorological data, strengthening scientific capability to predict the weather/climate, and communication of predictions to all the stakeholders including farmers, government departments, finance and insurance institutions, etc., in a well-structured, understandable and actionable manner is a basic requirement for effective drought monitoring and assessment. Considering that drought incidence is spatially highly variable, reverse information flow between the locales of drought incidence to the decision-making bodies should also happen in real time.

Over 50 drought indicators and indices are available in literature, but not a single one of these indices connects weather anomalies to socioeconomic vulnerabilities, coping capacities or impacts of previous droughts to estimate future risk and related impacts. Virtually all climate monitoring and forecasting efforts concentrate on hazards rather than on impacts (Enenkel et al., 2020). Just identification of drought-prone areas is not sufficient, social and economic vulnerabilities must also become part of drought management plan. Efforts have been made to find objective drought declaration criteria, however, the critics point out that universal standard criteria may not be applicable to all soil types, crops, crop growth phases and regions (Sharma, 2019). Different crops have different water requirements, and sensitivity to water stress also depends on growth stage, so fixed soil moisture values, rainfall deficits, etc. may sometimes give very misleading results while giving very good results in other situations. Clear indicators and thresholds for the onset, severity and end of drought are pre-requisite, but very elusive to find out.

Awareness about the contingency crop planning to cope with weather aberrations, along with needed support systems are critical components of drought preparedness. If drought occurs, prompt and appropriate relief measures are needed so that households do not have to resort to erosive coping measures such as selling household and farm assets, discontinuing children's education, reduction in food consumption, which only further aggravate the risk and make the recovery to normal life longer and harder. Public and private investments should be promoted for rural industrialization in drought-prone areas so that employment opportunities are strengthened.

Conservation and sustainable utilization of natural resources, especially soil and water, is a priority area. A number of soil and water conservation works have been done through programs such as DPAP, DDP, MGNREGS, watershed development, etc., but the upkeep of these structures is rarely satisfactory, and the intended benefits are not realized. Community participation and ownership is probably the key element to achieve long-term gains. The creation of irrigation facilities, including canals, should be complemented with enhanced water use efficiency, which calls for a combination of technological and social engineering.

Droughts affect the economic decisions of farmers, not just when they happen but also when they do not. Farmers become reluctant to make investments in costly inputs, and credit institutions become reluctant to lend to farmers in drought-prone areas. Several crop insurance schemes have been launched, but the desired outcome is still elusive. It needs serious thought and planning.

Timely and sufficient availability of funds can save people, especially the vulnerable section, from severe hardship. A composite index for computing the required amount of funds for drought relief is as much required as an objective criteria for drought declaration, to eliminate political interference, real or perceived. It may include the extent and intensity of drought, irrigated area, human and livestock population dependent on agriculture, economic status of affected people, number of very vulnerable people, etc.

The elaborate institutional infrastructure already in place should be geared to function in a pro-active and coordinated manner to implement drought management plans effectively. Cooperation among the central, state and local governments along with the involvement of farmers in planning and execution of drought management strategies is a key requirement. Sharing experiences of both successes and failures, among different state, national and international organizations/departments is probably the best way forward.

LIST OF ABBREVIATIONS

CCS	Cabinet Committee on Security
CDRC	Central Drought Relief Commissioner
CMG	Crisis Management Group
CMP	Crisis Management Plan
CRED	Center for Research on the Epidemiology of Disasters
CRF	Calamity Relief Fund
CWWG	Crop Weather Watch Group
CWWGDM	Crop Weather Watch Group for Drought Management
DAC	Department of Agriculture and Cooperation
DACFW	Department of Agriculture, Cooperation and Farmers Welfare
DACP	District Agriculture Contingency Plan
DAFW	Department of Agriculture and Farmers Welfare
DDMA	District Disaster Management Authority
DDP	Desert Development Programme
DM	Disaster Management
DMC	Drought Monitoring Centre
DMD	Disaster Management Division
DMP	Drought Management Plan
DPAP	Drought Prone Areas Programme
DRR	Disaster Risk Reduction
ENSO	El-Nino Southern Oscillation
FC	Famine Code
GoI	Government of India
GWDI	Groundwater Drought Index
GWP	Global Water Partnership
IFRC	International Federation of Red Cross and Red Crescent Societies
IMD	India Meteorological Department
INR	Indian Rupees
IOD	Indian Ocean Dipole
IPCC	Intergovernmental Panel on Climate Change
ISM	Indian Summer Monsoon
MAI	Moisture Adequacy Index
MGNREGA	Mahatma Gandhi National Rural Employment Guarantee Act
MGNREGS	Mahatama Gandhi National Rural Employment Guarantee Scheme
MHA	Ministry of Home Affairs
MI	Moisture Index
MoAFW	Ministry of Agriculture and Farmers Welfare
MoRD	Ministry of Rural Development
NAAS	National Academy of Agricultural Sciences
NADAMS	National Agricultural Drought Assessment and Monitoring System

NCCF	National Calamity Contingency Fund
NCMC	National Crisis Management Committee
NDMA	National Disaster Management Authority
NDMF	National Disaster Mitigation Fund
NDMP	National Disaster Management Plan
NDRF	National Disaster Response Fund
NDVI	Normalized Difference Vegetation Index
NDWI	Normalized Difference Wetness Index
NEC	National Executive Committee
NEM	Northeast Monsoon
NFCR	National Fund for Calamity Relief
NGOs	Non-Governmental Organizations
NPDM	National Policy on Disaster Management
NRAA	National Rainfed Area Authority
PASM	Percent Available Soil Moisture
PMKSY	Pradhan Mantri Krishi Sinchai Yojana
RSI	Reservoir Storage Index
SDGs	Sustainable Development Goals
SDMA	State Disaster Management Authority
SDMF	State Disaster Mitigation Fund
SDRF	State Disaster Response Fund
SEC	State Executive Committee
SFDI	Stream-Flow Drought Index
SPI	Standardized Precipitation Index
UN	United Nations
UNDP	United Nations Development Programme
UNFCCC	United Nations Framework Convention on Climate Change
UNICEF	United Nations Children's Fund
UNISDR	United Nations Office for Disaster Risk Reduction
VCI	Vegetation Condition Index
WMO	World Meteorological Organization

REFERENCES

Aadhar, S., and V. Mishra. 2021. On the occurrence of the worst drought in South Asia in the observed and future climate. *Environmental Research Letters* 16: 024050. doi: 10.1088/1748-9326/abd6a6.

Adger, W. N. 2006. Vulnerability. *Global Environmental Change* 16(3): 268–281.

Aggarwal, P. K. 2003. Impact of climate change on Indian agriculture. *Journal of Plant Biology* 30: 189–198.

Aiyar, S. 2012. From financial crisis to Great Recession: The role of globalized banks. *American Economic Association* 102: 225–230.

Alexander, S. 2019. Crop insurance schemes need better planning. *Livemint.* https://www.livemint.com/industry/agriculture/why-crop-insurance-needs-to-be-better-designed-1562741755465.html (assessed August 14, 2022).

Algur, K., D. S. K. Patel, and S. Chauhan. 2021. The impact of drought on the health and livelihoods of women and children in India: A systematic review. *Children and Youth Services Review* 122: 105909. DOI: 10.1016/j.childyouth.2020.105909.

Ashok, K., Z. Guan, and T. Yamagata. 2001. Impact of the Indian Ocean dipole on the relationship between the Indian monsoon rainfall and ENSO. *Geophysics Research Lettetter* 28: 4499–4502.

Bahinipati, C. S. 2020. Assessing the costs of droughts in rural India: A comparison of economic and non-economic loss and damage. *Current Science* 118: 1832–1841.

Barlow, M., H. Cullen, and B. Lyon. 2002. Drought in central and southwest Asia: La Nina, the warm pool, and Indian Ocean precipitation. *Journal of Climatology* 15: 697–700.

Bevere, L. 2021. Natural catastrophes in 2020. *Swiss RE Sigma*. https://www.swissre.com/institute/research/sigma-research/sigma-2021-01.html.

Bhalme, H. N., and D. A. Mooley. 1980. Large-scale droughts/floods and monsoon circulation. *Monthly Weather Review* 108: 1197–1211. DOI: 10.1175/1520-0493.

Bhatia, B. M. 1967. *Famines in India 1860–1965*. London: Asia Publishing House.

Bhatia, B. M. 1985. *Famines in India: A Study in Some Aspects of the Economic History of India with Special Reference to Food Problem*. Delhi: Konark Publishers Pvt. Ltd.

Bhattacharyya, T., P. Chandran, S. K. Ray, D. K. Pal, M. V. Venugopalan, C. Mandal, and S. P. Wani. 2007. Changes in levels of carbon in soils over years of two important food production zones of India. *Current Science* 93: 1854–1863.

Bhuiyan, C., R. P. Singha, and F. N. Koganc. 2006. Monitoring drought dynamics in the Aravalli region (India) using different indices based on ground and remote sensing data. *International Journal of Applied Earth Observation and Geoinformation* 8: 289–302.

Blaikie, P., T. Cannon, I. Davis, and B. Wisner. 1994. *At Risk: Natural Hazards, People's Vulnerability, and Disasters*. London: Routledge.

Boustan, L. P., M. E. Kahn, P. W. Rhode, and M. L. Yanguas. 2020. The effect of natural disasters on economic activity in US counties: A century of data. *Journal of Urban Economics* 118: 103257. DOI: 10.1016/j.jue.2020.103257.

Brennan, L. 1984. The Development of the India's famine codes: Personalities, policies and politics. In *Famine: As a Geographical Phenomenon*, B. Currey, and G. Hugo, eds. Frankfort: D. Reidel Publishing Company, 91–110.

Brown, J. F., B. D. Wardlow, T. Tadesse, M. J. Hayes, and B. C. Reed. 2008. The vegetation drought response index (VegDRI): A new integrated approach for monitoring drought stress in vegetation. *GIScience & Remote Sensing* 45: 16–46.

Cai, W., G. Wang, B. Dewitte, et al. 2018. Increased variability of eastern Pacific El Nino under greenhouse warming. *Nature* 564: 201–206. DOI: 10.1038/s41586-018-0776-9

Cain, M. 1981. Risk and insurance: Perspectives on fertility and agrarian change in India and Bangladesh. *Population and Development Review* 7: 435–474.

Carpena, F. 2019. How do droughts impact household food consumption and nutritional intake? A study of rural India. *World Development* 122: 349–369.

Chandrasekara, S. K. S., H. Kwon, M. Vithanage, J. Obeysekera, and T. Kim. 2021. Drought in South Asia: A review of drought assessment and prediction in South Asian countries. *Atmosphere* 12: 369. DOI: 10.3390/atmos12030369.

Chary, G. R., C. S. Rao, B. M. K. Raju, K. P. R. Vittal, and V. S. Babu. 2014. Integrated land use planning for sustainable rainfed agriculture and rural development: A rainfed agro-economic zone approach. *Agropedology* 24: 234–252.

Chatterjee, S., R. Chakraborty, and H. Banerjee. 2020. Economic impact assessment of conservation agriculture on small and marginal farm households in eastern India. *Agricultural and Resource Economics Review* 33(Conference issue): 127–138.

Chetty, V. K., and D. K. Ratha. 1987. *The Unprecedented Drought and Government Policies*. New Delhi: Indian Statistical Institute.

Congress.gov. 2020. Healthy Soils Healthy Climate Act of 2020. October 23, 2020. https://www.congress.gov/bill/116th-congress/senate-bill/4850. (accessed January 13, 2023).

Conway, G. 2008. *The Science of Climate Change in Africa: Impacts and Adaptation*. London: Department for International Development.

Cook, E. R., K. J. Anchukaitis, B. M. Buckley, R. D. D. Arrigo, G. C. Jacoby, and W. E. Wright. 2010. Asian monsoon failure and mega-drought during the last millennium. *Science* 486(April): 486–490. DOI: 10.1126/science.1185188.

Cook, E. R., R. Seager, M. A. Cane, and D. W. Stahle. 2007. North American drought: Reconstructions, causes, and consequences. *Earth Science Reviews* 81: 93–134.

Crausbay, S. D., A. R. Ramirez, S. L. Carter, et al. 2017. Defining ecological drought for the twenty-first century. *American Meteorology Society* 98: 2543–2550.

Curley, D. L. 1977. Fair grain markets and Mughal famine policy in late eighteenth-century Bengal. *Calcutta Historical Journal* 2(1): 1–26.

DAC (Department of Agriculture and Cooperation). 2009. *Manual for Drought Management*. New Delhi: Department of Agriculture and Cooperation, Ministry of Agriculture, Government of India.

DAC (Department of Agriculture and Cooperation). 2014. *National Mission for Sustainable Agriculture: Operational Guidelines*. New Delhi: Department of Agriculture and Cooperation, Ministry of Agriculture.

DACFW (Department of Agriculture, Cooperation and Farmers Welfare). 2016. *Manual for Drought Management*. New Delhi: Department of Agriculture, Cooperation and Farmers Welfare, Ministry of Agriculture and Farmers Welfare, Government of India.

DAFW (Department of Agriculture and Farmers Welfare). 2020a. *Manual for Drought Management*. New Delhi: Department of Agriculture and Farmers Welfare, Ministry of Agriculture and Farmers Welfare, Government of India.

DAFW (Department of Agriculture and Farmers Welfare). 2020b. *National Agriculture Disaster Management Plan*. New Delhi: Department of Agriculture and Farmers Welfare, Ministry of Agriculture and Farmers Welfare, Government of India.

DAFW (Department of Agriculture and Farmers Welfare). 2022. *National Crisis Management Plan for Drought*. New Delhi: Department of Agriculture and Farmers Welfare, Ministry of Agriculture and Farmers Welfare, Government of India.

Dai, A., 2011. Drought under global warming: A review. *Wiley Interdisciplinary Reviews Climate Change* 2(1): 45–65. DOI: 10.1002/wcc.81.

Das, B. S., S. P. Wani, D. K. Benbi, et al. 2022. Soil health and its relationship with food security and human health to meet the sustainable development goals in India. *Soil Security* 8: 100071. DOI: 10.1016/j.soisec.2022.100071.

Das, S. K., R. K. Gupta, and H. K. Varma. 2007. Flood and drought management through water resources development in India. *WMO Bulletin* 56(3): 2007. https://public.wmo.int/en/bulletin/flood-and-drought-management-through-water-resources-development-india.

Department of Land Resources. 2013. *Draft National land Utilisation Policy: Framework for Land Use Planning & Management*. New Delhi: Department of Land Resources, Ministry of Rural Development.

Deshpande, R. S. 2022. *Under the Shadow of Development: Rainfed Agriculture and Droughts in Agricultural Development of India - NABARD Research and Policy Series No. 2/2022*. Mumbai: NABARD. p. 114.

Diffenbaugh, N. S., D. L. Swain, and D. Touma. 2015. Anthropogenic warming has increased drought risk in California. Proceedings National Academy of Science 112: 3931–3936. DOI: 10.1073/pnas.1422385112.

Dimri, A. P., T. Yasunari, B. S. Kotlia, U. C. Mohanty, and D. R. Sikka. 2016. Indian winter monsoon: Present and past. *Earth-Science Review* 163: 297–322

Dixit, Y., D. A. Hodell, and C. A. Petrie. 2014. Abrupt weakening of the summer monsoon in northwest India ~4100 yrs ago. *Geology* 42(4): 339–342. DOI: 10.1130/G35236.

DM & Relief Department Rajasthan. 2022a. About DM relief. https://www.dmrelief.rajasthan.gov.in/index.php/organization/about-dm-relief (accessed August 14, 2022).

DM & Relief Department Rajasthan. 2022b. Position of drought since year 81–82. https://www.dmrelief.rajasthan.gov.in/index.php/irrigation-calender/position-of-drought-since-year-81-82 (accessed August 14, 2022).

DMD (Disaster Management Division). 2022. *State-Wise Release of Funds from CRF-NCCF and SDRF-NDRF from 2005–06 to 2020–2021*. Disaster Management Division. https://ndmindia.mha.gov.in/images/State-wise%20Release%20of%20funds%20from%20CRF-NCCF%20and%20%20SDRF-%20NDRF%20from%202005-06%20to%20-2021 (accessed August 14, 2022).

Dreze, J. 1991. Famine prevention in India. In *The Political Economy of Hunger: Volume 2: Famine Prevention*. J. Dreze, and A. Sen, eds. 13–122. Oxford: Oxford Academic. DOI: 10.1093/acprof:oso/9780198286363.003.0002, (accessed September 11, 2022).

Dupdal, R., R. Dhakar, C. A. Rama Rao, J. Samuel, B. M. K. Raju, P. Vijaya Kumar, and V. U. M. Rao. 2020. Farmers' perception and economic impact assessment of agromet advisory services in rainfed regions of Karnataka and Andhra Pradesh. *Journal of Agrometeorology* 22: 258–265.

Ebi, K. L., and K. Bowen. 2015. Extreme events as sources of health vulnerability: Drought as an example. *Weather and Climate Extremes* 11: 95–102.

Enenkel, M., M. E. Brown, J. V. Vogt, et al. 2020. Why predict climate hazards if we need to understand impacts? Putting humans back into the drought equation. *Climatic Change* 162: 1161–1176. DOI: 10.1007/s10584-020-02878-0.

Engle, N. L. 2013. The role of drought preparedness in building and mobilizing adaptive capacity in states and their community water systems. *Climatic Change* 118: 291–306.

Estrela, T., and T. A. Sancho. 2016. Drought management policies in Spain and the European Union: From traditional emergency actions to drought management plans. *Water Policy* 18: 153–176.

European Commission. 2022. Soil health-protecting, sustainably managing and restoring EU soils. https://ec.europa.eu/info/law/better-regulation/have-your-say/initiatives/13350-Soil-health-protecting-sustainably-managing-and-restoring-EU-soils_en. (accessed on January 13, 2023).

Fankhauser, S., S. Dietz, and P. Gradwell. 2014. *Non-Economic Losses in the Context of the UNFCCC Work Programme on Loss and Damage*. Policy paper, Centre for Climate Change Economics and Policy, Grantham Research Institute on Climate Change and the Environment, UK.

Finance Commission. 2020. *Finance Commission in Covid times: Report for 2021–26, Volume II Annexes*. New Delhi: XV Finance Commission of India.

Fisher, M., and E. R. Carr. 2015. The influence of gendered roles and responsibilities on the adoption of technologies that mitigate drought risk: The case of drought-tolerant maize seed in eastern Uganda. *Global Environmental Change* 35: 82–92.

Fomby, T., Y. Ikeda, and N. V. Loayza. 2013. The growth aftermath of natural disasters. *Journal of Applied Econometrics* 3: 412–434.

Gadgil, S. 2018. The monsoon system: Land-sea breeze or the ITCZ? *Journal of Earth System Science* 127: 1. DOI: 10.1007/s12040-017-0916-x.

Gadgil, S., and S. Gadgil. 2006. The Indian monsoon, GDP and agriculture. *Economic and Political Weekly* 41: 4887–4895. www.jstor.org/stable/4418949.

Gautam, R. C. 2012. Impacts of drought on crops and ways to mitigate it. *Indian Farming* 62(6): 13–19.

Gautam, R. C., and R. S. Bana. 2014. Drought in India: Its impact and mitigation strategies-a review. *Indian Journal of Agronomy* 59: 179–190.

GoI (Government of India). 2016. *National Disaster Management Plan*. New Delhi: National Disaster Management Authority.

Gray, C., and V. Mueller. 2012. Drought and population mobility in rural Ethiopia. *World Development* 40: 134–145.

Gudade, B. A., S. Babu, A. B. Aage, S. S. Bora, K. Dhanapal, S. Bhat, T. Bhutia, and S. Singh. 2016. Influence of in-situ soil water conservation practices on growth, yield and economics of large cardamom under rainfed condition at North East India. *Journal of Experimental Agriculture International* 14(3): 1–8.

Gulati, A., P. Terway, and S. Hussain. 2018. *Crop Insurance in India: Key Issues and Way Forward, ICRIER Working Paper No. 352*. New Delhi: Indian Council for Research on International Economic Relations.

Gupta, A. K., P. Tyagi, and V. K. Sehgal. 2011. Drought disaster challenges and mitigation in India: strategic appraisal. *Current Science* 100: 1795–1806.

Hallegatte, S. 2012. *A Cost Effective Solution to Reduce Disaster Losses in Developing Countries: Hydro-Meteorological Services, Early Warning, and Evacuation. World Bank Policy Research Working Paper (6058)*. Washington DC: World Bank.

Hallegatte, S. 2014. *Natural Disasters and Climate Change: An Economic Perspective*. Switzerland: Springer International Publishing.

Hallegatte, S., A. Vogt-Schilb, M. Bangalore, and J. Rozenberg. 2017. *Unbreakable: Building the Resilience of the Poor in the Face of Natural Disasters. Climate Change and Development Series*. Washington, DC: World Bank. DOI: 10.1596/978-1-4648-1003-9.

Hirsch, T., S. Minninger, and S. Wirsching. 2017. *Non-Economic Loss and Damage – With Case Examples from Tanzania, Ethiopia, El Salvador and Bangladesh*. Berlin: Bread for the World.

Hirway, I. 2001. Food security and disaster management. In *Paper Presented at the National Consultation on 'Towards hunger free India* (Eds. M. D. Asthana, and P. Medrano)', New Delhi. World Food Programme, Planning Commission, Government of India and M. S. Swaminathan Research Foundation, Chennai.

Howitt, R., J. Medellin-Azuara, D. MacEwan, J. Lund, and D. Sumner. 2015. *Economic Analysis of the 2015 Drought for California Agriculture*. Davis, CA: Center for Watershed Sciences, University of California.

IFRC (International Federation of Red Cross and Red Crescent Societies). 2006. *Eastern Africa: Regional Drought Response. DREF Bulletin No. MDR64001*. Geneva: IFRC. https://www.ifrc.org/docs/appeals/06/MDR64001.pdf.

IMD (India Meteorological Department). 2005. *Climatological Features of Drought Incidences in India - Meteorological Monograph Climatology No. 21/2005*. Pune: India Meteorological Department.

IPCC (Intergovernmental Panel on Climate Change). 2014. Climate change 2014: Impacts, adaptation, and vulnerability. Part A: Global and sectoral aspects. In: *Contribution of Working Group II to the Fifth Assessment Report of the Intergovernmental Panel on Climate Change*, C. B. Field, V. R. Barros, D. J. Dokken, et al. eds. Cambridge; New York: Cambridge University Press.

IPCC (Intergovernmental Panel on Climate Change). 2021. Climate change 2021: The physical science basis. In: *Contribution of Working Group I to the Sixth Assessment Report of the Intergovernmental Panel on Climate Change*, V. Masson-Delmotte, P. Zhai, A. Pirani, et al. eds. Cambridge: Cambridge University Press.

Jain, V. K., R. P. Pandey, M. K. Jain, and H. -R. Byun. 2015. Comparison of drought indices for appraisal of drought characteristics in the Ken River Basin. *Weather and Climate Extremes* 8: 1–11.

Kamepalli, L. B. 2019. Disaster relief financing: A journey from margin money to state disaster response fund, *Economic and Political Weekly* 54(16): 20–22.

Kaur, J., and A. Singh. 2017. Direct seeded rice: Prospects, problems/constraints and researchable issues in India. *Current Agriculture Research Journal* 5(1). https://dx. doi.org/10.12944/CARJ.5.1.03.

Kazianga, H., and C. Udry. 2006. Consumption smoothing? Livestock, insurance and drought in rural Burkina Faso. *Journal of Development Economics* 79: 413–446.

Khanna, B. K., and N. Khanna. 2011. Drought. In *Disaster: Strengthening Community Mitigation and Preparedness*, B. K. Khanna, and Neena Khanna, eds. New Delhi: New India Publishing Agency, 45–53.

Khera, R. 2006. Political economy of state response to drought in Rajasthan, 2000–03. *Economic and Political Weekly* 41: 5163–5172. https://www.epw.in/journal/2006/50/special-articles/political-economy-state-response-drought-rajasthan-2000-03.html.

Kinsey, B., K. Burger, and J. W. Gunning. 1998. Coping with drought in Zimbabwe: Survey evidence on responses of rural households to risk. *World Development* 26: 89–110.

Kripalani, R. H., and A. Kulkarni. 1997. Rainfall variability over South-East Asia-connections with Indian monsoon and ENSO extremes: New perspectives. *International Journal of Climatology* 17: 1155–1168.

Kripalani, R. H., and P. Kumar. 2004. Northeast monsoon rainfall variability over south peninsular India vis-a-vis the Indian Ocean dipole mode. *International Journal of Climatology* 24: 1267–1282.

Kumar, A., 2003. Variability and predictability of 200-mb seasonal mean heights during summer and winter. *Journal of Geophysical Research*. 108(D5): 4169. DOI: 10.1029/2002jd002728.

Kumar, K. K., B. Rajagopalan, and M. A. Cane. 1999. On the weakening relationship between the Indian monsoon and ENSO. *Science* 284(5423): 2156–2159. DOI: 10.1126/science.284.5423.2156.

Kumar, K. K., B. Rajagopalan, M. Hoerling, G. Bates, and M. A. Cane. 2006. Unraveling the mystery of Indian monsoon failure during El Nino. *Science* 314: 115–119. DOI: 10.1126/science.1131152.

Kumar, K. N., M. Rajeevan, D. S. Pai, A. K. Srivastava, and B. Preethi. 2013. On the observed variability of monsoon droughts over India. *Weather and Climate Extremes* 1: 42–50.

Kumar, P., K. Rupa Kumar, M. Rajeevan, and A. K. Sahai. 2007. On the recent strengthening of the relationship between ENSO and northeast monsoon rainfall over South Asia. *Climate Dynamics* 28: 649–660.

Lal, R. 2020. Managing soil quality for humanity and the planet. *Frontiers of Agricultural Science and Engineering* 7: 251–253. DOI: 10.15302/J-FASE-2020329.

Laskar, A. H., and A. Bohra. 2021. Impact of Indian summer monsoon change on ancient Indian civilizations during the Holocene. *Frontiers of Earth Science* 9: 709455. DOI: 10.3389/feart.2021.709455.

Linke, A. M., J. O'Loughlin, J. T. McCabe, J. Tir, and F. D. Witmer. 2015. Rainfall variability and violence in rural Kenya: Investigating the effects of drought and the role of local institutions with survey data. *Global Environmental Change* 34: 35–47.

Loayza, N. V., E. Olaberria, J. Rigolini, and L. Christiaensen. 2012. Natural disasters and growth: Going beyond the averages. *World Development* 40: 1317–1336.

Logar, I. and J. C. J. M. van den Bergh. 2011. Methods for assessment of the costs of droughts. CONHAZ Rep. WP05. https://climate-adapt.eea.europa.eu (accessed August 12, 2022).

Lohmann, S., and T. Lechtenfeld. 2015. The effect of drought on health outcomes and health expenditures in rural Vietnam. *World Development* 72: 432–448.

Lokesh, S. and R. S. Poddar. 2018. Impact of drought on water resources and agriculture in Karnataka. *International Journal of Pure and Applied Bioscience* 6: 1102–1107.

Mahajan, D. R., B. M. Dodamani, and G. Mannina. 2016. Spatial and temporal drought analysis in the Krishna river basin of Maharashtra, India. *Cogent Engineering* 3: 15.

Maharatna, A. 1996. The demography of famines: An Indian historical perspective. Doctoral Dissertation, London School of Economics and Political Science.

Marris, E. 2014. Two-hundred-year drought doomed Indus Valley Civilization. Monsoon hiatus that began 4,200 years ago parallels dry spell that led civilizations to collapse in other regions. *Nature* (2014). https://doi.org/10.1038/nature.2014.14800. DOI: 10.1038/nature.2014.14800.

Martin-Carrasco, F., L. Garrote, A. Iglesias, and L. Mediero. 2013. Diagnosing causes of water scarcity in complex water resources systems and identifying risk management actions. *Water Resources Management* 27: 1693–1705.

Mckee, T. B., N. J. Doesken, and J. Kleist. 1993. The relationship of drought frequency and duration to time scales. In: Preprint - Eighth Con*ference on Applied Climatology. Anaheim, January 17–22, 1993*, Boston: American Meterological Society, 179–184.

McKee, T. B., N. J. Doeskin, and J. Kleist. 1995. Drought monitoring with multiple time scales. In: Preprint - Ninth Conference on Applied Climatology. Dallas, TX, 15–20 January 1995: Boston: American Meteorological Society, 233–236.

MHA (Ministry of Home Affairs). 2019. *Disaster Risks and Resilience in India - An Analytical Study*. New Delhi: Ministry of Home Affairs, GoI.

Mirza, M. M. Q. 2010. A review on current status of flood and drought forecasting in South Asia. In: *Global Environmental Changes in South Asia*, A. P. Mitra, and C. Sharma, eds. Dordrecht: Springer, 233–243.

Mishra, A. K., and V. P. Singh. 2009. Analysis of drought severity-area-frequency curves using a general circulation model and scenario uncertainty. *Journal of Geophysical Research: Atmospheres*. 114(D6): https://doi.org/10.1029/2008JD010986. DOI: 10.1029/2008JD010986.

Mishra, A. K., and V. P. Singh. 2010. A review of drought concepts. *Journal of Hydrology* 391: 202–216.

Mishra, P., T. R. Sahoo, F. H. Rahman, L. M. Garnayak, A. Phonglosa, N. Mohapatra, R. Bhattacharya, and S. N. Mishra. 2020. Yield and economics of brinjal (*Solanum melongena*) as affected by different mulching types and its effect on soil moisture content and weed dynamics in post flood situation of coastal Odisha, India. *International Journal of Environment and Climate Change* 10: 264–270.

Mishra, V. 2020. Long-term (1870–2018) drought reconstruction in context of surface water security in India. *Journal of Hydrology* 580(1–16): 124228. DOI: 10.1016/j.jhydrol.2019.124228.

Mishra, V., S. Aadhar, A. Asoka, S. Pai, and R. Kumar. 2016. On the frequency of the 2015 monsoon season drought in the Indo-Gangetic plain. *Geophysical Research Letters* 43(23): 12102–12112. DOI: 10.1002/2016GL071407.

Mishra, V, K. A. Cherkauer, and S. Shukla. 2010. Assessment of drought due to historic climate variability and projected future climate change in the Midwestern United States. *Journal of Hydrometeorology* 11: 46–68.

Mishra, V., B. V. Smoliak, D. P. Lettenmaier, and J. M. Wallace. 2012. A prominent pattern of year-to-year variability in Indian summer monsoon rainfall. *The Proceedings of the National Academy of Sciences* 109(19): 7213–7217. DOI: 10.1073/pnas.1119150109.

Mishra, V., A. D. Tiwari, A. Aadhar, R. Shah, M. Xiao, D. S. Pai, and D. Lettenmaier. 2019. Drought and famine in India, 1870–2016. *Geophysical Research Letters* 46(4): 2075–2083. DOI: 10.1029/2018GL081477.

MoAFW (Ministry of Agriculture and Farmers Welfare). 2017. *Drought Management Plan*. New Delhi: Ministry of Agriculture and Farmers Welfare, Department of Agriculture, Cooperation and Farmers Welfare, Government of India.

MoAFW (Ministry of Agriculture and Farmers Welfare). 2022. https://agricoop.nic.in/sites/default/files/drought_management.pdf (accessed October 17, 2022).

Mohapatra, T, P. K. Rout, and H. Pathak. 2022. Indian agriculture: Achievements and aspirations. In: *Indian Agriculture after Independence*, H. Pathak, J. P. Mishra, and T. Mohapatra, eds. New Delhi: Indian Council of Agricultural Research, 1–26.

Mooley, D. A., and G. B. Pant. 1981. Droughts in India over the last 200 years, their socio-economic impacts and remedial measures for them. In: *Climate and History*, T. Wigley, M. J. Ingram, and G. Farmer, eds. 465–478. Cambridge: Cambridge University Press.

Mooley, D. A., and B. Parthasarathy. 1983. Droughts and floods over India in summer monsoon seasons 1871–1980. In: *Variations in the Global Water Budget* (Eds. Street-Perrott, M. Beran, and R. Ratcliffe). Dordrecht: Springer, 239–252.

MoRD (Ministry of Rural Development). 1994. *Report of the Technical Committee on Drought Prone Areas Programme and Desert Development Programme*. New Delhi: Ministry of Rural Development, Government of India.

Mukherjee, S., A. Mishra, and K. E. Trenberth. 2018. Climate change and drought: a perspective on drought indices. *Current Climate Change Reports* 4: 145–163.

Mundetia, N., and D. Sharma. 2015. Analysis of rainfall and drought in Rajasthan state, India. *Global Network of Environmental Science and Technology Journal* 17: 12–21.

Murthy, C. S., and M. V. R. Sesha Sai. 2010. Agricultural drought monitoring and assessment. In: *Remote Sensing Applications*. P. S. Roy, R. S. Dwivedi, and D. Vijayan, eds. Hyderabad: National Remote Sensing Centre, Indian Space Research Organization, Department of Space, Government of India, 303–330.

Murton, B. 1984. Spatial and temporal patterns of famine in Southern India, before the FCs. In: *Famine: As a Geographical Phenomenon*, B. Currey, and G. Hugo, eds. Frankfort: D. Reidel Publishing Company, 71–90.

NAAS (National Academy of Agricultural Sciences). 2011. *Drought Preparedness and Mitigation, Policy Paper No. 50*. New Delhi: National Academy of Agricultural Sciences.

Nalbantis, I., and G. Tsakiris. 2009. Assessment of hydrological drought revisited. *Water Resources Management* 23: 881–897.

National Commission on Agriculture. 1976. *Report of the National Commission on Agriculture, 1976. Part IV: Climate and Agriculture*. New Delhi: Ministry of Agriculture and Irrigation.

Naveendrakumar, G., M. Vithanage, H. H. Kwon, S. S. K. Chandrasekara, M. C. M. Iqbal, S. Pathmarajah, W. C. D. K. Fernando, and J. Obeysekara. 2019. South Asian perspective on temperature and rainfall extremes: A review. *Atmospheric Research* 225: 110–120.

NDMA (National Disaster Management Authority). 2019. *National Disaster Management Plan*. New Delhi: National Disaster Management Authority, Government of India.

NRAA (National Rainfed Area Authority). 2009. Drought management strategies – 2009. *Draft Paper of National Rainfed Area Authority*. New Delhi: Ministry of Agriculture, Government of India.

NRAA (National Rainfed Area Authority). 2013. *Contingency and Compensatory Agriculture Plans for Droughts and Floods in India-2012. Position Paper No. 6*. New Delhi: National Rainfed Area Authority.

NRAA (National Rainfed Area Authority). 2020. *Prioritization of Districts for Development Planning in India: A Composite Index Approach. Report Submitted by Task Force on Revisiting Prioritization of Rainfed Areas*. New Delhi: NRAA.

Pai, D. S., P. Guhathakurta, A. Kulkarni, and M. N. Rajeevan. 2017. Variability of meteorological droughts over India. In: *Observed Climate Variability and Change over the Indian Region*. M. N. Rajeevan, and S. Nayak, eds. Singapore: Springer, 73–87.

Palmer, W. C. 1965. *Meteorological Drought. U. S. Weather Bureau, Research Paper No. 45*. Washington, DC: US Weather Bureau.

Panda, D. K., A. Mishra, S. K. Jena, B. K. James, and A. Kumar. 2007. The influence of drought and anthropogenic effects on groundwater levels in Orissa, India. *Journal of Hydrology* 43(3): 140–153. DOI: 10.1016/j.jhydrol.2007.06.007.

Pandey, S., H. Bhandari, and B. Hardy. 2007. *Economic Costs of Drought and Rice Farmers' Coping Mechanisms: A Cross-Country Comparative Analysis from Asia*. Manila: International Rice Research Institute.

Passmore, R. 1951. Famine in India: An historical survey. *The Lancet* 2(6677): 303–307. DOI: 10.1016/s0140-6736(51)93295-3.

Patel, N. R., B. R. Parida, V. Venus, S. K. Saha, and V. K. Dadhwal. 2012. Analysis of agricultural drought using vegetation temperature condition index (VTCI) from Terra/MODIS satellite data. *Environmental Monitoring and Assessment* 184: 7153–7163.

Pelling, M., A. Ozerdem, and S. Barakat. 2002. The macro-economic impact of disasters. *Progress in Development Studies* 2: 283–305.

Prabhakar, S. V. R. K., and R. Shaw. 2018. Climate change adaptation implications for drought risk mitigation: A perspective for India. *Climatic Change* 88: 113–130.

Prakash, S. 2018. Capabilities of satellite-derived datasets to detect consecutive Indian monsoon droughts of 2014 and 2015. *Current Science* 114: 2361–2368.

Prasad, Y. G., M. Maheswari, S. Dixit, et al. 2014. *Smart Practices and Technologies for Climate Resilient Agriculture*. Hyderabad: ICAR-Central Research Institute for Dryland Agriculture.

Rajeevan, M., C. K. Unnikrishnan, J. Bhate, K. Niranjan Kumar, and P. P. Sreekala. 2012. Northeast monsoon over India: Variability and prediction. *Meteorological Applications* 19: 226–236.

Raju, B. M. K., K. V. Rao, B. Venkateswarlu, A. V. M. S. Rao, C. A. R. Rao, V. U. M. Rao, M. S. Rao, and K. Nagasree. 2013a. Change in climate in India during last few decades. *Journal of Agrometeorology* 15(Special Issue 1): 30–36.

Raju, B. M. K., K. V. Rao, B. Venkateswarlu, et al. 2013b. Revisiting climatic classification in India: A district-level analysis. *Current Science* 105: 492–495.

Rama Rao, C. A., B. M. K. Raju, A. Islam, et al. 2019a. *Risk and Vulnerability Assessment of Indian Agriculture to Climate Change*. Hyderabad: National Innovations in Climate Resilient Agriculture, ICAR-CRIDA.

Rama Rao, C. A., B. M. K. Raju, J. V. N. S. Prasad, et al. 2018. *Assessing Resilience of Agriculture to Climate Change and Variability. Technical Brief 02/2018*. Hyderabad: ICAR-Central Research Institute for Dryland Agriculture.

Rama Rao, C. A., K. V. Rao, B. M. K. Raju, J. Samuel, R. Dupdal, M. Osman, and R. N. Kumar. 2019b. Levels and determinants of economic viability of rainwater harvesting farm ponds. *Indian Journal of Agricultural Economics* 74: 539–551.

Rama Rao, C. A., M. S. Rao, Y. S. Ramakrishna, K. P. R. Vittal, and Y. V. R. Reddy. 2003. *Economics of Some Dryland Agricultural Technologies*. Hyderabad: ICAR-Central Research Institute for Dryland Agriculture.

Rathore, B. M. S., R. Sud, V. Saxena, L. S. Rathore, T. S. Rathore, V. G. Subrahmanyam, and M. R. Roy. 2014a. *Drought -Conditions and Management Strategies in India*. New York: United Nations Development Programme. 7 p. https://www.droughtmanagement.info/literature/UNW-DPC_NDMP_Country_Report_India_2014.pdf (accessed August 12, 2022).

Rathore, M. S. 2004. *Adaptive Strategies to Droughts in Rajasthan*. Jaipur: Institute of Development Studies.
Rathore, M. S. 2005. *State Level Analysis of Drought Policies and Impacts in Rajasthan, India. Working Paper 93: Drought Series Paper No. 6*. Colombo: IWMI.
Rawat, S., A. Ganapathy, and A. Agarwal. 2022. Drought characterization over Indian sub-continent using GRACE-based indices. *Nature* 12:15432. DOI: 10.1038/s41598-022-18511-2.
Ray, S., S. Jain, and V. Thakur. 2021. *Financing India's Disaster Risk Resilience Strategy. Working Paper 404*. New Delhi: Indian Council for Research on International Economic Relations.
Rossi, G., and A. Cancelliere. 2013. Managing drought risk in water supply systems in Europe: A review. *International Journal of Water Resources Development* 29: 272–289.
Rouse, J. W., R. H. Haas, J. A. Schell, and D. W. Deering. 1973. *Monitoring the Vernal Advancement and Retrogradation (Green Wave Effect) of Natural Vegetation*. Texas: Texas A&M University.
Roxy, M. K., K. Ritika, P. Terray, R. Murtugudde, K. Ashok, and B. N. Goswami. 2015. Drying of Indian subcontinent by rapid Indian Ocean warming and a weakening land-sea thermal gradient. *Nature Communications* 6: 1–10. DOI: 10.1038/ncomms8423.
Roy, A. K., and I. Hirway. 2007. *Multiple Impacts of Droughts and Assessment of Drought Policy in Major Drought Prone States in India*. Ahmedabad: Centre for Development Alternatives.
Samaniego, L., S. Thober, R. Kumar, et al. 2018. Anthropogenic warming exacerbates European soil moisture droughts. *Nature Climate Change* 8: 421–426. DOI: 10.1038/s41558-018-0138-5.
Samra, J. S. 2004. *Review and Analysis of Drought Monitoring, Declaration and Management in India. Working Paper 84*. Colombo: International Water Management Institute.
Sarkar, J. 2011. Monitoring drought risks in India with emphasis on agricultural drought. In: Agricultural Drought Indices. Proceedings of the WMO/UNISDR Expert Group Meeting on Agricultural Drought Indices, M. V. K. Sivakumar, R. P. Motha, D. A. Wilhite, and D. A. Wood, eds. Geneva: World Meteorological Organization, 50–59.
Schmidt, G., J. J. Benitez, and C. Benitez. 2012. *Working Definitions of Water Scarcity and Drought*. Madrid: TYPSA Intecsa-Inarsa.
Sen, A. 2000. *Development as Freedom*. New York: Anchor Books.
Sen, S. and S. Bera. 2016. Frame national drought policy: Supreme court. *Livemint*. https://www.livemint.com/Politics/K40mUpuaDEctugrfHpYyNJ/SC-verdict-on-drought-case-today.html (accessed September 2, 2022).
Serdeczny, O. M., S. Bauer, and S. Huq. 2018. Non-economic losses from climate change: opportunities for policy oriented research. *Climate and Development* 10(2): 97–101.
Shadman, F., S. Sadeghipour, M. Moghavvemi, and R. Saidur. 2016. Drought and energy security in key ASEAN countries. *Renewable and Sustainable Energy Reviews* 53: 50–58.
Sharma, A. 2019. Drought management policy of India: An overview. *Disaster Advances* 12: 51–62.
Sharma, S., and M. S. Shruthi. 2017. Thank you water! (water in Hindu scriptures). In: *Water and Scriptures: Ancient Roots for Sustainable Development*. K. V. Raju and S. Manasi, eds. Cham: Springer International Publishing, 89–172.
Shewale, M. P., and S. Kumar. 2005. *Climatological Features of Drought Incidences in India*. Pune: India Meteorological Department.
Shukla, S., and A. W. Wood. 2008. Use of a standardized runoff index for characterizing hydrologic drought. *Geophysical Research Letters* 35: L02405. DOI: 10.1029/2007GL032487.
Sigdel, M., and M. Ikeda. 2010. Spatial and temporal analysis of drought in Nepal using standardized precipitation index and its relationship with climate indices. *Journal of Hydrology and Meteorology* 7: 59–74.
Singh, D., S. Ghosh, M. K. Roxy, and S. McDermid. 2019. Indian summer monsoon: extreme events, historical changes, and role of anthropogenic forcings. *Wiley Interdisciplinary Reviews: Climate Change* 10: e571. DOI: 10.1002/wcc.571.
Sinha Ray, K.C. 2006. Role of drought early warning systems for sustainable agriculture research in India. In *Early Warning Systems for Drought Preparedness and Drought Management*, D. A. Wilhite, M. V. K. Sivakumar, and D. A. Wood, eds. Geneva: World Meteorological Organization.
Sivakumar, M. V. K., R. Stefanski, M. Bazza, S. Zelaya, D. Wilhite, and A. R. Magalhaes. 2014. High level meeting on national drought policy: Summary and major outcomes. *Weather and Climate Extremes* 3: 126–132. DOI: 10.1016/j.wace.2014.03.007.
Smakhtin, V. U., and E. L. F. Schipper. 2008. Drought: The impact of semantics and perceptions. *Water Policy* 10: 131–143.
Srinivas, I., Ch. Srinivasa Rao, K. Sammi Reddy, B. Sanjeeva Reddy, R. V. Adake, A. S. Dhimate, M. Osman, J. V. N. S. Prasad, and C. A. R. Rao. 2017. *Up-Scaling Farm Machinery Custom Hiring Centres in India: A Policy Paper. CRIDA Policy Paper No. 1/2017*. Hyderabad: ICAR-Central Research Institute for Dryland Agriculture.

Srivastava, H. S. 1968. *History of Indian Famines and Development of Famine Policy 1858–1918*. Agra: Sri Ram Mehra and Co.

Stanke, C., M. Kerac, C. Prudhomme, J. Medlock, and V. Murray. 2013. Health effects of drought: A systematic review of the evidence. *PLoS Currents* 5(2013). DOI: 10.1371/currents.dis.7a2cee9e980f91ad7697b570bcc4b004.

Tayade, N. H. 2017. Assessment of drum seeder: An improved technology in rice production system. *International Journal of Pure and Applied Bioscience* 5: 221–224. DOI: 10.18782/2320-7051.2785.

Timmermann, A., S. I. An, J. S. Kug, et al. 2018. El Nino-southern oscillation complexity. *Nature* 559: 535–545. DOI: 10.1038/s41586-018-0252-6.

Tsakiris, G. 2017. Drought risk assessment and management. *Water Resources Management* 31: 3083–3095.

Udmale, P. D., Y. Ichikawa, S. Manandhar, H. Ishidaira, A. S. Kiem, N. Shaowei, and S. N. Panda. 2015. How did the 2012 drought affect rural livelihoods in vulnerable areas? Empirical evidence from India. *International Journal of Disaster Risk Reduction* 13: 454–469.

UN (United Nations). 1990a. *Desertification and Drought: At a Glance. International Decade for Natural Disaster Reduction*. Geneva: IDNDR Secretariat.

UN (United Nations). 1990b. *What Can Be Done about Drought? International Decade for Natural Disaster Reduction*. Geneva: IDNDR Secretariat.

UNDP (United Nations Development Programme). 2002. A climate risk management approach to disaster reduction and adaptation to climate change. UNDP expert group meeting, integrating disaster reduction with adaptation to climate change, Havana, June 17–19, 2002. https://research.fit.edu/media/site-specific/researchfitedu/coast-climate-adaptation-library/global/un-unep-unesco-undp-reports/UNDP.-2002.-Global-CC-Risk-Management.pdf (accessed August 21, 2022).

UNFCCC (United Nations Framework Convention on Climate Change). 2013. Non-economic losses in the context of the work programme on loss and damage, Technical Paper. https://unfccc.int/resource/docs/2013/tp/02.pdf (accessed August 14, 2022).

UNICEF (United Nations Children's Fund). 2016. When coping crumbles - Drought in India 2015–16: A rapid assessment of the impact of drought on children and women in India. https://reliefweb.int/report/india/drought-india-2015-16-when-coping-crumples-rapid-assessment-impact-drought-children-and (accessed August 16, 2022).

UNISDR and CRED (United Nations Office for Disaster Risk Reduction and Centre for Research on the Epidemiology of Disasters). 2018. Economic losses, poverty and disasters: 1998–2017. https://www.undrr.org/publication/economic-losses-poverty-disasters-1998-2017 (accessed August 14, 2022).

Van Loon, A. F. 2015. Hydrological drought explained. *Wiley Interdisciplinary Reviews* 2: 359–392.

Venkateswarlu, B. 2010. Agricultural drought management in India. In: *Proceedings of the SAARC Workshop on Drought Risk Management in South Asia Organized at Kabul, Afghanistan on August 8–9, 2010*. New Delhi: SAARC Disaster Management Centre, 88–104.

Venkateswarlu, B., B. M. K. Raju, K. V. Rao, and C. A. R. Rao. 2014. Revisiting drought prone districts in India. *Economic and Political Weekly* 49(25): 71–75.

Vicente-Serrano, S. M., S. Begueria, and J. I. Lopez-Moreno. 2010. A multiscalar drought index sensitive to global warming: The standardized precipitation evapotranspiration index. *Journal of Climate* 23: 1696–1718.

Victora, C. G., L. Adair, C. Fall, P. C. Hallal, R. Martorell, L. Richter, and H. S. Sachdev. 2008. Maternal and child undernutrition: Consequences for adult health and human capital. *The Lancet* 371(9609): 340–357. DOI: 10.1016/S0140-6736(07)61692-4.

Webster, P. J., V. O. Magana, T. N. Palmer, et al. 1998. Monsoons: Processes, predictability, and the prospects for prediction. *Journal of Geophysical Research: Oceans* 103: 14451–14510. DOI: 10.1029/97JC02719.

Wilhite, D. A. 1992. *Preparing for Drought: A Guidebook for Developing Countries*. Nairobi: Climate Unit, United Nations Environment Programme.

Wilhite, D. A. 2000. Drought as a natural hazard: Concepts and definitions, In: *Drought: A Global Assessment*, D. A. Wilhite, ed. London: Routledge Publishers, 33–40.

Wilhite, D. A., and M. H. Glantz. 1985. Understanding the drought phenomenon: The role of definitions. *Water International* 10: 111–120. DOI: 10.1080/02508068508686328.

Wilhite, D. A., M. J. Hayes, and C. L. Knutson. 2005. Drought preparedness planning: Building institutional capacity. In: *Drought and Water Crises: Science, Technology, and Management Issues*. D. A. Wilhite, ed. Boca Raton, FL: CRC Press, 93–135.

Wilhite, D. A., M. V. Sivakumar, and R. Pulwarty. 2014. Managing drought risk in a changing climate: The role of national drought policy. *Weather and Climate Extremes* 3: 4–13.

Wilhite, D. A., and O. Vanyarkho. 2000. Drought: Pervasive impacts of a creeping phenomenon. In: *Drought: A Global Assessment (Volume I, Chapter 18)*. D. A. Wilhite, ed. London: Routledge Publishers, 245–255.

WMO (World Meteorological Organization). 2006. *Drought Monitoring and Early Warning: Concepts, Progress and Future Challenges. WMO No-1006*. Geneva: World Meteorological Organization.

WMO and GWP (World Meteorological Organization and Global Water Partnership). 2014. National drought management policy guidelines: A template for action. In: *Integrated Drought Management Programme (IDMP) Tools and Guidelines Series 1*. D. A. Wilhite, ed. Geneva: WMO and Stockholm: GWP.

WMO and GWP (World Meteorological Organization and Global Water Partnership). 2016. Handbook of drought indicators and indices. In: *Integrated Drought Management Programme (IDMP), Integrated Drought Management Tools and Guidelines Series 2*. M. Svoboda, and B. A. Fuchs, eds. Geneva: WMO.

WMO and GWP (World Meteorological Organization and Global Water Partnership). 2017. Benefits of action and costs of inaction: Drought mitigation and preparedness – A literature review. In: *Integrated Drought Management Programme (IDMP) Working Paper 1*. N. Gerber, and A. Mirzabaev, eds. Geneva: WMO and Stockholm: GWP.

World Bank. 2003. *Report on Financing Rapid Onset Natural Disaster Losses in India: A Risk Management Approach. Report No. 26844-IN*. Washington DC: World Bank.

Yadav, R. K. 2012. Why is ENSO influencing Indian northeast monsoon in the recent decades? *International Journal of Climatology* 32: 2163–2180.

Zhang, X., R. Obringer, C. Wei, N. Chen, and D. Niyogi. 2017. Droughts in India from 1981 to 2013 and implications to wheat production. *Scientific Reports* 7(2017): 44552. DOI: 10.1038/srep44552.

Zubair, L., and C. F. Ropelewski. 2006. The strengthening relationship between ENSO and northeast monsoon rainfall over Sri Lanka and southern India. *The Journal of Climate* 19: 1567–1575.

4 Soil, Water, and Nutrient Management in Drylands

S.K. Chaudhari, Priyabrata Santra, Deepesh Machiwal, Mahesh Kumar, V.K. Singh, K. Sammi Reddy, and S. Kundu

4.1 INTRODUCTION

Drylands are characterized by their limited water supply, low and highly variable rainfall, and recurrent drought. Globally, drylands cover about 41% of the earth's land area and are home to 38% of the world's population living in about 100 countries (Prâvâlie, 2016), most of which are developing countries. These lands are characterized by their limited water supply, low and highly variable rainfall, and recurrent drought. The increased population pressure aggravates the problems in various production systems in dry lands, affecting food, fodder, and livelihood security. Inhabitants of drylands have learned to cope with their harsh climatic conditions through appropriate practices. However, amid widespread poverty and increased human pressure on the fragile resource base, coping strategies are becoming insufficient to reduce peoples' vulnerability. It is widely believed that the poor will be hit hard due to climate change. This is especially true for those communities who live in dryland areas and rely largely or totally on dryland agriculture for their livelihoods. The most important issue facing the people who live in the dryland regions of the world is to ensure long-term agricultural productivity while conserving land, water, and biodiversity. In this context, sustainable agricultural development is very crucial. Lack of access to water supplies, overpopulation, migration, and conflicts overwhelm soil resources and reduce the ability of communities in providing food. Sustainable development of countries affected by drought and desertification can only come about through concerted efforts based on a sound understanding of the different factors that contribute to land degradation around the world. Therefore, in order to sustainably manage current dryland resources with respect to impacts of climate variability and to adequately address likely future climate change impacts, decision-makers at all levels need to be aware of all aspects of the climate system.

Climate change is no longer a distant reality; it is upon us now. The increasing frequency of recurrent droughts, unseasonal rains, and dry spells are already taking a toll on smallholder farmers across the country. The arid and semi-arid tropics across Asia and sub-Saharan Africa are where climate change is going to have the greatest impact. Climate change could cause yields of staple crops to drop, and to ensure food security, and farmers will have to switch to cultivating more climate-hardy crops and farming practices. The critical nexus of poverty, drought, and land degradation in the drylands can be broken by adopting a four-pronged science-based strategy that involves growing drought-tolerant and climate change-ready crops to match the available length of the growing season and low soil moisture, contingent action to replace affected crops with those that are more drought-tolerant, efficient management of natural resources, arresting land degradation, conserving soil moisture, and harvesting water in the rainy season for supplemental irrigation, and by empowering stakeholders through capacity building, enabling rural institutions, and formulating policies supportive of dryland agriculture.

Livestock production from the drylands is quite important, both in terms of the volume of product produced and with respect to the number of people dependent on livestock for their livelihood. Drylands support the full spectrum of production systems, from commercial ranching in

Australia, the western USA, Mexico, Argentina, and southern Africa through various gradations of crop-livestock systems in Africa and Asia to traditional nomadic and semi-nomadic subsistence systems. Almost everywhere, livestock production is undergoing rapid change in response to political, social, economic, environmental, and demographic pressures.

In India, net cultivated land area covers 143 million ha (M ha), of which about 97 M ha (68%) of that land is under dryland/rainfed agriculture with a total food production that meets 44% demand at the national level and supports 40% of human and 60% of livestock populations (NBSSLUP, 2001). Thus, drylands play a crucial role in country's food security. It is estimated that 75 M ha land would be under rainfed farming even if the country's full irrigation potential of 139.5 M ha is realized (Mazumdar, 2007). Geographically, dryland agriculture lands in India are located in the desert land in the northwest (Rajasthan), the plateau region (Central India), the alluvial plains of the Ganga-Yamuna River Basin, the Central Highlands (Gujarat, Maharashtra, and Madhya Pradesh), the rain-shadow region of Deccan (Maharashtra), the Deccan Plateau (Andhra Pradesh), and the highlands (Tamil Nadu) (Figure 4.1).

Dryland region is generally defined by aridity index (AI, ratio of mean annual precipitation, and mean annual evapotranspiration) of <0.65. Thus, dryland areas include hyper-arid areas with AI <0.05, arid areas of AI of 0.05–0.20, semi-arid areas with AI of 0.20–0.50, and dry sub-humid areas with AI of 0.50–0.65. Globally, the arid, semi-arid and dry sub-humid areas cover 42%, 37%,

FIGURE 4.1 Physiographic locations of dryland agricultural region in India. Source: Singh et al. (2004).

and 21% areas of drylands. In India, drylands are distributed in the following major rainfall zones (Kanwar, 1999): (i) less than 500 mm rainfall (15 M ha dryland of the arid region), (ii) 500–750 mm rainfall (15 M ha), (iii) 750–1,150 mm rainfall (42 M ha), and (iv) more than 1,150 m rainfall (25 M ha). In terms of production, dryland agriculture contributes approximately 80% of cereals, 50% of maize, 88% of soybean, 65% of pigeon pea, chickpea, etc. Overall, dry land areas contribute 40% of food production and support two-thirds of the livestock population (Srinivasarao et al., 2015). A major part of drylands in India is located in arid western India (AWI), which mainly comprises the western part of Rajasthan and the north-western part of Gujarata with some parts of Harynana and Punjab at its North East and East, respectively (Figure 4.2). It lies between 21°17¢–31°12¢N and 68°8¢–76°20¢E, covering an area of 32 M ha.

The southern, coastal part of the AWI is locally known as 'Kachch'. The central western and north-western parts of the region are dominantly covered with high and low dunes with an average height of 10–15 m, which are locally known as 'Marusthali'. The mean annual rainfall in the AWI is 400 mm. The 'Marusthali' region receives less rainfall (200–300 mm/year with 12–15 rainy days, mostly during July–September) than the 'Kachch' region (350–450 mm/year with 16–18 rainy days during July–September). The mean summer temperature in the 'Marusthali' region is as high as 49°C during the day and decreases to less than 20°C during the night. The mean day temperature in the coastal part of 'Kachch' is 36°C–38°C, which is low compared to the western and northern plains with dune areas. The arid landscape in India has witnessed some major changes in the past few decades, not all of which were conducive to the welfare of the arid ecosystem. A sharp growth in human and livestock population and the rapid developmental activities exert tremendous pressure

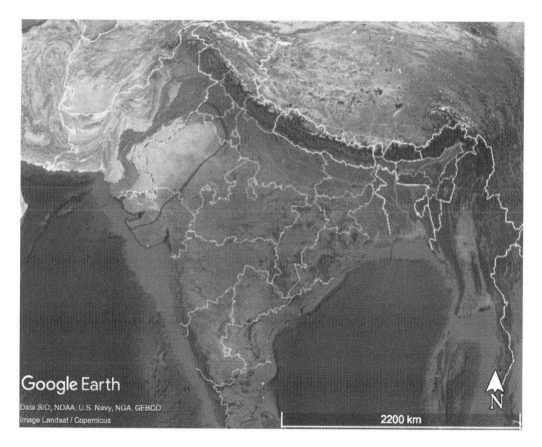

FIGURE 4.2 Arid western India (AWI) (marked in light gray colour) in Google Earth comprises a major part of drylands in India.

on the slender natural resource base of the region, with the apprehension of sustainability being at stake. All these necessitate farmers and farming practices in dryland areas to adapt to the predicted climate change.

4.2 CLIMATIC CONSTRAINTS

4.2.1 Low Rainfall

Rainfall constitutes as the only source for all kinds of water available in the drylands including surface water resources (Machiwal et al., 2020; Stewart, 1987). Hence, the crop growth gets largely affected from the low rainfall and failure of rains in the drylands under both rainfed and surface water irrigation conditions, and accordingly, lesser crop yields are obtained than the genetic potential of the crops (Ramakrishna and Rao, 2008). In addition, the occurrence of less frequent and highly intense rainy storms in the drylands has a worsening impact on crop production, making it less productive. The main characteristics of rainfall in the Indian dryland regions are its limited quantum ranging from 375 to 1,125 mm per annum, high temporal as well as spatial variability, uncertainty and unpredictability (Vijayan, 2016). In dryland regions, years receiving less than the mean annual rainfall are usually more in numbers that those getting below the mean rainfall, with the skewness value inversely related to rainfall magnitude. Under the recent rainfall scenario of high-intensity extreme events with enhanced frequency, vast quantities of runoff are generated even in response to the low quantum of rainfall. The substantial runoff water subsequently results in a huge amount of soil loss due to water erosion, which further aggravates the problem of reduced crop yields (Ramakrishna and Rao, 2008).

4.2.2 High Radiation

Among the entire drylands of the country, western Rajasthan is bestowed with abundant solar radiation throughout the year, as cloud cover is low during most of the year. It is estimated that the solar radiation in arid lands of western Rajasthan range from 15 to 18 MJ m^2/day in winter season, from 23 to 26.5 MJ m^2/day in summer season, and from 22 to 26.5 MJ m^2/day during monsoon season with the mean annual value of solar radiation as 22 MJ m^2/day. It is further seen that the solar radiation in drier regions of Jaisalmer and Barmer districts is relatively more (22.3 MJ m^2/day) than that in canal-irrigated belt of Hanumangarh district (20 MJ m^2/day).

4.2.3 Wide-Ranging Diurnal Temperatures

The drylands of the country are situated in sandy terrains with sparse vegetation, low soil moisture content, and low relative humidity. Thus, the diurnal, seasonal, and annual temperature ranges are high in the drylands due to their geographical location. In the summer season, the temperature may rise up to 50°C in the arid region, while in the winter, a temperature of −5.7°C has been recorded. The mean monthly maximum temperature in the winter season ranges from 22°C to 29°C and minimum temperature from 4°C to 14°C in the arid region where May is the hottest month with a mean maximum temperature of 40°C–42°C. The temperature declines by 3°C–5°C during the monsoon season but again increases slightly during September–October with the withdrawal of monsoon.

4.2.4 Low Humidity

Relative humidity in drylands, mostly situated in the northwest, central, and southern parts of India, remains relatively low (~35%–50%) in all the seasons in comparison to other parts. In the winter and pre-monsoon seasons, relative humidity values remain low in the northwest and southern drylands of the country. Relative humidity remains highly variable over the space during pre-monsoon season compared to other seasons. During the winter, low values of relative humidity are caused by the higher surface temperature prior to the monsoon. Recently, trend analysis of relative humidity reveals that a few stations in the northwest and central parts of the country depicted a strong increasing trend throughout the year (Khan et al., 2022). Rest of the stations in the northwest showed a decreasing trend.

4.2.5 High Evaporation

Evaporation is a vital process of the water cycle in the drylands that is responsible for the major loss of water from soil and water bodies to the atmosphere without any beneficial use. The evaporation depends on many atmospheric parameters, including air temperature, wind speed, vapor pressure deficit, solar radiation, atmospheric pressure, and the surrounding environment. Hence, it is difficult to estimate evaporation accurately, and its precise estimation further becomes complex in arid (mean monthly rainfall is less than potential evapotranspiration through the year) and semi-arid (mean monthly rainfall is less than potential evapotranspiration, except during the monsoon season) regions where accounting for water resources is a challenging task due to the relatively low amount of each water component (Ali et al., 2008).

4.2.6 High Wind Speed

The wind direction in the drylands of the country usually follows the wind pattern of a typical monsoon. In general, the wind blows in the southwest direction during the summer and rainy seasons and in the northeast direction during the winter season. The wind speed remains quite high during the summer (9–12 km/hour) and low (3–4 km/hour) during the winter season. It is experienced that strong winds of 15–18 km/hour speed often blow during June in summers. Sometimes, the wind speed reaches 60–80 km/hour especially during severe dust storms. It is reported that wind speed showed a significant decrease during 1971–2010 period, and this decline was more pertinent during summer months. Also, the decrease in wind speed was more prominent up to mid-1990s. In fact, the decrease in wind speed was observed in the entire northern hemisphere (Vautard et al., 2010), which was likely due to changing patterns of atmospheric circulation, increased vegetation due to afforestation, changing landscape management practices and increase in urban density.

4.3 DROUGHT IN DRYLANDS

4.3.1 Drought and Its Magnitude in Drylands

Drought is a recurring, complex, and extreme climatic phenomenon characterized by subnormal precipitation for months to years triggering negative impacts on agriculture (Cammalleri et al., 2017). Droughts have several definitions that are based on different schools of thought (Heim Jr, 2002), thus making it more difficult to quantify their impacts in terms of their magnitude, duration, intensity, and spatial extent. Drought is often categorized into four types, namely meteorological, agricultural, hydrological, and socio-economic drought (Wilhite and Glantz, 1985). Meteorological drought is caused by sub-normal precipitation for months to years (Carrão et al., 2014). Agricultural drought occurs due to lack of soil moisture to support crop production (Keyantash and Dracup, 2004) and it can occur at any crop growth stages (e.g., early, mid, and late) resulting in a reduction in crop yield (Leng and Hall, 2019). Hydrological drought takes place when river streamflow and water storages in water bodies (e.g., aquifers, lakes, or reservoirs) drop below long-term average levels (Van Loon, 2015). Socio-economic drought occurs when there is an excess demand for economic goods owing to a lack of water supply resulting in negative impacts on society, economy, and the environment (Guo et al., 2019).

There are several drought indices for monitoring drought events (Chatterjee et al., 2022). Conventional drought assessment and monitoring often focus on the meteorological aspects of drought and precipitation and/or evapotranspiration data are often used to build meteorological drought indices for monitoring over space and time (Heim Jr, 2002). The Palmer Drought Severity Index (PDSI; Palmer, 1965), the Standardized Precipitation Index (SPI; McKee et al., 1993), and the US Drought Monitor (USDM; Svoboda et al., 2002) are examples of popular meteorological drought indices. The PDSI relies on a water balance method that combines precipitation, evapotranspiration (ET), and soil moisture. The SPI is exclusively based on precipitation data, whereas the Standardized Precipitation Evapotranspiration Index (SPEI) combines the features in PDSI and SPI and is able to depict the effects of temperature variability on drought assessment (Vicente-Serrano et al., 2010; Beguería et al., 2014). Agricultural drought assessment is based on a deficit of soil moisture during plant/crop growing season.

Occurrence of drought in drylands of India is a highly frequent phenomena specifically in arid and semi-arid regions. Magnitude of drought shows large variation across locations and year. During last century, about 47%–62% of the years in arid region experienced drought (Rao and Singh, 1998). In general, drought occurs once in every 2.5–5 years in arid region of India and this recurrence cycle is even smaller in hyper arid situation in western part of Rajasthan. For example, number of drought events during 1901–2022 in Jaisalmer, Barmer, Bikaner and Jodhpur was 68, 48, 46 and 43, respectively. The mean annual rainfall, moisture index and length of growing period in different drought-prone areas in India are presented in Table 4.1.

4.3.2 Spatial and Temporal Variation in Occurrence of Drought across Drylands

Droughts are characteristics of drylands and can be defined as periods (1–2 years) where the rainfall is below average. A drought is a departure from average or normal conditions, in which shortage of water adversely impacts ecosystem functioning and the resident populations of people. Drought has been a recurring feature of agriculture in India. In the past, India experienced 24 large scale droughts in 1891, 1896, 1899, 1905, 1911, 1915, 1918, 1920, 1941, 1951, 1965, 1966, 1972, 1974, 1979, 1982, 1986, 1987, 1988, 1999, 2000, 2002, 2009 and 2012 with increasing frequencies during the periods 1891–1920, 1965–90 and 1999–2012 (NRA, 2014). The risk involved in successful cultivation of crops depends on the nature of drought (chronic and contingent), its duration, and frequency of occurrence within the season. In the arid region where the mean annual rainfall is less than 500 mm, drought is almost an inevitable phenomenon in most of the years (Ramakrishna, 1997). In semi-arid regions (mean annual rainfall 500–750 mm), droughts occur in 40%–60% of the years due to deficit seasonal rainfall or inadequate soil moisture availability between two successive rainfall events. Even in dry sub-humid regions (annual rainfall 750–1,200 mm), contingent drought situations occur due to breaks in monsoon.

Intermittent and prolonged droughts are major causes of yield reduction in most crops. Long-term data for India indicates that rainfed areas witness 3–4 drought years in every 10-year period. Of these, 2–3 are in moderate and one may be of severe intensity. However, so far, no definite trend has been seen in the frequency of droughts as a result of climate change. For any R&D and policy initiatives, it is important to know the spatial distribution of drought events in the country. Different drought scenarios indicate that most of the parts of the country are affected by moderate to severe droughts, which is a major concern for food security (Figure 4.3).

4.3.3 Impact of Drought in Dryland Farming

Dryland agriculture involves different components of agriculture and allied sectors such as cereal and horticultural crops, vegetables, livestock, etc. and is, thus, a complex and vulnerable system. The occurrence of drought has several impacts on dryland farming. It affects crop growth by limiting the available soil moisture in the root zone. In severe cases, standing crops are totally dried,

TABLE 4.1
Rainfed Production System, Climate, Soil Type, Growing Period and Mean Annual Rainfall in India (Srinivasrao et al., 2009)

Location	Agro-Climatic Zone (NARP)/Agro-Eco Subregion (AESR)	Climate	Mean Annual Rainfall (mm)	Length of Growing Period (days)	Dominant Soil Type	Major Rainfed Crop Based Production System
Agra	South-western semiarid zone in Uttar Pradesh (4.1)	Semiarid (Hot dry)	665	90–120	Inceptisols	Pearl millet
Akola	Western Vidarbha zone in Maharashtra (6.3)	Semiarid (Hot moist)	824	120–150	Vertisols	Cotton
Anantapur	Scarce rainfall zone (Rayalaseema) in Andhra Pradesh (3.0)	Arid (Hot)	544	90–120	Alfisols	Groundnut
Arjia	Southern zone in Rajasthan (4.2)	Semiarid (Hot dry)	656	90–120	Vertisols	Maize
B.Saunkhri	Kandi region in Punjab (9.1)	Sub humid (Hot dry)	1,011	120–150	Inceptisols	Maize
Bangalore	Central, eastern and southern dry zone in Karnataka (8.2)	Semi-arid (Hot moist)	926	120–150	Alfisols	Finger millet
Bellary	Northern dry zone in Karnataka (3.0)	Arid (Hot)	502	90–120	Vertisols	*Rabi* Sorghum
Bijapur	Northern dry zone in Karnataka (6.1)	Semi-arid (Hot dry)	595	90–120	Vertisols	*Rabi* Sorghum
BiswanathChariali	North Bank Plain zone in Assam (15.2)	Humid (Hot)	1,990		Alfisols	Rice
Chianki	Western plateau zone of Jharkhand (11.0)	Sub humid (Hot moist)	1,179	150–180	Inceptisols	Rice
Faizabad	Eastern plain zone in Uttar Pradesh (9.2)	Sub humid (Hot dry)	1,051	150–180	Inceptisols	Rice
Hisar	South-western dry zone in Haryana (2.3)	Arid (Hyper)	412	60–90	Inceptisols	Pearl millet
Indore	Malwa plateau in Madhya Pradesh (5.2)	Semi arid (Hot moist)	958	120	Vertisols	Soybean
Jagadalpur	Basthar Plateau zone in Chattisgarh (12.1)	Sub humid (Hot moist)	1,297		Inceptisols	Rice
Jhansi	Bundhelkhand zone in Uttar Pradesh (4.4)	Semi arid (Hot moist)	870	120	Inceptisols	Rainy Sorghum
Jodhpur	Arid Western zone of Rajasthan (2.1)	Arid (Hyper)	331		Aridisols	Pearl millet
Kovilpatti	Southern zone of TamilNadu (8.1)	Semi arid (Hot dry)	723	120	Vertisols	Cotton
Parbhani	Central Maharastra Plateau Zone in Maharashtra (6.2)	Semi-arid (Hot moist)	901		Vertisols	Cotton
Phulbani	Eastern Ghat Zone in Orissa (12.1)	Sub-humid (Hot moist)	1,580	180–210	Oxisols	Rice
Rajkot	North Saurashtra zones in Gujarat (5.1)	Semi-arid (Hot dry)	590	60–90	Vertisols	Groundnut
Rakh Dhiansar	Low altitude subtropical zone in Jammu and Kashmir (14.2)	Semi-arid (Moist dry)	860	150–210	Inceptisols	Maize
Rewa	Key more plateau and Satpura Hill zone in Madhya Pradesh (10.3)	Sub-humid (Hot dry)	1,088	150	Vertisols	Soybean
S.K. Nagar	Northern Gujarat in Gujarat (2.3)	Semiarid/Arid (Hot dry)	670	60–90	Entisols	Pearl millet
Solapur	Scarcity zone in Maharashtra (6.1)	Semi-arid (Hot dry)	732	90–120	Vertisols	*Rabi* Sorghum
Varanasi	Eastern Plain and Vindhyan Zone in Uttar Pradesh (4.3/9.2)	Semi-arid (Hot moist) Sub-humid (Hot dry)	1,049	150–180	Inceptisols	Rice

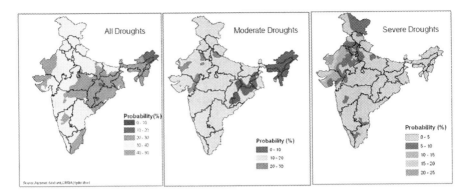

FIGURE 4.3 Frequency of different degree of droughts in India.

which lead to failure in crop production. The availability of fodder resources for livestock is drastically reduced during drought year since biomass production is limited due to water stress. Since livestock is the backbone of agricultural activity in dryland farming, non-availability of fodder has a serious impact on livestock population. Even the poor availability of drinking water for the livestock population during a severe drought year often forces farmers to migrate to suitable, distant places for the sustenance of their livelihoods. It has been observed that during drought years, the probability of occurrences of early, mid, and late-season drought was 26%, 9%, and 26%, respectively. Among these three categories of drought, the mid- and late-season drought caused a greater reduction in crop yields (72%–85% in pearl millet and 23%–61% in pulses) than the early-season drought (48% in pearl millet and 11% in pulses). However, the occurrence of drought at the very early stage of the cropping season often makes farmers unable to sow the crops in the field.

The drylands of arid and semi-arid areas lying towards the western and central parts of the country are more prone to droughts than those in other parts. Reduction in agricultural production due to drought poses great challenges in achieving national food production target since about 45% of the total agricultural production in the country is contributed from the drylands (Samuel et al., 2022). Since India's independence in 1947, many drought events occurred in the country. Among all these drought events, the one in 1987 with a rainfall deficiency of 19% was the worst, which affected 59%–60% of the normal cropped area and about 265 million people (GoI, 2017). Such high magnitude drought again occurred in 2002 with 19% overall deficiency of rainfall in the country as a whole. However, 2002 drought affected over 300 million people living in 18 states along with around 150 million cattle. Also, an exceptional reduction of 29 million tonnes in food grain production was recorded. In 2009, about 22% deficiency in rainfall of the country as a whole could decrease food grain production by 16 million tonnes. In addition, drought severity during 2014–15 and 2015–16 caused hardships to the affected population over large parts of the country encompassing the major agricultural regions of the country. The increased frequency of droughts is posing severe challenges to the productivity of drylands in the country, which will further aggravate the situation in future due to increasing trend of drought severity and frequency under the climate change scenario (Mallya et al., 2016; Kala, 2017). It is further evidenced from the Fifth Assessment Report of the Intergovernmental Panel on Climate Change (IPCC) that impact of the extreme events such as drought on livelihoods and poverty will increase in the coming decades, which will exacerbate rural poverty in Asia (Denton et al., 2014). It is estimated that by 2030, there will be a monetary loss of more than US$ 7.6 billion in Indian agriculture (Debaje, 2014) and the annual loss in agriculture from droughts is reported to increase. However, this loss could be reduced by 80% by implementing climate-resilient technologies and adaptation measures (Sutton et al., 2013). It has been reported that risks associated with agricultural production due to drought are the most crucial risks that farmers face frequently (Rao et al., 2020).

4.4 FRAGILE NATURAL RESOURCES IN DRYLANDS

Natural resources in terms of soil, water, and weather are highly limited in drylands in terms of their availability and quality. Most of the soils in drylands are coarse textured, and hence the retention of both nutrients and water is poor. Very high soil temperature regime in drylands has limited support for soil biota, thus sustaining the biological productivity of the soil. Surface soil temperature of the region reaches 60°C–70°C during summer months and remains >50°C for most periods in a year. Due to this high thermal regime, most biological processes are suppressed to maintain a congenial environment for plant growth. Sand content of the soils is 80%–90% in most cases and soil organic carbon (SOC) status also being low, soils are rather loose in structural attributes. Moreover, these loose surface soils are highly prone to erosion both by wind and water. The addition of organic matter as such does not lead to improvement in SOC because of low biological activity. The concern on very hot and dry soil physical environment in drylands is more serious under global climate change situations especially under warming scenarios. The availability of both surface water and groundwater resources is also limited in drylands. Groundwater depth below ground level is increasing at a very fast rate because of continuous withdrawal of it with less amount of rainwater recharge. The quality of scarce groundwater resources is again a limitation because of inherent salinity in most parts of the drylands. Rainfall is very limited (250–500 mm) in most part of drylands, which is again affected by climate change and a shifting pattern from low intensity distributed rainfall to few high intensity rainstorms. Evapotranspiration rate is very high with an annual value of 1,500–2,000 mm.

Despite limitations in natural resources, there is an increasing trend in the human and livestock populations in the region. For example, human population in arid western Rajasthan has increased by >250% over a period from 1961 to 2011, while the animal population has increased by about 125.2% between 1956 and 2012 census. Irrespective of frequent droughts, the region has a dominant agricultural economy. An estimate of changes in agricultural activity in arid Rajasthan between 1982–83 and 2005–06 indicated an increase in net-irrigated (128%) and double cropped area (70%), whereas a decline in culturable waste. This indicates magnitude of biotic pressure on limited natural resources in drylands. The rapid rate of land conversion in recent times has further aggravated the problem. Land use change to an arable farming system is obvious to sustain the livelihood of continuously increasing population pressure. Reversely, suitable on-farm technologies need to be developed to tackle the problem of sustainable production in the region.

4.4.1 Edaphic Constraints of Drylands

4.4.1.1 Soil Types and Its Spread

Alfisols, Vertisols, Inceptisols, Mollisols, Oxisols, and Aridisols/Entiols are the major soil types in drylands located in semi-arid and dry subhumid regions of India. About 30% of dryland areas are covered by Alfisols and associated soils while 35% by Vertisols and associated soils having vertice properties and 10% by Entisols of the alluvial soil regions (Virmani et al., 1991). Crusting in Alfisols, swelling and shrinking in Vertisols, low water and nutrient holding capacity in Aridisols and Inceptisols are the major productivity constraints. Major soil types of arid parts of India, defined according to United States Department of Agriculture (USDA) Soil Taxonomy include: Aridisols (37.8%), Entisols (50.1%), and Inceptisols (13.1%) (Figure 4.4). However in western Rajasthan covering 62% area of arid region in India, soils are classified under two soil orders: Entisol and Aridisol, which constitute about 61% and 38% of the total area of western Rajasthan, respectively. Aridisols are mainly observed in buried pediments, interdunal plains and old alluvial plains. Average depth of such profiles is 106 cm with well demarcated horizons; concretions of calcite below the soil profile are common. Psamments-Orthids are the major sub-order associations under Aridisols.

Torripsamments is the major great groups under Entisol covering 87.7% area, whereas Haplocambids are the major great group under Aridisol covering 71.1% area. Soils under Entisols

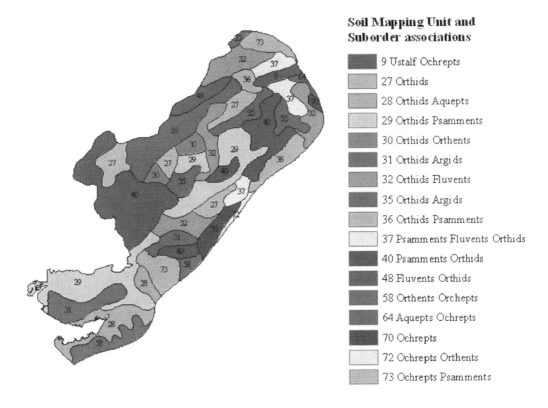

FIGURE 4.4 Soil suborder association map of hot Arid Western India. Numbers refer to mapping units as described in Table 4.2. Source: Santra et al. (2018).

are generally found in those parts of western Rajasthan where high aeolian activities are observed. In contrast to Entisols, Aridisols are dominant in those parts of arid regions where aeolian activity is comparatively less. Average depth of this type of soil is 100 cm with well-demarcated horizons. Concretions of calcite below the soil profile are a common feature of these soils.

4.4.1.2 Nutrient-Depleted Soils

In the hot arid regions of north western India the soils are usually sandy, deficient in several major and micro-nutrients, and there is large spatial variability in the plant available nutrients content of the soils (Gupta et al., 2000; Praveen et al., 2009; Mahesh et al., 2021). N status in the soils of arid Rajasthan varies from very low to low; P low to medium and K medium to high (Kathju et al., 1998). The SOC content in arid soils is low and ranges from 0.01% to 0.84%, with a mean of 0.16%. Overall, 99.37% of area contains low to very low SOC (Mahesh et al., 2021). Available phosphorus in these soils shows wide variability (1.19–96.64 kg/ha) with a mean value of 13.34 kg/ha. As the mean values across the region suggest, in general the quantity is low to medium. While 48.74% of the area has low P, 46.86% of the area has medium content. The available potassium (K) ranges from 56.00 to 974.00 kg/ha, with mean values of 250.60 kg/ha and the percent sample category under low, medium and high was 11.0, 60.5 and 28.5, with an overall medium fertility rating. By contrast the distribution of available K is adequate to more than sufficient in large parts of the region.

The Diethylenetriamine pentaacetate (DTPA)-extractable iron (Fe) ranged from 0.84 to 42.74 mg/kg, with a mean of 6.42 mg/kg, and could be considered as marginally adequate (Mahesh et al., 2021). Despite its adequacy over a large area, deficiency of Fe has been mapped in about 41% of areas (Mahesh et al., 2021) and needs to be addressed. The DTPA-extractable zinc (Zn) in the

TABLE 4.2
Area under Different Mapping Units and Suborder Associations in Arid Western India (Santra et al., 2018)

Map Unit	Suborder Associations	FAO Name	% Area	Major Suborder	% Area	Major Order	% Area
27	Orthids	Solonchaks/Yermosols	6.4	Orthids	31.5	Aridisols	36.8
37	Psamments-Fluvents-Orthids	Arenosols-Fluvisols	4.2				
40	Psamments-Orthids	Arenosols-Yermosols	18.9				
48	Fluvents-Orthids	Fluvisols-Yermosols	2.0				
31	Orthids-Argids	Yermosols-Yermosols	5.3	Argids	5.3		
35	Psamments	Arenosols	12.7	Psamments	33.5	Entisols	50.1
29	Orthids-Psamments	Solonchak/Yermosols-Arenosols	16.0				
73	Ochrepts-Psamments	Cambisols-Arenosols	4.8				
30	Orthids-Orthents	Yermosols-Regosols	2.0	Orthents	2.6		
72	Ochrepts-Orthents	Cambisols-Regosols	0.6				
32	Orthids-Fluvents	Yermosols-Fluvisols	10.5	Fluvents	14.0		
36	Psamments-Fluvents	Arenosols-Fluvisols	3.5				
70	Ochrepts	Cambisols	2.3	Ochrepts	7.6	Inceptisols	13.1
9	Ustalf-Ochrepts	Planosols/Luvisols/Nitosols-Cambisols	1.2				
64	Aquepts-Ochrepts	Gleysols-Cambisols	0.3				
58	Orthents-Ochrepts	Regosols-Cambisols	3.8				
28	Orthids-Aquepts	Solonchak-Solonchaks	5.5	Aquepts	5.5		

soils of the hot arid region of Rajasthan ranges from 0.05 to 5.20 mg/kg, with a mean of 0.76 mg/kg, which is marginally low. The DTPA-extractable copper (Cu) in soils varied from 0.04 to 4.86 mg/kg, with overall mean of 0.66 mg/kg, which suggests adequacy. In general, arid soils are not deficient in available Cu, and it varies from 0.21 to 4.28 mg/kg in arid soils of Rajasthan (Dhir, 1977). The DTPA-extractable manganese (Mn) in the soils ranged from 0.70 to 45.90 mg/kg, with mean value of 9.44 mg/kg, i.e., moderately high. Based on the threshold limit (2.0 mg/kg), only 2.5% samples were deficient in DTPA extractable Mn. Deficiency of Mn is not a serious problem in hot arid regions of Rajasthan.

4.4.1.3 Soil Physical Conditions

The soils of the region are predominantly sandy, and the sand content in the soils varies generally from 40% to 97%. The clay and silt contents vary from 2.0% to 35.8% and 1% to 33.4%, respectively, with the higher amounts being recorded in the alluvial plains while the lower amounts are found in the sand dunes. The range of moisture retention capacity values at 0.1 bar (field capacity) and 15 bars (permanent wilting point) tension is 8.0%–10.0% and 2.0%–3.0%, respectively. The soils predominantly have macro-porosity and, therefore, there is fast movement of water into and through the soil profile. The initial infiltration rate varies from 15 to 30 cm/ha and saturated hydraulic conductivity from 5 to 10 cm/ha. The average bulk density of these soils is 1.46 mg/m^3 whereas the average sand content is about 61%. Entisols are found in places where aeolian activity is dominant with Orthids-Psamments as major sub-order association. The average depth of Entisols is 105 cm and the bulk density is 1.54 mg/m^3. Surface horizons are richer in sand content than subsurface horizons; the average sand content is 78%. Soils under Inceptisols are mainly observed in western and southern borders of arid western India with Ochrepts as the dominating soil sub-order. The average soil depth of Inceptisols is 83 cm, with an average sand content of 51%, an average clay content of 34%; and an average bulk density of 1.51 mg/m^3. The average soil physical conditions in arid western India is depicted in Table 4.3 (Santra et al., 2018).

4.4.1.4 Low Concentration of Biota

Microbial populations in the Indian arid zone are reported to have relatively smaller population (1.5×10^2–$5 \times 10^4 g^{-1}$ soil) (Venkateswarulu and Rao, 1981; Khathuria, 1998). Gram-positive spore formers are dominant and the populations do not decline significantly even during summers (Rao and Venkateswarlu, 1983). Actinomycetes may constitute ~50% of the total microbial bacterial population in desert soils. Dominant microflora of desert soils is made up of coryneforms, i.e. *Archangium, Cystobacter, Myxococcus, Polyangium, Sorangium* and *Stigmatella*; sub-dominant forms comprise *Acinetobacter, Bacillus, Micrococcus, Proteus* and *Pseudomonas*. Cynobacteria also contribute significantly (0.02–$2.63 \times 10^4 g^{-1}$ soil) to the biota of hot arid regions in terms of primary productivity and nitrogen fixation (Bhatnagar et al., 2003). The dominant cynobacterial forms of Thar desert are *Chroococcus minutus, Oscillatoria pseudogeminata* and *Phormidium tenue*;

TABLE 4.3
Soil Physical Conditions in Arid Western India (Santra et al., 2018)

Soil Properties	Mean	Standard Deviation	Range
Sand (%)	64	23	4–97
Silt (%)	14	11	1–68
Clay (%)	23	14	1–68
Soil organic carbon (g/kg)	2.6	1.9	0.1–11.0
Bulk density (mg/m^3)	1.49	0.13	1.0–1.8
Field capacity (%, g/g)	18	9	3–48
Permanent wilting point (%, g/g)	7	4	

Nostoc sp. dominates amongst heterocystous forms. Fungal populations as viable propagules range from 0.5 to 14.7×10^3. The dominant genera include *Aspergillus, Curvularia, Fusarium, Mucor, Penicillium, Paecilomyces, Phoma* and *Stemphylium*. Xeric mushrooms such as *Coprinus, Fomes, Terfezia* and *Teramania* and arbuscular mycorrhizal fungi such as genera *Glomus, Gigaspora* and *Sclerocystis* have also been reported from desert (Trappe 1981; Pande and Tarafdar, 2004).

Rhizosphere is an important site of microbial activity in desert soils, since it provides ample carbon substrate in an otherwise organic matter poor soils of arid regions. Generally, the ratio of rhizosphere and soil is high in arid soils for nearly all metabolic types of bacteria (viz. heterotrophs, dizotrophs, cellulites and nitrifiers) and fungi in most plants studied (Khathuria, 1998; Singh and Tarafdar, 2002).

Crusts on soils are formed by microorganisms and microphytes. Diversity status of cryptobiotic crusts at Thar desert in India showed 43 morphotypes of diazotrophs in BG11-N enrichment and 71 of algae and cyanobacteria in the same medium supplemented with nitrogen (Bhatnagar et al., 2003). Most frequent form was *Phormidium tenue*. In the case of diazotrophs the most frequent forms were *Nostoc punctiforme, Nostoc commune* and *Nostoc polludosum*. Cynobacteria presence was influenced by their plant partners. *Alternaria* sp. was the dominant fungus in these crusts.

4.4.2 SCARCE WATER RESOURCES

4.4.2.1 Limited Surface Water Resources

The Indian hot arid region is known for very scarce water resources, a large part of the rainwater is lost as evapotranspiration. Western Rajasthan has been divided into three broad hydrological zones (Moharana et al., 2016).

> *Zone I:* This is the main canal irrigated zone in arid Rajasthan where major input of surface water comes from more humid region. About 60% area of Sri Ganganagar district, 50% area of Bikaner district, and 25% area of Jaisalmer district in the northwest lie in this zone.
> *Zone II:* This region in 52% area of arid Rajasthan has a system of repetitive micro-hydrology, with a primitive or no stream network. Churu, Jhunjhunun, Sikar, Nagaur, Jodhpur and parts of Bikaner, Jaisalmer ,and Barmer districts fall within this zone.
> *Zone III:* This region has an integrated stream network of Luni basin, occurring in Pali, Jalor and parts of Jodhpur and Barmer districts.

The total surface water resources excluding IGNP (Indira Gandhi Nahar Pariyojna) of arid zone of Rajasthan is $1361 \times 10^6 \, m^3$ which is equivalent to 7.2 mm in depth or 7,200 m^3/km in the region. Large numbers of tanks, reservoirs, minor irrigation dams, and check dams have been constructed to store runoff water during the monsoon period. In western Rajasthan, 550 storage tanks with the capacity ranging from less than $314.1 \times 10^6 \, m^3$ are functional with total utilizable capacity of nearly $1169.28 \times 10^6 \, m^3$ for providing irrigation on 0.102 M ha land. Out of these, six reservoirs (viz., Jaswantsagar, Sardar Samand, Jawai, Hemawas, Ora and Bankali) are the major irrigation tanks with capacity of irrigation of more than 4,000 ha each. Jawai is the main source of drinking water supply to many towns and villages. Table 4.4 presents surface water resources for the arid Rajasthan.

4.4.2.2 Groundwater Resources: Low Quantity and Poor Quality

After independence in 1947, the country's water resources were developed in a planned way by creating water storage projects along with renovation, extension and modernization of existing water projects (Das et al., 2007). Many areas of the country, especially those situated in the arid and semi-arid regions, suffer due to low and uncertain availability of surface water resources, and hence, these areas solely rely on groundwater resources. In fact, groundwater resources were key to the country's Green Revolution in the 1970s, though it was mostly occurred in water-rich northern plains where Punjab and Haryana are situated. Of the total irrigated area of the country, more

TABLE 4.4
Surface Water Resources in Arid Rajasthan (Moharana et al., 2016)

District	Zone-I	Zone-II	Zone-III	Total (mcm)
Sri Ganganagr[a]	6.49	11.19	-	17.68
Bikaner	7.14	26.16	-	33.3
Churu	-	22.82	-	22.82
Jhunjhunu	-	8.04	96	104.04
Sikar	-	10.48	96	106.48
Jaisalmer	5.03	38.38	-	43.41
Jodhpur	-	20.66	-	20.66
Nagaur	-	53.18	-	53.18
Pali	-	-	869	869
Barmer	-	19.64	-	19.64
Jalore	-	-	71	71
Total	18.66	210.55	1132	1361.21

[a] It includes Hanumangarh.

than 60% depends on groundwater supplies. Furthermore, about 60% of the cultivated area in the country is rainfed, and most of the rainfed area occurs in arid and semi-arid regions. In addition to surface water, groundwater use for irrigation purposes increased significantly, making the country self-reliant in food production. Since 1950, the country has seen a steady growth in development of groundwater resources reaching a stage of 62% with overdeveloped groundwater in many parts as in Punjab, Haryana, Rajasthan, and Delhi where groundwater draft exceeded the annual recharge and 74%–77% of groundwater development occurred in Tamil Nadu and Uttar Pradesh. Presently, of the total 6,584 blocks/taluks in the country, about 20% are categorized under overexploited or critical condition (Das, 2019). The overexploitation of groundwater resources has severe repercussion on the dependable water availability in the arid and semi-arid regions of the country. Uncontrolled development of groundwater has led to indiscriminate groundwater withdrawals, fast-depleting groundwater levels, increased energy and cost of lifting water, continuously diminishing well and aquifer yields, hampered agricultural production, and degraded groundwater quality, among others.

Decline in groundwater resource in the drylands induces the problems of poor groundwater quality due to chemical contamination making an aquifer vulnerable (Machiwal et al., 2018a). Groundwater in the arid and semi-arid areas of the country, receiving annual rainfall of less than 650 mm, usually has electrical conductivity (EC) of 1,000–1,500 μs/cm at 25°C, which is of fair quality and can be used for irrigation as well as drinking purposes. However, groundwater salinity is found quite high with an EC of more than 2,500–10,000 μs/cm in the arid areas of western Rajasthan, southern Haryana, NCT Delhi, southern Punjab, western Uttar Pradesh, and parts of Gujarat along with scattered patches in states of Peninsular India, i.e., Maharashtra, Madhya Pradesh, Andhra Pradesh, Karnataka, etc. Inland salinity in the country is reported in about 19.3 M ha area (CGWB, 1997; Das, 2019). Groundwater contamination in drylands has been recently emerging as a major concern due to three factors: (i) natural processes such as geology, groundwater movement, recharge water quality, and soil/rock interactions with water, (ii) anthropogenic activities including agricultural production, industrial growth, urbanization with increasing exploitation of water resources, and (iii) atmospheric input (Chan, 2001; Machiwal and Jha, 2015). Furthermore, the groundwater contamination in the drylands of the country is being exacerbated due to overexploitation of the groundwater resources. Future implications of climate change and global warming scenarios for groundwater alterations include changes in surface water availability with response to changes in runoff particularly in arid and semi-arid regions of drylands, fast depletion of groundwater levels due to heavy pumping and reduced recharge potential, which will increase degree of mineralization in groundwater. Contamination of fluoride in groundwater is a common hazard in semi-arid and arid areas of Rajasthan (5.0–20 mg/L) and southern Punjab (up to 11.7 mg/L), southern Haryana (up to 15 mg/L), Gujarat (up to 11 mg/L), Uttar Pradesh (10–12 mg/L), Karnataka (up to 4.7 mg/L), Odisha (up to 8 mg/L), Telangana (2–5 mg/L), among other states of the country (Das, 2011). The groundwater contamination threatens the reliability and sustainability of the groundwater resources in the drylands of the country.

4.4.2.3 Canal Water Resources

The major developments in irrigation facility in arid region of India started in the late 1960s with the introduction of the IGNP canal and later in 2008 with the introduction of Narmada Canal Project (NCP) established in the southern part of the Thar Desert covering Jalore and Barmer districts of Rajasthan with a canal command area of about 24.6 M ha. At present, about 61% of net irrigated area in the Thar Desert of Rajasthan is irrigated through wells and tube wells whereas 38% is irrigated through the IGNP and NCP canals (Rajasthan Agricultural Statistics, 2016–17).

IGNP is one of the largest irrigation projects in the world. It was conceived to transform Thar Desert into a land of plenty and had the objectives of "drought proofing, provision of drinking water, industrial and irrigation facilities, creation of employment opportunities, settlement of human population of thinly populated desert areas; improvement of fodder, forage and agriculture facilities, check spread of desert area and improve ecosystem through large-scale afforestation, develop road network and provide requisite opportunities for overall economic development" (IGNP, 2002). The project encompasses the districts of Ganganagar, Hanumangarh, Churu, Bikaner, Jaisalmer, Jodhpur, and Barmer with a culturable command area (CCA) of 1.963 M ha. The project has been divided into two stages. Stage I comprises a 204 km long feeder canal, with a discharge capacity of 460 m^3/second. Stage I also consists of a 189 km long main canal and 3,454 km long distribution system to serve 0.553 M ha of CCA. Stage II comprises a 256 km long main canal and 5,606 km long lined distribution system and serves 1.41 M ha of CCA.

The Narmada canal network in the desert of Rajasthan receives water from the Narmada River, which originates from the Amarkantak in the Shahdol district of Madhya Pradesh and runs a great distance (≈1,300 km) in Madhya Pradesh, Maharashtra and Gujarat states before entering into the Thar Desert. The length of the main Narmada canal in Rajasthan is 74 km with 12 main distributaries taking off from the main canal. The total length of the distribution system comprising distributaries, minors, sub-minors is about 1,719 km. This project was designed to provide irrigation and drinking water to Jalore and Barmer district of Rajasthan located in the southwest part of the state. For efficient utilization and ensured optimized regulation of canal water, irrigation with drip and sprinklers has been made mandatory in the command area. The water from the canal is being drawn into a surface water reservoir locally known as '*diggies*' and then water from '*diggies*' is conveyed to farmers' field through underground HDPE (high-density polyethylene) pipe network. Farmers at each end node of the irrigation network apply irrigation water in their field as per the allocated share decided by water user association committee for each '*diggi*' using pressurized irrigation network e.g. drippers and sprinklers. The command area of the Narmada canal is broadly grouped into three zones: (i) Flow area (ii) Lift area and (iii) Ned area. The command area has three physiographic units viz. alluvial plains, aeolian plains and duny complex. The intensive agricultural practices are now followed by farmers due to the availability of canal water in the region. Cropping system consisting of pearl millet (*Penisetum galucum*), mung bean (*Vigna radiata*), and cluster bean (*Cyamopsis tetragonoloba*) during *kharif* season (rainy) and wheat (*Triticum aestivum*), cumin (*Cuminum cyminum*), isabgol (*Plantago ovata*), mustard (*Brassica juncea*) during *rabi* season (winter) is the dominant in the canal command areas of NCP (Mahesh et al., 2022). It has been reported that introduction of irrigation facility in arid and semiarid regions improved the crop production by up to 400% and thus provided better economic returns; however, it also brought some unwanted environmental consequences such as water logging and secondary salinization (Cireli et al., 2009).

4.4.2.4 Non-Judicious Use of Water Resources

The introduction of irrigation in desert area brought considerable prosperity to the farmers. Some of the positive impacts of the introduction of irrigation in the desert include improvement in micro climate, change in land use/in cropping pattern, improvement of soil and moisture conditions and associated soil fertility and biological properties, but it has also brought in its wake the problems of water logging and secondary salinization. However, after a few years of irrigation with canal water, some negative effects emerged such as rise in the water table, waterlogging, formation of marshy lands and soil salinity at few places. Lack of proper drainage, excess irrigation, seepage from the canals and poor drainage planning under such situation have resulted in a rise in water table, followed by salinity build-up. Under these situations, two different sources of soluble salts are accumulated in irrigated soil; one is irrigation water itself, and the sub-soil or the parent rock is impregnated with salts before irrigation begins. The average rate of rise in water table in the command areas of IGNP is 0.88 m/year, while that in the Ganga Canal command is 0.53 m/year, and in the Bhakra canal command 0.66 m/year. Within the Ganga Canal command, the Ghaggar flood plain is experiencing a rise of 0.77 m/year (Kar et al., 2009). The problem of waterlogging in Sri Ganganagar and Hanumangarh districts under the IGNP has changed the environment of the districts, from dry and desertic tract to highly productive land. In about two decades now the Scarcity- Prosperity- Scarcity cycle seems to have become a full circle. The introduction of canal irrigation in the Thar Deserts of Rajasthan has completed the journey from one wasteland (water starved) to another wasteland (water soaked). At first instance, reduction in crop yield is observed which is followed by restrictions on the type of crop, and ultimately leads to the abandonment of previously productive land at few places. The current estimates indicate that about 0.208 M ha land is already affected by waterlogging and associated salinity in IGNP command area (Table 4.5). The salt-affected and waterlogged soils in this command are mainly located in Anupgarh Branch, Suratgarh Branch and Charanwala branch.

The NCP was extended up to Rajasthan to provide irrigation to the drought-prone areas in Jalore and Barmer districts. The NCP has been designed to utilize 0.50 million acre-feet (MAF) of

TABLE 4.5
Waterlogged Area (ha) in IGNP Commands of Rajasthan (Shrivastava et al., 2013)

	Years						
Category	2001–02	2002–03	2003–04	2004–05	2005–06	2006–07	2007–08
			Stage I				
Waterlogged area[a]	10,098	5,755	2,531	2,968	3,125	6,875	1,875
Critical area[b]	11,355	8,750	9,259	10,625	11,250	16,875	12,500
Potentially sensitive area[c]	179,170	164,375	195,000	168,750	196,875	202,150	181,250
			Stage II				
Waterlogged area	78	16	4	4	484	805	320
Critical area	1,261	453	317	476	1,129	2,576	1,120
Potentially sensitive area	18,304	24,572	13,481	16,018	18,548	15,906	11,840

[a] Waterlogged area (water table within 0–1 m).
[b] Critical area (water table within 1–2 m).
[c] Potentially sensitive area (water table within 2–6 m).

Narmada water for a total of 0.246 M ha CCA. Presently, the Narmada canal, through lift and flow systems, provides irrigation facilities in about 0.239 M ha area in both districts. With the available irrigation water from Narmada canal, cultivating crops in this area has become a reality in the command areas. The crop production in the canal command areas has been increased by manifold with the introduction of irrigation facility in arid areas, but it has also brought the problems of secondary salinization and water logging at some locations. The development of salinity in the newly developed command area of the Narmada canal in Rajasthan has arisen chiefly from the pre-existing salt deposits in the sub-stratum rather than from the irrigation water (Mahesh et al., 2016b). Sub-soil conditions, resulting in waterlogging, a rising water table and the resultant salinity have already rendered some parts of the Narmada canal command area in Sanchore tehsil of Jalore district. Saturation extract analysis of these soils revealed that the pH and EC of these soils varied from 7.7 to 9.5 and 1.6 to 41.5 dS/m. Among the cations, Na^+ was dominant followed by calcium+ magnesium. While chloride followed by sulphate were by and large the dominant anions and were present in the range of 16–156 and 12–126 me/L, respectively.

4.4.2.5 Poor Response of Rainfall to Runoff

Water pouring through the atmosphere during a rainy event partly contributes to interception and infiltration processes before generating surface runoff that flows over the land surface. It is observed that response of rainfall to the generated runoff quantities varies over the space and time depending upon a number of factors such as rainfall characteristics (intensity, duration, amount, antecedent moisture conditions), soil properties (texture, initial moisture content, temperature, etc.), land use (cultivated land, built-up land, waterbody, etc.) among others. In dryland areas of the country, rainfall shows a poor response to runoff as the process of runoff generation depends largely upon several parameters and some of them are listed above. The rainfall-runoff relationships are established for the semi-arid tropics of the country in a recent study (Garg et al., 2022), which revealed that the generated quantities of surface runoff varied largely depending upon geographic location, type of soil and variability of rainfall. The rainfall showed a linear relationship with runoff up to 120 mm of rainfall, and the relationship changed from linear to the exponential on exceeding rainfall threshold of 120 mm. Likewise, value of runoff ratio was found to be 0.33 (runoff coefficient ~ 33%) up to the mean rainfall of 120 mm with 40 mm runoff at 120 mm rainfall. Beyond the rainfall threshold value, value of the runoff ratio was found to be 0.90 with 230 mm runoff for 250 mm rainfall (runoff coefficient ~ 90%). The mean runoff quantities generated in low (<600 mm), medium (600–800 mm) and high (800–1,000 mm) rainfall zones were 40, 70, and 190 mm, respectively (Garg et al., 2022).

In arid regions of the country, value of the runoff ratio usually remains much lower (even less than 0.10) than that found in semi-arid regions. In sandy plains of Jodhpur district in the western Rajasthan, scanty runoff with runoff ratio of 0.05 (runoff coefficient ~ 5%) has been found in the studies reported during 1970s. The major reason behind the low runoff ratio value is very high infiltration rate in sandy tracts of the arid region, which causes less runoff generation. However, value of the runoff ratio was found higher in the years receiving high intensity rainfall. In the recent years, frequency of high intensity storms has increased in the arid regions of the country, which results in substantial runoff quantities with enhanced value of the runoff ratio.

4.5 ROLE OF SOIL, WATER AND NUTRIENT MANAGEMENT FOR SUSTAINABILITY IN DRYLANDS

The adverse impact of drought can be alleviated by sustainable management of soil, water and nutrients. In-situ soil moisture conservation by mulching may help to reduce high evapotranspiration losses of soil moisture and in turn provides prolonged supply of soil moisture to roots. Runoff water during high intensity rainfall events may be harvested in water reservoir or farm ponds for its efficient use during dry spells. Efficient irrigation systems (e.g., pressurized irrigation network using drippers and sprinklers) has the potential to improve water productivity and thus saves water resources. Addition of organic matter improves soil condition to retain more amount of water and nutrient in soil so as to supply them to crop roots for a prolonged period. Soil nutrients may be applied to soil as per the need basis and as per the availability of it in soil system. This will not only reduce the cost of cultivation but also improve the efficiency of nutrient use.

4.5.1 Interventions on Soil and Nutrient Management

4.5.1.1 Managing Soil Carbon

Dryland environments are characterized by a set of features that affect their capacity to sequester C. Lack of water and erratic nature of rainfall constraints plant productivity severely and therefore affect the accumulation of C in soils. In addition, SOC pool tends to decrease exponentially with temperature (Lal, 2002). However, dry soils are less likely to lose C than wet soils (Glenn et al., 1993) as lack of water limits soil mineralization and therefore the flux of C to the atmosphere. SOC stock has the largest contribution to total global carbon stocks contributing 1,550 Pg (1 Pg = 10^{15} g) of carbon to 1-m depth, which is about three times that of biotic and twice that of the atmospheric pools (Mishra et al., 2010). Presently, in the context of global warming scenarios and forcible land use changes under increased population pressure, soil carbon is continuously being lost to atmosphere (Davidson et al., 2000; Lettens et al., 2005). Intensive cultivation in dryland regions has resulted in the decline of its meagre SOC pool (~ less than 1 g/kg in most areas) at a faster rate and even more under climate change related desertification processes (Lal, 2001).

Soil carbon plays an important role in soil health. However, its content in arid soil is very poor. Most of the soils in the arid region have SOC content <0.2%. The clay content of soils in the arid region is also very less (5%–15%) and thus there is little scope to improve SOC content even after the long-term application of farmyard manure. This is because of the lack of retention of carbon in the soil matrix. It is evidenced from the high loss of soil carbon through eroded soil due to wind e.g. SOC content is eroded soil was found about 20–25 times higher in eroded soil than parent soil (Santra et al., 2010). This is mainly due to lack of intra-particulate organic matter in the soil.

However, there is scope to improve SOC content from very low (~0.1%) status to a desirable limit to sustain crop growth and yield. Looking into these scenarios, it is very important to use compost as an organic source of nutrients, which can be produced from farm wastes at a very low cost. One big advantage of compost preparation from farm waste is the efficient utilization of farm resources e.g. pre-monsoon grass vegetation, cow dung and farm labours, which are easily available. Dominant pre-monsoon vegetation commonly available in current fallow, permanent fallow, arable lands and permanent pasture lands in hot arid regions of Rajasthan includes *Crotolaria burhia* (*Saniya*), *Aerva javenica* (*Bui*), *Leptodenia pyrotechnica* (*Khimp*), and *Teprosia purpurea* (*Dhamasha*). Conventionally these native vegetations are removed from fields before cultivation of *kharif* crops. However, the above-ground biomass potential of these under shrub vegetations in the hot arid region of Rajasthan (~15 Tg/year) has great scope for in-situ preparation of compost in the region. Thus, preparation and use of compost can be a feasible solution to tackle the problem of poor fertility of the soils and also will make the small and marginal farmers self-reliant in terms of supplying nutrient resources in their fields.

4.5.1.2 Tackling Multi-Nutrient Deficiency

Declining soil fertility and nutrient imbalance is another major issue affecting agricultural productivity. Organic matter levels have declined sharply in intensively cropped areas. In addition to universal deficiency of nitrogen (N), deficiencies of potassium (K), sulphur (S) and micro nutrients are increasing. Zinc (Zn) deficiency has become most acute followed by sulphur (S) and boron (B) (Motsara, 2002). It is estimated that 29.4 M ha of soils in India are experiencing decline in fertility

TABLE 4.6
Macro and Micronutrient Percentage Deficiencies in Soils of Some States of India (Motsara, 2002; Takkar, 2006; Mahesh et al., 2021)

S. No	Location	Percentage Deficiency of Available Macro and Micronutrients							References
		N	P	K	Zn	Fe	Cu	Mn	
1	Madhya Pradesh	40	39	10	63	3	1	3	Motsara (2002)
2	Uttar Pradesh	80	71	12	45	6	1	3	Takkar (2006)
3	Maharashtra	67	86	8	-	-	-	-	
4	Andhra Pradesh	62	57	9	51	2	1	2	
5	Karnataka	29	31	7	78	39	5	19	
6	Orissa	60	59	33	-	-	-	-	
7	Tamil Nadu	75	24	12	53	15	3	8	
8	India	63	42	13	-	-	-	-	
9	Gujarat	-	-	-	24	8	4	4	
10	Bihar	-	-	-	54	6	3	2	
11	Arid Rajasthan	-	49	11	57	42	5	2.5	Mahesh et al. (2021)

with a net negative balance of 8–10 M Mg of nutrients per annum (Lal, 2004a,b). Poor nutrient use efficiency is another cause of concern. Indian soils are low in organic matter thus soils are showing 63% low, 26% medium and 11% high status. About 80% of soils tested were low to medium in available P. K status was low to medium in 50% soil samples. Similarly, among micronutrients, Zn is most deficient followed by Fe, Mn and Cu (Table 4.6). So far soil fertility issues have been mainly focused in irrigated agriculture, but recent studies indicated that drylands are not only thirsty but also hungry (Srinivasarao and Vittal, 2007). Most of the rainfed regions are low in organic carbon and available N and soils are multi-nutrient deficient (Table 4.7).

4.5.1.3 Control Mechanisms to Reduce Soil Erosion

4.5.1.3.1 Surface Cover Management for Controlling Wind Erosion

Use of surface cover to control wind erosion may be of two types, vegetative and non-vegetative. Protection of land surface through vegetative surface cover of grasses or crops is perhaps the most effective, easy and economical. Grasslands of sewan (*Lasuirus sindicus*) and foxtail buffalo grass (*Cenchrus ciliaris*) and maintenance of cover crops like colocynth (*Citrullus colocynthis*) have a large role in the reduction of soil loss from sandy plains due to wind erosion. In addition to the standing vegetation, crop residues often are placed artificially on the soil to provide temporary cover until the establishment of permanent vegetation. It was further reported that in the case of residue application, better control of wind erosion may be obtained if the residues are well anchored to the surface.

Other than vegetative covers, various surface films such as resin-in-water emulsion (petroleum origin), rapid curing cutback asphalt, asphalt-in-water emulsion, starch compounds, latex in water emulsion (elastomeric polymer emulsion), by-products of the paper pulp industry and wood cellulose fibre were used to control wind erosion (Woodruff et al., 1972), which mainly aim to decrease soil erodibility. Sand or pebble (>2 mm) mulch are often used to control erosion. However, these types of materials as surface cover are mainly used in non-agricultural lands where it is not feasible to obtain cover by growing and managing vegetation.

TABLE 4.7
Nutrient Deficiencies in Dry Land Soils in Various States of India (Srinivasarao and Vittal, 2007; Sahrawat et al., 2007; Mahesh et al., 2021)

Area	Deficient Nutrient	References
Varanasi	Nitrogen, Zinc, Boron	Srinivasarao and Vittal (2007)
Faizabad	Nitrogen	
Phulbani	Nitrogen, Calcium, Magnesium, Zinc, Boron	
Ranchi	Magnesium, Boron	
Rajkot	Nitrogen, Phosphorus, Sulphur, Zinc, Iron, Boron	
Anantapur	Nitrogen, Potassium, Magnesium, Zinc, Boron	
Indore	-	
Rewa	Nitrogen, Zinc	
Akola	Nitrogen, Phosphorus, Sulphur, Zinc, Boron	
Kovilpatti	Nitrogen, Phosphorus	
Bellari	Nitrogen, Phosphorus, Zinc, Iron	
Bijapur	Nitrogen, Zinc, Iron	
Jhansi	Nitrogen	
Solapur	Nitrogen, Phosphorus, Zinc	
Agra	Nitrogen, Potassium, Magnesium, Zinc, Boron	
Hisar	Nitrogen, Magnesium, Boron	
S.K. Nagar	Nitrogen, Potassium, Sulphur, Calcium, Magnesium, Zinc, Boron	
Bangalore	Nitrogen, Potassium, Calcium, Magnesium, Zinc, Boron	
Arjia	Nitrogen, Magnesium, Zinc, Boron	
Ballowal-Saunkri	Nitrogen, Potassium, Sulphur, Magnesium, Zinc	
Rakh Dhiansar	Nitrogen, Potassium, Calcium, Magnesium, Zinc, Boron	
Andhra Pradesh (Nalgonda, Mahabubnagar, Kurnool, Ananthapur and Prakasam)	Sulphur, Boron, Zinc	Sahrawat et al. (2007)
Karnataka (Kolar, Tumkur, Chitra durga, Haveri and Dharwad)	Sulphur, Boron, Zinc	Sahrawat et al. (2007)
Rajasthan (Bundi, Dungarpur and Udaipur)	Sulphur, Boron, Zinc	Sahrawat et al. (2007)
Districts of Madhya Pradesh (Vidisha and Dewas)	Sulphur, Boron, Zinc	Sahrawat et al. (2007)
Gujarat (Junagadh)	Sulphur, Boron, Zinc	Sahrawat et al. (2007)
Tamil Nadu (Tirunelveli, Salem, Kanchipuram, Vellore and Karur)	Sulphur, Boron, Zinc	Sahrawat et al. (2007)
Arid Rajasthan	Phosphorus, Zinc, Iron	Mahesh et al. (2021)

Permanent grass covers in rangelands are thought to be more effective than crop covers, which exist in the field for a short period. Among crops, dense row crops and creeping crops are highly favorable. After the plants complete their crop growth, residues become the primary cover. The decay of leguminous residues is faster than that of cereal or other crops and thus is less durable in the field to control wind erosion. Moreover, the more erect, finer and denser the residue, the smaller the risk of wind erosion. Maintenance of grass cover in rangelands is very important to control wind erosion and hence it is always better to adopt controlled grazing practices in rangelands so as to maintain the primary productivity as well as to provide sufficient protective grass covers. From a field experiment at two grazing situations in the Jaisalmer region of the Indian Thar desert revealed that the aeolian mass transport rate was almost three times higher at the overgrazed site than at the controlled grazing site during mid of June to mid of July (Mertia et al., 2010). Wind erosion also regulates the nutrient cycle in arid region. For example, soil nutrients lost through the wind erosion process are deposited in other places and enrich the soil fertility and thus the source area become unfertile while deposited areas become fertile. Thus, continuous overgrazing associated with severe drought situations in the desert may lead to loss of top fertile soils in quick time.

Eroded soil mass or commonly known as dust, emitted during wind erosion events over overgrazed rangelands remains in the atmosphere for a long time as suspended particles. These suspended particulate matters mix with other minute particles of the atmosphere e.g. carbon soot, smoke, salts etc. and produce a blanket of haze, which is known as aerosols. Aerosols have both direct and indirect effects on net radiative forcing and therefore play a key role in climate change of deserts as well as in the globe. Moreover, suspended particulate matters in the atmosphere have an adverse effect on respiratory and cardiovascular activity of people and thus are considered as health hazard.

Particle size analysis of collected aeolian samples during dust storm event of 15 June 2009 at Jaisalmer showed that particulate matter having size less than 10 μm (PM_{10}) was 30% in eroded aeolian mass, whereas the same for the desert surface was only 7.5% (Santra et al., 2010). Particle size distribution of eroded soil revealed that fine sized particles were more in eroded soils than on land surface and therefore have a great role to aerosol load in the atmosphere.

4.5.1.3.2 Sand Dune Stabilization

Sand dune stabilization by vegetation cover in checkerboard method is a popular wind erosion control technology in the arid region (Figure 4.5). Planting suitable vegetation on the denuded dune surface results in decreasing surface wind speed, prevention of scouring action and amelioration of soil conditions, which ultimately lead to improved micro-climatic condition of the area. In view of the limited water, high percolation, high ambient temperature and more evapotranspiration in arid regions, it is important to select plants that have an adaptive edge to survive in such demanding situations. Of many criteria, the ones to be taken care of in sand dune stabilization are that these should be able to survive in (i) extreme temperature condition, (ii) a variety of salinity conditions, (iii) variable speed and direction of wind, (iv) severe sand storm events, (v) very low soil moisture condition, and (vi) biotic stress situations.

4.5.1.3.3 Shelterbelt Plantations along Canals and Roads

Shelterbelts are barriers of trees or shrubs that are planted to reduce wind speed and, as a result, prevent wind erosion; they frequently provide direct benefits to agricultural crops, resulting in higher yields, and provide shelter to livestock, grazing lands, and farms. A few fundamental aspects of the shelterbelt are presented in Figure 4.6. Plant species of trees, shrubs and herbs to be used in shelterbelt need to be suitably chosen as per its height and canopy porosity for obtaining effective shelter effect.

Shelterbelt when planted across the wind direction and on the field boundaries effectively protects crops as well as controls sand drifting (Ganguli and Kaul, 1969). The effectiveness of the shelterbelt in reducing wind speed depends on wind speed and direction at the windward side and the shape, width, height and density of tree component. Tree species suitable for different purposes in the Indian Thar desert are mentioned in Table 4.8.

4.5.1.3.4 Wind Erosion Control in Agricultural Field

Wind erosion in agricultural field may also be controlled through micro shelterbelts of high crops (Figure 4.7). In this method, a few rows of high crops e.g. pearl millet (*Pennisetum glaucum*),

FIGURE 4.5 Wind erosion control on dunes through checkerboard method. Source: Santra et al. (2016).

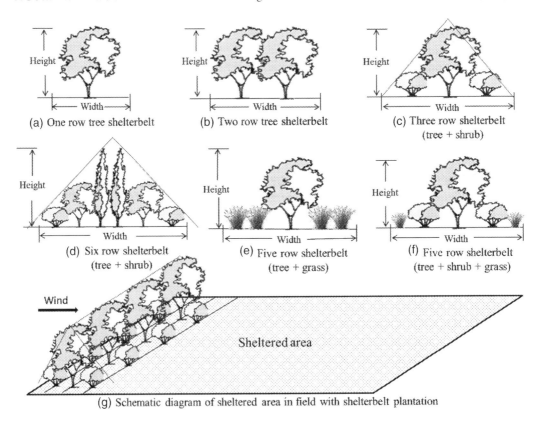

FIGURE 4.6 Basic design of shelterbelts composing of trees, shrubs and grass. Source: Santra et al. (2016).

sesame (*Sesamum indicum*) are sown in 15–20 m distance to give shelter effect to low height crops like mung bean (*Vigna radiata*), cluster bean (*Cyamopsis tetragonoloba*), groundnut (*Arachis hypogaea*) etc. The studies conducted at Central Arid Zone Research Institute, Jodhpur from 1976 to 1981 by Gupta et al. (1984) revealed that three rows of pearl millet as shelterbelt increased the yield

TABLE 4.8
Design and Species for Shelterbelt Plantation (Mertia et al., 2006)

Purpose	Design	Suitable Species
Road side	Three to five staggerred rows	*Acacia tortilis, Albizia lebbek, Azadirachta indica, Dalbergia sissoo, Prosopis juliflora, Tamarix articulate*
Railway side	Six rows	*Parkinsonia aculeata, P. juliflora, T. articulate*
Canal side	Rows	*Acacia nilotica, Eucalyptus spp., Tecomella. undulata, A. tortilis, P. juliflora, D. sissoo, P. cineraria, A. nubica*
Farm boundary (rainfed)	1/2/3 rows	*Acacia tortilis, A. lebbeck, A. indica, D. sissoo, P. aculeata, P. juliflora*
Farm boundary (irrigated)	Two rows	*A. tortilis, A. lebbeck, Dicrostachys cinerea, P. juliflora*

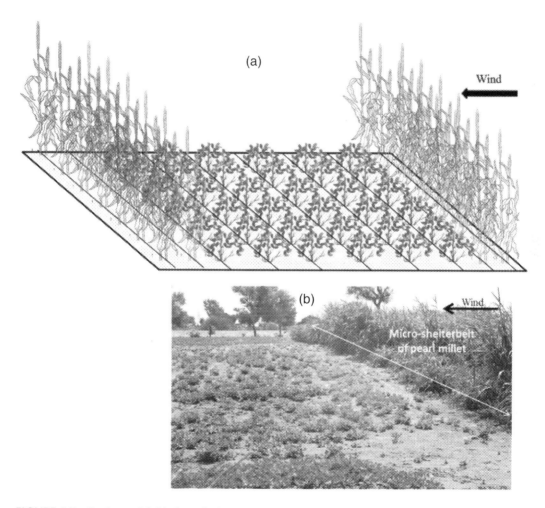

FIGURE 4.7 Design and field view of micro-shelterbelt plantation to control wind erosion. Source: Santra et al. (2016).

of cowpea and okra crops in order of 21% and 44% over the unsheltered crop in the summer season. In addition to increase in vegetable yield, the sheltered field provides an additional income from bajra fodder.

Wind strip cropping is another method of controlling wind erosion. Perennial grass or other plants are generally established in the field at the right angle to the prevailing wind direction and crops were grown in between these strips. Grasses like sewan (*Lasiurus scindicus*) and foxtail buffalo grass (*Cenchrus ciliaris*) are suitable for strip cropping. The purpose of these plantations is to create a mechanical obstacle to the free sweep of wind and to reduce its velocity, which ultimately helps in controlling erosion. A more acceptable way is to grow trees around the field boundaries. The suitable trees are mesquite (*Prosopis juliflora*), Indian jujube (*Ziziphus mauritiana*), gum Arabic tree (*Acacia senegal*), Indian blackwood (*Hardwickia bianata*), karonda (*Carissa carandas*) and Assyrian plum (*Cordia myxa*). Along with soothing of microclimate they provide some economical products e.g., gum from *Acacia senegal*, leaf fodder from *Hardwickia bianata*, fruits from *Cordia myxa* and *Ziziphus* mauritiana.

4.5.1.4 Managing Soil Salinity

Soil salinity and sodicity are major problems in irrigated areas specifically in canal command areas and in fields irrigated with poor quality groundwater. The use of poor-quality water for irrigation purpose leaves excessive salts in the soil matrix. Even excessive use of water helps the soluble salts to move up from deep soil profile to surface through the evaporation process and leaves the salts on the soil surface. Many of the groundwater resources in arid regions of Rajasthan are highly saline. Even, groundwater low in salinity often contains high residual sodium carbonate (RSC). Irrigation with such water results in sodification of land. As a result, sodicity with a pH 9.2–10.0 and exchangeable sodium percentage (ESP) 40%–50% has been developed. Even RSC water of 5 me/L has induced high sodicity in the rainfall zone of 200–300 mm (Joshi and Dhir, 1994). The soils in this situation acquire unusual hardness; water infiltration is reduced to a greater extent and workability of soils becomes very difficult.

There are several management practices followed by farmers to reclaim saline and sodic soil. Adoption of improved irrigation practices, deficit irrigation, change in crop selection, agroforesty and biological-drainage, application of soil amendments e.g. gypsum, mechanical reclamation, conjunctive use of water etc are some of the techniques available for management of saline and sodic soil. Most farmers are aware of the advantages of improved irrigation practices e.g. sprinkler irrigation and drip-irrigation, land levelling etc., but lack of exposure to and familiarity with such practices, along with existing constraints (labour, equipment, etc.), make them hesitant in trying them. Deficit irrigation is another technique for managing soil salinity, which refers to deliberate underirrigation of a crop as compared to its evaporative and leaching needs. The modelling study showed that in areas where water tables are shallow, irrigation requirements could be reduced to 80% of the total crop evapotranspiration without reducing crop yields. Growing salt-tolerant crops e.g. date palm, pomegranate etc. and low water requiring crops are another good option for improving land productivity from saline soil. In areas where undulated land topography does not permit gravity surface drains, and where groundwaters are saline, bio-drainage may be a potential option to control soil salinity. The potential of certain tree species to draw more water than the agricultural crops because of their deeper root systems, higher transpiration rates throughout the year and the ability to minimize recharge from rain by intercepting it on their foliage, provides a technique for keeping water table under control. *Eucalyptus camaldulensis, E. tereticornis, Atriplex lentformis, Acacia nilotica*, and *Acacia ampliceps* are species that offer a great potential to work as bio-pumps.

Since evaporation from the water table deposits salt on the soil surface, breaking the hydraulic connectivity of capillary up-flow by cultivating abandoned soil prior to and in between monsoon rains would lead to the reclamation of saline soil. In this strategy, monsoon rains provide leaching, while cultivation breaks up the hydraulic connectivity. Application of chemical amendments such as gypsum, calcium chloride dehydrate, sulfuric acid, hydrochloric acid and farmyard manure can reclaim sodic soil in the arid region. Soil amendments are usually broadcast across the field, and ploughed in. Thus, treatment is limited to surface layers and the quantity of amendment needed is high. Reclamation of sodic soil requires an increase in the infiltration rates, so that water can flow through the soil matrix and leach the sodium ions. Short-term increases in the water infiltration rate can be achieved mechanically by ploughing. However, medium- to long-term solutions require the replacement of sodium by divalent (calcium or magnesium) ions. Therefore, an ideal solution to reclaiming sodic soils requires a combination of mechanical and chemical measures.

The poor-quality of groundwater constrains its recycling and reuse as a means of crop irrigation, without proper management practices. Continuous recycling and reuse of saline-sodic groundwater cause an imbalance in the salt balance of an irrigation system, in general, and in the crop root zone, in particular (secondary salinization). Direct use of saline-sodic tube well water cannot be made for crop production without having a proper soil, water and crop management system in place. Under these conditions, frequent light irrigations, use of chemical amendments (gypsum, H_2SO_4, etc.), along with adequate leaching, growing salt-tolerant and moderately salt-tolerant crops in proper cropping sequence are essential requisites, if saline-sodic tube well water is to be used for irrigation on a sustainable long-term basis. Conjunctive use of good-quality and bad-quality waters through blending or cyclic application could be practiced to minimize the adverse effects of poor-quality waters on land and water resources. A blending strategy is useful under the conditions when fresh and saline water qualities are such that the mixed water would have less salinity than the threshold salinity of a given crop.

The negative impact of high RSC water on infiltration and nutrient availability could be mitigated if irrigation is done after the gypsum treatment @ 50% and 100% of soil requirement. The improvements were also reflected in terms of increased yields in loamy sand soils of Barmer and Jodhpur districts of Rajasthan (Joshi and Dhir, 1991; Mahesh et al., 2016a). The quantity of gypsum required by soils plus to neutralize RSC in excess of 3 me/L, resulted in an increase of 400–1,600 kg/ha grain yield of wheat (Mahesh et al., 2016a,b). The higher quantity of gypsum [100% gypsum requirement (GR)] is more effective in lowering soil pH by 0.1–0.8 units and decrease of sodium adsorption ration (SAR) by 6.4–10.7 and improvement in nutrient status could be attained (Mahesh et al., 2016a,b).

4.5.1.5 Improving Biological Activities in Soil Profile

Biological productivity refers to the quantity of organic matter or its equivalent in dry matter, carbon, or energy content which is accumulated during a given period of time, and expressed in terms of weight per unit area per unit time (g/m^2 year) (Park, 2007). In other words, the concept of biological productivity signifies the reproduction of the biomass of plants, microorganisms, and animals in an ecosystem. Though it is similar in many respects to that of soil fertility, it is, but, broader in content and scope.

Healthy soils provide sustained biological productivity, promote environmental quality, and maintain plant and animal health, by offering important ecosystem services like nutrient cycling, pathogen control, and improvement in soil structure. The microbial community within soil is responsible for transforming nutrients into a plant-available form from organic and inorganic sources, and for the release of nutrients near roots where the plants can access them. For maintaining this nutrient cycling in soil, it is important to add organic amendments like compost or manures which are a great source of labile carbon and associated macro- and micronutrients. Inoculation of soils with microbial communities such as diverse plant growth promoting rhizobacteria and mycorrhizal fungi which supply plants with essential nutrients and promote abiotic stress resilience secure the healthy functioning of the soil ecosystem. In such cases, fertilizer inputs should be calculated to complement nutrient cycling from organic matter and enhance the nutrient-sourcing activities of microbial inoculants. The use of microbial inoculants for biological control along with the implementation of practices like crop rotation, cover crops, mulching and healthy sanitary practices can improve soil health and reduce disease incidence in a sustainable manner, thereby enhancing the crop productivity. Biological processes in soil can also improve soil structure. Some bacteria and fungi produce substances like polysaccharides that can chemically and physically bind soil particles into micro-aggregates. The hyphal strands of fungi can cross-link soil particles helping to form and maintain aggregates. Sustenance of soil biological productivity requires the rebuilding of a healthy soil microbial community through the use of management practices that maintain or increase the organic matter content of soil which is an important source of carbon, energy and nutrients for soil organisms. Soil management should also involve the management of soil pH, the use of agricultural inputs which complement the activities of soil microorganisms, maintenance of rotational diversity, maintenance and conservation of ground cover, and adoption of appropriate grazing management strategies.

4.5.1.6 Integrated Nutrient Management

Soil fertility under arid conditions, is constrained by environmental extremes of hot and cold temperatures, as well as by low water availability. With some exceptions, these soils are inherently low fertile. Nitrogen is generally deficient, and availability of phosphorus is low to medium, while potassium is in adequately available (Mahesh et al., 2019; Praveen et al., 2009). Soils have low water-holding capacity, high pH, low SOC content (0.1%–0.5%); are shallow, stony, and at times have other specific limitations. It is unlikely that Indian arid lands can be competitive for grain production with the rest of India since arid areas have low production potential given their water and soil constraints. A few important low-input technologies and their impacts have been discussed below.

   a. *Application of optimum quantity:* A major step for achieving higher fertilizer use efficiency is to apply the right amount of fertilizer. But long-term estimations of N requirement of crops under arid rainfed arid systems are difficult as the yield levels vary from zero to three times of the average. Uncertainties in soil N supply can be reduced with site-specific N management approaches and selecting yield goals based upon (i) highest yield within the past 5 years under proper crop management (ii) yield goal set at 1.5 times of long-term average, and (iii) yield goal based on soil capabilities. This approach is very different from most extension services in India which provide a single, standard fertilizer recommendation for an entire district or region.
   b. *Internal nutrient circulation*: The approach comprises of practices like recycling of crop residues either through direct application or as manure, or growing legumes, often referred as the integrated nutrient management. The practice of using inorganic and organic sources of nutrients in a proper proportion not only reduces the requirement of inorganic fertilizers but also improves the physical conditions of soil, especially soil water retention and its availability. Biological properties of soil are also improved considerably and fertilizer use efficiency is increased.
   c. *Cereal-legume crop rotation*: Cereal-legume rotations, intercropping of cereals and legumes are important practices for restoring soil fertility on large land holdings. The amount of N returned from legume rotations depends on whether the legume is harvested for seed, used for forage, or incorporated as a green manure. Pearl millet in rotation with cluster bean has also been reported to yield higher than its continuous cultivation. The rotation of pearl millet with green gram or clusterbean was better than its rotation with moth bean. The beneficial effect of legumes to pearl millet also depended on the number of seasons of their cultivation before growing pearl millet.

d. *Nitrogen-fixing trees*: Leguminous trees can survive with arid soils' low levels of N due to their N-fixing capacity. These trees can fix 43–581 kg N/ha year, compared to about 15–210 kg N/ha year from grain legumes (Dakora and Keya, 1997). Nitrogen-fixing trees such as *Acacia* sp and *Prosopis* sp are some of the best sources for N enrichment in arid regions. In addition, most of these species are sources of highly nutritious fodder, fuel, food, charcoal, gums, fibre and timber.

e. *Conservation agriculture:* Conservation agriculture (CA) aims to conserve and improve the natural resource base while using the resources available for agricultural production more efficiently. It avoids or minimizes soil tillage, maintains a permanent soil cover of crops and/or residues, and utilizes efficient crop rotations. It has been successful in many parts of the world, but has been least successful to date in the arid areas where production of organic matter is too low for permanent soil cover with crop residues because of water shortages. But it has been found that even small amounts of crop residues can reduce wind erosion considerably and increase soil water storage. Since significant quantities of soil and nutrients are lost by wind erosion where the soil remains bare for most of the year, even small savings are worth pursuing. No dramatic increases in production and soil fertility can be expected by adopting conservation agriculture in arid areas in the short term, and farmers are unlikely to commit themselves to conservation agriculture without adequate incentives and policy support.

Integrated nutrient management (INM) approach is an effective way to deal with low productivity and poor soils. Recent literatures on the effect of INM practice involved various treatment combinations of organic and inorganic nutrient sources under different cropping systems shows the benefits in crop yields, net returns and B:C ratio. Strong evidence also presented that INM could be an effective and innovative practice for climate-resilient agriculture that brings sustainability in the ecosystem. Organic manures play a significant role in the INM package; increase SOC which proliferates the microbial activity in soil, helps retain soil moisture longer and reduces the leaching of plant nutrients besides imparting drought tolerance during dry spells. The advantages of INM practices have been reported in hundreds of researches in the Indian subcontinent (Bijay-Singh and Ali, 2020).

The basic concept of the INM model is the adjustment of soil fertility and supplement of plant nutrients in an available form at an optimum level in order to sustain the crop productivity and optimization of advantages of plant nutrients from all possible resources in an integrated approach (Srinivasarao et al., 2017a–e). The adoption of new interventions has developed new dimensions in the INM system. Improved farm mechanization, conservation agriculture system intensification (CASI) technologies, various organic amendments and renewed focus on recycling of existing natural organic nutrient flows further broadening the concept of INM to make it more context-specific for local environmental conditions (Wu and Ma, 2015). There are plenty of organic and inorganic nutrient source combinations adopted in the INM model as shown in Table 4.9. Integrated nutrient management options for various crops in rainfed regions of India under different rainfall regions are presented in Table 4.10.

4.5.1.7 Soil Test-Based Nutrient Management

Soils in most parts of India not only show deficiency of NPK but also of secondary nutrients (S, Ca and Mg) and micronutrients (B, Zn, Cu, Fe, Mn etc.). Balanced nutrient application to crops based on the nutrient requirement to produce a unit quantity of yield, the native nutrient supplying capacity of soil and specific targeted yield improves crop yields while minimizing nutrient losses and cost of cultivation. In Andhra Pradesh and Telangana, balanced nutrition was demonstrated in eight districts to address nutrient deficiencies which exist within farmers' fields, with promising results. For example, in Warangal, balanced nutrition improved cotton yields significantly in many farmers' fields. In some fields, cotton yields reached 1.6 mg/ha with balanced nutrition registering an increase in yield by

TABLE 4.9
Effective Integrated Nutrient Management (INM) Practices for Rainfed Crops Across the Country (Annual reports of All India Coordinated Research Projects On Dryland Agriculture, 2009-10, 2010-11, 2011-12, 2013-14 and 2015-16)

Location	Crop	Fertilizer (kg/ha)			Remarks
		N	P_2O_5	K_2O	
Jhansi	Cluster bean	15	60	0	Inoculation with *Rhizobium*
	Sorghum+ Dolichos	60	20	0	
Rajkot	Sorghum	90	30	0	FYM @ 6 mg/ha
	Pearl millet	80	40	0	FYM @ 6 mg/ha
	Groundnut	12	25	0	FYM @ 6 mg/ha
	Cotton	40	0	0	FYM @ 6 mg/ha
Solapur	Sorghum	50	0	0	9–10 mg/ha leucaena (*Leucaena leucocephala*) loppings can substitute 25 kg N/ha
Indore	General	(N plus P)			4–6 mg/ha FYM in alternate years
	Soybean	20	13	0	FYM @ 6 mg/ha
Bijapur		(NP or NPK)			Mulching with tree lopping @ 5 mg/ha
Arjia	Corn-Pigeon pea	50	30	0	50% N through organics.
	Safflower and Rapeseed-Mustard	30	15	0	Reduction in N by half if these crops follow legumes such as green gram/chickpea
Agra	Barley	60	30	0	Use of FYM plus *Azotobacter*
Ranchi	Soybean	20	80	40	Inoculation with *Rhizobium*
	Groundnut	25	50	20	Inoculation with *Rhizobium*
	Pulses	20	40	0	Inoculation with *Rhizobium*
Dantiwada	Green gram	0	20	0	Inoculation with *Rhizobium*
Jodhpur	Pearl millet	10	0	0	Addition of 10 mg/ha FYM
Hoshiarpur	Corn	80	40	20	Addition of FYM
	Wheat	80	40	0	
	Chickpea	15	40	0	
Akola	Cotton+Green gram	25	25	0	Along with FYM to meet 25 kg N/ha

TABLE 4.10
Integrated Nutrient Management Options for Various Crops in Rainfed Regions of India (Kumar et al., 2011)

S. No	Location/Crop/Cropping System/Soil Type/Rainfall (mm)	Technology
1	Hyderabad (Telangana); Sorghum; Alfisol; 770	Conjunctive use of inorganic N through urea and loppings of farm-based organics such as leucaena (*Leucaena leucocephala*) and gliricidia (*Gliricidia maculata*) to sorghum crop has been proved as potential options. Following these options, average yield level of about 1.7 Mg ha^{-1} of sorghum grain can be achieved. This way 50% of N requirement of sorghum can be substituted through organic sources of nutrients. Since these farm based materials have C:N ratios of less than 20, there is no risk of adverse effect of immobilization of native soil N during the process of decomposition.

(Continued)

TABLE 4.10 (Continued)
Integrated Nutrient Management Options for Various Crops in Rainfed Regions of India (Kumar et al., 2011)

S. No	Location/Crop/Cropping System/Soil Type/Rainfall (mm)	Technology
2	Hyderabad (Telangana); sorghum; Alfisol;770	Other potential options of INM technologies for sorghum crop in this region are i. Application of 2 mg glyricidia loppings+20 kg N through urea, ii. Application of 4 mg FYM+2 mg glyricidia loppings or iii. application of 4 mg FYM+20 kg N through urea.
3	Hyderabad (Telangana); mung bean (*Vigna radiata*); Alfisol; 770	Application of 2 mg FYM+10 kg N through urea under conventional tillage. Application of 2 mg FYM+1 mg Gliricidia loppings under reduced tillage.
4	Faizabad; Sorghum; Inceptisol; 980	Application of 15 kg N through compost+20 kg N through inorganic fertilizer
5	Ranchi; Upland rice; Oxisol; 1200	Application of 40-30-20 NPK kg/ha in rice Application of 30-20-0 NPK kg/ha in upland rice.
6	Varanasi; Upland rice; Entisol; 1080	Application of 40-30-20 kg NPK/ha in rice
7	Arjia; horse gram (*Macrotyloma uniflorum*), groundnut (*Arachis hypogea*) Maize+pigeon pea (*Cajanas cajan*) intercropping; Verisol; 660	Application of 45 kg P_2O_5 Rhizobium+PSB inoculation in groundnut (*Arachis hypogea*). Combination of 50% of inorganic and remaining of inorganic source to FYM+compost (60-30-0 NPK kg/ha)
8	RakhDhiansar; Maize; Entisol; 1180	Application of 60-40-20 kg NPK/ha+20 kg/ha zinc sulphate
9	Rewa; Soybean and Black gram Chick pea; Vertisol; 1080	Seed treatment of rhizobium bacteria and PSB in soybean and urad bean (*Vigna mungo*). Use of bio-fertiliser, Rhizobium +PSB in chickpea
10	Indore; Soybean-Safflower system Soybean-Pigeon pea system (2:2) Soy bean; Vertisol; 850	Application of 6 mg FYM/ha+N-20 kg/ha P-13 kg/ha Application of 45-40-25 kg NPK/ha in soybean (*Glycine max*)-pigeon pea (*Cajanas cajan*) system FYM @ 6 mg/ha with 20 kg N/ha and 30 kg P/ha in soybean
11	Anantapur; Groundnut; Alfisol; 610	Use of Rhizobium @ 500 g/ha+20–40-40 N, P_2O_5 and K_2O kg/ha
12	Kovilpatti; Cotton +Urdbean intercropping system; Vertisol; 740	20 kg N/ha through FYM-20 kg N/ha as urea
13	Bangalore; Fingermillet; Alfisol; 780	Application of 10 mg FYM/ha+50% NPK in groundnut-finger millet system Combination of organic (FYM 10 mg/ha) and inorganic (50-40-25 NPK kg/ha) in finger millet (*Eleusine coracana*)
14	Bijapur; Rabi sorghum Sunflower, Chickpea; Inceptisol; 360	Application of in-situ incorporation of Sunn hemp+application of FYM in *rabi* sorghum and sunflower (*Helianthus annuus*). Application of 50 kg N/ha through fertilizer and 5 mg compost in chickpea
15	Bellary; Sorghum and chickpea; Vertisol/Inceptisol; 500	15 kg N through green leaf+10 kg N through inorganic fertilizer
16	Agra; pearl millet; Entisol and Inceptisol; 630	40 kg N/ha

(*Continued*)

TABLE 4.10 (*Continued*)
Integrated Nutrient Management Options for Various Crops in Rainfed Regions of India (Kumar et al., 2011)

S. No	Location/Crop/Cropping System/Soil Type/Rainfall (mm)	Technology
17	Hisar; pearl millet + cowpea pearl millet, mustard; Aridisol; 350	40 kg N-30 kg P_2O_5/ha to pearl millet + cowpea (*Vigna unguiculata*). Seed inoculation of strain of MAC-68 and HT-54c in pearl millet Application of 20 kg N + Azotobacter culture in pearl millet Rhizobium + Azotobacter inoculation in mustard
18	Hoshiarpur; Wheat; Entisol; 1000	In *rabi* crops, fertilizer must be drilled at or before sowing so as to place it in the moist zone. In maize and sorghum, half of N and whole of P and K should be drilled at sowing and remaining half of N broadcast 1 month after. Incorporate this green manure (sunn hemp) crop in the field in the middle of August and raise wheat/wheat + gram mixture during *rabi*. Application of 80 kg N/ha in two equal splits
19	Rajkot; Groundnut; Vertisol; 500	Application of 6-12-0 NPK kg/ha + mulching of sunn hemp (*Crotolaria juncea*) in between rows + Rhizobium and PSB treatment for higher pod yields.
20	Akola; Pigeonpea; Vertisol; 810	Application of 5 mg/ha of FYM + 40 kg P_2O_5/ha + microbial culture @ 1.5 kg/ha to pigeon pea (*Cajanas cajan*)
21	Jhansi; Rabi crops; Alfisol; 930	In-situ incorporation of sunn hemp (*Crotolaria juncea*) at 45 DAS prior to *rabi* crops to improve the yield of *rabi* crops.
22	Solapur; *rabi* sorghum cowpea and *rabi* sorghum; Vertisol; 720	Application of 50% N through crop residue (sorghum/leucaena) + 50% N through fertilizer to *rabi* sorghum. Use of 12.5–25-0 NPK for cowpea and 25 kg N for *rabi* sorghum in sequence
23	Bhubaneswar; legume-finger millet; Red and lateritic; 1600	Legumes like cowpea, common mung bean and urad bean can be grown with the onset of monsoon. Pods harvested and rest of the biomass can be incorporated in soil to add 20–30 kg N/ha. Followed by this, a crop like ragi can be transplanted to produce 1.0–1.5 mg/ha without adding anymore N fertilizer.

10%–25% over farmers' practice. Further, farmers in this district achieved 30%–40% yield improvements in most of the crops due to balanced nutrition (Srinivasarao et al., 2010).

Need-based fertilizer applications: Need-based fertilizer application is a system where the amount and type of fertilizer applications to the soils and/or plants are based on soil testing. A Leaf colour chart (LCC: Ali et al., 2015), SPAD chlorophyll meter (Varinderpal-Singh et al., 2012) are used. The need-based N management through an LCC or chlorophyll meter reduces N requirement of rice from 12.5% to 25%, with no loss in yield and higher N use efficiency.

Site-specific nutrient management (SSNM): Site-specific nutrient management often results in efficient utilization of applied nutrients and higher crop productivity besides environmental benefits. In rainfed regions, most of the soils are N deficient and therefore N application is a must depending upon crop and soil conditions. Available P is building up in Alfisol regions of southern India due to continued DAP (di-ammonium phosphate) application. K application is concentrated on cash crops like cotton, tobacco, banana etc. Micronutrient application for rainfed crops is rather rare. Therefore, multiple nutrient deficiencies are emerging in these regions. Mg deficiency is emerging in Bt cotton-growing red soil regions of the country. Therefore, considering soil analysis and crop, SSNM strategies were developed in eight districts of Andhra Pradesh and Telangana. Several on farm trails on SSNM showed higher yields and higher B:C ratio over conventional recommendations. Bt cotton responded significantly to SSNM in cotton growing districts of Telangana (Adilabad, Khammam, Warangal, Nalgonda and Mahaboobnagar).

4.5.1.8 Fertigation Systems

Water soluble fertilizers can be used in sprinkler/drip irrigation systems/foliar spray which helps in improving crop yields and the quality of fruits and vegetable crops. These fertilizers are having low salt index that avoids burning of plant tissues, 100% soluble in water and suitable for foliar application or fertigation. Water soluble fertilizers are basically a combination of all essential nutrients with different ratios suit the matrix of type of crop, quality of water, status of soil fertility, and climatic conditions (Malhotra, 2016). Nutrient composition of different water-soluble fertilizers is presented in Table 4.11. Various researchers studied the effect of the application of water-soluble fertilizers on increasing productivity showing that much better nutrient use efficiency was observed with liquid and water-soluble fertilizers applied through fertigation in mango, litchi, grapevine, banana, kiwifruit, sweet cherry, citrus, apple, okra, garlic, squash, bell pepper, etc.

4.5.2 INTERVENTIONS ON WATER MANAGEMENT

4.5.2.1 Improving Water Retention

The application of water through irrigation and then its utilization by crops are the two major key processes to describe water use in the agriculture sector. Out of these two, the second sub-process is mostly dependent on the inherent soil characteristics, which are reflected in soil water retention behaviour and conductivity of water in the soil matrix. Soil water retention and other soil hydraulic

TABLE 4.11
FCO (Fertilizer Control Order) Approved 100% Water-Soluble Fertilizers in India (Malhotra, 2016)

Product (Grade)	Nutrient Composition (%)						
	N	P	K	S	Ca	Mg	Zn
NPK (13-40-13)	13	40	13				
NPK (18-18-18)	18	18	18				
NPK (13-5-26)	13	5	26				
NPK (6-12-36)	6	12	36				
NPK (20-20-20)	20	20	20				
NPK (19-19-19)	19	19	19				
NPK (12-30-15)	12	30	15				
NPK (12-32-14)	12	32	14				
Potassium nitrate (13-0-45)	13	0	45				
Mono potassium phosphate (0-52-34)	0	52	34				
Calcium nitrate	15.5				18.8		
Potassium magnesium sulphate			22	20		18	
Mono ammonium phosphate (12-61-0)	12	61	0				
Urea phosphate (17-44-0)	17	44	0				
Urea phosphate with SOP (18-18-18)	18	18	18	6.1			
NPK Zn (7.6-23.5-7.6-3.5)	17.6	23.5	7.6				

properties play a crucial role in water use efficiency in the agriculture sector. Soil water retention behaviour defines the relations between energy status and the amount of water in the soil. The energy level of water in the soil is governed by three major components: (i) gravity, (ii) pressure, and (iii) osmosis, which in together is denoted by soil water potential. Most of the time soil remains below its saturation state under field conditions, and hence, soil water potential remains negative. Sometimes, the term suction or tension is used instead of negative potential. In simple connotation, gravity potential indicates gravitational energy. Thus, the deeper the soil layer from the reference surface level, the lower the potential or the higher the suction or tension. Hence, water always flows from topsoil layers towards deeper layers. Pressure potential is also known as capillary potential or matric potential and mostly depends on soil particle's surface and pore space in soil matrix. In sandy soil, soil moisture content is more governed by pore space than the adsorption of water surrounding particles, which is otherwise dominant in clayey soil. Osmotic potential is governed by the presence of salt content in soil water, and thus, the higher is the soil salinity, the lower the soil water potential. Under such condition, plant's roots cannot be able to extract water from soil in spite of high soil moisture content but low soil water potential, and this is often referred as physiologically dry soil. In Figure 4.8, a typical relationship between the energy status of water and its content is depicted for three soil textural groups. It is also known as soil water retention curve and shows how the soil water content changes with changes in soil water potential. In field, we can measure either soil moisture content or soil water potential to describe the soil moisture status. The steepness of soil water retention curve indicates how tightly water is held in the soil matrix. It is visible from the graph that in light textured soils, which are prevalent in drylands, soil moisture content sharply decreases with a decrease in soil water potential or increase in soil water suction. Therefore, soil water content reduces drastically when it dries after saturation. It is also to be noted that soil water content at field capacity, corresponding to a potential of −0.33 bar or −330 mbar is higher in sandy loam soil and loamy sand soil than sandy soil. Thus, available soil water to be extractable by plant roots is higher in sandy loam soil than loam sand and sandy soil. Therefore, the major focus in water management of dryland system should be on improving the soil water retention or simply the pulling of soil water retention curve slightly upward between −0.33 and −15 bar potential levels.

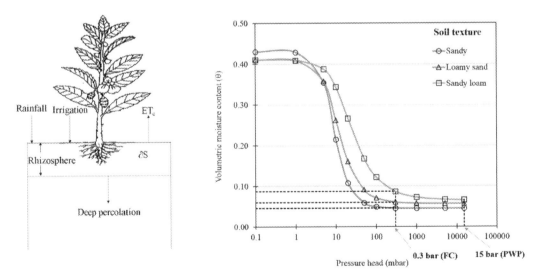

FIGURE 4.8 Water balance components in field and typical soil water retention curves for different soil textural groups. Source: Santra et al. (2020).

There are different ways to improve soil water retention of light-textured soil. Application of pond silt to agricultural field is a common practice in dryland areas to modify soil texture to some extent to improve soil water retention. However, it is not always feasible because of the limited availability of pond silt and heavy machinery works involved with it. Application of synthetic polymer as a water absorbent molecule may be another promising option for improving soil water retention e.g. synthetic hydrogel (Narjary et al., 2012) or guar gum-based superabsorbent molecules (Thombare et al., 2018). However, attention needs to be taken before its application on its degradability in soil and the adsorption and desorption pattern.

4.5.2.2 Improving Water Use Efficiency/Water Productivity

Shortage of water is the major constraint for limiting crop yield in arid and semi-arid areas, and improving effective utilization of water is an urgent need for sustainable crop production in these areas. Deficit irrigation (DI) has been emerging as an effective practice to improve water productivity (WP), and saving of water. It has been reported that moderate level of deficit irrigation e.g. applying 20% less water than full irrigation can improve water productivity as well as quality of peanut (Rathore et al., 2021). However, slight reduction in yield (6%) and nitrogen use efficiency (7%) was observed. Combination of moderate level of deficit irrigation and application of 30 kg N/ha promoted growth, yield components and biomass partitioning, all of which increased yield and water productivity. In addition, our results demonstrate that there is scope for increasing net farm income under both land and water -limiting conditions by adopting suitable deficit irrigation strategy. In regions where irrigation water is scarce relative to land, which is most common in most of the arid regions of the world, adoption of deficit irrigation by 20% reduction in irrigation water supply, is a suitable irrigation strategy for maximizing net income.

Khadin is a unique runoff farming system of the Thar Desert in India. The Jaisalmer district, lying at the centre of the desert, receives 100–200 mm rainfall annually. This runoff farming system involves collecting and storing runoff water from a high-elevation catchment area with shallow soil and underlying rocks in relatively low-elevation areas with deep soil. As the monsoon withdraws, the accumulated water starts receding due to seepage and evaporation. After the recession of accumulated water, *khadins* are cultivated to grow *kharif* or *rabi* season crops, depending on the depth of impounded water. In the *khadin* system, the bund at the lowest level is provided with a spillway and a sluice to regulate and drain out excess impounded water (Kolarkar et al., 1983). Delineated catchment and khadin system of a typic khadin from Bharamsar village of Jaisalmer is depicted in Figure 4.9.

4.5.2.3 Integrating Solar PV Pumping System with Pressurized Irrigation

For sustainable production of food from agricultural farms, irrigating crops at the right stages and efficient utilization of available water resources are highly important (Santra, 2021). Pressurized irrigation systems help in the application of irrigation water preciously in the agricultural field, however energy is required for this purpose. Fossil fuel-based irrigation pumps, generally grid-connected electric pumps and diesel pumps, meet the required energy demand for irrigation purpose. Fossil fuel-based energy sources have adverse effect on climate and it is depleting at a very fast rate. Therefore, alternative solutions to meet the energy demand in agriculture need to be looked for. Solar photovoltaic (PV) pumps are quite useful under such circumstances. Typically, a solar PV pumping system consists of PV modules, mounting structure, pump unit (AC/DC), and tracking system.

Groundwater resources are mostly used for irrigation purpose, however, it has been depleting at a very fast rate to meet the civilization targets and therefore needs due attention to optimally use this precious resource. Harvesting of runoff water in surface storage structures followed by pumping it for irrigation purpose may be a potential solution to achieve the set goal of 'crop per drop' mission (Abraham and Mathew, 2018). Therefore, a module for optimum use of harvested runoff water in farm ponds through solar PV pumping system for irrigation is presented in Figure 4.10. The module

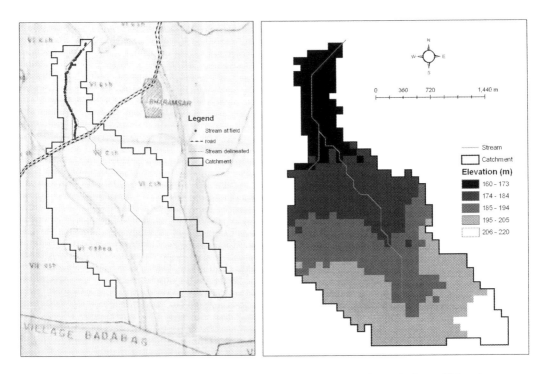

FIGURE 4.9 Delineated catchment area and digital elevation model of a typical runoff farming system (*khadin*) at Bharamsar, Jaisalmer.

consisted of two sub modules: (i) runoff generation potential calculation and (ii) use of harvested runoff water in farm ponds for irrigation purpose through a solar PV pumping system.

Solar PV pumping systems are economically more feasible (low life cycle cost) than corresponding grid-connected pumps and diesel-operated pumps, although initial investment for capital cost is quite high for the solar PV pumping system. Solar PV pumping system also helps in mitigating the greenhouse gas (GHG) emission potential. Carbon footprint of solar PV pumping systems is almost negligible as compared to grid-connected electric pumping system and diesel-operated pumping system. For example, carbon footprint per ha-mm irrigation is 1.214 kg CO_2-eq for grid connected electricity pump (1 HP) whereas for diesel operated pumping system, it is about 0.382 kg CO_2-eq. It is also to be noted here that grid-connected electric pumping system has the highest carbon footprint and is almost double of the carbon footprint from diesel-operated pumping systems. This is mainly because of use of coal as the major input in thermal power-based grid networks of the country and the carbon emission factor of coal is comparatively higher than diesel fuel.

4.5.2.4 Harvesting Rainwater

In earlier times, a general belief was that field-level rainwater harvesting in arid regions of the country cannot provide a dependable source of water supply in the agriculture sector, mainly due to the low amount and less frequent occurrence of rainfall (Machiwal et al., 2018b). However, rainfall in the Indian arid region has been significantly increasing since the year 2002 due to changing climate (Machiwal et al., 2016, 2018b). The increase is mainly in rainfall intensity as the number of rainy days is more or less the same as earlier. In the hot arid region of Gujarat, a large proportion of the annual rainfall ranging from 32% to 76% occurred in either a single or consecutive 2–4 days

Soil, Water, and Nutrient Management in Drylands

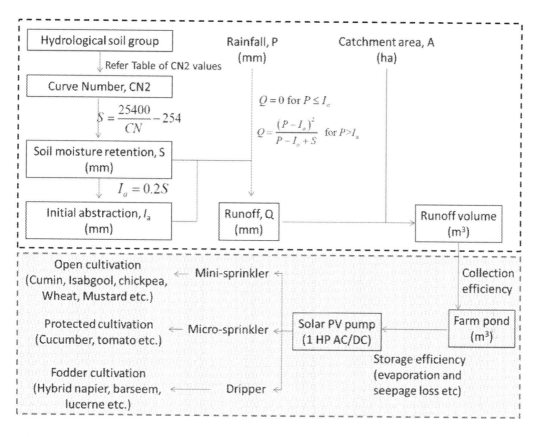

FIGURE 4.10 Module for solar PV pump based irrigation from harvested rainwater in farm pond. Source: Santra (2021).

period (Machiwal et al., 2015). Hence, it is now felt that the rainwater harvesting can be successfully adopted in arid regions of the country as the climate is favoring the same. A few of the case studies from the Indian arid regions proving scope of the rainwater harvesting are briefly described ahead.

a. *Rainwater harvesting for crop production:*

Looking at the scope of rainwater harvesting, a small reservoir of about 20,000 m^3 capacity was constructed in the research farm of Regional Research Station of the ICAR-Central Arid Zone Research Institute, Bhuj, Gujarat in the year 2001. The reservoir stored a large amount of runoff harvested from the catchment and satisfied domestic as well as irrigation requirements at the station. The reservoir storage capacity was enhanced by 9184.5 m^3 during 2008–09. The reservoir water was applied to irrigate wheat (1.5 ha) and mustard (2 ha) crops using two 5-HP diesel-operated centrifugal pumps. Daily monitoring of reservoir inflow and outflow components and computation of cost-economics revealed the unit cost of harvested rainwater as ₹1.51 m^{-3} for the 30-year average reservoir life, which seems to be fairly adequate for successful rainwater harvesting in the arid regions. Values of economic indicator criteria such as benefit-cost ratio (1.01), net present value (₹10,093) and internal rate of return (10.12%) prove the economic feasibility of the rainwater harvesting for crop production in arid regions of the country. Over a period of 22 years, it is realized that the reservoir storage fills up to its maximum capacity in response to 250–300 mm of annual rainfall in the area.

b. *Rainwater harvesting for groundwater recharge:*

Two small size rainwater harvesting-cum-percolation ponds, each of 486 m³ capacity, were constructed in the main campus of ICAR-CAZRI, Jodhpur in the year 2020. Both the ponds were connected through a surface drain such that water in the second pond enters when the first pond overflows. Both the ponds were kept unlined to allow percolation of the harvested rainwater through sides and bottom. In the first rainy season after construction of the ponds, both got filled up to their maximum storage capacity with the harvested runoff water from a single high-intensity (56 mm/hour) rainy storm of 45 minutes duration yielding 42 mm of rainfall. From the next day onwards, the pond water started depleting and groundwater recharge potentially began. Daily water budgeting of both the ponds revealed that evaporation loss was negligible in comparison to water percolated down from the ponds. The harvested rainwater depleted over a week with the average recharge rate of 47 cm/day or 74,658 L/day (first pond) and 52 cm/day or 78,240 L/day (second pond). Overall, more than 1 ML of rainwater received in less than 1 hour could be adequately harvested in two ponds and utilized for groundwater recharge in arid lands of the western Rajasthan. Moreover, the recharged freshwater may mitigate the groundwater salinity of the underlying aquifer system by having a positive impact on groundwater quality.

4.5.2.5 Using Salt-Affected Water for Specific Production System

Many of the groundwater resources in arid regions are highly saline, while waters low in salinity often contain high RSC. Groundwater irrigation is also causing deterioration of the land quality, especially where the water is rich in RSC. According to Minhas and Gupta (1992) about 84% groundwater in western Rajasthan is of poor quality. Often the water has EC >2.2 dS/m and up to 10 dS/m, coupled with high RSC (up to 20 me/L). Irrigation with such waters leads to the development of secondary salinization in the soils of the region (Gupta et al., 2000). Land degradation mapping in the districts of Barmer, Jalor, Pali, Jodhpur and Sikar revealed the problems of high soil pH (8.6–10.5), unusual hardness, restricted water infiltration and decreased nutrient availability to plants in the soils irrigated with high RSC water, so much so that even after frequent ploughing during the rainy season and application of higher doses of farmyard/organic manure land productivity could not be restored (Raina and Sen, 1991; Raina et al., 1991; Raina, 1999). The emergence of seedling, growth of crop and yield of harvest are severely affected under the described situation (Joshi, 1992). Large areas irrigated with high RSC water have gone out of cultivation. Even frequent ploughing during the rainy season and application of higher dose of farmyard/organic manure could not produce the desired yield in restoring the productivity.

However, the negative impact of high RSC water on infiltration and nutrient availability could be mitigated, if irrigation is done after the gypsum treatment @ 50% and 100% of soil requirement. The improvements were also reflected in terms of increased yields in loamy sand soils of Barmer and Jodhpur districts of Rajasthan (Mahesh et al., 2016a,b; Joshi and Dhir, 1991). The quantity of gypsum required by soils plus to neutralize RSC in excess of 5 me/L, resulted in an increase of 400–1,600 kg/ha grain yield of wheat (Mahesh et al., 2016a,b). The higher quantity of gypsum (100% GR) is more effective in lowering soil pH by 0.1–0.8 units and decrease of SAR by 6.4–10.7 and also improvement in nutrient status could be attained (Mahesh et al., 2016a,b). Inadequate and unreliable canal water supplies (especially at the tail end of distributaries and water courses) and change in cropping patterns forced farmers to use this marginal-quality water for irrigation.

4.6 DROUGHT PROOFING

4.6.1 Future Strategies of Crop Planning in case of Occurrence of Drought

The district agricultural contingency plans are technical documents aimed to be a ready reckoner for line deaprtments and farming community on prevailing farming systems and technological interventions for various weather aberrations such as droughts, floods, cyclones, hailstorms, heat and cold waves addressing different sectors of agriculture including horticulture, livestock, poultry, fisheries which could be used during weather aberrations and to sustain the production systems. In the country, the district contingency plans have been prepared by the Indian Council of Agricultural Research (ICAR), New Delhi along with State Agricultural Universities and Krishi Vigyan Kendras (KVKs) under the overall guidance and supervision of the ICAR and Department of Agriculture (DAC). For successful implementation of the developed contingency plans, orientation workshops are conducted for nodal officers of state agricultural universities to sensitize them about the standard template developed for the purpose. Vetting workshops are organized in different states to scrutinize and finalize the plans in the presence of ICAR institutes and respective university authorities. Up to now, contingency plans for 650 districts of the country have been prepared and their details are provided on ICAR/DAC websites (http://farmer.gov.in/, http://agricoop.nic.in/acp.html, http://crida.in/) and circulated to all state agriculture departments.

4.6.2 Agro-Advisory Services Based on Meteorological Forecasts

Agromet advisories are the measures devised based on the current and future meteorological or weather conditions to safeguard the standing crops from the likely weather aberrations. The components of agromet advisories include current crop conditions, meteorological forecasts, diagnose weather-related stresses (drought, flood, cold and heat waves etc.), weather-based farm management advisory, dissemination of agromet advisory bulletin, responding to specific queries and feedback from the farmers. In the country, a concept of micro-level agromet advisories (MAAS) at was developed block level under the ICAR's project on National Initiatives for Climate Resilient Agriculture (ICAR-NICRA), which has been implemented on pilot basis with the help of block level forecasts provided by the Indian Meteorological Department (IMD), agrometeorologists of All India Co-ordinated Research Project on Agrometeorology (AICRPAM) cooperating centers and subject matter specialists of KVKs in 25 selected blocks in 25 selected districts. The program was successfully started first by Bijapur in Belgaum District under ten talukas. Field Information Facilitators were introduced, and they help in collecting information to dissemination of agromet advisories to farmers and also advise the farmers in weather-based farm operations with the help of line departments.

4.6.3 Adoption of Water-Efficient Crops, Cropping Systems and Farming Systems

Traditionally, the farmers of drylands practiced only subsistence farming due to scarcity of water resources. Thus, mainly rainfed crops having the ability to sustain in absence of adequate rains such as pearl millet (*Pennisetum galucum*), mung bean (*Vigna radiata*), moth bean (*Vigna aconitifolia*), etc. were taken under cultivations in the drylands. In addition, other low water-requiring crops such as cumin and psyllium (*Plantago ovata*) have also been cultivated over a large area of drylands with limited irrigations depending upon the small quantities of water available in surface water bodies and shallow wells. The accessibility to water resources has been improved drastically over the years with the advent of technology such as electric-operated pumps, the development of policies for raising agricultural production such as subsidies on agricultural inputs, etc., which have an adverse impact of status of water resources availability in the drylands. For example, with the advent of canal irrigation and ease of access to groundwater resources through tube wells in the arid region of

Rajasthan, several new crops like cotton, groundnut, sugarcane, rice, wheat, castor, etc., have been introduced. Water demands of the newly introduced crops in drylands are very high and almost unsustainable in the long term. In the arid region of Rajasthan, areas under the traditional rainfed crops showed a significant decrease during the period from 2007–08 to 2016–17 with the negative values of compound annual growth rate (CAGR) for pearl millet (−4.3%), moth bean (−2.3%) and sesame (−1.8%). On the other hand, areas under the water-intensive crops have expanded at an alarming rate with positive CAGR values for groundnut (12.8%), rice (6.8%) and castor (4.0%). Therefore, efficient management of water resources is imperative to deal with the escalated future water demands of agriculture in the arid western Rajasthan. This can be achieved by emphasizing on production of water-efficient and high value crops suitable to the different agro-climatic conditions of this region by adopting appropriate strategies for the efficient management of irrigation water.

4.7 WAY FORWARD

4.7.1 ARTIFICIAL INTELLIGENCE AND IoT APPLICATIONS

Present day agriculture requires innovative approaches and practices to help farmers to increase production efficiency with simultaneous reduction in the required amount of natural resources (Santra et al., 2021). Machine learning (ML) tools and Internet of Things (IoT) have great scope to modernise the agriculture system to smart one. Few commonly used machine learning tools are multiple linear regression (MLR), random forest regression (RFR), support vector machine (SVM), artificial neural network (ANN) etc. Using these machine learning tools robust predictive models may be developed using large database on past experiences, which is otherwise not feasible with conventional computation algorithms. Possible ML-based applications in agriculture are digital soil maps, predictive models for yield prediction, identification of pest and disease infestation through image analysis, identification of weeds through hyperspectral image analysis, species recognition through digital image processing of leaves and floral structures, soil moisture prediction through modelling water balance components etc. IoT is an embedded system with sensors, software, actuators, electronics and computer to connect and exchange data. A few of the most commonly used IoT applications in agriculture are IoT-based operations of agricultural machineries, smart irrigation system, IoT smart greenhouse, etc. One of the big challenge of wide application of IoTs in agriculture is interference across networks especially with the IoT devices using the unlicensed spectrum, such as ZigBee, Wi-Fi, Sigfox, and LoRa. However, in recent times with digital India mission in the country and with the requirement of increasing water and other input use efficiency in agriculture, ML and IoT-based smart agriculture system has a great scope to improve agricultural productivity. This not only will fetch higher income for farmers but will make the agricultural production system a commercial and profitable venture specifically for the educated youth of the country.

4.7.2 SENSOR-BASED SMART FARMING

Sensor-driven agricultural operations starting from sowing to harvesting may lead to an increase in resource use efficiency in dryland farming system. Advances in sensors and machineries in recent times provide a good platform to shift from agrarian agriculture to a smart farming system (Bogue, 2017). With the involvement of IoTs in agricultural machineries, there is high scope to improve crop production and productivity. Some machineries such as drones and robots have a direct application of IoTs in agricultural machinery, which is operated in autopilot mode (Basri et al., 2021). The robotization of agricultural machineries e.g. auto rice transplanters, smart combine harvesters equipped with yield monitoring systems and smart spraying in the near future are expected to change the agriculture scenario in terms of work accuracy and efficiency. The mechatronics and its application in precision planter and seeder are also found to be a better way to achieve accurate seed spacing with higher efficiency (Gautam et al., 2019). In recent time, agricultural robots (Agribots) have been developed for chemical spraying, harvesting, picking fruits and crop monitoring (Gollakota and Srinivas, 2011). Agribots have big potential for multitasking activities, sensory acuity, operational consistency as well as suitability to perform operations in odd soil conditions. Smart irrigation systems have recently been adopted in farmers' field for improving water use efficiency, which includes a solar-powered automated IoTs-based drip/sprinkler irrigation system. Greenhouse cultivation or protected cultivation is another way of smart farming system, in which crops are grown inside structures made of polythene sheets or shade nets. The main purpose of greenhouse cultivation is to avoid the negative effects of unfavourable weather in open conditions. Therefore, controlling of inside environment is a major requirement of greenhouse cultivation. Since manual intervention is labor-intensive and prone to pest and disease infestation, automated monitoring and control of inside environment is advantageous for improved production in greenhouse (Rayhana et al., 2020).

4.7.3 Nano-Fertilizers

Nutrient management in the agricultural field is a difficult task since nutrient use efficiency still lies in the range of 25%–30%. Again, for supplying nutrients to soil for plant nutrition, farmers largely depend on synthetic chemical fertilizers which may not be environmentally friendly. Continuous application of chemical fertilizers may degrade the soil in future. Under such a situation, nano-fertilizers have the potential to enhance nutrient uptake and plant production by regulating the availability of fertilizers in the rhizosphere; extend stress resistance by improving nutritional capacity; and increase plant defense mechanisms (Subramanian and Tarafdar, 2011; Tarafdar et al., 2020). They may also substitute for synthetic fertilizers for sustainable agriculture. Nano-fertlizers also help in mitigating environmental stresses and enhancing tolerance abilities under adverse atmospheric conditions. Since nanoparticles are considered as the emerging cutting-edge agri-technology for agri-improvement in the near future, nano-fertilizer may play a big role to improve the nutrient use efficiency and also to increase the farm productivity (Qureshi et al., 2018).

4.7.4 Hyperspectra-Based Soil Assessment

Spectroscopy has been widely applied to identify the composition of materials. Hyperspectral soil reflectance spectrum (digital signature) is a proxy of inherent chemical composition of the soil, and thus may help in the quick assessment of soil properties (Santra et al., 2009, 2015). To cover the intensely large and diverse Indian soils, rapid and non-invasive techniques of hyperspectral spectroscopy and imaging might be the only solution to cater for the need of assessing and updating soil information of more than 100 million farmers. Although soil spectroscopy has attained a maturity, establishment of standard protocols of collecting, scanning and collating information of soils across distinct agro-cecological regions of the country is the need of the hour. The much ambitious soil health card scheme under the soil health management mission of the country has been launched to aid farmers with the required information of his soils and to recommend fertilizers and other inputs. With the aid of hyperspectral sensors, the ambitious goal of providing soil health card-based recommendations to a large number of farmers will be easily achievable. Moreover, the glitch in a huge load of soil testing with a limited workforce is facing the crunch of a short-timed extraction and analytical procedure. Therefore, finding a suitable multi-nutrient extractant and validation for a broader region for estimating plant available essential soil nutrients will be useful, since this will save time for their estimation. Therefore, there is need for research efforts with the following two major objectives: (i) The national level soil spectral library may be developed which may cater the need of futuristic digital application in soil resource assessment and (ii) development of soil type specific and machine learning-driven algorithms which may improve the accuracy of estimation of soil properties across different agroecological regions of India.

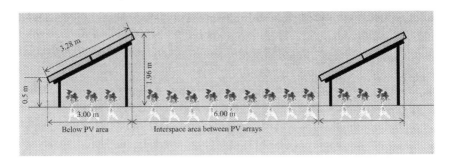

FIGURE 4.11 Design of agrivoltaic system. Source: Santra and Yadav (2020).

4.7.5 AGRIVOLTAIC SYSTEM

Competition for land in arid western Rajasthan may arise in future for agricultural use and PV-based electricity generation because of risk associated with rainfed cultivation under climate change. Agrivoltaic system (AVS) integrating both food production and PV generation from a single land use system was designed and developed at ICAR-Central Arid Zone Research Institute, Jodhpur (Santra et al., 2017, 2021). In the AVS, crops are cultivated at inter space areas between PV arrays as well as at below PV areas (Figure 4.11). Another important feature of the AVS is the rain water harvesting from the top of PV module to a water storage tank, which is recycled for cleaning of PV module and for providing supplemental irrigation to crops.

Height of crops is a key parameter for the selection of crops for agri-voltaic system because tall-growing crops may create shade on PV modules and thus reduce the PV-based electricity generation. Therefore, crops with low height (preferably shorter than 50 cm) and which tolerates a certain degree of shade and require less amount of water are most suitable for the solar farming system.

The AVS may bring huge opportunity to dryland farmers since rainfed-based crop production is risky because of uncertainty and scarcity of rainfall. Life cycle cost analysis of the system shows simple payback period of 5.87 years, discounted payback period of 10.40 years and internal rate of return of 16%. Ecological and Social benefits of agrivoltaic system are as follows:

i. Increased income from farmland and land productivity (land equivalent ratio may be increased up to 1.41)
ii. Reduction in soil erosion by wind and decreases dust load on PV module,
iii. Improvement in microclimate for crop cultivation and optimum PV generation
iv. Recycling of harvested rainwater for cleaning PV modules and irrigating crops (1.5 lakh litre per acre and can provide 40 mm irrigation in 1 acre land)
v. Soil moisture conservation by reducing the wind speed on ground surface
vi. Reduction in GHG emission (598.6 tons of CO_2 savings/year/ha)

4.8 CONCLUSIONS

Low productivity of soils and lack of sustainability often have a negative impact on livelihood development in drought-prone areas. Degraded soils with low moisture holding capacity, depleting groundwater table and multiple nutrient deficiencies leading to lower crop yields and further exacerbating the land degradation. Frequent drought in the form of intermittent dry spells, uneven distribution of rainfall, soil crusting in the red soil and swelling and shrinkage in the black soil, soil salinity, low investment capacity of the farmers lead to low productivity and returns in rainfed areas. In this document, different edaphic, water and climatic constraints of drylands in India are discussed in detail which include poor soil water and nutrient retention capacity, scarce water resources both in terms of rainfall and groundwater, extremely hot and dry environment etc. Thus, addressing a holistic approach of managing edaphic, hydrological and soil nutrient-related constraints is necessary to reduce the bottleneck. To overcome these inherent challenges, different interventions are discussed e.g. managing soil carbon, tackling multi-nutrient deficiency, soil test-based fertilizer recommendation, integrated nutrient management, control of wind erosion through vegetative barriers and shelterbelts, rainwater harvesting in farm ponds and its utilization through solar photovoltaic (PV) pump operated pressurized irrigation, deficit irrigation, etc. Increasing soil water storage, sequestering carbon, conservation agriculture, organic farming, mulching, suitable and region-specific land configuration, application of soil amendments like manures, tank silts, and some of the other novel materials viz., biochar, polymers, zeolites, bentonites can increase the water and nutrient retention and subsequently improve the use efficiency of water and nutrients. Climate-resilient intercropping systems, profitable integrated farming system models, alternate land use systems are some of the proven practices for increasing resilience against drought. In situ and ex situ rain water management, use of farm ponds for water harvesting and supplemental irrigation, improving water use efficiency through sprinkler or drip irrigation along with efficient nutrient management viz., foliar spraying of water-soluble fertilizers, nano fertilizers, site specific nutrient management, conjunctive use of chemical fertilizers and locally available organic manures can bridge the yield gaps dryland areas. A few innovative and advanced solutions for sustainable management of soil, water and nutrient resources in drylands are also suggested e.g., smart farming, artificial intelligence and IoT-based farming, utilization of nano-fertiliser, agrivoltaic system, etc. For implementing these interventions for imparting climate resilience in dryland areas, village-level agromet advisories and agro ecology specific real-time contingency planning is necessary.

LIST OF ABBREVIATIONS

AC	Alternating current
Agribots	Agricultural robots
AI	Aridity index
AICRPAM	All India Co-ordinated Research Project on Agrometeorology
ANN	Artificial neural network
AVS	Agrivoltaic system
AWI	Arid Western India
CA	Conservation agriculture
CAGR	Compound annual growth rate
CASI	Conservation agriculture system intensification
CAZRI	Central Arid Zone Research Institute
CCA	Culturable command area
DAC	Department of Agriculture and Cooperation
DAP	Di-ammonium phosphate
DC	Direct current
DI	Deficit irrigation
dS	deciSiemens
DTPA	Diethylenetriamine pentaacetate
EC	Electrical conductivity
ESP	Exchangeable sodium percentage
ET	Evapotranspiration
FCO	Fertilizer control order
GHG	Greenhouse gas
GR	Gypsum requirement
HDPE	High-density polyethylene
HP	Horse power
ICAR	Indian Council of Agricultural Research

IGNP	*Indira Gandhi Nahar Pariyojna*
IMD	Indian Meteorological Department
INM	Integrated nutrient management
IoT	Internet of things
KVK	*Krishi Vigyan Kendra*
LCC	Leaf colour chart
M ha	Million hectare
MAAS	Micro-level agromet advisories
ML	Machine learning
MLR	Multiple linear regression
NCP	Narmada canal project
NICRA	National Initiatives for Climate Resilient Agriculture
PDSI	Palmer drought severity index
PV	Photovoltaic
RFR	Random forest regression
RSC	Residual sodium carbonate
SAR	Sodium adsorption ratio
SOC	Soil organic carbon
SPEI	Standardized precipitation evapotranspiration index
SPI	Standardized precipitation index
SSNM	Site specific nutrient management
SVM	Support vector machine
USDA	United States Department of Agriculture
USDM	US drought monitor
WP	Water productivity

REFERENCES

Abraham, M., R.A. Mathew. 2018. Assessment of surface runoff for tank watershed in Tamil Nadu using hydrologic modelling. Int. J. Geophys. Article ID 2498648: 1–10. https://doi.org/10.1155/2018/2498648.

Ali, A.M., Thind, H.S., Sharma, S., and Y. Singh. 2015. Site-specific nitrogen management in dry direct-seeded rice using chlorophyll meter and leaf colour chart. Pedosphere 25(1): 72–81.

Ali, S., Ghosh, N.C., and R. Singh. 2008. Evaluating best evaporation estimate model for water surface evaporation in semi-arid region, India. *Hydrological Processes* 22: 1093–1106.

Annual Report 2009-10. All India Coordinated Research Project for Dryland Agriculture. ICAR - Central Research Institute for Dryland Agriculture, Indian Council of Agricultural Research, Hyderabad – 500 059, India, p 288

Annual Report 2010-11. All India Coordinated Research Project for Dryland Agriculture. ICAR - Central Research Institute for Dryland Agriculture, Indian Council of Agricultural Research, Hyderabad – 500 059, India, p 301

Annual Report 2011-12. All India Coordinated Research Project for Dryland Agriculture. ICAR - Central Research Institute for Dryland Agriculture, Indian Council of Agricultural Research, Hyderabad – 500 059, India, p 296.

Annual Report 2013-14. All India Coordinated Research Project for Dryland Agriculture. ICAR - Central Research Institute for Dryland Agriculture, Indian Council of Agricultural Research, Hyderabad – 500 059, India, p 305

Annual Report 2015-16. All India Coordinated Research Project for Dryland Agriculture. ICAR - Central Research Institute for Dryland Agriculture, Indian Council of Agricultural Research, Hyderabad – 500 059, India, p 312

Basri, R., Islam, F., Shorif, S.B., and M.S. Uddin. 2021. Robots and drones in agriculture-a survey. In Computer Vision and Machine Learning in Agriculture, eds. M.S. Uddin and J.C. Bansal (pp. 9–29). Springer, Singapore.

Beguería, S., Vicente-Serrano, S.M., Reig, F., and B. Latorre. 2014. Standardized precipitation evapotranspiration index (SPEI) revisited: Parameter fitting, evapotranspiration models, tools, datasets and drought monitoring. *International Journal of Climatology* 34(10): 3001–3023. DOI: 10.1002/joc.3887.

Bhatnagar, A., Bhatnagar, M., Makandar, Md.B. and M.K. Garg. 2003. *Satellite Centre for Microalgal Biodiversity in Arid Zones of Rajasthan. Project Completion Report*, Funded by Department of Biotechnology, New Delhi.

Bijay-Singh and A.M. Ali. 2020. Using hand-held chlorophyll meters and canopy reflectance sensors for fertilizer nitrogen management in cereals in small farms in developing countries. *Sensors* 20(4): 1127.

Bogue, R., 2017. Sensors key to advances in precision agriculture. *Sensor Review* 37(1): 1–6. DOI: 10.1108/SR-10-2016-0215.

Cammalleri, C., Vogt, J., and P. Salamon. 2017. Development of an operational low-flow index for hydrological drought monitoring over Europe. *Hydrological Science Journal* 62(3): 346–358. DOI: 10.1080/02626667.2016.1240869.

Carrão, H., Singleton, A., Naumann, G., Barbosa, P., and J.V. Vogt. 2014. An optimized system for the classification of meteorological drought intensity with applications in drought frequency analysis. *Journal of Applied Meteorology and Climatology* 53(8): 1943–1960. DOI: 10.1175/JAMC-D-13-0167.1.

CGWB. 1997. *Status of Groundwater Quality Including Pollution Aspects in India*. Central Ground Water Board (CGWB), Ministry of Water Resources, Government of India, Faridabad, 68p.

Chan, H.J. 2001. Effect of land use and urbanization on hydrochemistry and contamination of groundwater from Taejon area, Korea. *Journal of Hydrology* 253: 194–210.

Chatterjee, S., Desai, A.R., Zhu, J., Townsend, P.A., and J. Huang. 2022. Soil moisture as an essential component for delineating and forecasting agricultural rather than meteorological drought. *Remote Sensing of Environment* 269: 112833.

Cirelli, F.A., L.J. Arumi, D. Rivera, and W.P. Boochs. 2009. Environmental effects of irrigation in arid and semi arid regions. *Chilean Journal of Agricultural Research* 69: 27–40.

Dakora, F.D. and S.O. Keya. 1997. Contribution of legume nitrogen fixation to sustainable agriculture in Sub-Saharan Africa. Soil Biology and Biochemistry 29(5-6): 809–817.

Das, B.S., M.C. Sarathjith, P. Santra, R. Srivastava, R.N. Sahoo, A. Routray, and S.S. Ray. 2015. Hyperspectral remote sensing: Opportunities, status and challenges for rapid soil assessment in India. *Current Science* 108(5): 860–868.

Das, S. 2011. *Groundwater Resources of India*. National Book Trust, India, 248p.

Das, S. 2019. Water management in arid and semiarid areas of India. In *Ground Water Development-Issues and Sustainable Solutions*, ed. S. P. S. Ray (pp. 15–33). Springer Nature Singapore.

Das, S.K., Gupta, R.K., and H.K. Varma. 2007. Flood and drought management through water resources development in India. *WMO Bulletin*, 56(3): 179–188.

Davidson, E.A., Trumbore, S.E. and R. Amundson. 2000. Biogeochemistry: Soil warming and organic carbon content. *Nature* 408: 789–790.

Debaje, S. B. 2014. Estimated crop yield losses due to surface ozone exposure and economic damage in India. *Environmental Science and Pollution Research* 21(12): 7329–7338. DOI: 10.1007/s11356-014-2657-6.

Denton, F., Wilbanks, T.J., Abeysinghe, A.C., Burton, I., Gao, Q., Lemos, M.C., and K. Warner. 2014. Climate-Resilient pathways: Adaptation, mitigation, and sustainable development. In *Climate Change 2014: Impacts, Adaptation, and Vulnerability* eds C.B. Field and V.R. Barros (pp. 1101–1131). Cambridge University Press, Cambridge.

Dhir, R.P. 1977. Western Rajasthan soils: Their characteristics and properties. In *Desertification and Its Control* (pp. 102–115). ICAR, New Delhi.

Ganguli, J.K. and R.N. Kaul. 1969. *Wind Erosion Control*. ICAR: New Delhi.

Garg, K.K., Akuraju, V., Anantha, K.H., Singh, R., Whitbread, A.M. and S. Dixit. 2022. Identifying potential zones for rainwater harvesting interventions for sustainable intensification in the semiarid tropics. *Scientific Reports* 12: 3882. DOI: 10.1038/s41598-022-07847-4.

Gautam, P.V., Kushwaha, H.L., Kumar, A. and D.K. Kushwaha. 2019. Mechatronics application in precision sowing: A review. *International Journal of Current Microbiology and Applied Sciences* 8(4): 1793–1807.

Glenn, E., Squires, V., Olsen, M., and R. Frye. 1993. Potential for carbon sequestration in the drylands. *Water, Air and soil Plollution* 70: 341–355.

GoI. 2017. *Drought Management Plan*. Department of Agriculture, Cooperation & Farmers Welfare, Ministry of Agriculture and Farmers Welfare, Government of India, New Delhi.

Gollakota, A. and M.B. Srinivas. 2011. Agribot-A multipurpose agricultural robot. In 2011 Annual IEEE India Conference (pp. 1–4). BITS Pilani, Hyderabad Campus, Hyderabad, India.

Guo, Y., Huang, S., Huang, Q., Wang, H., Fang, W., Yang, Y., and L. Wang. 2019. Assessing socioeconomic drought based on an improved multivariate standardized reliability and resilience index. *Journal of Hydrology* 568: 904–918. DOI: 10.1016/j. jhydrol.2018.11.055.

Gupta, J.P., Joshi, D.C., and G.B. Singh. 2000. Management of arid agro-ecosystem. In Natural Resource Management for Agricultural Production in India, eds. J. S. P. Yadav and G. B. Singh (pp. 551–668). Secretary General, International Conference on Managing Natural Resources for Sustainable Agricultural Production in the 21st century, New Delhi.

Gupta, J.P., Rao, G.G.S.N., Ramakrishna, Y.S. and B.V.R. Rao. 1984. Role of shelterbelts in arid zone. *Indian Farming*, October 1984: 29–30.

Heim Jr., R.R., 2002. A review of twentieth-century drought indices used in the United States. *Bulletin of the American Meteorological Society* 83(8): 1149–1166. DOI: 10.1175/1520-0477-83.8.1149.

IGNP, 2002. *History of Indira Gandhi Nahar: A Venture to Turn Thar Desert into Granary*, Indira Gandhi Nahar Board, Bikaner.

Joshi, D.C. and R.P. Dhir. 1991. Rehabilitation of degraded sodic soils in an arid environment by using residual Na-carbonate water for irrigation. *Arid Soil Research and Rehabilitation* 5: 175–185.

Joshi, D.C. 1992. Relationship between the quantity and intensity parameters of labile potassium in Aridisols of Indian Desert. Journal of the Indian Society of Soil Science 40(3): 431–438.

Joshi, D.C. and R.P. Dhir. 1994. Amelioration and management of soils irrigated with sodic water in the arid region of India. *Soil Use and Management* 10: 30–34.

Kala, C.P. 2017. Environmental and socioeconomic impacts of drought in India: Lessons for drought management. *Applied Ecology and Environmental Sciences*, 5(2): 43–48.

Kar, A., Moharana, P.C., Raina, P., Kumar, M., Soni, M.L., Santra, P., Ajai, Arya, A.S., and Dhinwa, P.S. 2009. Desertification and its control measures, In Trends in Arid Zone Researches in India, eds A. Kar, B.K. Garg, M.P. Singh, S. Kathju (pp. 1-47), Central Arid Zone Research Institute, Jodhpur, India.

Kanwar, J.S. 1999. Need for a future outlook an mandate for dryland agriculture in India. In Fifty years of Dryland Agricultural Research in India, eds H.P. Singh et al. (pp. 11–19). Central Research Institute for Dryland Agriculture, Hyderabad.

Kathju, S., Joshi, N.L., Rao, A.V. and P. Kumar. 1998. Arable crop production. In *Fifty Years of Arid Zone research in India*, eds. A. S. Faroda and M. Singh (pp. 214–252). Central Arid Zone Research Institute, Jodhpur.

Keyantash, J.A., and J.A. Dracup. 2004. An aggregate drought index: assessing drought severity based on fluctuations in the hydrologic cycle and surface water storage. *Water Resource Research* 40(9): W09304, 1-13. DOI: 10.1029/2003WR002610.

Khan, P.I., Ratnam, D.V., Prasad, P., Basha, G., Jiang, J.H., Shaik, R., Ratnam, M.V. and P. Kishore. 2022. Observed climatology and trend in relative humidity, CAPE, and CIN over India. *Atmosphere* 13: 361, DOI: 10.3390/atmos13020361.

Khathuria, N. 1998. Rhizosphere microbiology of desert, M.Sc. dissertation, p. 52. Department of Microbiology, Maharshi Dayanand Saraswati University, Ajmer, Rajasthan

Kolarkar, A.S., Murthy, K.N.K. and N. Singh. 1983. Khadin-A method of harvesting water for agriculture in the Thar Desert. *Journal of Arid Environments* 6(1): 59–66.

Kumar, S., Sharma, K.L., Kareemulla, K., Chary, G.R., Ramarao, C.A., Rao, C.S. and B. Venkateswarlu. 2011. Techno-economic feasibility of conservation agriculture in rainfed regions of India. Current Science: 1171–1181.

Lal, R. 2001. Potemntial of dertificatiopn control to sequester carbon and mitigate the green house effect. *Climate Chnage* 51: 35–72.

Lal, R. 2002. Soil carbon dynamics in cropland and rangeland. Environmental Pollution 116: 353–362.

Lal, R. 2004a. Soil carbon sequestration impacts on global climate change and food security. *Science* 304: 1623–1627.

Lal, R. 2004b. Soil carbon sequestration to mitigate climate change. *Geoderma* 123: 1–22.

Leng, G. and J. Hall. 2019. Crop yield sensitivity of global major agricultural countries to droughts and the projected changes in the future. Science of the Total Environment 654: 811–821. DOI: 10.1016/j.scitotenv.2018.10.434.

Lettens, S., Orshoven, J.V., Wesemael, B., Vos, B.D., and B. Muys. 2005. Stocks and fluxes of soil organic carbon for landscape units in Belgium derived from heterogeneous data sets for 1990 and 2000. *Geoderma* 127: 11–23.

Machiwal, D., Dayal, D., and S. Kumar, S. 2015. Assessment of reservoir sedimentation in arid region watershed of Gujarat. *Journal of Agricultural Engineering, ISAE* 52(4): 40–49.

Machiwal, D. and M.K. Jha. 2015. Identifying sources of groundwater contamination in a hard-rock aquifer system using multivariate statistical analyses and GIS-based geostatistical modeling techniques. *Journal of Hydrology: Regional Studies* 4(Part A): 80–110.

Machiwal, D., Jha, M.K., and A. Gupta. 2020. Development of a rainfall stability Index using probabilistic indicators. *Ecological Indicators* 115: 106406. DOI: 10.1016/j.ecolind.2020.106406.

Machiwal, D., Jha, M.K., Singh, V.P. and C. Mohan 2018a. Assessment and mapping of groundwater vulnerability to pollution: Current status and challenges. *Earth-Science Reviews* 185: 901–927.

Machiwal, D., Kumar, S., and D. Dayal. 2018b. Evaluating cost-effectiveness of rainwater harvesting for irrigation in arid climate of Gujarat, India. *Water Conservation Science and Engineering* 3: 289–303.

Machiwal, D., Kumar, S., Dayal, D., and S. Mangalassery. 2016. Identifying abrupt changes and detecting gradual trends of annual rainfall in an Indian arid region under heightened rainfall rise regime. *International Journal of Climatology* 37(5): 2719–2733.

Mahesh, K., Kar, A., Raina, P., Singh, S.K., Moharana, P.C., and J.S. Chauhan. 2021. Assessment and mapping of available soil nutrients using GIS for nutrient management in hot arid regions of north-western India. *Journal of the Indian Society of Soil Science* 69(2): 119–132.

Mahesh, K., Moharana, P.C., Santra, P., Roy, S., Panwar, N.R., and C.B. Pandey. 2016a. Characterization of salt affected soils in Narmada canal command area of Rajasthan for reclamation and management. In *25th National Conference on "Natural Resource Management in Arid and Semi-Arid Ecosystem for Climate Resilient Agriculture and Rural Development"* (p. 19), Swami Kesavanand Rajathan Agricultural University (SKRAU), Bikaner.

Mahesh, K., Santra, P., Panwar, N.R., Moharana, P.C., and C.B. Pandey. 2022. Whether canal command irrigation through pressurized irrigation system deteriorates soil properties in hot arid ecosystem of India? *Geoderma Regional* 28: e00459. DOI: 10.1016/j.geodrs.2021.e00459.

Mahesh, K., Singh, R. and A. Kar. 2016b. Interventions of high residual sodium carbonate water-degraded soils amelioration technology in Indian Thar Desert and farmers' response. *National Science Academy Letters* 39(4): 245–249.

Mazumdar, A., 2007. Participatory approaches to sustainable rural water resources development and management: Indian Perspective. *Journal of Developments in Sustainable Agriculture* 2(1): 59–65.

Malhotra, S.K. 2016. Water soluble fertilizers in horticultural crops-An appraisal. *Indian Journal of Agricultural Sciences* 86(10): 1245–1256.

Mallya, G., Mishra, V., Niyogi, D., Tripathi, S., and Govindaraju, R.S. 2016. Trends and variability of droughts over the Indian monsoon region. *Weather and Climate Extremes*, 12: 43–68.

McKee, T.B., Doesken, N.J., and J. Kleist. 1993. The relationship of drought frequency and duration to time scales. In Proceedings of the 8th Conference on Applied Climatology (pp. 179–183. Vol. 17, No. 22), American Meteorological Society, Boston, Mass., USA.

Mertia, R.S., Prasad, R., Gajja, B.L., Samra, J.S. and P. Narain. 2006. *Impact of Shelterbelts in Arid Region of Western Rajasthan*. Central Arid Zone Research Institute, Jodhpur.

Mertia, R.S., Santra, P., Kandpal, B.K., and R. Prasad. 2010. Mass-height profile and total mass transport of wind eroded aeolian sediments from rangelands of Indian Thar Desert. *Aeolian Research* 2: 135–142.

Minhas, P.S. and R.K. Gupta. 1992. Quality of Irrigation Water-Assessment and Management. Indian Council of Agricultural Research, New Delhi. pp. 123.

Mishra, U., Lal, R., Liu, D. and M. Van Meirvenne. 2010. Predicting the Spatial Variation of the Soil Organic Carbon Pool at a Regional Scale. *Soil Science Society of America Journal* 74: 906–914.

Moharana, P.C., Santra, P., Singh, D.V., Kumar, S., Goyal, R.K., Machiwal, D. and O.P. Yadav. 2016. ICAR-Central Arid Zone Research Institute, Jodhpur: Erosion processes and desertification in the Thar Desert of India. *Proceedings of the Indian National Science Academy* 82(3): 1117–1140.

Motsara, M.R. 2002. Available nitrogen, phosphorus and potassium status of Indian soils as depicted by soil fertilizer maps. *Fertilizer News* 47(8): 15–21.

Narjary, B., Aggarwal, P., Singh, A., Chakraborty, D., and R. Singh. 2012. Water availability in different soils in relation to hydrogel application. *Geoderma* 187: 94–101.

NBSSLUP. 2001. *Classification of Dryland Regions in India*. National Bureau of Soil Survey and Land Use Planning (NBSSLUP), Nagpur.

NRAA. 2014. Expanded Rains, Drought, Floods, Cyclone, hail storms and Crop Production Prospects during 2013-14. *Position Paper No.7*. National Rainfed Area Authority, NASC Complex, DPS Marg, New Delhi – 110012, India : 142

Palmer, W.C., 1965. *Meteorological Drought* (Vol. 30). US Department of Commerce, Weather Bureau.

Pande, M. and J.C. Tarafdar. 2004. Arbuscular mycorrhizal fungal diversity in neem-based agroforestry systems in Rajasthan. *Applied Soil Ecology* 26: 233241.

Park, C. 2007. *A Dictionary of Environment and Conservation* (1st ed.). Oxford University Press, DOI: 10.1093/acref/9780198609957.001.0001.

Prâvâlie, R., 2016. Dryland extent and environmental issues: A global approach. *Earth-Science Reviews* 161: 259–278.

Praveen, K., Tarafdar, J.C., Painuli, D.K., Raina, P., Singh, M.P., Beniwal, R.K., Soni, M.L., M. Kumar, Santra, P., and M. Shamsuddin. 2009. Variability in arid soil characteristics. In *Trends in Arid Zone Research in India*, eds. A. Kar, B. K. Garg, M. P. Singh, S. Katju (pp. 78–112). Central Arid Zone Research Institute, Jodhpur.

Qureshi, A., Singh, D.K. and S. Dwivedi. 2018. Nano-fertilizers: A novel way for enhancing nutrient use efficiency and crop productivity. *International Journal of Current Microbiology and Applied Sciences* 7(2): 3325–3335.

Raina, P. and A.K. Sen. 1991. Soil degradation studies under different land use system in an arid environment. Annals of Arid Zone 30(1): 11–15.

Raina, P., Joshi, D.C. and A.S. Kolarkar. 1991. Land degradation mapping by remote sensing in the arid region of India. Soil use and management 7(1): 47–51.

Raina, P. 1999. Soil degradation assessment through remote sensing and its impact on fertility status of soils of western Rajasthan. Agropedology 9: 30–40.

Rajasthan Agricultural Statistics. 2016–17. *Directorate of Economics and Statistics*, Department of Planning, Government of Rajasthan, India.

Ramakrishna, R. and D.T. Rao. 2008. Strengthening Indian Agriculture through Dryland Farming: Need for reforms. *Indian Society of Agricultural Economics* 63(3): 1–17.

Rao, A.S. and R.S. Singh. 1998. Climatic features and crop production. In *Fifty Years of Arid Zone Research in India*, eds. A. S. Faroda and M. Singh (pp. 17–38). Central Arid Zone Research Institute, Jodhpur.

Rao, A.V. and B. Venkateswarlu. 1983. Microbial ecology of the soils of Indian desert. *Agricultural Ecosystem and Environment* 10: 361–369

Rao, C.A.R., Venkateswarlu, B., and R.G. Chary. 2020. Rainfed agriculture in India: Importance, challenges and imperatives. In *The Hindu Business Line's Handbook on Indian Agriculture* (pp. 36–40). THG Publishing Private Limited, Chennai.

Rao, K.V., Venkateswarlu, B., Sahrawat, K.L., Wani, S.P., Mishra, P.K., Dixit, S., Reddy, K.S., Kumar, M., and U.S. Saikia. 2009. *Proceedings of National Workshop-Cum-Brain Storming*. CRIDA, Hyderabad.

Rathore, V.S., Nathawat, N.S., Bhardwaj, S., Yadav, B.M., Kumar, M., Santra, P., P. Kumar, Reager, M.L., Yadav, N.D., and O.P. Yadav. 2021. Optimization of deficit irrigation and nitrogen fertilizer management for peanut production in an arid region. *Scientific Reports* 11: 5456.

Rayhana, R., Xiao, G. and Z. Liu. 2020. Internet of things empowered smart greenhouse farming. *IEEE Journal of Radio Frequency Identification* 4(3): 195–211.

Sahrawat, K.L., Wani, S.P., Rego, T.J., Pardhasaradhi, G. and K.V.S. Murthy. 2007. Widespread deficiencies of sulphur, boron and zinc in dryland soils of the Indian semi-arid tropics. *Current Science* 93(10): 1–6.

Samuel, J., Rao, C.A.R., Raju, B.M.K., Reddy, A.A., Pushpanjali, Reddy, A.G.K., Kumar, R.N., Osman, M., Singh, V.K. and J.V.N.S. Prasad. 2022. Assessing the impact of climate resilient technologies in minimizing drought impacts on farm incomes in drylands. *Sustainability*, 14: 382. DOI: 10.3390/su14010382.

Santra, P., 2021. Performance evaluation of solar PV pumping system for providing irrigation through micro-irrigation techniques using surface water resources in hot arid region of India. *Agricultural Water Management* 245: 106554. DOI: 10.1016/j.agwat.2020.106554.

Santra, P., Kumar, M., and B.L. Dhaka. 2020. Artificial intelligence and digital application in soil science: Potential options for sustainable soil management in future. *SATSA Mukhapatra-Annual Technical Issue* 24: 1–23.

Santra, P., Kumar, M., Kumawat, R.N., Painuli, D.K., Hati, K.M., Heuvelink, G., and N. Batjes. 2018. Pedotransfer functions to estimate soil water retention at field capacity and permanent wilting point in hot arid western India. *Journal of Earth System Science* 127: 35. DOI: 10.1007/s12040-018-0937-0.

Santra, P., Kumar, M., and R.S. Yadav. 2020. Soil water retention in dry sandy soil and its importance in crop water use efficiency. *CAZRI News Letter* 10(4): 3–5.

Santra, P., Kumar, R., Sarathjith, M.C., Panwar, N.R., Varghese, P., and B.S. Das. 2015. Reflectance spectroscopic approach for estimation of soil properties in hot arid western Rajasthan, India. *Environmental Earth Science* 74: 4233–4245

Santra, P., Kumar, S., Moharana, P.C., Pande, P.C., Roy, M.M., R.K. Bhatt, 2016. *Wind Erosion and Its Control in Western Rajasthan* (64 p). ICAR-Central Arid Zone Research Institute, Jodhpur.

Santra, P., Meena, H.M., and O.P. Yadav. 2021. Spatial and temporal variation of photosynthetic photon flux density within agrivoltaic system in hot arid region of India. *Biosystems Engineering* 209: 74–93. DOI: 10.1016/j.biosystemseng.2021.06.017

Santra, P., Mertia, R.S., and H.L. Kushawa. 2010. A new wind erosion sampler for monitoring dust storm events in the Indian Thar desert. *Current Science* 99(8): 1061–1067.

Santra, P., Pande, P.C., Kumar, S., Mishra, D., and R.K. Singh. 2017. Agri-voltaics or Solar farming: the concept of integrating solar PV based electricity generation and crop production in a single land use system. *International Journal of Renewable Energy Research* 7(2): 694–699.

Santra, P., Sahoo, R.N., Das, B.S., Samal, R.N., Pattanaik, A.K., and V.K. Gupta. 2009. Estimation of soil hydraulic properties using proximal spectral reflectance in visible, near-infrared, and shortwave-infrared (VIS-NIR-SWIR) region. *Geoderma* 152: 338–349.

Santra, P. and O.P. Yadav. 2020. Agri-voltaic (AVS) system at CAZRI exploring the possibility of cultivating crops, generating electricity and harvesting water. In *Compendium on Solar Powered Irrigation Systems in India*, eds. Shirsath PB, Saini S, Durg N, Senoner D, Ghose N, Verma S, Sikka A (pp. 58–60). CGIAR Research Program on Climate Change, Agriculture and Food Security (CCAFS), New Delhi.

Shrivastava, M., Sharma, I.K., and D.D. Sharma. 2013. Ground water scenario in Indira Gandhi Nahar Pariyojana (IGNP) in parts of Sri Ganganagar, Hanumangarh, Churu, Bikaner, Jaisalmer, Jodhpur and Barmer districts, Rajasthan. *Memoir Geological Society of India* 82: 16–35.

Singh, H.P., Sharma, K.D., Reddy, G.S. and K.L. Sharma. 2004. Dryland Agriculture in India. In *Challenges and Strategies for Dryland Agriculture*, eds Srinivas C Rao and John Ryan (pp. 67–92, Special Publication No. 32, Chapter 6). Crop Science Society of America and American Society of Agronomy, Madison, WI.

Singh, J.P. and J.C. Tarafdar. 2002. Rhizospheric microflora as influenced by sulphur application, herbicide and Rhizobium inoculation in summer mung bean (*Vigna radiata* L.). *Journal of the Indian Society of Soil Science* 50: 127129.

Srinivasarao, C. and K.P.R. Vittal. 2007. Emerging Nutrient Deficiencies in Different Soil Types under Rainfed Production Systems of India. *Indian Journal of Fertilisers*, 3(5): 37.

Srinivasarao, C., Girija Veni, V., Prasad, J.V.N.S., Sharma, K.L., Chandrasekhar, Ch., Rohilla, P.P. and Y.V. Singh. 2017a. Improving carbon balance with climate-resilient management practices in tropical agro-ecosystems of Western India. *Carbon Management* 8(2): 175–190.

Srinivasarao, C., Gopinath, K.A., Ramarao, C.A. and B.M.K. Raju. 2017b. Dryland agriculture in South Asia: experiences, challenges and opportunities. In *Innovations in Dryland Agriculture*. Springer.

Srinivasarao, C., Indoria, A.K. and K.L. Sharma. 2017c. Effective management practices for improving soil organic matter for increasing crop productivity in rainfed agroecology of India. *Current Science* 112(7):1497–1504

Srinivasarao, C., Lal, R., Prasad, J.V.N.S., Gopinath, K.A., Singh, R., Jakkula, V.S., Sahrawat, K.L., Venkateswarlu, B., Sikka, A.K. and S.M. Virmani. 2015. Potential and challenges of rainfed farming in India. *Advance in Agronomy* 133: 113–181 https://dx. doi.org/10.1016/bs.agron.2015.05.004

Srinivasarao, C., Prasad, J.V.N.S., Osman, M., Prabhakar, M., Kumara, B.H., and A.K. Singh. 2017d. *Farm Innovations in Climate Resilient Agriculture*. Central Research Institute for Dryland Agriculture, Hyderabad.

Srinivasarao, C., Rejani, R., Rama Rao, C.A., Rao, K.V., Osman, M., Srinivasa Reddy, K., M. Kumar and P. Kumar. 2017e. Farm ponds for climate-resilient rainfed agriculture. *Current Science* 112(3): 471–477.

Srinivasarao, C., Wani, S.P., Sahrawat, K.L., Krishnappa, K. and B.K. Rajasekhara Rao. 2010. Effect of balanced nutrition on yield and economics of vegetable crop in participatory watersheds in Karnataka. *Indian Journal of Fertilizers* 6(3): 39–42.

Stewart, B.A. 1987. *Advances in Soil Science* (Vol. 7), Springer-Verlag, New York.

Subramanian, K.S. and J.C. Tarafdar. 2011. Prospects of nanotechnology in Indian farming. *The Indian Journal of Agricultural Sciences* 81(10): 887–893.

Sutton, W.R., Srivastava, J.P., and E.N. James. 2013. *Looking Beyond the Horizon: How Climate Change Impacts and Adaptation Responses Will Reshape Agriculture in Eastern Europe and Central Asia*. World Bank Publications, Washington, DC.

Svoboda, M., Le Comte, D., Hayes, M., Heim, R., Gleason, K., Angel, J., Rippey, B., Tinker, R., Palecki, M., Stooksbury, D., and D. Miskus. 2002. The drought monitor. *Bulletin of the American Meteorological Society* 83(8): 1181–1190. DOI: 10.1175/1520-0477-83.8.1181.

Tarafdar, C., Daizy, M., Alam, M.M., Ali, M.R., Islam, M.J., Islam, R., Ahommed, M.S., Aly Saad Aly, M., and M.Z.H. Khan. 2020. Formulation of a hybrid nanofertilizer for slow and sustainable release of micronutrients. *ACS Omega* 5(37): 23960–23966.

Takkar, P.N., 2006. *Handbook of Agriculture*. Indian Council of Agricultural Research, New Delhi.

Thombare, N., Mishra, S., Siddiqui, M.Z., Jha, U., Singh, D., and G.R. Mahajan. 2018. Design and development of guar gum based novel, superabsorbent and moisture retaining hydrogels for agricultural applications. *Carbohydrate Polymers* 185: 169–178.

Trappe, J.M. 1981. Mycorrhizae and productivity of arid and semi arid range lands. In *Advances in Food Production System for Arid and Semi Arid Lands*, eds. J. J. Manassha, and E. J. Briskey (pp. 581–599). Academic Press, New York.

Van Loon, A.F., 2015. Hydrological drought explained. *Wiley Interdiscipilinary Reviews: Water* 2(4): 359–392. DOI: 10.1002/wat2.1085.

Varinderpal-Singh, V., Kaur, R., Singh, B., Singh, B. B., and A. Kaur. 2012. Precision Nutrient Management: A Review. *Indian Journal of Fertilisers* 12(11): 48–62.

Vautard, R., Cattiaux, J., Yiou, P., Thépaut, J.N., and P. Ciais. 2010. Northern Hemisphere atmospheric stilling partly attributed to an increase in surface roughness. *Nature Geoscience* 3(11): 756–761.

Venkateswarulu, B. and A.V. Rao. 1981. Distribution of microorganisms in stabilized and unstabilised sand dunes of Indian desert. *Journal of Arid Environment* 4:203–208.

Vicente-Serrano, S.M., Beguería, S., and J.I. López-Moreno. 2010. A multiscalar drought index sensitive to global warming: the standardized precipitation evapotranspiration index. *Journal of Climate* 23(7): 1696–1718. DOI: 10.1175/2009JCLI2909.1.

Vijayan, R. 2016. Dryland agriculture in India - problems and solutions. *Asian Journal of Environmental Science* 11(2): 171–177.

Virmani, S.M., Pathak, P. and R. Singh. 1991. Soil related constraints in dryland crop production in Vertisols, Alfisols and Entisols of India. *Soil Related Constraints in Crop Production. Bulletin* 15: 80–95.

Wilhite, D.A. and M.H. Glantz. 1985. Understanding: the drought phenomenon: the role of definitions. *Water International* 10(3): 111–120. DOI: 10.1080/ 02508068508686328.

Woodruff, N.P., Lyles, L., Siddoway, F.H., and D. W. Fryrear. 1972. How to control wind erosion. *USDA ARS Agriculture Information Bulletin No. 354*.

Wu, W. and B. Ma. 2015. Integrated nutrient management (INM) for sustaining crop productivity and reducing environmental impact: A review. *Science of Total Environment* 512: 415–427.

5 Water Storage in the Rest Zone by Enhancing Soil Organic Matter Content in the Rest Zone

*Ch. Srinivasarao, S. Rakesh, G. Ranjith Kumar,
M. Jagadesh, K.C. Nataraja, R. Manasa, S. Kundu,
S. Malleswari, K.V. Rao, JVNS Prasad, R.S. Meena,
G. Venkatesh, P.C. Abhilash, J. Somasundaram, and R. Lal*

5.1 INTRODUCTION

With the rise in global population, food demand is increasing particularly in densely populated south Asia. In India, the rainfed area covers about 55% of the net sown area (139.42 M ha) and about 61% of the farmers are cultivating crops under the rainfed region. Almost 80% of small and marginal farmers of the country depended on rainfed farming for their livelihoods. Rainfall is the main source of water in drylands. Coping with the extreme variability in rainfall, high-intensity storms, and high frequency of dry spells are the key challenges in rainfed agriculture as it is complex, highly diverse and risk prone. Soil organic carbon (SOC) has a critical role in soil plant water relationships and contributes to drought mitigation. Increasing SOC by 1% may increase the available water holding capacity (AWHC) by 2% to >5%. Soil organic matter content (SOM) is also vital to soil processes like nutrient dynamics, water interactions, and for maintaining the biological and physical health of soil. The low level of SOC content in rainfed drylands is due to rapid decomposition of added organic matter, loss of carbon through soil erosion and use of inappropriate crop management practices. Thus, improving SOC in drylands contributes to productivity enhancement and stability due to higher available water retention during mid-season droughts while improving other soil productivity factors. Therefore, intensive implementation of site-specific available C enrichment technologies in different agro-ecosystems can maintain the overall productivity functions of soil even under adverse conditions. Thus, the present chapter aims to cover the SOC status of rainfed drylands, drought management with improving SOC technologies along with various national programs that address to improve SOC content and stocks in agroecosystems of India.

The population of India increased from 451 million to 1.39 billion from 1960 to 2021, which reflects the growth of 209.3% over the past 61 years. The highest annual increase in the population of India was recorded in 1974 at 2.36%, and the lowest increase of 0.97% was observed in 2021. India accounts for 2.4% of the global land surface but hosts more than 17.73% of the world population (ICAR- Data book 2022). Due to population pressure, India has faced several problems like food insecurity, low income, and vulnerability to environmental and biotic factors. The total geographical area of the country is 328.73 million ha (M ha), in which the net area sown is 139.35 M ha and the net irrigated area is 71.55 M ha. The cropping intensity of the country increased from 131% in 2000–01 to 144.9% in 2018–19. Fertilizer consumption increased from 26.75 million tons in 2015–16 to 32.54 million tons in 2020–21. As per the 2020–21 statistics, food grain, horticultural, nutri-cereals, pulses and oilseeds production were 308.6, 331, 51.15, 25.72 and 36.1 million tons (Mt),

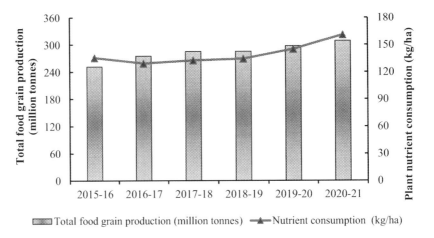

FIGURE 5.1 Total food grain production and nutrient (NPK) consumption of India.

respectively (ICAR-Data Book, 2022). Total food grain production increased from 251.6 (2015–16) to 308.6 Mt (2020–21) in conjunction with the nutrient consumption of 136 kg/ha (2015–16) to 161 kg/ha (2020–21); showing a direct cause-effect relation (Figure 5.1).

In the 2021–22 crop year, total food grain production of 315.7 Mt has been the highest till now, and the Government of India (GOI) has set a food grain production target of 328 Mt for the 2022–23 crop year which is 4% higher than production for 2020–21. Due to population pressure, by 2050 the total food grain production should be increased to 377 Mt (Srinivasarao et al., 2021a). As per the 2019–20 estimates, per capita availability of milk and eggs was 406 g/day and 86/annum, respectively. Globally, the country's milk and egg production has achieved first and third places. However, despite phenomenal growth in food grain production, the country still needs more food grains to feed the growing population, especially women and children.

In India, small and marginal farmers constituted 86.1% of the total landholdings in 2015–16 against 85.0% in 2010–11 and contributed about 60% of the total food grain production and over 50% of the country's fruits and vegetables production (Agricultural Census, 2015–16). The average size of the land holdings of small and marginal farmers is about 0.38 ha when compared to large farmers (17.37 ha) and the country's total average size of the land holding declined from 2.28 ha (1970–71) to 1.08 ha (ICAR-Data Book, 2022) which cannot efficiently contribute to employment and income of the nation. Nonetheless, small and marginal farmers are highly efficient in food production as compared to large farmers (Chand et al. 2011); their contribution to total food grain production and poverty reduction is rather high compared with that of large farmers (Gururaj et al. 2017).

Rainfed farming is complex, highly diverse and risk prone. In India, the rainfed area covers almost 55% of the net sown area (139.42 M ha) and about 61% of the farmers are cultivating crops under rainfed conditions. Thus, rainfed farming is crucial for food security and economy. At present, rainfed farming contributes around 40% of the total food grain production (85%, 83%, 70% and 65% of nutri-cereals, pulses, oilseeds, and cotton, respectively). Further, rainfed farming impacts the livelihoods of 80% of small and marginal farmers in the country (NRAA, 2022). Alfisols and Vertisols are predominant soil orders in the peninsular plateau of India. Aridisols exist in hot dry climates along with Entisols and Inceptisols. Alluvial (Inceptisols) soils are most dominant (93.1 M ha) when it comes to agriculture production and land use and management, followed by red (Alfisols, 79.7 M ha), black (Vertisols, 55.1 M ha), desert (Entisols, Aridisols, 26.2 M ha), and lateritic (Plinthic horizon, 17.9 M ha) soils. However, in the rainfed regions, Inceptisols are dominant followed by Entisols, Alfisols, Vertisols, Mixed soils, Aridisols, Mollisols, Ultisols, and Oxisols (Srinivasarao et al., 2013a). In general, most of the soils belonging to the rainfed areas are coarse

Water Storage in the Rest Zone

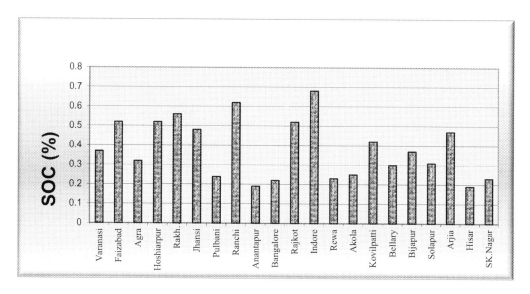

FIGURE 5.2 Organic carbon status of various soil types under diverse rainfed production systems of India

textured, except Vertisols. Accordingly, their ability to retain water and nutrients is low, crops cultivated on these soils are susceptible to drought and nutrient deficiencies compared to irrigated soils. Rainfed soils have multiple nutrient deficiencies. The SOC levels are as low as 5 g/kg, whereas the desired level should be more than 11 g/kg (Srinivasarao et al., 2009a, b; NRAA, 2022). Inherent fertility is low in most rainfed soils except in some Vertic group which are rich in bases. Low aggregate stability, insufficient SOM content and high erosion rates are serious problems in rainfed soils which decline the crop yields and threaten the nation's food security (Srinivasarao et al., 2009a, b, 2015, 2019b) (Figure 5.2). Nonetheless, large yield gaps remain in several crops grown under rainfed farming. There was a modest increase in productivity from 0.6 tons in the 1980s to 1.1 tons at present. In general, the productivity of rainfed areas is around 1.1 tons/ha, compared with 3 tons/ha in irrigated areas (NRAA 2022). Thus, there is a strong need for the uplifting of the rainfed farming community to meet the food demand of the country.

5.2 CHALLENGES IN ACHIEVING WATER PRODUCTIVITY IN RAINFED SOILS OF INDIA

India is the world 12th most water-stressed country. As per "Prioritization of Districts for Development Planning in India – A Composite Index Approach (2020)" report by NRAA, about 15 M ha of rainfed (arid, semi-arid, dry, sub-humid and humid regions) cropped area in the arid region receives less than 500 mm rainfall per annum. Another 15 M ha receives 500–750 mm, and about 42 M ha receives 750–1,100 mm. About 73 out of 127 agroclimatic zones in India are predominantly rainfed. This variability requires the identification of location-specific cultivation approaches. Climate change can influence rainfed farming. Long-term data indicate that rainfed areas experience 3–4 drought years per decade which are moderate to severe in intensity. Rainfed crops are often affected by irregular patterns of precipitation and high temperatures which result in frequent modifications in sowing time and harvesting dates.

The Indian economy is severely affected by variability in monsoons. Rainfall reduction causes drought, which is a climatic disaster affecting farming activities. Prolonged dry spells in rainfed regions initially hamper nutrient uptake, diminish crop yield, and can eventually collapse the country's agrarian economy (Singh et al., 2021). Rainfed areas are more vulnerable to extreme climatic events. Thus, water resource management and mitigation of hydrological disasters are highly dependent on climate variability in rainfed regions (Neog et al., 2019). Rainfall deficiency causes depletion of surface and ground water levels which leads to deleterious effects on cultivation practices because of inadequate availability of water for the crops, particularly during the critical stages of plant growth. Water scarcity impacts in rainfed regions are determined by climate change, soil and hydrological profiles, soil moisture availability, crop selection, agricultural operations, etc. Poor rainfall in sequential years adversely affects the ground water table by reducing the scope of surface and ground water recharge, and replenishment of soil moisture. Increased depletion rates of groundwater and declined recharge may deplete aquifers, adversely affect water quality (e.g., salt concentration, acidity etc.) and jeopardize soil's biological productivity. Too much exploitation of groundwater resources by tube wells has depleted this finite resource. In rainfed areas, crops are affected by delays in the onset of monsoons which narrows the sowing window. A wider sowing window reduces the risks of any major yield loss. Delayed monsoons reduce the crop growth duration and hamper agronomic yield. Delays in the onset of monsoon may result in poor crop growth and development due to lack of soil moisture during critical growth stages. Similarly, an early withdrawal of monsoon may adversely affect the reproductive stage of crop, which is crucial to attaining the desired yield. Mid-season drought occurring during the vegetative phase may result in stunted growth. It adversely affects crop yield when it occurs at flowering or early reproductive stage. Nonetheless, nutrient top-dressing, plant protection, and intercultural operations are most useful during mid-season drought. Among all types of droughts, terminal drought is critical to farmers due to contingency plans do not work out and grain yield is strongly related to water availability during the reproductive stage. Sometimes terminal drought leads to forced maturity which declines the quality of the yield (Srinivasarao et al., 2019a, b). Most of the rainfall is received during the south-west monsoon, and India is already witnessing extreme weather events. In the year 2021, rainfed regions of central and northwest India received 6% above normal rain throughout the country and received 1% below normal rainfall despite the longer duration. Then there was month-to-month variation in the 2021 monsoon (Nayar, 2021). The changing pattern of monsoon in rainfed regions of the country has taken a toll on the lives of people, livestock and crops due to the high intensity of rains in short durations. Due to undulating topography and the low moisture retention capacity of the soil, a major portion of the rainwater is lost through runoff, accelerating losses by soil erosion. Deficiency/uncertainty in rainfall of high intensity causes excessive loss of soil through erosion which leads to the loss of carbon and nutrients. Insufficient soil moisture is left in the profile to support plant growth and grain production from rainfed drylands. Due to erratic behavior and inappropriate rainfall distribution, agriculture is risky, especially in soils of low carrying capacity like rainfed drylands, where most of the resource poor smallholder farmers exist. There occurs a widespread problem of the lack of resources and tools which are inefficient and are the cause of low productivity.

5.3 WATER RESOURCES AVAILABILITY IN INDIAN RAINFED DRYLANDS

The principal source of water in rainfed regions of India is precipitation (rainfall and snow stocks from the Himalayan mountains). Out of total rainfall received only a part of it is stored as groundwater and the remaining is lost as runoff and evaporation. On average, India receives a total annual precipitation of around 4,000 km^3. Out of this, the majority share of rainfall is from the southwest monsoon, and it contributes 3,000 km^3. The rainfed regions in India comprise predominantly arid, semi-arid and dry sub-humid (Figure 5.3) with an annual rainfall ranging from 100 to 1,000 mm whereas, moist sub-humid, humid and per humid regions receive rainfall ranging from 1,000 to >2,500 mm annually (Sharma and Kumar, 2014).

The distribution of water resources in India is highly uneven and skewed and it varies from dry and semi-arid Rajasthan in western India to wet and water-excess states of West Bengal and Orissa (Kumar et al., 2005). Yet, there has been a paradigm shift in water resources used in India since the 1950s from communities (tanks and small water structures) to government (major and medium

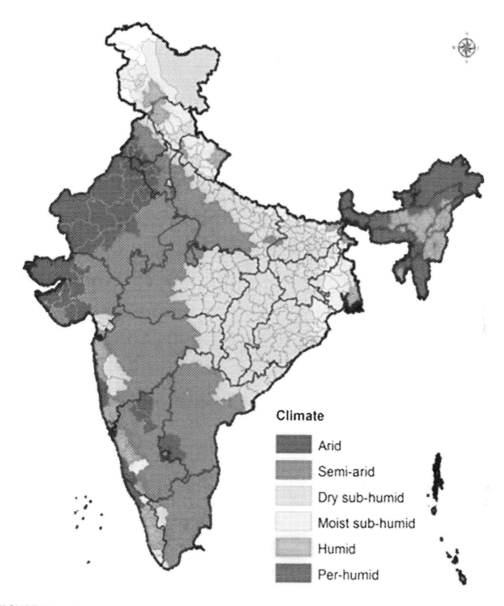

FIGURE 5.3 Predominant rainfed dryland regions in India (arid, semi- arid, and dry sub-humid). Source: Adopted from Srinivasarao et al. (2020a).

irrigation projects), and the private domain (groundwater). Groundwater is the source for 70% of the irrigation and 80% of the drinking water. The average annual per capita availability of water for a major part of India is around 2,251 m^3/year. About 40% of the population in India has 70% of India's water resources while the remaining 60% has only 30% indicating extreme disparities in distribution (Majumdar, 2008). Out of the total annual precipitation of around 3,880 BCM (10^9 m^3) in India, the groundwater and surface water are about 184.56 BCM, comprising 690 BCM (37%) of surface water, and 436 BCM of groundwater (CWC, 2022). By 2050, the total per annum consumption of national water may exceed the utilizable water resources, unless significant changes occur through potential water management. Thus, available water resources need judicious management through the artificial recharge of groundwater, desalination of brackish water, rainwater harvesting etc.

5.3.1 Surface Water Resources

India's average annual surface runoff through rainfall and snowmelt is estimated to be 1,869 km³. Yet, only about 213 km³ (11.4%) of surface water can be harnessed as 90% of the annual flow of the Himalayan rivers occurs for only 4 months period and there is no potential to capture due to limited suitable storage reservoir sites. Soon (by 2050), this capacity may double to about 21%. The utilization in the peninsular basins (e.g., Godavari, Krishna, Cauvery, Mahanadi, Tapti, and Narmada) is more than 70% of the present capacity. The Krishna River basin has the highest storage capacity (49 km³) and can store 64% of the mean annual river flow, or about 220 days of average flow (CWC, 2022).

5.3.2 Ground Water Resources

The annual replenishable groundwater resource is contributed by rainfall (280 km³) and other sources (156.1 km³) from canal seepage return flow from irrigation, seepage from water bodies, and artificial recharge. Rainfall contributes about 67% to the country's annual replenishable groundwater. The central groundwater board has drafted a model bill for ensuring sustainable and equitable development and use of groundwater resources, that can be adopted by different states of India (Srinivasarao et al., 2015).

5.3.3 Rainfall Trends in Rainfed Drylands

Rainfall is the ultimate source of water in drylands (Srinivasarao et al., 2014a, b). Coping with the extreme variability in rainfall, high-intensity storms, and high frequency of dry spells are the key challenges for rainfed agriculture. The water retention and length of crop growing period in diverse production systems under varied soil types and climatic conditions of India is presented in Table 5.1.

5.3.4 Rainwater Management

5.3.4.1 In-Situ Moisture Conservation

This technique is more practical and feasible as a large part of rainfed agriculture consists of marginal and smallholdings. Hence, location-specific in-situ moisture conservation practices were developed based on rainfall, soil types and overall agro-ecological conditions (Table 5.2) (Srinivasarao et al., 2014b). Measures such as ridge/furrow, sowing across the slope and paired row sowing are observed to be important water conservation measures for effective water conservation besides draining out the excess rainwater.

5.3.4.2 Ex-Situ Rainwater Harvesting through Farm Ponds

India receives about 4,000 km³ rainfall per annum, of which, about 1,869 km³ flows as run-off every year in the country. Due to geographical limitations, we are unable to utilize a good amount of surface water. Farm ponds constructed in coarse-textured soils (Alfisols) require lining to minimize seepage, but clayey soils (Vertisols) having negligible seepage may not require lining. The unlined farm ponds which have higher seepage can be utilized to recharge the aquifers (Srinivasarao et al., 2017). In this context, the importance of rainwater harvesting has increased in recent years because of the depletion of groundwater levels and increased rainfall variability. Watershed management is the flagship program in India to enhance water resource availability and improve agricultural production in a sustainable manner (Ravindra Chary and Gopinath, 2022). As per the projections, about 27.5 M ha of Indian rainfed area can contribute an amount of 114 km³ of water for water harvesting that will be adequate to supply water for one supplementary irrigation of 100 mm depth to 20 M ha during drought years and 25 M ha during normal years (Sharma et al., 2010; Srinivasarao and Gopinath, 2016). However, harvesting the runoff water and storing it in farm ponds in rainfed

TABLE 5.1
Mean Annual Rainfall, Climate, Soil Type, Production System Length of Growing Period and Water Retention (% wt. Basis) (Mean of Seven Layers of Soil Profile) of Rainfed Drylands in India

Location/State	Mean Annual Rainfall (mm)	Climate	Soil Type	Production System	Water Retention at (bar) 1/3 (%)	15 (%)	Available (%)	Growing Period (Days)
Hissar, Haryana	412	Arid	Alluvial Deep -Aridisols	Pearl millet	15.9	5.7	10.3	60–90
Bellary Karnataka	500	Arid	Black deep Vertisols	Rabi sorghum	35.8	23.0	12.8	90–120
Sardar Krushinagar Gujarat	550	Arid	Desert Deep - Aridisols	Pearl millet	3.1	1.7	1.4	60–90
Anantapur, Andhra Pradesh	560	Arid	Red Shallow- Alfisols	Groundnut	18.3	8.3	10.0	90–120
Rewa, Madhya Pradesh	590	Semi - arid	Black medium deep Vertisols	Soybean	29.6	16.0	13.6	150
Rajkot, Gujarat	615	Arid	Black Deep -Vertisols	Groundnut	35.1	23.5	11.6	60–90
Arija, Rajasthan	656	Semi-arid	Black shallow deep Vertsols	Maize	6.6	2.5	4.0	90–120
Agra, Uttar Pradesh	665	Semi-arid	Alluvial deep Inceptisol	Pearl millet	21.1	8.4	13.0	90–120
Bijapur Karnataka	680	Semi-arid	Black medium deep Vertisols	Rabi sorghum	45.7	24.7	21.0	90–120
Solapur, Maharashtra	723	Semi-arid	Black medium deep vertic/Vertisols	Rabi sorghum	42.6	30.5	12.1	90–120
Kovilpatti Tamilnadu	743	Semi-arid	Black deep Vertisols	Cotton	40.4	26.7	13.8	120
Akola, Maharashtra	825	Semi-arid	Black medium deep vertic/Vertisols	Cotton	36.8	25.3	11.5	120–150
Bengaluru, Karnataka	926	Semi-arid	Red deep Alfisols	Finger millet	14.1	8.6	5.5	120–150
Indore Madhya Pradesh	944	Semi-arid	Black deep Vertisols	Soybean	33.0	19.9	13.1	120
Ballowal Sunkari, Punjab	1,000	Semi-arid	Alluvial deep Inceptisol	Maize	9.3	4.1	5.2	120–150
Jhansi, Uttar Pradesh	1,017	Semi-arid	Alluvial deep Inceptisol	Rabi Sorghum	20.3	3.8	16.4	120
Faizabad, Uttar Pradesh	1,057	Sub humid	Alluvial deep Inceptisol	Rice	24.6	10.4	14.2	90–120
Varnasi, Uttar Pradesh	1,080	Sub humid	Alluvial deep Inceptisol	Rice	21.1	7.5	13.6	150–180
Rakh Dhinsar, J&K	1,180	Sub humid	Alluvial deep Inceptisol	Maize	5.1	2.0	3.0	150–210
Ranchi, Jharkhand	1,299	Sub humid	Red shallow Alfisols	Rice	15.8	10.9	4.9	150–180
Phulbani, Orissa	1,378	Sub humid	Red yellow deep Alfisols	Rice	18.2	9.8	8.4	180–210

Source: Complied and modified from Srinivasarao et al. (2009a, 2013a).

TABLE 5.2
Recommended Soil Moisture Conservation Measures/Practices Based on Rainfall Received in Rainfed Regions of India

Mean Annual Rainfall (mm) Received in Rainfed Region			
< 500 mm	500–750	750–1,000	>1,000
• Conservation furrows	• Conservation furrows	• Broad bed furrow (for Vertisols)	• Broad bed furrow (for Vertisols)
• Contour bunds	• Contour cultivation	• Conservation furrows	• Field bunds
• Contour cultivation	• Ridging	• Sowing across slopes	• Vegetative bunds
• Ridging	• Sowing across slopes	• Tillage	• Graded bunds
• Sowing across slopes	• Scoops	• Lock and spill drains	• Level terrace
• Mulching	• Tied ridges	• Small basins	
• Off season tillage	• Mulching	• Field bunds	
• Scoops	• Zing terrace	• Vegetative bunds	
• Tied ridges	• Off season tillage	• Graded bunds	
• Inter row water harvesting systems	• Broad bed furrow	• Zing terrace	
• Small basins	• Inter row water harvesting systems		
• Field bunds	• Small basins		
	• Field bunds		

areas depends on the amount of rainfall received per annum, topography, and soil type. The harvested water during the rainy season can be utilized for supplemental irrigation during dry spells coinciding with critical growth stages in the rainy season or for the establishment of winter crops. Half of Indian agriculture is rain-dependent and therefore, in-situ and ex-situ rainwater conservation in terms of farm or community ponds is highly prioritized (Srinivasarao et al., 2019a, b).

5.4 IMPROVING SOIL MOISTURE STORAGE THROUGH SOC BUILDUP

SOM has great potential for water retention, thus becoming a key factor in contributing to available water holding capacity (AWHC) (Figure 5.4). SOM has a practical implication for water management in agriculture. Increasing 1% SOM increases the AWHC range to varying degrees (Hudson, 1994). SOM could potentially contribute to available water by 2.2%–12.5% (Emerson, 1995). Soils from the USDA Natural Resources Conservation Service (NRCS) soil survey reported that 1% increase in SOM resulted in 2% to >5% increase in AWHC (soil water retention at −1,500 and −33 kPa), however, the magnitude of increase depended on soil textural properties (Olness and Archer, 2005). Several old studies reported that enhancement of AWHC under SOM increments was observed to be more prominent in lighter than in heavy soils (Rawls et al., 2003). In the case of fine-textured soils, SOM improves soil structure and aggregation and decreases bulk density (BD) that improves water storage closer to field capacity than at the wilting point (Rawls et al., 2003). Libohova et al. (2018) evaluated the effect of SOM on AWHC using the National Cooperative Soil Survey (NCSS) database. In their study, it was observed that AWHC correlated positively with silt content ($r=0.56$). Decreased BD values increased the AWHC ($r=-0.34$), but this relation was observed to be extremely variable with the SOM and soil textural properties. Porosity is another important factor that governs the relationship between SOC and soil water storage. A unit increase of SOC would enable a large increase in porosity (0.2–5 and 480–720 µm diameter classes) that ultimately enhances the AWHC and unsaturated hydraulic conductivity (Fukumasu et al., 2022). A positive correlation recorded between SOC and mean weight diameter of water-stable aggregates may be ascribed to the fact that decreased water wettability of aggregates with larger SOC content (Chenu et al., 2000). Also, the variations between SOC and soil water storage (SWS) are more highly influenced by the land use system than the local environmental conditions. Chen et al.

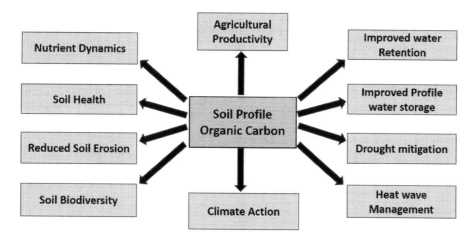

FIGURE 5.4 Critical role of SOC in improved profile water storage and drought mitigation.

(2022) observed that the interaction of SOC and SWS was maximum at surface depth (0–20 cm) and the topography was a predominant factor that influenced the SOC and SWS in the deep soil layers (100–200 cm). Soil texture was a stable driving factor which influenced these in the whole soil profile (0–200 cm). Some of the recent studies (Zhou et al., 2020) also reported that factors such as soil texture, BD, SOM, and plant roots and litters determine the soil water retention.

In addition, while positive correlations between SOC and AWHC were reported by Olness and Archer (2005), recent global and continental-scale studies found limited effects of SOC on AWHC(Minasny and McBratney, 2018). The relationships between SOC and porosity have implications for soil water dynamics. A 10 g/kg (1%) increase in SOC increased the AWHC of about 3 mm to 100 mm^{-1} soil depth (Fukumasu et al., 2022) and this was larger than the results of a meta-analysis (1.1–1.9 mm 100 mm^{-1} soil depth) (Minasny and McBratney 2018). Nevertheless, the farmland practices like crop residue management and tillage practices also influence the SOC, SWS and soil structure (Zhao et al., 2020). Amalgamation of improved SOC and SWS and soil structure should be promoted to ensure sustainable food production and national food security. A good soil structure generally has better water-holding capacities, hydraulic conductivity, and adequate aeration which are extremely helpful for better plant growth and development (Karami et al., 2012). The critical role of SOC in improved profile water storage and drought mitigation is illustrated in Figure 5.4.

Hence, it is paramount to increase the SOC storage, in order to significantly increase the soil water storage in the soil profile. However, there are several important factors that are highly influencing the SOC storage (Figure 5.5) and therefore, it is important to critically consider all these factors to efficiently manage and enhance the SOC in the agriculture system.

5.5 SOIL ORGANIC MATTER AND KEY CONTRIBUTING FACTORS

SOM has an impact on several factors that are essential for maintaining soil productivity. As the "lifeline" of the soil, it significantly affects qualities like aggregate stability, water-holding capacity, buffering capacity, cation exchange capacity (CEC), acidification, solidification, solidification, solidification, sodification, salinization, etc. However, soils in rainfed regions are highly diverse and include Vertisols and Vertic sub-groups, Inceptisols, Entisols, Alfisols, Oxisols, Aridisols, etc. in different agroecological zones of India. The main constraints on soil health in rainfed regions are moisture stress, unfavorable permeability, nutrient P-fixation, poor nutrient retention, erosion, and slope. In addition, there is a significant variation in annual rainfall (between 400 and 1,500 mm). In rainfed agroecological regions, the SOM status of soil in various locations was found to be low which is an emerging issue (Srinivasarao et al., 2009b). In most cases, the SOM concentration

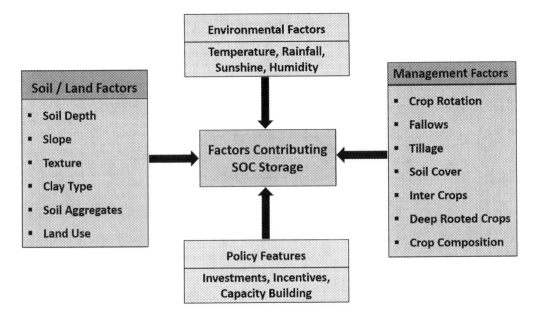

FIGURE 5.5 Factors contributing to SOC storage in particular agro-ecosystem.

measured under various crop production techniques in rainfed (non-irrigated) agro-ecological zones was estimated to be <5 g/kg soil. In addition, surface soils (0–30 cm) in many Indian regions under rainfed crop production systems are significantly low in SOC, which may be as low as 0.15% in some of the rainfed agroecological locations (Srinivasarao, 2011a; Srinivasarao and Vittal, 2007). Therefore, the reduction in SOM content can be slowed and crop yield can be greatly increased by adopting appropriate soil-crop management strategies. Overall, SOC building up in dryland agriculture depends on soil factors (clay, depth of soil, slope), environment (temperature, rainfall, humidity, and sunshine) as well as the management practices in agriculture (Jayaraman et al., 2021; Ramesh et al., 2019).

5.6 SOIL WATER STORAGE THROUGH BUILDING SOM

Soils in dry regions are degraded and have low quantities of SOM, which is vital for key soil processes: nutrient dynamics, water interactions, and biological and physical soil health. Due to the quick oxidation process that occurs in the dry regions, most dryland soils have low quantities of SOM (Srinivasarao, 2011a; Srinivasarao et al., 2008, 2009b; Srinivasarao and Vittal, 2007). The limited biomass production and loss of the top layer of soil rich in SOM during heavy rains are two important factors that govern the SOM status in tropical regions. Low water infiltration and porosity affect local and regional water cycles, agroecosystem resilience, plant productivity, and global carbon cycles, which have significant negative implications on water use efficiency (Wani et al., 2009). In drylands, where there are climatic extremes and unpredictability, SOM performs a critical role in maintaining optimum WHC and infiltration rates, hence restricting yield losses. To prevent further harm and improve soil health, carbon sequestration, greenhouse gas (GHG) mitigation, and the country's food security, management strategies that enhance SOM and its maintenance at a threshold level are urgently needed at this hour (Srinivasarao, 2011a). Improved management suggests sustainable productivity along with improved soil quality (physical, chemical, and biological parameters), increased carbon sequestration, and other benefits as shown by long-term research at the International Crops Research Institute for the Semi-Arid Tropics (ICRISAT), India (Wani et al., 2003). To maintain or boost SOM, amendments such as farmyard manure, plant residues and farm compost must be applied continuously. However, it is challenging to apply these to soil at the appropriate rates due to a lack of organic manures and enormous uses of farm waste as feed and fuel. There are several technologies that significantly improve the SOC and water storage in the soil profile.

5.6.1 MULCH WITH MANURING OF SOILS

Evaporation in mulched soil is minimal when compared to bare soil. Total porosity, texture, and soil structure are the critical factors affecting the overall and accessible soil moisture storage capacity (Prem et al., 2020). It can be achieved by using organic mulching material. Soil wetting depth increases with an increase in mulching materials. Mulching with straw has the capacity to store more soil water from light precipitation (Ji and Unger, 2001). Using straw mulch can increase soil moisture storage by 55% compared to that under control (Sood and Sharma, 1996). The application of wheat residue mulch at a rate of 6,730 kg/ha has been shown to improve AWHC to a depth of 1.5 m of soil in comparison to bare soil (Black, 1973). In Rabi (winter) sorghum, total soil moisture loss from planting to harvest was 9.14%, 11.33%, and 11.92%, respectively, at 15, 30, and 45 cm of soil depth. The percentage increases in soil moisture for the sugarcane bagasse mulch, soybean straw mulch, wheat straw mulch, and intercultural operation were 7.54, 12.26, 17.81, and 28.19, respectively over the control (without mulch). Average soil temperatures were 19.58°C, 20.04°C, 20.37°C, 20.73°C, and 21.33°C, respectively, in the intercultural operation, sugarcane waste mulch, wheat straw mulch, soybean straw mulch, and control (without mulch) (Ranjan et al., 2017). In contrast, fields with mulched grass have the lowest observed soil temperature.

Agricultural wastes used as mulching material (e.g., grass clippings, bark chips, wheat or paddy straw, rice hulls, plant dry leaves, compost, and sawdust) are examples of materials with an organic origin that can decay naturally. It decomposes over time, increasing the soil's ability to store water (Unger, 1974). As it decomposes, it also adds nutrients to the soil. In addition, it indirectly increases water use efficiency (Ossom et al., 2001). There are about 38 trillion Mt of organic waste produced annually due to various human activities across the world. In India alone, there are about 600–700 Mt of agricultural waste each year (along with 272 Mt of crop residues), but most of these materials are not utilized (Suthar, 2009).

5.6.2 DRIP IRRIGATION WITH ORGANIC MULCH

Drip or trickle irrigation is the frequent, slow distribution of water to soils by mechanical emitters or applicators located at certain points along delivery lines. A small amount of water can discharge from the emitters because they dissipate the pressure from the distribution system via orifices, vortices, and long or tortuous flow routes. Sub-surface drip irrigation is a unique and efficient method of irrigation which provides flexible and light irrigation, especially in dryland or arid conditions. It eliminates surface water evaporation and avoids potential surface run-off thereby improving soil moisture availability. With the use of this technique, a small amount of soil may be kept moist by applying small amounts of water frequently (Du et al., 2015), thereby decreasing the drainage of water from the root zone soil and restricting the rooting zone to moist soil (Du et al., 2010). Subsurface drip irrigation significantly reduces evaporation from topsoil in comparison to surface drip irrigation and enhances irrigation WUE (Lamm and Trooien, 2003). Micro irrigation like drip and sprinkler along with organic mulches improve SOC levels and WUE further. The frequency of critical irrigation required in rainfed dryland crops is reduced by one-third with micro irrigation with organic mulches.

5.6.3 GREEN MANURING FOR SOC AND MOISTURE STORAGE

Regardless of the cropping pattern and the climate, it is generally known that adding biomass from green manuring crops to soil improves its SOC and other nutrients, particularly nitrogen (N),

(Walia and Kler, 2007). Cluster beans, cowpea, green gram, sesbania, sunhemp, dhaincha, etc. are common green manuring crops in various rainfed ecosystems since they are leguminous in nature and have a high ability to fix atmospheric N in the root nodules. Considerable research on this topic, done under the umbrella of irrigated ecology (Rao et al., 2017), showed that various green manuring crops, especially dhaincha and sun hemp, had a significant ability to augment SOC content and enhance other related soil functions. Green manuring with cowpea, green gram, and sun hemp crops significantly increased the activities of various enzymes (dehydrogenase, and phosphatase activity) when compared to the control (Yogesh and Hiremath, 2014; Mrunalini et al., 2022).

Dryland soils with light sandy loam are ideal for growing green manuring crops viz., lobia, guar, green gram, and black gram (Meena, 2019). Cowpea and sun hemp green manure (GM) plots recorded significantly higher dehydrogenase activity (5.58 and 4.79 µg TPF/g of soil/day, respectively) followed by the green gram GM (3.55 µg TPF/g of soil/day) as compared to fallow land (2.78 µg TPF/g of soil/day). The higher dehydrogenase activity in cowpea and sun hemp was due to the addition of a higher quantity of easily mineralizable phytomass and biomass when compared to green gram (GM). In addition, the former GM crops had a higher leaf-to-stem ratio, which may have aided in the soil microbes' quicker multiplication. The cowpea GM plots had higher levels of phosphatase activity (40.02 mg of PNP/g of soil/h) than the sun hemp and green gram GM plots (34.70 mg of PNP/g of soil/h, 31.45 mg of PNP/g of soil/h respectively). In fallow, phosphatase activity was noticeably reduced (Yogesh and Hiremath, 2014). The plants that produce green leaf manure acclimatize well to diverse soil types and are crucial for increasing the fertility and productivity of the soil. The leaves of some tree species, including *Azadirachta indica*, *Pongamia glabra*, *Delonix regia*, and *Peltophorum ferrugenum*, can be utilized as mulching material in the rainfed (non-irrigated) agroecology as a source of SOC. These species are easily cultivated in different parts of India (Srinivasarao, 2011a; Vineela et al., 2008). Tree green leaf manuring with the lopping of *Gliricidia sepium* has been proven to be a cost-effective and environment-friendly method (Figure 5.6). Another key characteristic of Gliricidia is its ease of growth in a variety of soil types, including acidic, dry, and even marginally damaged fields. Adopting green leaf manuring is one of the key strategies for improving SOC content. Interestingly, when biomass is used as mulching material, it aids in the conservation of moisture and lowers soil erosion losses.

5.6.4 Intercropping as Efficiency Production System under Droughts

Numerous studies across the world have shown how intercropping can increase the effectiveness of resource utilization (Martin-Guay et al., 2018). Intercropping is considered a vital strategy for ensuring food security, diversifying cropping systems, advancing agricultural progress in a sustainable manner, and making the most of smallholder farms' scarce labor resources (Ouma and Jeruto, 2010). Intercropping has the potential to significantly boost the primary yield of land per unit area by raising the use efficiency of resources like water, light, heat, and fertilizer. This would significantly improve global food security (Foley et al., 2011; Jensen et al., 2015) and raise water use efficiency and economic profit for farmers (Yin et al., 2018). The goal of intercropping research is to increase resource use efficiency as agricultural production gradually shifts from a resource-intensive to a technology-efficient mode. The best ways to optimize the benefits of intercropping systems and generate stable yields while conserving water should continue to be the focus of research.

Intercropping can use less water if the intercrop strips are laid out optimally. As an illustration, strip intercropping of maize (*Zea mays* L.) and pea (*Pisum sativum* L.) (4:4 model, four rows of maize and four rows of pea) decreased water usage by 10.2%–13.7% in comparison to solo cropping, (Mao et al., 2012). When using an alternative irrigation technique to flood irrigation, intercropping's water consumption was decreased by 16.1% and 15.3% at high and low water supply levels, respectively (Yang et al., 2011). Water consumption is reduced by 4.4%–8.5% (save 25.1–70.9 mm of soil water) and irrigation water is saved by 15% with limited water supply after the booting stage of wheat (*Triticum aestivum* L.) in wheat-maize intercropping without significantly reducing crop

FIGURE 5.6 Gliricidia nursery for green manuring to improve SOC and moisture storage of soil.

productivity (Wang et al., 2015). Because wheat and maize are intercropped, the amount of evaporation from the soil in the maize strips is decreased because wheat is more competitive than maize throughout the co-growth period and competes for soil water from the maize strips (Yin et al., 2018). Contrarily, intercropped maize receives compensatory soil water from the wheat strips after the wheat harvest, reducing soil evaporation in wheat strips (Yin et al., 2018). The overlapping of root spatial distribution is a natural result of diverse root morphologies of the numerous intercrops (Gao et al., 2010) Crop species with rapid water absorption and rapid growth have an advantage over those with efficient soil water usage but slow growth in the zone where crop roots overlap (Bramley et al., 2007). Legumes can obtain the water below the root zone of maize and increase the water supply of maize by water lifting during the intercropping between maize and legumes (Sekiya and Yano, 2004). The ability of intercropping systems to use water depends on soil water availability. Cowpea (*Vigna unguiculata*) and maize intercropping have higher WUE than the comparable solo cropping when soil water availability is high; however, when soil water availability is low, intercropping has higher WUE than sole cowpea but lower WUE than sole maize (Droppelmann et al., 2000).

5.6.5 CONTOUR FARMING FOR EROSION CONTROL

One of the biggest challenges to efficient land management is limiting surface runoff and sediment loss. Soil loss along with carbon and nutrients are key determinants for productivity loss in rainfed drylands. To enhance crop production while also efficiently protecting the soil, practical strategies are required. Terracing and contouring are two such techniques that aid with this. Today's improvements in precision conservation allow for better terrace location and field-based contour grade maintenance. Contour farming is one of the simplest and most efficient sustainable farming techniques used to minimize erosion. It is also known as strip farming, contour cropping, contour cultivation, terrace farming, or terracing (Dorren and Rey, 2004). The effects of erosion are dramatically reduced when agricultural techniques are carried out along the contours of a sloping land area. According to some estimates, contour farming can minimize erosion by up to 50% (Babubhai, 2018). Terracing is used as one of the ways of soil conservation because of its ability to control erosion. Reducing the chance of environmental damage through soil loss and land deformation is also advisable. Organic matter in soil increases because of contour farming (Abiye, 2022). This happens because of the conservative method, which enables organic matter to build up on (and in) soil without being lost to erosion or leaching. Contour cropping improves soil health by enhancing carbon concentration and lowering soil disturbance. Moreover, the regional water quality is also improved because contour farming reduces the runoff of pollutants into water bodies (Gathagu et al., 2018; Srinivasarao et al., 2019a).

5.6.6 CONSERVATION AGRICULTURE FOR CARBON STORAGE AND WATER USE EFFICIENCY

Conservation agriculture (CA) with zero-tillage (ZT) involves three principles that include no-tillage, residue retention and diversified crop rotations. Crop residue biomass is rich in carbon. Returning this C-rich biomass into soil or placing it on the surface would provide multiple benefits like soil surface protection, temperature moderation, restricting the evaporation loss at the initial stages of the farming etc. Later it will be converted into SOC, then plays a greater role in maintaining soil health like activating the soil microbial activity, reducing BD, soil water storage, enhancing the plant available nutrients etc. Indeed, there are several benefits from the adoption of CA in the existing farming practice (Figure 5.7). CA practice plays effectively in sustainable agriculture, helps in mitigating GHG emissions, protects the soil from erosion, and improves SOC and soil water storage (Singh et al., 2020). A large data meta-analysis on the impact of the adoption of CA practices on soil C stock revealed the increase in soil C stocks at the surface soil depth (0–15 cm) with a potential of 15 Mg C/ha (Jat et al., 2022). The maximum total organic carbon (TOC) and its fractions at the surface layers in the soils under ZT practice are due to the crop residue retention with higher biomass residue and a slow decomposition rate due to minimum soil disturbance (Roy et al., 2022). Sinha et al. (2019) to evaluate the effect of adoption of CA on the key soil parameters in the Indo-Gangetic plains of West Bengal. They observed that the soils under ZT had greater amounts of SOC among all sites studied as compared to conventional tillage. Enhancement of SOC in soil certainly has a positive impact on water storage. Residue retention on the soil surface or return to the

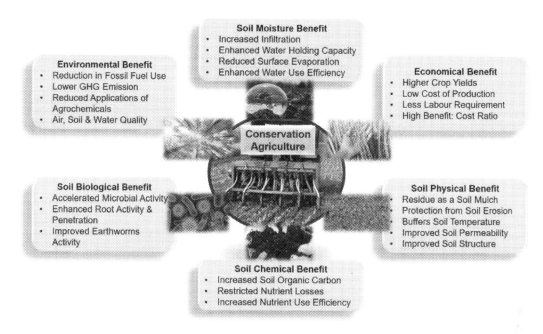

FIGURE 5.7 Multi-benefits obtained from the adoption of Conservation Agriculture.

soil significantly increased soil water storage (SWS) and decreased soil water consumption (Zhao et al., 2022). About 11.7% increments in SWS are noticed in case of residue return in the rainy years when compared with that under no residue in dry year (Zhao et al., 2022). Thus, leaving the crop residues on the soil surface reduces the C footprints, and helps in achieving sustainability from an environmental perspective (Rakesh et al., 2021). Adoption of ZT in the wheat crop helped in conserving more soil residual moisture that supported better wheat growth than that without residue retention (Sahoo et al., 2022).

5.6.7 INTEGRATED NUTRIENT MANAGEMENT (INM): WIN-WIN FOR SOC AND WATER STORAGE IN SOIL PROFILE

Though organic fertilizers are slow in releasing plant available nutrients, they have a greater residual effect on subsequent crops; combining these with inorganic fertilizers would provide multiple benefits to farmers. Recently, farmers and agriculture specialists changing their mindset to substitute a part of inorganic fertilizers with eco-friendly, sustainable, and economic natural nutrients like farmyard manures (FYM), crop residues, green manures, vermicompost, soil amendments, and agroforestry (Selim, 2018). The key goals of integrated nutrient management (INM) are holistic and balanced management of existing natural resources with fertilizers, optimizing nutrient-use efficiency, minimizing nutrient losses, enhancing WUE (Wu and Ma, 2015), sustainability, grain superiority, and high economic returns (Selim, 2020). Using biofertilizers along with organic and inorganic fertilizers helps in enhancing SOC content, aggregate stability, and moisture-retention capacity (Kumari et al., 2017).

Srinivasarao et al. (2011b, 2012a,–f, 2020b, 2021b) have investigated the impact of various INM treatments on the SOC stocks under different cropping systems and showed that INM practice in pearl-millet cropping improved the profile SOC stock by 110%–112% over the 100% inorganic treatment. The increase in SOC stock was 32% in pearl millet-cluster bean-castor, 43%–47%, in sole groundnut, 16%–22% in finger millet, 16%–22%, in post monsoon sorghum, 9.6%–9.7%, in groundnut–finger millet, 7.2%–6%, in soybean-safflower, 3.5%, in rice-lentil, and 3.5% in sole groundnut in comparison with the 100% inorganic treatment. The improved productivity of dryland crops with various location-specific INM practices is due to improved water storage in the soil profile with enhanced SOC content along with soil fertility improvement. The enhanced storage of water due to SOC enrichment was due to higher water retention capacity parameters of soil estimated during the mid-season droughts. The improved water storage estimated with enhanced SOC content was equivalent to the water required for one or two critical irrigation micro irrigation system (Srinivasarao et al., 2014a). The relation of water conserving, or water use is largely dependent on the soil structure, quality, and quantity of SOM as it is considered the primary control among different soil properties. Thus SOC becomes a basis of soil physical (structure), chemical (plant nutrients) and biological (microbial population) attributes. Brady and Weil (2005) observed that 1% of SOM (i.e., 1 g SOM per 100 kg of dry soil) has the potential to hold 30 kg of water. Many investigators have also reported similar values (Bastida et al., 2008) and highlighted the importance of the relationship between soil structure and soil productivity (Yadav and Meena, 2009). Thus, improvement of soil structure through improvement in SOC facilitates the water absorption capacity of the soil. The INM can significantly improve the SOC which is and indictor of soil fertility and soil structure along with its greater role in nutrient availability, acceleration of microbial population and its activity, reducing BD, cation-exchange capacity, soil pH, and soil aeration (Mohammad et al., 2012). Such an improvement in overall soil quality leads to improvement in water infiltration, increase in soil field capacity, and WHC that ultimately results in achieving economic water use and WUE despite attaining the desirable crop yields (Nazli et al., 2015).

5.6.8 Organic Amendments

Most of the organic amendments are bulky in nature, eco-friendly and cost-effective. Such amendments have a greater potential in holding soil moisture and help in saving time, energy, and money in crop production. Improved water storage as contributed by the application of OM or FYM can solve the problems of water limitations during the crop growth and development. Several researchers have documented the percent increase in SOM that can hold the percent amount of water in the soil. A field experiment revealed that for every 1% increment in SOM can hold 16,500 gallons of plant-available water per acre of soil; that is equal to 1.5 quarts of water per cubic foot of soil (Gould, 2015). Application of vermicompost in agricultural fields results in improving crop productivity and curtailing the water stress problems as the vermicomposts are porous with excellent water storage capacity (Abaranji et al., 2021). The use of vermicompost in agriculture helps in improving soil physical structure including some macro and micronutrients (Azarmi et al., 2008); while combining this with other organic or inorganic fertilizers effectively increases the growth and yield of crops (Javaad and Panwar, 2013). The usage of beneficial microbes in farming started during the 1960s. In recent decades, bio-fertilizers played a commendable role in enhancing agricultural productivity through biotic and abiotic stress management such as water and nutrient deficiency and heavy metal contamination (Wu et al., 2005). However, combining bio-fertilizers with organic amendments would further enhance its performance in overall soil quality buildup. The application of poultry manure and filter mud cake resulted in improving the status of N, P and K but the inclusion of bio-fertilizers significantly increased the decomposition of the organic waste (Bakr, 2016) which is the key mechanism behind the release of plant available nutrients. Application of *Azolla-caroliniana* compost significantly improved soil carbon sequestration by 1.1–1.4 folds and reduced the global warming potential by 1.2–1.4 folds when compared to cow dung and green manure treatments. The addition of *Azolla-caroliniana* and rice husk dust was also observed to suppress methane (CH_4^+) by 30%–36% through the enhancement of porosity, C-storage, and recalcitrant C fractions in soil (Bharali et al., 2021). Input of compost has some advantages because of its rich C biomass which is the key element used as an energy for their growth promotion activities (Ullah et al., 2021); consequently, it helps in improving soil quality and WHC (Alzamel et al., 2022). Compost coupled with bio-fertilizer and filter mud cake are best amendments for improving soil porosity, soil fertility and harvesting healthy crops. The addition of filter mud cake into the soil also significantly enhanced nutrient and water movement (Alzamel et al., 2022).

5.6.9 Biochar

The use of biochar is getting more attention in India for its role in improving SOC content and soil health (Samra and Srinivasarao, 2021). Biochar is a C - rich product produced from crop residues through controlled pyrolysis at temperatures of 400°C–600°C under anaerobic conditions (Sinha et al., 2021). The concentration of total C and total N in different biochar materials shows that agro-forestry-based biochar has higher levels (Wood and Eucalyptus) compared to paper mill, green waste, and poultry litter (Srinivasarao et al., 2013b). Thus, biochar can be effectively used for agricultural purposes (Lehmann and Joseph, 2009). It helps in soil carbon sequestration, reduces farm waste, and improves soil quality (Srinivasarao et al., 2012g, 2013b). It has highly concentrated carbon chains, hydrophilic characteristics, and high surface area (Qian et al., 2020; Wang et al., 2019) as it is produced by thermal decomposition (Beesley and Marmiroli, 2011). Biochar has numerous advantages in agriculture; most importantly increasing carbon sequestration, improving soil WHC, increasing drought mitigation, accelerating the microbial population and their activity, leading to an overall improvement of soil physical condition, and increasing crop yields (Lal, 2008).

Increased porosity of biochar helps in maximizing the soil's WHC (Singh et al., 2010) and ultimately results in dissolving the soil nutrients in water that provides large amounts of nutrients for plant uptake (Lehmann and Joseph, 2009). The application of biochar at 2% mixture rate increased the WHC of a loamy sand soil (Novak et al., 2009). Yet, temperature also plays a crucial role in biochar performance. Varying pyrolysis temperature from 250°C to 750°C, showed that WHC of the soil ranged from 7% to 16% (Yu et al., 2013). Biochar can effectively minimize external fertilizer additions, enhance the WHC, improve crop productivity and reduce the nutrient leaching losses (Li et al., 2021). In fact, biochar fills the empty space between soil particles increases total porosity and improves water retention parameters such as FC and PWP (Alghamdi et al., 2020). In rainfed maize production systems on Alfisols of southern India, the addition of biochar significantly improved biomass and economic yield during the season with mid-season droughts (Srinivasarao et al., 2013a,b).

5.6.10 Tank Silt

"Tank silt" (TS) is rich in SOC and other plant nutrients, with greater potentiality in enhancing the WHC, is highly suitable to cope with the mid-season droughts under rainfed agriculture. SOC helps in soil aggregation that restores the degraded lands (Rakesh et al., 2022). The addition of TS, which is rich in microbial biomass, would enhance crop production and restore degraded soils (Tiwari et al., 2014). Application of TS in the farmlands improves soil physicochemical and biological properties that ultimately increase crop productivity (Indoria et al., 2018). The addition of TS in the Alfisol significantly enhances the profitability of rainfed agriculture (Osman et al., 2007). TS-amended plots showed a greater residual impact on Horse gram biomass production besides mitigating N_2O emissions in the degraded semi-arid Alfisol (Sharan et al., 2022). However, the direct and residual influence of TS application in *Kharif(summer)* and *Rabi(winter)* crops is highly influenced by soil texture and moisture availability and management practices (Sharma et al., 2015).

5.6.11 Cover Crops and Market Wastes

Cover cropping is highly effective in preventing nutrient losses through leaching and percolation, restricting weed growth and enhancing soil C sequestration (Srinivasarao et al., 2021c). A field study conducted by Mohanty et al. (2015), to evaluate the impact of combining tillage with different cropping systems, revealed that involving CA practice that included tillage (conventional and minimum) with cropping systems (sole maize and maize+cowpea intercrop) and followed by cover

FIGURE 5.8 Crop residue converted into biochar for SOC improvement and water storage in the soil profile.

crops [fallow, horsegram, and toria (rapeseed)] significantly improved SOC storage under the inclusion of cover crops when compared with no – cover crop. Venkateswarlu et al. (2007) reported that incorporation of crop biomass into soil improved SOC, MBC, and available nutrients over 10 years. Market-waste materials are rich in plant nutrients. Market wastes contain a mixture of vegetables, fruits, flowers, animal wastes etc. that are easily decomposable and carbon -rich compost. Using such wastes as manure enables the effective utilization of wastes in agriculture and helps in enhancing SOC and other available nutrients that ultimately maximize the crop yields and soil health.

5.7 NATIONAL AND STATE GOVERNMENT INITIATIONS FOR SOIL CARBON AND WATER STORAGE

The National Mission for Sustainable Agriculture (NMSA) was launched under the eight Missions outlined under National Action Plan on Climate Change (NAPCC) in 2010 to sustain agricultural productivity through the conservation of natural resources like soil and water in conjunction with the development of rainfed agriculture in India. The Mahatma Gandhi National Rural Employment Guarantee Act (MGNREGA) was launched in 2005 and aimed to improve soil fertility, carbon sequestration, rainwater harvesting, and restoration of degraded lands etc. targeting the rural communities of India. The Soil Health Card (SHC) Mission was initiated in 2015 to improve soil fertility status through soil test-based fertilizer recommendations. This mission made a revolution in India

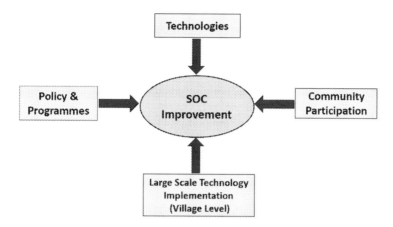

FIGURE 5.9 Critical requirements for enhancing SOC storage in Indian agriculture.

and distributed about 10.48 crores of soil health cards. The National Policy for Management of Crop Residues (NPMCR) programme came into existence in 2014 and widely restricted crop residue burning and promoted its utilization through innovative farm mechanization and this scheme also helped in monitoring the residue burning through satellite-based technologies. The Paris Climate Change Agreement committed by India also contributes to the mitigation of GHG emissions and enhances carbon sink. The Neem Coated Urea (NCU) launched in 2015 promoted widely to use this fertilizer that helps maintain soil health besides protecting the ground water from excess nitrate nutrient pollution. The Rashtriya Krishi Vikas Yojana (RKVY) scheme was launched in 2007 for the holistic development of agriculture and allied sectors through soil and water conservation and to strengthen the farmer's effort, risk mitigation and promoting agri-business entrepreneurship.

The Integrated Watershed Management Programme (IWMP) was implemented by the Department of Land Resources of the Ministry of Rural Development. This programme aimed to restore ecological balance by harnessing, conserving, and developing degraded soil and water while increasing overall biodiversity. The Desert Development Programme (DDP) was started during the year 1977–78 with the objectives to restore the natural resources like land, water, vegetative cover and implement watershed approach, water resource development etc. The scheme Watershed Development Project in Shifting Cultivation Areas (WDPSCA) was implemented during the year 1995–96 to protect the hill slopes of jhum (shifting cultivation) areas through soil and water conservation measures on a watershed basis. The Accelerated Irrigation Benefits Programme (AIBP) was launched during 2009–10 and sponsored by the Ministry of Water Resources to increase the area under irrigation for increasing the crop productivity and socio-economic condition of the people. The NABARD Loan- Soil & Water Conservation Scheme under the Rural Infrastructure Development Fund (RIDF) from 2000 to 01 onwards to enhance the agricultural productivity in small river valleys of rural areas. The synergy requirements of technologies, policy, and programmes along with community participation for enhancing SOC storage in Indian agriculture have been illustrated in Figure 5.7.

5.8 CONCLUSION

The barriers to achieving the potential productivity of rainfed farming in India can be broadly categorized as being connected to knowledge and institutional, technology/resource, and socio-economic factors. Of these, the technological and resource-related constraints can be efficiently handled using the technologies already in use, which can greatly boost productivity in rainfed conditions. The secret to increasing productivity and narrowing or even eliminating the production gaps is effective soil and water management. Building the SOM for soil health restoration should be the focus. Water is a precious natural resource, and rainfed farming can only be sustained by managing rainfall in-situ or by collecting runoff and recycling it. Effective use of water, soil, and farm management practices in an integrated strategy is both necessary and a requirement for making rainfed farming more economic and sustainable. To achieve gains in productivity and widespread effects, it is necessary to scale these demand-driven approaches through farm science centers, agricultural technology management agencies, and various national and state government initiatives, which are present in every region of developing nations. Yet, there is no "one-size-fits-all" solution to the complex problems of water scarcity in diverse agricultural systems. Considering the current scenario, agriculture waste management technologies like CA, mulching, organic amendments, novel approaches like biochar, TS, green manuring, integrated nutrient management options etc. are some of the "win–win" strategies to unlock the potential of rainfed dryland soils without jeopardizing the quality SOM content. As these areas are crucial for India's food security, publishing this chapter will further motivate and enhance efforts to eliminate yield gaps and to unleash the potential of rainfed agriculture.

5.9 WAY FORWARD

- Coordinating policies of integrating soil carbon with national climate commitments to achieve more carbon sequestration in India,
- Stopping crop residue burning to restrict further carbon emissions,
- Training programs to educate farmers about the importance of carbon and its greater role in agriculture and food security,
- Developing innovative methodologies for a quick, cheaper, and authentic measurement of soil carbon changes,
- Remapping is critical to find out the actual SOC depleted areas at regional and national levels for diverse land use and management for the year 2021–22,
- Integrating of water, agriculture and forest policies is vital to bring coordination and collective achievements,
- Involving the community is critical in collecting and sharing the information,
- Promoting carbon neutrality policy at the national level to strengthen carbon finance,
- Encouraging farmers to shift to organic agriculture/carbon smart agriculture by adopting diversified cropping systems and minimizing fertilizer usage,
- Providing strong support is needed to R & D activities to bring innovations and dissemination of feasible technologies related to carbon sequestration,
- Focusing is also needed on the groundwater recharge and water management programmes.
- Provisioning of knowledge and skills on soil carbon improvement and soil condition monitoring for land managers and planners,
- Converging of inter-ministries with a group of secretaries for the effective monitoring of national programmes.
- Developing national mission on SOC sequestration and a cross-learning platform for SOC sequestration at micro level, and
- Financing the farmers who adopt climate-resilient agricultural technologies.

LIST OF ABBREVIATIONS

%	Percent
°C	Centigrade
AIBP	Accelerated irrigation benefits programme
AICRPDA	All India Co-Ordinated Research Project for Dryland Agriculture

AWHC	Available water holding capacity
BCM	Billion cubic
BD	Bulk density
C	Carbon
C:N ratio	Carbon:nitrogen ratio
CA	Conservation agriculture
CEC	Cation exchange capacity
cm	Centimeter
CWC	Central Water Commission
DDP	Desert Development Programme
FAO	Food and Agriculture Organization
FYM	Farmyard manure
g/kg	Gram per kilogram
GHG	Greenhouse gas
GHI	Global hunger index
GM	Green manure
GOI	Government of India
ha	Hectare
ICAR	Indian Council of Agricultural Research
ICRISAT	International Crops Research Institute for the Semi-Arid Tropics
INM	Integrated Nutrient Management
IWMP	Integrated Watershed Management Programme
K	Potassium
kg/ha	Kilogram per hectare
km3	Kilometer cube
M ha	Million hectare
Mg C/ha	Mega gram carbon per hectare
MGNREGA	Mahatma Gandhi National Rural Employment Guarantee Act
mm	Millimeter
Mt	Million tons
N	Nitrogen
N2O	Nitrogen dioxide
NABARD	National Bank for Agriculture and Rural Development
NAPCC	National Action Plan on Climate Change
NCSS	National Cooperative Soil Survey
NCU	Neem Coated Urea
NMSA	National Mission for Sustainable Agriculture
NPMCR	National Policy for Management of Crop Residues
NRAA	National Rainfed Area Authority
NRCS	Natural Resources Conservation Service
P	Phosphorus
PAWC	Plant Available Water Capacity
RIDF	Rural Infrastructure Development Fund
RKVY	Rashtriya Krishi Vikas Yojana
SHC	Soil Health Card
SOC	Soil organic carbon
SOM	Soil organic matter
SWS	Soil water storage
t	tons
TS	Tank silt
USDA	United States Department of Agriculture

WDPSCA	Watershed Development Project in Shifting Cultivation Areas
wt.	Weight
WUE	Water use efficiency
ZT	Zero-tillage

REFERENCES

Abaranji, S., Panchabikesan, K., & Ramalingam, V. (2021). Experimental study on the direct evaporative air-cooling system with vermicompost material as the water storage medium. *Sustainable Cities and Society*, 71, 102991.

Abiye, W. 2022. Soil and water conservation nexus agricultural productivity in Ethiopia. *Advances in Agriculture*, Article ID 8611733, 1–10. https://doi.org/10.1155/2022/8611733.

Alghamdi, A.G., Alkhasha, A., & Ibrahim, H.M. 2020. Effect of biochar particle size on water retention and availability in a sandy loam soil. *Journal of Saudi Chemical Society*, 24(12), 1042–1050. https://doi.org/10.1016/j.jscs.2020.11.003.

All India Coordinated Research Project for Dryland Agriculture (AICRPDA), Annual Reports 2019–20, Agricultural Research Station, Anantapur.

All Report on Agricultural Census 2015–16, Agriculture Census Division, Department of Agriculture, Cooperation and Farmers Welfare, Ministry of Agriculture and Farmers Welfare, GoI.

Alzamel, N.M., Taha, E.M., Bakr, A.A., & Loutfy, N. 2022. Effect of organic and inorganic fertilizers on soil properties, growth yield, and physiochemical properties of sunflower seeds and oils. *Sustainability*, 14(19), 12928. https://doi.org/10.3390/su141912928.

Azarmi, R.M., Giglou, T., & Taleshmikail, R.D. 2008. Influence of vermicompost on soil chemical and physical properties in tomato (*Lycopersicum esculentum*) field. *African Journal of Biotechnology*, 7, 2397–2401. https://www.academicjournals.org/AJB.

Babubhai, S.H. 2018. Cover cropping and contour cultivation. *Biotech Articles*. https://biotecharticles.com/Agriculture-Article/Cover-Cropping-and-Contour-Cultivation-4326.html

Bakr, A.A. 2016. Dynamic of some plant nutrients in soil under organic farming conditions. Ph.D. Thesis, Faculty of Agriculture, Assiut University, Assiut.

Bastida, F., Kandeler, E., Hernandez, T., & Garcıa, C. 2008. Long term effect of municipal solid waste amendment on microbial abundance and humus-associated enzyme activities under semiarid conditions. *Microbial Ecology*, 55(4), 651–661. https://doi.org/10.1080/03650340.2013.766721.

Beesley, L., & Marmiroli, M. 2011. The immobilisation and retention of soluble arsenic, cadmium and zinc by biochar. *Environmental Pollution* 159(2), 474–480. https://doi.org/10.1016/j.envpol.2010.10.016

Bharali, A., Baruah, K.K., Bhattacharya, S.S., & Kim, K.H. 2021. The use of Azolla caroliniana compost as organic input to irrigated and rainfed rice ecosystems: Comparison of its effects in relation to CH4 emission pattern, soil carbon storage, and grain C interactions. *Journal of Cleaner Production*, 313, 127931.

Black, A.L. 1973. Crop residue, soil water, and soil fertility related to spring wheat production and quality after fallow. *Soil Science Society of America Journal*, 37, 754–758.

Brady, N.C. & Weil, R.R. 2005. *Nature and Properties of Soil*, 13th edition, MacMillan Publishing Co. Ltd., New York, NY. https://doi.org/10.1155/2020/2821678.

Bramley, H., Turner, D.W., Tyerman, S.D., Turner, N.C. 2007. Water flow in the roots of crop species: the influence of root structure, aquaporin activity, and waterlogging. *Advances in Agronomy*, 96, 133–196. https://doi.org/10.1016/S0065-2113(07)96002-2.

Chand, R., Prasanna, L.P., & Singh, A. 2011. Farm size and productivity: Understanding the strengths of smallholders and improving their livelihoods. *Economic and Political Weekly*, 5, 11.

Chen, Y., Wei, T., Ren, K., Sha, G., Guo, X., Fu, Y., & Yu, H. 2022. The coupling interaction of soil organic carbon stock and water storage after vegetation restoration on the Loess Plateau, China. *Journal of Environmental Management*, 306, 114481. https://doi.org/10.1016/j.jenvman.2022.114481.

Chenu, C., Le Bissonnais, Y., & Arrouays, D. 2000. Organic matter influence on clay wettability and soil aggregate stability. *Soil Science Society of America Journal*, 64, 1479–1486.

CWC. 2022. *Water Resources at a Glance-2022*. Central Water Commission (CWC), Ministry of Jal Shakti, Government of India. cwc.gov.in.

Dorren, L., & Rey, F. 2004. A review of the effect of terracing on erosion. In: *Briefing Papers of the 2nd SCAPE Workshop*. Soil Conservation and Protection for Europe, Citeseer, pp. 97–108. https://www.ecorisq.org/docs/Dorren_Rey.pdf

Droppelmann, K.J., Lehmann, J., Ephrath, J.E., & Berliner, P.R. 2000. Water use efficiency and uptake patterns in a runoff agroforestry system in an arid environment. *Agroforestry Systems*, 49, 223–243.

Du, T., Kang, S., Sun, J., Zhang, X., & Zhang, J. 2010. An improved water use efficiency of cereals under temporal and spatial deficit irrigation in north China. *Agricultural Water Management*, 97, 66–74.

Du, Z., Ren, T., Hu, C., & Zhang, Q. 2015. Transition from intensive tillage to no-till enhances carbon sequestration in microaggregates of surface soil in the North China Plain. *Soil Tillage & Research*, 146, 26–31.

Emerson, W.W. 1995. Water retention, organic C and soil texture. *Australian Journal of Soil Research*, 33, 241–251. https://doi.org/10.1071/SR9950241.

Foley, J.A., Ramankutty, N., Brauman, K.A., Cassidy, E.S., Gerber, J.S., Johnston, M., Mueller, N.D., O'Connell, C., Ray, D.K., & West, P.C. 2011. Solutions for a cultivated planet. *Nature*, 478, 337–342.

Fukumasu, J., Jarvis, N., Koestel, J., Kätterer, T., & Larsbo, M. 2022. Relations between soil organic carbon content and the pore size distribution for an arable topsoil with large variations in soil properties. *European Journal of Soil Science*, 73(1), e13212. https://doi.org/10.1111/ejss.13212.

Gao, Y., Duan, A., Qiu, X., Liu, Z., Sun, J., Zhang, J., & Wang, H. 2010. Distribution of roots and root length density in a maize/soybean strip intercropping system. *Agricultural Water Management* 98, 199–212.

Gathagu, J.N., Mourad, K.A., & Sang, J. 2018. Effectiveness of contour farming and filter strips on ecosystem services. *Water (Basel)*, 10, 1312.

Gould, C.M., 2015. *Compost Increases Water Holding Capacity of Droughty Soils*, Michigan State University: East Lansing, MI.

Gururaj, B., Hamsa, K.R., & Mahadevaiah, G.S. 2017. Doubling of small and marginal farmer's income through rural non-farm and farm sector in Karnataka. *Economic Affairs*, 62(4), 581–587.

Hudson, B.H. 1994. Soil organic matter and available water capacity. *Journal of Soil and Water Conservation*, 49(2), 189–194.

ICAR Data Book. 2022. *Indian Council of Agricultural Research. Agricultural Research Data Book 2022*. Available Online: https://apps.iasri.res.in/agridata/22data/home.html.

Indoria, A.K., Sharma, K.L., Reddy, K.S., Srinivasarao, Ch., Srinivas, K., Balloli, S.S., Osman, M., Pratibha, G., & Raju, N.S. 2018. Alternative sources of soil organic amendments for sustaining soil health and crop productivity in India-impacts, potential availability, constraints and future strategies. *Current Science*, 115, 2052–2062. https://krishi.icar.gov.in/jspui/handle/123456789/32401.

Jat, M.L., Chakraborty, D., Ladha, J.K., Parihar, C.M., Datta, A., Mandal, B., Nayak, H., Maity, P., Rana, D.S., Chaudhary, S.K., & Gerard, B. 2022. Carbon sequestration potential, challenges, and strategies towards climate action in smallholder agricultural systems of South Asia. *Crop and Environment*. 1(1): 86–101. https://doi.org/10.1016/j.crope.2022.03.005.

Javaad, S., & Panwar, A, 2013. Effect of biofertilizer, vermicompost and chemical fertilizer on different biochemical parameters of Glycine max and Vigna mungo. *Recent Research in Science and Technology*, 5, 40–44.

Jayaraman, S., Sinha, N.K., Kumar, S., & Patra A.K. 2021. Sustaining Soil Carbon to Enhance Soil Health, Food, Nutritional Security, and Ecosystem Services. *Frontiers in Sustainable Food Systems*, 5, 777495. https://doi.org/10.3389/fsufs.2021.777495.

Jensen, E.S., Bedoussac, L., Carlsson, G., Journet, E.-P., Justes, E., Hauggaard-Nielsen, H. 2015. Enhancing yields in organic crop production by eco-functional intensification. *Sustainable Agricultural Research*, 4, 42–50.

Ji, S., & Unger, P.W. 2001. Soil water accumulation under different precipitation, potential evaporation, and straw mulch conditions. *Soil Science Society of America Journal*, 65, 442–448.

Karami, A., Homaee, M., Afzalinia, S., Ruhipour, H., & Basirat, S. 2012. Organic resource management: Impacts on soil aggregate stability and other soil physico-chemical properties. *Agriculture Ecosystem. Environmrnt*, 148, 22–28. https://doi.org/10.1016/j.agee.2011.10.021.

Kumar, R., Singh, R.D., & Sharma, K.D. 2005. Water resources of India. *Current Science*, 89(5): 794–811.

Kumari, R., Kumar, S., Kumar, R. et al. 2017. Effect of long-term integrated nutrient management on crop yield, nutrition and soil fertility under rice-wheat system. *Journal of Applied and Natural Science*, 9(3), 1801–1807. https://doi.org/10.31018/jans.v9i3.1442.

Lal, R. 2008. Carbon sequestration. *Philosophical Transactions of the Royal Society B: Biological Sciences*, 363(1492), 815–830. https://doi.org/10.1098/rstb.2007.2185.

Lamm, F.R., & Trooien, T.P. 2003. Subsurface drip irrigation for corn production: A review of 10 years of research in Kansas. *Irrigation Science*, 22, 195–200.

Lehmann, J., & Joseph, S. 2009. *Biochar for Environmental Management: Science and Technology*. Earthscan/James & James, London.

Li, L., Zhang, Y-J., Novak, A., Yang, Y., & Wang, J. 2021. Role of biochar in improving sandy soil water retention and resilience to drought. *Water*, 13(4), 407. https://doi.org/10.3390/w13040407.

Libohova, Z., Seybold, C., Wysocki, D., Wills, S., Schoeneberger, P., Williams, C & Owens, P.R. 2018. Reevaluating the effects of soil organic matter and other properties on available water-holding capacity using the National Cooperative Soil Survey Characterization Database. *Journal of Soil and Water Conservation*, 73(4), 411–421. https://doi.org/10.2489/jswc.73.4.411.

Mao, L., Zhang, L., Li, W., van der Werf, W., Sun, J., Spiertz, H., Li, L. 2012. Yield advantage and water saving in maize/pea intercrop. *Field Crops Research*, 138, 11–20.

Martin-Guay, M.-O., Paquette, A., Dupras, J., & Rivest, D. 2018. The new green revolution: sustainable intensification of agriculture by intercropping. *Science of the Total Environment*, 615, 767–772.

Meena, R. 2019. Green manuring. An approach to improve soil fertility and crop production, Munich, GRIN Verlag, https://www.grin.com/document/468117.

Minasny, B., & McBratney, A.B. 2018. Limited effect of organic matter on soil available water capacity. *European Journal of Soil Science*, 69, 39–47. https://doi.org/10.1111/ejss.12475.

Mohammad, W., Shah, S.M., Shehzadi, S., & Shah, S.A. 2012. Effect of tillage, rotation and crop residues on wheat crop productivity, fertilizer nitrogen and water use efficiency and soil organic carbon status in dry area (rainfed) of north-west Pakistan. *Journal of Soil Science and Plant Nutrition*, 12(4), 715–727. https://doi.org/10.4067/S0718-95162012005000027.

Mohanty, A., Mishra, K.N., Roul, P.K., Dash, S.N., & Panigrahi, K.K. 2015. Effects of conservation agriculture production system (CAPS) on soil organic carbon, base exchange characteristics and nutrient distribution in a tropical rainfed agro-ecosystem. *International Journal of Plant, Animal and Environmental Sciences*, 5, 310–314.

Mrunalini, K., Behera, B., Jaraman, S., Abhilash, P.C., Dubey, P.K., Narayanaswamy, G., Prasad, J.V.N.S., Rao, K.V., Krishnan, P., Pratibha, G., & Srinivasarao, Ch. 2022. Nature based solutions in soil restoration for improving agricultural productivity. *Land Degradation and Development*. 33(8): 1269–1289 https://doi.org/10.1002/ldr.4207.

Mujumdar, P.P. 2008. Implications of climate change for sustainable water resources management in India. *Physics and Chemistry of the Earth, Parts A/B/C*, 33(5), 354–358.

National Rainfed Area Authority (NRAA). 2022. *Accelerating the Growth of Rainfed Agriculture – Integrated Farmers Livelihood Approach*. Department of Agriculture and Farmers' Welfare, Ministry of Agriculture & Farmers' Welfare, New Delhi.

Nayar, L. 2021. Outlook India magazine. https://www.outlookindia.com/author/lola-nayar-40.

Nazli, R.I., Inal, I., Kusvuran, A., Demirbas, A., & Tansi, V. 2015. Effects of different organic materials on forage yield and nutrient uptake of silage maize (*Zea mays* L.). *Journal of Plant Nutrition*, 39(7), 912–921. https://doi.org/10.3906/tar-1302-62.

Neog, P., Sarma, P.K., Saikia, D., Borah, P., Hazarika, G.N., Sarma, M.K., Sarma, D., Ravindra Chary, G., & Srinivasarao, Ch. 2019. Management of drought in Sali rice under increasing rainfall variability in the North Bank Plains Zone of Assam, North East India. *Climatic Change*, 158: 473–484. https://doi.org/10.1007/s10584-019-02605-4.

Novak, J.M., Lima, I., Xing, B., Gaskin, J.W., Steiner, C., Das, K., Ahmedna, M., Rehrah, D., Watts, D.W., Busscher, W.J., 2009. Characterization of designer biochar produced at different temperatures and their effects on a loamy sand. *Annals of Environmental Science*, 3(1), 195–206. https://www.aes.northeastern.edu/.

Olness, A., & Archer, D. 2005. Effect of organic carbon on available water in soil. *Soil Science*, 170(2), 90–101. https://doi.org/10.1097/01.ss.0000155496.63323.35.

Osman, M., Ramakrishna, Y.S., & ShaikHaffis. 2007. Rejuvenating tanks for self-sustainable rainfed agriculture in India. *Agriculture Situation India*, 64, 67–70.

Ossom, E.M., Pace, P.F., Rhykerd, R.L., & Rhykerd, C.L. 2001. Effect of mulch on weed infestation, soil temperature, nutrient concentration, and tuber yield in *Ipomoea batatas* (L.) Lam. in Papua New Guinea. *Tropical Agriculture*, 78, 144.

Ouma, G., & Jeruto, P. 2010. Sustainable horticultural crop production through intercropping: The case of fruits and vegetable crops: A review. *Agriculture and Biology Journal of North America*, 1, 1098–1105.

Prem, M., Ranjan, P., Seth, N., & Patle, G.T. 2020. Mulching techniques to conserve the soil water and advance the crop production-A review. *Current World Environment*, 15, 10–30.

Qian, Z., Tang, L., Zhuang, S., Zou, Y., Fu, D., & Chen, X. 2020. Effects of biochar amendments on soil water retention characteristics of red soil at south China. *Biochar* 2(4):479–488. https://doi.org/10.1007/s42773-020-00068-w.

Rakesh, S., Sarkar, D., Sinha, A.K., Shikha, Mukhopadhyay, P., Danish, S., Fahad, S., & Datta, R. 2021. Carbon mineralization rates and kinetics of surface-applied and incorporated rice and maize residues in Entisol and Inceptisol soil types. *Sustainability*, 13(13), 7212. https://doi.org/10.3390/su13137212.

Rakesh, S., Sinha, A.K., Juttu, R., Sarkar, D., Jogula, K., Reddy, S.B., Raju, B., Danish, S., & Datta, R. 2022. Does the accretion of carbon fractions and their stratification vary widely with soil orders? A case study in Alfisol and Entisol of sub-tropical eastern India. *Land Degradation & Development*, 33(12), 2039–2049. https://onlinelibrary.wiley.com/doi/10.1002/ldr.4291.

Ramesh, T., Bolan, NS., Kirkham, M.B., Wijesekara, H., Manjaiah, K.M., Srinivasarao, Ch., Sandeep, S., Rinklebe, J., Ok, YS., Choudhury, B.U., Want, H., Tang, C., Song, Z., & Freeman II, O.W. (2019). Soil organic carbon dynamics: Impact of land use changes and management practices: A review. *Advances in Agronomy*, 156, 1–125.

Ranjan, P., Patle, G.T., Prem, M., & Solanke, K.R. 2017. Organic mulching-a water saving technique to increase the production of fruits and vegetables. *Current Agriculture Research Journal*, 5(3), 371–380.

Rao, S., Indoria, A.K., & Sharma, K.L. 2017. Effective management practices for improving soil organic matter for increasing crop productivity in rainfed agroecology of India. *Current Science*, 112, 1497. https://doi.org/10.18520/cs/v112/i07/1497-1504.

Ravindra Chary, G. & Gopinath, K.A. 2022. Agro-ecology specific rainwater management interventions for higher productivity and income in rainfed areas. In B. Krishna Rao, S. Annapurna, B. Renuka Rani, Z. Srinivasa Rao, K. Sunitha, M. SchinDutt, S.K. Jamanal, & V. Ramesh. *Soil and Water Conservation Techniques in Rainfed Areas (e-book)*. National Institute of Agricultural Extension Management (MANAGE) & Water and Land Management Training and Research Institute (WALAMTARI), Hyderabad.

Rawls, W.J., Y.A. Pachepsky, J.C. Ritchie, T.M. Sobecki, & H. Bloodworth. 2003. Effect of soil organic carbon on soil water retention. *Geoderma*, 116, 61–76. https://doi.org/10.1016/S0016-7061(03)00094-6.

Roy, D., Datta, A., Jat, H.S., Choudhary, M., Sharma, P.C., Singh, P.K., & Jat, M.L. 2022. Impact of long-term conservation agriculture on soil quality under cereal based systems of North West India. *Geoderma*, 405, 115391. https://doi.org/10.1016/j.geoderma.2021.115391.

Sahoo, S., Mukhopadhyay, P., Sinha, A.K., Bhattacharya, P.M., Rakesh, S., Kumar, R., … Kumar, U. 2022. Yield, nitrogen-use efficiency, and distribution of nitrate-nitrogen in the soil profile as influenced by irrigation and fertilizer nitrogen levels under zero-till wheat in the eastern Indo-Gangetic plains of India. *Frontier Environmental Science*, 10, 970017. https://doi.org/10.3389/fenvs.2022.970017.

Samra, J.S., & Srinivasarao, Ch. 2021. *Circular Carbon Economy in India: Efficient Crop Residue Management for harnessing Carbon, Energy and Manure with Co-benefits of Greenhouse Gases (GHGs) Emissions Mitigation*. Policy Paper, ICAR-National Academy of Agricultural Research Management, Hyderabad, p. 20.

Sharan, B.R., Srinivasarao, Ch., Chandrasekhar, R.P., Lal, R., Rakesh, S., Kundu, S., Singh, R.N., Dubey, P.K., Abhilash, P.C., Rao, K.V., Abrol, V., & Somasundaram, J. 2022. Greenhouse Gas Emission and Agronomic Productivity as Influenced by Varying Levels of N Fertilizer and Tanksilt in Degraded Semi-Arid Alfisol of Southern India. *Land Degradation and Development*, 34(4), 943-955. https://doi.org/10.1002/ldr.4507.

Sekiya, N., & Yano, K. 2004. Do pigeon pea and sesbania supply groundwater to intercropped maize through hydraulic lift?-Hydrogen stable isotope investigation of xylem waters. *Field Crops Research*, 86, 167–173.

Selim, M. 2018. Potential role of cropping system and integrated nutrient management on nutrients uptake and utilization by maize grown in calcareous soil. *Egyptian Journal of Agronomy*, 40(3), 297–312. https://doi.org/10.21608/AGRO.2018.6277.1134.

Selim, M.M. 2020. Introduction to the integrated nutrient management strategies and their contribution to yield and soil properties. *International Journal of Agronomy*. 2020. https://doi.org/10.1155/2020/2821678.

Sharma, B.R., Rao, K.V., Vittal, K.P.R., Ramakrishna, Y.S., & Amarasinghe, U. 2010. Estimating the potential of rainfed agriculture in India: Prospects for water productivity improvements. *Agricultural Water Management*, 97(1), 23–30.

Sharma, P.K. & Kumar, M., 2014. Status and management of water in rainfed agriculture. *Efficient Water Management for Sustainable Agriculture*, 41–57.

Sharma, S.K., Sharma, R.K., Kothari, A.K., Osman, M., & Chary, G.R. 2015. Effect of tank silt application on productivity and economics of maize-based production system in southern Rajasthan. *Indian Journal of Dryland Agriculture Research and Development*, 30(2), 24–29. https://doi.org/10.5958/2231-6701.2015.00021.4.

Singh, B., Singh, B.P., & Cowie, A.L. 2010. Characterisation and evaluation of biochars for their application as a soil amendment. *Soil Research*, 48(7), 516–525. https://doi.org/10.1071/SR10058.

Singh, D., Lenka, S., Lenka, N.K., Trivedi, S.K., Bhattacharjya, S., Sahoo, S., et al. 2020. Effect of reversal of conservation tillage on soil nutrient availability and crop nutrient uptake in soybean in the vertisols of central India. *Sustainability*, 12(16), 6608.

Singh, N.P., Anand, B., Singh, S., Srivastava, S.K., Srinivasarao, Ch., Rao, K.V., & Bal, S.K. 2021. Synergies and trade-offs for climate-resilient agriculture in India: an agro-climatic zone assessment. *Climatic Change*, 164, 11. https://doi.org/10.1007/s10584-021-02969-6.

Sinha, A.K., Ghosh, A., Dhar, T., Bhattacharya, P.M., Mitra, B., Rakesh, S., Paneru, P., Srestha, S.R., Manandhar, S., Beura, K., Dutta, S., Pradha, A.K., Rao, K.K., Hossain, A., Siddquie, N., Molla, M.S.H., Chaki, A.K., Gathala, M.K., Islam, M.S., Dalal, R.C., Gaydon, D.S., Laing, A.M. & Menzies, N.W. 2019. Trends in key soil parameters under conservation agriculture- ased sustainable intensification farming practices in the Eastern Ganga Alluvial Plains. *Soil Research*, 57(8), 883–893. https://doi.org/10.1071/SR19162.

Sinha, A.K., Rakesh, S., Mitra, B., Roy, N., Sahoo, S., Saha, B.N., Dutta, S., & Bhattacharya, P.M. 2021. Agricultural waste management policies and programme for environment and nutritional security. In R. Bhatt, et al. (eds) *Input Use Efficiency for Food and Environmental Security*. Springer Nature. Chapter 21.

Sood, B.R., & Sharma, V.K. 1996. Effect of intercropping and planting geometry on the yield and quality of forage maize. *Forage Research*, 24, 190–192.

Srinivasarao, Ch. 2011a. Soil health improvement with gliricidia green leaf manuring in rainfed agriculture: On farm experiences. Central Research Institute for Dryland Agriculture, Santoshnagar, PO. Saidabad, Hyderabad 500 059, Andhra Pradesh, p.16

Srinivasarao, Ch., Chary, G.R., Raju, B.M.K., Jakkula, V.S., Rani, Y.S. & Rani, N., 2014a. Land use planning for low rainfall (450–750 mm) regions of India. *Agropedology*, 24(2), 197–221.

Srinivasarao, Ch., Deshpande, A.N., Venkateswarlu, B., Lal, R., Singh, A.K., Kundu, S., Vittal, K.P.R., Mishra, P.K., Prasad, J.V.N.S., Mandal, U.K., & Sharma, K.L. 2012a. Grain yield and carbon sequestration potential of post monsoon sorghum cultivation in Vertisols in the semi-arid tropics of central India. *Geoderma*, 175–176, 90–97.

Srinivasarao, Ch. & Gopinath, K.A. 2016. Resilient rainfed technologies for drought mitigation and sustainable food security. *Mausam*, 67(1), 169–182.

Srinivasarao, Ch., Gopinath, K.A., Venkatesh, G., Dubey, A.K., Wakudkar, H., Purakayastha, T.J., Pathak, H., P. Jha, Lakaria, B.L., Rajkhowa, D.J., Mandal, S., Jeyaraman, S., Venkateswarlu, B., & Sikka, A.K. 2013b. *Use of Biochar for Soil Health Management and Greenhouse Gas Mitigation in India: Potential and Constraints*. Central Research Institute for Dryland Agriculture, Hyderabad. 51 p.

Srinivasarao, Ch., Kareemulla, K., Krishnan, P., Murthy, G.R.K., Ramesh, P., Ananthan, P.S. & P.K. Joshi. 2019b. Agro-ecosystem based sustainability indicators for climate resilient agriculture in India: A conceptual framework. *Ecological Indicators*, 105(2019), 621–633.

Srinivasarao, Ch., Kundu, S., Rakesh, S., Lakshmi, C.S., Kumar, G.R., Manasa, R., … Prasad, J.V.N.S. 2021c. Managing soil organic matter under dryland farming systems for climate change adaptation and sustaining agriculture productivity. In *Soil Organic Carbon and Feeding the Future: Basic Soil Processes*. R. Lal (ed). First edition, Chapter 10, pp. 219–251. *Advances in Soil Science*, CRC Press. ISBN 9781032150673. https://doi.org/10.1201/9781003243090-10.

Srinivasarao, Ch, Kundu, S., Yashavanth, B.S., Rakesh, S., Akbari, K.N., Sutaria, G.S., Vora, V.D., Hirpara, D.S., Gopinath, K.A., Chary, G.R., Prasad, J.V.N.S., Bolan, N.S., Venkateswarlu, B. 2020b. Influence of 16 years of fertilization and manuring on carbon sequestration and agronomic productivity of groundnut in vertisol of semi-arid tropics of Western India. *Carbon Management*, 12(1), 13–24. https://doi.org/10.1080/17583004.2020.1858681.

Srinivasarao, Ch., Lal, R., Prasad, J.V., Gopinath, K.A., Singh, R., Jakkula, V.S., Sahrawat, K.L., Venkateswarlu, B., Sikka, A.K., & Virmani, S.M. 2015. Potential and challenges of rainfed farming in India. In *Advances in Agronomy*, D.L. Sparks (ed). Vol. 133, pp. 113–181. https://doi.org/10.1016/bs.agron.2015.05.004.

Srinivasarao, Ch., Lal, R., Sumanta Kundu, M.B.B. Prasad Babu, B. Venkateswarlu, & A.K. Singh. (2014b) Soil carbon sequestration in rainfed production systems in the semiarid tropics of India. *Science of the Total Environment*, 487, 587–603.

Srinivasarao, Ch., Prasad, R.S. & Mohapatra, T. 2019a. *Climate Change and Indian Agriculture: Impacts, Coping Strategies, Programmes and Policy. Technical Bulletin/Policy Document*. Indian Council of Agricultural Research, Ministry of Agriculture and Farmers' Welfare and Ministry of Environment, Forestry and Climate Change, Government of India, New Delhi. p25.

Srinivasarao, Ch, Rakesh, S., Kumar, G.R., Manasa, R., Somashekar, G., Lakshmi, C.S. & Kundu, S. 2021a. Soil degradation challenges for sustainable agriculture in tropical India. *Current Science*, 120(3), 492–500.

Srinivasarao, Ch., Rao, K.V., Chary, G.R., Vittal, K.P.R., Sahrawat, K.L. & Kundu, S. 2009a. Water retention characteristics of various soil types under diverse rainfed production systems of India. *Indian Journal of Dryland Agricultural Research and Development*, 24(1), 1–7.

Srinivasarao, Ch, Rejani, R, Rama Rao, C.A, Rao, K.V, Osman, M., Srinivasa Reddy, K., Kumar, M. & Kumar, P. 2017. Farm pond for climate-resilient rainfed agriculture. *Current Science*, 112(3), 471–477.

Srinivasarao, Ch, Singh, S.P., S. Kundu, V. Abrol, Lal, R., Abhilash, P.C., Chary, G.R., Pravin, B.T., Prasad, J.V.N.S., & Venkateswarlu, B. 2021b. Integrated nutrient management improves soil organic matter and agronomic sustainability of semiarid rainfed Inceptisols of the Indo-Gangetic Plains. *Journal of Plant Nutrition and Soil Science*, 184(5), 562–572.

Srinivasarao, Ch, Subha Lakshmi, C., Sumanta Kundu, S., Ranjith Kumar, G., Manasa, R. & Rakesh, S. 2020a. Integrated nutrient management strategies for rainfed agro-ecosystems of India. *Indian Journal of Fertilizers*, 16(4), 344–361.

Srinivasarao, Ch., Venkateswarlu, B., Lal, R., Singh, A.K. & Kundu, S. 2013a. Sustainable management of soils of dryland ecosystems of India for enhancing agronomic productivity and sequestering carbon. *Advances in Agronomy*, 121, 253–329.

Srinivasarao, Ch., Venkateswarlu, B., Lal, R., Singh, A.K., Kundu, S., Vittal, K.P.R., Balaguruvaiah, G., Vijaya Shankar Babu, M., Ravindra Chary, G., Prasadbabu, M.B.B. & Yellamanda Reddy, T. 2012g. Soil carbon sequestration and agronomic productivity of an Alfisol for a groundnut-based system in a semiarid environment in southern India. *European Journal of Agronomy*, 43, 40–48.

Srinivasarao, Ch., Venkateswarlu, B., Lal, R., Singh, A.K., Kundu, S., Vittal, K.P.R., Balaguravaiah, G., VijayaShankarBabu, M., RavindraChary, G., Prasadbabu, M.B.B., YellamandaReddy, T. 2012e. Soil carbon sequestration and agronomic productivity of an Alfisol for a groundnut-based system in a semi-arid environment in South India. *European Journal of Agronomy*, 43, 40–48. https://doi.org/10.1016/j.eja.2012.05.001.

Srinivasarao, Ch., Venkateswarlu, B., Lal, R., Singh, A.K., Kundu, S., Vittal, K.P.R., Basavapura, K.R., Narayanaiyer, G. 2012d. Yield sustainability and carbon sequestration potential of groundnut-finger millet rotation in Alfisols under semi-arid tropical India. *International Journal of Agricultural Sustainability*, 10(3), 1–15.

Srinivasarao, Ch., Venkateswarlu, B., Lal, R., Singh, A.K., Kundu, S., Vittal, K.P.R., Patel, J.J., & Patel, M.M. 2011b. Long-term manuring and fertilizer effects on depletion of soil organic carbon stocks under pearl millet cluster bean-castor rotation in western India. *Land Degradation & Development*. 25(3), 173–183. https://doi.org/10.1002/ldr.1158.

Srinivasarao, Ch., Venkateswarlu, B., Lal, R., Singh, A.K., Kundu, S., Vittal, K.P.R., Sharma, S.K., Sharma, R.A., Jain, M.P., Chary, G.R. 2012f. Sustaining agronomic productivity and quality of a Vertisolic Soil (Vertisol) under soybean-safflower cropping system in semi-arid central India. *Canadian Journal of Soil Science*, 92, 771–785.

Srinivasarao, Ch., Venkateswarlu, B., Lal, R., Singh, A.K., Vittal, K.P.R., Kundu, S., Singh, S.R., & Singh, S.P. 2012c. Long-term effects of soil fertility management on carbon sequestration in a rice-lentil cropping system of the Indo-Gangetic plains. *Soil Science Society of America Journal*, 76(1), 168–178.

Srinivasarao, Ch., Venkateswarlu, B., Singh, A.K., Vittal, K.P.R., Kundu, S., Gajanan, G.N., Ramachandrappa, B., & Chary, G.R. 2012b. Critical carbon inputs to maintain soil organic carbon stocks under long term finger millet (*Eleusine coracana* (L.) Gaertn) cropping on Alfisols in semi-arid tropical India. *Journal of Plant Nutrition and Soil Science*, 175(5), 681–. https://doi.org/10.1002/jpln.201000429.

Srinivasarao, Ch., & Vittal, K.P.R. 2007. Emerging nutrient dificiencies in different soil types under rainfed production systems of India. *Indian Journal of Fertilisers*, 3, 37.

Srinivasarao, Ch., Vittal, K.P.R., Gajbhiye, P.N., Kundu, S., & Sharma, K.L. 2008. Distribution of micronutrients in soils in rainfed production systems of India. *Indian Journal of Dryland Agricultural Research and Development*, 23, 29–35.

Srinivasarao, Ch., Vittal, K.P.R., Venkateswarlu, B., Wani, S.P., Sahrawat, K.L., Marimuthu, S., & Kundu, S. 2009b. Carbon stocks in different soil types under diverse rainfed production systems in tropical India. *Communication in Soil Science and Plant Analysis*, 40, 2338–2356. https://doi.org/10.1080/00103620903111277.

Suthar, S. 2009. Impact of vermicompost and composted farmyard manure on growth and yield of garlic (*Allium stivum* L.) field crop. *International Journal of Plant Production*, 3(1), 1735–6814. https://doi.org/10.22069/IJPP.2012.629.

Tiwari, R., Ramakrishna Parama, V.R., Murthy, I.K. & Ravindranath, N.H. 2014. Irrigation tank silt application to croplands: Quantifying effect on soil quality and evaluation of nutrient substitution service. *International Journal of Agricultural Science Research*, 3, 1–10. https://academeresearchjournals.org/journal/ijasr.

Ullah, N., Ditta, A., Imtiaz, M., Li, X., Jan, A.A., Mehmood, S., Rizwan, M.S., & Rizwan, M. 2021. Appraisal for organic amendments and plant growth promoting rhizobacteria to enhance crop productivity under drought stress: A review. *Journal of Agronomy and Crop Sciences*, 207, 1–20. https://doi.org/10.1111/jac.12502.

Unger, P. 1974. Crop residue management. In JL Hatfield and BA Stewart (eds), Proceedings. CRC Press. pp. 45–56. https://www.taylorfrancis.com/chapters/edit/10.1201/9781351071246-3/residue-management-strategies-great-plains-unger

Venkateswarlu, B., Srinivasarao, Ch., Ramesh, G., Venkateswarlu, S. & Katyal, J.C. 2007. Effects of long-term legume cover crop incorporation on soil organic carbon, microbial biomass, nutrient build-up and grain yields of sorghum/sunflower under rain-fed conditions. *Soil Use and Management* 23l, 100–107.

Walia, S.S., & Kler, D.S. 2007. Ecological studies on organic vs inorganic nutrient sources under diversified cropping systems. *Indian Journal of Fertilisers*, 3, 55.

Wang, D., Li, C., Parikh, S.J., & Scow, K.M. 2019. Impact of biochar on water retention of two agricultural soils - a multi-scale analysis. *Geoderma*, 340, 185–191. https://doi.org/10.1016/j.geoderma.2019.01.012.

Wang, X., Wang, J., Xu, M., Zhang, W., Fan, T., & Zhang, J. 2015. Carbon accumulation in arid croplands of northwest China: pedogenic carbonate exceeding organic carbon. *Scientific Reports*, 5, 1–12.

Wani, S.P., Pathak, P., Jangawad, L.S., Eswaran, H., & Singh, P. 2003. Improved management of Vertisols in the semiarid tropics for increased productivity and soil carbon sequestration. *Soil Use and Management*, 19, 217–222.

Wani, S.P., Sreedevi, T.K., Rockström, J., & Ramakrishna, Y.S. 2009. Rainfed agriculture-past trends and future prospects. *Rainfed Agriculture: Unlocking the Potential*, 7, 1–33.

Wu, S.C., Cao, Z.H., Li, Z.G., Cheung, K.C., & Wong, M.H. 2005. Effects of biofertilizer containing N-fixer, P and K solubilizers and AM fungi on maize growth: a greenhouse trial. *Geoderma*, 125(1–2), 155–166. https://doi.org/10.1016/j.geoderma.2004.07.003.

Wu, W. & Ma, B. 2015. Integrated nutrient management (INM) for sustaining crop productivity and reducing environmental impact: a review. *Science of the Total Environment*, 512–513, 415–427. https://doi.org/10.1016/j.scitotenv.2014.12.101.

Yadav, R.L. & Meena, M.C. 2009. Available micronutrient status and their relation with soil properties of Degana soil series of Rajasthan. *Journal of the Indian Society of Soil Science*, 57(1), 90–92. https://www.researchgate.net/publication/287778539.

Yang, C., Huang, G., Chai, Q., & Luo, Z. 2011. Water use and yield of wheat/maize intercropping under alternate irrigation in the oasis field of northwest China. *Field Crops Research*, 124, 426–432.

Yin, W., Guo, Y., Hu, F., Fan, Z., Feng, F., Zhao, C., Yu, A., & Chai, Q. 2018. Wheat-maize intercropping with reduced tillage and straw retention: A step towards enhancing economic and environmental benefits in arid areas. *Frontiers in Plant Science*, 9, 1328.

Yogesh, T.C., & Hiremath, S.M. 2014. Incorporation of green manure crops on soil enzymatic activities under rainfed condition. *Karnataka Journal of Agricultural Sciences*, 27, 300–302.

Yu, O., Raichle, B., & Sink, S. 2013. Impact of biochar on the water holding capacity of loamy sand soil. *International Journal of Energy and Environmental Engineering*, 4(1), 1–9. https://www.journal-ijeee.com/content/4/1/44.

Zhao, H., Qin, J., Gao, T., Zhang, M., Sun, H., Zhu, S & Ning, T. 2022. Immediate and long-term effects of tillage practices with crop residue on soil water and organic carbon storage changes under a wheat-maize cropping system. *Soil and Tillage Research*, 218, 105309. https://doi.org/10.1016/j.still.2021.105309.

Zhao, X., Virk, A.L., Ma, S.T., Kan, Z.R., Qi, J.Y., Pu, C & Zhang, H.L. 2020. Dynamics in soil organic carbon of wheat-maize dominant cropping system in the North China Plain under tillage and residue management. *Journal of Environmental Management*, 265, 110549. https://doi.org/10.1016/j.jenvman.2020.110549.

Zhou, H., Chen, C., Wang, D., Arthur, E., Zhang, Z., Guo, Z & Mooney, S.J. 2020. Effect of long-term organic amendments on the full-range soil water retention characteristics of a Vertisol. *Soiland Tillage Research*, 202, 104663. https://doi.org/10.1016/j.still.2020.104663.

6 Drought Management in Soils of the Semi-Arid Tropics

Kaushal K. Garg, K.H. Anantha, M.L. Jat, Shalander Kumar, Gajanan Sawargaonkar, Ajay Singh, Venkataradha Akuraju, Ramesh Singh, Md. Irshad Ahmed, Ch Srinivas Rao, R.S. Meena, Martin M. Moyo, Bouba Traore, Gizaw Desta, Rebbie Harawa, Bruno Gerard, Y.S. Saharawat, Alison Laing, and Mahesh K. Gathala

6.1 INTRODUCTION

The drylands of semi-arid tropics face multiple challenges of land degradation, water scarcity, climate change, low agricultural productivity, food insecurity and migration. Soils of the drylands in these landscapes are highly stressed with very low organic carbon and are deficient in nutrients essential for sustainable production. Among multiple challenges in the drylands of the semi-arid tropics, drought is a common phenomenon impacting humanity in these regions. Globally, 1.4 to 4 billion people face water scarcity for 1 to 12 months and most of those live in arid and semi-arid tropics. Climate change having impacts on the precipitation in these regions, further aggravates the challenges, especially through intensifying the hydrological and agricultural droughts. Therefore sustainable management of soils and other natural resources is critical for minimizing the impacts of droughts. Landscape approaches for resource conservation coupled with climate resilient technologies hold enormous opportunities to address the drought in soils of the semi-arid tropics for sustainable food security. In this chapter, we provide the synthesises of the learnings from science-based evidence generated over decades in the drylands of Asia and Africa. Integration of landscape and field scale conservation technologies together holds the key to mitigating drought, sustainable intensification and strengthening ecosystem services. This strategy can potentially help in addressing poverty and malnutrition while contributing to land degradation neutrality and reducing water-energy-carbon footprints towards achieving UN sustainable development goals.

Water shortages and land degradation are the main issues in the drylands of the semi-arid tropics, which are characterised by irregular rainfall, poor soil health, high temperatures, and high evaporative demand. A massive change in land use, such as converting forests and rangelands into agricultural and grazing lands and urbanization in a recent century, has negatively affected the landscape retention ability, including organic carbon, biodiversity and altered hydrological processes at local, regional and planetary scales. As a result, several planetary boundary parameters have crossed or reached permissible thresholds causing a threat to the sustainability of available ecosystem services (Rockstrom et al., 2009). Climatic conditions, one of the planetary boundaries have crossed the safe threshold limit influencing the uncertainty in rainfall amounts and its distribution; volatility in temperature poses serious threats to a range of ecosystem services, including agricultural systems, especially for small and marginal farmers.

Land degradation limits crop productivity and affects the system's sustainability by reducing the landscape's carrying capacity, producing several negative externalities both at upstream and downstream ecologies (Brandolini et al., 2023; Brottrager et al., 2023). Generally, these areas are inhabited by poor and vulnerable groups and face these externalities' brunt. In such scenarios, people living in these ecologies follow shifting cultivation or migrate to urban centres for livelihoods facing huge socio-economic challenges (Mukul et al., 2016). Land degradation and poverty levels are strongly correlated, especially in low-income nations in South Asia and Sub-Saharan Africa (Barbier and Hochard, 2018). Crop productivity, particularly in rainfed systems, ranges from 0.6 to 2.1 t/ha (Singh et al., 2014). The problems already present are further made worse by climate change. Uncertainty in rainfall amounts and distribution due to climate change impacts poses major concerns to agriculture systems, especially for small and marginal farmers who depend mainly on farming and related sectors for their livelihood.

Despite the above challenges, large potential and ample opportunities exist if science is integrated with development pathways. This chapter discusses global challenges and a range of opportunities to address these challenges based on the scaling-up experience of ICRISAT and partners in drylands of semi-arid tropics.

6.2 STATUS OF LAND DEGRADATION, WATER SCARCITY AND NUTRITION

Globally, the quality and quantity of natural resources available for food production are declining due to land degradation, freshwater shortages, and soil nutrient loss. Figure 6.1 (top panel) shows the land degradation status globally, indicating that most agriculture-dominated landscapes are degraded/deteriorated. However, ecologies with rainforests and grasslands also indicate land degradation due to deforestation, soil erosion and nutrient losses. Figure 6.1 (bottom panel) summarises the land degradation status of major land uses across the globe. About 50% of croplands are either degraded or under deterioration. Further, cropland has been divided into irrigated and rainfed croplands in which 46% of rainfed ecologies are degraded/deteriorated, whereas 62% of irrigated ecologies are in the same category. Similar to croplands, trees and shrubs are also under threat as 10%–15% of these lands are categorised under degradation and 35%–40% under the deterioration stage.

Despite significant regional diversity, agriculture accounts for over 70% of freshwater withdrawals (Hoekstra and Mekonnen, 2011). Globally, it is thought that water demand has been increasing at an up that is more than double that of population growth, and there is increased competition for freshwater resources among different businesses, home uses, and ecological uses (Gleick and Palaniappan, 2010; Stenzel et al., 2021). In addition, it is anticipated that by 2030, a further 15% more water will be needed to produce the necessary amount of food due to population growth and a continued shift in dietary preferences towards foods that require much water. Irrigated land raises 40% of the world's food (Schewe et al., 2014; FAO, 2018). This shows how important rainfed agriculture is to ensuring food security for the world's steadily growing population.

Figure 6.2a and Figure 6.2b showed the historical change in per capita availability of cropland for major regions, including the entire World. The highest per capita cropland availability is in Europe and the least in Asia due to high population density. Globally, cropland availability has declined per capita from 0.43 ha in 1961 to 0.20 ha in 2020. Whereas in Asia, the per capita cropland availability was 0.26 in 1961, which declined to 0.12 ha in 2020; similarly, in Africa, this declined from 0.57 ha in 1961 to 0.21 ha in 2020. A similar trend was also observed for other regions.

Figure 6.2b shows the per cent crop land equipped with irrigation for major regions worldwide. Globally, per cent cropland with irrigation has increased from 12% in 1961 to 22% in 2020, which indicates a 0.2% rate at which the irrigated area is expanding annually. The expansion in irrigated areas is found highest in Asia per cent irrigated croplands in 1961 was 23% which increased to 41% in 2020. On average, the rate of expansion is 0.32% per year. On the other hand, this expansion in Africa is negligible as total irrigated cropland increased from 4.42% in 1961 to 6% in 2020.

Though there is an expansion of irrigated areas, the water scarcity level has yet to come down due to increasing demand by different sectors. Mekonnen and Hoekstra (2016) mapped that about 1.4 to 4 billion people worldwide face water scarcity from 1 to 12 months in a year at different severity levels (Figure 6.3a and Figure 6.3b). Figure 6.3b clearly explains the number of months people

Drought Management in Soils of the Semi-Arid Tropics

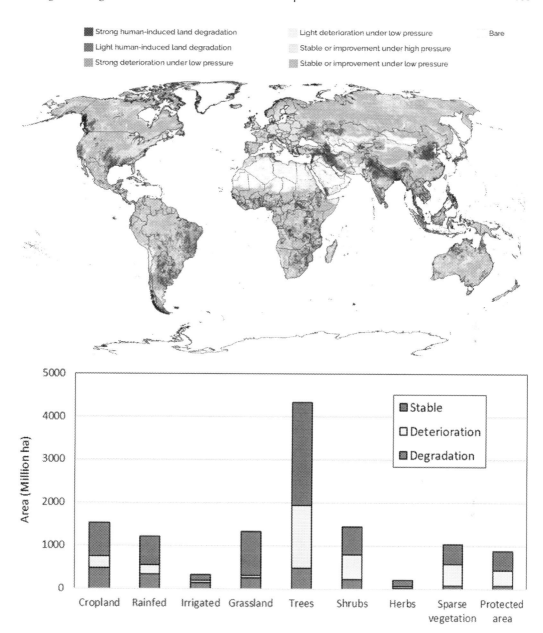

FIGURE 6.1 Land degradation status across the globe. (Source: FAO, 2021.)

face water scarcity globally. For example, 3.97 billion people face water scarcity minimum for one month a year, whereas 0.5 billion people face water scarcity across 12 months in a year, which is a matter of concern.

To fulfil the rising need for food, cultivated land is being expanded, and technical advancement is resulting in higher agricultural yields (FAO, 2018). However, the high dependence on fertilisers, pesticides, and water extraction required for this yield rise has strained the environment. Although N and P are often replaced, other nutrients must be supplemented, according to Fyles and Madramootoo (2016). So, it is necessary to balance the loss of nutrients from agricultural soils through crop absorption, leaching, and erosion through external inputs. Parallel to this, Bossio et al. (2010)

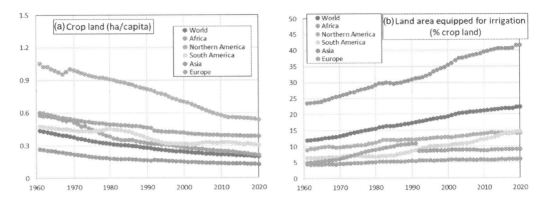

FIGURE 6.2 (a) Change in cropland across major regions (ha/capita) and (b) land area equipped for irrigation (per cent croplands). (Data source: FAO, 2021.)

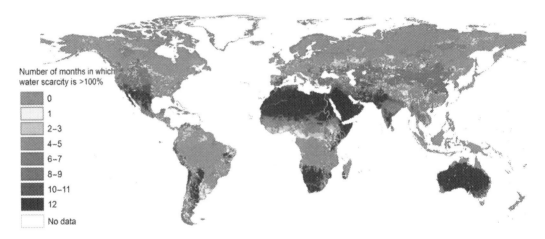

FIGURE 6.3a Spatial variability of water scarcity (number of months) over the globe. (Source: Mekonnen and Hoekstra, 2016.)

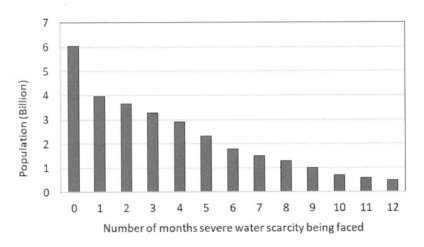

FIGURE 6.3b *n* number of months people across the globe are facing water scarcity. (Source: Mekonnen and Hoekstra, 2016.)

assessed several studies and found that the current rates of soil nitrogen depletion in many Asian and Latin American countries are unsustainable. The deficiency of soil nutrients in India has significantly reduced the effectiveness of N and P fertilisers and decreased agricultural output. The use efficiency of added fertiliser nutrients reduced over the decades due to falling soil organic carbon (Srinivasarao et al., 2021a, 2022a). According to Jones et al. (2013), there was a significant decline in the amount of grain grown for every kg of fertiliser used, from 13.4 in 1970 to 3.7 kg in 2013. Estimates show that 12–22 mt of inorganic phosphorus and 23–42 mt of nitrogen are lost to soil erosion annually from agricultural regions. When nutrients are lost due to erosion, expensive fertiliser nutrients are also lost (Quinton et al., 2010; Brandolini et al., 2023).

Climate change affects food security on a national and regional scale. Various anticipated effects on crop productivity are due to regional changes in growing seasons, minimum and maximum temperatures, precipitation frequency and intensity, and the emergence of agricultural pests and diseases (IPCC, 2014; Ren et al., 2023). According to a range of biophysical and socioeconomic factors, including soil texture, nutrient and organic matter levels, and farmers' ability to adjust to changes in temperature and precipitation, the consequences of climate change on agricultural production are expected to change extensively. Utilising agrochemicals, additional irrigation, better crop types, and changing farm management practises, including the timing of field activities and the usage of conservation agriculture, are some of the elements that can help agriculture resist the hazards associated with climate change. For instance, many studies indicate that during the past 50 years in India, drought and extremely heavy rains have lowered rainfed rice yields by roughly 6%, while wheat yields have not improved in ten years (Lobell et al., 2011). The amount of food available daily per person will decrease from 2425 to 2242 kcal by 2050, according to estimates of crop production reductions in South Asia due to climate change (Nelson et al., 2010). Figure 6.4 shows the changes in major nutritional indicators such as undernourishment, stunted, obese and anaemia women during reproductive age (between 15 and 49).

Though there are several challenges simultaneously, there are ample opportunities to mitigate the above challenges by adopting natural resource management strategies at local, regional and national scales. The following section outlines best management practices for building system-level resilience and mitigating the impact of droughts.

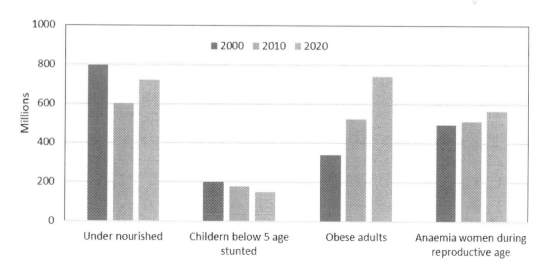

FIGURE 6.4 Population of undernourished, stunted children, obese adults and anaemia women during reproductive age across the globe. (Source: FAO, 2021.)

6.3 CHARACTERISTIC FEATURES OF THE SEMI-ARID TROPICS: AGRICULTURE AND LIVELIHOODS, FOOD AND NUTRITION SECURITY, EMPLOYMENT, ETC

Semiarid regions are a subtype of dry land with an aridity index (i.e., ratio of total annual precipitation to potential evapotranspiration) between 0.20 and 0.50 (Lal, 2004). The semi-arid tropics (SAT) contain parts of 48 developing nations, including the majority of India, certain regions of Southeast Asia, much of Southern, Eastern and Western Africa, and a small portion of Latin America. Degradation of the natural resource base, coupled with high rates of population growth and food insecurity, is a major development problem in the semi-arid rainfed areas of sub-Saharan Africa and Asia. Most of the poor and food insecure are concentrated in rural areas, where their livelihoods depend on smallholder agriculture, rural labour markets, and livestock production. Alleviating poverty, managing agricultural development, and ensuring food security for fast growing populations in SAT will increasingly depend on land-use intensification, as much of the land suitable for agriculture has already been used. Sustainable intensification of agricultural production (Sawargaonkar et al., 2016, 2018) in the less favoured and marginal environments therefore continues to pose enormous challenges to researchers, development practitioners, and policy makers. Poor soil fertility and scarcity of water (low and variable rainfall), accompanied by underdevelopment of infrastructure, institutions, and markets, make the rainfed areas of the semi-arid tropics inherently risky. This means that the poor inhabiting such areas will have to adjust and adapt their livelihood strategies in ways that smoothen their livelihoods in such a risky environment. Risk-reducing adaptive strategies also influence agricultural technology choices, including investments in natural resource management (NRM) innovations. The high degree of abiotic and biotic constraints in the system complicates and hinders scientific breakthroughs and slows progress in designing and developing technologies suitable to these locations. A multiplicity of abiotic stresses such as extreme temperatures, drought, flash floods, salinity and radiation have detrimental effects on plant growth and yield, especially when several occur together with the biotic stresses caused by plant pathogens (Shiferaw, 2002; Mittler, 2006). Climate models show that semiarid parts around the globe are likely to experience increased variability in rainfall and more extended drought periods in the coming decades (IPCC, 2014). Endogenic (e.g. land size and quality, water shortage, cultural and demographic) and exogenic (e.g. climate change, international market, migration) factors are most often cited as drivers of change under drylands (van Ginkel et al., 2013). High poverty rates, food insecurity, the triple burden of malnutrition, rapid population expansion, and environmental uncertainty are intractable challenges in the vulnerable SAT ecosystems and require long-term systemic solutions (Bantilan et al., 2001). Despite several challenges, the SAT regions also present a few opportunities available to intensify agricultural production and generate employment if the natural resources are managed sustainably aligning with socio-economic conditions and preferences.

6.4 SOILS OF SEMI-ARID TROPICS AND VEGETATION INDEX

Soils in the semi-arid tropics are facing several challenges and experience severe land degradation. With changes in land use and reduced organic carbon, water retention ability has declined. Table 6.1 shows major soil types and their extent in semi-arid tropics across the globe. The majority of the soils in semi-arid tropics are covered with Lithosols (15%), Kastanozems (15%), Luvisols (11%) and Xerosols (9%) (Figure 6.5).

Figure 6.6 shows the Normalized Difference Vegetation Index (NDVI) variations across Africa and Asia continents. NDVI values range from +1.0 to −1.0 representing the vegetation condition and density. The more the NDVI value is closer to +1, the more it is related to vegetation cover and its vigour. The greater the value of the NDVI, the higher the density, and vice versa for a lower value. Seven classes were classified based on NDVI values which represent snow, waterbodies, bare soil/no vegetation cover, sparse vegetation, moderate vegetation, dense vegetation, and high dense

Drought Management in Soils of the Semi-Arid Tropics

TABLE 6.1
Major Soil Types and Their Extent in Semi-arid Tropics

Sl No	DOMINATING SOIL TYPES	Area in Million ha	% Area
1	CAMBISOLS	130.2	5
2	GLEYSOLS	77.5	3
3	LITHOSOLS	363.5	15
4	FLUVISOLS	67.9	3
5	KASTANOZEMS	368.5	15
6	LUVISOLS	265.0	11
7	AREMOSOLS	189.8	8
8	REGOSOLS	129.9	5
9	SLONETZ	50.2	2
10	VERTISOLS	167.6	7
11	PLANOSOLS	48.6	2
12	XEROSOLS	207.0	9
13	YERMOSOLS	133.2	6
14	Others	180.9	8

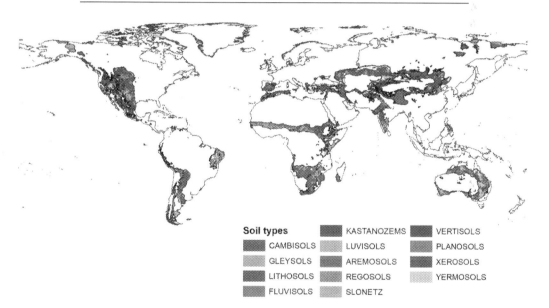

FIGURE 6.5 Major soil types across the world. (Source: FAO, 2021.)

vegetation cover as shown in the figure. Clouds, water, and non-vegetation objects have an NDVI value of less than zero. Areas of barren rock, sand, or snow usually show very low NDVI values. NDVI values between 0 and 0.2 are classified as bare soil/no vegetation cover in arid regions of the world. The values between 0.2 and 0.4 represent sparse vegetation areas. Sparse vegetation such as shrubs and grasslands. NDVI values from 0.4 to 0.6 and 0.6 to 0.8 represent moderate to high dense vegetation areas of agriculture. High NDVI values (approximately 0.8 to 1) correspond to dense vegetation which can be found in tropical forests or crops at their peak growth stage. This analysis shows the NDVI variations across the Africa and Asia continents. In arid as well as semi-arid regions, it has been observed that significantly less NDVI (ranging from 0 to 0.2) indicates less vegetation throughout the year.

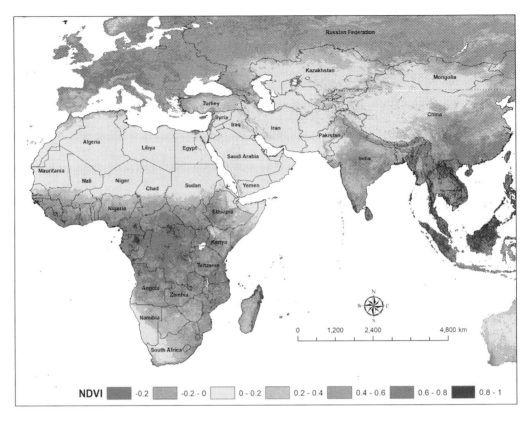

FIGURE 6.6 Normalised Difference Vegetation Index (NDVI) variations under major soil types across the world.

6.5 IMPACT OF CLIMATE CHANGE ON SEMI-ARID AGROECOSYSTEM

The IPCC's Sixth Assessment Report indicates that Africa and Asia, particularly the semi-arid drylands, are at the highest risk from climate change, given the magnitude of existing stresses in the continents (IPCC, 2021). In the coming years, significant areas of the drylands will likely see changing rainfall patterns with more frequent and more intense extreme events such as droughts and floods. Climate change impacts are (1) direct, on crops and livestock productivity domestically, (2) indirect, on the availability or prices of food domestically and in international markets, and (3) indirect, on income from agricultural production at both the farm and country levels (IPCC, 2014; Srinivasarao et al., 2019).

South Asia: Climate change will likely to cause environmental and social stress in many of Asia's rangelands and drylands. Surface and ground water resources in the arid and semi-arid Asian countries play vital roles in forestry, agriculture, fisheries, livestock production and industrial activity. The water and agriculture sectors are likely to be most sensitive and hence vulnerable to climate change-induced impacts in arid and semi-arid tropical Asia (Ratnakumar et al., 2011; Srinivasarao et al., 2015). Climate change impacts are increasingly visible in South Asia (SA) with greater variability of the monsoon, noticeably a declining trend with more frequent deficits (IPCC, 2014). There has also been an increase in the occurrence of extreme weather events, such as heat waves and intense precipitation, that affect agricultural production and thereby the food security and livelihoods of many small and marginal farmers, particularly in the more stress-prone regions of the central and eastern Indo-Gangetic Plain (IGP) (Rosenzweig et al., 2021). Warming trends and increasing temperature extremes have been observed across most of the Asian region over the past century (Hijioka et al., 2014). Because of the current marginality of soil-water and nutrient reserves, some ecosystems in semi-arid regions may be among the first to show the effects of climate change. In semi-arid areas, both positive and negative, rainfed agriculture is sensitive to climate change (Ratnakumar et al., 2011). Generally, climatic variability and change will impact sustained agricultural production in tropical Asia in the coming decades.

Stress on water availability in Asia is likely to be exacerbated by climate change (Rosenzweig et al., 2021). The elevated temperature and decreased precipitation in arid and semi-arid rangelands could cause a manifold increase in potential evapotranspiration, leading to severe water-stress conditions. In India for instance, a changing climate was projected to reduce monsoon sorghum grain yield by 2 to 14% by 2020, with worsening yields by 2050 and 2080. In the Indo-Gangetic Plains, a large reduction in wheat yields is projected. A systematic review and meta-analysis of data in 52 original publications projected mean changes in yield by the 2050s across South Asia of 16% for maize and 11% for sorghum (Knox et al., 2012). No mean change in yield was projected for rice. Climate change can potentially exacerbate the loss of biodiversity in this region. An increase in temperature could lead to an increase in the survival rate of pathogens (Skendžić et al., 2021). Damage from diseases may be more serious because heat-stress conditions will weaken the disease-resistance of host plants and provide pathogenic bacteria with more favourable growth conditions. The growth, reproduction, and spread of disease bacteria also depend on air humidity; some diseases-such as wheat scab, rice blast, and sheath and culm blight of rice will be more widespread in tropical regions of Asia if the climate becomes warmer and wetter (Skendžić et al., 2021).

sub-Saharan Africa: SSA is regarded as the region most vulnerable to climate change, affecting the livelihoods of the millions of people who directly depend on agriculture (Ascott et al., 2022). The high levels of dependence on precipitation for the viability of sub-Saharan African agriculture, in combination with observed crop sensitivities to maximum temperatures during the growing season (Lobell et al., 2011) indicate significant risks to the sector from climate change. Recurrent climate extremes are indicators of change in the patterns, timing, and amount of precipitation. They are major drivers of vulnerability in many agro-ecological zones of SSA, as these may lead to changes in crop yields and higher food prices. Climate change impacts may lead to loss of vegetation and biodiversity, and destruction of wildlife ecosystems.

Climate change projections for sub-Saharan Africa point to a warming trend, particularly in the inland subtropics; frequent occurrence of extreme heat events; increasing aridity; and changes in rainfall—with a particularly pronounced decline in Southern Africa and an increase in East Africa. Impacts of climate variability and change in the arid and semi-arid tropics of Africa can be described as those related to projected temperature increases, the possible consequences to water balance of the combination of enhanced temperatures and changes in precipitation and sensitivity of different crops/cropping systems to projected changes (Serdeczny et al., 2017). The observed impacts of extreme climate events have intensified in recent years, including prolonged dry spells, abnormal rainfall patterns, consequent shortage of water, and heat stress (Ayanlade et al., 2022).

Extreme weather events resulting from climate change are likely to negatively impact the yields of major cereal crops in the SSA region, with a reduction in crop yields and crop quality (IPCC, 2014, 2021). In many SSA regions, recent studies have established a warming trend through extreme heat events, with changes in rainfall (Ayanlade et al., 2022). Climate change is also likely to have direct and indirect impacts on the socioeconomic sectors of SSA. For example, Arndt and Thurlow (2015) estimated climate change impacts on Mozambique's economy, with about 13% reduction in GDP by 2050. Similar results were obtained in different southern African countries: Namibia, Angola, Malawi, Mozambique, and Zambia (Arndt et al., 2019). As agricultural livelihoods in the SSA region become riskier due to extreme events, the rural-urban migration rate may be expected to grow, adding to the already significant urbanization trend in the region (Serdeczny et al., 2017). This is because extreme events affect the rainfed agricultural systems on which the livelihoods of a large proportion of the region's population currently depend.

The overall impacts of climate change on agriculture are expected to be negative, particularly in the Sahelian countries, threatening regional food security. Schlenker and Lobell (2010) note that a "worst-case" projection indicates losses of 27%–32% for maize, sorghum, millet and groundnut for a warming of about 2°C above pre-industrial levels by mid-century. In the Southern Sahelian zone of West Africa, where the predominant soils are sandy, increased mean temperature could substantially affect the maximum temperatures at the soil surface. It is conceivable that surface soil temperatures could exceed even 60°C. Under higher temperatures, enzyme degradation will limit photosynthesis and growth. Increased temperatures will result in increased rates of potential evapotranspiration. In the long term, the very establishment and survival of species in this region's managed and unmanaged ecosystems may be threatened, resulting in a change in the community structure.

Livestock, an important enterprise in the drylands of SSA farming systems will also be negatively affected by climate change (Thornton et al., 2009; Thornton and Herrero, 2015). For example, the pastoral systems of the Sahel are highly dependent on natural resources, including pasture, fodder, forest products and water, all of which are directly affected by climate variability. Livestock is vulnerable to drought, particularly where it depends on local biomass production, with a strong correlation between drought and animal death (Godde et al., 2021).

6.6 DROUGHTS IN SEMI-ARID ECOSYSTEMS

A synthesis of frequency and occurrence of drought events in semi-arid tropics and economic losses due to droughts are discussed in this section. Drought is a natural calamity brought on by climate change that causes severe social, environmental, and economic harm. Extreme events such as drought and floods have been projected to increase globally in frequency and intensity under future climate scenarios (Hirabayashi et al., 2013; Trenberth et al., 2013; Gao et al., 2019). Out of all extreme events, drought occurs frequently and lasts long. However, drought may occur in any region of the world, but it is more intense and frequent in arid and semi-arid regions due to the large variability of climate in such regions (Huang et al., 2016; Ramkar and Yadav, 2018; Adejuwon et al., 2019; Srinivasarao et al., 2020). The farmers of this region (arid and semi-arid) are the ones who are more prone to drought-related effects. Drought at any stage of crop growth has profoundly affected them (Chand and Biradar, 2017). Around 15% of the surface of the Earth is covered by semi-arid regions, which in 2000 were home to 14.4% of all people on the planet (Safriel and Adeel, 2005). Such locations have delicate ecosystems that are highly susceptible to interactions between human activity and climate change (Xue, 1996; Zeng et al., 1999; Huang et al., 2010; Rotenberg and Yakir, 2010). Water shortage, which varies greatly in terms of time, place, quantity, and duration, is typically a feature of arid and semi-arid countries. In the past fifty years, droughts have occurred with a significant rate of recurrence and severity in the majority of the world (Vicente-Serrano et al., 2014; Wilhite et al., 2014). The increased impact of drought is brought on by pressure from human population growth, overgrazing, and ongoing land exploitation (Abdi et al., 2013; Adejuwon et al., 2021).

In the tropical semi-arid regions, numerous studies on drought have been conducted, including those in India (Singh, 2014; Jayasree et al., 2015; Ramkar and Yadav, 2018; Wable et al., 2019), Chile (Leon, 2002; Masih et al., 2014), Peru (Dzavo et al., 2019), Argentina (Rivera et al., 2017; Casali et al., 2018), Brazil (Marengo et al., 2017) and Mexico (Moreno and Huber-Sannwaid, 2011; Mavhura et al., 2015). Nearly 37.6 per cent (123.4 Mha) of India's land area is semiarid, and about 15.8 per cent (50.8 Mha) of it is arid. India experiences regular droughts that cause devastation in various areas of the nation, especially in the northeast and west central (Sharma and Majumdar, 2017). According to Mishra et al. (2019), seven significant drought periods—1876–82, 1895–1900, 1908–24, 1937–45, 1982–90, 1997–2004, and 2011–15 occurred in India between 1870 and 2016. According to climate change modelling research, the frequency and severity of droughts in the tropics of Asia and Africa may significantly vary (IPCC, 2007). The Intergovernmental Panel on The IPCC (2013) anticipated that droughts would occur more frequently in India's semi-arid regions due to climate change. For instance, Maharashtra has a semiarid climate, which covers around 83 per cent of its land area (World Bank, 2003; Kalamkar, 2011) and the projected rise in temperature (by 1.58°C–3.8°C) over the region is very likely to intensify the drought in Maharashtra, which will lead to higher agrarian stress. The highest frequency of droughts was recorded in Maharashtra's chronically dry region (Deosthali, 2002). A study conducted on the duration and severity of the drought in the semi-arid region of Karnataka reported that the precipitation is showing declining trends because of climatic change, hence drought severity will increase in that region in the future (Jayasree et al., 2015).

Several experts from around the world have evaluated various drought indices to discover the best drought indicator for the semi-arid regions (Pandey et al., 2008; Roudier and Mahe, 2010). The most accurate drought index is the Effective Drought Index (EDI) across semi-arid, sub-humid, temperate, and dry regions of India and Western Africa (Jain et al., 2015). In a different study, the effectiveness of five drought indices was tested for a semi-arid area in western India. They concluded that the Standardized Precipitation Evapotranspiration Index (SPEI) is the most appropriate drought index for assessing the drought severity in that area (Wable et al., 2019). Leon (2002) conducted a study at the Limarí River basin (a semi-arid area in Chile) and reported that from 1950 to 1999 there were 15 severe droughts observed in this region. Studies reveal that droughts tend to be more frequent, longer and more severe in the boreal spring and summer in East Africa, as the overall precipitation and water storage abruptly decline (Haile et al., 2019). Another study on geospatial and temporal variation of droughts in Africa indicated that droughts have become more frequent, intense and widespread during the last 50 years in the Africa region (Masih et al., 2014). The extreme droughts are 1972–73, 1983–84, 1991–92, 1999–2002 (in northwest Africa) and 2010–11 (in eastern Africa). The Horn of Africa (HoA) region comprises—Djibouti, Eritrea, Ethiopia, Kenya, Somalia, South Sudan, Sudan, and Uganda is one of the most drought-prone regions in the world with many arid and semiarid areas. The frequency, duration, and intensity of droughts increased in the HoA region over the past decades (Han et al., 2022).

In terms of several socioeconomic factors, such as agriculture, human health, sea level rise, labour shortages, disease prevalence, etc., droughts have an extensive impact on humanity (Adger, 1999). Droughts are predicted to affect people's ability to make a living, and their prevalence will worsen poverty and the sustainability of available means of subsistence in the years to come. Moreover, there is a huge economic loss due to drought. For instance, Gupta et al. (2011) reported a loss of US$ 588 million and US$ 910.72 million due to the drought of 2000 and 2002 respectively in India. According to Mishra et al. (2007), a drought can have a wide range of negative effects on the country's financial system. These effects extend well beyond the area that is experiencing the drought. Moreover, the dramatic drought of 1982–84 led to a yield loss of approximately 18% in West Africa for millet compared to the more humid year 1985 (Sultan et al., 2019).

6.6.1 Classification of Drought

Numerous studies on drought, three types of physical droughts—meteorological, agricultural, and hydrological—as well as one type of socioeconomic drought (Wilhite, 2000). The physical settings and sectors that are prone to drought form the basis for this classification (Wilhite et al., 2007; Alam et al., 2014; Mekonen et al., 2020; Spinoni et al., 2020). Figure 6.7 depicts the development of various droughts across time, including changes to various physical characteristics and agricultural, hydrological, and socio-economic droughts that affected a specific region.

Four fundamental approaches—meteorological, hydrological, agricultural, and socioeconomic—were used by Wilhite and Glantz (1985) to categorise the drought. The last method looks at drought in terms of supply and demand, following how a water shortage affects socioeconomic systems. The first three approaches focus on measuring drought as a physical occurrence. Due to little rainfall and excessive evapotranspiration rates, metrological drought first manifests itself. Agricultural drought results from the meteorological drought in terms of decreased crop output and soil moisture. Afterwards because of

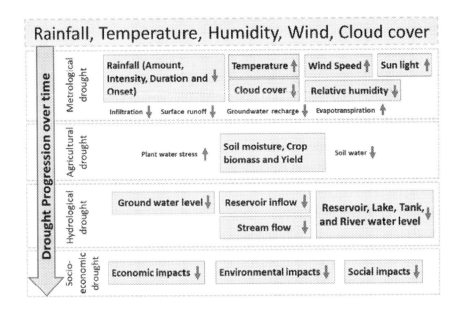

FIGURE 6.7 Schematic representation of drought process. (Source: Zargar et al., 2011.)

depletion of ground water and surface water resources, it is converted to hydrological drought. Finally, socioeconomic drought is formed due to huge economic loss and social distress. Different types of droughts are described as follows:

6.6.2 Meteorological Drought

The basis for meteorological drought, (also known as climatological drought) is the deviation of precipitation from the average value during a specific period and region. It is typically defined in terms of how dry something is and how long it has been dry. Since the atmospheric conditions that lead to a lack of precipitation vary greatly from region to region, definitions of meteorological drought must be considered region-specific.

6.6.3 Hydrological Drought

Lack of surface or subsurface water due to insufficient precipitation defines a hydrological drought. The main focus of a hydrological drought is how this deficit affects elements of the hydrological system including soil streamflow, moisture, groundwater, and reservoir levels. Hydrological system components (like soil moisture, stream flow, groundwater levels, and reservoir levels) take longer to show signs of deficits in rainfall. As a result, there is a time lag between these effects. For example, a lack of precipitation may cause a rapid loss in soil moisture that is almost immediately apparent to farmers. At the same time, its impacts on reservoir levels may not be felt for several months, postponing any consequences on the generation of hydroelectric power or recreational activities. Water usage in hydrologic storage systems (such as reservoirs and rivers) frequently conflicts with one another, making it difficult to predict when an effect will occur and how to assess its impact.

The hydrological drought is associated with low rainfall and resulted in declined blue water availability (surface and groundwater). Recurring hydrological droughts occur in several regions largely due to climate change, creating huge water scarcity including domestic and industrial sectors. To understand this effect, long-term data collected from micro watersheds of Alfisols and Vertisols at the International Crops Research Institute for the Semi-Arid Tropics (ICRISAT), Hyderabad is depicted in Figure 6.8.

Drought Management in Soils of the Semi-Arid Tropics

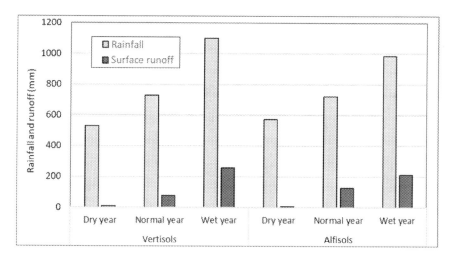

FIGURE 6.8 Runoff response with change in rainfall amount in Vertisols and Alfisols at ICRISAT research station, Hyderabad. (Source: Garg et al., 2022a.)

Runoff is found as one of the most sensitive parameters. Generally, runoff generated against 500 mm rainfall or below was found negligible in both Alfisols and Vertisols landscapes. The received amount is captured in the form of soil moisture but does not generate surface runoff or a negligible amount. In normal years with rainfall of 725 mm, the runoff generated from Vertisols was found about 10% and Alfisols 17%. Further, runoff generated in wet years from both landscapes was 21%–23%. This indicates blue water, which is surplus after satisfying the initial requirement, is the most sensitive component. Under the changing climatic conditions, slight changes in rainfall can create hydrological drought or flooding in downstream areas.

This is evident in one of the regions of the Central Indian landscape called Bundelkhand region where rainfall pattern has changed significantly. Rainfall in 1960–90 ranged between 750–1050 mm and declined to 550–850 mm in 1991–2020 resulting in frequent hydrological droughts, outmigration, water scarcity including domestic purposes (Figure 6.9). The region receives about 85% of the annual rainfall from July to September. Long-term data (between 1950 and 2017) of 23 rainfall stations covering all seven districts of Uttar Pradesh's Bundelkhand region showed declining rainfall status. Figure 6.9 shows spatial rainfall variability patterns between 1950–83 and 1984–2017. It revealed that districts such as Jalaun, Hamirpur, Mahoba are affected more in terms of reduction in rainfall amount in recent decades (1984–2017) compared to previous decades (1950–83). The average annual rainfall of the study region was 867 mm in 1950–83, which declined to 683 mm in 1984–2017. Out of 23 stations, two stations showed no reduction; whereas, rainfall reduced by 10%–20% for 10 stations, 20%–30% for 3 stations, and more than 30% for 7 stations during 1984–2017 compared to 1950–83. Such a declining trend is due to the reduction in number of rainfall events from 47 in 1950–83 to 39 in 1984–2017 in different categories (medium: 10–30 mm, high: 30–50 mm and very high: >50 mm) which clearly indicates evidence of climate change in the region. This hurt the water balance at the regional scale, especially on groundwater recharge (Singh et al., 2014).

6.6.4 Agricultural Drought

Agricultural drought relates numerous traits of both hydrological and meteorological drought to effects on agriculture. It concentrates on the lack of precipitation, discrepancies between actual and potential evapotranspiration, soil water shortages, and decreased groundwater or reservoir levels. A plant's water requirements are influenced by the weather, the sort of plant it is, where it is in its growth cycle, the biological and physical characteristics of the soil, and other factors. The varying vulnerability of crops at various stages of crop development, from emergence to maturity, must be

174 Managing Soil Drought

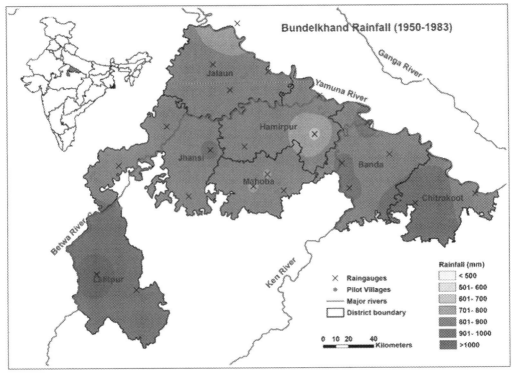

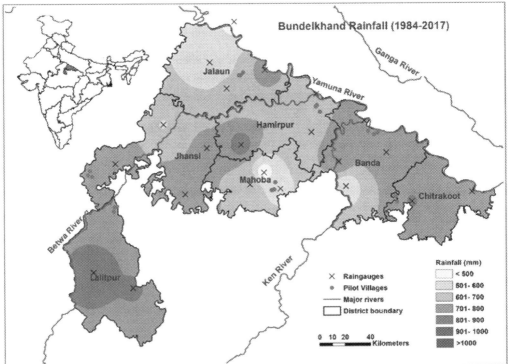

FIGURE 6.9 Location of pilot villages along with rainfall variability in Bundelkhand region, Uttar Pradesh, Central India.

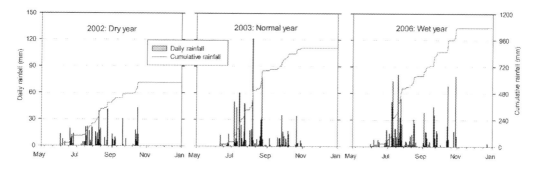

FIGURE 6.10 Rainfall distribution in selected dry, normal and wet years at ICRISAT research station, Hyderabad, India.

considered in any definition of agricultural drought. Germination may be hampered by insufficient surface soil moisture at planting, which could result in small plant populations and a lower yield overall.

Agricultural drought is a situation where the landscape cannot supply enough moisture required for plant growth at different agronomical stages, resulting in yield penalty. Recurrence of agricultural drought may lead to crop failure resulting in food insecurity in the region. Rainfall distribution and infrastructure availability are important in defining the severity of agricultural drought. Supplemental irrigation from any water source can mitigate the effects of agricultural droughts.

Figure 6.10 shows the rainfall distribution in selected dry, normal and wet year. The year 2002 was categorised as a dry year with a total rainfall of 522 mm between June and December. Runoff recorded during this year was negligible (< 20 mm). While the highest rainfall event during 2002 was about 40 mm, its distribution was relatively uniform as the longest dry spell recorded during the year was less than 12 days. The year 2003 was a normal year with 887 mm of rainfall of which about 70% (~644 mm) was received from July to August. Further, from the last week of August to the third week of September (25 days), there was a long dry spell except for one meagre rainfall event (<8 mm). Total runoff volumes generated during 2003 were in the order of about 181–198 mm, respectively. The year 2005 was a wet year with a total rainfall of 1072 mm, which was well distributed with a combination of high to low-intensity showers throughout the season.

Table 6.2 shows the yields from different cropping systems during dry, normal and wet years in Alfisols landscapes. Between 2002 and 2006, 2 years were normal with 750 mm rainfall, 2 were wet with 1070 mm rainfall and 1 year was dry with 573 mm rainfall. Yields from sole sorghum were the highest in wet years (2.8 t/ha) followed by dry (2.47 t/ha) and normal years (2.16 t/ha). In the groundnut/pigeon pea intercrop, the average groundnut yield in the normal year was 0.4 t/ha and pigeon pea yielded was 0.87 t/ha; however, yields obtained during dry and wet year were relatively higher. Rainfall distribution played an important role in fulfilling crop-water requirements. The agricultural drought explains this phenomenon. For example, the year 2002 was a dry year with 573 mm rainfall, but better rainfall distribution led to maximum yields. The highest productivity was achieved in groundnut/pigeon pea intercrop (3.53–9.12 t/ha) followed by sole sorghum (2.16–2.80 t/ha). Similarly, other parameters such as economic water productivity and net income are also higher in dry and wet years than the normal years (Anantha et al., 2023).

6.6.5 Socio-economical Drought

Socioeconomic definitions of drought link meteorological, hydrological, and agricultural drought features with the supply and demand of some economic good. The major cause of this kind of drought is a shortage of water caused by weather when the demand for an economic commodity exceeds the supply. As opposed to the other types of droughts mentioned above, their occurrence

TABLE 6.2
Crop Yield, Economic Water Productivity and Net Return during Dry, Normal and Wet Years in ICRISAT Research Station, Hyderabad, India

Particulars		Dry Years ($n=1$)	Normal Years ($n=2$)	Wet Years ($n=2$)
Landform treatment		Raised bed	Raised bed	Raised bed
Rainfall (mm)		573	752	1070
Runoff (mm)		9	115	257
Effective rainfall (mm)		563	637	812
Measured crop yields (t/ha)				
1. Sole sorghum	Sorghum	2.47	2.16	2.80
2. Groundnut/pigeon pea intercrop	Groundnut	1.87	0.40	1.23
	Pigeon pea	1.51	0.87	1.03
Sorghum equivalent yield (t/ha)				
1. Sole sorghum		2.47	2.16	2.80
2. Groundnut/pigeon pea intercrop		9.12	3.53	6.11
Sorghum equivalent WP (kg/m^3)				
1. Sole sorghum		0.44	0.39	0.34
2. Groundnut/pigeon pea intercrop		1.62	0.66	0.75
Net return (US$/ha)				
1. Sole sorghum		377	308	452
2. Groundnut/pigeon pea intercrop		1821	553	1139
Economic WP (US$/m^3)				
1. Sole sorghum		0.07	0.06	0.06
2. Groundnut/pigeon pea intercrop		0.32	0.11	0.14

Source: Anantha et al. (2023).

is dependent on the dynamics of supply and demand across time and space. The weather impacts numerous economic commodities, such as water, pasture, food grains, fish, and hydroelectric power. Due to the natural unpredictability of the climate, water availability is adequate in some years but insufficient to meet human and environmental needs in other years. As power plants' reliance on streamflow, a drought in Uruguay in 1988–89 drastically decreased hydroelectric power generation. To meet the country's energy needs, the government had to rely more heavily on more expensive petroleum and put in place strict energy-saving measures. This shows the wider impact of drought.

6.7 DROUGHT MANAGEMENT STRATEGIES/OPTIONS IN SOILS OF SEMI-ARID TROPICS

6.7.1 Enhance the Soil Functions and Soil Ecosystem Services

Through the land-atmosphere feedback loop, soils contribute to the emergence and persistence of agricultural drought and, to a lesser extent, hydrological and meteorological drought. Soils have been identified as a key ecosystem component that may offer critical hazard regulating functionality. The ability of soils to store water, biomass, and energy has long been recognised as having a regulatory effect on natural disasters (Andréassian, 2004). The linked ecology and soils can control water dynamics. For example, water can be stored in the soil profile in the wet season may be utilised during drought in summer. Strategic soil management has the potential to increase an ecosystem's resistance to the effects of local or regional droughts by concentrating on managing the dynamic biotic and abiotic processes of soils, such as drainage capacity or organic matter (Carminati et al., 2010).

Drought Management in Soils of the Semi-Arid Tropics

There are several measures in agricultural systems that can be utilised to lessen the consequences of local or regional drought. These solutions mostly centre on soil management. For instance, keeping the soil surface covered with mulching and cover crops will prevent runoff and decrease evaporation (Bodner et al., 2015; Srinivasarao et al., 2022b). Mulch has been shown to increase soil water storage by 10% by reducing soil evaporation (Mrabet et al., 2003). Moreover, no-till operations report a higher soil water content than conventional tillage (Verhulst et al., 2011). A review of several studies under different soil types and agroecologies concluded higher water use efficiencies in conservation agricultural practices than in conventional practices in South Asia (Jat et al., 2020). The percolation of the soil improves three-fold under conservation agriculture managed fields than in conventional managed fields (Jat et al., 2017). It has been demonstrated that using hedgerows, both separately and in conjunction with no-till techniques, helps to reduce the risk of drought (Frank et al., 2014). Additionally, a recent study found that organic fertilisation increases the soil microbial communities' resilience and resistance to extreme drought (Sun et al., 2022).

6.7.2 Sustainable Intensification/Diversification of Cropping/Farming Systems

Diversification of cropping/farming systems is important in mitigating drought in semi-arid regions (Wani and Sawargaonkar, 2018a). When faced with water scarcity issues, crop variety has been praised as a technique to control insect outbreaks, maintain biodiversity, and improve water management. In crop diversification, the existing monoculture cropping system combines with the crop which requires less water that can withstand deficit irrigation, in this way, it helps in mitigation of drought (De Boni et al., 2022). Another study concluded that crop rotation diversification improved maize yield drought resistance (better-maintained yield under drought) (Renwick et al., 2021). Moreover, the adoption of drought-tolerant crops (Evans et al., 2008; Nikolaou et al., 2020) may contribute to improving biodiversity, solving the problems of water scarcity and making the agro-ecosystems more resilient towards drought.

6.7.2.1 Case Study from Mali

In Mali, farming is a low-input system characterised by limited resources, low technology, low funds, and limited information (Kaya et al., 2000). Cotton (*Gossypium hirsutum* L.) is the main cash crop. It is often grown in rotation alongside cereals such as sorghum (*Sorghum bicolor* (L.) Moench), millet (pearl millet (*Pennisetum glaucum* L.) R.Br.), maize (*Zea mays* L.), as well as legumes such as groundnut (*Arachis hypogaea* L.), and cowpea (*Vigna unguiculata* L.) Walp.). Cotton and maize are the major crops in the region that receive nutrient input additions in the form of manure and/or mineral fertilisers, whereas other cereal crops rarely receive fertilisers (Autfray et al., 2012; Traore et al., 2014). These practices eventually contribute to the nutrient depletion of the low-fertile soils in the fields. Soil fertility depletion negatively affects up to 40% of the annual agricultural revenue of farmers in southern Mali (DeRidder et al., 2004). To reverse this trend, farmers in southern Mali collect and store biomass not only for animal feed but also to produce organic manure (Blanchard, 2010), resulting in nutrient flows from biomass, organic household waste, compost and animal manure to crop fields. Therefore, the flow of organic matter/biomass from the production sources is a major factor in soil fertility management on most agricultural fields in southern Mali. Despite the remarkable promising results regarding crop biomass management and soil fertility management (Autfray et al., 2012), there is a knowledge gap on nutrient flow dynamics from households to farm fields particularly for farms of different resource ownerships including livestock and production orientation and intensities.

Previous studies have shown negative nutrient balances in the ferruginous tropical soils located in the subhumid zone of West Africa (Kanté et al., 2007; Ramisch, 2014; Audouin et al., 2015). Furthermore, N and K balances were negative (-27 to $-34\,kg\,N$, -18 to $-28\,kg\,K\,ha^{-1}$) (Roy et al., 2005) while P content was close to zero (Cobo et al., 2010; Smaling et al., 2013) in Southern Mali. The study's objective was to characterise the dynamics of nutrient flows from households to farm fields using the Nutmon (nutrient monitoring for tropical farming systems) model. The specific objectives were to: (i) quantify biomass and manure flows from different organic sources, (ii) evaluate nutrient flows concerning organic sources and farm typologies, and (iii) determine the nutrient balances associated with each farm typology and crop type.

6.7.2.1.1 Variations in Nutrient Management from Biomass and Manure

Despite significant efforts to collect biomass, the produced stocks were still low for all farm types and only covered 46% of fodder needs for both HRE and LRE farms compared to 42% for MRE farms. Lack of biomass results in the departure of animals especially with regard to the conversion of large farm typologies to transhumant practices in areas with high forage potential (Umutoni et al., 2016; Ayantunde et al., 2019), thus leading to a transfer of fertility (Coulibaly et al., 2009; Andrieu et al., 2015) and strong pressures around resources (Brottem, 2016). At the farm scale, nutrient management depends on farmers' decisions of soil fertility management which are also influenced by both socio-economic and biophysical environments, resource endowment and production objectives (i.e. land use and crop selection). Thus, HRE and MRE farm typologies provide more nutrients based-manure to cotton and maize crops with high monetary income, unlike LRE which prioritises millet and sorghum to address household food self-sufficiency.

Furthermore, nutrient management varies with distance from the homestead, leading to strong soil fertility gradients. Fields that are distant from the homestead or more difficult to access due to steep slopes are managed as remote fields and reserved for millet and sorghum crops which utilize less mineral nutrients (Tittonell et al., 2005). Lack of biomass demonstrates the available opportunities regarding forage-based crop management practices not only for improving agricultural productivity but also for improving the sustainable management of local ecosystems (Ayantunde et al., 2014). Perennial non-cultivated fodder, such as *Andropogon gayanus*, can yield up to 22 t ha^{-1} of biomass (Fall et al., 2005). This production can facilitate animal feeding and increase manure production. Forage production from maize, sorghum (Sanon et al., 2007), and millet (Pasternak et al., 2012) must be scaled up for farmers in order to alleviate feed shortages. As happens with mineral fertilisers, subsidies be granted to small farmers, and access of all farmers to fodder seed crops such as A. gayanus and Brachiaria ruziziensis R., as well as dual-purpose varieties such as Soubatimi and Tiendougou-coura, should be facilitated. In addition, the promotion of fodder tree/shrub legumes such Leucaena leucocephala, Piliostigma reticulatum, etc. will alleviate feed shortages and address soil nutritional deficiencies. Furthermore, livestock manure can be effectively managed by adopting animal corralling systems in the field (Rahman et al., 2019), which could result in significantly positive impacts on soil nutrient recycling and texture and fertility improvements.

Results obtained for manure indicate that the HRE and MRE farms produced 60 t/year and 34 t/year of manure respectively, when compared to the 10.3 t/year produced by LRE farms. Similarly, the NPK amounts produced from HRE farms were 50% and 80% higher than those produced by MRE and LRE farms, respectively. The differences in manure production are due in part to the differences in biomass access depending on the farm typology with regards to technical means such as animals, available workers, and cropland area (Autfray et al., 2012; Coly et al., 2013; Bonaudo et al., 2017).

Overall, 25% of manure came from compost at the HRE and MRE farms compared to only 8% at the LRE farms, indicating opportunities for increasing compost production, especially for the LRE farm typology. The main challenge, however, involves access to biomass and worker, and addressing these issues would promote the use of fodder crops and the processing of residues from communal pasture areas. LRE farmers could organise themselves into groups for collective compost production.

Regarding the distribution of manure per crop, we found that on HRE farms cotton fields can be fertilised with 5 t farmyard manure per hectare, while maize fields receive only 3 t farmyard manure per hectare corresponding to 65% of the fields in HRE and 47% in MRE. These findings can be attributed to farmers placing greater value on cotton and maize by supplying more fertiliser to these crops for economic purposes (Hussein, 2004). In addition, cotton is the main cash crop and the Compagnie Malienne buys the production pour le Developpement des Textiles (CMDT), which provides farmers with credit facilities for fertiliser on cotton and maize (Falconnier, 2016). The income generated in this way can support critical family needs, such as children's health and education costs (Anderson and Valenzuela, 2006; Baquedano et al., 2010).

For millet and sorghum grown at MRE and LRE farms, the produced manure covers only 35% and 21% of the fields, respectively, versus barely 12% of these fields at HRE farms, in which these crops are primarily integrated into rotation to benefit residual fertiliser from cotton and maize (Tounkara, 2018). This result indicates that millet and sorghum receive less manure or seldom receive any mineral fertiliser. This indicates the need to increase mineral fertiliser or manure application on millet and sorghum. Moreover, millet/sorghum yields are known to increase significantly when intercropped with a legume than when grown as a sole crop (Trail et al., 2016) through soil moisture improvement due to soil covering, which limits evapotranspiration (Sissoko et al., 2013).

6.7.2.1.2 Sustainability of the Current Agricultural Production System

Early studies on nutrient balances in sub-Saharan Africa indicate serious nutrient deficiencies in the farming systems (Hengl et al., 2017) especially in the cotton zone located in southern Mali, where 30% of farmers' income is derived from soil nutrients that have been mined and not replaced (van der Pol, 1992). Since that study, possible solutions have been applied through integrated crop and livestock systems (Ayantunde et al., 2020), including cattle corralling systems and systems in which crop residues are transformed into compost that is then mainly incinerated directly in the fields (Gandah et al., 2003).

The pessimistic results of the 1990s have evolved into the relatively positive results ($\pm$ 3.2 kg N ha^{-1}) obtained around the 2000s (Kanté, 2001) and reflect the findings of this study. Phosphorus has long been considered the main limiting nutrient for cereal production in the Sahel (Bationo et al., 2015). However, our findings, indicated that the phosphorus scenario has improved due to continuous application of phosphorus in the region via NPK fertiliser (Paul and Annicet Hugues, 2018).

In addition, this positive development might be attributable to the advancements in small agricultural mechanization systems, which have improved the overall crop production system management efficiency (Aune et al., 2020; Peyraud et al., 2020). Moreover, the recent popularization of rock phosphate application (Traoré et al., 2007), which enhances the soil pH and leads to an increase in P nutrition, might have played a significant role in the mitigation of past P deficiencies (Bagayoko et al., 2000; Buerkert et al., 2000).

The nutrient balances obtained herein reflect a positive nitrogen balance for cotton, which can be explained by the fact that cotton is the main cash crop and benefits up to 70% of farm-scale manure production in addition to mineral fertiliser inputs. At the same time, sorghum and millet receive little or no organic or mineral fertilisers (Gaborel et al., 2006). It can be inferred that the residual effects of cotton fertilization are not sufficient for maintaining nutrient balances under the current sorghum and millet cropping systems.

Although farmers do not have sufficient financial means to purchase mineral fertilisers, it is still recommended that these farmers adopt the microdose approach (Ibrahim et al., 2016; Traore et al., 2019), which consists of hill placements of 6 g of NPK and 200 g of manure protected from erosion and corresponds to 60 kg ha^{-1} and 2 t ha^{-1}, respectively. This approach is relatively inexpensive, maximises the residual effects of fertilization, and improves millet and sorghum growth and production rates (Ibrahim et al., 2015; Coulibaly et al., 2019). This presents an opportunity for policymakers and extension workers to promote the microdosing technique under cereal cropping, especially in the regions where cotton is driving the system (Sogoba et al., 2020).

Despite current efforts associated with organo-mineral application, our results showed generalised potassium deficiencies for all studied crops. Previous research results also confirmed that leaf deficiencies (Gaborel et al., 2006) led to photosynthesis dysfunction and poor grain quality

(Hafsi et al., 2014). The negative partial potassium balance observed herein can be explained by removing crop residues from agricultural fields which is a current practice in the study area. In the last two decades, livestock rearing has become a secondary activity that has been integrated into the practices of agriculturalists residing in the analysed zones (Ayantunde et al., 2014). Crop residues are no longer left in the fields and thus account for potassium loss (Gerardeaux, 2009). Therefore, to meet the demands of feeding animals and manure production, this imbalance can be addressed by revising mineral fertiliser formulas and increasing the K contents in available fertilisers.

6.7.3 Reducing Carbon-Energy Footprint: Solar-Based Irrigation Systems (SBIS) in sub-Saharan Africa

In many sub-Saharan African (SSA) countries, much of the agricultural development is constrained by the gap in knowledge on improved agricultural practices (Attia et al., 2022). This has resulted in few options for smallholder rural farm households to improve their livelihoods. The problem is aggravated by the increased frequency of rainfall variability and depleted soil nutrients from many farm fields. As a result, crop and livestock productivity has remained low for many years in most SSA countries (Birhanu et al., 2019). Much of the agricultural land in Mali has been degraded and is less fertile because of rigorous cultivation over the years along with wind and water erosion. This has resulted in increased food shortages as the land has not been able to support the food demands of the ever-increasing population. Lack of reliable access to available water resources in the sub-surface hydrological system further hampers better agricultural and nutritional security options. Sustainable and efficient water management practices are key to improved smallholders' agricultural productivity and natural resource management in rainfed agricultural systems.

Farmers in rural Mali cultivate vegetable gardens during the dry season using traditional irrigation systems. The vegetable gardens, though limited in scope, allow food diversification in the household, leading to an increased household income. Traditional irrigation uses shallow wells with depths ranging from 6.5 to 14.5 meters (Birhanu and Tabo, 2016). A recently conducted survey on water availability and access in rural Mali revealed that while 39% of rural Malians always lack water, periodic water shortages are experienced by the majority of communities (61%) (Sanogo et al., 2021) sometimes or the other. In the traditional system, water for irrigation is collected manually from shallow wells through a bucket connected to a rope.

6.7.3.1 Prospects for SBISs Development in Mali

The agricultural system in Mali heavily relies on rainfed agriculture which suffers from land degradation and a limited contribution from the small-scale irrigation system. Rising population growth and increased fragmentation of household farmland necessitate special and urgent attention to the use of CSA practices that include SBISS (IWMI, 2019; Attia et al., 2022). In communities with large household sizes (for example the average in southern Mali is 27), increasing agricultural practices with irrigation-based technologies would ensure better household food security. Most rural Malians practice traditional irrigation systems except for a few donor-funded projects. Due to the limited water lifting systems and scarce availability of water in most shallow wells, the SBISs are practised once a year in most farm fields (69%).

Rural farmers still consider irrigation practices using solar energy as complementary to rainfed agriculture. However, the result of our study highlighted that ~20% of the available land area in the two districts is suitable for solar-based irrigation investment. This potential, together with the untapped groundwater resource in Mali (Birhanu and Tabo, 2016), if properly managed and appropriate investments are in place, would be a game changer for the Malian agricultural system. Groundwater reserve in Mali is an untapped resource requiring due consideration and other management practices in the changing climate conditions.

However, the promotion and scaling of SBISs are limited by the low rate of literacy level among most rural communities. The technology requires skills in implementation, maintenance, and use.

However, 46% of the studied districts' rural communities do not have formal education (Sanogo et al., 2021). When there are no industries or equivalent employers to provide local people with income-generating opportunities, implementing small-scale SBISs with the provision of required skills could be a useful option to enhance the livelihood of rural communities. This needs to include introducing low-cost soil water sensors that are useful to determine the minimum amount of water and frequency of irrigation (Adimassu et al., 2020).

The other limitation is the availability of suitable land. Land requirements, considered environmental factors, have been identified as one of the most critical factors for irrigation investments (Rabia et al., 2013). Despite being a major constraint in most developmental projects (), land use type is the foundation to plan and allocate land for diverse investment options (Tahri et al., 2015). For example, land with appropriate climate conditions for solar energy investments may have a lower value if the land use factor is considered (Carrión et al., 2008). Additionally, the dominant soil types in both districts, i.e., Regosols (in Bougouni), and Lixisols (in Koutiala) are characterised by low nutrient status and storage capacity and are susceptible to erosion. In this case, areas with Haplic Lixisols, Gleysols, and Ferric Luvisols were found to be suitable for solar-based irrigation development with the additional application of organic manure and mulches (Jalloh et al., 2011). In particular, Gleysols that are found in low-lying landscape positions with shallow groundwater reserves are better sites as they contain relatively higher organic matter and available nutrients (Jalloh et al., 2011).

Areas receiving annual rainfall >800mm were also found to be suitable. However, the spatial and temporal rainfall variability as evidenced in southern Mali (Ebi et al., 2011; Akinseye et al., 2020; Sanogo et al., 2021) could be a limiting factor to recharge groundwater aquifers. This could also impact the sustainability of SBISs. As smallholder farmers cannot afford hydrocarbon-energised motor pumps or electrical pumps, the affordability of solar panels in many rural places as explained by Schmitter et al. (2018) makes SBISs an emerging climate-smart technology for most rural Malian populations. Hence, for a better result, the output of the study needs to be integrated with other intervention measures such as good agronomic practices (Traore et al., 2017) and landscape-based soil and water conservation practices (Traore and Birhanu, 2019; Birhanu et al., 2022).

6.7.3.2 SBISs as a Game Changer for Smallholder Agriculture in sub-Sahara Africa

In SSA, innovative irrigation systems are essential to secure smallholder farmers' year-round food production to contribute to the increased demand for food. SBISs are potentially a game changer for SSA smallholder agriculture from several aspects. SBISs are alternatives enabling smallholder farmers to grow more crops in a year by utilizing abundant sunlight and groundwater, mitigating climate change by reducing CO_2 emissions (Brunet et al., 2018). SBISs can provide clean irrigation to millions of farmers, empowering their adaptive and resilient capacity by raising agricultural productivity and their incomes (Ockwell et al., 2018). SBISs can open new avenues of opportunity and potential for agricultural growth while transforming SSAs' entire range of farming systems.

However, activating SBISs' game-changer potential requires a long-term commitment and comprehensive ingenuity of thought and action, time, and determination across scales. Several attempts are being undertaken in SSA countries to install electric pumps fed by solar energy and modern irrigation system to promote renewable energy and water use efficiency in agriculture (Noubondieu et al., 2018; Mugisha et al., 2021). Investments in west African countries (e.g., Ghana, Senegal, Mali, Gambia) focus on multiple initiatives for testing, de-risking, subsidizing, and analysing policy reform to sustainably scale solar-powered pumps (Brunet et al., 2018; Lefore et al., 2021). SBISs as game-changers must have the ability to manage uncertainties and overcome obstacles. Enabling this ability requires adaptive approaches to overcome systemic barriers related to the lack of contextually relevant innovation bundles, appropriate end-user financing, policy frameworks biased toward large-scale irrigation and rain-fed agriculture, weak market linkages, nascent private sector investment and increasing competition for water among sectors (IWMI, 2021a). It also needs to integrate public, research, and private sector actors to respond to diverse incentives and environmental trade-offs with the underground water depletion and e-wastes (Minh et al., 2020; Lefore et al., 2021) and improve stakeholder coordination, enact more effective policies, and facilitate integration within the value chains and across sectors (Izzi et al., 2021).

In the case of Mali, enabling SBISs as a game-changer is driven by viable business and investment opportunities, appropriate finance tools, and market integration for solar entrepreneurs and irrigators. There is a growing presence of private sector solar technologies suppliers. Still, the solar technology market can grow to reach many farmers directly when the demand for and supply of SBISs are matched. Understanding the diversity of farmers' SBISs demands, coupled with the solar irrigation suitability map (IWMI, 2021b) helps the suppliers to prioritise geographical areas for their marketing activities and business investments.

Farmers' demands differ in terms of the amount of water needed, land and water access, pump preferences, and capacity to pay for the SBISs. Tailoring the supply business models to different demands and abilities to invest is one of the necessary conditions to unpack the market bottlenecks. For example, in the areas where resource-limited and resource-poor farmers can access shallow groundwater like in areas of Dogo, Kokele, and Sibirila of Bougouni district, and N'golonianasso and M'Pessoba communes of Koutiala district, the business models should focus on supplying the solar-powered pumps with low capacity to match these farmers' ability to invest.

Finally, accelerating the SBISs requires an enabling environment where domestic manufacturers, irrigation and input suppliers, and small processing businesses can grow. Sustainable financing models help de-risk private sector investments in irrigation markets, especially products and services that support gender and youth inclusion. Win-win partnerships between entrepreneurs, farmer groups, cooperatives, and private and public sector actors help optimise the engagement of private sector companies that supply different equipment along the continuum of agricultural water management, creating a more robust irrigation market for farmers. Multi-stakeholder dialogues and platforms are ways to engage diverse business actors to increase market density and integration (Minh et al., 2020). They also encourage collaboration and learning to drive responsive innovations to address social and gender inequality, economic empowerment, water governance, and multi-sector/stakeholder coordination to accommodate local contexts, diverse partners and stakeholders, and emerging needs for feasible and sustainable SBISs (Lefore et al., 2021).

6.7.4 Landscape Rejuvenation Approaches

A growing number of research shows that dryland concerns may be solved by using integrated natural resource management interventions after landscape-based conservation of resources initiatives (Anantha et al., 2022; Anantha et al., 2021b, c; Garg et al., 2022b). Landscape-based resource conservation techniques largely concentrate on rainwater gathering by combining in-situ, and ex-situ practices with biological measures that follow the hydrological unit. Nature-based solutions offer the sustainable management of natural resources in rainfed ecosystems at the local level (Mrunalini et al., 2022). In this method, the landscape is altered starting at the crest using a combination of earthen and masonry structures, allowing for the prolonged retention of freshwater in the form of surface and groundwater. These interventions and structures are designed with field size hydrology in mind, and the terrain's topography is heavily utilised to save costs. Managing nutrient application in semi-arid Alfisols can be rationalised along with tank silt application in small holder farmers, providing improved nutrient and water retention for longer period (Sharan-Bhopal Reddy, 2022).

Increased retention in the landscape not only enhances water resource availability but also strengthens a range of ecosystem services, such as flood control and enhanced base flow. In addition, biological measures, including agroforestry further improve the sustainability of these structures. Without these interventions, the landscape remains unattended leading to excessive runoff generation and permanent fallows with high opportunity costs. Further, the landscape approach provides an opportunity for regenerating fragile farming systems into profitable farming systems in which various interventions (agriculture and allied sectors) could be integrated.

Several public welfare programs in Asia and Africa have been implemented by focusing on natural resource management to achieve UN SDGs (Anantha et al., 2021b). However, in the absence of a cluster approach, the full potential of these programs has yet to be realised. Under the landscape approach, various landscape and field scale technologies are bundled together to enhance resource availability and efficiency. The landscape approach ensured that the most degraded upland was given higher priority, and it was possible by following a decentralised rainwater harvesting approach. A fraction of runoff generated from individual plots was harvested by constructing earthen field bunds, farm ponds, and deepening and widening of stream networks. These low-cost interventions generate pockets of recharge zone even in landscapes with highly impermeable clay layers or hard rock geology. In these landscapes, decentralised rainwater harvesting largely generates opportunities for groundwater recharge through preferential flow. Otherwise, conventional methods could be more effective. While designing landscape conservation measures, it is important to understand land resource inventory and hydrology. This science approach helps quantify the availability of water resources, current freshwater demand and scope for bridging the demand-supply gap.

6.7.4.1 Case Study of Bundelkhand Region, Central India

The Drylands in the semi-arid tropics are a hotspot for water shortages, land degradation, poverty, and poor socioeconomic status. A higher level of agricultural production (0.6–1.6 t/ha) might be achieved in these areas due to limited groundwater recharge, high temperatures, and little to no regular rainfall. In particular, the Bundelkhand area of Central India, which is single-cropped and heavily dependent on rainfall, confronts several difficulties (Garg et al., 2020; Singh et al., 2014). In terms of quantity and distribution across time, rainfall is very variable. In contrast to the monsoon season, where dry spells of over 12–16 days or longer are also possible, more than 6–8 days are standard and happen multiple times (6–7) in a season. A hot summer precedes the tropical monsoonal climate of the area, with minimum air temperatures of 17°C to 29°C and maximum air temperatures of 31°C to 47°C in May. After that comes a chilly winter with the lowest air temperatures between 2°C and 19°C and maximum air temperatures in January between 20°C and 31°C. Most of the region's soils are deteriorated and have low levels of organic matter and nutrients. Hard rocks and aquifers, which make up the majority of the region's geology, are primarily the result of weathering and have formed a two-layered system: (1) unconsolidated fractured layers created by prolonged weathering of bedrock within 10–15 m, depending on topography, drainage, and vegetation; and (2) relatively impermeable basement beginning at a depth of 15 to 20 m. The sole significant water source for home and agricultural use in this area is shallow-dug wells that are 5 to 15 m deep due to the hard rock aquifers' poor transmissibility (Singh et al., 2014). The area's hard rock geology contributes to the groundwater aquifer's low specific yield. After the monsoon season, many dug wells begin to dry up, creating severe water shortages later in the year. The intensification of rainfed agriculture, which frequently experiences water scarcity, and problems to human well-being, is prevalent in this region.

6.7.4.1.1 Innovations

a. *Field bunding with field drainage structures*
   Laser land levelling technology was introduced in project villages to enhance water use efficiency. Farms with slopes less than one per cent were identified, and levelling operations were undertaken using laser-guided machines. In addition, for farms with undulated topography (2%–5% slope), there were better options than laser land levelling as a lot of soil mass had to be cut and filled to level it. Therefore, in-situ conservation measures such as field bunding were introduced. About 1.0 ha farm was divided into 4 to 5 parcels such that runoff and soil sedimentation is checked within the field and a piped outlet or masonry drainage structure was constructed to dispose-off excess water from the field.

The impact of earthen field bunding on crop yield of different crops and net income is described in Table 6.3. Barley, field pea and wheat yield increased by 10 to 20%; mustard yield increased by 40–50% compared to the non-intervention stage. This has translated into an increase in net income ranging from US$ 100 to US$ 400 per ha. Increased moisture availability and controlled water stagnation at the lower portion of the field were the main reasons for increased productivity.

b. *Rejuvenation of traditional rainwater harvesting structures*

In the Bundelkhand area, haveli agriculture is one of the ancient rainwater-gathering technologies that date back between 400 and 550 years (Sahu et al., 2015; Singh et al., 2022). According to catchment, farmers used to build 100–300 m-long earthen embankments (5–12 m broad and 2–3 m high) across the land's slope (Garg et al., 2012; Metre et al., 2014, 2016; Sahu et al., 2015; Garg et al., 2020; Garg et al., 2021; Singh et al., 2021). This system has a spillway capability for discharging extra runoff during the monsoon season and a conduit option for draining away all the water. Communities have typically repaired earthen embankments, water outlets, and scheduled water discharge to preserve these structures (Metre et al., 2016). Typically, a haveli system's catchment area is between 100 and 250 acres. These structures capture catchment runoff during the monsoon, allowing shallow-dug wells and borehole wells across the hamlet to be recharged. The impounded water is drained and rained out once the monsoon has subsided (by the end of October), and the haveli bed is then ready to cultivate rabi crops using the remaining soil moisture (Sahu et al., 2015; Garg et al., 2020). Farmers downstream utilise the haveli's drained water for pre-sowing irrigation as well. During the monsoon season, the haveli system serves as a reservoir (shallow water bodies with a depth of 1–3 metres). During the post-monsoon season, it transforms into agricultural fields (Figure 6.11). As a result, farmers maintain their

TABLE 6.3
Impact of Field Bunding on Crop Yield and Net Income in Central Indian Landscape

Crops	Yield (Kg/ha)		Net Income (US$/ha)	
	Before	After	Before	After
Barley	2560	3030	280	375
Field pea	2700	2950	1860	2050
Wheat	3125	3750	540	710
Mustard	1310	1940	610	990

FIGURE 6.11 (a) Harvested water in rejuvenated haveli structure during the monsoon period and (b) crop cultivated in the post-monsoon period on the haveli bed during the post-monsoon season using residue moisture.

land through haveli farming. The production of crops supplied by the haveli in the rabi season is 15%–25% greater than that of surrounding regular fields due to decomposed organic matter and extra humus (Sahu et al., 2015). Additionally, the need for irrigation is cut in half, drastically lowering cultivation costs and increasing net revenue by between 60 and 70 per cent. These motivating elements encourage farmers to engage in haveli agriculture even while their fields are drowned during the rainy season (Dev et al., 2022). However, haveli cultivation, which once supported a variety of ecosystem services and helped the area meet its freshwater needs, gradually ceased to exist for the following reasons: (i) Most of the earthen embankments collapsed, especially during heavy intensity rainfall events, largely because of burrowing and inadequate structure strength; (ii) There weren't enough outlets for the region's excess runoff to be safely disposed of; and (iii) the dissolution of water user groups and rural institutions (Garg et al., 2020).

ICRISAT and collaborators revitalised existing systems by developing an original core wall idea to overcome the difficulties mentioned above. Instead of using earthen embankments, a masonry core wall of the right height was built (together with the necessary foundation) to submerge the intended haveli fields fully. After the core wall was built, it was covered with soil to provide a sufficient cross-section for the necessary structural strength and to prevent thermal hysteresis for a longer lifespan. Additionally, adhering to the proper engineering designs, a masonry outlet was built at an appropriate location (at raised spots).

c. *Diversion drains, deepening and widening of drainage network for decentralised rainwater harvesting*

Degraded landscapes have high runoff-generating potential at the uppermost topo-sequence during monsoon season. The runoff, thus, generated can flood agricultural lands causing damage to crops and inundation in low-lying areas during kharif. The degraded uplands also suffer from acute water scarcity due to poor groundwater recharge and hence need to be cultivated in the rabi season (Figure 6.12). Farmers tend to leave such lands as fallows and migrate elsewhere to earn their livelihoods.

Diversion drainage channels with specific technical specifications per hydrology and plugs at suitable intervals are potential measures. Diversion drains along with surface channels along the slope facilitate water harvesting cum safe disposal of runoff without flooding the agricultural fields. Further, widening and deepening of stream network with plugs at suitable intervals is an important activity that offers the opportunity to create large storage without losing farmers' land and helps recharge groundwater and improve baseflow. The rainwater harvesting structures constructed in the ridgeline will drain excess water to these drainage networks.

FIGURE 6.12 Deepening and widening of stream networks for rainwater harvesting in decentralised approach (example from Jhansi, Central India).

6.7.4.1.2 Impact of Landscape Interventions on Drought Mitigation and Crop Intensification

Renovating the defunct *havelis* and various landscape-based NRM interventions offer an opportunity to overcome the challenges (mentioned above) typical to this region. Building a masonry core wall and a suitable masonry outlet has helped to sustain the *haveli* structures. The per unit cost of water harvesting by *haveli* renovation system is much lower (< 0.1 US$/m^3) compared to those of check dams and farm ponds (1–10 US$/m^3). Also, treating the landscape from the upper most ridge point by *in-situ* interventions and agroforestry has ensured sustainable intensification and diversification of the system. Field bunding and agroforestry have stabilised the landscape at the upland. During the field bunding, a provision for field drainage structures (a range of outlet systems) helps to dispose-off the excess runoff from the field, which helps control land degradation and enhance soil moisture availability. Otherwise, in the absence of field drainage structures, these field bunds ended up with short life by breaching it during heavy rainfall events, and its purpose has yet to be realised in most of the development programs.

Compared to the baseline scenario, landscape restoration interventions have increased base flow by 40%–60%, increased surface runoff by 50–80 mm/year, and decreased soil loss by 80%–85%. These increases have increased groundwater availability by 60%–160%. The water table in the wells was raised, often by 2 to 10 metres, to allow for additional irrigation. Increased soil moisture allowed for the resumption of profitable farming, improving land and water use efficiency.

From 120% (before the intervention) to 190% (after the intervention), cropping intensity rose as a result. Field bundling and laser land levelling are two examples of landform management techniques that have improved soil moisture availability (by 60 to 90 mm each season) and decreased the need for irrigation. Integrating landscape and field size interventions has increased crop productivity by 20%–80% in several cropping systems. Compared to the baseline scenario, the combined effects of field-scale interventions and landscape restoration have increased income from agriculture by 50%–150%. The income of the farmers in project villages increased from US$ 650/hh/year (before) to US$ 1800/hh/year (after) for smallholder farmers (having 1–2 ha land holdings). Many tribal families and vulnerable farmers, especially those who had migrated to the nearby urban centres for livelihoods, have returned to their villages and started practising agriculture and making a reasonable living for themselves.

6.7.4.2 Water Spreading Weirs in Drought-Affected Areas in Ethiopia

The importance of multiple economic, social and environmental benefits derived from land-based resources has increased in recent years. Sound management of these resources is therefore prerequisite to sustainable resource-based production systems. Landscape management, which is the application of land resource management systems through collaborative means, is considered by many to be the most appropriate approach to ensuring the protection, conservation and sustainability of all land-based resources and improving the living conditions of people in the uplands and lowlands. Integrated landscape management has become widely accepted as the approach best suited for the sustainable management of natural resources in upland areas and unlocking the potential of natural resources conservation for resilient production systems in lowland agropastoral systems and driving intensification for improved livelihoods.

Research and development projects strive to develop integrated solutions in dry and degraded farming systems where inappropriate management of water, nutrient, and genetic resources for a sustainable management of natural resources and livelihood improvement are important bottlenecks. Major issues of landscapes under dry environment conditions that require integrated solutions include nutrient depletion, water scarcity, low intensified production systems, resilient capacity to cope with shocks, and cross-scale biophysical and socio-economic issues. Thus, the focus of integrated landscape management activities is to enhance water productivity and nutrient efficiency through the rehabilitation of natural resources and the application of landscape-targeted water and nutrient management technologies and tools.

The efforts exerted on landscape management research anticipated to select and apply appropriate sustainable natural resources management and intensification options, investigate pathways of water, sediment and nutrients in response to landscape management, and collective approaches to collaborative land and water management. This chapter highlights the experiences and outcomes achieved on the application of flood spreading weirs for drought management in agropastoral system and present as case study to illustrate the potential of flood farming.

6.7.4.2.1 Practices of Flood Farming in Agro-Pastoral Systems

Managing flood water for productive use is one of the flood mitigation strategies that reduce flood hazards and simultaneously addresses food security issues in dry lowland environments. Many research findings stated that flood-based farming is an entry point to efficiently utilizing hazardous flood to productive use through various forms of agricultural development. Flood-based farming, according to Van Steenbergen et al. (2010, 2011), contributes to food security and provides numerous environmental benefits across a wide range of geographies by making flood water available for agricultural use. Flood-based farming is important for nutritional security and household coping mechanisms during the dry season when other food sources are depleted (Singh, et al., 2021). Annual flood regimes of rivers are important in flood-affected areas, where flood-based farming could be an effective solution to meet food security and sustain livelihoods (Motsumi et al., 2012; Sidibe et al., 2016; Balana et al., 2019).

Flood-based farming is a rainfed farming system that occurs in dryland areas and relies on supplementary water derived from various types of floods. It is a practice that depends on the residual soil moisture and soil nutrient deposits that remain after flood recedes (Nederveen, 2012; Balana et al., 2019). Flood-based farming usually occurs in relatively low-lying areas with gentle topography. Various forms of flood-based farming are found across the world's drylands (Van Steenbergen et al., 2010; Varisco et al., 1983). To determine the extent and duration of flood-based farming, water supply in flood farming is often difficult to predict due to uncertainties in the timing, duration, size, and frequency of floods from ephemeral and perennial streams (Van Steenbergen, et al., 2011). Furthermore, to utilize riverine floods, the river courses change from season to season leading to changes in riverbed levels and sediment accumulations.

In Afar region of Ethiopia, located near the base of the eastern escarpment of the Ethiopian highlands, is largely drought-prone. Annual rainfall ranges from 200 to 500 mm per year, spreading from March to September. Most of these rangelands depend on the flood stemming from the mountain highlands of Wollo and Tigray provinces which receive sufficient annual rainfall. Drought in the pastoral areas hits this region hard. ICRISAT, in collaboration with local partners, have been engaged to create a farming system that converts flood water harvested through flood water spreading weirs (WSWs) into productive land uses. The weirs reduced the velocity of the flash flood water and distributed it to the wider plain rangelands. The captured water was then used for producing dryland food and feeding crops. The intervention was implemented through a participatory approach with the community and local government. The local community is actively engaged and capacitated through training on land preparation, choice of crops and varieties, agronomic management, biomass management, crop-livestock integration, postharvest management and so on. A land use suitability analysis was conducted at the intervention site. The entire area was tracked using GPS to characterise soil-water distribution and soil fertility gradients created by the WSWs, which was effective in capturing sediment emerging from the highlands along with the flood, enriching the flat plains. The soil moisture regimes of the flood recession areas were plotted, and different parcels were assigned into different land use categories. Different food crops (maize, sorghum, teff, mung bean, cowpea, pigeon pea and sesame) and forage crops (elephant grass, Napier grass and natural grass) have been recommended for the different land use categories (Figures 6.13 and 6.14).

FIGURE 6.13 Appearance of flood spreading weirs immediately after a flash flood and the residual sediment accumulated in the flood recession plain. (Photo: Gizachew Legesse, ICRISAT.)

6.7.4.2.2 Impacts of Applying Flood Spreading Weirs

The WSWs have been implemented with communities in which their livelihoods are purely relying on the pastoral farming system. The WSWs intervention attempted to introduce and intensify an agro-pastoral farming system. The WSW (Water Spreading Weirs) created different farming zones following moisture and sediment distribution regimes. The flood water spreading weirs and demonstrations of food and fodder crop technologies resulted in production and environmental outcomes.

The water spreading weirs have created fields with varying sediment deposition where sands appear in upstream fields and silt and clay in downslope fields, and a moisture gradient between 10% and 22% across fields (Getnet et al., 2020). This in turn affects differences in crop yields and allows local lost grasses and trees to rejuvenate. For example, one of the learning sites has benefited 52 households with about 360 household members, providing food for them and fodder for their livestock. Regarding productivity, the communities manged to produce about 280 tonnes of biomass in a season from 35 ha of land. The high moisture zones provided 5 tonnes per hectare maize grain yield, while in low moisture zones, there is a total failure in grain. By increasing the plant population to 80,000 plants per hectare, maize biomass yield boosts to 15 tonnes per hectare. Sorghum biomass yield was 3.5 tonnes per hectare. Teff and mung bean performed very well in the systems yielding 1.5 and 2.0 tonnes per hectare, respectively. The performance of dryland legumes such as mung beans, cowpea and pigeon pea were promising. In one season, about 11.2 tonnes of fodder biomass per hectare was produced.

The water spreading weirs changed the landscape characteristics, enhanced soil moisture, reduced torrential floods, and enabled a fast recovery of degraded and dry landscapes into a productive and greening of abandoned and degraded rangeland. Using flood-spreading weirs, harnessing flash floods emerging from neighbouring highlands helped convert degraded dry rangelands to be productive and support diversified crop and fodder production (Getnet et al., 2020). The integrated solutions increased cultivated land by 44%, improved vegetation cover from 13% to 29%, reduced degraded rangelands from 87% to just 28%, improved land quality (soil moisture and nutrients), and enabled a gradual change in land allocation from communal use covered by grasses and shrubs to individual occupation. New dryland crops and fodder varieties were integrated into flood-based farming (Getnet et al., 2020). The environmental and socio-economic impact has been huge contributing to shift from pure pastoralist and dryland system to agro-pastoral system. Three learning sites have been implemented in Afar Regional State where more than 200 agropastoral households have directly benefitted. Flood-spreading weirs could potentially be scaled to 720,000 ha in the *Meher* season and 550,000 ha in the dryer *Belg* season in Afar alone to support crop and fodder production (Gumma et al., 2020). Although the dryland restoration practices show the greatest promise to enhance ecosystem services, the extent of implementation of integrated agronomic, mechanical, and vegetative management practices

Drought Management in Soils of the Semi-Arid Tropics

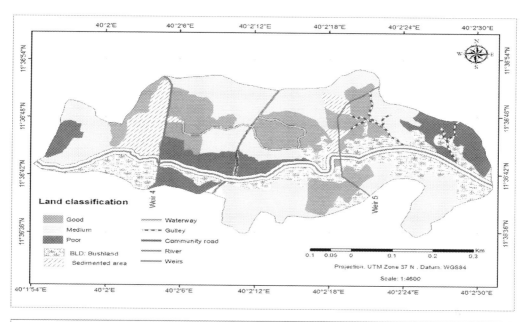

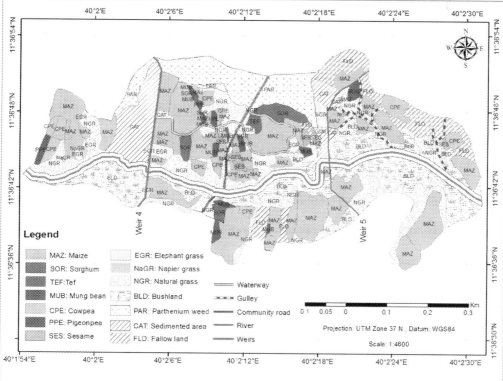

FIGURE 6.14 Mapping of homogenous farming zones (upper picture) and land use suitability plan (below picture) based on soil moisture gradient at the start of Meher season in 2017 at Chifra in Afar. (Source: Getnet et al., 2020.)

has not been well strategised by the extension service system. Thus, addressing land degradation risks, particularly soil erosion and extreme flooding, is not only a matter of implementing land management practices but also a concern on how to increase the efficiency and sustained use of these practices at the landscape scale (Figure 6.15).

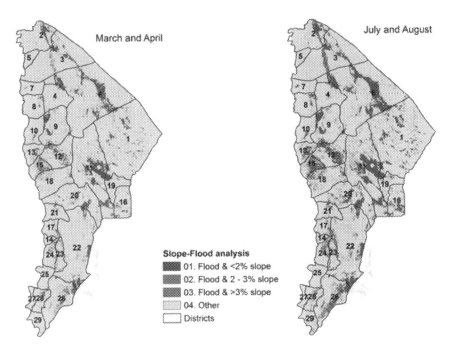

FIGURE 6.15 Potential flood areas for scaling at different slope categories in Afar. (Source: Gumma et al., 2020.)

6.7.5 Conservation Agriculture

Conservation agriculture (CA) is an ecosystem approach to regenerative agriculture and soil management systems based on three interlinked principles: (i) continuous minimum mechanical soil disturbance using no-till or reduced tillage-based crop establishment, (ii) maintenance of permanent soil cover using crop residues and cover crops and (iii) diversification of cropping system using economically, environmentally and socially adapted rotations including legumes and cover crops coupled with other complementary good agronomic management practices. Globally, CA has been adopted with less than 1 million ha in 8 countries in 1970 to 205 million ha in 102 countries in 2019 which is 15% of the world's cropland area (Kassam et al., 2022). Over the past couple of decades, a considerable attention has been given to conservation agriculture (CA) as a 'sustainable intensification' strategy in the semi-arid tropics of Asia and Africa (Jat et al., 2020; Jat and Chaudhari, 2022) and to optimise soil health and resilience to climate change (FAO, 2015; Farooq and Siddique, 2015; Kar et al., 2021). A meta-analysis of large datasets on CA research across South Asia (Jat et al., 2020) shows science-based evidence on the performance of CA on key parameters (yield, protein, water, profits, GHGs, soil health, adaptive capacity to climatic risks, etc.) which indicates that CA not only has multiple benefits but also has potential to contribute to the UN Sustainable Development Goals through alleviating the multiple stresses including emerging social issues currently faced by agriculture in South Asia. The results of meta-analysis of CA-based practices in South Asia show a mean yield advantage of 5.8% with 12.6% higher water use efficiency, an increase in net economic return of 25.9% and a reduction of 12%–33% in global warming potential while building soil carbon, reducing the climatic risks (Jat et al., 2020). Further, Jat et al. (2023) using an intensive review of the literature on the impact of CA on soil health reported that under the majority of the studies, implementation of CA practices has led to improvements in soil health which resulted in building resilience. In this section, various best management practices are described, which showed promising results.

6.7.5.1 Raised Bed Along with Intercropping for Enhancing Resource Use Efficiency

Due to the more effective and better utilisation of resources like green water and soil nutrients, intercropping or mixed cropping systems are more robust than monocropping systems in rainfed conditions (Sawargaonkar et al., 2008). These systems are steady amid unfavourable weather and pest/disease conditions. Basic land and water management elements for conservation and the controlled removal of surplus water include field drain formation and ground smoothing. For clayey soils with low infiltration rates, as the soil profile becomes saturated and waterlogged as the rainy season progresses, the broad bed and furrow (BBF) system is an enhanced in-situ soil and moisture conservation and drainage method (Garg et al., 2022a).

According to data from long-term research trials conducted at ICRISAT, controlling Vertisols with better management choices enhanced the soil's physical, chemical, and biological characteristics in micro-watersheds. Crops are sown on graded wide bed and furrow (BBF) of 45 cm practice for in-situ soil and water conservation. Extra runoff during a heavy rainstorm is safely disposed of as part of field-scale intervention of enhanced management. CA systems in rainfed Alfisols of Southern India show improved agronomic productivity besides mitigating greenhouse gas (GHGs) (Pratibhal et al., 2015).

Before the monsoon rains, the rainy season crops (sole and intercrops), along with pigeonpea, maize, sorghum, soybean, and green gramme, were seeded in the dry bed. Two crops were farmed each year in a rotation. Application of 80 kg N and 40 kg P per hectare was required for fertiliser management. The seedbed was kept flat per custom, and one crop—sorghum or chickpea—was produced in the post-rainy season using the profile's stored soil moisture. Farmyard manure was applied at a rate of 10 t/ha every two years, and no mineral fertilisers were used. According to the findings, fields with better management had considerably higher soil porosity, infiltration rate, and carbon content than areas with conventional management. Due to decreased runoff in BBF fields and increased storage of rainfall in green water form, such changes in bio-physical qualities also affected the hydrological cycle. For various dry, typical, and rainy years, Table 6.4 compares the runoff produced by Vertisols and Alfisols in raised bed and flatbed situations. In both soils, a runoff was found to be insignificant during dry years. In contrast to Vertisols, produced runoff during typical years is higher in Alfisols. Furthermore, it was virtually discovered that the runoff produced by both soils in wet years was same. In addition, it was shown that elevated beds harvested an extra 40–50 mm of runoff during the monsoon season. Importantly, raised bed farms experienced 30%–60% less soil loss than flatbed fields. Furthermore, despite multiple water shortages and surplus years, crop output in BBF fields was consistently higher than 4.5 t/ha (Pathak et al., 2005; Wani et al., 2011). This is because a sizeable portion of total rainfall is utilised in productive transpiration. On the other hand, it was discovered that the conventionally maintained field had an average crop yield of 1 t/ha. In BBF fields, the average crop water productivity was found to be 0.7 Kg m^{-3} as opposed to 0.2 Kg m^{-3} in conventionally maintained fields.

TABLE 6.4
Effects of the Raised Bed (BBF) Technology on Soil Loss and Surface Runoff in ICRISAT Research Stations, Hyderabad, India

Site	Year	Rainfall (mm)	Surface Runoff		Soil Loss	
			Raised Bed	Flat Bed	Raised Bed	Flat Bed
Vertisols	Dry	530	2 (0%)	9 (2%)	0.0	0.0
	Normal	729	53 (7%)	77 (11%)	1.1	2.6
	Wet	1101	215 (20%)	258 (23%)	3.1	5.2
Alfisols	Dry	573	9 (2%)	6 (1%)	0.1	0.2
	Normal	722	115 (16%)	126 (17%)	3.0	4.5
	Wet	986	186 (19%)	213 (22%)	7.1	9.1

In-field studies on the management of Vertisols in central India showed that, compared to farmers' practises, the BBF method increased soybean output by 35% during the rainy season and chickpea yield by 22% during the post-rainy season. A comparable yield benefit was noted in the rotation of wheat and maize using the BBF method. At the Haveri, Dharwad, and Tumakuru watersheds in Karnataka, yield advantages of 25% above farmers' practices were seen in maize, soybean, and peanuts with conservation furrows on Alfisols. In cropping systems comprising soybean-chickpea, maize-chickpea, and soybean/maize-chickpea under enhanced land management techniques, yield advantage in rainfall-use efficiency (RUE) was also exhibited.

6.7.5.2 Direct Seeded Rice (DSR)

The inefficient use of inputs (fertiliser, water, and labour), the increasing scarcity of resources, particularly water and labour, the changing climate, the emerging energy crisis and rising fuel prices, the rising cost of cultivation, and emerging socioeconomic changes like urbanisation, labour migration, the preference for non-agricultural work, and worries about farm-related p DSR offer chances to reduce labour costs, water use, and energy use. Dry and Wet-DSR use less water than transplanted rice and benefit from it. In contrast, Dry-DSR with zero tillage has the greatest potential due to the growing water crisis since it has a significantly larger potential for labour and water savings (Anantha et al., 2021). A global meta and mixed model analysis using a dataset involving 323 on-station and 9 on-farm studies (a total of 3878 paired data) comparing transplanted and direct seeded rice showed that shifting from transplanting to direct-seeding produced similar yields with a significant reduction in water input, lower greenhouse gas emissions, and higher net returns (Chakraborty et al., 2017).

A zero-till multi-planter is used to sow paddy seeds under the DSR method (Figure 6.16). Using a zero-till multi-planter makes it possible to sow seeds without any seedbed preparation or puddling, significantly lowering the cost of labour and energy. The zero-till machine produces a 5–6 cm sharp cut in the top layer of soil, then evenly distributes seed and fertiliser before covering it with dirt. Because the topsoil is not disturbed, any available moisture is shielded from wasteful evaporation losses. Pre-emergent weedicide must be used at the specified dosage to manage the weeds.

Crop cutting examines carried out in various scaling-up research trials revealed that in peninsular India, the crop yield attained via DSR was between 2.8 t ha^{-1} and 5.8 t ha^{-1} as opposed to 2.6 t ha^{-1} and 5.4 t ha^{-1} in the TPR system. Since there is no need for nursery rearing, seedling uprooting, or transferring, direct sowing requires less labour overall. DSR also does away with puddling processes, which further reduces labour costs. Since land preparation is mostly automated, machine labour is more cost-effective than human labour in this process. Studies conducted over a short to medium period of time on-station found that DSR required 35–47% less machine labour than TPR (Saharawat et al., 2010; Wani et al., 2016; Anantha et al., 2021). In addition to conserving labour,

FIGURE 6.16 Sowing direct seeded rice using zero till multi-crop planter in Southern India.

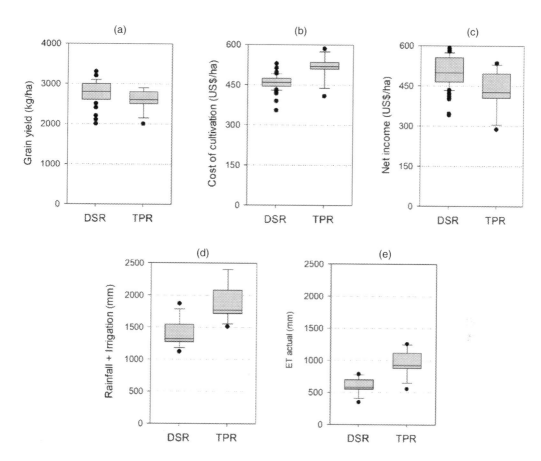

FIGURE 6.17 A comparison of the productivity of direct seeded rice (DSR) and transplanted rice (TPR) based on 110 field demonstrations in Southern India, including (a) crop yield, (b) cultivation cost, (c) net income, (d) measured irrigation application, and (e) real ET. (Data source: Anantha et al., 2021.)

DSR requires less labour over a longer period of time than transplanted rice. The crucial transplanting procedure necessitates a lot of labour, sometimes leading to a labour shortage. Utilising family labour to its fullest and reducing reliance on hired labour is the benefit of the spread-out labour need. With a benefit-cost ratio of more than 1, which denotes the profitability of the technology, the net return under DSR was afterwards much higher.

Results from extensive research experiments in Southern India are displayed in Figure 6.17. The crop yield was equivalent to that of the standard approach. In addition, compared to transplanted rice (TPR), there was a reduction in the cost of cultivation by US$ 160/ha, the quantity of irrigation saved by 350–425 mm, and the corresponding energy consumption (550 kWh/ha). Comparing DSR to TPR, the economic gain was $100/ha higher. The findings also revealed that the real evapotranspiration in the DSR was 600–750 mm as opposed to 1050–1250 mm in the TPR. When compared to TPR, DSR has been demonstrated to be more resistant to severe occurrences like flash floods due to its greater establishment of root development (Jat et al., 2020).

6.7.5.3 Use of Zero Tillage Multi-Crop Planter

Farmer needs help to buy many types of machinery for planting various crops due to fragmented and tiny land holdings and varying farmer typologies. The multi-crop planter offers a straightforward solution since it can plant several crops with varying seed sizes, seed rates, depths, spacing, etc. The multi-crop planters have accurate seed metering systems employing inclined rotating plates with changeable groove numbers and sizes for varying seed sizes and spacing for diverse crops, in addition to row spacing, depth, and gears for power transfer modifications to seed and fertiliser metering systems, and so on. The long-term data shows that the use of zero tillage helps in savinf residue moisture by 30–50 mm and enhances the protection of soil carbon (Cooper et al., 2020; Anantha et al., 2021).

6.7.5.4 Laser Land Levelling

Laser land levelling technology is one of the promising interventions to enhance water use efficiency. Laser-assisted precision land levelling considered as a precursor technology for Conservation Agriculture based management practices has been reported to improve crop yields and input-use efficiency including water and nutrients and lowering GHG emissions (Jat et al. 2009, 2011, 2015). Hill et al. (1991) rated the development of laser technology for precision land levelling as second only to breeding of high yielding crop varieties. The laser land levelling technology in South Asia has bed one of the most significant interventions that led to a water revolution. Farms with a slope of less than one per cent are generally suitable for land levelling using laser-guided machines. For farms with undulated topography (2%–5% slope), it is better to treat the landscape by field bunding as it is not technically feasible as much soil mass had to be cut and filled to level it (Figure 6.18).

Table 6.5 summarises the impact of laser land levelling on the cost of cultivation, crop yield, and net income from research trials undertaken in Central India. Land levelling has significantly impacted on wheat yield from 2290 to 4330 kg/ha. With the introduction of laser land levelling,

FIGURE 6.18 Fields before and after laser land levelling in Central Indian Landscape.

TABLE 6.5
Impact of Laser Land Levelling on Crop Yield, Irrigation Application, Energy Consumption and Net Income in Different Districts, Bundelkhand Region, Uttar Pradesh

Description	Without Treatment	With Laser Levelling
Wheat yield (Kg/ha)	2290 ($\sigma=220$)	4330 ($\sigma=150$)
No. of irrigation (−)	5 ($\sigma=1$)	3 ($\sigma=1$)
Diesel consumption (Liter/ha)	80 ($\sigma=15$)	50 ($\sigma=8$)
Number of man-days/ha for irrigation	45 ($\sigma=12$)	20 ($\sigma=7$)
Cost of cultivation (US$/ha)	310 ($\sigma=50$)	210 ($\sigma=30$)
Net income (US$/ha)	350	1040

Note: Based on data collected from 60 farmers' fields

Drought Management in Soils of the Semi-Arid Tropics

the moisture across the field was uniformly available, which helped better crop establishment and growth. This also has influenced a reduction in 1–2 irrigations and reduced energy consumption (diesel consumption reduced from 80 to 50 lit/ha). Before the intervention, irrigation application was labour intensive as farmers had to manually direct the water flow and about 40–50-man days were invested per ha for wheat cultivation. Whereas after the laser land levelling, only 15–25 man-days were sufficient enough for irrigation application. Three-folds increased overall net income from US$ 350 to US$ 1040/ha has made a significant contribution to household income.

6.7.6 Organic Amendments/Live Mulching

Live mulch and organic amendment enhance the soil characteristics, which can control its function. The breakdown of agricultural leftovers makes it easier for soil to aggregate, which leads to the development of soil organic matter (SOM), which reduces soil bulk density when no-till (NT) is used in conjunction with live mulch (Yang et al., 2020). There are more pores than with conservation tillage, enhancing the porosity in semi-arid tropics soils. Compared to conventional practices live mulch and organic amendment increase moisture content up to 60 cm dept of the soil (Muzangwa et al., 2020). Live mulching using cover crops and leftovers improves soil organic carbon, soil water holding capacity, raising matric potential in the semi-arid tropics' soils (Srinivasarao et al., 2021a, 2016). Maintaining living mulch coupled with the subsequent increase in soil organic carbon (SOC) stocks and improved soil aggregation opens faunal pores. It increases soil macro porosity, which improves infiltration and soil moisture content in semi-arid tropics soils (Heitkamp, 2017). Organic amendments/live mulching role in semi-arid tropics soils presented in Figure 6.19.

FIGURE 6.19 Organic amendments/live mulching role in semi-arid tropics soils.

6.7.7 TECHNIQUES FOR IMPROVING WATER USE EFFICIENCY

The first issue water scarcity is brought on by meteorological dryness, while the second issue—plants' incapacity to utilise the available water is caused by soil drought. Lengthy dry periods, inadequately lengthy growth seasons, or complete lack of rain are all symptoms of meteorological droughts. Climate change and abnormal rainfall are the causes. Farmers can use various indigenous and cutting-edge strategies to combat this drought (Araya and Stroosnijder, 2011). Deficient irrigation and furrow-diking are two instances. Contrarily, dryness is caused by deteriorated physical soil characteristics or a lack of (balanced) soil nutrients. While some infiltrating water can be held in the root zone and used for plant development, the majority is lost through deep drainage. Non-productive losses of water can be quite large if the soil is not sufficiently covered with plants. Desertification and a lack of agricultural intensification are the main causes (Nyakudya and Stroosnijder, 2011). Due to increased non-productive water losses brought on by desertification, water usage efficiency is directly impacted. According to a recent study, farmers may reduce soil dryness and improve water usage efficiency using cutting-edge techniques and technology. The desire to feed the globe necessitates water-efficient land management techniques. There has been a lot of progress in finding ways to do this. Here are some examples of methods to boost infiltration and water storage, lessen evaporation losses, collect and store water in semi-arid tropical regions, and improve water usage efficiency (Figure 6.20).

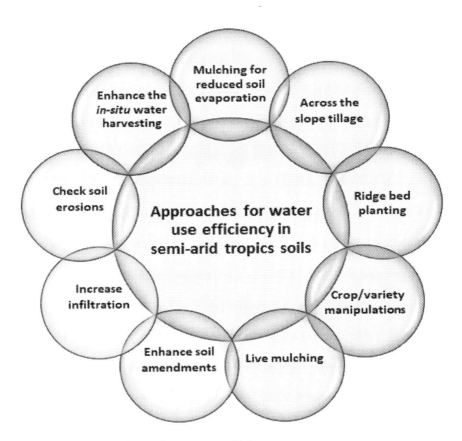

FIGURE 6.20 Techniques for improving water use efficiency.

6.8 ROADMAP FOR DROUGHT MANAGEMENT IN SOILS OF THE SEMI-TROPICS

Drought management is key to sustainable intensification for achieving food and nutritional security and reducing water-energy-carbon footprints. With growing climate change scenarios, occurrence of extreme events and drought is a common phenomenon resulting in uncertainty of resource availability, yield penalty, crop failure and intensifying the issue of food insecurity, poverty and malnutrition. Further, recurring drought events create several socio-economic challenges such as mass migration, increasing cost of cultivation and inflation. Large-scale public welfare programs in Asia and Africa are targeting to mitigate the impact of drought by devising various social safety nets, incentive mechanisms, creating job opportunities and provisioning services. These mechanisms have helped millions of people to mitigate the effects of droughts. However, there is a need to improve the efficacy of such investments by integrating science-led interventions and approaches. A range of technologies developed by knowledge-generating institutes holds the potential to mitigate the effects of droughts which need to be leveraged on these welfare programs. Below are the key strategies for managing drought in soils of the semi-arid tropics:

6.8.1 INTEGRATION OF LANDSCAPE RESOURCE CONSERVATION MEASURES AND FIELD SCALE CLIMATE RESILIENT TECHNOLOGIES

Annual rainfall of semi-arid tropics of Asia and Africa ranges from 400 to 1100 mm which is generally received within 20 to 75 rainy days annually. Rainfall is the major source of freshwater for which all sectors compete which often lead to a demand-supply imbalance with a deficit which keeps accumulating. Therefore, rainwater should be harvested using the decentralised approach to enhance the longevity of freshwater resources especially in drylands. Integration of engineering and biological measures enhances moisture retention ability and facilitates to enhance surface and groundwater recharge which in turn helps in the sustainable intensification of farming systems and strengthening ecosystem services. The engineering measures are helpful to conserve and store surface runoff; whereas the biological measures further support minimizing land degradation thereby ensuring the sustainability of the system. Landscape resource conservation approach which captures multi-sectoral aspects of resource mapping, demand estimation, and optimum utilization of resources, is needed. With this approach, it is possible to convert degraded landscape including fallow lands (permanent and seasonal fallows) into productive cultivation in the drylands.

6.8.2 SYSTEMS APPROACH FOR DECOMPOSING YIELD GAP

Systems approaches integrating genetics, social and ecological aspects of the farming are critical to realise the full potential of the technologies and innovations. In the absence of such integration, crop cultivars developed for drylands often underperform in many situations. Therefore, breeding crop varieties adapted to specific production environments, resource availability and socio-ecological settings ($G \times E \times M$) using systems approaches is critical. In this respect, the advancement in geospatial and big data science provides ample opportunities for the multiplier effect of $G \times E \times M$. Also, the geospatial and digital tools can help in providing safety nets through designing insurance products, yield estimation, contingency planning and taking corrective measures timely with minimum human interference for better transparency. Similarly, there are emerging technologies such as diffused reflectance spectroscopy (DRS) for rapid soil nutrient analysis hold a huge potential to minimise the sample analysis cost and precision fertiliser application.

6.8.3 REDUCING WATER-ENERGY-CARBON FOOTPRINTS

To better understand the synergies and trade-offs between the ecological boundaries of the production systems and the demand for food, scenario analysis could help in targeting systems and prioritizing technologies for sustainable and resilient farming systems. Market-driven water budget-based cropping/farming system design is an innovative approach for sustainable food production within planetary boundaries. Using such an approach, suitable cropping systems can be designed considering moisture supplying capacity of the landscape and identifying permissible irrigation threshold such that utilization of freshwater should be within the limit of the planetary boundaries for a given agroecological region. The reduction in carbon footprints can provide opportunities for incentivizing farmers for carbon credits which will create a pull factor for the adoption of resilient farming practices.

6.8.4 ESTABLISHING LIVING LABS FOR BRIDGING THE KNOWLEDGE GAP AND EVIDENCE-BASED POLICY INFORMING

For generating science-based evidence, establishment of innovative platforms/living labs at farm to landscape scale through integrating a range of validated technologies across agroecological regions is an important step towards adoption, knowledge dissemination and scaling up process. Such living labs at farm and landscape scale need to be encouraged to understand the system dynamics including techno-economic, social feasibility and generate continued evidence on system performance for guiding policy.

6.9 CONCLUSION

This chapter discusses the challenges of drylands of the semi-arid tropics in Asia and Africa. Drought-induced water scarcity is the major challenge which limits moisture availability thereby affecting crop production. In the absence of supplemental irrigation, agriculture is risk-prone and non-remunerative. A decentralised approach to landscape resource conservation measures is crucial for addressing these challenges. This chapter synthesises best management practices to mitigate the impact of droughts in the semi-arid tropics. There is clear evidence that a combination of *in-situ* and *ex-situ* rainwater harvesting interventions helped in enhancing freshwater availability by increasing infiltrability and facilitating groundwater recharge. Science evidence shows that a fraction of surface runoff harvested in the landscape, particularly during the normal and wet years remains available in succeeding seasons and alleviates the effects of droughts. This also has reduced the cost of cultivation by reducing irrigation requirements and energy consumption at least by 30%. In addition, field scale conservation agriculture technologies helped to conserve available residual moisture by 50–80 mm. Non-productive evaporation can be converted into productive transpiration by the effective implementation of conservation agricultural technologies. This chapter clearly shows the importance of the integration of landscape resource conservation technologies for moisture augmentation and field scale conservation agriculture technologies for enhancing resource use efficiency. These together can lead to sustainable intensification even in drylands of semi-arid tropics which is crucial for addressing the challenges of food security, poverty and malnutrition in the drylands.

ACKNOWLEDGEMENTS

We sincerely acknowledge the support received from the Governments of Uttar Pradesh and Odisha, India and Indian Council of Agricultural Research (ICAR), International Crops Research Institute for the Semi-Arid Tropics (ICRISAT) and JSW Foundation for this synthesis.

REFERENCES

Abdi, O.A., Glover, E.K., and Luukkanen, O. 2013. Causes and impacts of land degradation and desertification: CAS+B2:B189e study of the Sudan. *International Journal of Agriculture and Forestry*, 3(2), 40–51.

Adejuwon, J.O., and Dada, E. 2021. Temporal analysis of drought characteristics in the tropical semi-arid zone of Nigeria. *Scientific African*, 14, e01016.

Adejuwon, J.O., and Olaniyan, S.B. 2019. Drought occurrence in the sub-humid eco-climatic zone of Nigeria. *Theoretical and Applied Climatology*, 137, 1625–1636. https://doi.org/10.1007/s00704-018-2670-7

Adger, W.N. 1999. Social vulnerability to climate change and extremes in coastal Vietnam. *World Development*, 27(2), 249–269.

Adimassu, Z., Tamene, L., and Degefie, D.T. 2020. The influence of grazing and cultivation on runoff, soil erosion, and soil nutrient export in the central highlands of Ethiopia. *Ecological Processes*, 9, 23. https://doi.org/10.1186/s13717-020-00230-z

Aditi, K., Abbhishek, K., Chander, G., Singh, A., Falk, T., Mequanint, M.B., and Nagaraji, S. 2023. Assessing residue and tillage management options for carbon sequestration in future climate change scenarios. *Current Research in Environmental Sustainability*, 5, 100210.

Akinseye, F.M., Ajeigbe, H.A., Traore, P.C.S., Agele, S.O., Zemadim, B., and Whitbread, A. 2020. Improving sorghum productivity under changing climatic conditions: A modelling approach. *Field Crops Research*, 246, 107685. https://doi.org/10.1016/j.fcr.2019.107685

Alam, N.M., Ranjan, R., Jana, C., Singh, R.J., Patra, S., Ghosh, B.N., and Sharma, N.K. 2014. Drought classification for policy planning and sustainable agricultural production in India. Popular Kheti. 2. 53–59.

Amede, T., Van den Akker, E., Berdel, W., Keller, C., Tilahun, G., Dejen, A., Legesse, G., and Abebe, H., 2022. Facilitating livelihoods diversification through flood-based land restoration in pastoral systems of Afar, Ethiopia. *Renewable Agriculture and Food Systems*, 37(S1), S43–S54. doi:10.1017/S1742170520000058.

Anantha, K.H., Garg, K.K., Barron, J., Dixit, S., Venkataradha, A., Singh, R., and Whitbread, A.M. 2021a. Impact of best management practices on sustainable crop production and climate resilience in smallholder farming systems of South Asia. *Agricultural Systems*, 194, 103276.

Anantha, K.H., Garg, K.K., Moses S.D., Patil, M.D., Sawargaonkar, G.L., Kamdi, P.J., Malve, S., Sudi, R., Raju, K.V., and Wani, S.P. 2021b. Impact of natural resource management interventions on water resources and environmental services in different agroecological regions in India. *Groundwater for Sustainable Development*, 13(1): 100574.

Anantha, K.H., Garg, K.K., Petrie, C.P., Dixit, S. 2021c. Seeking sustainable pathways for fostering agricultural transformation in peninsular India. *Environmental Research Letters*, 16, 044032. https://doi.org/10.1088/1748-9326/abed7b

Anantha, K.H., Garg, K.K., Singh, R., Venkataradha, A., Dev, I., Petrie, C.A., Whitbread, A.M., and Dixit, S. 2022. Landscape resource management for sustainable crop intensification. *Environmental Research Letters*, 17, 014006. https://doi.org/10.1088/1748-9326/ac413a

Anantha, K.H., Garg, K.K., Venkataradha, A., Sawargaonkar, G., Purushothaman, N.K., Das, B.S., Singh, R., and Jat, M.L. 2023. Sustainable intensification opportunities for Alfisols and Vertisols landscapes of the semi-arid tropics. *Agricultural Water Management*. 284(C) In press.

Anderson, K., and Valenzuela, E. 2006. WTO's Doha Cotton Initiative: A Tale of Two Issues. Development Research Group. The World Bank, Washington, DC.

Andréassian, V. 2004. Waters and forests: From historical controversy to scientific debate. *Journal of Hydrology*, 291(1–2), 1–27.

Andrieu, N., Vayssières, J., Corbeels, M., Blanchard, M., Vall, E., and Tittonell, P. 2015. From farm scale synergies to village scale trade-offs: Cereal crop residues use in an agro-pastoral system of the Sudanian zone of Burkina Faso. *Agricultural Systems*, 134, 84–96. https://doi.org/10.1016/j.agsy.2014.08.012

Araya, A., and Stroosnijder, L. 2011. Assessing drought risk and irrigation need in northern Ethiopia, *Agricultural and Forest Meteorology*, 151, 425–436.

Arndt, C., Chinowsky, P., Fant, C., Paltsev, S., Schlosser, A., Strzepek, K., Tarp, F., and Thurlow, J. 2019. Climate change and developing country growth: The cases of Malawi, Mozambique, and Zambia. *Climatic Change*, 154(3–4), 335–349. https://doi.org/10.1007/s10584-019-02428-3

Arndt, C., and Thurlow, J. 2015. Climate uncertainty and economic development: Evaluating the case of Mozambique to 2050. *Climatic Change*, 130, 63–75. https://doi.org/10.1007/s10584-014-1294-x

Arya, A.S., Dhinwa, P.S., Pathan, S.K., and Raj, K.G. 2009. Desertification/land degradation status mapping of India. *Current Science*, 97(10), 1478–1483.

Ascott, M., Macdonald, D.M.J., Sandwidi, W.J.P., Black, E., Verhoef, A., Zongo, G., Tirogo, J., and Cook, P. 2022. Time of emergence of impacts of climate change on groundwater levels in sub-Saharan Africa. *Journal of Hydrology*, 612, 128107. https://doi.org/10.1016/j.jhydrol.2022.128107

Attia, A., Qureshi, A.S., Kane, A.M., Alikhanov, B., Kheir, A.M.S., and Ullah, H., et al. 2022. Selection of potential sites for promoting small-scale irrigation across Mali using remote sensing and GIS. *Sustainability*, 14, 12040. https://doi.org/10.3390/su141912040

Audouin, E., Vayssières, J., Odru, M., Masse, D., Dorégo, S., Delaunay, V., and Lecomte, P. 2015. Réintroduire l'élevage pour accroître la durabilité des terroirs villageois d'Afrique de l'Ouest : le cas du bassin arachidier au Sénégal. *Les sociétés rurales face aux changements climatiques et environnementaux en Afrique de l'Ouest*, 403–427.

Aune, J.B., Tadesse, B.A., Coulibaly, A., and Borgvang, S. 2020. L'intensification agricole au Mali et au 'Soudan à travers l'amélioration de la fertilité du sol et la mécanisation. In IER, 2020. *Adaptation de l'Agriculture et de l'Élevage au Changement Climatique au Mali - Résultats et leçons apprises au Sahel*. Bamako, Mali, 404p.

Autfray, P., Sissoko, F., Falconnier, G., Ba, A., and Dugué, P. 2012. Usages des résidus de récolte et gestion intégrée de la fertilité des sols dans les systèmes de polyculture élevage: étude de cas au Mali-Sud. *Cahiers Agricultures*, 21, 225–234.

Ayanlade, S., Oluwaranti, A., Ayanlade, O., Borderon, M., Sterly, H., Sakdapolrak, P., Jegede, M., Weldemariam, L., and Ayinde, F. 2022. Extreme climate events in sub-Saharan Africa: A call for improving agricultural technology transfer to enhance adaptive capacity. *Climate Services*, 27, 100311. https://doi.org/10.1016/j.cliser.2022.100311

Ayantunde, A.A., Asse, R., Said, M.Y., and Fall, A. 2014. Transhumant pastoralism, sustainable management of natural resources and endemic ruminant livestock in the sub-humid zone of West Africa. *Environment, Development and Sustainability*, 16, 1097–1117.

Ayantunde, A.A., Oluwatosin, B.O., Yameogo, V., and Van Wijk, M. 2020. Perceived benefits, constraints and determinants of sustainable intensification of mixed crop and livestock systems in the Sahelian zone of Burkina Faso. *International Journal of Agricultural Sustainability*, 18, 84–98.

Ayantunde, A.A., Umutoni, C., Dembele, T., Seydou, K., and Samake, O. 2019. Effects of feed and health interventions on small ruminant production in mixed crop-livestock systems in Southern Mali. *Revue d'Elevage et de Médecine Vétérinaire des Pays Tropicaux*, 72, 65. https://doi.org/10.19182/remvt.31747

Bado, B.V., 2002. Rôle des légumineuses sur la fertilité des sols ferrugineux tropicaux des zones guinéenne et soudanienne du Burkina Faso, Thèse, Ph.D., Faculté des études supérieures de l'Université Laval Québec. Décembre 2002, 184p. Ph.D. thesis.

Bagayoko, M., Alvey, S., Neumann, G., and Bürkert, A. 2000. Root-induced increases in soil pH and nutrient availability to field-grown cereals and legumes on acid sandy soils of Sudano-Sahelian West Africa. *Plant Soil*, 225, 117–127.

Balana, B., Bizimana, J.-C., Richardson, J., Lefore, N., Adimassu, Z., and Herbst, B. 2019. Economic and food security effects of small-scale irrigation technologies in northern Ghana. *Water Resources and Economics*, 29, 100141. https://doi.org/10.1016/j.wre.2019.03.001

Bantilan, M.C.S., Rao, P.P., and Padmaja, R. 2001. Future of agriculture in the semi-arid tropics: Proceedings of an international symposium on future of agriculture in semi-arid tropics, international crops research institute for the semi-arid tropics, Patancheru, India, 14 Nov 2000. International Crops Research Institute for the Semi-Arid Tropics.

Baquedano, F.G., Sanders, J.H., and Vitale, J. 2010. Increasing incomes of Malian cotton farmers: Is elimination of US subsidies the only solution? Agricultural Systems, 103, 418–432.

Barbier, E.B., and Hochard, J.P. 2018. Land degradation and poverty. *Nature Sustainability*, 1(11), 623–631. https://doi.org/10.1038/s41893-018-0155-4

Bationo, A., Fairhurst, T., Giller, K., Kelly, V., Lunduka, R., Mando, A., Mapfumo, P., Oduor, G., Romney, D., Vanlauwe, B., Wairegi, L., and Zingore, S. 2015. Manuel de gestion intégrée de la fertilité des sols. Consortium Africain pour la Santé des Sols, Nairobi, Edité par Thomas Fairhurst (CTA), 179p.

Birhanu, B., and Tabo, R. 2016. Shallow wells, the untapped resource with a potential to improve agriculture and food security in southern Mali. *Agriculture & Food Security*, 5, 5. https://doi.org/10.1186/s40066-016-0054-8

Birhanu, B., Traoré, K., Sanogo, K., Tabo, R., Fischer, G., and Whitbread, A. 2022. Contour bunding technology-evidence and experience in the semiarid region of southern Mali. *Renewable Agriculture and Food Systems*, 37, S55–S63. https://doi.org/10.1017/S1742170519000450

Birhanu, L., Hailu, B.T., Bekele, T., and Demissew, S. 2019. Land use/land cover change along elevation and slope gradient in highlands of Ethiopia. *Remote Sensing Applications: Society and Environment*, 16, 100260. https://doi.org/10.1016/j.rsase.2019.100260

Blanchard, M., 2010. Gestion de la fertilité des sols et rôle du troupeau dans les systèmes coton-céréales-élevage au Mali- sud, savoirs techniques locaux et pratiques d'intégration agriculture élevage. Thèse de Doctorat, Sciences de l'Univers et Environnement, 298p.

Bodner, G., Nakhforoosh, A., and Kaul, H.P. 2015. Management of crop water under drought: A review. *Agronomy for Sustainable Development*, 35, 401–442.

Bonaudo, T., Billen, G., Garnier, J., Barataud, F., Bognon, S., Dupré, D., and Marty, P. 2017. Analyser une transition agro-alimentaire par les flux d'azote: Aussois un cas d'étude du découplage progressif de la production et de la consommation. *Revue d'Économie Régionale Urbaine*, Décembre, 967–991. https://doi.org/10.3917/reru.175.0967

Bossio, D., Geheb, K., and Critchley, W. 2010. Managing water by managing land: Addressing land degradation to improve water productivity and rural livelihoods. *Agricultural Water Management*, 97(4), 536–542. https://doi.org/10.1016/j.agwat.2008.12.001

Brandolini, F., Kinnaird, T.C., and Srivastava, A. et al. 2023. Modelling the impact of historic landscape change on soil erosion and degradation. *Scientific Reports*, 13, 4949. https://doi.org/10.1038/s41598-023-31334-z

Brottem, L.V. 2016. Environmental change and farmer-herder conflict in agro-pastoral West Africa. *Human Ecology*, 44, 547–563.

Brottrager, M., Crespo Cuaresma, J., and Kniveton, D. et al. 2023. Natural resources modulate the nexus between environmental shocks and human mobility. *Nature Communications*, 14, 1393. https://doi.org/10.1038/s41467-023-37074-y

Brunet, C., Savadogo, O., Baptiste, P., and Bouchard, M.A. 2018. Shedding some light on photovoltaic solar energy in Africa-A literature review. *Renewable and Sustainable Energy Reviews*, 96, 325–342. https://doi.org/10.1016/j.rser.2018.08.004

Buerkert, A., Bationo, A., and Dossa, K. 2000. Mechanisms of residue mulch-induced cereal growth increases in West Africa. *Soil Science Society of America Journal*, 64, 346–358.

Carminati, A., Moradi, A.B., Vetterlein, D., Vontobel, P., Lehmann, E., Weller, U., and Oswald, S.E. 2010. Dynamics of soil water content in the rhizosphere. *Plant and Soil*, 332, 163–176.

Carrión, J.A., Estrella, A.E., Dols, F.A., Toro, M.Z., Rodríguez, M., and Ridao, A.R. 2008. Environmental decision-support systems for evaluating the carrying capacity of land areas: Optimal site selection for grid-connected photovoltaic power plants. *Renewable and Sustainable Energy Reviews*, 9, 2358–2380. https://doi.org/10.1016/j.rser.2007.06.011

Casali, L., Rubio, G., and Herrera, J.M. 2018. Drought and temperature limit tropical and temperate maize hybrids differently in a subtropical region. *Agronomy for Sustainable Development*, 38, 1–12.

Chakraborty, D., Ladha, J.K., Rana, D.S., Jat, M.L., Gathala, M.K., Yadav, S., Rao, N.A, Ramesha, M.S., and Raman, A. 2017. A global analysis of alternative tillage and crop establishment practices for economically and environmentally efficient rice production. *Scientific Reports*, 7, 9342.

Chand, K., and Biradar, N. 2017. Socio-economic impacts of drought in India. In: *Drought Mitigation and Management* (eds. S. Kumar, S.P.S Tanwar and A. Singh), Scientific Publishers, New Delhi. Pp. 245–226.

Charney, J. 1975. Dynamics of deserts and drought in the Sahel. *Quarterly Journal of the Royal Meteorological Society*, 101(428), 193–202.

Cobo, J.G., Dercon, G., and Cadisch, G. 2010. Nutrient balances in African land use systems across different spatial scales: A review of approaches, challenges and progress. *Agriculture, Ecosystems & Environment*, 136, 1–15. https://doi.org/10.1016/j.agee.2009.11.006

Coly, I., Diop, B., and Akpo, E. 2013. Transformation locale des résidus de récolte en fumier de ferme dans le terroir de la Néma au Saloum (Sénégal) *Journal of Applied Biosciences*, 70, 5640–5651.

Coulibaly, A., Woumou, K., and Aune, J.B. 2019. Sustainable Intensification of Sorghum and Pearl Millet Production by Seed Priming, Seed Treatment and Fertilizer Microdosing under Different Rainfall Regimes in Mali. Agronomy, 9(10), 664

Coulibaly, D., Poccard-Chapuis, R., Ba, A. 2009. Dynamiques territoriales et changements des modes de gestion des ressources pastorales au Mali Sud (Mali) [Territorial dynamics and changes of pastoral management resources in Mali Sud (Mali)]. In: *Seizièmes rencontres autour des recherches sur les ruminants, Paris les 2 et 3 décembre 2009*. INRA. Paris: Institut de l'élevage, pp. 357–360. (Rencontres autour des recherches sur les ruminants, 16).

CSIRO. 2021. Centre for Scientific and Industrial Research Organization. Chameleon Soil Water Sensor. https://www.csiro.au/en/research/plants/crops/farming-systems/chameleon-soil-water-sensor.

De Boni, A., D'Amico, A., Acciani, C., and Roma, R. 2022. Crop diversification and resilience of drought-resistant species in semi-arid areas: An economic and environmental analysis. *Sustainability*, 14(15), 9552. https://doi.org/10.3390/su14159552

De Ridder, N., Breman, H., Van Keulen, H., and Stomph, T.J. 2004. Revisiting a "cure against land hunger": Soil fertility management and farming systems dynamics in the West African Sahel. *Agricultural Systems*, 80, 109–131. https://doi.org/10.1016/j.agsy.2003.06.004

Deosthali, V., 2002. Dry farming in Maharashtra. Geography of Maharashtra, J. Diddee et al., Eds, Rawat Publications, 180–196. Weather, Climate, and Scoiety Volume 11. https://books.google.co.in/books/about/Geography_of_Maharashtra.html?id5Ey1uAAAAMAAJ&redir_esc5y.752

Dev, I., Singh, R., Garg, K.K., Ram, A., Singh, D., Kumar, N., Dhyani, S.K., Singh, A., Anantha, K.H., VenkataRadha, A., Dixit, S., Tewari, R.K., Dwivedi, R.P., and Arunachalam, A. 2022. Transforming livestock productivity through watershed interventions: A case study of Parasai-Sindh watershed in Bundelkhand region of Central India. *Agricultural Systems*, 196, 103346.

Dzavo, T., Zindove, T.J., Dhliwayo, M., and Chimonyo, M. 2019. Effects of drought on cattle production in sub-tropical environments. *Tropical Animal Health and Production*, 51, 669–675.

Evans, R.G., and Sadler, E.J. 2008. Methods and technologies to improve efficiency of water use. *Water Resources Research*, 44(7), 1–15.

Falconnier, G.N., Descheemaeker, K., Mourik, T.A.V., and Giller, K.E. 2016. Unravelling the causes of variability in crop yields and treatment responses for better tailoring of options for sustainable intensification in southern Mali. *Field Crops Research*, 187, 113–126. https://doi.org/10.1016/j.fcr.2015.12.015

Fall, S.T., Rippstein, G., and Corniaux, C. 2005. Les fourrages et les aliments du bétail. In: Amadou, T.G., Taïb, D., Ababacar, N., Jacques, D., Benoît, G., ean-Pascal, P., Emile Victor, C., Amadou, B., Jean-Pierre, N., Emmanuel, S., Mour, G., Adama, N., Massamba, N., Ousmane Timéra, T., and Awa, D., Bilan de la recherche agricole et agroalimentaire au Sénégal (pp. 267–279). Institut Sénégalais de Recherches Agricoles/Institut de Technologie Alimentaire/Centre de Coopération Internationale en Recherche Agronomique pour le Développement.

FAO, IFAD, UNICEF, WFP and WHO. 2018. The State of Food Security and Nutrition in the World 2018. Building climate resilience for food security and nutrition. Rome, FAO. Licence: CC BY-NC-SA 3.0 IGO

FAO, ITPS, F., 2015. Status of the world's soil resources (SWSR)-Main report. Food and Agriculture Organization of the United Nations and Intergovernmental Technical Panel on Soils, Rome, Italy. https://www.fao.org/3/a-i5199e.pdf

FAO. 2021. The state of the world's land and water resources for food and agriculture - Systems at breaking point. Synthesis report 2021. Rome. https://doi.org/10.4060/cb7654en

Farooq, M., and Siddique, K.H.M. 2015. Conservation agriculture: Concepts, brief history, and impacts on agricultural systems. In Conservation agriculture (M. Farooq, and K.H. Siddique, K.H, eds). Springer, Cham. p. 317.

Field, C.B., Barros V.R., Dokken D.J., Mach K.J., Mastrandrea M.D., Bilir T.E., Chatterjee M., Ebi K.L., Estrada Y.O., Genova R.C., Girma B., Kissel E.S., Levy A.N., MacCracken S., Mastrandrea P.R., and White L.L. 2014. IPCC. Climate change 2014: Impacts, adaptation, and vulnerability - summary for policy makers. In: Field editors. Part A: Global and Sectoral Aspects. Contribution of Working Group II to the Fifth Assessment Report of the Intergovernmental Panel on Climate Change. Cambridge University Press; Cambridge, United Kingdom and New York, NY, USA: 2014. pp. 1–32. https://www.ipcc.ch/report/ar5/wg2/summary-for-policymakers/

Frank, S., Fürst, C., Witt, A., Koschke, L., and Makeschin, F. 2014. Making use of the ecosystem services concept in regional planning-trade-offs from reducing water erosion. *Landscape Ecology*, 29, 1377–1391.

Gaborel, C., Crétenet, M., and Guibert, H. 2006. La fertilisation du cotonnier en Afrique sub saharienne. Cirad, France.

Gandah, M., Bouma, J., Brouwer, J., Hiernaux, P., and Van Duivenbooden, N. 2003. Strategies to optimize allocation of limited nutrients to sandy soils of the Sahel: A case study from Niger, West Africa. *Agriculture, Ecosystems & Environment*, 94, 311–319.

Gao, Y., Hu, T., Wang, Q., Yuan, H., and Yang, J. 2019. Effect of drought-flood abrupt alternation on rice yield and yield components. *Crop Science*, 59(1), 280–292.

Garg, K.K., Anantha, K.H., Dixit, S., Nune, R., Venkaradha, A., Wable, P., Budama, N., and Singh R. 2022a. Impact of raised beds on surface runoff and soil loss in Alfisols and Vertisols. *Catena*, 211, 105972. https://doi.org/10.1016/j.catena.2021.105972

Garg, K.K., Anantha, K.H., Venkataradha, A., Dixit, S., Singh, R., and Ragab, R. 2021. Impact of rainwater harvesting on hydrological processes in a fragile watershed of South Asia. *Groundwater*, 59, 839–855. https://doi.org/10.1111/gwat.13099

Garg, K.K., Karlberg, L., Barron, J., Wani, S.P., and Rockstrom, J. 2012. Assessing impacts of agricultural water interventions in the Kothapally watershed, Southern India. *Hydrological Processes*, 26, 387–404.

Garg, K.K., Singh, R., Anantha, K.H., Singh, A.K., Akuraju, V.R., Barron, J., Dev, I., Tewari, R.K., Wani, S.P., Dhyani, S.K., and Dixit, S. 2020. Building climate resilience in degraded agricultural landscapes through water management: A case study of Bundelkhand region, Central India. *Journal of Hydrology*, 591, 125592.

Garg, K.K., Venkataradha, A., Anantha, K.H., Singh, R., Whitbread, A.M., and Dixit, S. 2022b. Identifying potential zones for rainwater harvesting interventions for sustainable intensification in the semi-arid tropics. *Scientific Reports*, 12: 3882. https://doi.org/10.1038/s41598-022-07847-4

Gerardeaux, E., 2009. Ajustement de la Phenologie, de la croissance et de la production de biomasse du Cotonnier (Gossypium Hirsutum L.) Face à des carences en potassium. Thèse de Doctorat, Université de Boreaux I, 100p.

Getnet, M., Amede, T., Tilahun, G., Legesse, G., Gumma, M.K., Abebe, H., Gashaw, T., Ketter, C., and Akker, E.V. 2020. Water spreading weirs altering flood, nutrient distribution and crop productivity in upstream-downstream settings in dry lowlands of Afar, Ethiopia. *Renewable Agriculture and Food Systems*, 37(S1) 1–11. https://doi.org/10.1017/S1742170519000474

Gleick, P.H., and Palaniappan, M. 2010. Peak water limits to freshwater withdrawal and use. *Proceedings of the National Academy of Sciences*, 107(25), 11155–11162. https://doi.org/10.1073/pnas.1004812107

Godde, C.M., Mason-D'Croz, D., Mayberry, D.E., Thornton, P.K., and Herrero, M. 2021. Impacts of climate change on the livestock food supply chain; a review of the evidence. *Global Food Security*, 28, 100488. https://doi.org/10.1016/j.gfs.2020.100488

Gumma, M., Amede, T., Getnet, M., Pinjarla, B., Panjala, P., Legesse, G., Tilahun, G., Van den Akker, E., Berdel, W., Ketter, C., Siambi, M., and Anthony, W. 2020. Assessing potential locations for flood-based farming using satellite imagery: A case study in Afar region, Ethiopia. *Renewable Agriculture and Food Systems*, 37(S1), 1–15. https://doi.org/10.1017/S1742170519000516

Gupta, A., Tyagi, P., and Sehgal, V.K. 2011. Drought disaster challenges and mitigation in India: Strategic appraisal. *Current Science*, 100(12), 1795–1806.

Hafsi, C., Debez, A., and Abdelly, C. 2014. Potassium deficiency in plants: Effects and signaling cascades. *Acta Physiologiae Plantarum*, 36, 1055–1070. https://doi.org/10.1007/s11738-014-1491-2

Haile, G.G., Tang, Q., Sun, S., Huang, Z., Zhang, X., and Liu, X. 2019. Droughts in East Africa: Causes, impacts and resilience. *Earth-Science Reviews*, 193, 146–161.

Haileslassie, A., Priess, J., Veldkamp, E., Teketay, D., and Lesschen, J.P. 2005. Assessment of soil nutrient depletion and its spatial variability on smallholders' mixed farming systems in Ethiopia using partial versus full nutrient balances. *Agriculture, Ecosystems & Environment*, 108(1), 1–16. https://doi.org/10.1016/j.agee.2004.12.010

Han, X., Li, Y., Yu, W., and Feng, L. 2022. Attribution of the extreme drought in the horn of Africa during short-rains of 2016 and long-rains of 2017. *Water*, 14(3), 409. https://doi.org/10.3390/w14030409

Hengl, T., Leenaars, J., Shepherd, K., Walsh, M., Heuvelink, G., Mamo, T., Tilahum, H., Berkhout, E., Cooper, M., Fegraus, E., Wheeler, I., and Kwabena, N. 2017. Soil nutrient maps of Sub-Saharan Africa: Assessment of soil nutrient content at 250m spatial resolution using machine learning. *Nutrient Cycling in Agroecosystems*, 109, 77–102. https://doi.org/10.1007/s10705-017-9870-x

Hijioka, Y., E. Lin, J.J., Pereira, R.T., Corlett, X., Cui, G.E., Insarov, R.D., Lasco, E., Lindgren, and A. Surjan, 2014. Asia. In: Climate Change 2014: Impacts, Adaptation, and Vulnerability. Part B: Regional Aspects. Contribution of Working Group II to the Fifth Assessment Report of the Intergovernmental Panel on Climate Change.

Hill, J.E, Bayer, D.E, Bocchi, S., and Clampett, W.S. 1991. Direct Seeded Rice in the Temperate Climates of Australia," Direct Seeded Flooded Rice in the Tropics, IRRI, Manila, 1991, pp. 91–102.

Hirabayashi, Y., Mahendran, R., Koirala, S., Konoshima, L., Yamazaki, D., Watanabe, S., Kim, H., and Kanae, S. 2013. Global flood risk under climate change. *Nature Climate Change*, 3(9), 816–821. https://doi.org/10.1038/nclimate1911

Huang, J., Minnis, P., Yan, H., Yi, Y., Chen, B., Zhang, L., and Ayers, J. 2010. Dust aerosol effect on semi-arid climate over Northwest China detected from A-Train satellite measurements. *Atmospheric Chemistry and Physics*, 10(14), 6863–6872.

Huang, Y., Gerber, S., Huang, T., and Lichstein, J.W. 2016. Evaluating the drought response of CMIP5 models using global gross primary productivity, leaf area, precipitation, and soil moisture data. *Global Biogeochemical Cycles*, 30(12), 1827–1846.

Hussein, K. 2004. Importance of cotton production and trade in West Africa, In: Proceedings of WTO african Regional Workshop on Cotton, Cotonou Benin, 18p.

Ibrahim, A., Abaidoo, R.C., Fatondji, D., and Opoku, A. 2016. Fertilizer micro-dosing increases crop yield in the Sahelian low-input cropping system: A success with a shadow. *Soil Science and Plant Nutrition*, 62, 277–288. https://doi.org/10.1080/00380768.2016.1194169

Ibrahim, A., Pasternak, D., and Fatondji, D. 2015. Impact of depth of placement of mineral fertilizer micro-dosing on growth, yield and partial nutrient balance in pearl millet cropping system in the Sahel. *The Journal of Agricultural Science*, 153, 1412–1421. https://doi.org/10.1017/S0021859614001075

International Water Management Institute (IWMI). 2022. IWMI Annual report 2021. Colombo, Sri Lanka: International Water Management Institute (IWMI) 58p.

IWMI. 2021. Assessing the Potential for Sustainable Expansion of Small-Scale Solar Irrigation in Segou and Sikasso, Mali. Colombo: International Water Management Institute, 8.

Izzi, G., Denison, J., and Veldwisch, G.J. 2021. The Farmer-led Irrigation Development Guide: A What, Why and How-to for Intervention Design. World Bank. https://pubdocs.worldbank.org/en/751751616427201865/FLID-Guide-March-2021-Final.pdf

Jain, V.K., Pandey, R.P., Jain, M.K., and Byun, H.R. 2015. Comparison of drought indices for appraisal of drought characteristics in the Ken River Basin. *Weather and Climate Extremes*, 8, 1–11.

Jalloh, A., Nelson, G.C., Thomas, T.S., Zougmore, R., and Roy-Macauley, H. (Eds.). 2013. Overview. In: West African Agriculture and Climate Change: A Comprehensive Analysis. International Food Policy Research Institute, Washington, DC 20006-1002, USA, pp. 1–52.

Jat, M.L., and Chaudhari, S.K. 2022. Conservation Agriculture and Sustainable Development Goals in South Asia. In: Souvenir, National Seminar on Agrophysics for Smart Agriculture, February 22-23, 2022, Indian Society of Agricultural Physics, New Delhi, India

Jat, M.L., Chakraborty, D., Ladha, J.K., Rana, D.S., Gathala, M.K., McDonald, A., and Gerard, B. 2020. Conservation agriculture for sustainable intensification in South Asia. *Nature Sustainability*, 3(4), 336–343.

Jat, M.L., Gathala, M.K., Choudhary, M., Sharma, S., Jat, H.S., Gupta, N., and Singh, Y. 2023. Conservation agriculture impacts on regenerating soil health, carbon sequestration, and climate change mitigation in South Asia. *Advances in Agronomy*, 181, 183–277. In Press.

Jat, M.L., Gathala, M.K., Ladha, J.K., Saharawat, Y.S., Jat, A.S., Kumar, V., Sharma, S.K., Kumar, V., and Gupta, R. 2009. Evaluation of precision land leveling and double zero-till systems in rice-wheat rotation: Water use, productivity, profitability and soil physical properties. *Soil and Tillage Research*, 105, 112–121.

Jat, M.L., Gupta, R., Saharawat, Y.S., and Khosla, R. 2011. Layering precision land leveling and furrow irrigated raised bed planting: Productivity and input use efficiency of irrigated bread wheat in indo-gangetic plains. *American Journal of Plant Sciences*, 2, 578–588.

Jat, H.S., Datta, A., Sharma, P.C., Kumar, V., Yadav, A.K., Choudhary, M., et al. 2017. Assessing soil properties and nutrient availability under conservation agriculture practices in a reclaimed sodic soil in cereal-based systems of North-West India. *Archives of Agronomy and Soil Science*, 64(4), 531–545.

Jayasree, V., and Venkatesh, B. 2015. Analysis of rainfall in assessing the drought in semi-arid region of Karnataka State, India. *Water Resources Management*, 29, 5613–5630.

Jones, M.R., Singels, A., and Ruane, A. 2013. Simulated impacts of climate change on water use and yield of irrigated sugarcane in South Africa. *Proceedings South African Sugar Technologists' Association*, 86, 184–189.

Kalamkar, S.S. 2011. *Agricultural Growth and Productivity in Maharashtra: Trends and Determinants*. Allied Publishers, New Delhi, India. 218 pp.

Kanté, S. 2001. Gestion de la fertilite des sols par classe d'exploitation au Mali-Sud. Tropical resource Manegement Papers, No. 38 (2001); ISSN 0926-9495, also published as thesis (2001), Wageningen University and Research Centre, Departement of plant Science, ISBN 90.

Kanté, S., Smaling, E.M.A., and Van Keulen, H. 2007. Nutrient balances for different farm types in Southern Mali. In A. Bationo (Ed.), Advances in Integrated Soil Fertility Management in Sub-Saharan Africa: Challenges and Opportunities. 557–568. Springer.

Kar, S., Pramanick, B., Brahmachari, K., Saha, G., Mahapatra, B.S., Saha, A., and Kumar, A. 2021. Exploring the best tillage option in rice based diversified cropping systems in alluvial soil of Eastern India. *Soil and Tillage Research*, 205, 104761.

Kassam, A., Friedrich, T., and Derpsch, R. 2022. Successful experiences and lessons from conservation agriculture worldwide. *Agronomy*, 12, 769. https://doi.org/10.3390/agronomy12040769

Kaya, B., Hildebrand, P.E., and Nair, P.K.R. 2000. Modeling changes in farming systems with the adoption of improved fallows in southern Mali. *Agricultural Systems*, 66, 51–68.

Knox, J., Hess, T., Daccache, A., Wheeler, T. 2012. Climate change impacts on crop productivity in Africa and South Asia. *Environmental Research Letters*, 7, 1–8.

Lal, R. 2004. Soil carbon sequestration impacts on global climate change and food security. *Science*, 304(5677), 1623–1627. https://doi.org/10.1126/science.1097396

Lefore, N., Closas, A., and Schmitter, P. 2021. Solar for all: A framework to deliver inclusive and environmentally sustainable solar irrigation for smallholder agriculture. *Energy Policy*, 154, 112313. https://doi.org/10.1016/j.enpol.2021.112313

León, A. 2002. Drought frequency and its social impacts in the semi-arid region of chile. *Investigaciones Marinas*, 30(1), 89–90.

Liman, I., Whitney, C.W., Kungu, J., and Luedeling, E. 2017. Modelling risk and uncertainty in flood-based farming systems in East Africa. In Tropentag Bonn "Future Agric. Socio-Ecological Transitions Bio-Cultural Shifts"; Academia: San Francisco, CA, USA, 2017; p. 289.

Lobell, D.B., Schlenker, W., and Costa-Roberts, J. 2011. Climate trends and global crop production since 1980. *Science*, 333(6042), 616–620. https://doi.org/10.1126/science.1204531

Marengo, J.A., Alves, L.M., Alvala, R., Cunha, A.P., Brito, S., and Moraes, O.L. 2017. Climatic characteristics of the 2010-2016 drought in the semiarid Northeast Brazil region. *Anais da Academia Brasileira de Ciências*, 90, 1973–1985.

Masih, I., Maskey, S., Mussá, F.E.F., and Trambauer, P. 2014. A review of droughts on the African continent: A geospatial and long-term perspective. *Hydrology and Earth System Sciences*, 18(9), 3635–3649.

Masson-Delmotte, V., Zhai, P., Pirani, A., Connors, S.L., Péan, C., Berger, S., Caud, N., Chen, Y., Goldfarb, L., Gomis, M.I., Huang, M., Leitzell, K., Lonnoy, E., Matthews, J.B.R., Maycock, T.K., Waterfield, T., Yelekçi, O., Yu, R., and Zhou, B. (eds.) 2014. IPCC, 2021: Summary for policymakers. In: Climate Change 2021: The Physical Science Basis. Contribution of Working Group I to the Sixth Assessment Report of the Intergovernmental Panel on Climate Change, Cambridge University Press, Cambridge, United Kingdom and New York, NY, USA, pp. 3–32, https://doi.org/10.1017/9781009157896.001

Mavhura, E., Manatsa, D., and Mushore, T. 2015. Adaptation to drought in arid and semi-arid environments: Case of the Zambezi Valley, Zimbabwe. *Jàmbá: Journal of Disaster Risk Studies*, 7(1), 1–7.

McCartney, M., Foudi, S., Muthuwatta, L., Sood, A., Simons, G., Hunink, J., Vercruysse, K., and Omuombo, C. (eds.) 2019. International Water Management Institute (IWMI). Quantifying the services of natural and built infrastructure in the context of climate change: The case of the Tana River Basin, Kenya. Colombo, Sri Lanka: International Water Management Institute (IWMI) 61p. (IWMI Research Report 174)

Mekonen, A.A., Berlie, A.B., and Ferede, M.B. 2020. Spatial and temporal drought incidence analysis in the northeastern highlands of Ethiopia. *Geoenvironmental Disasters*, 7, 1–17.

Mekonnen, M.M., and Hoekstra, A.Y. 2011. The green, blue and grey water footprint of crops and derived crop products. *Hydrology and Earth System Sciences*, 15(5), 1577–1600. https://doi.org/10.5194/hess-15-1577-2011

Mekonnen, M.M., and Hoekstra, A.Y. 2016. Four billion people facing severe water scarcity. *Science Advances*, 2, e1500323.

Minh, T., Cofie, O., Lefore, N., and Schmitter, P. 2020. Multi-stakeholder dialogue space on farmer-led irrigation development in Ghana: An instrument driving systemic change with private sector initiatives. *Knowledge Management for Development Journal*, 15, 98–118.

Mishra, A.K., Desai, V.R., and Singh, V.P. 2007. Drought forecasting using a hybrid stochastic and neural network model. *Journal of Hydrologic Engineering*, 12(6), 626–638.

Mishra, V., Tiwari, A.D., Aadhar, S., Shah, R., Xiao, M., Pai, D.S., and Lettenmaier, D. 2019. Drought and famine in India, 1870-2016. *Geophysical Research Letters*, 46, 2075–2083. https://doi.org/10.1029/2018gl081477

Mittler, R. 2006. Abiotic stress, the field environment and stress combination. *Trends in Plant Science*, 11(1), 15–19.

Moreno, T.A., and Huber-Sannwald, E. 2011. Impacts of drought on agriculture in Northern Mexico. In Coping with Global Environmental Change, Disasters and Security: Threats, Challenges, Vulnerabilities and Risks (pp. 875–891). Berlin, Heidelberg: Springer Berlin Heidelberg.

Motsumi, S., Magole, L., and Kgathi, D. 2012. Indigenous knowledge and land use policy: Implications for livelihoods of flood recession farming communities in the Okavango Delta, Botswana. *Physics and Chemistry of the Earth*, 50–52, 185–195. https://doi.org/10.1016/j.pce.2012.09.013

Mrabet, R., El-Brahli, A., Anibat, I., and Bessam, F. 2003. No-tillage technology: Research review of impacts on soil quality and wheat production in semiarid Morocco. *Options Méditerranéennes*, 60, 133–138.

Mrunalini, K., Behera, B., Jaraman, S., Abhilash, P.C., Dubey, P.K., Narayanaswamy, G., Prasad, J.V.N.S., Rao, K.V., Krishnan, P., Pratibha, G. and Srinivasarao, Ch. 2022. Nature based solutions in soil restoration for improving agricultural productivity. *Land Degradation and Development*, 33, 1269–1289, https://doi.org/10.1002/ldr.4207

Mugisha, J., Ratemo, M.A., Keza, B.C.B., and Kahveci, H. 2021. Assessing the opportunities and challenges facing the development of off-grid solar systems in Eastern Africa: The cases of Kenya, Ethiopia, and Rwanda. *Energy Policy*, 150, 112131. https://doi.org/10.1016/j.enpol.2020.112131

Mukul, S., Herbohn, J., and Firn, J. 2016. Tropical secondary forests regenerating after shifting cultivation in the Philippines uplands are important carbon sinks. *Scientific Reports*, 6, 22483. https://doi.org/10.1038/srep22483

Muzangwa, L., Mnkeni, P.N.S., and Chiduza, C. 2020. The use of residue retention and inclusion of legumes to improve soil biological activity in maize-based No-till systems of the eastern cape province, South Africa. *Agricultural Research*, 9, 66–76. [CrossRef] 31. Shahbaz, M., Kuzyakov, Y., and Heitkamp, F. 2017. Decrease of soil organic matter stabilization with increasing inputs: Mechanisms and controls. *Geoderma*, 304, 76–82.

Nederveen Pieterse, J. 2012. Periodizing globalization: Histories of globalization. *New Global Studies*, 6, 1–25. https://doi.org/10.1515/1940-0004.1174

Nederveen, S. 2012. Flood recession farming: An overview and case study from the upper Awash catchment, Ethiopia. MSc Thesis Hydrology. Code 450122. Wageningen University, The Netherlands.

Nelson, G.C., Valin, H., Sands, R.D., Havlík, P., Ahammad, H., Deryng, D., and Willenbockel, D. 2013. Climate change effects on agriculture: Economic responses to biophysical shocks. *Proceedings of the National Academy of Sciences*, 111(9), 3274–3279. https://doi.org/10.1073/pnas.1222465110

Nikolaou, G., Neocleous, D., Christou, A., Kitta, E., and Katsoulas, N. 2020. Implementing sustainable irrigation in water-scarce regions under the impact of climate change. *Agronomy*, 10(8), 1120.

Noubondieu, S., Flammini, A., and Bracco, S. 2018. Costs and benefits of solar irrigation systems in Senegal. Dakar: FAO, 28.

Nyakudya, I.W., and Stroosnijder, L. 2011. Water management options based on rainfall analysis for rainfed maize (Zea mays L.) production in Rushinga district, Zimbabwe. *Agricultural Water Management*, 98, 1649–1659.

Ockwell, D., Sagar, A., and de Coninck, H. 2014. Collaborative research and development (R&D) for climate technology transfer and uptake in developing countries: Towards a needs driven approach. *Climatic Change*, 131(3), 401–415. https://doi.org/10.1007/s10584-014-1123-2

Pandey, R.P., Dash, B.B., Mishra, S.K., and Singh, R. 2008. Study of indices for drought characterization in KBK districts in Orissa (India). *Hydrol Process*, 22, 1895–1907.

Parry, M.L., Canziani, O.F., Palutikof, J.P., van der Linden, P.J. and Hanson, C.E. (eds.) IPCC, 2007: Climate Change 2007: Impacts, Adaptation and Vulnerability. Contribution of Working Group II to the Fourth Assessment Report of the Intergovernmental Panel on Climate Change, Cambridge University Press, Cambridge, UK, 976pp.

Pasternak, D., Ibrahim, A., and Augustine, A., 2012. Evaluation of five pearl millet varieties for yield and forage quality under two planting densities in the Sahel. *African Journal of Agricultural Research*, 7, 4526–4535.

Pathak, P., Sahrawat, K.L., Rego, T.J., and Wani, S.P. 2005. Measurable biophysical indicators for impact assessment: Changes in soil quality. In *Natural resource management in agriculture: Methods for assessing economic and environmental impact*, ed. B. Shiferaw, H.A. Freeman, and S.M. Swinton, 53–74. Wallingford, UK: CAB International.

Paul, A.K., and Annicet Hugues, N. 2018. Effets de l'engrais sur la fertilité, la nutrition et le rendement du maïs: Incidence sur le diagnostic des carences du sol Paul Kouadjo Akanza, et Annicet Hugues N ' Da. *Journal de la Société Ouest-Africaine de Chimie*, 45, 54–66. https://doi.org/www.researchgate.net/publication/327653123

Peyraud, J., Cellier, P., Aarts, F., Béline, F., Bourblanc, M., Delaby, L., Donnars, C., and Dupraz, P. 2020. Rapport d'expertise scientifique collective Les flux d'azote liés aux élevages: réduire les pertes, rétablir les équilibres. rapport d'expertise, INRA, Paris, HAL Id: hal-01198315, 527p. https://hal.archives-ouvertes.fr/hal-01198315

Pratibha, G., Srinivas, I., Rao, K.V., Raju, B.M.K., Thyagaraj, C.R., Korwar, G.R., and Srinivasarao, C. 2015. Impact of conservation agriculture practices on energy use efficiency and global warming potential in rainfed pigeonpea-castor systems. *European Journal of Agronomy*, 66, 30–40. https://doi.org/10.1016/j.eja.2015.02.001

Pratibha, G., Srinivas, I., Rao, K.V., Raju, B.M.K., Thyagaraj, C.R., Korwar, G.R., Venkateswarlu, B., Shanker, A.K., Choudhary, D.K., Rao, K.S., and Srinivasarao, C. 2015. Impact of conservation agriculture practices on energy use efficiency and global warming potential in rainfed pigeonpea- castor systems. *European Journal of Agronomy*, 66, 30–40.

Quinton, J., Govers, G., and Van Oost, K. et al. 2010. The impact of agricultural soil erosion on biogeochemical cycling. *Nature Geoscience*, 3, 311–314. https://doi.org/10.1038/ngeo838

Rabia, A., Figueredo, H., Huong, T., Lopez, A., Solomon, H., and Alessandro, V. 2013. Land suitability analysis for policy making assistance: A GIS based land suitability comparison between surface and drip irrigation systems. *International Journal of Environmental Science and Development*, 4, 1–6. https://doi.org/10.7763/IJESD.2013.V4.292

Rahman, N.A., Larbi, A., Opoku, A., Tetteh, F.M., and Hoeschle-Zeledon, I. 2019. Crop-livestock interaction effect on soil quality and maize yield in Northern Ghana. *Agronomy Journal*, 111, 907. https://doi.org/10.2134/agronj2018.08.0523

Ramisch, J. 2014. La longue saison sèche: Interaction agriculture-élevage dans le sud du Mali. Agricultural Ecosystems Research Group, Agronomy departement, University of Wisconsin, 1575 Linden Drive, Madison, WI 53706, US. Fax: +1 608 265 3437: jjramisch@facstaff.wwisc.e.

Ramkar, P., and Yadav, S.M. 2018. Spatiotemporal drought assessment of a semi-arid part of middle Tapi River Basin, India. *International Journal of Disaster Risk Reduction*, 28, 414–426.

Ratnakumar, P., Vadez, V., Krishnamurthy, L., and Rajendrudu, G. 2011. Semi-arid crop responses to atmospheric elevated CO2. *Plant Stress*, 5(1), 42–51.

Reddy, S.B., Srinivasarao, C., Rao, P.C., Lal, R., Rakesh, S., Kundu, S., Singh, R.N., Dubey, P.K., Abhilash, P.C., Rao, K.V., Abrol, V. and Somasundaram, J. 2022. Greenhouse gas emission and agronomic productivity as influenced by varying levels of N fertilizer and tanksilt in degraded semi-arid alfisol of Southern India. Land Degradation and Development, https://doi.org/10.1002/ldr.4507

Ren, C., Zhang, X., and Reis, S. et al. 2023. Climate change unequally affects nitrogen use and losses in global croplands. *Nature Food*. https://doi.org/10.1038/s43016-023-00730-z

Renwick, L.L., Deen, W., Silva, L., Gilbert, M.E., Maxwell, T., Bowles, T.M., and Gaudin, A.C. 2021. Long-term crop rotation diversification enhances maize drought resistance through soil organic matter. *Environmental Research Letters*, 16(8), 084067.

Rivera, J.A., Araneo, D., Penalba, O., and Villalba, R. 2017. Linking climate variations with the hydrological cycle over the semi-arid Central Andes of Argentina. Past, present and future, with emphasis on streamflow droughts. In EGU General Assembly Conference Abstracts, p. 995.

Rockström, J., W. Steffen, K. Noone, Å. Persson, F.S. Chapin, III, E. Lambin, T.M. Lenton, M. Scheffer, C. Folke, H. Schellnhuber, B. Nykvist, C.A. De Wit, T. Hughes, S. van der Leeuw, H. Rodhe, S. Sörlin, P.K. Snyder, R. Costanza, U. Svedin, M. Falkenmark, L. Karlberg, R.W. Corell, V.J. Fabry, J. Hansen, B. Walker, D. Liverman, K. Richardson, P. Crutzen, and J. Foley, 2009. Planetary boundaries: Exploring the safe operating space for humanity. *Ecology and Society*, 14(2), 32.

Rosenzweig, C., Carolyn Z. Mutter, Alex C. Ruane, Erik Mencos Contreras, Kenneth J. Boote, Roberto O., Valdivia, Joske Houtkamp, Dilys, S., MacCarthy, Lieven Claessens, Roshan Adhikari, Wiltrud Durand, Sabine Homann-Kee Tui, Ashfaq Ahmad, Nataraja Subash, Geethalakshmi Vellingiri, and Nedumaran, S. 2021. AgMIP Regional Integrated Assessments: High-level Findings, Methods, Tools, and Studies (2012-2017). In Handbook of Climate Change and Agroecosystems: Climate Change and Farming System Planning in Africa and South Asia: AgMIP Stakeholder-driven Research Part 1 (pp. 123–142). World Scientific Publishing Co. https://doi.org/10.1142/9781786348791_0005

Rotenberg, E., and Yakir, D. 2010. Contribution of semi-arid forests to the climate system. *Science*, 327(5964), 451–454.

Roudier, P., and Mahe, G. 2010. Study of water stress and droughts with indicators using daily data on the Bani river (Niger basin, Mali). *International Journal of Climatology*, 30(11), 1689–1705.

Roy, N., Misra, R.V., Lesschen, J., and Smaling, K. 2005. Evaluation du bilan en éléments nutritifs du sol. Approches et méthodologies, Organisation des nations Unies pour l'alimentation et l'agriculture (FAO), Rome, 2005, Bulletin FAO engrais et nutrition végétale n°14, https://www.fao.org, Approches Méthodologiques.

Safriel, U., and Adeel, Z. 2005. Dryland systems. In: Hassan R, Scholes R, Ash N (eds.), *Ecosystems and human well-being, current state and trends*, vol. 1. Island Press, Washington, pp. 625–658.

Sahu, S., Kumar, Sg., Bhat, Bv., Premarajan, K., Sarkar, S., Roy, G., and Joseph, N. 2015. Malnutrition among under-five children in India and strategies for control. *Journal of Natural Science, Biology and Medicine*, 6(1), 18. https://doi.org/10.4103/0976-9668.149072

Samaké, O., Smaling, E.M.A., Kropff, M.J., Stomph, T.J., and Kodio, A. 2005. Effects of cultivation practices on spatial variation of soil fertility and millet yields in the Sahel of Mali. *Agriculture, Ecosystems & Environment*, 109, 335–345.

Sanogo, D., Ndour, B.Y., Sall, M., Toure, K., Diop, M., Camara, B.A., and Thiam, D. 2017. Participatory diagnosis and development of climate change adaptive capacity in the groundnut basin of Senegal: Building a climate-smart village model. *Agriculture & Food Security*, 6(1). https://doi.org/10.1186/s40066-017-0091-y

Sanogo, S., Peyrillé, P., Roehrig, R., Guichard, F., and Ouedraogo, O. 2021. Heavy precipitating events in satellites and rain-gauge products over the Sahel, EGU General Assembly 2021, online, 19-30 Apr 2021, EGU21-8261, https://doi.org/10.5194/egusphere-egu21-8261, 2021

Sanon, H.O., Kaboré-Zoungrana, C., and Ledin, I. 2007. Edible biomass production from some important browse species in the Sahelian zone of West Africa. *Journal of Arid Environments*, 71, 376–392.

Sawargaonkar, G.L., Girish Chander, G.C., Wani, S.P., Dasgupta, S.K., and Pardhasaradhi, G. 2018. Increasing agricultural productivity of farming systems in parts of central India - Sir Ratan Tata Trust initiative. In CABI publication: Corporate social responsibility: Win-win propositions for communities, corporates and agriculture, pp. 161–179. https://doi.org/10.1079/9781786394514.0161

Sawargaonkar, G., Rao, S.R., and Wani, S.P. 2016. An integrated approach for productivity enhancement, In CABI Publication: Harnessing dividends from drylands: Innovative scaling up with soil nutrients, pp. 201–235.

Sawargaonkar, G., Shelke, D.K., and Shinde, S.A. 2008. Influence of cropping systems and fertilizer doses on dry matter accumulation and nutrient uptake by maize (Zea mays L.). *International Journal of Agricultural Science*, 4(1), 45–50.

Schewe, J. et al. 2014. Multimodel assessment of water scarcity under climate change. *Proceedings of the National Academy of Sciences of the United States of America*, 111, 3245–3250.

Schlenker, W., and Lobell, D.B. 2010. Robust negative impacts of climate change on African agriculture. *Environmental Research Letters*, 5, 014010. https://iopscience.iop.org/article/10.1088/1748-9326/5/1/014010

Schmitter, P., Kibret, K.S., Lefore, N., and Barron, J. 2018. Suitability mapping framework for solar photovoltaic pumps for smallholder farmers in sub-Saharan Africa. *Applied Geography*, 94, 41–57. https://doi.org/10.1016/j.apgeog.2018.02.008

Serdeczny, O., Adams, S., Baarsch, F., Coumou, D., Robinson, A., Hare, B., Schaeffer, M., Perrette, M., and Reinhardt, J. 2017. Climate change impacts in Sub-Saharan Africa: From physical changes to their social repercussions. *Regional Environmental Change*, 17, 1–16. https://doi.org/10.1007/s10113-015-0910-2

Sharma, S. and Majumdar, P. 2017. Increasing frequency and spatial extent of concurrent meteorological droughts and heat waves in India. *Scientific Reports*, 7, 1–9.

Shiferaw, B. 2002. Poverty and natural resource management in the semi-arid tropics: Revisiting challenges and conceptual issues. Working Paper Series no. 14. Working Paper. International Crops Research Institute for the Semi-Arid Tropics, Patancheru, Andhra Pradesh, India.

Sidibe, Y., Williams, T.O., and Kolavalli, S. 2016. Flood recession agriculture for food security in Northern Ghana: Literature review on extent, challenges, and opportunities. *Ghana Strategy Support Program Working Paper*, 42, 1–18, https://doi.org/10.13140/RG.2.1.3250.8405

Singh, N.P., Bantilan, C., and Byjesh, K. 2014. Vulnerability and policy relevance to drought in the semi-arid tropics of Asia-A retrospective analysis. *Weather and Climate Extremes*, 3, 54–61.

Singh, R., Garg, K.K., Anantha, K.H., Venkataradha, A., Dev, I., Dixit, S., and Dhyani, S.K., 2021. Building resilient agricultural system through groundwater management interventions in degraded landscapes of Bundelkhand region, Central India. *Journal of Hydrology: Regional Studies*, 37, 100929.

Singh, R., Garg, K.K., Wani, S.P., Tewari, R.K., and Dhyani, S.K. 2014. Impact of water management interventions on hydrology and ecosystem services in Garhkundar-Dabar watershed of Bundelkhand region, Central India. *Journal of Hydrology*, 509, 132–149.

Singh, R., Venkataradha, A., Anantha, K.H., Garg, K.K., Barron, J., Whitbread, A.M., Dev, I., and Dixit, S. 2022. Traditional Rainwater Management (Haveli cultivation) for Building System Level Resilience in a Fragile Ecosystem of Bundelkhand Region, Central India. *Frontiers in Sustainable Food Systems*, 6, 826722. https://doi.org/10.3389/fsufs.2022.826722

Sissoko, F., Affholder, F., Autfray, P., Wery, J., and Rapidel, B. 2013. Wet years and farmers' practices may offset the benefits of residue retention on runoff and yield in cotton fields in the Sudan-Sahelian zone. *Agricultural Water Management*, 119, 89–99.

Skendžić, S., Zovko, M., Živković, I.P., Lešić, V., and Lemić, D. 2021. The impact of climate change on agricultural insect pests. *Insects*, 12(5), 440. https://doi.org/10.3390/insects12050440

Smaling, E.M.A., Lesschen, J.P., Van Beek, C.L., De Jager, A., Stoorvogel, J.J., Batjes, N.H., and Fresco, L.O. 2013. Where do we stand 20 Years after the assessment of soil nutrient balances in sub-Saharan Africa? openscience.library@wur.nl. *University research, Wageningen, World Soil Resource Food Security*, 499–537. https://doi.org/10.1201/b11238-15

Sogoba, B., Traoré, B., Safia, A., Samaké, O.B., Dembélé, G., Diallo, S., Kaboré, R., Benié, G.B., Zougmoré, R.B., and Goïta, K., 2020. On-farm evaluation on yield and economic performance of Cereal-Cowpea intercropping to support the smallholder farming system in the Soudano-Sahelian zone of Mali. *Agriculture*, 10, 1–15. https://doi.org/10.3390/agriculture10060214.

Spinoni, J., Barbosa, P., Bucchignani, E., Cassano, J., Cavazos, T., Christensen, J.H., Christensen, O., Coppola, E., Evans, J., Geyer, B., Giorgi, F., Hadjinicolaou, P., Jacob, D., Katzfey, J., Koenigk, T., Laprise, R., Lennard, C., Kurnaz, L., Li, D. and Giorgi, F. 2020. Future global meteorological drought hotspots: A study based on CORDEX data. *Journal of Climate*, 33(9), 3635–36561.

Srinivasarao, Ch., Gopinath, K.A., Prasad, J.V.N.S., Prasannakumar, and Singh, A.K. 2016. Climate resilient villages for sustainable food security in tropical India: Concept, process, technologies, institutions, and impacts. *Advances in Agronomy*, 140(3), 101–214.

Srinivasarao, Ch., Prasad R.S. and Mohapatra, T. 2019. Climate Change and Indian Agriculture: Impacts, Coping Strategies, Programmes and Policy. Technical Bulletin/Policy Document. Indian Council of Agricultural Research, Ministry of Agriculture and Farmers' Welfare and Ministry of Environment, Forestry and Climate Change, Government of India, New Delhi, p25.

Srinivasarao, Ch., Prasad, J.V.N.S., Rao, K.V., Kiran, B.V.S., Ranjit, M., Girija Veni, V., Priya, P., Abhilash, P.C. and Chaudhari, S.K. 2022a. Land and water conservation technologies for building carbon positive villages in India. *Land Degradation & Development*, 33(3), 395–412. https://doi.org/10.1002/ldr.4160

Srinivasarao, Ch., Ramesh Naik, M., Ranjit Kumar, G., Manasa, R., Kundu, S., Narayana Swamy, G., Nataraj, K.C. and Prasad, J.V.N.S. 2022b. Cover-crop technology for soil health improvement, land degradation neutrality, and climate change adaptation. *Indian Journal of Fertilisers*, 18(5), 440–460.

Srinivasarao, Ch., Rao, K.V., Gopinath, K.A., Prasad, Y.G., Arunachalam, A., Ramana, D.B.V., Ravindra Chary, G., Gangaiah, B., Venkateswarlu, B. and Mohapatra T. 2020. Agriculture contingency plans for managing weather aberrations and extreme climatic events: Development, implementation and impacts in India. *Advances in Agronomy*, 159, 35–91. https://doi.org/10.1016/bs.agron.2019.08.002.

Srinivasarao, Ch., Rattan Lal, Prasad, J.V.N.S., Gopinath, K.A., Rajbir Singh, Vijay S. Jakkula, Sahrawat, K.L., Venkateswarlu, B., Sikka, A.K., and S.M. Virmani. 2015. Potential and challenges of rainfed farming in India. *Advances in Agronomy*, 133, 113–181.

Srinivasarao, Ch., Singh, S.P., Sumanta Kundu, Vikas Abrol, Rattan Lal, Abhilash, P.C., Chary, G.R., Pravin B. Thakur, Prasad, J.V.N.S. and Venkateswarlu, B. 2021b. Integrated nutrient management improves soil organic matter and agronomic sustainability of semiarid rainfed Inceptisols of the Indo-Gangetic Plains. *Journal of Plant Nutrition and Soil Science*, 1–11. https://doi.org/10.1002/jpln.202000312

Srinivasarao, Ch. and Singh, A.K. (Conveners). 2021a. Strategies for Enhancing Soil Organic Carbon for Food Security and Climate Action. Policy Paper No. 100, National Academy of Agricultural Sciences, New Delhi, p16.

Stenzel, F., Greve, P., and Lucht, W. et al. 2021. Irrigation of biomass plantations may globally increase water stress more than climate change. *Nature Communications*, 12, 1512. https://doi.org/10.1038/s41467-021-21640-3

Stocker, T.F., et al., Eds., 2013. IPCC, 2013: Climate Change 2013: The Physical Science Basis. Cambridge University Press, 1535 pp.

Sultan, B., Defrance, D. and Iizumi, T. 2019. Evidence of crop production losses in West Africa due to historical global warming in two crop models. *Scientific Reports*, 9, 1283. https://doi.org/10.1038/s41598-019-49167-0

Sun, Y., Tao, C., Deng, X., Liu, H., Shen, Z., Liu, Y., and Geisen, S. 2022. Organic fertilization enhances the resistance and resilience of soil microbial communities under extreme drought. *Journal of Advanced Research*, 47, 1–12.

Tahri, M., Hakdaoui, M., and Maanan, M. 2015. The evaluation of solar farm locations applying Geographic Information System and Multi-Criteria Decision-Making methods: Case study in southern Morocco. *Renewable and Sustainable Energy Reviews*, 51, 1354–1362. https://doi.org/10.1016/j.rser.2015.07.054

Thornton, P.K., van de Steeg, J., Notenbaert, A., and Herrero, M. 2009. The impacts of climate change on livestock and livestock systems in developing countries: A review of what we know and what we need to know. *Agricultural Systems*, 101, 113–127. https://doi.org/10.1016/j.agsy.2009.05.002

Tittonell, P., Vanlauwe, B., Leffelaar, P.A., Shepherd, K.D., and Giller, K.E. 2005. Exploring diversity in soil fertility management of smallholder farms in western Kenya II. Within-farm variability in resource allocation, nutrient flows and soil fertility status. *Agriculture, Ecosystems & Environment*, 110, 166–184.

Tounkara, A. 2018. Legacy effects of cropping systems on soil carbon stock, millet yield and NPK fertilizer efficiency in Sahel. https://hdl.handle.net/20.500.11766/9470

Trail, P., Abaye, O., Thomason, W.E., Thompson, T.L., Gueye, F., Diedhiou, I., Diatta, M.B., and Faye, A. 2016. Evaluating intercropping (living cover) and mulching (desiccated cover) practices for increasing millet yields in Senegal. *Agronomy Journal*, 108, 1742–1752. https://doi.org/10.2134/agronj2015.0422

Traore, A., Ouattara, B., Ouedraogo, S., Yabi, A.J., and Lompo, F. 2019. Mineral fertilisation by microdose: Incentives for widespread adoption in Burkina Faso. *African Crop Science Journal*, 27, 29–43.

Traore, B., Corbeels, M., van Wijk, M.T., Rufino, M.C., and Giller, K.E. 2013. Effects of climate variability and climate change on crop production in southern Mali. *European Journal of Agronomy*, 49, 115–125.

Traore, B., Van Wijk, M.T., Descheemaeker, K., Corbeels, M., Rufino, M.C., and Giller, K.E. 2014. Evaluation of climate adaptation options for Sudano-Sahelian cropping systems. *Field Crops Research*, 156, 63–75. https://doi.org/10.1016/j.fcr.2013.10.014

Traoré, O., Traoré, K., Bado, V.B., and Lompo, D.J. 2007. Crop rotation and soil amendments: Impacts on cotton and maize production in a cotton-based system in western Burkina Faso. *International Journal of Biological and Chemical Sciences*, 1, 143–150.

Trenberth, K.E., Anthes, R.A., Belward, A., Brown, O.B., Habermann, T., Karl, T.R., Running, S., Ryan, B., Tanner, M., and Wielicki, B. 2013. Challenges of a sustained climate observing system. In *Climate Science for Serving Society* (pp. 13–50). New York: Springer.

Umutoni, C., Ayantunde, A.A., and Sawadogo, G.J. 2016. Connaissance locale des pratiques de la transhumance dans la zone soudano-sahélienne du Mali. *d'Elevage et de Médecine Vétérinaire des Pays Tropicaux*, 69, 53–61.

van der Pol, F. 1992. Soil mining, an unseen contributor to farm income in southern Mali. Royal Tropical Institute (KIT), Amsterdam. Bulletin 325.

Van Ginkel, M., Sayer, J., Sinclair, F., Aw-Hassesn A., Bossieo, D., Crufurd, P., El Mourid M., Haddad, N., Hoisington, D., Johnson, N., Velarde, C.L., Mares, V., Mude, A., Nefzaoui, A., Noble, A., Rao, K.P.C., Serraj, R., Tarawali, S., Vodouhe, R., and Ortiz, R. 2013. An integrated agro-ecosystem and livelihood systems approaches for the poor and vulnerable in dry areas. Food Security. https://doi.org/10.1007/s1257-013-0305-5

Van Meter, K.J., Basu, N.B., Veenstra, J.J., and Burras, C.L. 2016. The nitrogen legacy: Emerging evidence of nitrogen accumulation in anthropogenic landscapes. *Environmental Research Letters*, 11(3), 035014. https://doi.org/10.1088/1748-9326/11/3/035014

Van Steenbergen F., Abrham M.H., Taye A., Tena A., and Yohannes G. (2011). Status and potential of spate irrigation in Ethiopia. *Water Resources Management*. https://doi.org/10.1007/s11269-011-9780-7

van Steenbergen, F., Lawrence, P., Mehari, A.; Salman, M., and Faurès, J.M. 2010. Guidelines on Spate Irrigation; FAO: Rome, Italy, 2010.

Varisco, D.M. 1983. Sayl and Ghayl: The ecology of water allocation in Yemen. *Human Ecology*, 11, 365–383. https://doi.org/10.1007/BF00892245

Verhulst, N., Nelissen, V., Jespers, N., Haven, H., Sayre, K.D., Raes, D., and Govaerts, B. 2011. Soil water content, maize yield and its stability as affected by tillage and crop residue management in rainfed semi-arid highlands. *Plant and Soil*, 344, 73–85.

Vicente-Serrano, S.M., Lopez-Moreno, J.I., Beguería, S., Lorenzo-Lacruz, J., Sanchez-Lorenzo, A., García-Ruiz, J.M., and Espejo, F. 2014. Evidence of increasing drought severity caused by temperature rise in southern Europe. *Environmental Research Letters*, 9(4), 044001.

Wable, P.S., Jha, M.K. and Shekhar, A. 2019. Comparison of Drought Indices in a Semi-Arid River Basin of India. *Water Resour Manage*, 33, 75–102. https://doi.org/10.1007/s11269-018-2089-z

Wani, S.P., Anantha, K.H., Sreedevi, T.K., Sudi Raghavendra, Singh, Sn., and Souza Marcella. 2011. Assessing the Environmental Benefits of Watershed Develop- ment: Evidence from the Indian Semi-Arid Tropics. *Assessing the Environmental Benefits of Watershed Development: Evidence from the Indian Semi-Arid Tropics*, 1(1). 10–20. https://doi.org/10.5147/jswsm/2011/0036.

Wani, S.P., and Sawargaonkar Gajanan. 2018. Future smart crops for paddy fallow agri-food systems in Southeast Asia. Future smart food-rediscovering hidden treasures of neglected and underutilized species for zero hunger in Asia. FAO, Bangkok pp. 61–78.

Wilhite, D.A., Sivakumar, M.V., and Pulwarty, R. 2014. Managing drought risk in a changing climate: The role of national drought policy. *Weather and Climate Extremes*, 3, 4–13.

Wilhite, D.A. 2000. Drought as a natural hazard: Concepts and Definitions. Drought, a Global Assessment, 1.

Wilhite, D.A. and Glantz, M.H. 1985. Understanding the drought phenomenon: The role of definitions. *Water International*, 10(3), 111–120.

Wilhite, D.A., Svoboda, M.D., and Hayes, M.J. 2007. Understanding the complex impacts of drought: A key to enhancing drought mitigation and preparedness. *Water Resources Management*, 21, 763–774.

World Bank. 2008. Climate change impacts in drought and flood affected areas: Case studies in India. Sustainable Development Department.

World Bank. 2003. World Development Report 2003: Sustainable Development in a Dynamic World–Transforming Institutions, Growth, and Quality of Life. (c) World Bank. License: CC BY 3.0 IGO. https://hdl.handle.net/10986/5985

Wu, J., Gu, Y., Sun, K., Wang, N., Shen, H., Wang, Y., and Ma, X. 2023. Correlation of climate change and human activities with agricultural drought and its impact on the net primary production of winter wheat. *Journal of Hydrology*, 620, 129504.

Xue, Y. 1996. The impact of desertification in the Mongolian and the Inner Mongolian grassland on the regional climate. *Journal of Climate*, 9(9), 2173–2189.

Yang, H., Wu, G., Mo, P., Chen, S., Wang, S., Xiao, Y., Hongli, M., Tao, W., Xiang, G., and Gaoqiong, F. 2020. Conservation tillage and residue management improves soil properties under a upland rice-rapeseed system in the subtropical eastern Himalayas. *Land Degradation & Development*, 31, 1775–1791.

Zargar, A., Sadiq, R., Naser, B., and Khan, F.I. 2011. A review of drought indices. *Environmental Reviews*, 19(NA), 333–349. https://doi.org/10.1139/a11-013

Zeng, N., Neelin, J.D., Lau, K.M., Tucker, C.J. 1999. Enhancement of interdecadal climate variability in the Sahel by vegetation interaction. *Science* 286 (5444):1537–1540

7 Managing Drought in Semi-Arid Regions through Improved Varieties and Choice of Species

Faisal Nadeem, Abdul Rehman, Aman Ullah, Muhammad Farooq, and Kadambot H.M. Siddique

7.1 INTRODUCTION

Geographically dry lands (hyper-arid, arid, semi-arid, and dry sub-humid ecosystems) are demarcated by a water deficit with precipitation to potential evaporation ratio (also called the aridity index; AI)<0.65 (Hassan et al. 2005; Cima et al. 2015; Plaza et al. 2018). The AI of humid drylands is >0.65, arid areas are ≥ 0.05 and <0.2, and semi-arid regions are ≥ 0.2 and <0.5 (Abatzoglou et al. 2018; Figure 7.1).

The aridity, in meteorology, is "the degree to which a climate lacks effective, life-promoting moisture" (https://glossary.ametsoc.org/wiki/Aridity). Extreme drought events in semi-arid regions depend on pre-existing anthropic stressors (de Melo et al. 2022) that influence ecosystem dynamics. Agriculture is limited to livestock grazing in arid areas, with a large part of semi-arid areas used for animal grazing. The semi-arid and sub-humid areas account for 27% of the earth's area (4 billion ha), with only 10% farmed each year as most of the arable land has long-term crop–fallow rotations. These areas house around 2.5 billion people and are thus crucial for global food security despite their challenging soil and climatic conditions (Stewart and Thapa 2016). Semi-arid grasslands are among the most extensive global ecosystems, spread across continental to temperature zones and parts of some subtropical zones.

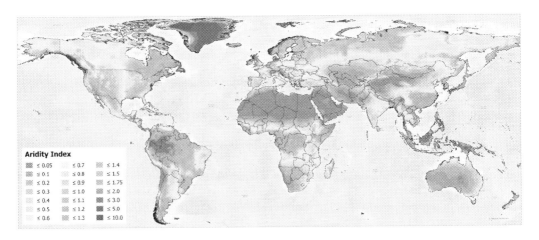

FIGURE 7.1 Global aridity index (AI-v3) based on the FAO-56 Penman–Monteith equation for reference evapotranspiration. Green and blue colors indicate environments with high humidity; red, brown, and yellow colors reflect regions with high aridity.

Drought is "a period of abnormally dry weather sufficiently long enough to cause a serious hydrological imbalance" (https://glossary.ametsoc.org/wiki/Drought). Soils in semi-arid regions have been degraded by historical land use, resulting in low soil organic carbon (SOC) content, erosion, salinity, and degradation (FAO 2016), linked to losses in soil water storage capacity and greenhouse gas (GHG) emissions. Despite these sensitivities, semi-arid soils are considered 'hot spots' due to their importance for crop production and contribution to food security. Semi-arid ecosystems contribute the most to internal variability in the land CO_2 sink (57% or 0.04 Pg C year^{-1}) relative to the global land CO_2 sink (0.07 Pg C year^{-1}) (Ahlstrom et al. 2015).

Rainfall variations associated with climate change pose a risk to the water- and temperature-sensitive semi-arid ecosystems (Kottek et al. 2006; Sommer et al. 2011). In semi-arid regions, surface SOC is more sensitive to climate change scenarios than subsoil SOC (Albaladejo et al. 2013). Nevertheless, the mechanisms controlling C dynamics differ between near-surface soils and subsoils (Hobley et al. 2016b), and subsoils respond rapidly to land use change (Hobley et al. 2016a). As such, whole soil profile management is important, as subsoils are reservoirs for soil nutrients and SOC and have a high-water storage capacity, providing resilience to drought.

Drought is a key stress factor with a high impact on crop yields (Bennett et al. 2014; Table 7.2). The projected high drought spells and variation in interannual rainfall with climate change will further exacerbate the vulnerability of drought- and temperature-sensitive semi-arid ecosystems (Kottek et al. 2006; Sommer et al. 2011), impacting the global food supply. For instance, the impact of drought on cereal crop productivity is significant and widespread, with consistent reductions in average grain yield occurring in many regions around the world, including Europe, Africa, Asia, Australia, South America, Central America, and North America (Daryanto et al. 2016; Zampieri et al. 2017; Poggi et al. 2023). The principal source of livelihood in arid and semi-rid Sub-Saharan Africa is agriculture, currently comprising around 43% of the total area and is projected to increase due to long drought spells and climate change. Water shortage is the major limitation to crop production in these areas (Hadebe et al. 2017).

Introducing staple crops with lower water requirements and high water productivity is needed to improve crop production and food security in semi-arid areas. Wheat (*Triticum aestivum* L.), rice (*Oryza sativa* L.), and maize (*Zea mays* L.) currently dominate global cropland areas, supplying monotonous diets. The yields of these crops are already affected by water shortages, and stability or increased production is not certain due to the rapidly changing climate. For instance, climate change is expected to worsen drought conditions in large wheat-growing areas worldwide (Poggi et al. 2023). Introducing highly nutritious drought-resilient alternate crops can help improve crop production and food security in semi-arid environments (Jamalluddin et al. 2021), such as millets, quinoa (*Chenopodium quinoa*), amaranth (*Amaranthus caudatus*), and teff (*Eragrostis tef*). Likewise, cowpea and bambara groundnuts can thrive in water deficit and poor soil conditions (Tadele 2017). The inherent tolerance of these crops makes them suitable for cultivation in semi-arid environments (Chivenge et al. 2015). Likewise, developing drought-tolerant crop varieties is needed to keep pace with rising food demands. Modern breeding, molecular, and transgenic approaches can help develop drought-tolerant, high-yielding crop cultivars of staple crops. This review highlights drought management strategies, including crop diversification with climate-resilient crops and crop improvement, using molecular breeding and transgenic approaches in semi-arid regions.

7.2 DROUGHT SCENARIOS FOR SEMI-ARID REGIONS

Semi-arid and arid regions regularly encounter drought, with far-reaching consequences for ecosystem services, land sustainability, and economic development. Climate models have predicted that arid and semi-arid regions will experience increased drought intensity and severity as these regions encounter more aridity and less precipitation (Schwabe et al. 2013). Projected surface temperature means and extremes will be 2°C higher compared to 1.5°C by 2050 in most land regions. The projected increase in temperature will aggravate drought events (through irregular rainfall) and further reduce vegetative productivity (Maurer et al. 2020). Dryland regions are sensitive to precipitation; thus, the projected increase in potential evapotranspiration will expand the global dryland area (Huang et al. 2016).

In the past two decades, frequent drought episodes in Northeast China, the United States of America (USA), and southeastern Australia have forewarned the widespread surge in climatic events (Sheffield and Wood 2008). Drought spells in recent decades—including those in Pakistan (Balochistan province; 1997–2003), China (1994, 1997, 2002), the Indian subcontinent (2002), the USA [Texas (2012); California (2012–2015)], Tibetan Plateau (1997–2002), Australia (1997–2010), and East Africa (2010–2011) (McVicar and Biewirth 2001; Prabhakar and Shaw 2008; Sheffield et al. 2009; van Dijk et al. 2013; Zhang et al. 2013; Hao et al. 2018)—have caused considerable damage to agricultural productivity and water supply (Peterson et al. 2012; Hoerling et al. 2014).

India has faced drastic water scarcity in the last 100 years and has become more sensitive to droughts with implications for all sectors (Goldin 2016). Reduction in freshwater availability and precipitation are threatening agricultural, urban and environmental interests in many arid and semi-arid regions of Europe, Africa, and Australia (Shindell et al. 2006; Seager et al. 2007). For instance, drought in 2009 resulted in 285,000 acres fallowed, $340 million of revenue loss in California's San Joaquin valley (Howitt et al. 2011); while in Australia it caused a 20% reduction in the value of irrigated agriculture (Kirby et al. 2012). Spain's 2004–2005 drought resulted in agricultural production losses of US$670 million and hydro-electric generation losses of US$123 million (Schwabe et al. 2012).

Asia and Africa are highly vulnerable to climatic extremes, with most rural people dependent on agriculture (Hoddinott and Quisumbing 2008; Nhemachena et al. 2008). Drought-sensitive regions have been profiled in India, northeastern India, and the Himalayas, with most of the vulnerable masses in the semi-arid tropics unable to cope with the adversities of drought due to the lack of access to adequate resources and weak institutional mechanisms (Jodha 2005; Ravindranath et al. 2011; Pandey and Jha 2012; Singh et al. 2013). Miyan et al. (2015) reported that Asian and Southeast Asian countries in the monsoon climatic zone, including Bangladesh, Nepal, Bhutan, and Cambodia, also suffer from frequent droughts arising from delayed and variable distribution patterns. In arid and semi-arid regions, fallow land is increasing considerably due to the lack of alternative irrigation options and unreliable monsoon periods. Moreover, increased crop failure and fluctuating product prices force farmers to abandon cultivation.

From early 2000 onwards, extreme dry spells affected large areas of South Asia, including southern and central Pakistan, western India, Sri Lanka, and Afghanistan. In the past five decades, these drought-prone regions have faced dry spells at least once every three years. For instance, Pakistan imposed a state of emergency in 2012 in Mirpur Khas and Tharparkar districts due to the severity of drought (Tareq 2012), and extreme water scarcity in Cambodia affected long-duration and late-season genotypes (Tsubo et al. 2009). From 2007 to 2009, severe droughts in Yemen displaced 22 million people (www.irinnews.org). During this time, Yemen received 200 m^3 of water (per person/year), below the international poverty line of 1,000 m^3 (www.theguardian.com). Extreme drought in Bangladesh made 2 million small farmers and 2.4 million rural wage laborers vulnerable to disaster, and 90% of perennial streams stopped flowing in the dry monsoon resulting in a serious water crisis (Rahman 2011).

7.3 CROP DIVERSIFICATION USING ALTERNATE AND DROUGHT-STRESS-TOLERANT CROPS

Crop diversification includes crop cultivation practices, such as legume–cereal intercropping, perennial forage crops, relay cropping, and agroforestry, aimed at introducing alternate crops or climate-resilient crop cultivars into existing cropping systems (Figure 7.2).

Crop diversification offers diverse options for the farmers' economic benefit, by combatting yield losses from monocropping and cultivating marginal lands (Hussain et al. 2020). Including

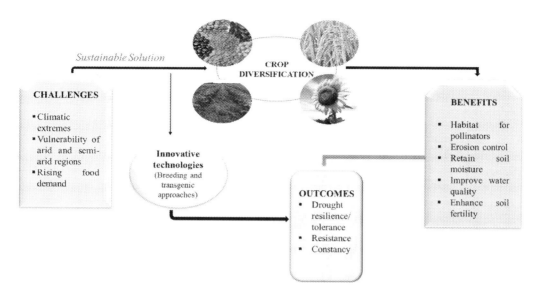

FIGURE 7.2 Crop diversification benefits and outcomes in semi-arid regions.

drought-tolerant crop cultivars such as quinoa, pearl millet, sorghum, barley, perennial grasses, safflower, and mustard, with low water requirements and wide food and industrial uses can improve crop production under water deficit. In addition, increasing agricultural crop diversification reduces biodiversity losses by increasing agroecosystem resilience (Altieri et al. 2015; Van Zonneveld et al. 2020). Crop diversification can suppress pest outbreaks, preserve biodiversity, and optimize water management when facing water scarcity.

7.3.1 Cereals

Most cereal crops, including wheat (*Triticum aestivum* L.), maize (*Zea mays* L.), sorghum (*Sorghum bicolor* L.), barley (*Hordeum vulgare* L.), and pearl millet (*Pennisetum glaucum* L.), are grown in arid and semi-arid regions, where they often endure exposure to recurrent drought episodes throughout their growth cycles. Implementing crop management practices such as using improved varieties can potentially alleviate the harmful effects of drought (Lamaoui et al. 2018). For example, Haley et al. (2007) developed a drought-tolerant variety of wheat called 'Ripper' that had superior grain yields and milling and bread-baking qualities under non-irrigated conditions in Colorado. Badu-Apraku and Yallou (2009) developed maize varieties with superior yields under drought conditions compared to the control varieties. Shamsudin et al. (2016) developed the drought-tolerant elite Malaysian rice cultivar MR219 by pyramiding three quantitative trait loci (QTL): qDTY2.2, qDTY3.1, and qDTY12.1. Dixit et al. (2017) developed the rice variety TDK1 for high yield under drought by incorporating three QTLs (qDTY3.1, qDTY6.1, and qDTY6.2).

Among cereals, barley has excellent growth and adaptation in water-limited and marginal soils. Mansour et al. (2017) reported that drought-tolerant barley genotypes had higher water use efficiency and grain yield than drought-sensitive genotypes. Barley can grow successfully in water-deficit soil supplied with irrigation management (drip irrigation) at the booting and heading stages. The synchronization of reproductive development with water availability is crucial for the success of crops in arid regions. Abu-Elenein et al. (2021) evaluated the variations in flowering and drought tolerance in barley genotypes and found that drought tolerant genotype (Rum) was superior in agronomic performance and stability under dry conditions, and there was a negative correlation between heading date and yield in both genotypes. They further observed that drought stress had a pronounced effect on the drought-sensitive genotype, but in the drought-tolerant genotype, it led to an advanced meristem development stage and earlier heading. This suggests that integrating vernalization and photoperiod signals in drought-tolerant barley genotypes may lead to higher productivity in dry environments. A recent study by Su et al. (2022) demonstrated that the use of drought-tolerant maize hybrids with deficit irrigation resulted in an increase in yield and water savings of 15%–17% compared to conventional hybrids. This highlights the potential of promoting the adoption of drought-tolerant maize hybrids with deficit irrigation as a viable strategy for increasing yields in semi-arid environments.

Millets are small-grain annual cereal crops, and millets excluding pearl millet and sorghum are called small or minor millets. Sorghum is an important cereal crop for food and fodder in China, Central and South America, Africa, South Africa, India, and Indonesia (Hussain et al. 2020). Sorghum can grow successfully in semi-arid environments with <100 mm annual precipitation (Assefa et al. 2010). Early maturing sorghum varieties are encouraged to overcome drought effects in semi-arid regions where seasonal rainfall has a short or erratic distribution. Hybrid cultivars of sorghum are often preferred due to their higher yields and more stable performance than open-pollinated varieties under a wide range of environmental conditions (Belete 2018). In Kenya, hybrid sorghum produced a 50% yield advantage over open-pollinated varieties under extreme drought (Chane 2018).

Triticale is a hybrid cereal developed by crossing rye and wheat that can grow under water deficit. Grzesiak et al. (2012) evaluated 11 triticale genotypes for drought tolerance and dry weight; drought-sensitive genotypes had higher grain yields and biomass under well-watered conditions, while drought-tolerant genotypes (CHD 247, CHD 220, Migo and Wanad) had minimal reductions in yield under drought stress compared to drought-sensitive genotypes. Quinoa is a highly nutritious alternative cereal crop with the potential to grow under drought stress due to its intrinsically low water requirements and ability to resume photosynthetic activities and leaf area growth after drought (Jacobsen et al. 2009).

González et al. (2012) demonstrated that quinoa accumulated high sugar and proline contents, biomarkers for environmental stresses such as drought, salinity, and low temperature (Vera-Hernández et al. 2018). Recently, Valdivia-Cea (2021) evaluated four quinoa genotypes for their response to deficit irrigation, concluding that quinoa can be successfully grown under water deficit conditions as drought-tolerant genotypes (Regalona, Cahuil and Morado) under severe drought stress (0% available water in the root zone) produced average yields of 2.90 t ha^{-1} compared to 3.68 t ha^{-1} under well-watered conditions (100% available water in the root zone). Teff is another cereal crop, mostly cultivated in Africa; teff variety DZ-Cr-37 is drought tolerant (Cannarozzi et al. 2014; Abraha et al. 2016).

7.3.2 LEGUMES

Legumes have at least two ways to resist drought: (i) drought avoidance via efficient stomatal regulation and (ii) drought tolerance via osmotic adjustment, allowing root growth to proceed (Chaves et al. 2002; Vadez et al. 2008). Legumes such as common bean, cowpea, and lupin can maintain leaf water content and avoid tissue dehydration during mild drought spells by controlling stomatal conductance and closure (Campos et al. 1999; Pinheiro et al. 2001), consequently decreasing internal CO_2 concentrations that eventually limit photosynthesis and shoot growth.

The second mechanism—osmotic regulation through increased solute concentration—demands less energy than drought avoidance (Amede and Schubert 2003). Solutes, mostly organic substrates such as sugars, sugar alcohols, and amino compounds, are allocated to the roots to lower their osmotic potential (Chaves et al. 2003; Streeter 2003) and thus continue extracting water at low soil water potentials. Maintaining turgor and plant water content by lowering epidermal conductance is an important trait for several legumes (e.g., chickpea, cowpea, common bean, pigeon pea) while lowering osmotic potential has been reported in common bean, faba bean, and cowpea in response to water deficit (Amede and Schubert 2003).

Chickpea (*Cicer arietinum* L.) is the second most important grain legume globally after common bean (*Phaseolus vulgaris*). It is cultivated predominantly on low fertile sandy soils in arid and semi-arid regions with little rainfall and high vulnerability to climate change (Kumar and Abbo 2001). Chickpea has low water and input requirements and can grow in climatic and soil conditions where many commercial crops fail (Hussain et al. 2020). Cowpea (*Vigna unguiculata* L.) Walp.) is another legume crop that can improve crop diversification in marginal soil and environments, with great genetic diversity for drought stress tolerance. Ishiyaku and Aliyu (2013) screened 22 cowpea genotypes for drought tolerance indices and found two genotypes—IT93K-452-1 and IT98K-412–13—with high drought tolerance indexes. In another drought tolerance index screening of 30 cowpea cultivars, Belko et al. (2014) identified five short-duration (IT85F-3139, IT93K-693-2, IT97K-499-39, KVx-61-1, Mouride) and five medium-duration genotypes (KVx-421–25, KVx-403, IT93K-503-1, IT97K-207–15 and IT96D-610) that were drought tolerant.

Faba bean (*Vicia faba* L.) is a highly nutritious grain legume that can be grown under low water and soil fertility conditions. It requires less moisture than many cereal crops, such that any carryover moisture will enhance N availability for the non-leguminous succeeding crop (Miller et al. 2002). Maalouf et al. (2015) observed significant variation among 11 faba bean genotypes under different water regimes, with FLIP06–010FB producing the highest yield under well-watered and drought-stressed conditions. Faba bean under limited water availability can produce comparable yields with well-watered crop (Theib et al. 2005). In an arid Mediterranean climate, 14 different faba bean genotypes were evaluated for their response to drought stress and were classified into five groups (A–E) based on their drought-response indices, with A being the most tolerant and E being the most sensitive. The drought-tolerant genotypes produced more yield with less water compared to the drought-sensitive genotypes. These drought-tolerant genotypes are recommended for faba bean production in water-limited environments (Mansour et al. 2021). Additionally, certain physiological parameters such as photosynthetic pigment, net photosynthetic rate, transpiration rate, leaf nutrient status, relative water content, and membrane stability index were positively associated with seed yield and yield-contributing traits.

Moth bean (*Vigna aconitifolia* (Jacq.) Maréchal) is an extraordinarily hardy legume crop, offering great potential for crop diversification in arid and semi-arid regions (Sharma et al. 2014). It can produce economic yields in water-scarce soils and survive evaporative stresses (Kumar and Singh 2002). Moth bean's genetic makeup and high buffering capacity help it adapt to very low moisture (desert) and nutrient-deficient conditions. A random amplified polymorphic DNA (RAPD) genetic diversity analysis of ten moth bean genotypes identified Jwala, PLMO-7, IC-39749, PLMO-2, and IC-39754 as drought-tolerant accessions.

Legumes are unique in their capacity to resist drought due to their interactions with N-fixing (i.e., rhizobia) bacteria and arbuscular mycorrhiza (Lodeiro et al. 2000). While some studies have suggested that water deficit inhibits N_2 fixation, there is evidence that genetic variation exists among species, which may explain their variable resistance to water stress. Furthermore, this N-fixing trait could be an important determining factor of yield potential since legumes combine biomass accumulated from photosynthesis and fixed N to form essential grain components (Sinclair et al. 2004).

Variety in nodule typology could also be responsible for the higher N_2 fixation of some legumes. For example, nodules formed in the endodermis (i.e., indeterminate nodules), such as in faba bean and groundnut, resist water stress better than those that are superficially attached (i.e., determinate nodules), such as in cowpea, black gram and green gram (Sprent 2008). While indeterminate nodules can proliferate after periods of adverse conditions, determinate nodules are short-lived and must be replaced when plant growth resumes. This replacement process can sometimes be incomplete (Sprent 2008), limiting N_2 fixation efficiency.

Legumes species that exhibit relatively high N_2 fixation during drought (e.g., groundnut and faba bean) also tend to produce higher yields than species with limited N_2 fixation (e.g., green gram, black gram, and cowpea) (Lodeiro et al. 2000). Thus, some legume species may benefit from symbiosis more than others since the investment to maintain nodules is about the same for most plants (i.e., about 20% of net photosynthate) (Sprent 2008). Therefore, species with less sensitivity to N_2 fixation under water deficits could be selected for regions where drought is a recurring phenomenon.

7.3.3 OILSEEDS

Oilseed crops, such as soybean (*Glycine max* L.), rapeseed/canola (*Brassica napus* L.), sunflower (*Helianthus annuus* L.), peanut (*Arachis hypogaea* L.), safflower (*Carthamus tinctorius* L.), flax (*Linum usitatissimum* L.), cotton (*Gossypium hirsutum* L.), castor (*Ricinus communis* L.), and sesame (*Sesamum indicum* L.), are predominantly grown for the oil contained in their seeds. The major oilseed-producing countries are the USA, Brazil, China, Argentina, and India (Rahman and de Jimenez 2016). Brassica species with large taproots can reduce soil compaction and increase water infiltration (Clark 2007; Chen and Weil 2011). Switching from cereals to drought-tolerant pulse and oilseed cultivars will help reduce crop failure.

Safflower (*Carthamus tinctorius* L.) has a very low water requirement compared to other oilseed crops and showed less decline in grain yield under drought stress (Movahhedy-Dehnavy et al. 2009; Hussain et al. 2020). Among four safflower genotypes grown in sandy and clay soils under water deficit, IAPAR and IMA-4409 tolerated drought better than the other tested accessions (Santos et al. 2017). Likewise, a drought-tolerant safflower genotype (L.R.V.5151) produced high oil yields (14.4%) even after exposure to drought stress at the vegetative stage (Movahhedy-Dehnavy et al. 2009). Mustard (*Brassica juncea* L.) can also grow under limited water availability. Chauhan et al. (2007) evaluated 14 Indian mustard genotypes for their performance under semi-arid and irrigated conditions; 6PSR-20, RH-819, JMMWR-941, PRO-97024, PCR-7, IS-1787, and RC-1446 had lower drought susceptibility indexes and higher yields than the other tested genotypes. Desert gourd (*Citrullus colocynthis* L.) Schrad.) is a xerophytic perennial creeper with high oil seed content (Bande et al. 2012). It often grows in sandy soils as a wild plant and can survive hyper-arid desert environments with very low precipitation (50 mm) and coastal areas with brackish water (Qasim et al. 2011).

Peanut is a widely grown food crop in tropical and semi-arid regions where water scarcity is a major challenge to production. In regions with limited water availability, upright cultivars are an ideal option for farmers as they have shorter growth cycles and require less water during cultivation (Painawadee et al. 2009). In a 2-year field study under a semi-arid environment in Brazil, four peanut genotypes were evaluated for their yield and drought tolerance response. The top two genotypes showed the physiological ability to tolerate drought, with LBM Branco, an early and upright line, being classified as drought tolerant and performing similarly to BR 1. This genotype had low losses in pod and seed yield compared to its runner parent (LViPE-06). Based on its physiological and agronomic performance, LBM Branco could be recommended for use in semi-arid environments (de Lima Pereira et al. 2016). The root architecture plays a crucial role in peanut productivity in dry conditions. Jongrungklang et al. (2014) demonstrated that the root surface area of a specific genotype (ICGV 98305) among six tested genotypes was greater in deeper soil layers under pre-flowering drought and this corresponded to an increase in pod yield. Additionally, the drought tolerance index was positively correlated with root surface area and pod yield under limited water conditions, suggesting that peanut plants with better root architecture can produce good yields under arid environments.

Sesame is a cash crop that is commonly grown by small and marginal farmers, primarily in regions with limited water resources. Recently, Gopika et al. (2022) evaluated 25 sesame genotypes for their response to drought stress in an arid/semi-arid environment. Seed yield decreased by 54% under drought stress compared to irrigation conditions. Genotypes SI 1802 and SI 9823 produced the highest seed yield under both irrigation and drought stress conditions and also had higher values for stress tolerance indices, suggesting that these genotypes can be useful in breeding programs as well as for cultivation in semi-arid environments.

7.3.4 COVER CROPS

Cover crops are those used to cover the ground surface to protect the soil from erosion and prevent nutrient losses in deep layers through leaching and surface runoff (Pieters and McKee 1938; Kaye and Quemada 2017). Ideal cover crops should germinate and emerge quickly, tolerate adverse climatic conditions, fix atmospheric nitrogen, absorb soil nutrients by developing deep roots and produce high amounts of biomass in short periods (Reddy 2016; Sharma et al. 2018). In addition, using cover crops in semi-arid regions reduces the summer fallow while improving soil health (Jones et al. 2020).

Crop species with fibrous root systems (e.g. ryegrass, rye, and oats) can control soil erosion, while cover crops with thick roots (e.g. white mustard and fodder radish) are less effective in preventing soil erosion (De Baets et al. 2011; Sharma et al. 2018). Cover crops help reduce surface soil evaporation, conserve moisture from irrigation and rainfall, and provide soil moisture to subsequent crops. For instance, winter rye as a cover crop enhanced water storage in soil over wet and dry years. In top soil, rye cover crops increased the water content than field capacity (10%–11%) and plant available water (21%–22%) (Basche et al. 2016).

Cover crops act as a barrier between the soil surface and precipitation, especially rainfall, reducing the intensity of rainfall hitting the ground. Water drops slowly trickle into the soil through soil pores formed by soil macrofauna and cover crop root growth. Thus, the slow water infiltration recharges soil water storage (Joyce et al. 2002; Sharma et al. 2011, 2012; Sammis et al. 2012). Cover crops can decrease added-in fertilizer requirements, particularly nitrogen, for the subsequent crop; nitrogen is captured by roots, preventing nitrate from leaching into groundwater and downward movement into the soil profile (Gabriel et al. 2013; Sharma et al. 2018). Crops such as pea and oat had higher earthworm populations than bare fallow plots or a cover crop–spring barley rotation, while Brassica species such as mustard had large aboveground biomass but small earthworm populations (Roarty et al. 2017).

Cover crops increase soil water content with crop biomass and infiltration but decrease soil water content through transpiration. Cover crops also help decrease surface drainage and maintain soil quality (Qi and Helmers 2010; Sharma et al. 2018).

7.4 DROUGHT STRESS TOLERANCE THROUGH CROP IMPROVEMENT

Drought tolerance is a complex trait that results from various mechanisms plants use to adapt to water scarcity. This trait encompasses a wide range of morphological, physiological, biochemical and molecular processes that are activated at different stages of plant development (Rampino et al. 2006; Nezhadahmadi et al. 2013). All these adaptations can confer drought tolerance through drought escape (completion of life cycle before onset of drought), avoidance (maintaining a better water status under drought) or tolerance (ability to withstand dehydration) (Kooyers 2015). Crop improvement using various strategies such as breeding, molecular, and transgenic approaches can potentially mitigate the drastic effects of drought stress.

7.4.1 BREEDING AND MOLECULAR APPROACHES

Improving drought tolerance in plants using conventional breeding methods has been challenging. Despite developing numerous field crop varieties using conventional breeding methods such as pedigree breeding, inbreeding, hybridization, and backcrossing (Ashraf 2010), conventional breeding is time-consuming, expensive and labor-intensive, requiring numerous selection and breeding cycles (Ahmar et al. 2020). The effects of drought can vary greatly depending on the environment. The relationship between plant characteristics and yield under drought conditions is often unclear. An approach that takes into account the specific environment may be the best way to improve crop resistance to drought (Poggi et al. 2023). Plant breeders endeavor to use intra-specific variation to improve crops at the genetic level, eventually enhancing drought tolerance. Crop wild relatives have high intra-specific variation and are thus highly valuable for crop variety adaptation to climate change (Mammadov et al. 2018). Crossbreeding between exotic varieties and local landraces resulted in higher biomass, grain yield, and stover yield compared to the exotic varieties alone (Yadav 2008). For instance, a diploid sunflower species of hybrid origin (*Helianthus anomalus*), resistant to desert dune environments, has been successfully used in drought tolerance (Ludwig et al. 2004). In rice, interspecific backcrossing of *Oryza glaberrima* has been used to improve the drought resistance of *Oryza sativa* (Ndjiondjop et al. 2010). Similarly, the wild emmer wheat (*Triticum dicoccoides*) is more drought tolerant than its domesticated counterpart (Budak et al. 2013). Wild relatives are considered a valuable resource for diversifying primary gene pools of pearl millet. Pre-breeding a method used to utilize gene pool diversity from wild relatives for crop improvement by linking important traits with variety development programs (Ogbonnaya et al. 2013), is crucial for incorporating traits from wild relatives into primary gene pools of crops like pearl millet (Sharma et al. 2013). Yadav (2010) found that landrace populations of pearl millet had the highest drought Resistance Index (DRI), and grain yield compared to Elite composites, which were found to be more sensitive to drought. The pearl millet cultivar 'Okashana-1' (ICMV 88,908) is a popular and drought-resistant variety that is grown in Namibia and Zimbabwe and was developed by pearl millet landraces (Daisuke 2005). Likewise, Pusa Composite-443, a drought-resistant pearl millet cultivar, was developed for cultivation in drought-prone areas of India using the Jakhrana landrace of Rajasthan. Additionally, 20 hybrids and 5 OPVs were released and recommended for use in drought-prone A1 zone of pearl millet cultivation in India through a coordinated research project (Satyavathi et al. 2018).

Conventional breeding approach has been successful in creating crop varieties that can withstand drought conditions (Baenziger et al. 2008; Kumar et al. 2014), but the process is time-consuming and requires significant effort, taking several years to move from identifying plant characteristics and genetic traits to developing commercial varieties through crossbreeding. Speed breeding is a method that helps to speed up crop research by shortening the breeding cycle. This can be done in many ways, such as increasing the duration of plants to light exposure, and harvesting seeds early, which helps to reduce the time it takes to go from one generation to the next in some long-day or day-neutral crops (Ghosh et al. 2018). Speed breeding by using specific and highly controlled environmental conditions such as long photoperiods (e.g. 22 h) can help in achieving six generations per year for certain crops like spring wheat, durum wheat, barley, and chickpea, and 4 generations for canola, in contrast to the 2–3 generations that can be achieved under normal glasshouse conditions. This technique can be utilized for the development of drought-tolerant crop cultivars. This technique can be accomplished by using LED lights to shorten the length of the daylight hours that the plants receive, allowing for faster breeding cycles. This approach is effective in developing new cultivars for short-day crops like rice (Jähne et al. 2020). In Australia, a speed breeding technique was used to create a drought-tolerant wheat variety (Christopher et al. 2015). This technique allowed the researchers to generate up to the F5 generation of stay-green inbred lines in just 18 months. This study produced over 40,000 molecular markers, which can be used to identify new genes responsible for stay-green traits.

Advanced genomics-assisted breeding techniques such as genomic selection (GS), marker-assisted selection (MAS), marker-assisted backcrossing (MABC), and marker-assisted recurrent selection (MARS) offer novel opportunities to impart drought tolerance in crops. Most breeding programs focus on grain yield under water deficit when selecting drought-tolerant genotypes. MAS requires the identification of molecular markers and QTL/genes that explain the phenotypic variance of drought tolerance. This method focuses on identifying candidate genes that express under moisture stress conditions using genomic technologies such as microarray and transcriptomic analysis (Daskowska-Golec et al. 2018; Ghorbani et al. 2019). QTL linked to drought tolerance has been identified in field crops (wheat, barley; Table 7.1) and introgressed from crop wild relatives to include new favorable alleles in the breeding population (Merchuk-Ovnat et al. 2016).

TABLE 7.1
Quantitative Trait Loci (QTL) Identified in Different Field Crops for Drought Tolerance

Crop	QTLs	Traits	Chromosomal Linkage Group (LG)	Reference
Wheat	qGYWD.3B.2	Grain yield	–	Gupta et al. (2017)
	4A		–	
	4A–a		–	
	3B	1000-grain weight	–	
	QTgw		–	
	QSrm-ipk-2D	Stem reserve mobilization	–	
	QSrm-ipk-5D		–	
	QSrm-ipk-7D		–	
Rice	qHt1	Plant height	–	Sabar et al. (2019)
	qTill11	Tillers per plant	–	
	qDRL3	Deep root length	–	
	qDRL9		–	
	qDRL11		–	
	qDRSA3.1	Deep root surface area	–	
Maize	qGW-J1-1	Grain weight per ear	–	Zhao et al. (2019)
	qGW-J2-1		–	
	qGW-Ch-4-1		–	
	qEHPH-J1-1	Ear height to plant ratio	–	
	qEHPH-Ch.1-2		–	
Chickpea	Q3-1	Drought tolerance	LG3	Rehman et al. (2011)
	Q1-1		LG1	
	QR3rld01	Root length density	CaLG04	Varshney et al. (2014)
	QR3rsa01	Root surface area	CaLG04	
	QR3df01	Days to 50% flowering	CaLG08	
	QR3pod01	Pods per plant	CaLG04	
	QR3100sdw01	100-seed weight	CaLG01	Varshney et al. (2014)
	QR3100sdw03		CaLG04	
	QR3100sdw04		CaLG01	
	QR3100sdw03		CaLG04	
	QR3100sdw02		CaLG04	
	QR3100sdw04		CaLG01	
	QR3100sdw03		CaLG04	
	QR3yld01	Grain yield	CaLG04	Varshney et al. (2014)
	QR3yld02		CaLG01	
	QR3yld03		CaLG01	
	QR3yld04		CaLG04	
	QR3yld05		CaLG04	
	QR3yld06		CaLG06	
	QR3yld07		CaLG03	
	QR3yld08		CaLG04	
	QR3yld09		CaLG04	
	QR3dti01	Drought tolerance indices	CaLG01	Varshney et al. (2014)
	QR3dti02		CaLG04	
	QR3dti03		CaLG01	
	QR3dti04		CaLG04	
	QR3dti03		CaLG01	

(Continued)

TABLE 7.1 (*Continued*)
Quantitative Trait Loci (QTL) Identified in Different Field Crops for Drought Tolerance

Crop	QTLs	Traits	Chromosomal Linkage Group (LG)	Reference
	QR3dti04		CaLG04	
	QR3dsi01	Drought susceptibility indices	CaLG07	Varshney et al. (2014)
	QR3dsi02		CaLG07	
	QR3dsi03		CaLG04	
	Ca_04551	Drought tolerance indices	hotspot_a	Kale et al. (2015)
	Ca_04552			
	Ca_04561		hotspot_b	
	Ca_04564			
	Ca_04569			
	Qncl.Sw1	Grain weight	LG1	Radhika et al. (2007)
	NCPGR-50		3	Hamwieh et al. (2013)
	TR-50		3	
	SCEA19		7	
	TAA-58		–	
	H6C-07	Drought resistance score	3	Hamwieh et al. (2013)
Cowpea	Dro-1	Stay green	–	Muchero et al. (2009)
	Dro-3		–	
	Dro-7		–	
	Dro-1	Drought-induced senescence	GL1	Muchero et al. (2009)
	Dro-2		–	
	Mat-1	Days to maturity	LG7	
	Mat-2		LG8	
Lentil	QRSratioIX-2.30	Root shoot ratio	–	Idrissi et al. (2016)
	QLRNIII-98.64	Lateral root number	–	
	QSRLIV-61.63	Specific root length	–	
Pigeon pea	QTL-RF-1	Fertility restoration	–	Saxena et al. (2011)
	QTL-RF-2		–	
	QTL-RF-3		–	
	QTL-RF-4		–	
Soybean	Q_Root_Gm01	Fibrous roots	Gm01	Abdel-Haleem et al. (2011)
	Q_Root_Gm03		Gm03	
	Q_Root_Gm04		Gm04	
	Q_Root_Gm08		Gm08	
	GLR1	Lateral root number	–	Prince et al. (2019)

* Drought tolerance and susceptibility indexes; root traits (root length density, root dry weight, root depth, root surface area), morphological traits (shoot dry weight, plant height, primary branches), phenological (days to 50% flowering and days to maturity), and yield-related (pods per plant, seeds per pods, 100-seed weight, biomass, harvest index, and yield).

TABLE 7.2
Yield Losses in Different Field Crops with Drought Stress Imposed at Different Growth Stages

Crop	Growth Stages	Yield Reduction (%)	References
Rice	Reproductive (mild stress)	53–92	Lafitte et al. (2007)
Rice	Reproductive (severe stress)	48–94	
Rice	Grain filling	60	Basnayake et al. (2006)
Maize	Vegetative	25–60	Atteya et al. (2003)
Maize	Reproductive	63–87	Kamara et al. (2003)
Maize	Grain filling	79–81	Monneveux et al. (2006)
Potato	Flowering	13	Kawakami et al. (2006)
Barley	Seed filling	49–57	Samarah et al. (2005)
Chickpea	Reproductive	45–69	Nayyar et al. (2006)
Common beans	Reproductive	58–87	Martínez et al. (2007)
Cowpea	Reproductive	11–60	Ogbonnaya et al. (2003)
Soybean	Reproductive	46–71	Samarah et al. (2006)
Canola	Reproductive	30	Sinaki et al. (2007)
Sunflower	Reproductive	60	Mazahery-Laghab et al. (2003)

Molecular marker application has significantly improved the efficiency of breeding drought-tolerant crops (Mwamahonje et al. 2021). Ribaut and Ragot (2007) introduced five QTLs linked to flowering and grain yield traits into a drought-sensitive maize line using MABC, resulting in higher yield and lower anthesis silking interval. The QTL-sequence approach has been applied to identify specific regions or genes in the genome which confer tolerance to various biotic and environmental stresses in crops (Chen et al. 2021). Only large-effect QTL, identified and validated across a wide range of genetic backgrounds, are introduced into elite varieties using MABC (Bernardo 2008). Unlike MABC, MARS uses a subset of markers significantly related to target traits to accumulate medium-effect QTL in a given population (Bernardo 2008; Ribaut et al. 2010).

Another marker-based approach, GS uses the available significant or insignificant marker information simultaneously to envisage the genetic value of progenies for selection (Lorenz 2013). In maize, GS was an effective method for enhancing genetic gains in water-deficit environments compared with pedigree-based traditional phenotypic selection (Beyene et al. 2015). Understanding genes and gene networks for stomatal patterning, size, and density regulation in wheat helped modulate the stomatal index and improve transpiration efficiency under drought tolerance (Kulkarni et al. 2017). To overcome the low heritability of drought tolerance, plant breeders have integrated DNA molecular markers into their programs, improving drought tolerance in cereals.

7.4.2 Transgenic Approaches

Advances in biotechnology with genome sequencing offer diverse techniques for manipulating and identifying genes involved in drought tolerance. Contemporary research on developing drought-tolerant plants is based primarily on the transfer of target gene/s involved in signaling and regulatory pathways, or encoding enzymes involved in pathways leading to the synthesis of functional and structural protectants, including stress tolerance conferring proteins, antioxidants, and osmolytes (Vinocur and Altman 2005; Khan et al. 2019). Several drought-responsive genes have been identified, including *ZmNAC111* (Mao et al. 2015), *ZmTIP1* (Zhang et al. 2020), *ZmVPP1* (Wang et al. 2016), *TsVP* (encoding V-H$^+$-PPase from *Thellungiella halophila*) and *BetA* (encoding choline dehydrogenase from *Escherichia coli*) (Wei et al. 2011) from maize, *SbER1-1*, *SbER2-1* from sorghum (Li et al. 2019) and *TsCBF1* from *T. halophila* (Zhang et al. 2010) whose overexpression resulted development of drought stress tolerance (Wei et al. 2011; Jia et al. 2020; Muppala et al. 2021). Shou et al. (2004) incorporated Nicotiana Protease Kinase (NPK1) through genetic transformation to develop transgenic maize, which induced glutathione-S-transferases to protect photosynthetic machinery during dehydration stress.

Drought tolerance in transgenic crops can be induced by manipulating ABA pathways through the upregulation or downregulation of genes and respective proteins involved in ABA synthesis, signaling, and degradation (Villa et al. 2005). In barley, farnesyltransferase ERA1 (enhanced response to ABA1) participates in ABA signaling, and a transgenic barley mutant developed with antisense ERA1 exhibited higher drought stress tolerance (Daszkowska-Golec et al. 2018). Late embryogenesis abundant protein expression helps alleviate osmotic stress in plants by stabilizing cellular organelles, water status, and membrane stability. In transgenic Indian mustard, overexpression of *AtLEA4-1* enhanced drought and salt stress tolerance by reducing lipid membrane peroxidation and reactive oxygen species generation and improving water status and antioxidant activities (Saha et al. 2016). In rice, *OsESG1* (S-domain receptor-like kinase) has been found to play a role in the development of crown roots and the response to drought. The *OsESG1* mutant rice plants have been observed to have fewer crown roots and shorter shoots compared to wild plants, and there are disruptions in the way auxin signaling and pollen transport (Pan et al. 2020). Transgenic brassica plants, overexpressing *BnKCS1-1*, *BnKCS1–2*, and *BnCER1–2* showed improved drought tolerance due to higher density of wax crystals on the leaf surface, increased level of aldehydes, alkanes, secondary alcohols, and significantly reduced ketone level (Wang et al. 2020).

Dehydration-responsive element binding transcription factors enhanced drought tolerance in transgenic plants of rice, wheat, barley, peanut, and tomato (Bhatnagar-Mathur et al. 2007; Wang et al. 2008; Xiao et al. 2009). Vadez et al. (2007) reported that DREB1A improved groundnut water efficiency by modifying the root/shoot ratio under drought stress. In wheat seedlings, overexpression of the DREB1A transcription factor delayed leaf wilting symptoms compared to the control ten days after the cessation of watering.

7.5 CONCLUSION AND FUTURE PERSPECTIVE

Drought stress is one of the most intricate natural hazards causing substantial losses to agricultural productivity, food security, and land sustainability. Semi-arid regions are more prone to water insufficiency due to low rainfall and high aridity. This complex phenomenon is a significant threat to global food security due to uncultivated fallow fields and/or crop failure and low water supplies. Over the past two decades, severe drought spells have been recorded worldwide, affecting millions of small farmers due to huge yield losses. Implementing appropriate crop management practices and crop diversification by including cereals, legumes, oilseeds, and cover crops are effective strategies for curbing drought-induced adversities.

Therefore, shifting cropping patterns from cereals to pulses and growing drought-tolerant and early maturing cultivars in semi-arid regions will help reduce drought-induced yield losses. In addition, advanced breeding and molecular approaches such as speed breeding, GWS, MAS, MABS, and MARS are crucial for developing genotypes with high drought tolerance. Moreover, transgenic methods can identify genes of interest, which can be transferred into target genotypes to enhance drought stress tolerance. Research priorities need to be developed, including developing a suitable methodology for calculating the drought index, an important criterion for planning effective adaptation and mitigation strategies for drought vulnerability in field crops. Future research should also include legumes and cover crops in existing cropping systems to enhance soil capacity against water deficit conditions (Table 7.3).

TABLE 7.3
Drought Indexes Used in South Asia

Index Name	Input Parameters	Calculation	Applied Area	Strength	References
Standardized precipitation evapotranspiration index (SPEI)	Precipitation and potential evapotranspiration	Monthly climatic water balance (difference between rainfall and PET)	Monitor and point out drought conditions linked with various drought impacts	Temperature inclusion and standardization of index increase application of to all climatic variations	Vicente-Serrano et al. (2010)
Normalized difference vegetation index (NDVI)	NOAA AVHRR satellite data	Global vegetation index data and value of radiance measured in near-infrared channels and visible	Monitor drought spells affecting agriculture	Very high resolution and spatial covering	Svoboda et al. (2016)
Palmer drought severity index	Available soil moisture contents, temperature, and rainfall	Difference in water contents from average conditions following bucket-type water balance model	Monitor droughts in agriculture	Estimates water demand and supply and provides availability	Palmer (1965)
Percent of normal precipitation (PNPI)	Rainfall	Actual rainfall divided by normal rainfall for the required period	Assess and monitor different drought impacts	Less time and basic maths required	Svoboda et al. (2016)
Standardized precipitation index (SPI)	Rainfall	Precipitation probability of 1–48 h or long period using historical rainfall data	Monitor agricultural drought, meteorological drought	Easy calculation, uses rainfall data, 3-month timescale beneficial for stations with intermittent missing rainfall	McKee et al. (1993)

REFERENCES

Abatzoglou, J.T., A.P. Williams, and R. Barbero. 2019. Global emergence of anthropogenic climate change in fire weather indices. *Geophysical Research Letters* 46:326–336.

Abdel-Haleem, H., G. Lee, and H.R. Boerma. 2011. Identification of QTL for increased fibrous roots in soybean. *Theoretical and Applied Genetics* 122:935–946.

Abraha, M.T., H. Shimelis, M. Laing, and K. Assefa. 2016. Performance of tef [*Eragrostis tef* (Zucc.) Trotter] genotypes for yield and yield components under drought-stressed and non-stressed conditions. *Crop Science* 56:1799–1806.

Abu-Elenein, J., R. Al-Sayaydeh, Z. Akkeh, Z. Al-Ajlouni, A.A. Al-Bawalize, S. Hasan, T. Alhindi, R.N. Albdaiwi, J.Y. Ayad, and A.M. Al-Abdallat. 2021. Agronomic performance and flowering behavior in response to photoperiod and vernalization in barley (*Hordeum vulgare* L.) genotypes with contrasting drought tolerance behavior. *Environmental and Experimental Botany* 192:104661.

Ahlstrom, A., M.R. Raupach, G. Schurgers, B. Smith, A. Arneth, M. Jung, M. Reichstein, J.G. Canadell, P. Friedlingstein, A.K. Jain, E. Kato, B. Poulter, S. Sitch, B.D. Stocker, N. Viovy, Y.P. Wang, A. Wiltshire, S. Zaehle, and N. Zeng. 2015. Carbon cycle. The dominant role of semi-arid ecosystems in the trend and variability of the land CO_2 sink. *Science* 348:895–899.

Ahmar, S., R.A. Gill, K.H. Jung, A. Faheem, M.U. Qasim, M. Mubeen, and W. Zhou. 2020. Conventional and molecular techniques from simple breeding to speed breeding in crop plants: Recent advances and future outlook. *International Journal of Molecular Sciences* 21:2590

Albaladejo, J., R. Ortiz, N. Garcia-Franco, A.R. Navarro, M. Almagro, J.G. Pintado, and M. Martínez-Mena. 2013. Land use and climate change impacts on soil organic carbon stocks in semi-arid Spain. *Journal of Soils and Sediments* 13:265–277.

Altieri, M.A., C.I. Nicholls, A. Henao, and M.A. Lana. 2015. Agroecology and the design of climate change-resilient farming systems. *Agronomy for Sustainable Development* 35:869–890.

Amede, T., and S. Schubert. 2003. Mechanisms of drought resistance in grain legumes: 1. Osmotic adjustment. *Ethiopian Journal of Journal of Science* 26:37–46.

Ashraf, M. 2010. Inducing drought tolerance in plants: Recent advances. *Biotechnology Advances* 28:169–183.

Assefa, Y., S.A. Staggenborg, and V.P. Prasad. 2010. Grain sorghum water requirement and responses to drought stress: A review. *Crop Management* 9:1–11.

Atteya, A.M. 2003. Alteration of water relations and yield of corn genotypes in response to drought stress. *Bulgarian Journal of Plant Physiology* 29:63–76.

Badu-Apraku, B., and C.G. Yallou. 2009. Registration of Striga-resistant and drought-tolerant tropical early maize populations TZE-W Pop DT STR C4 and TZE-Y Pop DT STR C4. *Journal of Plant Registrations* 3:86–90.

Baenziger, P.S., B. Beecher, R.A. Graybosch, A.M.H. Ibrahim, D.D. Baltensperger, L.A. Nelson, Y. Jin, S.N. Wegulo, J.E. Watkins, J.H. Hatchett and M.S. Chen. 2008. Registration of 'NE01643'wheat. *Journal of Plant Registrations* 2:36–42.

Bande, Y., N. Adam, Y. Azmi, and O. Jamarei. 2012. Determination of selected physical properties of egusi melon (*Citrullus lanatus*) seeds. *Journal of Basic & Applied Sciences* 8:257–265.

Basche, A.D., T.C. Kaspar, S.V. Archontoulis, D.B. Jaynes, T.J. Sauer, T.B. Parkin, and F.E. Miguez. 2016. Soil water improvements with the long-term use of a winter rye cover crop. *Agricultural Water Management* 172:40–50.

Basnayake, J., S. Fukai, and M. Ouk. 2006. Contribution of potential yield, drought tolerance and escape to adaptation of 15 rice varieties in rainfed lowlands in Cambodia. *Proceedings of the Australian Agronomy Conference*, Australian Society of Agronomy, Brisbane, Australia.

Belete, T. 2018. Breeding for resistance to drought: A case in sorghum (*Sorghum bicolor* (L.) Moench). *Journal of Agriculture and Forest Meteorology Research* 1.1:1–10.

Belko, N., N. Cisse, N.N. Diop, G. Zombre, S. Thiaw, S. Muranaka, and J. Ehlers. 2014. Selection for post-flowering drought resistance in short-and mediumduration cowpeas using stress tolerance indices. *Crop Science* 54:25–33.

Bennett, L.T., C. Aponte, T.G. Baker, and K.G. Tolhurst. 2014. Evaluating long-term effects of prescribed fire regimes on carbon stocks in a temperate eucalypt forest. *Forest Ecology and Management* 328:219–228.

Bernardo, R. 2008. Molecular markers and selection for complex traits in plants: Learning from the last 20 years. *Crop Science* 48:1649–1664.

Beyene, Y., K. Semagn, S. Mugo, A. Tarekegne, R. Babu, B. Meisel, P. Sehabiague, D. Makumbi, C. Magorokosho, S. Oikeh, and J. Gakunga. 2015. Genetic gains in grain yield through genomic selection in eight bi-parental maize populations under drought stress. *Crop Science* 55:154–163.

Bhatnagar-Mathur, P., V. Vadez, and K.K. Sharma. 2008. Transgenic approaches for abiotic stress tolerance in plants: Retrospect and prospects. *Plant Cell Reports* 27:411–424.

Budak, H., M. Kantar, and K.Y. Kurtoglu. 2013. Drought tolerance in modern and wild wheat. *The Scientific World Journal* 2013:1–16.

Campos, P.S., J.C. Ramalho, J.A. Lauriano, M.J. Silva, and M.D.C. Matos. 1999. Effects of drought on photosynthetic performance and water relations of four Vigna genotypes. *Photosynthetica* 36:79–87.

Cannarozzi, G., S. Plaza-Wuthrich, K. Esfeld, S. Larti, Y.S. Wilson, D. Girma, E. de Castro, S. Chanyalew, R. Blösch, L. Farinelli, and E. Lyons. 2014. Genome and transcriptome sequencing identifies breeding targets in the orphan crop tef (Eragrostistef). *BMC Genomics* 15:581.

Chauhan, J.S., M.K. Tyagi, A. Kumar, N.I. Nashaat, M. Singh, N.B. Singh, and S.J. Welham. 2007. Drought effects on yield and its components in Indian mustard (*Brassica juncea* L.). *Plant Breeding* 126:399–402.

Chaves, M.M., J.P. Maroco, and J.S. Pereira. 2003. Understanding plant responses to drought-from genes to whole plant. *Functional Plant Biology* 30:239–264.

Chaves, M.M., J.S. Pereira, J. Maroco, M.L. Rodrigues, C.P.P. Ricardo, M.L. Osorio, I. Carvalho, T. Faria, and C. Pinheiro. 2002. How plants cope with water stress in the field. Photosynthesis and growth. *Annals of Botany* 89:907–916.

Chen, G., and R.R. Weil. 2011. Root growth and yield of maize as affected by soil compaction and cover crops. *Soil and Tillage Research* 117:17–27.

Chen, L., Q. Wang, M. Tang, X. Zhang, Y. Pan, X. Yang, G. Gao, R. Lv, W. Tao, L. Jiang, and T. Liang. 2021. QTL mapping and identification of candidate genes for heat tolerance at the flowering stage in rice. *Frontiers in Genetics* 11:621871.

Chivenge, P., T. Mabhaudhi, A. Modi, and P. Mafongoya. 2015. The potential role of neglected and under utilised crop species as future crops under water scarce conditions in Sub-Saharan Africa. *International Journal of Environmental Research and Public Health* 12:5685–5711.

Christopher, J., C. Richard, K. Chenu, M. Christopher, A. Borrell, and L. Hickey. 2015. Integrating rapid phenotyping and speed breeding to improve stay-green and root adaptation of wheat in changing, water-limited, Australian environments. *Procedia Environmental Sciences*, 29:175–176.

Cima, D., A. Luik, and E. Reintam. 2015. Organic farming and cover crops as an alternative to mineral fertilizers to improve soil physical properties. *International Agrophysics* 29:405–412.

Clark, A. 2007. *Managing Cover Crops Profitably*, 3rd ed. National SARE Outreach Handbook Series Book 9. National Agricultural Laboratory, Beltsville, MD.

Daisuke, U.N.O. 2005. Farmers' selection of local ad improved pearl millet varieties in Ovamboland, Northern Namibia. *African Studies Monographs* 30:107–117

Daryanto, S., L. Wang, and P.A. Jacinthe. 2016. Global synthesis of drought effects on maize and wheat production. *PloS One* 11:e0156362.

Daszkowska-Golec, A., A. Skubacz, K. Sitko, M. Słota, M. Kurowska, and I. Szarejko. 2018. Mutation in barley ERA1 (enhanced response to ABA1) gene confers better photosynthesis efficiency in response to drought as revealed by transcriptomic and physiological analysis. *Environmental and Experimental Botany* 148:12–26.

De Baets, S., J. Poesen, J. Meersmans, and L. Serlet. 2011. Cover crops and their erosion-reducing effects during concentrated flow erosion. *Catena* 85:237–244.

de Lima Pereira, J.W., M.B. Albuquerque, P.A. Melo Filho, R.J.M.C. Nogueira, L.M. de Lima, and R.C. Santos. 2016. Assessment of drought tolerance of peanut cultivars based on physiological and yield traits in a semiarid environment. *Agricultural Water Management* 166:70–76.

de Melo, D.B., M. Dolbeth, F.F. Paiva, and J. Molozzi. 2022. Extreme drought scenario shapes different patterns of Chironomid coexistence in reservoirs in a semi-arid region. *Science of The Total Environment* 821:153053.

Dixit, S., A. Singh, N. Sandhu, A. Bhandari, P. Vikram, and A. Kumar. 2017. Combining drought and submergence tolerance in rice: Marker-assisted breeding and QTL combination effects. *Molecular Breeding* 37:143.

FAO. 2016. *The State of Food and Agriculture. Climate Change, Agriculture and Food Security*. Food and Agriculture Organization, United Nations, Rome.

Gabriel, J.L., A. Garrido, and M. Quemada. 2013. Cover crops effect on farm benefits and nitrate leaching: Linking economic and environmental analysis. *Agricultural Systems* 121:23–32.

Ghorbani, R., A. Alemzadeh, and H. Razi. 2019. Microarray analysis of transcriptional responses to salt and drought stress in Arabidopsis thaliana. *Heliyon* 5:e02614.

Ghosh, S., A. Watson, O.E. Gonzalez-Navarro, R.H. Ramirez-Gonzalez, L. Yanes, M. Mendoza-Suárez, J. Simmonds, R. Wells, T. Rayner, P. Green, and A. Hafeez. 2018. Speed breeding in growth chambers and glasshouses for crop breeding and model plant research. *Nature Protocols* 13:2944–2963.

Goldin, T. 2016. India's drought below ground. *Nature Geoscience* 9:98–98.

González, J.A., Y. Konishi, M. Bruno, M. Valoy, and F.E. Prado. 2012. Interrelationships among seed yield, total protein and amino acid composition of ten quinoa (*Chenopodium quinoa*) cultivars from two different agroecological regions. *Journal of the Science of Food and Agriculture* 92:1222–1229.

Gopika, K., P. Ratnakumar, A. Guhey, C.L. Manikanta, B.B. Pandey, K.T. Ramya, and A.L. Rathnakumar. 2022. Physiological traits and indices to identify tolerant genotypes in sesame (*Sesamum indicum* L.) under deficit soil moisture condition. *Plant Physiology Reports* 27:744–754.

Grzesiak, M.T., I. Marcińska, F. Janowiak, A. Rzepka, and T. Hura. 2012. The relationship between seedling growth and grain yield under drought conditions in maize and triticale genotypes. *Acta Physiologiae Plantarum* 34:1757–1764.

Gupta, P.K., H.S. Balyan, and V. Gahlaut. 2017. QTL analysis for drought tolerance in wheat: Present status and future possibilities. *Agronomy* 7:5.

Hadebe, S.T., A.T. Modi, and T. Mabhaudhi. 2017. Drought tolerance and water use of cereal crops: A focus on sorghum as a food security crop in sub-Saharan Africa. *Journal of Agronomy and Crop Science* 203:177–191.

Haley, S.D., J.J. Johnson, F.B. Peairs, J.S. Quick, J.A. Stromberger, S.R. Clayshulte, J.D. Butler, J.B. Rudolph, B.W. Seabourn, G. Bai, Y. Jin, and J. Kolmer. 2007. Registration of 'Ripper' wheat. *Journal of Plant Registrations* 1:1–6.

Hamwieh, A., M. Imtiaz, and R.S. Malhotra. 2013. Multi-environment QTL analyses for drought related traits in a recombinant inbred population of chickpea (*Cicer arientinum* L.). *Theoretical and Applied Genetics* 126:1025–1038.

Hao, Z., F. Hao, V.P. Singh, and X. Zhang. 2018. Changes in the severity of compound drought and hot extremes over global land areas. *Environmental Research Letters* 13:124022.

Hassan, R., R. Scholes, and N. Ash. 2005. *Ecosystems and Human Well-Being: Current State and Trends: Findings of the Condition and Trends Working Group* (Millennium Ecosystem Assessment Series). Island Press, Washington DC. 1:623–662.

Hobley, E.U., J. Baldock, and B. Wilson. 2016a. Environmental and human influences on organic carbon fractions down the soil profile. *Agriculture, Ecosystems & Environment* 223:152–166.

Hobley, E.U., A.J. Le Gay Brereton, and B. Wilson. 2016b. Forest burning affects quality and quantity of soil organic matter. *Science of The Total Environment* 575:41–49.

Hoddinott, J., and A. Quisumbing. 2008. Methods for micro econometric risk and vulnerability assessments. In: Deressa, T., R.M. Hassan, and C. Ringler (eds.), *Measuring Ethiopian Farmers' Vulnerability to Climate Change Across Regional States*, IFPRI Discussion Paper 00806. International Food Policy Research Institute (IFPRI), Washington.

Hoerling, M., J. Eischeid, A. Kumar, R. Leung, A. Mariotti, K. Mo, S. Schubert, and R. Seager. 2014. Causes and predictability of the 2012 Great Plains drought. *Bulletin of the American Meteorological Society* 95:269–282.

Howitt, R., D. MacEwan, and J. Medellin-Azuara. 2011. Drought, Jobs, and Controversy: Revisiting 2009. University of California Giannini Foundation. Available online: https://giannini.ucop.edu/media/are-update/files/articles/V14N6_1.pdf

Huang, J., M. Ji, Y. Xie, S. Wang, Y. He, and J. Ran. 2016. Global semi-arid climate change over last 60 years. *Climate Dynamics* 46:1131–1150.

Hussain, M.I., M. Farooq, A. Muscolo, and A. Rehman. 2020. Crop diversification and saline water irrigation as potential strategies to save freshwater resources and reclamation of marginal soils-A review. *Environmental Science and Pollution Research* 27:28695–28729.

Idrissi, O., S.M. Udupa, E. De Keyser, R.J. McGee, C.J. Coyne, G.C. Saha, F.J. Muehlbauer, P. Van Damme, and J. De Riek. 2016. Identification of quantitative trait loci controlling root and shoot traits associated with drought tolerance in a lentil (Lens culinaris Medik.) recombinant inbred line population. *Frontiers in Plant Science* 7:1174.

Ishiyaku, M F., and H. Aliyu. 2013. Field evaluation of cowpea genotypes for drought tolerance and Striga resistance in the dry savanna of the North-West Nigeria. *International Journal of Plant Breeding and Genetics* 7:47–56.

Jacobsen, S.E., F. Liu, and C.R. Jensen. 2009. Does root-sourced ABA play a role for regulation of stomata under drought in quinoa (*Chenopodium quinoa* Willd.). *Scientia Horticulturae* 122:281–287.

Jähne, F., V. Hahn, T. Würschum, and W.L. Leiser. 2020. Speed breeding short day crops by LED controlled light schemes. *Theoretical and Applied Genetics* 133:2335–2342.

Jamalluddin, N., R.C. Symonds, S. Mayes, W.K. Ho, and F. Massawe. 2021. Diversifying crops for food and nutrition security: A case of vegetable amaranth, an ancient climate-smart crop. In: Galanakis, C.M., (eds.), *Food Security and Nutrition*. Academic Press, pp. 125–146.

Jia, T.J., L.I. Jing-Jing, L.F. Wang, Y.Y. Cao, M.A. Juan, W.A.N.G. Hao, D.F. Zhang, and H.Y. Li. 2020. Evaluation of drought tolerance in ZmVPP1-overexpressing transgenic inbred maize lines and their hybrids. *Journal of Integrative Agriculture* 19:2177–2187.

Jodha, N.S., N.P. Singh, and M.C.S. Bantilan. 2012. Enhancing farmers' adaptation to Climate Change in Arid and Semi-Arid Agriculture of India: Evidences from Indigenous Practices: Developing International Public Goods from Development-Oriented Projects. Working Paper Series No.32.International Crops Research Institute for the Semi-Arid Tropics, Patancheru 502324, Andhra Pradesh, India, p. 28.

Jones, C., K. Olson-Rutz, P. Miller, and C. Zabinski. 2020. Cover crop management in semi-arid regions: Effect on soil and cash crop. *Crop Soils* 53:42–51.

Jongrungklang, N., S. Jogloy, T. Kesmala, N. Vorasoot, and A. Patanothai. 2014. Responses of rooting traits in peanut genotypes under pre-flowering drought stress. *International Journal of Plant Production* 8:335–352.

Joyce, B.A., W.W. Wallender, J.P. Mitchell, L.M. Huyck, S.R. Temple, P. Brostrom, and T.C. Hsiao. 2002. Infiltration and soil water storage under winter cover cropping in California's Sacramento Valley. *Transactions of the ASAE* 45:315–326.

Kale, S.M., D. Jaganathan, P. Ruperao, C. Chen, R. Punna, H. Kudapa, M. Thudi, M. Roorkiwal, M.A. Katta, D. Doddamani, and V. Garg. 2015. Prioritization of candidate genes in "QTL-hotspot" region for drought tolerance in chickpea (*Cicer arietinum* L.). *Scientific Reports* 5:1–14.

Kamara, A.Y., A. Menkir, B. Badu-Apraku, and O. Ibikunle. 2003. The influence of drought stress on growth, yield and yield components of selected maize genotypes. *The Journal of Agricultural Science* 141:43–50.

Kawakami, J., K. Iwama, and Y. Jitsuyama. 2006. Soil water stress and the growth and yield of potato plants grown from microtubers and conventional seed tubers. *Field Crops Research* 95:89–96.

Kaye, J.P., and M. Quemada. 2017. Using cover crops to mitigate and adapt to climate change. A review. *Agronomy for Sustainable Development* 37:1–17.

Khan, S., S. Anwar, S. Yu, M. Sun, Z. Yang, and Z.Q. Gao. 2019. Development of drought-tolerant transgenic wheat: Achievements and limitations. *International Journal of Molecular Sciences* 20:3350.

Kirby, M., J. Connor, R. Bark, E. Qureshi, and S. Keyworth. 2012. The economic impact of water reductions during the Millennium Drought in the Murray-Darling Basin. *Presented at the Australian Agricultural and Resource Economics Society Annual Conference 2012*, Freemantle, Australia. Available online: https://ageconsearch.

Kooyers, N.J. 2015. The evolution of drought escape and avoidance in natural herbaceous populations. *Plant Science* 234:155–162.

Kottek, M., J. Grieser, C. Beck, B. Rudolf, and F. Rubel. 2006. World map of the Köppen-Geiger climate classification updated. *Meteorologische Zeitschrif* 15:259–263.

Kulkarni, M., R. Soolanayakanahally, S. Ogawa, Y. Uga, M.G. Selvaraj, and S. Kagale. 2017. Drought response in wheat: Key genes and regulatory mechanisms controlling root system architecture and transpiration efficiency. *Frontiers in Chemistry* 5:106.

Kumar, A., S. Dixit, T. Ram, R.B. Yadaw, K.K. Mishra, and N.P. Mandal. 2014. Breeding high-yielding drought-tolerant rice: Genetic variations and conventional and molecular approaches. *Journal of Experimental Botany* 65:6265–6278.

Kumar, J., and S. Abbo. 2001. Genetics of flowering time in chickpea and its bearing on productivity in semi-arid environments. *Advances in Agronomy* 72:107–138.

Lafitte, H.R., G. Yongsheng, S. Yan, and Z.K. Lil. 2007. Whole plant responses, key processes, and adaptation to drought stress: The case of rice. *Journal of Experimental Botany* 58:169–175.

Lamaoui, M., M. Jemo, R. Datla, and F. Bekkaoui. 2018. Heat and drought stresses in crops and approaches for their mitigation. *Frontiers in Chemistry* 6:26.

Li, H., X. Han, X. Liu, M. Zhou, W. Ren, and B. Zhao. 2019. A leucine-rich repeat-receptor-like kinase gene SbER2-1 from sorghum (*Sorghum bicolor* L.) confers drought tolerance in maize. *BMC Genomics* 20:737.

Lodeiro, A.R., P. González, A. Hernández, L.J. Balague, and G. Favelukes. Comparison of drought tolerance in nitrogen-fixing and inorganic nitrogen-grown common beans. *Plant Science* 154:31–41.

Lorenz, A.J., 2013. Resource allocation for maximizing prediction accuracy and genetic gain of genomic selection in plant breeding: A simulation experiment. *G3 Genes Genomes Genetics* 3:481–491.

Ludwig, F., D.M. Rosenthal, J.A. Johnston, N.C. Kane, B.L. Gross, C. Lexer, S.A. Dudley, L.H. Rieseberg, and L.A. Donovan. 2004. Selection on leaf ecophysiological traits in a desert hybrid helianthus species and early-generation hybrids. *Evolution* 58:2682–2692.

Maalouf, F., M. Nachit, M.E. Ghanem, and M. Singh. 2015. Evaluation of faba bean breeding lines for spectral indices, yield traits and yield stability under diverse environments. *Crop and Pasture Science* 66:1012–1023.

Mammadov, J., R. Buyyarapu, S.K. Guttikonda, K. Parliament, I.Y. Abdurakhmonov, and S.P. Kumpatla. 2018. Wild relatives of maize, rice, cotton, and soybean: Treasure troves for tolerance to biotic and abiotic stresses. *Frontiers in Plant Science* 9:886.

Mansour, E., E.S.M. Desoky, M.M. Ali, M.I. Abdul-Hamid, H. Ullah, A. Attia, and A. Datta. 2021. Identifying drought-tolerant genotypes of faba bean and their agro-physiological responses to different water regimes in an arid Mediterranean environment. *Agricultural Water Management* 247:106754.

Mansour, E., M.I. Abdul-Hamid, M.T. Yasin, N. Qabil, and A. Attia. 2017. Identifying drought-tolerant genotypes of barley and their responses to various irrigation levels in a Mediterranean environment. *Agricultural Water Management* 194:58–67.

Mao, H., H. Wang, S. Liu, Z. Li, X. Yang, J. Yan. 2015. Qin A transposable element in a NAC gene is associated with drought tolerance in maize seedlings. *Nature Communications* 6:8326.

Martínez, J.P., H. Silva, J.F. Ledent, and M. Pinto. 2007. Effect of drought stress on the osmotic adjustment, cell wall elasticity and cell volume of six cultivars of common beans (*Phaseolus vulgaris* L.). *European Journal of Agronomy* 26:30–38.

Maurer, G.E., A.J. Hallmark, R.F. Brown, O.E. Sala, and S.L. Collins. 2020. Sensitivity of primary production to precipitation across the United States. *Ecology Letters* 23:527–536.

Mazahery-Laghab, H., F. Nouri, and H.Z. Abianeh. 2003. Effects of the reduction of drought stress using supplementary irrigation for sunflower (*Helianthus annuus*) in dry farming conditions. *Pajouhesh-Va-Sazandegi* 59:81–86.

Mckee, T.B., N.J. Doesken, and J. Kleist. 1993. The relationship of drought frequency and duration to time scales. In: *Proceedings of the 8th Conference on Applied Climatology*, Anaheim, CA, USA, pp. 179–184.

Merchuk-Ovnat, L., V. Barak, T. Fahima, F. Ordon, G.A. Lidzbarsky, T. Krugman, and Y. Saranga. 2016. Ancestral QTL alleles from wild emmer wheat improve drought resistance and productivity in modern wheat cultivars. *Frontiers in Plant Science* 7:452.

Miller, P.R., J. Waddington, C.L. McDonald, and D.A. Derksen. 2002. Cropping sequence affects wheat productivity on the semiarid northern Great Plains. *Canadian Journal of Plant Science* 82:307–318.

Miyan, M.A., 2015. Droughts in Asian least developed countries: Vulnerability and sustainability. *Weather and Climate Extremes* 7:8–23.

Monneveux, P., C. Sánchez, D. Beck, and G.O. Edmeades. 2006. Drought tolerance improvement in tropical maize source populations: Evidence of progress. *Crop Science* 46:180–191.

Movahhedy-Dehnavy, M., S.A.M. Modarres-Sanavy, and A. Mokhtassi-Bidgoli. 2009. Foliar application of zinc and manganese improves seed yield and quality of safflower (*Carthamus tinctorius* L.) grown under water deficit stress. *Industrial Crops and Products* 30:82–92.

Muchero, W., N.N. Diop, P.R. Bhat, R.D. Fenton, S. Wanamaker, M. Pottorff, S. Hearne, N. Cisse, C. Fatokun, J.D. Ehlers, and P.A. Roberts. 2009. A consensus genetic map of cowpea [*Vigna unguiculata* (L) Walp.] and synteny based on EST-derived SNPs. *Proceedings of the National Academy of Sciences* 106:18159–18164.

Muppala, S., P.K. Gudlavalleti, K.R. Malireddy, S.K. Puligundla, and P. Dasari. 2021. Development of stable transgenic maize plants tolerant for drought by manipulating ABA signaling through agrobacterium-mediated transformation. *Journal of Genetic Engineering and Biotechnology* 19:1–14.

Mwamahonje, A., J.S.Y. Eleblu, K. Ofori, S. Deshpande, T. Feyissa, and P. Tongoona. 2021. Drought tolerance and application of marker-assisted selection in sorghum. *Biology* 10:1249.

Nayyar, H., S. Kaur, S. Singh, and H.D. Upadhyaya. 2006. Differential sensitivity of Desi (small-seeded) and Kabuli (large-seeded) chickpea genotypes to water stress during seed filling: Effects on accumulation of seed reserves and yield. *Journal of the Science of Food and Agriculture* 86:2076–2082.

Ndjiondjop, M.N., B. Manneh, M. Cissoko, N. Dramé, R.G. Kakaï, R. Bocco, H. Baimey, and M. Wopereis. 2010. Drought resistance in an interspecific backcross population of rice (*Oryza spp.*) derived from the cross WAB56-104 (*O. sativa*)×CG14 (*O. glaberrima*). *Plant Science* 179:364–373.

Nhemachena, J., and B.G. Glwadys. 2008. Vulnerability to climate change and adaptive capacity in South African agriculture. In: Deressa, T., R.M. Hassan, and C. Ringler (eds.), *Measuring Ethiopian Farmers' Vulnerability to Climate Change Across Regional States*, IFPRI Discussion Paper 00806. International Food Policy Research Institute (IFPRI), Washington.

Ogbonnaya, C.I., B. Sarr, C. Brou, O. Diouf, N.N. Diop, and H. Roy-Macauley. 2003. Selection of cowpea genotypes in hydroponics, pots, and field for drought tolerance. *Crop Science* 43:1114–1120.

Ogbonnaya, F.C., O. Abdalla, A. Mujeeb-Kazi, A.G. Kazi, S.S. Xu, N. Gosman, E.S. Lagudah, D. Bonnett, M.E. Sorrells, and H. Tsujimoto. 2013. Synthetic hexaploids: Harnessing species of the primary gene pool for wheat improvement. *Plant Breeding Reviews* 37:35–122.

Painawadee, M., S. Jogloy, T. Kesmala, C. Akkasaeng, and A. Patanothai. 2009. Heritability and correlation of drought resistance traits and agronomic traits in peanut (*Arachis hyogaea* L.). *Asian Journal of Plant Sciences* 8:325–334.

Palmer, W.C. 1965. Meteorological drought Research Paper No. 45. U.S. Department of Commerce, Washington, DC, p. 58.

Pandey, R., and S. Jha. 2012. Climate vulnerability index - measure of climate change vulnerability to communities: A case of rural Lower Himalaya, India. *Mitigation and Adaptation Strategies for Global Change* 17:487–506.

Peterson, T.C., P.A. Stott, and S. Herring. 2012. Explaining extreme events of 2011 from a climate perspective. *Bulletin of the American Meteorological Society* 93:1041–1067.

Pieters, A., and R. McKee. 1938. The use of cover and green-manure crops. In: *Soils and Men, the Yearbook of Agriculture*, United States Department of Agriculture Printing Office, Washington, DC. pp. 431–444.

Pinheiro, C., M.M. Chaves, and C.P. Ricardo. 2001. Alterations in carbon and nitrogen metabolism induced by water deficit in the stems and leaves of *Lupinus albus* L. *Journal of Experimental Botany* 52:1063–1070.

Plaza, C., D. Hernández, J.C. García-Gil, and A. Polo. 2018. Microbial activity in pig slurry-amended soils under semiarid conditions. *Soil Biology and Biochemistry* 36:1577–1585.

Poggi, G.M., S. Corneti, I. Aloisi, and F. Ventura. 2023. Environment-oriented selection criteria to overcome controversies in breeding for drought resistance in wheat. *Journal of Plant Physiology* 280:153895.

Prabhakar, S.V.R.K., and R. Shaw. 2008. Climate change adaptation implications for drought risk mitigation: A perspective for India. *Climatic Change* 88:113–130.

Prince, S.J., B. Valliyodan, H. Ye, M. Yang, S. Tai, W. Hu, M. Murphy, L.A. Durnell, L. Song, T. Joshi, and Y. Liu. 2019. Understanding genetic control of root system architecture in soybean: Insights into the genetic basis of lateral root number. *Plant, Cell & Environment* 42:212–229.

Qasim, M., S. Gulzar, and M.A. Khan. 2011. Halophytes asmedicinal plants. In: Ozturk, M., A.R. Mermut, and A. Celik (eds.), *Urbanization, Land Use, Land Degradation and Environment*. Daya Publishing House, Turkey. pp. 330–343.

Qi, Z., and M.J. Helmers. 2010. Soil water dynamics under winter rye cover crop in Central Iowa. *Vadose Zone Journal* 9:53–60.

Radhika, P., S.J.M. Gowda, N.Y. Kadoo, L.B. Mhase, B.M. Jarnadagni, M.N. Sainani, S. Chandra, and V.S. Gupta. 2007. Development of an integrated intraspecific map of chickpea (*Cicer arietinum* L.) using two recombinant inbred line populations. *Theoretical and Applied Genetics* 115:209–216.

Rahman, M., and M.M. de Jiménez. 2016. Designer oil crops. In: Gupta, S.K. (eds.), *Breeding Oilseed Crops for Sustainable Production*. Academic Press, pp. 361–376.

Rahman, M.A. 2011. Study on the changes of Coastal Zone: Chittagong to Cox's Bazar along the Bay of Bengal. *Global Summit on Coastal Seas*. EMECS 9:28–31.

Ravindranath, N.H., S. Rao, N. Sharma, M. Nair, R. Goplakrishanan, A.S. Rao, S. Malaviya, R. Tiwari, A. Sagadevan, M. Munsi, N. Krishna, and G. Bala. 2011. Climate change vulnerability profiles for North East India. *Current Science* 101:384–394.

Reddy, P.P. 2016. Cover/green manure crops. In: Reddy, P.P., (ed.), *Sustainable Intensification of Crop Production*. Springer, Singapore, pp. 55–67.

Rehman, A.U., R.S. Malhotra, K. Bett, B. Tar'An, R. Bueckert, and T.D. Warkentin. 2011. Mapping QTL Associated with Traits Affecting Grain Yield in Chickpea (*Cicer arietinum* L.) under terminal drought stress. *Crop Science* 51:450–463.

Ribaut, J.M., and M. Ragot. 2007. Marker-assisted selection to improve drought adaptation in maize: The backcross approach, perspectives, limitations, and alternatives. *Journal of Experimental Botany* 58:351–360.

Ribaut, J.M., M.C. De Vicente, and X. Delannay. 2010. Molecular breeding in developing countries: Challenges and perspectives. *Current Opinion in Plant Biology* 13:213–218.

Roarty, S., R.A. Hackett, and O. Schmidt. 2017. Earthworm populations in twelve cover crop and weed management combinations. *Applied Soil Ecology* 114:142–151.

Sabar, M., G. Shabir, S.M. Shah, K. Aslam, S.A. Naveed, and M. Arif. 2019. Identification and mapping of QTLs associated with drought tolerance traits in rice by a cross between Super Basmati and IR55419-04. *Breeding Science* 69:169–178.

Saha, B., S. Mishra, J.P. Awasthi, L. Sahoo, and S.K. Panda. 2016. Enhanced drought and salinity tolerance in transgenic mustard [*Brassica juncea* (L.) Czern & Coss.] overexpressing Arabidopsis group 4 late embryogenesis abundant gene (AtLEA4-1). *Environmental and Experimental Botany* 128:99–111.

Samarah, N.H. 2005. Effects of drought stress on growth and yield of barley. *Agronomy for Sustainable Development* 25:145–149.

Samarah, N.H., R.E. Mullen, S.R. Cianzio, and P. Scott. 2006. Dehydrin-like proteins in soybean seeds in response to drought stress during seed filling. *Crop Science* 46:2141–2150.

Sammis, T., P. Sharma, M. Shukla, J. Wang, and D. Miller. 2012. A water-balance drip-irrigation scheduling model. *Agricultural Water Management* 113:30–37.

Santos, R.F., D. Bassegio, and M. de Almeida Silva. 2017. Productivity and production components of safflower genotypes affected by irrigation at phenological stages. *Agricultural Water Management* 186:66–74.

Satyavathi, T.C., V. Khandelwal, B.S. Rajpurohit, A. Supriya, B.R. Beniwal, K. Kamlesh, B. Sushila, S. Shripal, C.K. Mahesh, and S.L. Yadav. 2018. Pearl millet hybrids and varieties. In: *ICAR-All India Coordinated Research Project on Pearl Millet*, Mandor, Jodhpur, India, p. 142.

Saxena, K.B., R. Sultana, R.K. Saxena, R.V. Kumar, J.S. Sandhu, A. Rathore, P.B. KaviKishor, and R.K. Varshney. 2011. Genetics of fertility restoration in A4 based diverse maturing hybrids in pigeonpea [*Cajanus cajan* (L.) Millsp.]. *Crop Science* 51:574–578.

Schwabe, K., J. Albiac, J. Connor, R. Hassan, and L. Meza-González. 2012. *Drought in Semi-Arid and Arid Environments: A Multi-Disciplinary and Cross-Country Perspective*. Springer Publishing, Dordrecht, The Netherlands.

Seager, R., M. Ting, I. Held, Y. Kushnir, J. Lu, G. Vecchi, H. Huang, N. Harnik, A. Leetmaa, N. Lau, C. Li, J. Velez, and N. Naik. 2007. Model projections of an imminent transition to a morearid climate in southwestern North America. *Science* 316:1181–1184.

Shamsudin, N.A.A., B.P.M. Swamy, W. Ratnam, M.T.S. Cruz, A. Raman, and A. Kumar. 2016. Marker assisted pyramiding of drought yield QTLs into a popular Malaysian rice cultivar, MR219. *BMC Genetics* 17:30.

Sharma, P., A. Singh, C. Kahlon, A. Brar, K. Grover, M. Dia, and R. Steiner. 2018. The role of cover crops towards sustainable soil health and agriculture-A review paper. *American Journal of Plant Sciences* 9:1935–1951.

Sharma, P., M.K. Shukla, and J.G. Mexal. 2011. Spatial variability of soil properties in agricultural fields of southern New Mexico. *Soil Science* 176:288–302.

Sharma, P., M.K. Shukla, T.W. Sammis, and P. Adhikari. 2012. Nitrate-nitrogen leaching from onion bed under furrow and drip irrigation systems. *Applied and Environmental Soil Science* 2012:650206.

Sharma, R., M. Jain, S. Kumar, and V. Kumar. 2014. Evaluation of differences among *Vigna aconitifolia* varieties for acquired thermotolerance. *Agricultural Research* 3:104–112.

Sharma, S., H.D. Upadhyaya, R.K. Varshney, and C.L.L. Gowda. 2013. Pre-breeding for diversification of primary gene pool and genetic enhancement of grain legumes. *Frontiers in Plant Science* 4:309.

Sheffield, J., and E.F. Wood. 2008. Projected changes in drought occurrence under future global warming from multi-model, multi-scenario, IPCC AR4 simulations. *Climate Dynamics* 31:79–105.

Sheffield, J., K.M. Andreadis, E.F. Wood, and D.P. Lettenmaier. 2009. Global and continental drought in the second half of the twentieth century: Severity-area-duration analysis and temporal variability of large-scale events. *Journal of Climate* 22:1962–1981.

Shindell, D.T., G. Faluvegi, R.L. Miller, G.A. Schmidt, J.E. Hansen, and S. Sun. 2006. Solar and anthropogenic forcing of tropical hydrology. *Geophysical Research Letters* 33:L24706.

Shou, H., P. Bordallo, and K. Wang. 2004. Expression of the Nicotiana protein kinase (NPK1) enhanced drought tolerance in transgenic maize. *Journal of Experimental Botany* 55:1013–1019.

Sinaki, J.M., E.M. Heravan, A.H.S. Rad, G. Noormohammadi, and G. Zarei. 2007. The effects of water deficit during growth stages of canola (*Brassica napus* L.). *American-Eurasian Journal of Agricultural & Environmental Sciences* 2:417–422.

Sinclair, T.R., L.C. Purcell, and C.H. Sneller. 2004. Crop transformation and the challenge to increase yield potential. *Trends in Plant Science* 9:70–75.

Singh, N.P., C. Bantilan, and K. Byjesh. 2013. Vulnerability to climate change: Adaptation strategies and layers of resilience. Quantifying vulnerability to climate change in SAT India. International Crop Research Institute for the semi-arid tropics (ICRISAT). Research Report, Patancheru 502324, Andhra Pradesh, unpublished.

Sommer, R., J. Ryan, S. Masri, M. Singh, and J. Diekmann. 2011. Effect of shallow tillage, moldboard plowing, straw management and compost addition on soil organic matter and nitrogen in a dryland barley/wheat-vetch rotation. *Soil and Tillage Research* 115:39–46.

Sprent, J.I. 2008. Evolution and diversity of legume symbiosis. In: Dilworth, M.J., E.K. James, and J.I. Sprent (eds.), *Nitrogen Fixing Leguminous Symbiosis, Nitrogen Fixation: Origin, Applications and Research Progress*. Springer, Dordrecht, The Netherlands, pp. 1–18.

Stewart, B.A., and S. Thapa. 2016. Dryland farming: Concept, origin and brief history. In: Farooq, M., and K.H.M. Siddique (eds.), *Innovations in Dryland Agriculture*. Springer, Cham, pp. 3–29.

Streeter, J.G. 2003. Effects of drought on nitrogen fixation in soybean root nodules. *Plant, Cell and Environment* 26:1199–1204.

Su, Z., J. Zhao, T.H. Marek, K. Liu, and M.T. Harrison, and Q. Xue. 2022. Drought tolerant maize hybrids have higher yields and lower water use under drought conditions at a regional scale. *Agricultural Water Management* 274:107978.

Svoboda, M.D., B.A. Fuchs. 2016. Integrated Drought Management Programme (IDMP), "Handbook of Drought Indicators and Indices". Drought Mitigation Center Faculty Publications. 117.

Tadele, Z. 2017. Raising crop productivity in Africa through intensification. *Agronomy* 7:22.

Tareq, S.M. 2012. Integrated Drought Management in South Asia - A Regional Proposal. Global Water Partnership South Asia. https://www.slideshare.net/Global water partnership/

Theib, O., H. Ahmed, and M.P. Mustafa. 2005. Faba bean productivity under rainfed and supplementing irrigation in northern Syria. *Agricultural Water Management* 73:57–72.

Tsubo, M., S. Fukai, J. Basnayake, and M. Ouk. 2009. Frequency of occurrence of various drought types and its impact on performance of photoperiod-sensitive and insensitive rice genotypes in rainfed lowland conditions in Cambodia. *Field Crops Research* 113:287–296.

Vadez, V., S. Rao, J. Kholova, L. Krishnamurthi, J. Kashiwagi, P. Ratnakumar. 2008. Root research for drought tolerance in legumes: Quo vadis? *Journal of Food Legumes* 21:77–85.

Vadez, V., S. Rao, K.K. Sharma, M. Bhatnagar, and J.M. Devi. 2007. DREB1A allows for more water uptake in groundnut by a large modification in the root/shoot ratio under water deficit. *International Arachis Newsletter* 27:27–31.

Valdivia-Cea, W., L. Bustamante, J. Jara, S. Fischer, E. Holzapfel, and R. Wilckens. 2021. Effect of soil water availability on physiological parameters, yield, and seed quality in four quinoa genotypes (*Chenopodium quinoa* Willd.). *Agronomy* 11:1012.

Van Dijk, A.I., H.E. Beck, R.S. Crosbie, R.A. De Jeu, Y.Y. Liu, G.M. Podger, B. Timbal, and N.R. Viney. 2013. The Millennium drought in southeast Australia (2001-2009): Natural and human causes and implications for water resources, ecosystems, economy, and society. *Water Resources Research* 49:1040–1057.

Van Zonneveld, M., M.S. Turmel, and J. Hellin. 2020. Decision-making to diversify farm systems for climate change adaptation. *Frontiers in Sustainable Food Systems* 4:32.

Varshney, R.K., M. Thudi, S.N. Nayak, P.M. Gaur, J. Kashiwagi, L. Krishnamurthy, D. Jaganathan, J. Koppolu, A. Bohra, S. Tripathi, and A. Rathore. 2014. Genetic dissection of drought tolerance in chickpea (*Cicer arietinum* L.). *Theoretical and Applied Genetics* 127:445–462.

Vera-Hernández, P., M. Ramírez, M. Núñez, M. Ruiz-Rivas, and F. Rosas-Cárdenas. 2018. Proline as a probable biomarker of cold stress tolerance in sorghum (*Sorghum bicolor*). *Mexican Journal of Biotechnology* 3:77–86.

Vicente-Serrano, S.M., S. Beguería, and J.I. López-Moreno. 2010. A multiscalar drought index sensitive to global warming: The standardized precipitation evapotranspiration index. *Journal of Climate* 23:1696–1718.

Villa, T.C.C., N. Maxted, M. Scholten, and B. Ford-Lloyd. 2005. Defining and identifying crop landraces. *Plant Genetic Resources* 3:373–384.

Vinocur, B., and A. Altman. 2005. Recent advances in engineering plant tolerance to abiotic stress: Achievements and limitations. *Current Opinion in Biotechnology* 16:123–132.

Wang, Q., Y. Guan, Y. Wu, H. Chen, F. Chen, and C. Chu. 2008. Overexpression of a rice OsDREB1F gene increases salt, drought, and low temperature tolerance in both Arabidopsis and rice. *Plant Molecular Biology* 67:589–602.

Wang, X., B.-B. Li, T.-T. Ma, L.-Y. Sun, L. Tai, C.-H. Hu, W.T. Liu, W.Q. Li, and K.M. Chen. 2020. The NAD kinase *OsNADK1* affects the intracellular redox balance and enhances the tolerance of rice to drought. *BMC Plant Biology* 20:11.

Wang, X., H. Wang, S. Liu, A. Ferjani, J. Li, J. Yan, X. Yang, and F. Qin. 2016. Genetic variation in ZmVPP1 contributes to drought tolerance in maize seedlings. *Nature Genetics* 48:1233–1241.

Wang, Y., S. Jin, Y. Xu, S. Li, S. Zhang, Z. Yuan, J. Li, and Y. Ni. 2020. Overexpression of *bnkcs1-1*, *bnkcs1-2*, and *bncer1-2* promotes cuticular wax production and increases drought tolerance in *Brassica napus*. *The Crop Journal* 8:26–37.

Watson, A., S. Ghosh, M.J. Williams, W.S. Cuddy, J. Simmonds, M.D. Rey, M. Asyraf Md Hatta, A. Hinchliffe, A. Steed, D. Reynolds, and N.M. Adamski. 2018. Speed breeding is a powerful tool to accelerate crop research and breeding. *Nature Plants* 4:23–29.

Wei, A., C. He, B. Li, N. Li, and J. Zhang. 2011. The pyramid of transgenes TsVP and BetA effectively enhances the drought tolerance of maize plants. *Plant Biotechnology Journal* 9:216–229.

Xiao, B., X. Chen, C. Xiang, N. Tang, Q. Zhang, and L. Xiong. 2009. Evaluation of seven function-known candidate genes for their effects on improving drought resistance of transgenic rice under field conditions. *Molecular Plant* 2:73–83.

Yadav, O.P. 2010. Drought response of pearl millet landrace-based populations and their crosses with elite composites. *Field Crops Research* 118:51–57.

Zampieri, M., A. Ceglar, F. Dentener, and A. Toreti. 2017. Wheat yield loss attributable to heat waves, drought and water excess at the global, national and subnational scales. *Environmental Research Letters* 12:064008.

Zhang, S., N. Li, F. Gao, A. Yang, and J. Zhang. 2010. Over-expression of TsCBF1 gene confers improved drought tolerance in transgenic maize. *Molecular Breeding* 26:455–465.

Zhang, X., J. Cai, B. Wollenweber, F. Liu, T. Dai, W. Cao, and D. Jiang. 2013. Multiple heat and drought events affect grain yield and accumulations of high molecular weight glutenin subunits and glutenin macropolymers in wheat. *Journal of Cereal Science* 57:134–140.

Zhang, X., Y. Mi, H. Mao, S. Liu, L. Chen, and F. Qin. 2020. Genetic variation in ZmTIP1 contributes to root hair elongation and drought tolerance in maize. *Plant Biotechnology Journal* 18:1271–1283.

Zhao, X., J. Zhang, P. Fang, and Y. Peng. 2019. Comparative QTL analysis for yield components and morphological traits in maize (*Zea mays* L.) under water-stressed and well-watered conditions. *Breeding Science* 69:621–632.

8 Engineering Abiotic Stresses in Crops by Using Biotechnological Approaches

*Aladdin Hamwieh, Naglaa A. Abdallah,
Nourhan Fouad, Khaled Radwan, Tawffiq Istanbuli,
Sawsan Tawkaz, and Michael Baum*

8.1 INTRODUCTION

It is well known that abiotic stresses, such as extreme temperatures, high salinity, and drought, can have a detrimental impact on crops all over the world. Abiotic stress is caused by distressing environmental conditions that make it hard for plants to grow and develop. Plants may face abiotic stresses like a lack of water from drought or osmotic stress, too much salt in the soil, which causes salt stress, or high or low temperatures (heat or cold stress). The environmental conditions are widely acknowledged as posing significant threats to crop productivity. Plants have developed diverse adaptive mechanisms at the cellular and metabolic levels to counteract and recuperate from the deleterious consequences of abiotic stress. These strategies are activated promptly upon the onset of stress. At cellular and metabolic levels, plants have developed several ways to adapt to and recover from the harmful effects of abiotic stress. These strategies are triggered upon the onset of stress.

Changes in the water potential gradients between the soil and the plant are caused by drought. The phenomenon poses a challenge to the plant's water absorption process, resulting in a reduction of cell turgor, an alteration in cell volume, protein denaturation, and modifications in various physiological, molecular, and membrane constituents. The plant responds to this circumstance by initiating various cellular and metabolic mechanisms, such as impeding cell expansion, cell wall production, protein synthesis, stomata conductance, and photosynthetic function. Moreover, the flora elicits adaptive mechanisms, encompassing the synthesis of some phytohormones, such as abscisic acid (ABA), the generation of compatible solutes, and the expression of gene products implicated in osmotic regulation, cellular safeguarding, and mending (Bartels & Sunkar, 2005; Bhatnagar-Mathur et al., 2008; Janská et al., 2010).

Salt stress happens when there is a lot of salt in the soil, mostly in the form of sodium chloride (NaCl). When this occurs, the water potential of the soil goes down. This causes osmotic stress, which causes the same kinds of responses as when there is a drought. After that, the salt that has gotten into the root is carried to the shoot by the xylem, which causes more toxic effects that are specific to ions. These effects impede growth and hasten the process of leaf senescence. To address the subsequent stage, the plant initiates processes related to ion absorption and segregation, with the aim of restoring ionic equilibrium (Munns & Tester, 2008).

Heat stress is often caused by temperatures that are too high and out of the normal range physiologically for the plant. This is often accompanied by salinity stress and drought stress. Under stressful conditions, numerous proteins undergo denaturation and dysfunction, resulting in alterations to membrane fluidity and permeability. This disturbance to cellular homeostasis can lead to significant impediments in growth, development, and potentially fatal outcomes (Hasanuzzaman et al., 2013).

DOI: 10.1201/b23132-8

Plants have developed diverse physiological and biochemical mechanisms to manage stress; however, these mechanisms may not invariably prove efficacious. The use of genetic engineering presents a prospective strategy for augmenting crops' ability to withstand abiotic stress, thereby elevating their productivity. This chapter aims to investigate diverse approaches to inducing abiotic stresses in agricultural crops through engineering techniques.

Various strategies can be employed to engineer crops to withstand abiotic stresses in the field of agriculture. Several organisms have evolved various adaptations to cope with abiotic stresses. The implementation of biotechnology methodologies presents novel prospects for investigating the regulatory pathways of various stress-responsive genes and regulatory elements, including non-coding RNAs. Additionally, it enables the identification of genes that confer tolerance to single or multiple abiotic stressors. Numerous genes associated with stress have been identified and utilized in the development of crops that can adapt to rapidly changing environments while maintaining high productivity levels in the face of severe environmental stressors. Contemporary biotechnological methodologies, such as genetic engineering, genome editing, epigenetics, and biomolecules, enable scientists to manipulate the expression of these stress-associated genes in crops, which can lead to the development of new varieties that are more resilient to environmental stressors and can contribute to global food security. In addition, these biotechnological approaches can also reduce the use of harmful pesticides and herbicides, resulting in a more sustainable agricultural system.

8.2 TRANSGENIC APPROACH TO GENERATE ABIOTIC TOLERANCE IN CROPS

The intricate nature of abiotic stresses and the varying degrees of plant susceptibility to such stresses throughout their life cycle pose challenges to the selection of enhanced stress tolerance through traditional breeding methods. Hence, it is imperative to embrace an alternative methodology that can be utilized to ameliorate abiotic stress resilience and augment agricultural productivity and caliber. Genetic engineering represents a viable approach for producing transgenic crops that exhibit resilience in response to rapidly fluctuating environmental conditions. Transgenic plants have been produced through the stable integration and expression of foreign gene(s) from different species (Kumar et al., 2020). The plants produced modifications in the expression of distinct genes that are accountable for a particular characteristic through diverse techniques of transformation (such as Biolistic and Agrobacterium). Over the last 27 years, the commercialization of biotech crops has been proven to offer benefits on a global scale in a variety of sectors, including the environment, economy, health, and social spheres, for both small- and large-scale farmers. In 2019, the primary genetically modified crops that were embraced by nations cultivating GMOs included soybean, maize, cotton, and canola (ISAAA, 2019). Soybeans, maize, cotton, and canola are the most extensively utilized biotech crops. The International Service for the Acquisition of Agri-biotech Applications (ISAAA) has developed a database containing information on genetically modified (GM) crop events that have received approval for commercial cultivation, planting, and importation (food and feed).

8.2.1 Enhanced Salt Tolerance in Transgenic Crops

The implementation of the overexpression of multiple genes related to stress was carried out with the aim of mitigating the uptake of toxic ions in the cytosol and enhancing the salt tolerance of plants. To enhance the salt tolerance and yield of various crops, researchers have implemented the overexpression of vacuolar H^+-pyrophosphatase (V-PPase) genes in rice, sugarcane, wheat, and other crops, as demonstrated in studies conducted by Kim et al. (2020), Kumar et al. (2014), and Lv et al. (2015). He et al. (2005) conducted a study wherein they overexpressed the plant vacuolar Na^+/H^+ antiporter gene from Arabidopsis (*AtNHX1*) in cotton. The study's findings suggest that the transgenic lines demonstrated enhanced salt tolerance, elevated photosynthetic rate, and improved fiber production that follows exposure to salt treatment. Furthermore, the augmentation of salt tolerance was observed in tomato through the overexpression of SOS2 (salt overly sensitive), which encodes a calcineurin-interacting protein kinase, and in tobacco through the overexpression of SOS1 (Huertas et al., 2012; Yue et al., 2012). The genes known as salt overly sensitive (SOS), which include SOS1, SOS2, and SOS3, play a crucial role in the cellular exclusion of Na+ ions and the maintenance of ion homeostasis in root cells. This, in turn, leads to an improvement in salt tolerance in plants (Shi et al., 2000). In a study conducted by Huertas et al. (2012), a salt-tolerant variety of tomato was produced through the process of overexpressing the SOS2 gene, which encodes a protein kinase that interacts with calcineurin and is derived from *Solanum lycopersicum*. Transgenic soybean plants exhibited improved salt stress tolerance through the enhancement of chloride channel protein genes *GmCLC1* and *GsCLC-c2*, as reported by (Wei et al., 2019). Similarly, Liu et al. (2021) observed improved salt stress tolerance in cotton through the mutation of *GhCLCg*-1. The excessive expression of transcription factors that participate in salt signaling pathways has been observed to negatively impact the salt tolerance of plants. The overexpression of the *DREB1B* and *DREB1F* genes, which encode for dehydration-responsive element binding proteins, has been observed in various crops, including rice and soybean. This genetic modification has been found to enhance the crops' ability to tolerate salt stress (Datta et al., 2012; Nguyen et al., 2019). The MYB genes are responsible for encoding transcription factors that are associated with the responses of plants to abiotic stress. According to Tang et al. (2019), an increased amount of proline was produced by overexpressing the OsMYB6 gene, resulting in the development of salt-tolerant rice.

8.2.2 Enhanced Drought Tolerance in Transgenic Crops

In 2009, the initial commercially available variant of drought-resistant maize (DroughtGard, MON87460) was introduced. This variant expresses the cold shock protein B gene sourced from Bacillus subtilis (Liang, 2016). Phytohormones, namely abscisic acid (ABA), auxins (IAA), gibberellins (GAs), jasmonic acid (JA), and salicylic acid (SA), are crucial in the plant's reaction to abiotic stressors. The upregulation of NCED genes, which encode the 9-cis-epoxycarotenoid dioxygenase enzyme, in rice and soybean, resulted in elevated levels of ABA and a notable enhancement in drought resistance in genetically modified plants (Huang et al., 2019; Molinari et al., 2020). The transcription factors known as DRE (drought-responsive elements) facilitate the induction of dehydration-responsive genes in response to environmental stresses. According to reports by (Bhatnagar-Mathur et al., 2007; Ravikumar et al., 2014), the expression of *DREB1A* increased the tolerance to drought stress in transgenic rice and peanut plants. Increased biomass and boll number in the cotton show that the overexpression of the potato DREB2 gene, StDREB2, enhanced drought tolerance (El-Esawi & Alayafi, 2019). According to El-Esawi et al. (2019), the introduction of the transcription factor gene AtWRKY30 into wheat resulted in the upregulation of various stress-related genes, including *DREB1*, *DREB3*, *WRKY19*, and *TIP2*. This genetic modification led to an increase in the plant's ability to tolerate drought and heat stress. According to Liu et al. (2014), the augmentation of *SNAC1* gene expression, which governs the closure of stomata and the efficiency of water use, in cotton plants resulted in an enhancement of their ability to tolerate salt and drought stress. Additionally, the transgenic plants exhibited an improvement in their rooting system and a decrease in their transpiration rate. Based on what Li et al. (2018a,b) and Vendruscolo et al. (2007) reported, wheat and sugarcane that have undergone genetic modification to express the P5CS gene exhibit enhanced drought stress tolerance and increased proline production.

The HaHB4 transcription factor, which is classified as a homeodomain-leucine zipper, has been found to increase the transcript levels of multiple genes that are associated with defense-related processes and jasmonic acid biosynthesis. The expression of drought stress tolerance in wheat and soybean was achieved through the transformation of the *HaHB4* gene (González et al., 2019; Ribichich et al., 2020). The expression of ACO2 and LOX2 genes involved in JA biosynthesis is induced by transgenic wheat that expresses HaHB4, resulting in drought tolerance (González et al., 2019). The HB4 trait has been observed to enhance wheat yields by as much as 20%. Argentina has been the pioneer in implementing HB4 drought tolerance technology for wheat cultivation, which has subsequently been adopted by Brazil, Australia, and New Zealand.

8.2.3 ENHANCED HEAT TOLERANCE IN TRANSGENIC CROPS

The heat shock protein (HSP) acts as a molecular chaperone to stabilize, fix, and refold proteins that have become denatured. This keeps cells from being damaged by heat stress. Overexpression of *HSP* genes improves heat stress tolerance in plants. Transgenic soybeans expressing the heat shock transcription factor gene *GmHsfA1* showed increased tolerance to heat stress (Zhu et al., 2006). Overexpression of HSP70 in rice and cotton improved heat stress tolerance (Batcho et al., 2021; Qi et al., 2011). Overexpression of the transcription factor *LeAN2* gene in tomatoes increases the transcripts of several anthocyanin biosynthetic-related genes and enhances heat stress tolerance (Meng et al., 2015). The enhanced endurance to high-temperature stress in rice and wheat was achieved through the upregulation of the betaine aldehyde dehydrogenase gene *BADH*, resulting in heightened membrane stability (Kishitani et al., 2000; Wang et al., 2010).

8.3 GENOME EDITING APPROACH TO GENERATE ABIOTIC STRESS TOLERANCE IN CROPS

The development of modern biotechnology has led to the emergence of novel plant breeding methodologies that enable a precise and predictable modification of elite genetic backgrounds. The technique of genome editing enables plant breeders to precisely manipulate crop genomes at the nucleotide level. The technique of genome editing, which is also referred to as the new plant breeding technique, enables accurate and foreseeable alterations to be made within a superior genetic framework. This approach also diminishes the expenses linked to arbitrary gene transfer (transgenic plants) and traditional plant breeding. In recent years, genome editing has emerged as a promising technology with the potential to become the dominant and essential tool for managing abiotic and biotic stress in plants. The utilization of Clustered Regularly Interspaced Short Palindromic Repeats (CRISPR) associated protein (Cas) systems is considered a highly effective and suitable approach for the manipulation of targeted genes in plants. The Cas9 endonuclease, which has been modified through engineering, initially associates with a guide RNA of limited size (sgRNA). Subsequently, the CRISPR/Cas9 complex undertakes a thorough examination of the genomic DNA and attaches itself to regions located upstream of the protospacer adjacent motif (NGG-PAM) sequences that are rich in G. The Cas9 protein, acting as an endonuclease, initiates a double-stranded break at the designated site. This is followed by a DNA repair process, which can occur through either the non-homologous end joining (NHEJ) pathway or homology-directed repair (HDR). Ultimately, this results in the modulation of the genome. The use of the CRISPR/Cas methodology has been a pivotal milestone in the advancement of novel crop breeding techniques. The technique of genome editing, which was initially acquired in 2012, enables the precise modification of crops at the molecular level while circumventing the intricate regulations associated with genetically modified organisms (GMOs) by avoiding the introduction of genes from external organisms.

CRISPR technology has undergone remarkable advancements in enhancing genome editing applications. To reduce the frequency of off-target cleavage, a strategy involving the use of nickase mutants, which break only one strand of DNA, was employed. Specifically, paired gRNAs were utilized to achieve this goal by lowering the off-target editing rates while simultaneously increasing the on-target editing rates. Truncated guide RNA and Cas9 enzymes with high fidelity were employed to exhibit reduced off-target effects.

CRISPR editing technologies utilize editing mechanisms that are independent of Non-Homologous End Joining (NHEJ) or Homology-Directed Repair (HDR). Base editors have the ability to perform base conversions without inducing a double-strand break. This process involves the conversion of cytidine to thymidine or adenosine to guanosine. The fusion of Cas13 with ADAR2 deaminase is utilized for RNA editing, specifically for the conversion of adenosine to inosine on RNA. To facilitate gene activation or repression, transcriptional activators or repressors are fused to dCas9 (dead Cas9). To focus on epigenetic alterations, catalytically inactive dCas9 is used to target epigenetic modifiers, including p300 for histone acetylation, LSD1 for histone demethylation, DNMT3A or MQ1 for cytosine methylation, and Tet1 for cytosine demethylation. Develop a guide RNA (gRNA) with the aim of selectively targeting a particular promoter or enhancer region for the gene of interest.

8.4 NON-GM GENOME EDITING APPROACHES

To achieve DNA-free editing, four approaches for Cas9/gRNA delivery have been developed. Non-GM genome editing approaches are a set of techniques that can be used to modify the genetic modification of an organism without introducing foreign DNA. Nanoparticles and virus-like particles are examples of non-viral gene delivery systems that can be used to deliver gene-editing materials into cells. Agrobacterium-mediated plant transformation involves using a natural process in which soil bacterium Agrobacterium transfers DNA into plant cells to introduce desired DNA sequences into plants. Graft-mobile gene editing, on the other hand, involves using a modified form of CRISPR-Cas9 to edit genes in one part of a plant and then allowing the edited cells to move through the plant via the vascular system, resulting in edited cells throughout the plant. These approaches have shown promise in treating genetic diseases, producing genetically modified crops, and addressing other challenges in genetic engineering, without the need for introducing foreign DNA.

8.4.1 Nanoparticles

The utilization of nanoparticles facilitates the transportation of protein-RNA complexes and the integration of mRNA and gRNA, resulting in the effective manifestation of Cas9. This is followed by the formation of Cas9/gRNA within the plant cell, leading to the accomplishment of DNA-free editing (Wang et al., 2019). Long RNAs including Cas9 mRNA and sgRNAs were delivered using the synthesis and development of zwitterionic amino lipids (ZALs). DNA-loaded nanoparticles using polyethyleneimine (PEI)-coated Fe_3O_4 magnetic nanoparticles were implemented to carry DNA plasmids into the pollen grains of several dicot plants (Zhao et al., 2017). This process "known as magnetofection" allows for targeted gene delivery and has potential applications in gene therapy and drug delivery. The use of pollen as a carrier for these loaded particles could provide a more natural and sustainable alternative to traditional methods. Pollen combined with loaded particles needs a magnetic field to enhance magnetofection.

8.4.2 Virus-Like Particles

The CRISPR/Cas system and engineering viruses have been implemented for developing transgene-free plant genome editing. The utilization of the negative-strand sonchus yellow net rhabdovirus (SYNV) was explored to generate DNA-free genome-edited plants with a high degree of efficiency in tobacco (Ma et al., 2020). In a study conducted by Hu et al. (2019), a vector based on Barley Stripe Mosaic Virus (*BSMV*) was utilized to transport gRNA in wheat. Additionally, Potato Virus X was employed to deliver both Cas9 and gRNA in potatoes (Ariga et al., 2020).

8.4.3 Agrobacterium-Mediated Plant Transformation

Cas9 and the gRNA cloned at the T-DNA of the *Agrobacterium* Ti plasmid have been employed successfully in many plant species (Abdallah et al., 2022). Genetic segregation could be used to eliminate transgenes from edited transgenic plants and develop transgene-free plants. The transgene-free plants can only work for sexually propagated plants. The first generation would include transgenic plants containing the transgenes (non-edited), untransformed plants, and the transgenic plants of the edited target. Intensive methods are used for the selection of DNA-free plants in the following generation (Chen et al., 2018).

8.4.4 Graft-Mobile Gene Editing

The transportation of protein-encoding transcripts across graft junctions can be permitted by the presence of tRNA-like sequence (TLS) motifs and their respective variations, which are capable of inducing transcript mobility (Yang et al., 2023). Recently, edited plants were developed through a graft-mobile gene editing system for developing transgene-free offspring without the need for transgene elimination, culture recovery, selection, or the use of viral editing vectors (Yang et al., 2023). Transcript mobility was introduced by tRNA-like sequence (TLS) motifs to license the transport of protein-encoding transcripts over graft junctions. Adding TLS to Cas9 and gRNA transcripts resulted in their root-to-shoot movement and caused editing in recipient-grafted transgene-free tissues grafted on transgenic rootstocks. Mobile of Cas9/gRNA through TLS motifs expressed in the rootstocks produces seeds with edited genomes.

8.5 ENHANCING ABIOTIC STRESS TOLERANCE USING CRISPR/CAS9 APPROACHES

The utilization of the CRISPR/Cas9 editing system for genome editing has led to the development of various crops that exhibit improved environmental stress tolerance, including drought, salt, and heat tolerance (Table 8.1).

8.5.1 CRISPR for Enhancing Salt Tolerance

In a recent study conducted by Zhang et al. (2019), it was demonstrated that the CRISPR/Cas9 system can be utilized to enhance the salt stress tolerance of rice during the seedling stage. This was achieved through the suppression of the transcription factor gene OsRR22, which encodes the response regulator. The study conducted by Tran et al. (2021) found that the disruption of the SlHyPRP1 gene in tomato plants led to an enhancement in their ability to tolerate salt stress during both the germination

TABLE 8.1
Recent CRISPR/Cas9 Gene Editing in Crops for Improved Abiotic Stress Tolerance

Species	Trait Targeted	Gene(s) Edited	References
Arabidopsis	Stomata closure response and ABA	SlNPR1	Osakabe et al. (2016)
Rice	Herbicide tolerance	OSALS	Wang et al. (2021)
Rice	Cold tolerance	OsMYB30	Zeng et al. (2020)
Rice	Drought adaptive	OsDST	Santosh Kumar et al. (2020)
Rice	Drought tolerance	OsEBP89	Zhang et al. (2020)
Rice	Salinity tolerance	OsRR22	Zhang et al. (2019)
Rice	Root structure for saline soils	qSOR1	Kitomi et al. (2020)
Rice	Decreased Cd accumulation	OsLCT1, OsNramp5	Songmei et al. (2019)
Rice	Decreased Cd accumulation	OsNramp5	Yang et al. (2019)
Rice	Thermo-sensitive genic male-sterile	TMS5	Barman et al. (2019)
Rice	Drought tolerance	OsDST	Santosh Kumar et al. (2020)
Rice	Low cadmium content	TGW6, GW2 and GW5	Tang et al. (2017)
Tomato	Plant chilling tolerance	SlCBF1	Li et al. (2018a, b)
Tomato	Drought tolerance, as evidenced by higher levels of electrolytic leakage and malondialdehyde.	SBEIIb	Li et al. (2019)
Wheat	Drought resistance	TaDREB2 and TaERF3	Kim et al. (2018)

and vegetation stages. According to Wang et al. (2021), the enhancement of soybean salinity tolerance was achieved through the utilization of CRISPR/Cas9 editing to mutate the gene GmAITR, which encodes ABA-induced transcription repressors. The gene SlARF4 is involved in the regulation of both stomatal morphology and vascular bundle development. The gene SlARF4 was targeted using CRISPR/Cas9 in tomatoes, resulting in enhanced plant resilience to water deficits (Chen et al., 2021). The utilization of CRISPR-Cas9 technology to suppress the salt and drought tolerance (DST) gene in rice led to an increase in leaf water retention. This was attributed to the expansion of leaf width and a decrease in stomatal density. Consequently, the rice plant exhibited improved resistance to salt and drought stress and a higher yield of grains (Kumar et al., 2020).

8.5.2 CRISPR for Enhancing Heat/Cold Tolerance

The abscisic acid receptor genes, namely *OsPYL1*, *OsPYL4*, and *OsPYL6*, were found to be associated with heat stress susceptibility in rice. Knocking out these genes resulted in enhanced heat stress tolerance, which in turn led to a rise in yield production (Zhang et al., 2017). CRISPR/Cas9 technique was used to mutate one of the MAPK family members (SlMAPK3) in tomatoes, resulting in an improvement in heat stress tolerance (Yu et al., 2019). In addition, knocking out the SlAGL6 gene using CRISPR-Cas9-mediated genome editing improved the fruit setting of tomatoes under heat stress (Klap et al., 2017). The enhancement of cold tolerance was observed through the utilization of CRISPR/Cas9 technology to disrupt the rice *osmyb30* gene, which is a cold-responsive R2R3-type *MYB* gene (Zeng et al., 2020). The knockout of *OsMYB30*, a gene in rice known to respond to cold, was achieved using CRISPR/Cas9 technology. The resulting rice plants exhibited a greater degree of cold tolerance compared to wild-type rice (Zeng et al., 2020). The CBF1 mutants employed by CRISPR/Cas9 in tomatoes could increase cold-stress tolerance by reducing membrane damage (Li et al., 2019). The utilization of CRISPR/Cas9 technology to induce mutations in the *OsAnn5* gene resulted in a notable enhancement in the survival rates of rice seedlings when exposed to cold stress (Que et al., 2020).

8.5.3 CRISPR for Enhancing Drought Tolerance

The utilization of CRISPR/Cas9 technology to suppress the expression of regulatory genes *DERF1*, *PMS3*, *MSH1*, *MYB5*, and *SPP* has been employed to confer drought resistance in rice (Rahman et al., 2022). Mutation induced in *Arabidopsis*, CRISPR/Cas9 was deployed to mutate the *OST2* gene, resulting in the development of plant-tolerant plants (Osakabe et al., 2016). CRISPR/Cas9 was developed to improve drought tolerance through the activation of regulatory genes such as the vacuolar H+-pyrophosphate (AVP1) regulatory gene in Arabidopsis (Park et al., 2017) and the abscisic acid-responsive element binding gene (AREB1) in *Arabidopsis* (Roca Paixão et al., 2019). The utilization of CRISPR/Cas9 technology to generate SAPK2 mutants has revealed that these mutants exhibit ABA-insensitive phenotypes during germination and reduced tolerance to drought stress. These findings suggest that SAPK2 plays a crucial role in the drought stress response in rice (Lou et al., 2017). The downregulation of the abscisic acid hydroxylase 2 (abh2) gene in maize resulted in enhanced drought tolerance (Liu et al., 2020). Recently, the silencing of the six *TaSal1* genes (encoding 3′ (2′), 50-bisphosphate nucleotidase) in wheat using multiplex sgRNA-CRISPR/Cas9 genome editing enhanced stomatal closure and conferred drought tolerance at the seedling stage (Abdallah et al., 2022).

8.6 EPIGENETIC MODIFICATIONS

Epigenetic modifications are changes in gene expression that are passed down from generation to generation but don't change the DNA sequence. These modifications can alter gene expression patterns in response to abiotic stresses and can be targeted to improve abiotic stress tolerance in plants. Here are some examples of epigenetic modifications that have been used to improve abiotic stress tolerance in plants (Table 8.2). Epigenetic alterations have the potential to be transmitted to

TABLE 8.2
More Recent Research on Epigenetic Modifications to Improve Stress Tolerance

Title	Authors	Main Findings
Understanding the modification of plant epigenetics in response to abiotic stresses	Singh et al. (2022)	Reviewed several aspects of epigenetic modification in plants subjected to abiotic stresses, including DNA methylation, histone modification, the RdDM pathway, chromatin remodeling, and MSAP.
The underlying mechanisms, relevance, and implications of epigenetics and epigenomics for crop improvement	Agarwal et al. (2020)	Described the function of histone modifications and DNA methylation in regulating biotic and abiotic stresses and crop improvement. Indicated the potential of epigenome engineering and epigenome-based predictive models for enhancing agronomic traits.
Changes in epigenetic features in legumes under abiotic stresses	Yung et al. (2022)	Summarized the current knowledge on the changes in DNA methylation and histone modifications in legumes under salt, drought, and elevated temperatures. Discussed the potential applications of epigenetic engineering for legume improvement.
Epigenetic regulation of plant responses to abiotic stress	Liu et al. (2022)	Reviewed the recent advances in understanding the molecular mechanisms of epigenetic regulation of plant responses to abiotic stress, such as drought, salinity, heat, cold, and nutrient deficiency. Highlighted the challenges and future perspectives for epigenetic research in plant stress biology.
Epigenetic regulation of plant adaptation to environmental stress	Kaur et al. (2020)	Discussed the role of epigenetic mechanisms such as DNA methylation, histone modifications, chromatin remodeling, and non-coding RNAs in plant adaptation to environmental stress. Provided examples of crop plants that have been improved by manipulating epigenetic factors.
Epigenetics: a key player for crop improvement under abiotic stress	Kaur et al. (2019)	Reviewed the present understanding of the role of epigenetics in plant responses to diverse abiotic stresses, including drought, salinity, heat, cold, heavy metals, and ultraviolet radiation. Suggested some strategies for exploiting epigenetics for crop improvement under stress conditions.
Epigenetic modifications associated with abiotic stress tolerance	Amin et al. (2019)	Reviewed the recent advances in understanding the role of DNA methylation and histone modifications in regulating plant responses to different abiotic stresses such as drought, salinity, heat, cold, flooding, and heavy metals. Discussed the potential applications of epigenetics for enhancing stress tolerance in crops.

subsequent generations of dividing cells and are used indirectly to activate genes or repress them through the modification of the local chromatin state or DNA (Abdallah et al., 2022).

In DNA methylation, a methyl group is added to a cytosine residue in the DNA sequence. DNA methylation has been shown to play a role in regulating gene expression in response to abiotic stresses, such as drought and salt stress. Researchers have used genome-wide DNA methylation analysis to identify differentially methylated regions (DMRs) associated with stress tolerance in crops such as rice and wheat. By manipulating DNA methylation patterns in these regions, researchers have been able to improve stress tolerance in these crops (Garg et al., 2015; Sun et al., 2022). The results of correlation analysis suggest that DNA methylation exerts diverse impacts on gene expression in the context of drought stress, implying that it modulates gene expression via various regulatory pathways, either directly or indirectly. The research conducted by Ackah et al. (2022), Wang et al. (2016), and Zhao et al. (2022) revealed that drought-tolerant plants exhibit a more consistent methylome under drought conditions. The genes associated with differentially methylated regions (DMRs) were primarily linked to stress response, programmed cell death, and other pathways in rice, maize, mulberry, and mungbean, respectively.

Epigenetic **modifications of histone** proteins entail the addition or removal of chemical groups, thereby affecting chromatin structure and gene expression. The regulation of gene expression in response to abiotic stresses, such as salt stress, has been demonstrated to be influenced by histone modifications (Yuan et al., 2013). Researchers have used histone deacetylase inhibitors (HDACIs) to alter histone acetylation patterns in crops such as rice and maize, which has led to improved stress tolerance (He et al., 2020; Patanun et al., 2017).

Noncoding RNAs for enhancing drought tolerance, RNAi can be considered as an epigenetic modification when it involves the regulation of gene expression by small RNA molecules. RNAi has been used to improve abiotic stress tolerance in plants by targeting specific stress-responsive genes (Khare et al., 2018). For example, researchers have used RNAi to downregulate genes involved in salt stress responses in Arabidopsis, tomato, and wheat, which has led to improved salt tolerance (Borsani et al., 2005).

Chromatin remodeling is a type of epigenetic change that involves changing the structure of chromatin to control how genes are expressed. It has been shown to help control how genes are expressed in response to abiotic stresses like salt stress and drought. Researchers have used chromatin remodeling agents to change the structure of chromatin and the way genes are expressed in plants like rice, maize, and wheat. This has led to the plants being able to handle stress better (Luo et al., 2012).

8.7 BIOMOLECULES

Understanding the roles of these biomolecules in how plants respond to abiotic stress can teach us a lot about how plants can be bred to be better able to survive in harsh environments, which can improve agricultural productivity and food security. Biomolecules can be used to enhance abiotic stress tolerance in plants through a variety of mechanisms.

8.7.1 OVERPRODUCTION OF FUNCTIONAL PROTEINS

The cell undergoes several physiological alterations in response to abiotic stress, including the production of protective metabolites and proteins. To obtain plants with increased stress tolerance, it has been necessary to modulate the expression of genes involved in the biosynthesis of these metabolites, as well as those coding for proteins that play a direct role in the mechanisms of the protective function of the plant cell (Chinnusamy et al., 2005; Zhou et al., 2020).

To obtain plants with increased stress tolerance, researchers have focused on identifying and modulating the expression of genes involved in the biosynthesis of these metabolites and proteins. For example, overexpression of genes involved in proline biosynthesis has been shown to enhance drought and salt tolerance in plants. Similarly, overexpression of genes encoding ROS scavenging enzymes and heat shock proteins has been shown to improve plant stress tolerance (Yang et al., 2020). In addition to overexpression, the suppression of genes involved in the synthesis of stress-inducing compounds such as ethylene and abscisic acid has also been explored to enhance plant stress tolerance. These strategies, aimed at modulating the expression of genes involved in protective functions, have the potential to increase the stress tolerance of plants and help to ensure food security in the face of changing environmental conditions.

8.7.2 ENZYMES FOR THE PRODUCTION OF PROTECTIVE METABOLITES

Plants employ a shared physiological mechanism as a survival strategy in response to diverse stressors. This mechanism involves the buildup of water-soluble organic compounds with low molecular weight that exhibit non-toxic properties even at high concentrations. The substances referred to as osmoprotectants or compatible solutes are known to safeguard plants against stress through their contribution to cellular osmotic adjustment. Additionally, they act as ROS scavengers and/ or low-molecular-weight chaperones that aid in preserving membrane integrity and stabilizing proteins, thereby preventing damage and denaturation of cell structures (Chen & Murata, 2002). Several metabolites, such as amino acids (e.g., proline), quaternary and other amines [e.g., glycine betaine (GB), and polyamines (PAs)], as well as various sugars and sugar alcohols, have been recognized as osmoprotectants. (e.g., mannitol and trehalose). The implementation of a technique involving the excessive production of osmoprotectants has been employed as a means to enhance resistance to non-living environmental factors (Singh et al., 2015).

8.7.3 Proline

The increase in proline levels observed in plants under stress conditions can be attributed to the activation of genes involved in proline biosynthesis or the inhibition of genes responsible for proline degradation. In plants, the conversion of glutamate to proline is facilitated by pyrroline-5-carboxylate synthase (P5CS) and pyrroline-5-carboxylate reductase (P5CR), which act in tandem to catalyze two consecutive reductions (Delauney & Verma, 1993). Proline has the potential to be obtained from ornithine (Orn) through a transamination process that converts it to *P5C*. This conversion is facilitated by the mitochondrial enzyme Orn-aminotransferase (*OAT*) (Anwar et al., 2018). The augmentation of proline production through transgenic methods has been accomplished using comparable techniques, including the promotion of proline synthesis via overexpression of *P5CS, P5CR, or OAT* genes, or the hindrance of proline degradation via *PDH* silencing through antisense or knockout mutants. Consequently, numerous research endeavors have employed transgenic methodologies in diverse crop species (Kavi Kishor & Sreenivasulu, 2014).

8.7.4 Polyamines (PAs)

Putrescine (Put), spermidine (Spd), and spermine (Spm) are the most common PAs found in plants, and they have been shown to play a role in enhancing stress tolerance. Exogenous application of PAs such as Put, Spd, and Spm has been shown to improve plant growth, increase photosynthetic capacity, and enhance plant tolerance to a variety of abiotic stresses, including salinity, drought, and extreme temperatures (Bhattacharya & Rajam, 2007; Groppa & Benavides, 2008). In addition to the exogenous application, genetic engineering approaches that increase endogenous PA biosynthesis or alter PA metabolism have also been used to improve plant stress tolerance (Alcázar et al., 2010). For example, overexpression of genes involved in PA biosynthesis or transport has been shown to enhance plant growth and increase stress tolerance.

Many different plants produce the quaternary ammonium compound glycine betaine (GB) when under stress from high salinity, drought, or high temperatures. GB has been shown to play a crucial role in abiotic stress tolerance in plants. One of the primary functions of GB in plants is as an osmoprotectant. Under stress conditions such as drought or high salinity, GB accumulates in plant cells and helps to maintain cell turgor pressure and prevent cellular dehydration. Additionally, GB preserves protein structure and shields cellular membranes from harm from high salt or extreme temperatures.

8.7.5 Sugars and Sugar Alcohols

Sugars and sugar alcohols can help plants deal with abiotic stress by acting as osmoprotectants. Osmoprotectants are chemicals that protect cells from damage caused by things like drought, high salinity, extreme temperatures, and high light intensity. When plants accumulate sugars and sugar alcohols in response to stress like heat, drought, or salinity, these compounds can help to maintain cellular turgor pressure, stabilize membranes, scavenge reactive oxygen species, and protect proteins from denaturation (Bhattacharya & Kundu, 2020).

One way that sugars and sugar alcohols can help enhance abiotic stress tolerance is by regulating the expression of stress-responsive genes. For example, the sugar trehalose has been shown to upregulate genes involved in stress tolerance in Arabidopsis thaliana, a model plant species. In addition, sugars and sugar alcohols can act as signaling molecules that activate pathways involved in stress responses, such as the abscisic acid (ABA) pathway (Trivedi et al., 2016). The results indicate that the buildup of sugars and sugar alcohols in flora is a vital adaptive reaction to non-living environmental factors, which enables them to uphold cellular equilibrium and endure in unfavorable circumstances. Additional investigation is required to comprehensively comprehend the mechanisms that underlie these processes and their potential utilization to enhance crop resilience in the context of climate change.

8.7.6 Enzymes for Membrane Lipid Biosynthesis

Enzymes involved in membrane lipid biosynthesis can enhance abiotic stress tolerance in plants by altering the composition and properties of the plant cell membrane. When plants are under abiotic stress, their lipid metabolism changes, which causes their cell membranes to change in what they are made of and how fluid they are. Changes like these can hurt the growth and survival of plants if they interfere with membrane functions like transport, signaling, and structural support.

Enzymes that help make membrane lipids can help plants deal with abiotic stress by changing how the membrane's lipids are made in response to stress signals. For example, the enzyme phosphatidylglycerol (PG) synthase plays a key role in the synthesis of PG, a membrane phospholipid that is important for the stability and function of photosynthetic membranes (Luévano-Martínez & Kowaltowski, 2015). Overexpression of PG synthase has been shown to enhance abiotic stress tolerance in plants by increasing the accumulation of PG and other stress-responsive lipids (Li et al., 2020).

Other enzymes involved in membrane lipid biosyntheses, such as desaturases and acyltransferases, can also modulate membrane properties to enhance stress tolerance (Rogowska & Szakiel, 2020). Desaturases introduce double bonds into fatty acids, increasing the fluidity of the membrane and allowing it to better adapt to changes in temperature and other stress factors. Acyltransferases add fatty acid chains to lipids, which can alter the hydrophobicity and stability of the membrane (Rogowska & Szakiel, 2020).

8.7.7 Anti-Oxidant Enzymes

Antioxidant enzymes can enhance plant abiotic stress tolerance by helping to mitigate oxidative damage caused by stress-induced reactive oxygen species (ROS) accumulation (Choudhury et al., 2017). ROS are generated as a typical outcome of cellular metabolism. However, their concentration may escalate under unfavorable abiotic stressors such as extreme temperatures, high salinity, drought, and intense light. Reactive oxygen species (ROS) have the potential to induce oxidative damage to various cellular constituents, including proteins, lipids, and DNA, thereby resulting in cellular malfunction and eventual cell demise (Hasanuzzaman et al., 2019).

Antioxidant enzymes, such as superoxide dismutase (SOD), catalase (CAT), and peroxidase (POD), help to neutralize ROS and prevent oxidative damage (Gao et al., 2010; Saibi & Brini, 2018; Szőllősi, 2014; Vighi et al., 2016). Superoxide radicals undergo a conversion by SOD into hydrogen peroxide, which CAT and POD then break down into water and oxygen. Other enzymes, such as ascorbate peroxidase (APX) and glutathione peroxidase (GPX), also play important roles in detoxifying ROS. Under abiotic stress conditions, the expression and activity of antioxidant enzymes can increase in response to stress signals. This upregulation of antioxidant enzymes can help to maintain cellular redox balance and prevent oxidative damage. For example, overexpression of SOD has been shown to enhance abiotic stress tolerance in various plant species.

In addition to their direct role in getting rid of ROS, antioxidant enzymes can also indirectly improve a plant's ability to deal with abiotic stress by interacting with pathways that control how it responds to stress. For example, the antioxidant ascorbate has been shown to regulate the expression of genes involved in abiotic stress responses through its role in the ascorbate-glutathione cycle.

8.7.8 Transporters

Under abiotic stress conditions such as drought, high salinity, or nutrient deficiency, plants can experience changes in the availability and distribution of nutrients and ions, which can lead to cellular dysfunction and stress-induced damage. Transporters play a critical role in maintaining the balance of these essential molecules across cell membranes, which can help plants cope with stress.

For example, transporters such as aquaporins regulate the transport of water across cell membranes, which is crucial for maintaining cellular turgor pressure and hydration under drought stress (Yepes-Molina et al., 2020). In the same way, transporters like potassium channels and sodium transporters help control the movement of ions like potassium and sodium, which are important for maintaining membrane potential, osmotic balance, and other physiological processes (Francini et al., 2022). Transporters do more than just move nutrients and ions. They can also indirectly improve a plant's ability to deal with abiotic stress by interacting with pathways that control how it responds to stress. For example, the transport of the stress-response hormone abscisic acid (ABA) is controlled by specific transporters, which can change how much ABA builds up and how it sends signals in response to stress (Rehman et al., 2021).

8.7.9 Transcription Factors

Several families of transcription factors have been identified as key regulators of abiotic stress responses in plants. These include AP2/ERF family, MYB family, bZIP family, and WRKY family, among others. Each family of transcription factors has a unique set of target genes and regulatory mechanisms that contribute to abiotic stress tolerance in plants.

The DREB/CBF family of AP2/ERF transcription factors has been demonstrated to govern the transcriptional activity of genes implicated in plant dehydration and cold stress tolerance (Agarwal et al., 2013; Hussain et al., 2011). The augmentation of abiotic stress tolerance in diverse plant species has been observed through the overexpression of DREB/CBF transcription factors. In contrast to the *DREB1/CBF* genes, the enhancement of stress tolerance is not observed upon overexpression of *DREB2* in Arabidopsis, indicating that the regulation of *DREB2* proteins is subject to stringent control by upstream mechanisms, which include posttranscriptional modifications (Hayat et al., 2022; Liu et al., 1998). The transcription factor DREB2A harbors a domain with negative regulatory properties, which can be abrogated through deletion, resulting in the generation of a form that is constitutively active (DREB2A-CA). Arabidopsis plants that were genetically modified to overexpress *DREB2A-CA* exhibited stunted growth and enhanced resistance to both drought and heat stress (Chaudhari et al., 2022). Comparable outcomes are achieved upon overexpression of maize *ZmDREB2A* (Qin et al., 2007) and rice *OsDREB2B* (Matsukura et al., 2010) in a single plant. The MYB family of transcription factors has been demonstrated to regulate genes that are associated with drought and salt stress tolerance. Similarly, the bZIP family of transcription factors has been linked to the regulation of genes that are involved in oxidative stress responses.

8.7.10 MicroRNAs (miRNAs)

miRNAs are small, non-coding RNA molecules that play a key role in regulating gene expression in plants. They can help plants tolerate abiotic stresses by fine-tuning the expression of stress-responsive genes and controlling various physiological and biochemical pathways involved in stress adaptation. Under drought, high salinity, or extreme heat conditions, plants activate a complex network of signaling pathways that leads to changes in gene expression and cellular responses. MiRNAs play a critical role in regulating these stress-responsive genes by binding to specific mRNA molecules and repressing their translation or promoting their degradation.

Numerous microRNAs (miRNAs) have been demonstrated to participate in the plant's response to abiotic stress. MiRNAs have been identified as having various biological functions in plants, including regulation of growth and development, as well as responses to abiotic and biotic stress (Begum, 2022; Khraiwesh et al., 2012; Lima et al., 2012; Sunkar et al., 2012). TF genes have been identified as primary targets for various stress-regulated miRNAs, thereby presenting promising prospects for novel transgenic strategies aimed at modifying stress response. Tomato plants that exhibit overexpression of Sly-miR169 demonstrate increased tolerance to drought stress (Zhang et al., 2011). Furthermore, miR159 has been implicated in the regulation of genes associated with drought and salinity stress tolerance, whereas miR396 has been demonstrated to regulate genes involved in nutrient uptake and abiotic stress responses.

Recent studies have shown that plants can also produce stress-specific miRNAs in response to particular types of abiotic stress. For example, specific miRNAs have been identified in response to drought, salt, heat, and cold stresses, among others (Begum, 2022; Shriram et al., 2016). These stress-specific miRNAs can help plants fine-tune their responses to different types of stress by regulating the expression of specific sets of stress-responsive genes. In addition to their role in regulating gene expression, miRNAs can also interact with signaling pathways that regulate stress responses, such as the abscisic acid (ABA) signaling pathway (Jiang et al., 2019). ABA is a stress-responsive hormone that plays a key role in plant responses to drought, high salinity, and other types of abiotic stress.

8.8 ENHANCEMENT OF GENETIC GAIN UNDER DROUGHT STRESS

Biotechnology has enabled the identification and manipulation of genes involved in drought tolerance, such as transcription factors, signaling molecules, and stress-responsive genes. By introducing these genes into crops, genetic engineering has produced transgenic plants with improved drought tolerance (Liang, 2016). For example, prominent drought-resistant genotype, CIMBL55 of maize, the gene ZmRtn16 encoded a reticulon-like protein and contributed to drought resistance by facilitating the vacuole H+-ATPase activity, which highlights the role of vacuole proton pumps in maize drought resistance showed an increase of 6.7% in grain yield under drought conditions compared to the control (Tian et al., 2023). One of the success stories of using modern breeding and biotechnology approaches was reported in Mazie where the yield increase was by about 20%–30% in comparison to conventional varieties, even in the presence of moderate drought conditions (Fisher et al., 2015). According to CIMMYT (2022), the increased grain production in drought-affected regions of sub-Saharan Africa resulted in a gain of $160–200 million and benefited approximately 30–40 million individuals. Similarly, rice with overexpression of a transcription factor gene exhibited a higher survival rate and biomass after drought stress (Ahmed et al., 2020). The development and commercialization of drought-tolerant crops face many challenges, such as regulatory hurdles, public acceptance, and field performance (Liang, 2016; Nuccio et al., 2018). Therefore, with the continued advancement of biotechnology, the potential for enhancing drought tolerance in crops is immense, paving the way for a more resilient and productive agricultural sector.

8.9 CONCLUSION

Engineering abiotic stress tolerance in crops offers a promising approach to increase productivity and addressing food security challenges. Transgenic approaches, genome editing techniques, epigenetic modifications, and other biotechnological approaches are among the different strategies that have been used to enhance abiotic stress tolerance in crops. Each strategy has its advantages and limitations, and further research is needed to fully realize its potential. Nonetheless, the progress made so far provides hope that we can develop crops that can withstand harsh environmental conditions and help feed a growing global population.

REFERENCES

Abdallah, N. A., Elsharawy, H., Abulela, H. A., Thilmony, R., Abdelhadi, A. A., & Elarabi, N. I. (2022). Multiplex CRISPR/Cas9-mediated genome editing to address drought tolerance in wheat. *GM Crops & Food*, 12(2), 1–17.

Ackah, M., Guo, L., Li, S., Jin, X., Asakiya, C., Aboagye, E. T., Yuan, F., Wu, M., Essoh, L. G., & Adjibolosoo, D. (2022). DNA methylation changes and its associated genes in mulberry (*Morus alba* L.) Yu-711 response to drought stress using MethylRAD sequencing. *Plants*, 11(2), 190.

Agarwal, G., Kudapa, H., Ramalingam, A., Choudhary, D., Sinha, P., Garg, V., Singh, V. K., Patil, G. B., Pandey, M. K., & Nguyen, H. T. (2020). Epigenetics and epigenomics: Underlying mechanisms, relevance, and implications in crop improvement. *Functional & Integrative Genomics*, 20, 739–761.

Agarwal, P. K., Shukla, P. S., Gupta, K., & Jha, B. (2013). Bioengineering for salinity tolerance in plants: State of the art. *Molecular Biotechnology*, 54, 102–123.

Ahmed, R. F., Irfan, M., Shakir, H. A., Khan, M., & Chen, L. (2020). Engineering drought tolerance in plants by modification of transcription and signalling factors. *Biotechnology & Biotechnological Equipment*, 34(1), 781–789.

Alcázar, R., Altabella, T., Marco, F., Bortolotti, C., Reymond, M., Koncz, C., Carrasco, P., & Tiburcio, A. F. (2010). Polyamines: Molecules with regulatory functions in plant abiotic stress tolerance. *Planta*, 231, 1237–1249.

Amin, A. B., Rathnayake, K. N., Yim, W. C., Garcia, T. M., Wone, B., Cushman, J. C., & Wone, B. W. M. (2019). Crassulacean acid metabolism abiotic stress-responsive transcription factors: A potential genetic engineering approach for improving crop tolerance to abiotic stress. *Frontiers in Plant Science*, 10, 129.

Anwar, A., She, M., Wang, K., Riaz, B., & Ye, X. (2018). Biological roles of ornithine aminotransferase (OAT) in plant stress tolerance: Present progress and future perspectives. *International Journal of Molecular Sciences*, 19(11), 3681.

Ariga, H., Toki, S., & Ishibashi, K. (2020). Potato virus X vector-mediated DNA-free genome editing in plants. *Plant and Cell Physiology*, 61(11), 1946–1953.

Barman, H. N., Sheng, Z., Fiaz, S., Zhong, M., Wu, Y., Cai, Y., Wang, W., Jiao, G., Tang, S., & Wei, X. (2019). Generation of a new thermo-sensitive genic male sterile rice line by targeted mutagenesis of TMS5 gene through CRISPR/Cas9 system. *BMC Plant Biology*, 19, 1–9.

Bartels, D., & Sunkar, R. (2005). Drought and salt tolerance in plants. *Critical Reviews in Plant Sciences*, 24(1), 23–58.

Batcho, A. A., Sarwar, M. B., Rashid, B., Hassan, S., & Husnain, T. (2021). Heat shock protein gene identified from Agave sisalana (As HSP70) confers heat stress tolerance in transgenic cotton (*Gossypium hirsutum*). *Theoretical and Experimental Plant Physiology*, 33, 141–156.

Begum, Y. (2022). Regulatory role of microRNAs (miRNAs) in the recent development of abiotic stress tolerance of plants. *Gene*, 821, 146283.

Bhatnagar-Mathur, P., Devi, M. J., Reddy, D. S., Lavanya, M., Vadez, V., Serraj, R., Yamaguchi-Shinozaki, K., & Sharma, K. K. (2007). Stress-inducible expression of At DREB1A in transgenic peanut (*Arachis hypogaea* L.) increases transpiration efficiency under water-limiting conditions. *Plant Cell Reports*, 26, 2071–2082.

Bhatnagar-Mathur, P., Vadez, V., & Sharma, K. K. (2008). Transgenic approaches for abiotic stress tolerance in plants: Retrospect and prospects. *Plant Cell Reports*, 27, 411–424.

Bhattacharya, E., & Rajam, M. V. (2007). Polyamine biosynthetic pathway: A potential target for enhancing alkaloid production. In R. Verpoorte, A. Alfermann, & T. Johnson (Eds.), *Applications of Plant Metabolic Engineering* (pp. 129–143). Springer, Dordrecht.

Bhattacharya, S., & Kundu, A. (2020). Sugars and sugar polyols in overcoming environmental stresses. In A. Roychoudhury & D. K. Tripathi (Eds.), *Protective Chemical Agents in the Amelioration of Plant Abiotic Stress: Biochemical and Molecular Perspectives* (pp. 71–101). John Wiley & Sons Ltd. Chichester, West Sussex, PO19 8SQ, UK.

Borsani, O., Zhu, J., Verslues, P. E., Sunkar, R., & Zhu, J.-K. (2005). Endogenous siRNAs derived from a pair of natural cis-antisense transcripts regulate salt tolerance in Arabidopsis. *Cell*, 123(7), 1279–1291.

Chaudhari, R. S., Jangale, B. L., Krishna, B., & Sane, P. V. (2022). Improved abiotic stress tolerance in Arabidopsis by constitutive active form of a banana DREB2 type transcription factor, MaDREB20, than its native form, MaDREB20. *Protoplasma*, 260(3), 671–690.

Chen, M., Zhu, X., Liu, X., Wu, C., Yu, C., Hu, G., Chen, L., Chen, R., Bouzayen, M., & Zouine, M. (2021). Knockout of auxin response factor SlARF4 improves tomato resistance to water deficit. *International Journal of Molecular Sciences*, 22(7), 3347.

Chen, R., Xu, Q., Liu, Y., Zhang, J., Ren, D., Wang, G., & Liu, Y. (2018). Generation of transgene-free maize male sterile lines using the CRISPR/Cas9 system. *Frontiers in Plant Science, 9,* 1180.

Chen, T. H. H., & Murata, N. (2002). Enhancement of tolerance of abiotic stress by metabolic engineering of betaines and other compatible solutes. *Current Opinion in Plant Biology, 5*(3), 250–257.

Chinnusamy, V., Jagendorf, A., & Zhu, J. (2005). Understanding and improving salt tolerance in plants. *Crop Science, 45*(2), 437–448.

Choudhury, F. K., Rivero, R. M., Blumwald, E., & Mittler, R. (2017). Reactive oxygen species, abiotic stress and stress combination. *The Plant Journal, 90*(5), 856–867.

CIMMYT. (2022). Drought Tolerant Maize for Africa (DTMA). International Maize and Wheat Improvement Center. Retrieved from https://www.cimmyt.org/projects/drought-tolerant-maize-for-africa-dtma/

Datta, K., Baisakh, N., Ganguly, M., Krishnan, S., Yamaguchi Shinozaki, K., & Datta, S. K. (2012). Overexpression of Arabidopsis and rice stress genes' inducible transcription factor confers drought and salinity tolerance to rice. *Plant Biotechnology Journal, 10*(5), 579–586.

Delauney, A. J., & Verma, D. P. S. (1993). Proline biosynthesis and osmoregulation in plants. *The Plant Journal, 4*(2), 215–223.

El-Esawi, M. A., & Alayafi, A. A. (2019). Overexpression of StDREB2 transcription factor enhances drought stress tolerance in cotton (*Gossypium barbadense* L.). *Genes, 10*(2), 142.

El-Esawi, M. A., Al-Ghamdi, A. A., Ali, H. M., & Ahmad, M. (2019). Overexpression of AtWRKY30 transcription factor enhances heat and drought stress tolerance in wheat (*Triticum aestivum* L.). *Genes, 10*(2), 163.

Fisher, M., Abate, T., Lunduka, R. W., Asnake, W., Alemayehu, Y., & Madulu, R. B. (2015). Drought tolerant maize for farmer adaptation to drought in sub-Saharan Africa: Determinants of adoption in eastern and southern Africa. *Climatic Change, 133,* 283–299.

Francini, A., Toscano, S., Romano, D., & Ferrante, A. (2022). An overview of potassium in abiotic stress: Emphasis on potassium transporters and molecular mechanism. In N. Iqbal & S. Umar (Eds.), *Role of Potassium in Abiotic Stress* (pp. 249–262). Springer, Singapore.

Gao, S., Ou-yang, C., Tang, L., Zhu, J., Xu, Y., Wang, S., & Chen, F. (2010). Growth and antioxidant responses in *Jatropha curcas* seedling exposed to mercury toxicity. *Journal of Hazardous Materials, 182*(1–3), 591–597.

Garg, R., Narayana Chevala, V. V. S., Shankar, R., & Jain, M. (2015). Divergent DNA methylation patterns associated with gene expression in rice cultivars with contrasting drought and salinity stress response. *Scientific Reports, 5*(1), 1–16.

González, F. G., Capella, M., Ribichich, K. F., Curín, F., Giacomelli, J. I., Ayala, F., Watson, G., Otegui, M. E., & Chan, R. L. (2019). Field-grown transgenic wheat expressing the sunflower gene HaHB4 significantly outyields the wild type. *Journal of Experimental Botany, 70*(5), 1669–1681.

Groppa, M. D., & Benavides, M. P. (2008). Polyamines and abiotic stress: recent advances. *Amino Acids, 34,* 35–45.

Hasanuzzaman, M., Bhuyan, M. H. M. B., Anee, T. I., Parvin, K., Nahar, K., Mahmud, J. Al, & Fujita, M. (2019). Regulation of ascorbate-glutathione pathway in mitigating oxidative damage in plants under abiotic stress. *Antioxidants, 8*(9), 384.

Hasanuzzaman, M., Nahar, K., Fujita, M., Ahmad, P., Chandna, R., Prasad, M. N. V., & Ozturk, M. (2013). Enhancing plant productivity under salt stress: Relevance of poly-omics. Salt Stress in Plants: Signalling, Omics and Adaptations, pp. 113–156. In: Ahmad, P., Azooz, M.M., Prasad, M.N.V. (eds) Salt Stress in Plants. Springer, New York, NY.

Hayat, F., Sun, Z., Ni, Z., Iqbal, S., Xu, W., Gao, Z., Qiao, Y., Tufail, M. A., Jahan, M. S., & Khan, U. (2022). Exogenous melatonin improves cold tolerance of strawberry (Fragaria× ananassa Duch.) through modulation of DREB/CBF-COR pathway and antioxidant defense system. *Horticulturae, 8*(3), 194.

He, C., Yan, J., Shen, G., Fu, L., Holaday, A. S., Auld, D., Blumwald, E., & Zhang, H. (2005). Expression of an Arabidopsis vacuolar sodium/proton antiporter gene in cotton improves photosynthetic performance under salt conditions and increases fiber yield in the field. *Plant and Cell Physiology, 46*(11), 1848–1854.

He, S., Hao, Y., Zhang, Q., Zhang, P., Ji, F., Cheng, H., Lv, D., Sun, Y., Hao, F., & Miao, C. (2020). Histone deacetylase inhibitor SAHA improves high salinity tolerance associated with hyperacetylation-enhancing expression of ion homeostasis-related genes in cotton. *International Journal of Molecular Sciences, 21*(19), 7105.

Hu, J., Li, S., Li, Z., Li, H., Song, W., Zhao, H., Lai, J., Xia, L., Li, D., & Zhang, Y. (2019). A barley stripe mosaic virus-based guide RNA delivery system for targeted mutagenesis in wheat and maize. *Molecular Plant Pathology, 20*(10), 1463–1474.

Huang, Y., Cao, H., Yang, L., Chen, C., Shabala, L., Xiong, M., Niu, M., Liu, J., Zheng, Z., & Zhou, L. (2019). Tissue-specific respiratory burst oxidase homolog-dependent H2O2 signaling to the plasma membrane H+-ATPase confers potassium uptake and salinity tolerance in Cucurbitaceae. *Journal of Experimental Botany*, *70*(20), 5879–5893.

Huertas, R., Olias, R., Eljakaoui, Z., Gálvez, F. J., Li, J. U. N., De Morales, P. A., Belver, A., & Rodríguez-Rosales, M. P. (2012). Overexpression of SlSOS2 (SlCIPK24) confers salt tolerance to transgenic tomato. *Plant, Cell & Environment*, *35*(8), 1467–1482.

Hussain, S. S., Kayani, M. A., & Amjad, M. (2011). Transcription factors as tools to engineer enhanced drought stress tolerance in plants. *Biotechnology Progress*, *27*(2), 297–306.

ISAAA. (2019). International Service for the Acquisition of Agri-biotech Applications. GM Approval Database. Retrieved from https://www.isaaa.org/gmapprovaldatabase/default.asp.

Janská, A., Maršík, P., Zelenková, S., & Ovesná, J. (2010). Cold stress and acclimation-what is important for metabolic adjustment? *Plant Biology*, *12*(3), 395–405.

Jiang, D., Zhou, L., Chen, W., Ye, N., Xia, J., & Zhuang, C. (2019). Overexpression of a microRNA-targeted NAC transcription factor improves drought and salt tolerance in Rice via ABA-mediated pathways. *Rice*, *12*, 1–11.

Kaur, N., Alok, A., Kumar, P., Kaur, N., Awasthi, P., Chaturvedi, S., Pandey, P., Pandey, A., Pandey, A. K., & Tiwari, S. (2020). CRISPR/Cas9 directed editing of lycopene epsilon-cyclase modulates metabolic flux for β-carotene biosynthesis in banana fruit. *Metabolic Engineering*, *59*, 76–86.

Kaur, N., Kaur, G., & Pati, P. K. (2019). Deciphering strategies for salt stress tolerance in rice in the context of climate change. In: Mirza Hasanuzzaman, Masayuki Fujita, Kamrun Nahar, Jiban Krishna Biswas (Eds). Advances in Rice Research for Abiotic Stress Tolerance, pp. 113–132. Elsevier. Woodhead Publishing, Sawston, Cambridge.

Kavi Kishor, P. B., & Sreenivasulu, N. (2014). Is proline accumulation per se correlated with stress tolerance or is proline homeostasis a more critical issue? *Plant, Cell & Environment*, *37*(2), 300–311.

Khare, T., Shriram, V., & Kumar, V. (2018). RNAi technology: The role in development of abiotic stress-tolerant crops. In S. H. Wani (Ed.), Biochemical, *Physiological and Molecular Avenues for Combating Abiotic Stress Tolerance in Plants* (pp. 117–133). Elsevier. Academic Press, Cambridge, Massachusetts, United States.

Khraiwesh, B., Zhu, J.-K., & Zhu, J. (2012). Role of miRNAs and siRNAs in biotic and abiotic stress responses of plants. *Biochimica et Biophysica Acta (BBA)-Gene Regulatory Mechanisms*, *1819*(2), 137–148.

Kim, D., Alptekin, B., & Budak, H. (2018). CRISPR/Cas9 genome editing in wheat. *Functional & Integrative Genomics*, *18*(1), 31–41.

Kim, J.-J., Park, S.-I., Kim, Y.-H., Park, H.-M., Kim, Y.-S., & Yoon, H.-S. (2020). Overexpression of a proton pumping gene OVP1 enhances salt stress tolerance, root growth and biomass yield by regulating ion balance in rice (*Oryza sativa* L.). *Environmental and Experimental Botany*, *175*, 104033.

Kishitani, S., Takanami, T., Suzuki, M., Oikawa, M., Yokoi, S., Ishitani, M., Alvarez-Nakase, A. M., Takabe, T., & Takabe, T. (2000). Compatibility of glycinebetaine in rice plants: Evaluation using transgenic rice plants with a gene for peroxisomal betaine aldehyde dehydrogenase from barley. *Plant, Cell & Environment*, *23*(1), 107–114.

Kitomi, Y., Hanzawa, E., Kuya, N., Inoue, H., Hara, N., Kawai, S., Kanno, N., Endo, M., Sugimoto, K., & Yamazaki, T. (2020). Root angle modifications by the DRO1 homolog improve rice yields in saline paddy fields. *Proceedings of the National Academy of Sciences*, *117*(35), 21242–21250.

Klap, C., Yeshayahou, E., Bolger, A. M., Arazi, T., Gupta, S. K., Shabtai, S., Usadel, B., Salts, Y., & Barg, R. (2017). Tomato facultative parthenocarpy results from Sl AGAMOUS-LIKE 6 loss of function. *Plant Biotechnology Journal*, *15*(5), 634–647.

Kumar, K., Gambhir, G., Dass, A., Tripathi, A. K., Singh, A., Jha, A. K., Yadava, P., Choudhary, M., & Rakshit, S. (2020). Genetically modified crops: current status and future prospects. *Planta*, *251*, 1–27.

Kumar, T., Khan, M. R., Abbas, Z., & Ali, G. M. (2014). Genetic improvement of sugarcane for drought and salinity stress tolerance using Arabidopsis vacuolar pyrophosphatase (AVP1) gene. *Molecular Biotechnology*, *56*, 199–209.

Li, J., Liu, L.-N., Meng, Q., Fan, H., & Sui, N. (2020). The roles of chloroplast membrane lipids in abiotic stress responses. *Plant Signaling & Behavior*, *15*(11), 1807152.

Li, J., Phan, T.-T., Li, Y.-R., Xing, Y.-X., & Yang, L.-T. (2018a). Isolation, transformation and overexpression of sugarcane SoP5CS gene for drought tolerance improvement. *Sugar Tech*, *20*, 464–473.

Li, R., Liu, C., Zhao, R., Wang, L., Chen, L., Yu, W., Zhang, S., Sheng, J., & Shen, L. (2019). CRISPR/Cas9-Mediated SlNPR1 mutagenesis reduces tomato plant drought tolerance. *BMC Plant Biology*, *19*(1), 1–13.

Li, R., Zhang, L., Wang, L., Chen, L., Zhao, R., Sheng, J., & Shen, L. (2018b). Reduction of tomato-plant chilling tolerance by CRISPR-Cas9-mediated SlCBF1 mutagenesis. *Journal of Agricultural and Food Chemistry*, 66(34), 9042–9051.

Liang, C. (2016). Genetically modified crops with drought tolerance: Achievements, challenges, and perspectives. In M. Hossain, S. Wani, S. Bhattacharjee, D. Burritt, & LS. Tran (Eds.), *Drought Stress Tolerance in Plants, Vol. 2: Molecular and Genetic Perspectives* (pp. 531–547). Springer, Cham. Springer International Publishing Switzerland.

Lima, J. C. de, Loss-Morais, G., & Margis, R. (2012). MicroRNAs play critical roles during plant development and in response to abiotic stresses. *Genetics and Molecular Biology*, 35, 1069–1077.

Liu, G., Li, X., Jin, S., Liu, X., Zhu, L., Nie, Y., & Zhang, X. (2014). Overexpression of rice NAC gene SNAC1 improves drought and salt tolerance by enhancing root development and reducing transpiration rate in transgenic cotton. *PLoS One*, 9(1), e86895.

Liu, Q., Kasuga, M., Sakuma, Y., Abe, H., Miura, S., Yamaguchi-Shinozaki, K., & Shinozaki, K. (1998). Two transcription factors, DREB1 and DREB2, with an EREBP/AP2 DNA binding domain separate two cellular signal transduction pathways in drought-and low-temperature-responsive gene expression, respectively, in Arabidopsis. *The Plant Cell*, 10(8), 1391–1406.

Liu, S., Li, C., Wang, H., Wang, S., Yang, S., Liu, X., Yan, J., Li, B., Beatty, M., & Zastrow- Hayes, G. (2020). Mapping regulatory variants controlling gene expression in drought response and tolerance in maize. *Genome Biology*, 21, 1–22.

Liu, W., Feng, J., Ma, W., Zhou, Y., & Ma, Z. (2021). GhCLCg-1, a vacuolar chloride channel, contributes to salt tolerance by regulating ion accumulation in upland cotton. *Frontiers in Plant Science*, 12, 765173.

Liu, X., Quan, W., & Bartels, D. (2022). Stress memory responses and seed priming correlate with drought tolerance in plants: An overview. *Planta*, 255(2), 45.

Lou, D., Wang, H., Liang, G., & Yu, D. (2017). OsSAPK2 confers abscisic acid sensitivity and tolerance to drought stress in rice. *Frontiers in Plant Science*, 8, 993.

Luévano-Martínez, L. A., & Kowaltowski, A. J. (2015). Phosphatidylglycerol-derived phospholipids have a universal, domain-crossing role in stress responses. *Archives of Biochemistry and Biophysics*, 585, 90–97.

Luo, M., Liu, X., Singh, P., Cui, Y., Zimmerli, L., & Wu, K. (2012). Chromatin modifications and remodeling in plant abiotic stress responses. *Biochimica et Biophysica Acta (BBA)-Gene Regulatory Mechanisms*, 1819(2), 129–136.

Lv, S., Jiang, P., Nie, L., Chen, X., Tai, F., Wang, D., Fan, P., Feng, J., Bao, H., & Wang, J. (2015). H+-pyrophosphatase from *Salicornia europaea* confers tolerance to simultaneously occurring salt stress and nitrogen deficiency in Arabidopsis and wheat. *Plant, Cell & Environment*, 38(11), 2433–2449.

Ma, X., Zhang, X., Liu, H., & Li, Z. (2020). Highly efficient DNA-free plant genome editing using virally delivered CRISPR-Cas9. *Nature Plants*, 6(7), 773–779.

Matsukura, S., Mizoi, J., Yoshida, T., Todaka, D., Ito, Y., Maruyama, K., Shinozaki, K., & Yamaguchi-Shinozaki, K. (2010). Comprehensive analysis of rice DREB2-type genes that encode transcription factors involved in the expression of abiotic stress-responsive genes. *Molecular Genetics and Genomics*, 283, 185–196.

Meng, X., Wang, J.-R., Wang, G.-D., Liang, X.-Q., Li, X.-D., & Meng, Q.-W. (2015). An R2R3- MYB gene, LeAN2, positively regulated the thermo-tolerance in transgenic tomato. *Journal of Plant Physiology*, 175, 1–8.

Molinari, M. D. C., Fuganti-Pagliarini, R., Marin, S. R. R., Ferreira, L. C., Barbosa, D. de A., Marcolino-Gomes, J., Oliveira, M. C. N. de, Mertz-Henning, L. M., Kanamori, N., & Takasaki, H. (2020). Overexpression of AtNCED3 gene improved drought tolerance in soybean in greenhouse and field conditions. *Genetics and Molecular Biology*, 43(3), e20190292. https://doi.org/10.1590/1678-4685-gmb-2019-0292.

Munns, R., & Tester, M. (2008). Mechanisms of salinity tolerance. *Annual Review of Plant Biology*, 59, 651–681.

Nguyen, Q. H., Vu, L. T. K., Nguyen, L. T. N., Pham, N. T. T., Nguyen, Y. T. H., Le, S. Van, & Chu, M. H. (2019). Overexpression of the GmDREB6 gene enhances proline accumulation and salt tolerance in genetically modified soybean plants. *Scientific Reports*, 9(1), 19663.

Nuccio, M. L., Paul, M., Bate, N. J., Cohn, J., & Cutler, S. R. (2018). Where are the drought tolerant crops? An assessment of more than two decades of plant biotechnology effort in crop improvement. *Plant Science*, 273, 110–119.

Osakabe, Y., Watanabe, T., Sugano, S. S., Ueta, R., Ishihara, R., Shinozaki, K., & Osakabe, K. (2016). Optimization of CRISPR/Cas9 genome editing to modify abiotic stress responses in plants. *Scientific Reports*, 6(1), 26685.

Park, J.-J., Dempewolf, E., Zhang, W., & Wang, Z.-Y. (2017). RNA-guided transcriptional activation via CRISPR/dCas9 mimics overexpression phenotypes in Arabidopsis. *PLoS One*, 12(6), e0179410.

Patanun, O., Ueda, M., Itouga, M., Kato, Y., Utsumi, Y., Matsui, A., Tanaka, M., Utsumi, C., Sakakibara, H., & Yoshida, M. (2017). The histone deacetylase inhibitor suberoylanilide hydroxamic acid alleviates salinity stress in cassava. *Frontiers in Plant Science*, 7, 2039.

Qi, Y., Wang, H., Zou, Y., Liu, C., Liu, Y., Wang, Y., & Zhang, W. (2011). Over-expression of mitochondrial heat shock protein 70 suppresses programmed cell death in rice. *FEBS Letters*, 585(1), 231–239.

Qin, F., Kakimoto, M., Sakuma, Y., Maruyama, K., Osakabe, Y., Tran, L. P., Shinozaki, K., & Yamaguchi-Shinozaki, K. (2007). Regulation and functional analysis of ZmDREB2A in response to drought and heat stresses in Zea mays L. *The Plant Journal*, 50(1), 54–69.

Que, Z., Lu, Q., Liu, T., Li, S., Zou, J., & Chen, G. (2020). *The Rice Annexin Gene OsAnn5 Is a Positive Regulator of Cold Stress Tolerance at the Seedling Stage*. Research Square, Durham, NC.

Rahman, M., Zulfiqar, S., Raza, M. A., Ahmad, N., & Zhang, B. (2022). Engineering abiotic stress tolerance in crop plants through CRISPR genome editing. *Cells*, 11(22), 3590.

Ravikumar, G., Manimaran, P., Voleti, S. R., Subrahmanyam, D., Sundaram, R. M., Bansal, K. C., Viraktamath, B. C., & Balachandran, S. M. (2014). Stress-inducible expression of AtDREB1A transcription factor greatly improves drought stress tolerance in transgenic indica rice. *Transgenic Research*, 23, 421–439.

Rehman, A., Azhar, M. T., Hinze, L., Qayyum, A., Li, H., Peng, Z., Qin, G., Jia, Y., Pan, Z., & He, S. (2021). Insight into abscisic acid perception and signaling to increase plant tolerance to abiotic stress. *Journal of Plant Interactions*, 16(1), 222–237.

Ribichich, K. F., Chiozza, M., Ávalos-Britez, S., Cabello, J. V, Arce, A. L., Watson, G., Arias, C., Portapila, M., Trucco, F., & Otegui, M. E. (2020). Successful field performance in warm and dry environments of soybean expressing the sunflower transcription factor HB4. *Journal of Experimental Botany*, 71(10), 3142–3156.

Roca Paixão, J. F., Gillet, F.-X., Ribeiro, T. P., Bournaud, C., Lourenço-Tessutti, I. T., Noriega, D. D., Melo, B. P. de, de Almeida-Engler, J., & Grossi-de-Sa, M. F. (2019). Improved drought stress tolerance in Arabidopsis by CRISPR/dCas9 fusion with a histone acetyltransferase. *Scientific Reports*, 9(1), 8080.

Rogowska, A., & Szakiel, A. (2020). The role of sterols in plant response to abiotic stress. *Phytochemistry Reviews*, 19(6), 1525–1538.

Saibi, W., & Brini, F. (2018). Superoxide dismutase (SOD) and abiotic stress tolerance in plants: An overview. Superoxide Dismutase: Structure, Synthesis and Applications, Magliozzi, S. (ed.), pp. 101–142. Nova science publishers, Inc. Hauppauge, New York, United states.

Santosh Kumar, V. V, Verma, R. K., Yadav, S. K., Yadav, P., Watts, A., Rao, M. V, & Chinnusamy, V. (2020). CRISPR-Cas9 mediated genome editing of drought and salt tolerance (OsDST) gene in indica mega rice cultivar MTU1010. *Physiology and Molecular Biology of Plants*, 26(6), 1099–1110.

Shriram, V., Kumar, V., Devarumath, R. M., Khare, T. S., & Wani, S. H. (2016). MicroRNAs as potential targets for abiotic stress tolerance in plants. *Frontiers in Plant Science*, 7, 817.

Singh, S., Rahangdale, S., Pandita, S., Saxena, G., Upadhyay, S. K., Mishra, G., & Verma, P. C. (2022). CRISPR/Cas9 for insect pests management: A comprehensive review of advances and applications. *Agriculture*, 12(11), 1896.

Singh, T., Pun, K., Bhagat, K., Lal, B., Satapathy, B., Sadawarti, M., Katara, J., & Gautam, S. L. (2015). Abiotic stress in rice: Mechanism of adaptation. *Challenges and Prospective of Plant Abiotic Stress*, 4436(7), 259–310.

Songmei, L., Jie, J., Yang, L., Jun, M., Shouling, X., Yuanyuan, T., Youfa, L., Qingyao, S., & Jianzhong, H. (2019). Characterization and evaluation of OsLCT1 and OsNramp5 mutants generated through CRISPR/Cas9-mediated mutagenesis for breeding low Cd rice. *Rice Science*, 26(2), 88–97.

Sun, M., Yang, Z., Liu, L., & Duan, L. (2022). DNA Methylation in plant responses and adaption to abiotic stresses. *International Journal of Molecular Sciences*, 23(13), 6910.

Sunkar, R., Li, Y.-F., & Jagadeeswaran, G. (2012). Functions of microRNAs in plant stress responses. *Trends in Plant Science*, 17(4), 196–203.

Szőllősi, R. (2014). Superoxide dismutase (SOD) and abiotic stress tolerance in plants: An overview. In P. Ahmad (Ed.), *Oxidative Damage to Plants* (pp. 89–129). Academic Press. San Diego.

Tang, X., Cao, X., Xu, X., Jiang, Y., Luo, Y., Ma, Z., Fan, J., & Zhou, Y. (2017). Effects of climate change on epidemics of powdery mildew in winter wheat in China. *Plant Disease*, 101(10), 1753–1760.

Tang, Y., Bao, X., Zhi, Y., Wu, Q., Guo, Y., Yin, X., Zeng, L., Li, J., Zhang, J., & He, W. (2019). Overexpression of a MYB family gene, OsMYB6, increases drought and salinity stress tolerance in transgenic rice. *Frontiers in Plant Science*, 10, 168.

Tian, T., Wang, S., Yang, S., Yang, Z., Liu, S., Wang, Y., Gao, H., Zhang, S., Yang, X., & Jiang, C. (2023). Genome assembly and genetic dissection of a prominent drought-resistant maize germplasm. *Nature Genetics*, 55 (3), 1–11.

Tran, M. T., Doan, D. T. H., Kim, J., Song, Y. J., Sung, Y. W., Das, S., Kim, E., Son, G. H., Kim, S. H., & Van Vu, T. (2021). CRISPR/Cas9-based precise excision of SlHyPRP1 domain(s) to obtain salt stress-tolerant tomato. *Plant Cell Reports, 40*, 999–1011.

Trivedi, D. K., Gill, S. S., & Tuteja, N. (2016). Abscisic acid (ABA): Biosynthesis, regulation, and role in abiotic stress tolerance. In N. Tuteja & S. S. Gill (Eds.), *Abiotic Stress Response in Plants* (pp. 315–326). Wiley Online Library.

Vendruscolo, E. C. G., Schuster, I., Pileggi, M., Scapim, C. A., Molinari, H. B. C., Marur, C. J., & Vieira, L. G. E. (2007). Stress-induced synthesis of proline confers tolerance to water deficit in transgenic wheat. *Journal of Plant Physiology, 164*(10), 1367–1376.

Vighi, I. L., Benitez, L. C., do Amaral, M. N., Auler, P. A., Moraes, G. P., Rodrigues, G. S., Da Maia, L. C., Pinto, L. S., & Braga, E. J. (2016). Changes in gene expression and catalase activity in *Oryza sativa* L. under abiotic stress. *Genetics and Molecular Research, 15*, 1–15.

Wang, F., Xu, Y., Li, W., Chen, Z., Wang, J., Fan, F., Tao, Y., Jiang, Y., Zhu, Q.-H., & Yang, J. (2021). Creating a novel herbicide-tolerance OsALS allele using CRISPR/Cas9-mediated gene editing. *The Crop Journal, 9*(2), 305–312.

Wang, G. P., Zhang, X. Y., Li, F., Luo, Y., & Wang, W. (2010). Overaccumulation of glycine betaine enhances tolerance to drought and heat stress in wheat leaves in the protection of photosynthesis. *Photosynthetica, 48*, 117–126.

Wang, H., Tang, H., Yang, C., & Li, Y. (2019). Selective single molecule nanopore sensing of microRNA using PNA functionalized magnetic core-shell Fe3O4-Au nanoparticles. *Analytical Chemistry, 91*(12), 7965–7970.

Wang, W., Qin, Q., Sun, F., Wang, Y., Xu, D., Li, Z., & Fu, B. (2016). Genome-wide differences in DNA methylation changes in two contrasting rice genotypes in response to drought conditions. *Frontiers in Plant Science, 7*, 1675.

Wei, P., Che, B., Shen, L., Cui, Y., Wu, S., Cheng, C., Liu, F., Li, M.-W., Yu, B., & Lam, H.-M. (2019). Identification and functional characterization of the chloride channel gene, GsCLC-c2 from wild soybean. *BMC Plant Biology, 19*, 1–15.

Yang, C., Zhang, Y., & HUANG, C. (2019). Reduction in cadmium accumulation in japonica rice grains by CRISPR/Cas9-mediated editing of OsNRAMP5. *Journal of Integrative Agriculture, 18*(3), 688–697.

Yang, L., Machin, F., Wang, S., Saplaoura, E., & Kragler, F. (2023). Heritable transgene-free genome editing in plants by grafting of wild-type shoots to transgenic donor rootstocks. *Nature Biotechnology, 41*(7), 1–10.

Yang, R., Yu, G., Li, H., Li, X., & Mu, C. (2020). Overexpression of small heat shock protein LimHSP16.45 in Arabidopsis hsp17.6II mutant enhances tolerance to abiotic stresses. *Russian Journal of Plant Physiology, 67*, 231–241.

Yepes-Molina, L., Carvajal, M., & Martínez-Ballesta, M. C. (2020). Detergent resistant membrane domains in broccoli plasma membrane associated to the response to salinity stress. *International Journal of Molecular Sciences, 21*(20), 7694.

Yu, W., Wang, L., Zhao, R., Sheng, J., Zhang, S., Li, R., & Shen, L. (2019). Knockout of SlMAPK3 enhances tolerance to heat stress involving ROS homeostasis in tomato plants. *BMC Plant Biology, 19*(1), 1–13.

Yuan, L., Liu, X., Luo, M., Yang, S., & Wu, K. (2013). Involvement of histone modifications in plant abiotic stress responses. *Journal of Integrative Plant Biology, 55*(10), 892–901.

Yue, Y., Zhang, M., Zhang, J., Duan, L., & Li, Z. (2012). SOS1 gene overexpression increased salt tolerance in transgenic tobacco by maintaining a higher K+/Na+ ratio. *Journal of Plant Physiology, 169*(3), 255–261.

Yung, W., Huang, C., Li, M., & Lam, H. (2022). Changes in epigenetic features in legumes under abiotic stresses. *The Plant Genome, 16*(4), e20237.

Zeng, Y., Wen, J., Zhao, W., Wang, Q., & Huang, W. (2020). Rational improvement of rice yield and cold tolerance by editing the three genes OsPIN5b, GS3, and OsMYB30 with the CRISPR-Cas9 system. *Frontiers in Plant Science, 10*, 1663.

Zhang, A., Liu, Y., Wang, F., Li, T., Chen, Z., Kong, D., Bi, J., Zhang, F., Luo, X., & Wang, J. (2019). Enhanced rice salinity tolerance via CRISPR/Cas9-targeted mutagenesis of the OsRR22 gene. *Molecular Breeding, 39*, 1–10.

Zhang, G., Lu, T., Miao, W., Sun, L., Tian, M., Wang, J., & Hao, F. (2017). Genome-wide identification of ABA receptor PYL family and expression analysis of PYLs in response to ABA and osmotic stress in Gossypium. *PeerJ, 5*, e4126.

Zhang, X., Zou, Z., Gong, P., Zhang, J., Ziaf, K., Li, H., Xiao, F., & Ye, Z. (2011). Over- expression of microRNA169 confers enhanced drought tolerance to tomato. *Biotechnology Letters, 33*, 403–409.

Zhang, Y., Li, J., Chen, S., Ma, X., Wei, H., Chen, C., Gao, N., Zou, Y., Kong, D., & Li, T. (2020). An APETALA2/ethylene responsive factor, OsEBP89 knockout enhances adaptation to direct-seeding on wet land and tolerance to drought stress in rice. *Molecular Genetics and Genomics*, 295(4), 941–956.

Zhao, W., Wang, X., Zhang, Q., Zheng, Q., Yao, H., Gu, X., Liu, D., Tian, X., Wang, X., & Li, Y. (2022). H3K36 demethylase JMJ710 negatively regulates drought tolerance by suppressing MYB48-1 expression in rice. *Plant Physiology*, 189(2), 1050–1064.

Zhao, X., Meng, Z., Wang, Y., Chen, W., Sun, C., Cui, B., Cui, J., Yu, M., Zeng, Z., & Guo, S. (2017). Pollen magnetofection for genetic modification with magnetic nanoparticles as gene carriers. *Nature Plants*, 3(12), 956–964.

Zhou, Y., Zhang, Y., Wang, X., Han, X., An, Y., Lin, S., Shen, C., Wen, J., Liu, C., & Yin, W. (2020). Root-specific NF-Y family transcription factor, PdNF-YB21, positively regulates root growth and drought resistance by abscisic acid-mediated indoylacetic acid transport in Populus. *New Phytologist*, 227(2), 407–426.

Zhu, B., Ye, C., Lü, H., Chen, X., Chai, G., Chen, J., & Wang, C. (2006). Identification and characterization of a novel heat shock transcription factor gene, GmHsfA1, in soybeans (*Glycine max*). *Journal of Plant Research*, 119, 247–256.

9 Groundwater's Geochemical Status in Agricultural and Sustainable Use (Western Mitidja, Algeria)

Ahcène Semar, Hakim Bachir, and Rattan Lal

9.1 INTRODUCTION

As known, Algeria is a dry country that belongs to the arid-semi-arid climate (Smadhi et al., 2022). These vast interior spaces are characterized by a dry and cold climate in winter, and hot and dry in summer (Bachir et al., 2022). A deficient and irregularly distributed rainfall characterizes the climatic conditions that exert a very strong influence on agricultural growth (Bachir et al., 2016). They are an obstacle to the adoption of intensive agricultural practices without irrigation (Bachir et al., 2021). A judicious use of water in the main sectors of agriculture, industry and drinking water supply, should be based on the knowledge of its hydrochemical status (Hosseinifard and Mirzaei, 2015). The latter is a function of the mineralization acquisition process and pollution (Sayeh et al., 2022). Many factors are involved in the chemistry of water, such as climate, geology, the presence of a sea, and human action (Peters and Maybeck, 2000). Thus, it is prudent to link the changes in the chemical character of waters to natural conditions and anthropogenic activities. Due to their location on the surface, surface waters are relatively less loaded with minerals than ground waters, which are subject to geochemical, natural processes resulting from water-rock interactions. Numerous research studies have elucidated these geochemical reactions through case studies (Gupta et al., 2008; Al-Ahmadi and El-Fiky, 2009; Senthilkumar and Elango, 2013; Barbieri et al., 2017; Yi et al., 2017). Several approaches were used to assess water geochemistry. By using the determination of chemical facies or factors approach, it is possible to identify, the mixing processes of waters (Selvakumar et al., 2017), the recharge zones (Dragon and Gorski, 2015) or the chemical evolution of waters due to the human activities which impact in a direct or indirect way in the modification of hydrochemical behaviour (Eloïse, 2017). Geochemical indicators are widely discussed in the literature and have used the following parameters: Saturation and Chloro-alkalines indices, Gibbs diagram and some characteristic ion ratios. Human, through diverse activities, participates in a direct or indirect way in the modification of hydrochemical behavior.

Numerous studies have also been proposed and conducted on Mitidja, in the form of academic theses and scientific articles (Sekkal, 1986; Messaoud Nacer, 1987; Hadjoudj et al., 2014; Sbargoud et al., 2017). The western alluvial plain part of the Mitidja involves a vivid agricultural-based economy, it is relatively less affected by urbanization compared to the eastern part. The work of Sekkal (1986) focused, after presenting the natural environment, on a hydrodynamic study of the waters of the alluvial aquifer. This was followed by a hydrochemical study (emphasis on pollution based on analyses of metallic elements and nutrients) and groundwater modelling approach hydrodynamic aspect (piezometry and pumping tests) adopted by Messaoud Nacer (1987) on the western Mitidja. Indeed, the evaluation, management and preservation of water resources in Algeria, and of the Mitidja plain, is the responsibility of the NWRA which is affiliated with the Ministry of Water Resources created to follow the evolution of the water resource both from a quantitative and qualitative point of view. Hydrogeological and geophysical studies are undertaken by this institution and are available in the form of reports.

The present research focuses on the study of the alluvial groundwater of Sidi Rached region circumscribed in an agricultural perimeter that is an integral part of the Western Mitidja. Groundwater. This latter is widely used in irrigation but also in the supply of drinking water to peri-urban populations where 33 water points were sampled across the agricultural perimeter. The objective of this research is to try to explain the major mechanisms that contribute to its current geochemical status.

9.2 GEOGRAPHICAL FRAMEWORK, CLIMATE, GEOLOGY AND HYDROGEOLOGY

9.2.1 Geographical Framework

The study area is located about 70 km from the capital Algiers. Locally, it is bounded in the north by the Sahel reliefs at altitudes of around 200 m and in the south by the Blidean Atlas, with an average altitude of 1,400 m (Figure 9.1). It falls within the following coordinates: longitude between 2° 26′ E and 2° 37′ E and latitude between 36° 27′ N and 36° 34′ N with an estimated surface area of 150 km^2. The main localities in the study area are Sidi Rached, Hmar El Ain and Bourkika. From a geomorphological point of view, it corresponds to the western part of the Mitidja plain extended according to the general direction ENE - WSW. The topography of the site, according to a north-south profile, varies from 500 m of the southern relief, passing through the Sidi Rached basin, to some 60 m then reaching 200 m on the northern relief of the Sahel. The cluse of Mazafran allows the wadi Djer and Bouroumi originating in the south, to reach the sea, after bypassing the coastal relief. The main economic activity is related to agriculture, it covers an area of 156.4 km^2 of which 62.36% (97.53 km^2) is irrigated, occupied by cereals (45%), vegetables (25%), fruits trees (20%) and other crops (10%). The non-irrigated land accounts for 37.64% of the study area, consisting of forests, mountains, lands with hydromorphic soils (ex. The Halloula region located at the North-East of the Sidi Rached watershed) and finally roads and agglomerations (Sbargoud et al., 2017).

9.2.2 Climate

The study region comes under a Mediterranean climate, characterized by a mild, wet winter and a hot, dry summer. As one moves away from the Sahel, the climate becomes more and more continental with strong inter-monthly thermal amplitudes with colder winters and hotter summers.

According to the EEA (2012), the Mediterranean region has been subject to significant impacts in recent decades due to decreasing rainfall and increasing temperature, and these are expected to increase as the climate continues to change. The main impacts are the reduction of water and hence of crop availability and yields, the risk of droughts, loss of biodiversity, forest fires and heat waves. Analysis of rainfall data (period: 1975–2017) shows irregularity on a monthly and annual scale. Annual rainfall reaches 718 mm at the Blida station. On a monthly scale, rainfall increases from September (28 mm) to reach a maximum in January (94 mm) and then decreases to reach a minimum in August (8 mm). Monthly temperatures evolve differently from the rainfall and are relatively more regular. They are low in winter with a minimum of 11°C in January and increase gradually to reach peaks in summer mainly during July (33°C) and August (32°C).

9.2.3 Geology and Hydrogeology

Belonging to the alpine system, characterized by active tectonics, the Mitidja alluvial plain is an asymmetrically shaped depression, intra-montane and littoral, oriented ENE - WSW, about 100 km long and 5–20 km wide (Glangeaud, 1952). The synclinal shape of the basin is confirmed by deep drilling data and geophysical surveys. Bordering flexures to the north and south delineate the basin. The northern flexure corresponds to the fault contact between the geological layers of the basin and the Plio-Quaternary anticline of the Sahel (Meghraoui, 1991). Towards the south, the contact between the plain and the Atlas is underlined by an overlap (Boudiaf, 1996). The western and eastern limits are respectively the Menaceur relief and the Thenia fault line.

Like the Tellian domain to which the basin belongs, and as Figure 9.1 highlights, sedimentary rocks are widespread and dominate the geological outcrops of magmatic and metamorphic rocks. The succession of geological formations in the Western Mitidja alluvial basin is structured, from bottom to top, as follows:

- the Piacenzian, a thick marly layer, deposited in deep, grey, or blue water, sometimes sandy, constitutes the impermeable substratum. It outcrops in the hills of the Sahel and its thickness varies from one place to another and exceeds 240 m at the level of the Mazafran cluse (Glangeaud, 1952);
- the Astian attributed to the Middle Pliocene, of marine origin, is marked by a diverse lithology. Sandstone, limestone or sandstone limestone, molasse or marl facies are encountered. The thickness of this formation is around 100 m and is found at depths close to 300 m. The Astian is a locally exploited aquifer reservoir;
- the El Harrach formation, from the continental Pliocene, is predominantly clayey, alternating with yellow marls, gravelly clays and shallow sand and gravel beds. This formation is considered impermeable;
- the continental Quaternary Mitidja formations are made of sedimentary deposits of various types and sizes. Important wadi deposits as well as dejection cones at the foothills of the Atlas Mountains are mixed with fine elements in quantities. The latter form a silty or clayey surface layer that covers coarser formations. The maximum thickness is 200 m at the Mazafran cluse. In the Halloula region, which is limited to our study site, the existence of silty and clayey-silt sediments containing gravel is highlighted. Recent deposits of dune sand, from the foothills of the Atlas Mountains and current wadis are also present in the Mitidja basin. These formations are widely exploited at the level of the Mitidja plain.

From a hydrogeological point of view, two main aquifers can be distinguished in the Mitidja plain. The first aquifer is shallow and consists of alluvial deposits, while the second aquifer is relatively deep and is essentially sandstone in nature. The alluvial aquifer, the object of the present study, is widely exploited throughout the Mitidja due to its shallow depth. The lithological nature is relatively variable from one place to another, both horizontally and vertically. At the scale of the study zone, the drilling data capturing the quaternary aquifer shows a lithological heterogeneity marked by a succession of permeable and impermeable levels, particularly in the centre of the Sidi Rached perimeter. The geographical location of the boreholes shows that those located at the northern and southern extremities are relatively characterised by a higher rate of permeable formations and thicker layers. This vertical variability is indicated by the maximum depth of investigation estimated at 200 m. The permeable horizons are gravelly, sandy and often mixed with clay. Impermeable clayey and sometimes marly interlayers can be very frequent and thick. The general flow of groundwater is from South to North but is often influenced by intensive pumping. Recharging of the water table occurs from the impluvium and the wadis that cross the plain.

The second aquifer, known as the Astian, is deeper and captive. It is highlighted by geophysical prospecting by electric method which has located it under the stony yellow marls known as Maison Carrée. This second reservoir, nearly 100 m thick, is found at depths of 200 m in the centre of the Sidi Rached perimeter (Figure 9.2). It represents an important aquifer both at the local level and at the level of the Mitidja plain. The lithology of the (Astian) is of sandstone, sandy and calcareous type evolving towards the calcairo-sandstone or argilo-sandstone facies. The substratum of the

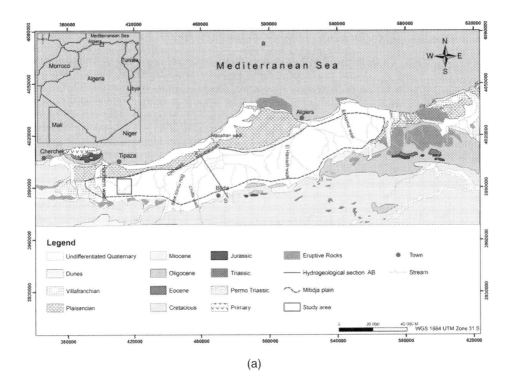

(a)

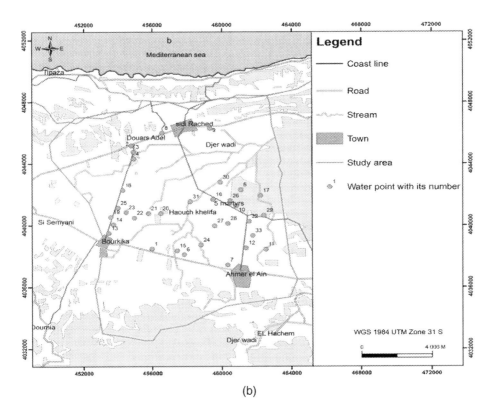

(b)

FIGURE 9.1 General and local framework. (a) Geographical and geological presentation of the Mitidja Plain. (b) Delimitation of the study area.

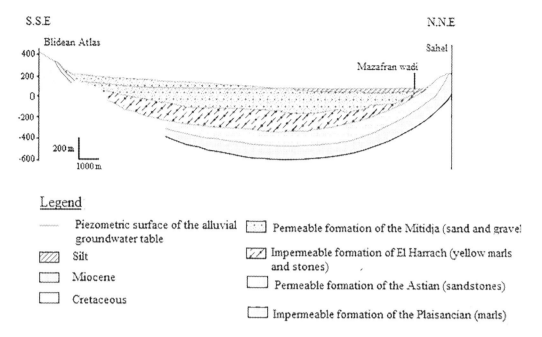

FIGURE 9.2 Hydrogeological section in the north-south direction.

Astian aquifer is made up of blue marls and clays from the Plaisancien period. The depth decreases towards the northern edge, while towards the south the Miocene and Cretaceous marl formations predomiate.

9.3 SAMPLING, WATER ANALYSIS AND GEOCHEMICAL APPROACH 6

9.3.1 Sampling

Water sampling focused on the groundwater of the alluvial aquifer of the Western Mitidja at the Sidi Rached site. This reservoir is widely exploited for the irrigation of different crops and for the domestic needs of a few scattered dwellings. The groundwater sampling was done from 16 to 23 May 2017, when 33 samples were collected through boreholes and artisanal wells. Sampling procedures are carried out according to standardised methods (APHA, 2005). A 1 L sample of water was collected in pre-cleaned polyethylene bottles and rinsed with the water collected. The samples were acidified with HNO_3 at pH = 2.

9.3.2 Water Analysis

All samples were analysed for major ions using standard methods. Field measurements (pH, T, TDS and EC) were carried out by a portable multi 340i multifunction device of the WTW brand. Calcium and magnesium were determined by the compleximetric titration method with respect to the EDTA standard solution, while sodium and potassium were evaluated by flame photometry. Chloride concentrations are determined by titrimetric titration against the $AgNO_3$ standard solution. Nitrate and sulphate contents were determined by spectrophotometry. The quality of the hydrochemical data was evaluated by the ionic balance establishing the relationship between all the cations (Ca^{2+}, Mg^{2+}, Na^+, K^+) and the total anions (Cl^-, SO_4^{2-}, HCO_3^- et NO_3^-) for each complete analysis. The validation of the analytical results was based on the calculation of the error and a threshold of 10% is tolerated (Domenico and Schwartz, 1998).

9.3.3 Geochemical Approach

9.3.3.1 Water Hardness

Water hardness is an indicator of water quality and is caused by the concentration of calcium and magnesium ions. It is expressed in mg of $CaCO_3$. Water with high hardness prevents the cooking of certain foodstuffs, does not easily foam soap and leads to limescale deposits in the pipes. Hardness is calculated by Equation (9.1) (Todd and Mays, 2005):

$$TH(CaCO_3) mg/L = 2.5Ca^{2+} + 4.1Mg^{2+} \tag{9.1}$$

where, Ca^{2+} and Mg^{2+} are expressed in mg/L

A classification of waters, according to TH values, is proposed by Sawyer and McCarty (1967) which qualifies waters as soft if TH < 75, moderately hard if TH is between 75 and 150, hard if TH is between 150 and 300 and finally very hard if TH > 300.

9.3.3.2 Chemical Types

Determination of chemical types allows for identifying the order of the dominant cations and anions present in water. The study used the Piper diagram because it allows projecting several water points, comparing them with each other, to determine the possible groups formed and to follow the evolution of the chemical types' function to the mineralization. The realization of this diagram was done by using the Diagram software (Simler, 2012) which is composed of two triangles, representing the cationic and anionic types, and a diamond synthesizing the global facies.

9.3.3.3 Saturation Index

The saturation index is an essential parameter in hydrogeological and geochemical studies because it allows the identification of the behaviour of certain key minerals in groundwater. In this study, the saturation indices retained are those relating to calcite, dolomite, aragonite, gypsum, anhydrite and halite. The saturation index is calculated according to Equation 9.2.

$$SI = \log(IAP / Ks) \tag{9.2}$$

where, IAP is "ion activity product of the mineral-water reaction" and Ks "is the thermodynamic equilibrium constant adjusted to the temperature of the given sample." The geochemical code Phreeqc (Parkhurst and Appelo, 1999) was used to evaluate the saturation indices. When groundwater is saturated with minerals, SI values are zero; positive SI values indicate over-saturation and negative values indicate under-saturation (Appelo and Postma, 2005; Drever, 1997).

9.3.3.4 Chloro-Alkaline Indices

Schoeller (1965) proposed chloro-alkaline indices to highlight cation exchanges between groundwater and the solid matrix of the hydrogeological reservoir. These indices, widely used in hydrochemistry, make it possible to identify changes in the chemical composition of water as it flows through the aquifer.

The chloro-alkaline called CAI-I and CAI-II are calculated according to Equations 9.3 and 9.4 where all ions are expressed in meq/L:

$$CAI\text{-}I = \frac{Cl^- - (Na^+ + K^+)}{Cl^-} \tag{9.3}$$

$$CAI\text{-}II = \frac{Cl^- - (Na^+ + K^+)}{HCO_3^- + SO_4^{2-} + CO_3^{2-} + NO_3^-} \tag{9.4}$$

When the indices are negative, the Ca^{2+} or Mg^{2+} cations of the groundwater are exchanged against the Na^+ or K^+ of the surrounding rock. Thus, the water in the aquifer is enriched in Na^+ and K^+. On the other hand, if the values of the chloro-alkali indices are positive, the reverse reaction, called Reverse Ion Exchange, occurs, i.e., the Na^+ or K^+ cations in the groundwater are exchanged against the Ca^{2+} or Mg^{2+} ions in the reservoir rock. As a result, the groundwater is enriched with alkaline earth.

9.3.3.5 Gibbs Diagram

The representation of hydrochemical data according to the Gibbs diagram makes it possible to understand and distinguish the influences of rock-water interaction, evaporation and precipitation (Gibbs, 1970). The Gibbs diagram is widely used in hydrogeological studies to identify the explanatory processes of water salinization. Gibbs (1970) recommends two diagrams to evaluate the dominant phenomenon between precipitation, weathering or rock weathering and evaporation in the geochemical evolution of groundwater. The Gibbs diagrams are based on graphs of the ratios $Na^+ + K^+ / [Na^+ + K^+ + Ca^{2+}]$ et $Cl^- / [Cl^- + HCO_3^-]$ as a function of log TDS.

9.4 HYDROCHEMICAL PROPERTIES

9.4.1 Chemical Composition

The pH of the water moderates the physico-chemical equilibrium, in particular the calcium-carbonate equilibria and therefore the action of the water on the carbonates. The minimum and maximum values of the pH of groundwater vary respectively between 7.2 and 8.2. These data reflect a pH ranging from neutral to strong alkalinity. The latter can be linked to strong evaporation. The average pH value is 7.58±0.22. The electrical conductivity (EC) of the groundwater varies between 1,150 and 4,600 µS/cm with an average of 1724.33±659.81 µS/cm. This wide variation in EC data is due to the different geochemical processes that prevail in this field of study.

TDS values range from 713 to 2,852 mg/L with an average of 1069.09±409.08 mg/L. These values reflect relatively soft and mineralized waters (Freeze and Cherry, 1979). The total hardness (TH) of the waters, in $CaCO_3$ equivalent, ranges from 260 to 1,410 mg/L with an average of 479.70±214.61 mg/L. These results reflect relatively hard waters (Sawyer and McCarty, 1967; Rodier et al., 2009).

Calcium (Ca^{2+}) concentrations range from 48 to 398 mg/L and magnesium (Mg^{2+}) concentrations range from 19 to 100 mg/L. The average concentrations are 124.51±66.45 mg/L and 40.48±18.09 mg/L respectively. These values are to be related to the alluvial nature of the aquifer containing limestone fragments and the clay matrix responsible for the basic exchanges. The minimum Na^+ and K^+ contents evolve respectively between 64 and 1 mg/L and the highest reach 315 and 4.2 mg/L, respectively. The average Na^+ content is 132.88±64.48 mg/L. The high Na^+ contents come from cationic exchanges with clays and probably from pollution and sea spray from the Mediterranean Sea due to its proximity. Cl^- concentrations range from 37 to 875 mg/L and with an average of 224.38±148.25 mg/L. The Cl^- would come from a marine origin by aerosols insofar as the geology has not revealed the existence of saline rocks of evaporitic type. The Cl^- can be of anthropic origin due to wastewater discharges or from livestock farming.

Anthropogenic sources of Cl^- in groundwater could also be due to the influence of irrigation return flows and chemical fertilizers (Rao et al., 2012). Sulphates range from 66 to 326 mg/L (mean 119.82±56.55 mg/L). They come from various origins. They are mainly due to evaporitic rocks (gypsum, anhydrite), pyrite oxidation, atmospheric deposition, and decomposition of organic matter (Appelo and Postma, 2005). HCO_3^- with an average of 364.67±78.18 mg/L, have values ranging from 168 to 554 mg/L. The sources of bicarbonates are the dissolution of carbonate minerals, alteration of silicates, and pollution. According to Venkatramanan et al. (2016), they may come from the dissolution of CO_2 from the air or soil in water, and reach the aquifer during irrigation return flows, especially in arid and semi-arid agricultural regions.

The partial pressures in CO_2 (pCO_2) of the groundwater of the alluvial aquifer vary between −0.266 and −0.095 MPa and have an average of −0.171 MPa. The latter value is well above the atmospheric reference set at −0.354 MPa (Table 9.1). This high value suggests that the groundwater is confined in a hydrogeological system open to CO_2 from the soil. When pCO_2 values are high in agricultural soil, the potential for CO_2 degassing during runoff leads to a relative increase in pH and consequently to supersaturation of carbonate minerals such as calcite (Singh et al., 2011). Summary statistics of the physical and chemical parameters of groundwater are shown in Table 9.1.

TABLE 9.1
Summary Statistics of Groundwater Hydrochemical Data

Parameters	Units	Minimum	Maximum	Mean	Standard Deviation
pH	–	7.20	8.20	7.58	0.22
EC	µS/cm	1150.00	4600.00	1724.33	659.81
TDS	mg/L	713.00	2852.00	1069.09	409.08
Ca^{2+}	mg/L	48.00	398.00	124.51	66.45
Mg^{2+}	mg/L	19.00	100.00	40.48	18.09
Na^+	mg/L	64.00	315.00	132.88	64.48
K^-	mg/L	1.00	4.20	2.12	0.67
Cl^-	mg/L	37.00	875.00	224.38	148.25
SO_4^{2-}	mg/L	66.00	326.00	119.82	56.55
HCO_3^-	mg/L	168.05	553.88	364.67	78.18
NO_3^-	mg/L	18.00	500.70	70.82	79.74
TH ($CaCO_3$)	mg/L	260.00	1410.00	479.70	214.61
pCO_2	MPa	−0.266	−0.095	−0.171	0.51
CAI-I	–	−4.12	0.65	−0.14	0.85
CAI-II	–	−0.46	1.05	0.05	0.32
SI anhydrite	–	−2.23	−1.32	−1.76	0.22
SI aragonite	–	−0.12	1.05	0.44	0.25
SI calcite	–	0.02	1.19	0.59	0.25
SI dolomite	–	−0.09	2.38	1.05	0.51
SI gypsum	–	−2.01	−1.10	−1.54	0.22
SI halite	–	−6.95	−5.40	−6.24	0.33

The correlation matrix (Table 9.2) highlights some relationships that could explain certain mineralogical associations. For example, the EC of groundwater in the alluvial aquifer is significantly controlled in decreasing order by Cl^-, Mg^{2+}, NO_3^-, Ca^{2+} and Na^+. Ca^{2+} is correlated respectively to NO_3^- (0.783), Cl^- (0.728) and Mg^{2+} (0.512) while Mg^{2+} is related respectively to Cl^-, Na^+, NO_3^- and SO_4^{2-} with respective values of (0.797), (0.629), (0.627) and (0.429). Na^+ is also significantly correlated with HCO_3^-, SO_4^{2-} and Cl^- with values of (0.452), (0.449) and (0.440) respectively. The only significant and negative correlation encountered is that between K^+ and HCO_3^- (−0.409).

TABLE 9.2
Correlation Matrix

Variables	pH	EC	Ca^{2+}	Mg^{2+}	Na$^+$	K$^+$	Cl$^-$	SO$_4^{2-}$	HCO$_3^-$	NO$_3^-$
pH	1									
EC	0.125	1								
Ca^{2+}	−0.098	0.769	1							
Mg^{2+}	0.226	0.882	0.512	1						
Na$^+$	0.195	0.525	−0.104	0.629	1					
K$^+$	0.117	−0.125	0.028	−0.239	−0.244	1				
Cl$^-$	0.218	0.941	0.728	0.797	0.440	−0.020	1			
SO$_4^{2-}$	0.018	0.286	0.064	0.428	0.449	−0.209	0.044	1		
HCO$_3^-$	−0.145	0.242	0.057	0.279	0.452	−0.409	0.039	0.273	1	
NO$_3^-$	−0.050	0.817	0.783	0.627	0.202	−0.056	0.790	0.041	−0.096	1

9.4.2 Hydrochemical Types

The distribution of major ions in the groundwater of the alluvial aquifer is varied. Among all the chemical representations, three formulas, with the same percentage, dominate the ion procession and correspond to about 60% of the total. The characteristic formulas are: Ca^{2+}>Na$^+$>Mg^{2+}>K$^+$ et Cl$^-$>HCO$_3^-$>SO$_4^{2-}$>NO$_3^-$; Ca^{2+}>Na$^+$>Mg^{2+}>K$^+$ et HCO$_3^-$>Cl$^-$>SO$_4^{2-}$>NO$_3^-$; Na$^+$>Ca^{2+}>Mg^{2+}>K$^+$ et HCO$_3^-$>Cl$^-$>SO$_4^{2-}$>NO$_3^-$.

The projection of the water points according to the Piper diagram reveals that in the cation triangle, the sodium type and calcium type facies are present but the non-dominant cation factors (mixed type) are more abundant. As for the triangle representing the anions, we observe mainly the chloride-type factors and the non-dominant anion factors (mixed type) widely represented (Figure 9.3). Consequently, the chemical factors found in the groundwater of the alluvial aquifer at the place of the agricultural perimeter of Sidi Rached are (mixed) Ca - Mg - Cl type with 64%, Na - Cl type with 21%, Ca - HCO$_3$ type with 9% and Ca - Cl type with 6%. The first factor, the most widespread, indicates a mixture of high salinity, to sources of contamination from the surface and especially to irrigation returns, followed by the phenomenon of reverse ion exchange (Selvakumar et al., 2017). The second factor suggests an interaction with brackish water, the effect of evaporation, ion exchanges and significant contact time between water and rock (Ravikumar and Somashekar, 2017; Zaidi et al., 2015; Mahaqi et al., 2018). The third factor is attributed to freshwater recharge (Jalali, 2007) while the fourth is linked to ionic exchanges and salinisation (Walraevens and Van Camp, 2004; Zaidi et al., 2015).

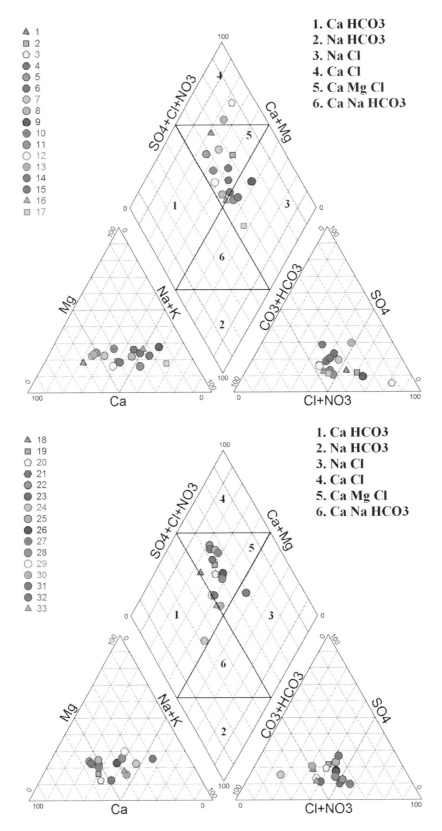

FIGURE 9.3 Groundwater facies according to the Piper diagram.

9.4.3 Marine Influence

Given the proximity of the Mediterranean Sea to the study area, it would be interesting to know its influence on groundwater chemistry. The geology has shown that the Sahel has an anticlinal configuration with a core of impermeable marls from the Piacenzian. It follows that marine intrusion by freshwater-seawater contact is unlikely. The explanation lies in the fact that as soon as the Sahel anticline was formed, it isolated the plain from the sea at the end of the Pliocene (Glangeaud, 1952; Rivoirard, 1952). The influence of marine aerosols is rather omnipresent and largely dependent on winds coming from the Mediterranean Sea. In the study of this marine influence, Cl and EC levels are the simplest indicators (Mercado, 1985; El Moujabber et al., 2006). Other authors recommend referring to rainwater chemistry or Cl-based reports (Sarin et al., 1989; Zhang et al., 1995; Singh et al., 2013; Khaska et al., 2013). In the absence of rainwater chemistry, we have referred to marine aerosol ratios, which are as follows: Na/Cl=0.85; Mg/Cl=0.1; K/Cl=0.02; Ca/Cl=0.02; SO_4/Cl=0.05 et NO_3/Cl=10^{-9} (Figure 9.4).

Applying the Na/Cl ratio gives values between 0.35 and 5.08. The average value is 1.13. Comparing the value of the ratio (0.85) with our data, the data presented herein show that 18 samples have a value lower or slightly higher than this reference (Figure 9.4). It follows that these lower values compared to seawater are the result of the cation exchange that occurs when seawater contaminates freshwater aquifers, with a deficit of Na^+ compensated by an excess of Ca^{2+}. This contamination is probably due to the current sea or the presence of deep saline water or sea spray. In the present case study, sea spray is responsible for this contamination. In situations where the Na/Cl ratio is greater than 1, this indicates that the groundwater is contaminated by an anthropogenic source. There are 14 water points in the study area, i.e., 42.42%, with a value greater than unity.

The minimum and maximum values of the Ca/Cl, Mg/Cl, SO_4/Cl et NO_3/Cl, ratios for the study area are (0.23–3.3; 0.31–2.44; 0.08–1.7; 0.05–0.83) respectively. The minimum values of these ratios are all in excess of the respective guide data by 0.02; 0.1; 0.05 and 10^{-9}. These results suggest an origin of Ca^{2+}, Mg^{2+}, SO_4^{2-} and NO_3^- different from the marine origin.

Furthermore, the fact that SO_4^{2-} is far from being correlated to Cl^- means that they do not come from the same source. They are to be linked to a geological origin and above all an exogenous origin from fertilisers and wastewater discharges. It is observed that SO_4^{2-} is significantly correlated to Na^+ and Mg^{2+} cations which reinforces our hypothesis.

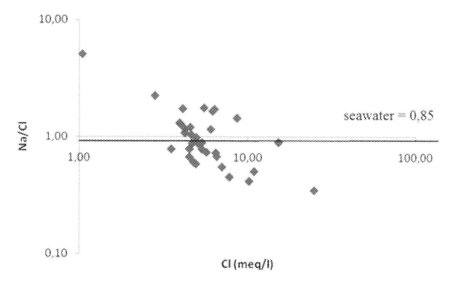

FIGURE 9.4 Molar ratio Cl vs. Na/Cl concentrations.

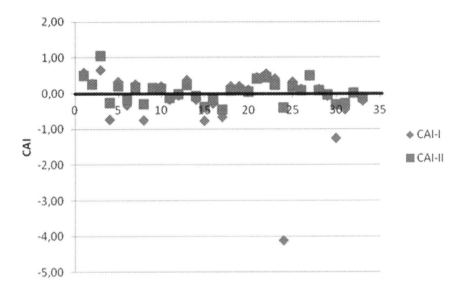

FIGURE 9.5 Representation of the chloro-alkaline indices.

El Moujabber et al. (2006) and Al Farrah et al. (2011) explained the Simpson report as $Cl/(HCO_3+CO_3)$. The report is divided into five classes: the first is of good quality (<0.5); the second is slightly contaminated (0.5–1.3); the third is moderately contaminated (1.3–2.8); the fourth is highly contaminated (2.8–6.6); and the fifth is highly contaminated (6.6–15.5). Applying this report to the data reveals that 25 samples, or 75.8%, belong to the lightly contaminated class. The moderately contaminated class is represented by 15.2%. The remainder is divided between the good (6%) and highly contaminated (3%) class.

9.4.4 Chloro-Alkaline Indices

The chloro-alkaline indices (CAI-I and CAI-II) obtained in this study show contrasting values. The minimum and maximum data for CAI-I and CAI-II vary between −4.12 and 0.65 and between −0.46 and 1.05 respectively. The average values are −0.14 for CAI-I and 0.05 for CAI-II. The majority of the indexes range between +1 and −1 with a slight dominance of positive values over negative ones (Figure 9.5). These results underline that in the alluvial aquifer, which is heterogeneous in nature, cation exchange takes place according to both reactions. Groundwater is sometimes enriched in Ca^{2+}, Mg^{2+} and sometimes in Na^+, K^+ with a slight dominance of the reverse ion exchange.

9.4.5 Saturation Index

The saturation index (SI) obtained shows that globally two types of minerals differ from zero (Figure 9.6). Indeed, the carbonate minerals (calcite, dolomite, and aragonite) have, in their majority, values higher than zero indicating oversaturation. The carbonate minerals show that the SI of calcite, dolomite and aragonite, range respectively between 0.02 and 1.19, between −0.09 and 2.38, and between −0.12 and 1.05 with respective averages of 0.59; 1.05 and 0.44. Of the 33 water samples analysed, almost all are in a state of supersaturation (SI>0) with respect to calcite with only one case close to zero (0.02). Dolomite appears supersaturated (SI>0) in 97% of the cases and only one sample is close to zero (−0.09) which represents 3% of the total. As for aragonite, 31 water points, or 94%, have a positive saturation index (SI>0) and two samples, or 6%, have negative values close to zero.

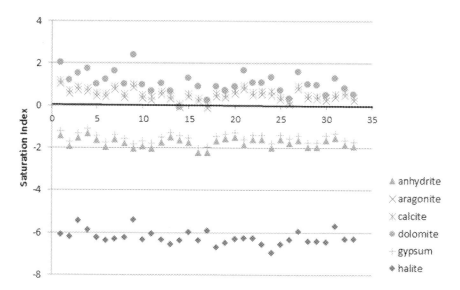

FIGURE 9.6 Groundwater saturation index.

In general, the majority of the samples analysed have slightly supersaturated carbonate mineral saturation indices, such as calcite, whose SI is mainly between 0 and 1.8, dolomite between 0 and 1.7 and aragonite between 0 and 1.2. This slight supersaturation could be explained by CO_2 degassing during runoff in agricultural soils, as pointed out by Singh et al. (2011). The SI values of evaporitic minerals show that anhydrite, gypsum and halite vary respectively between −2.23 and −1.32, −2.01 and −1.10 and −6.95 and −5.4 with respective averages of −1.76, −1.54 and −6.24 (Table 9.1). These results clearly show that the groundwater in the region of Sidi Rached is in a state of under saturation with respect to these minerals and consequently the possibility of dissolution will be more.

9.5 MECHANISM CONTROLLING GROUNDWATER CHEMISTRY

The sample projection on the Gibbs diagram (Figure 9.7) shows two phenomena controlling the geochemical processes and groundwater evolution in the study area. The dots overlap on two zones which are rock weathering and evaporation. The diagram highlights the influence of Ca^{2+}, Na^+ and Cl^- ions on TDS. The correlation matrix showed very significant statistical relationships (Table 9.2) of the EC, equivalent to TDS, with these ions. Concerning the rock weathering phenomenon, Figure 9.7 shows that rock weathering is an important source that controls groundwater chemistry and its evolution. Rock weathering facilitates the process by which salts and soluble minerals are incorporated into groundwater. According to Marandia and Shand (2018), unlike surface water, groundwater has a long residence time and therefore the rock-water interaction facilitates the dissolution of minerals. The sedimentary nature where clays predominate, the relatively flat relief, high temperatures, are all parameters that promote the degradation of rocks by chemical means. The near horizontal disposition of a large number of samples, with a relatively small variation in TDS could be explained by the phenomenon of cation exchange. The second phenomenon relating to evaporation is explained by the severity of the semi-arid climate marked by high temperatures generating strong evaporation. Although the water depth is relatively far from the soil surface, we believe that the return flows of irrigation water are affected by evaporation. The latter increases salinity by increasing Na^+, Cl^- and consequently TDS. The anthropogenic factor could also intervene through agricultural practices, domestic discharges or leaks in sewerage networks (Srinivasamoorthy et al., 2014; Selvakumar et al., 2017) by increasing Na^+ and Cl^- and therefore TDS, with the spreading of cation and anion ratios (Rao and Rao, 2009). The semi-arid climate, gentle slope, lack of good drainage conditions and longer groundwater residence time contribute directly or indirectly to groundwater quality (Rao, 2006).

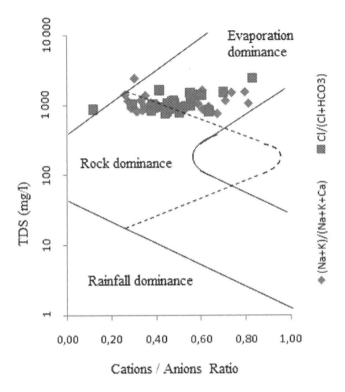

FIGURE 9.7 Projection of the samples on the Gibbs diagram.

9.6 INTERACTION AND GEOCHEMICAL EVOLUTION

The interaction and geochemical evolution can be identified by the establishment of characteristic ion ratios and graphs. The Na/Cl relationship is widely used, moreover, it has been subject to the influence of the sea. Its use can be extended to explain the effect of evaporation, listed at the level of the Gibbs diagram. If evaporation and evapotranspiration processes are dominant, assuming that no mineral species precipitate, the Na/Cl ratio would be unchanged (Jankowski and Acworth, 1997). Therefore, the plot of Na/Cl relative to EC would give a horizontal line. The relationship between EC and Na/Cl in groundwater was plotted and gave evidence of the evaporation phenomenon. The plot of EC versus Na/Cl (Figure 9.8a) shows that the trend line is slightly inclined with increasing

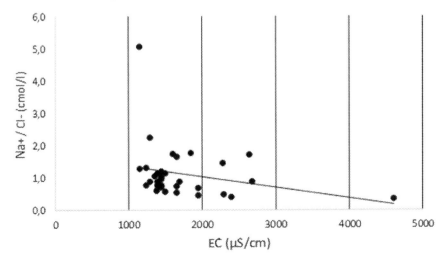

FIGURE 9.8a Scatter plot of Na^+/Cl^- vs EC.

EC, indicating that evaporation and evapotranspiration are not the dominant process. In the opinion of Zaidi et al. (2015), the trend line is related to the reverse ion exchange where groundwater becomes depleted in Na^+ but enriched in Ca^{2+} and Mg^{2+}.

It is admitted that the dissolution of halite in water releases a concentration of Na^+ equal to that of Cl^- and consequently their projection on a graph gives a 1:1 slope line. The data of the Sidi Rached perimeter projected on the graph Cl=function (Na) reveal points deviating from the 1:1 slope line (Figure 9.8b), moreover the correlation coefficient is only 0.44 (p=0.05). These results assume anthropogenic contamination (agricultural pollution, domestic waste, wastewater discharges). Cationic exchange between Ca, Mg and Na is also at the origin of the high Na contents. The minimum and maximum values are 0.35 and 5.08 respectively. The average is 1.13, while the percentages of samples above and below 1 are 42.42 and 57.57 respectively. According to Marghade et al. (2012), samples with a Na/Cl ratio higher than 1 indicate a significant deficit in Ca^{2+} and Mg^{2+} consistent with the cation exchange process. Nevertheless, about 58% of the samples have a ratio lower than 1, which reflects a dominance of Cl^- over Na^+, thus a deficit that can be compensated for by the alkaline earths. The cation exchanges are in both directions.

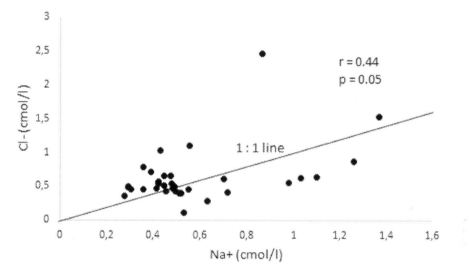

FIGURE 9.8b Scatter plot of Cl^- vs Na^+.

The Ca/Mg ratio shows values ranging from a minimum of 0.5 to a maximum of 4.4. The average is 2.02. In percentage terms, 57.57% (19 samples) have a ratio greater than 2; 27.27% (9 water points) between 1 and 2, and 12.12% (4 samples) have ratios less than unity. There is 1 water point close to 1 (1.01). From these figures, it can be seen that the Ca^{2+} content largely dominates the Mg^{2+}. The presence of $CaCO_3$ is undeniable, all the more so as fragments of limestone and marl are present in the lithology of the alluvial aquifer. If the Ca/Mg ratio is equal to 1, the dissolution of the dolomite should occur, whereas a higher ratio would result in a large contribution from calcite (Maya and Loucks, 1995). In the opinion of Prasanna et al. (2010), a Ca/Mg ratio between 1 and 2 is related to the dissolution of calcite. If the ratios are less than unity (12.12%), they are probably attributed to the dissolution of the dolomite. The Ca/Mg ratios are related to cationic exchanges and to various origins (agricultural inputs, livestock waste, domestic waste, etc.). The relationship between Ca+Mg and HCO_3 (Figure 9.8c) shows that the majority of points are below the 1:1 slope line although some points are close to the line which means overall an excess of bicarbonates over alkaline earths. The groundwater has become enriched in HCO_3. In the opinion of Sayeh et al. (2022), the latter would probably come from CO_2 resulting from the decomposition of organic matter contained in the soil rhizosphere on the one hand and from the alteration of silicate minerals of the Astian formation on the other hand. Anthropic action is very probable in cationic exchange in this case.

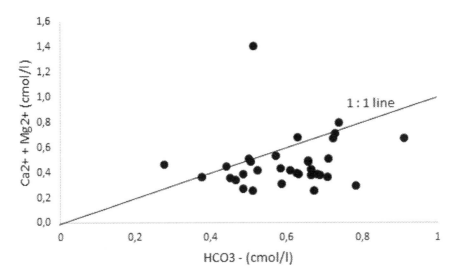

FIGURE 9.8c Scatter plot of $(Ca^{2+}+Mg^{2+})$ vs (HCO_3^-).

The Na+K/Ca+Mg straight line is often used as was done by Vasu et al. (2017). The plot of the line (Figure 9.8d) shows the consequent influence of silicate weathering. According to Drever (1997), a Na+K/Ca+Mg ratio of more than 0.5 is attributed to silicate mineral weathering. The application of this ratio to the data of the alluvial aquifer of Sidi Rached shows that all the points are related to this form of alteration.

The dissolution of gypsum can cause the presence of the cation Ca^{2+} in the groundwater. The plot of the Ca^{2+} vs. SO_4^{2-} line ($r=0.064$, $p=0.05$) shows that most of the water points sampled are not close to 1:1 (Figure 9.8e), indicating that the groundwater has a deficit of SO_4^{2-} and an excess of Ca^{2+}. According to the saturation indices, the waters are sub-saturated with respect to gypsum, whereas they are saturated to over-saturated with respect to carbonate minerals.

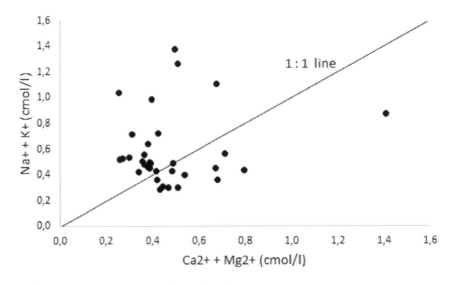

FIGURE 9.8d Scatter plot of $(Ca^{2+}+Mg^{2+})$ vs (Na^++K^+).

Groundwater's Geochemical Status

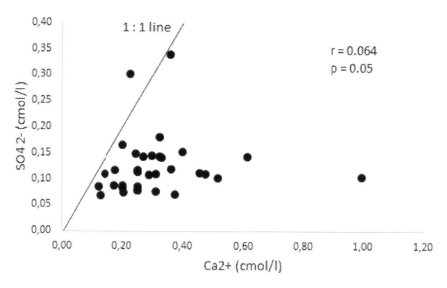

FIGURE 9.8e Scatter plot of Ca^{2+} vs SO_4^{2-}.

9.7 SUSTAINABLE WATER MANAGEMENT IN AGRICULTURE

Agriculture is an important element of rural development in Algeria and is considered one of the major components of the national economy, representing 25% of the labor force and contributing approximately 10% of the GDP (Laoubi and Yamao, 2012). For about 30 years, drastic socioeconomic changes have occurred in the Mitidja region, and the population has doubled. Agriculture must face problems of crops reconversion, crops intensification, food security and water requirements for industry and drinking water supply (Figure 9.9). Meeting national food security needs remains a challenge as the demand for food and water increases at an even faster rate than population growth.

Groundwater chemistry data were used to clarify the geochemical behavior of the alluvial groundwater in the areas characterized by human-intensive activities. The Sidi Rached basin situated in the western Mitidja represents the most emblematic case because it includes all the factors mentioned. The Mitidja rural area is one of the main agricultural plains in the Mediterranean basin where water supplies are mainly provided by groundwater resource (Laoubi and Yamao, 2009), where the strategic plan for sustainable development is a more than urgent alternative. Sustainable management of water for agriculture and human use.

9.7.1 Option for Water Management in Condition of Scarcity

9.7.1.1 Enhancing Crop Water Productivity

Water-related plant breeding is also the subject of research, like resistance to dry spells, reduced respiration rates, increase of the harvest ratio, enhancement of the photosynthetic efficiency or change in routing growing cycles to better match training season.

9.7.1.2 Instrument for Water Scarcity Management Irrigation

Tree instruments are available to improve water management or "supply management" consisting of developing the tree and necessary infrastructure to allow for safe and reliable use of the resources.

In the Mitidja basin characterized by water scarce, demand management is the immediate step following supply. Traditional irrigation in the region consumes only a fraction of the water it withdraws, while not all that all water is lost, inefficient management of water in irrigation usually translate into inequitable water distribution irrigation schemes, ineffective water delivery

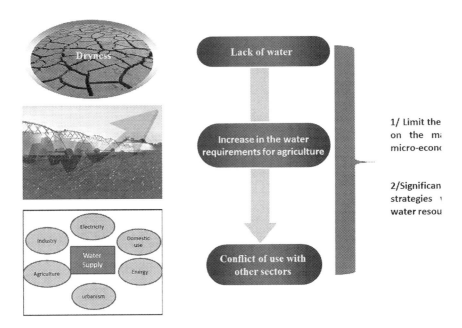

FIGURE 9.9 Current situation of water supply.

waterlogging in lower area and problems of groundwater mineralization in the most region part. All these factors affect the average yield and restrain the overall productivity of the irrigation infrastructures.

9.7.1.3 Water Productivity

The last, and necessary step in conditions of water scarcity, is to increase the productivity of irrigated water, here, The objective is not only to control the use of explosives losses but make sure that the return on water is optimized. This is usually obtained through truth institutional schemes, with an increase in utility authority given to the water users, coupled with technological upgrading allowing for a higher level of flexibility in water delivery. Access technologies are also a major condition for improved demand management and increased water productivity.

9.8 LAND URBANIZATION, AGRICULTURE AND WATER QUALITY

The Mitidja Basin has shown a rapid urbanization development trend. The analysis shows that their impact on water quality variables was obvious. Through the analysis of land use change, water quality change and their relationship, the main culprit of water quality deterioration is the transformation mode from agricultural land to urban land (Lounsi et al., 2022). Agricultural activities are still important sources of water pollution, so the proportion of agricultural land is positively related to the concentration of water pollutants. On the other hand, in fast urbanizing watersheds, agricultural activities may be a negligible pollution source (Chen et al., 2020). In this case, urban land may produce many non-point source pollution, which is the main pollution source.

In this case, a good land-use plan that takes into consideration aspects related to the problems of pollution of agricultural and urban origin is necessary. The solutions provided in this type of plan must essentially address the problem of urban sprawl by proposing integrated actions to reduce this pollution. An urban area with a reliable and downstream sewerage system to reduce the use of septic tanks, with a wastewater treatment plant will considerably reduce the pollution of groundwater and rivers.

For agriculture, good management of water and fertilizers is more than necessary. A surplus of irrigation leads the water loaded in excess to percolate in the groundwater. Irrigation doses must be given to the crops set up according to a scheme based on a knowledge of the climate, water and soil, which will allow a good management of irrigation.

Finally, regarding water resources sustainability, our thought should be directed to the sustainable use and management of the resources and the way they should be governed.

The solution is the adoption of a holistic integrated natural resources (soil and water) management approach with the promotion of the Mitidja basin level management and participatory approach. An appropriate water policy and economic framework under a good land-use plan are needed to promote natural conservation, social equity, food security to allocate efficiency of scarce water resources as well as environmental condition and human resources development.

9.9 CONCLUSIONS

The results of this study, undertaken in a peri-urban area within an agricultural perimeter close to the coastline subject to a semi-arid climate, reveal an insight into the knowledge of the geochemical processes to which groundwater is subject. The aquifer is characterized by a heterogeneous lithology marked by alternating gravelly, sandy and clay beds. Water mineralisation is significantly influenced by the Ca^{2+}, Mg^{2+}, Na^+, Cl^- and NO_3^- contents. The dominant ions are Ca^{2+} and Na^+ for cations and Cl^- and HCO_3^- for anions. The chemical facies of the waters of the alluvial aquifer are of four types: (mixed) Ca - Mg - Cl type (64%); Na - Cl type (21%); Ca - HCO_3 type (9%) and Ca - Cl type (6%). The use of a few marine aerosol ratios showed that for the Na/Cl ratio, about half of the samples are under the influence of the sea while the other ratios are not. Simpson's report highlighted a dominant class (75%) called slightly contaminated followed by the moderately contaminated class (15%). The chloro-alkaline indices have highlighted that the groundwater of the alluvial groundwater is sometimes enriched in Na^+, K^+ by ion exchange, sometimes in Ca^{2+}, Mg^{2+} by reverse ion exchange and the latter slightly predominates.

The study of saturation indices has shown that carbonate minerals (calcite, dolomite and aragonite) are in a state of saturation to supersaturation while evaporite minerals (gypsum, anhydrite, halite) are in a state of under saturation. Gibbs diagrams indicate that the mechanisms controlling groundwater chemistry have highlighted the importance of rock weathering, evaporation and possibly anthropogenic action. The interaction and geochemical evolution evaluated from ion ratios and graphs confirmed the effect of evaporation and transpiration, ion exchange in particular inverse ion exchange, the influence of lithology and anthropogenic action.

The results obtained can be further investigated by other approaches, such as isotopes. For the time being, they may be useful for undertaking sustainable management of groundwater resources, particularly in the current climate change situation. These waters, although acceptable for irrigation, must be used on crops with high water productivity and on tolerant crops to the chemical specificity of these waters. The use of soil drainage techniques is an option in this type of situation. The future of the agricultural basin of the Mitidja depends strongly on the effective, efficient and even sustainable management of natural resources (soil and water). The future of the quality of its resources is closely linked to the capacity managers of stakeholders and practices used by farmers and finally, it depends on the entropic human action responsible in part for the degradation of these resources.

LIST OF ABBREVIATIONS

APHA   American Public Health Association
CAI   Chloro-alkaline indices
EC   Electrical conductivity
EDTA   Ethylenediamine tetraacetic acid
EEA   European Environment Agency

GDP	Gross domestic product
IAP	Ion activity product
MPa	Megapascal
NWRA	National Water Resources Agency
pCO$_2$	Partial pressure of carbon dioxide
SI	Saturation index
T	Temperature
TDS	Total dissolved solids
TH	Total hardness

REFERENCES

Al-Ahmadi, M.E., and El-Fiky, A.A. (2009) Hydrogeochemical evaluation of shallow alluvial aquifer of Wadi Marwani, western Saudi Arabia. *J King Saud Univ* 21, 179–190.

Al Farrah, N., Martens, K., and Walraevens, K. (2011) Hydrochemistry of the Upper Miocene-Pliocene-Quaternary aquifer complex of Jifarah plain, NW-Libya. *Geol Belgica* 14(3–4), 159–174. https://popups.uliege.be/1374-8505/index.php?id=3325.

APHA. (2005) *Standard Methods for Examination of Water and Wastewater*, 21st ed. American Public Health Association, Washington, DC.

Appelo, C., and Postma, D. (2005) *Geochemistry, Groundwater and Pollution*, 2nd ed. Balkema, Rotterdam. https://doi.org/10.1201/9781439833544.

Bachir, H., Etsouri, S., Smadhi, D., and Semar, A. (2022) Representation of rainfall in regions with a low distribution of rain gauging stations. In International Conference on Natural Resources and Sustainable Environmental Management, pp. 262–271. Springer, Cham. https://doi.org/10.1007/978-3-031-04375-8_30.

Bachir, H., Kezouh, S., Semar, A., Smadhi, D., and Ouamer-ali, K. (2021) Improvement of interpolation using information from rainfall stations and comparison of hydroclimate changes (1913–1938)/(1986–2016). *Al-Qadisiyah J Agric Sci* 11(1), 54–67.

Bachir, H., Semar, A., and Mazari, A. (2016) Statistical and geostatistical analysis related to geographical parameters for spatial and temporal representation of rainfall in semi-arid environments: The case of Algeria. *Arab J Geosci* 9(7), 1–12. https://doi.org/10.1007/s12517-016-2505-8.

Barbieri, M., Nigro, A., and Petitta, M. (2017) Groundwater mixing in the discharge area of San Vittorino Plain (Central Italy): Geochemical characterization and implication for drinking uses. *Environ Earth Sci* 76, 393. https://doi.org/10.1007/s12665-017-6719-1.

Boudiaf, A. (1996) Etude sismotectonique de la région d'Alger et de Kabylie (Algérie) : Utilisation des modèles numériques de terrains (MNT) et de la télédétection pour la reconnaissance des structures tectoniques actives; contribution à l'évaluation de l'aléa sismique. Thèse Doctorat, Univ. Sciences et Technique du Languedoc, France, 268 p.

Chen, D., Elhadj, A., Xu, H., Xu, X., and Qiao, Z. (2020) A study on the relationship between land use change and water quality of the Mitidja watershed in Algeria based on GIS and RS. *Sustainability* 12(9), 3510.

Domenico, P.A., and Schwartz, F.W. (1998) *Physical and Chemical Hydrogeology*, 2nd ed. Wiley, New York, 506 p

Dragon, K., and Gorski, J. (2015) Identification of groundwater chemistry origins in a regional aquifer system (Wielkopolska region, Poland). *Environ Earth Sci* 73, 2153–2167. https://doi.org/10.1007/s12665-014-3567-0.

Drever, J.I. (1997) *The Geochemistry of Natural Waters: Surface and Groundwater Environments*. Prentice-Hall, Hoboken, NJ, p. 436

El Moujabber, M., Bou Samra, B., Darwish, T., and Atallah, T. (2006) Comparison of different indicators for groundwater contamination by seawater intrusion on the Lebanese coast. *Water Res Manag* 20, 161–180. https://doi.org/10.1007/s11269-006-7376-4.

Eloïse, C (2017) Pollution agricole des ressources en eau: approches hydrogéologique et économique. Optimisation et contrôle. Thèse Doct. 268p.

European Environment Agency (2012) Climate Change, Impacts and Vulnerability in Europe 2012 – An Indicator-Based Report. EEA Report No 12/2012.

Freeze, R.A., and Cherry, J.A. (1979) *Groundwater*. Prentice-Hall, Englewood Cliffs, NJ, 604 p.

Gibbs, R.J. (1970) Mechanisms controlling world water chemistry. *Sci N Ser* 170(3962), 1088–1090.

Glangeaud, L. (1952) Histoire géologique de la Province d'Alger. *XIX Congrès International. Monographies régionales. Serie n°01. N°01. Alger.*

Gupta, S., Mahato, A., Roy, P. et al. (2008) Geochemistry of groundwater, Burdwan District, West Bengal, India. *Environ Geol* 53, 1271–1282. https://doi.org/10.1007/s00254-007-0725-7.

Hadjoudj, O., Bensemmane, R., Saoud, Z., and Reggabi, M. (2014) Pollution des eaux souterraines de la Mitidja par les nitrates: État des lieux et mesures correctives. *Eur J Water Qual* 45, 57–68.

Hosseinifard, S.J., and Mirzaei Aminiyan, M. (2015) Hydrochemical characterization of groundwater quality for drinking and agricultural purposes: A case study in Rafsanjan plain. *Iran Water Qual Expos Health* 7(4), 531–544.

Jalali, M. (2007) Hydrochemical identification of groundwater resources and their changes under the impacts of human activity in the chah basin in Western Iran. *Environ Monit Assess* 130, 347–364. https://doi.org/10.1007/s10661-006-9402-7.

Jankowski, J., and Acworth, R.I. (1997) Impact of depris-flow deposits on hydrogeochemical processes and the development of dryland salinity in the Yass River catchment, New South Wales, Australia. *Hydrogeol J* 5(4), 71–88.

Khaska, M., Le Gal La Salle, C., Lancelot, J., Team, A., Mohamad, A., Verdoux, P., Noret, A., and Simler, R. (2013) Origin of groundwater salinity (current seawater vs. saline deep water) in a coastal karst aquifer based on Sr and Cl isotopes. Case study of the La Clape massif (southern France). *Appl Geochem* 37, 212–227. https://doi.org/10.1016/j.apgeochem.2013.07.006.

Laoubi, K., and Yamao, M. (2009) Irrigation schemes management in Algeria: An assessment of water policy impact and perspectives on development. *Wat. Res. Manag.* V.C.A. Brebbia & V. Popov (eds.), WIT Press: Southampton, UK, pp. 503–514.

Laoubi, K., and Yamao, M. (2012) The challenge of agriculture in Algeria: Are policies effective. *Bull Agric Fisher Econ* 12(1), 65–73.

Lounsi, A., Aidaoui, A., and Hartani, T (2022) Irrigation water quality in the western Mitidja, Algeria: A case study of Sidi Rached watershed. *Farm Manag* 7(2), 72–79.

Mahaqi, A., Moheghi, M.M., Mehiqi, M., et al. (2018) Hydrogeochemical characteristics and groundwater quality assessment for drinking and irrigation purposes in the Mazar-i-Sharif city, North Afghanistan. *Appl Water Sci* 8, 133. https://doi.org/10.1007/s13201-018-0768-9.

Marandia, A., and Shand, P. (2018) Groundwater chemistry and the Gibbs Diagram. *Appl Geochem* 97, 209–212.

Marghade, D., Malpe, D.B., and Zade, A.B. (2012) Major ion chemistry of shallow groundwater of a fast growing city of Central India. *Environ Monit Assess* 184, 2405–2418. https://doi.org/10.1007/s10661-011-2126-3.

Maya, A.L., and Loucks, M.D. (1995) Solute and isotopic geochemistry and groundwater flow in the Central Wasatch Range, Utah. *J Hydrol* 172, 31–59.

Meghraoui, M. (1991) Blind reverse faulting system associated with the Mont Chenoua Tipasa earthquake of the 29th October 1989 (North Central Algeria). *Terra Nova* 3, 84–93.

Mercado, A. (1985) The use of hydrogeochemical patterns in carbonate sand and sandstone aquifers to identify intrusion and flushing of saline waters. *Ground Water* 23, 635–645

Messaoud Nacer, N. (1987) Hydrogéologie et pollution des eaux: exemple du bassin versant du Mazafran, Mitidja (Algérie). Thèse Doctorat 3ème cycle en Géologie appliquée. Université Scientifique et Médicale de Grenoble, p. 240.

Parkhurst, D.L., and Appelo, C.A.J. (1999) User's guide to PHREEQC (Version 2) A computer program for speciation, batch-reaction, one-dimensional transport, and inverse geochemical calculations: U.S. Geological Survey Water-Resources Investigations Report 99-4259, 312.

Peters, N. E. and Meybeck, M. (2000) Water quality degradation effects on freshwater availability: Impacts of human activities. *Water Intern* 25(2), 185–193.

Prasanna, M.V., Chidambaram, S., Senthil Kumar, G., Ramanathan, A.L., and Nainwal, H.C. (2010) Hydrogeochemical assessment of groundwater in Neyveli Basin, Cuddalore District, South India. *Arab J Geosci* 4(1–2), 319–330.

Rao, N.S. (2006) Seasonal variation of groundwater quality in a part of Guntur district, Andhra Pradesh, India. *Environ Geol* 49, 413–429.

Rao, N.S. and Rao, P.S. (2009) - Major ion chemistry of groundwater in a river basin: A study from India. *Environ Earth Sci* 61(4), 757–775.

Rao, N.S., Rao, P.S., Reddy, G.V. et al. (2012) Chemical characteristics of groundwater and assessment of groundwater quality in Varaha River Basin, Visakhapatnam District, Andhra Pradesh, India. *Environ Monit Assess* 184, 5189–5214. https://doi.org/10.1007/s10661-011-2333-y.

Ravikumar, P. and Somashekar, R.K. (2017) Principal component analysis and hydrochemical facies characterization to evaluate groundwater quality in Varahi river basin, Karnataka state, India. *Appl Water Sci* 7, 745–755. https://doi.org/10.1007/s13201-015-0287-x.

Rivoirard, R. (1952) Aperçu sur l'hydrogéologie de la Mitidja. In *XIX Congrès Géologique International. La géologie et les problemes de l'eau. Tome II, Données sur l'hydrogeologie Algérienne*. Alger.

Rodier, J., Legube, B., and Merlet, N. (2009) *L'analyse de l'eau*, 9ème ed. Dunod, DL Paris.

Sarin, M.M., Krishnaswamy, S., Dilli, K., Somayajulu, B.L.K., and Moore, W.S. (1989) Major ion chemistry of the Ganga-Brahmaputra river system: Weathering processes and fluxes to the Bay of Bengal. *Geochim Cosmochim Acta* 53, 997–1009. https://doi.org/10.1016/0016-7037(89)90205-6.

Sawyer, C.N. and McCarty, P.L. (1967) *Chemistry for Sanitary Engineers*. McGraw-Hill, New York.

Sayeh, Y., Semar, A., and Bachir, H. (2022) Hydrochemistry and nitrate pollution of groundwater in the alluvial aquifer of the Eastern Mitidja (Algeria). *Al-Qadisiyah J Agric Sci* 12(1), 1–12. https://doi.org/10.33794/qjas.2022.132235.1018. https://jouagr.qu.edu.iq/.

Sbargoud, S., Hartani, T., Aidaoui, A., Herda, F., and Bachir, H. (2017) Assessment of groundwater vulnerability to nitrate based on the optimised drastic models in the GIS environment (case of Sidi Rached Basin, Algeria). *Geosciences* 7(2), 20.

Schoeller, H. (1965) Qualitative evaluation of groundwater resources. In Schoeller, H., Ed., *Methods and Techniques of Ground-Water Investigations and Development*. Water Resource Series No. 33, UNESCO, Paris, 44–52. pp. 54–83.

Sekkal, R. (1986) Hydrologie de la nappe de la Mitidja (Algérie): étude hydrodynamique des champs captants de la ville d'Alger. In *Géologie Appliquée*. Université Scientifique et Médicale de Grenoble. Ed .Hall open scienceeè. 207p.

Rached Sekkal. Hydrologie de la nappe de la Mitidja (Algérie) : étude hydrodynamique des champs captants de la ville d'Alger. Géologie appliquée. Université Scientifique et Médicale de Grenoble, 1986. Français.

Selvakumar, S., Chandrasekar, N., and Kumar, G. (2017) Hydrogeochemical characteristics and groundwater contamination in the rapid urban development areas of Coimbatore, India. *Water Res Indus* 17, 26–33. https://doi.org/10.1016/j.wri.2017.02.002.

Senthilkumar, M., Elango, L. (2013) Geochemical processes controlling the groundwater quality in lower Palar river basin, southern India. *J Earth Syst Sci* 122, 419–432. https://doi.org/10.1007/s12040-013-0284-0.

Simler, R. (2012) *Logiciel Diagramme*. Laboratoire d'hydrogéologie, Université Avignon France.

Singh, A.K., Tewary, B.K., and Sinha, A. (2011) Hydrochemistry and quality assessment of groundwater in Part of NONDA Metropolitan City, Uttar Pradesh. *J Geol Soc India* 78(6), 523–540.

Singh, A.K., Raj, B., Tiwari, A.K., et al. (2013) Evaluation of hydrogeochemical processes and groundwater quality in the Jhansi district of Bundelkhand region, India. *Environ Earth Sci* 70, 1225–1247. https://doi.org/10.1007/s12665-012-2209-7.

Smadhi, D., Zella, L., Amirouche, M., Bachir, H., and Semiani, M. (2022) Monthly rainfall variability and vulnerability of rainfed cereal crops in the tellian highlands of Algeria. In International Conference on Natural Resources and Sustainable Environmental Management. Springer, Cham, pp. 240–251.

Srinivasamoorthy, K, Gopinath, M., Chidambaram, S., Vasanthavigar, M., and Sarma, V.S. (2014) Hydrochemical characterization and quality appraisal of groundwater from Pungar sub basin, Tamilnadu, India. *J King Saud Uni Sci* 26(1), 37–52.

Todd, D.K., and Mays, L.M. (2005) *Groundwater Hydrology*, 3rd ed. Wiley, Hoboken, NJ.

Vasu, D., Singh, S.K., Tiwary, P., Sahu, N., Ray, S.K., Butte, P., and Duraisami, V.P. (2017) Influence of geochemical processes on hydrochemistry and irrigation suitability of groundwater in part of semi-arid Deccan Plateau, India. *Appl Water Sci* https://doi.org/10.1007/s13201-017-0528-2.

Venkatramanan, S., Chung, S.Y., Ramkumar, T., Rajesh, R., and Gnanachandrasamy, G. (2016) Assessment of groundwater quality using GIS and CCME WQI techniques: A case study of Thiruthuraipoondi city in Cauvery deltaic region, Tamil Nadu, India. *Desal Water Trea* 57(26), 12058–12073. https://doi.org/10.1080/19443994.2015.1048740.

Walraevens, K., and Van Camp, M. (2004) Advances in understanding natural groundwater quality controls in coastal aquifers. In *Proceedings of the 18th SaltWater IntrusionMeeting (SWIM), Cartagena, Spain, 31 May–3 June 2004*, pp. 451–460.

Yi, L., JiuJimmy, J., Wenzhao, L., and Xingxing, K. (2017) Hydrogeochemical characteristics in coastal groundwater mixing zone. *App Geoch* 85(A), 49–60. https://doi.org/10.1016/j.apgeochem.2017.09.002.

Zaidi, F.K., Nazzal, Y., Jafri, M.K. et al. (2015) Reverse ion exchange as a major process controlling the groundwater chemistry in an arid environment: A case study from northwestern Saudi Arabia. *Environ Monit Assess* 187, 607. https://doi.org/10.1007/s10661-015-4828-4.

Zhang, J., Huang, W.W., Letolle, R., and Jusserand, C. (1995) Major element chemistry of the Huanghe (Yellow river), China-weathering processes and chemical fluxes. *J Hydrol* 168, 173–203. https://doi.org/10.1016/0022-1694(94)02635-O.

10 Saharan Agriculture

Strategic Choice, Environmental Issues in a Perspective of Sustainable and Resilient Agriculture. Case of Northeastern Sahara of Algeria

Hakim Bachir, Ahcène Semar, Nour el houda Abed, Wassima Lakhdari, Dalila Smadhi, and Rattan Lal

10.1 INTRODUCTION

Algeria, as well as developing countries, has reoriented the concept of food security through institutional and administrative measures and policies aimed at improving the access of populations to abundant and healthy food. Thus, it is the availability of food, the accessibility to this food, the proper use, and the stability and safety of supplies (Amrani, 2021). Ambitious public policies put in place in terms of irrigation, energy and infrastructure development, also associated with the emergence of local economic dynamics, have allowed an extraordinary development of agriculture, especially in the Saharan regions, thus aiming to modernize agriculture with the purpose of increasing and stabilizing agricultural and food production (Bencharif, 2018). Benefiting from these incentives, in favor of Saharan agricultural projects, the increasing aggregation of the population in urban centers and, subsequently, the growth of consumer markets, stimulate the economic development of the entire region (Kouzmine et al., 2009).

In Algeria, the Sahara represents about 85% of the total area of Algeria, estimated at 2,381,741 km². Known mainly for its oil and gas resources, the Sahara also has enormous potential in terms of underground water resources, which are considered to be huge but poorly renewable or even fossilized (Dubost and Moguedet, 2002; Margat et al., 2006). The Saharan climate is characterized by an acute deficit in the water balance (Mohamed et al., 2022). Indeed, rainfall is low and temperatures are high, which leads to a very intense evaporative power of the atmosphere. Consequently, the water needs for drinking water supply, industry and agriculture can only be ensured by the intense exploitation of surface and groundwater (Chabour et al., 2021). As for the drinking water supply, industry and agriculture can only be ensured by the exploitation of groundwater. This groundwater has two major constraints, namely its low or even non-renewal on the one hand and its quality due to its saline load. The environment is harsh with large quantities of irrigation water and the presence of soils that are mainly coarse-textured and poor in organic matter. Soil salinization is a widespread land degradation phenomenon in arid regions that threatens agricultural production (Côte, 2002). Irrigation with mineralized water worsens the salinization of soils due to the intense evaporation to which the water is subjected. Because of its mineralization, its use, poorly controlled, leads to secondary salinization of soils (Aragüés et al., 2011; Qureshi et al., 2013; Singh, 2015; Li et al., 2018). The research conducted by Wang and Jia (2012) showed that climate change would favor the processes of soil salinization through the lack of precipitation and the increase of aridity. The other problem encountered, linked to uncontrolled irrigation, is that of the rise of water. Indeed, following excessive irrigation water inputs, the excess water feeds the water table whose topography is depressed and in the absence of a natural drainage system, the piezometric levels rise to the surface of the soil. The major consequences of this phenomenon are palm groves flooding (hydromorphy of the soils) and the extension of salinization at the level of the soils or the chotts (Remini, 2006). This phenomenon is suffered by the oases of El Oued and Ouargla, especially exacerbated by the combination of these waters with wastewater. This situation is still present even if the public authorities have made enormous efforts to deal with this excess water by pumping to lower the surface water table. Sustainable management of groundwater is particularly relevant in regions with low renewable water resources such as the Sahara. However, human activities, particularly agricultural, have had significant negative impacts on water and the environment. It is true that oasis agriculture alongside intensive agriculture in the Sahara constitutes spaces conducive to anthropic and economic activities (Saidani et al., 2022). Creating real attractive poles, they also constitute at the same time, a risk of potential anthropic pressure for the sustainability of oasis agrosystems and also on the future of intensive agricultural poles. It is thus a question of finding the right balance between the advantages provided by the attractiveness of the territories, the preservation of natural resources and the maintenance of agricultural and economic activity. Indeed, the non-adapted irrigation techniques, the unjustified quantities of water brought, the lack of drainage or its bad maintenance, are as many major factors which are missing to lead a good agricultural exploitation in this arid environment. Moreover, the symptoms of this unreflected management are the secondary salinization of soils (Semar et al., 2019; Abdennour et al., 2020), which tends to become widespread in several agricultural areas, significant drawdowns of piezometric levels, rising water levels in depressional areas and contamination of the water table by various pollutants.

The objective of this chapter is to highlight the current state of Saharan agriculture in its northeastern part, representing nearly 305,490 km^2 (Figure 10.1), the environmental problems and to introduce new approaches for the sustainable development of water and soil resources. To do this, we relied on observations made on some oases of the northern Sahara, results of researchers who conducted their studies in this Saharan space, represented by Biskra, Touggourt, Ouagla and Souf whose intensive practice of irrigated agriculture is taken into account in the assessment of these impacts on the environment.

10.2 AGRICULTURAL PRODUCTION AND CROPPING SYSTEMS

10.2.1 Plans and Programs Promoting Agricultural Development

In 2000, the NADP (National Agricultural Development Program) was already targeting the improvement of food security through better coverage of consumption by national production and the development of production capacities for agricultural inputs and reproductive material. However, the global food crisis has revealed even more clearly than in the past the extent of Algeria's vulnerability in terms of food security. The worst-case scenarios for 2030, which combine extensive liberalization and severe climate change, predict a drop in production of one-third compared to the 2000/2007 average (Omari et al., 2012).

As with previous reforms, the 2008 reforms failed to reduce the country's food dependency. With the persistent problem of the amount of fallow land still maintained at 3 Mha, a new land law was enacted in 2010. It includes both a plan to reduce fallow land and replace a right of perpetual ownership of plots of land, a system of concession up to 40 years for state land, and promoting the practice of agriculture in new areas of extension including in the Saharan zone. This development is reflected over the years (2000–14), by an increase in gross domestic agricultural production (GDP) which rose as a percentage of GDP from 8.3% in 2000 to 9.2% in 2010 and 11.2% in 2014 (Figure 10.2),

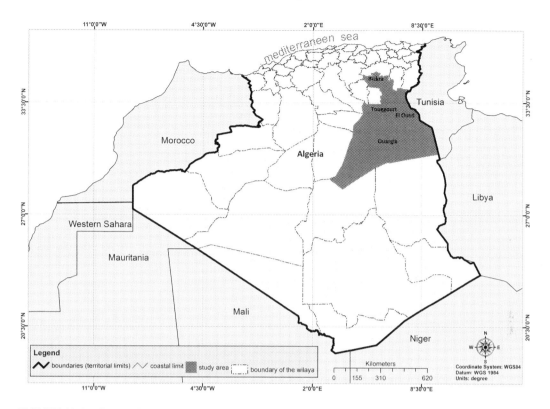

FIGURE 10.1 Geographic location of North East Sahara.

FIGURE 10.2 Contribution (in %) of agriculture to Algeria's GDP. Source: Adair et al. (2022).

an increase attributed to an increase in market garden and animal production (Bessaoud et al., 2019). The last period (2015–19) experienced the freezing of a large part of the programs of the rural renewal policy in 2017 to encourage returns on investment and also due to the consumption of budgets allocated to the program to 85.9% (Adair et al., 2022).

10.2.1.1 Farming Activities

Saharan agriculture is no longer at the experimental or project level, it is a concrete reality. Species that were once intended for self-subsistence are increasingly being replaced by other species for marketing purposes. The old cultivation systems characterized by high planting density (tiered cultivation) are regressing in the face of spaced and structured intensive planting under greenhouses, which better meets the new objectives of farmers who are still adapting to socio-economic and environmental changes, as well as the national objectives to achieve food security. Currently, there are 16.2 million productive palm trees of all varieties, offering a little more than 1.14 million tons of dates (FAO, 2022), ranking Algeria as the fourth largest producer in the world with 12.5% of world production in 2019; in addition to a market crop production of full season and off-season. Moreover, the introduction of these irrigated crops in these areas has transformed the desert land into a new agricultural Eldorado. This agricultural dynamic was initially driven by the law of accession to agricultural land ownership through land development by organizing, for the first time, the transfer (for a symbolic dinar) of ownership of public land to private operators and promoting agriculture in arid and semi-arid as a priority axis of the Algerian agriculture's development strategy. This transfer is organized in two forms: through the development of perimeters, then the installation of beneficiaries by the public authorities, or at the request of candidates who commit to develop, by their own resources, lands located outside the perimeters. According to Bouammar and Bakhti (2008), two types of agriculture can be distinguished in these new areas: the first type through the extension of palm groves which led to "peri-urban" agriculture or small-scale land development for small and medium-sized farms. Farmers operate individually or as cooperatives, generally of family structure, to access the State's assistance. The second type is described as large-scale development through vast concession programs. It is based on capitalistic agriculture on medium and large farms, sometimes based on cereal production systems. The large investments required for land development, the construction of forages, electrification and the construction of tracks require State assistance. This also reflects the difficult conditions under which farmers attempt to develop agricultural lands.

10.2.1.1.1 Oasian Agriculture

Houichiti et al. (2020) indicate that the old palm groves are represented by the marginal oasis production system, while the new palm groves correspond to the production system based on date production. The oases of the Biskra region differ from the oases of other regions by the good alignment of palm trees and by the merchant family character of almost all the oasis farms. According to Amichi et al. (2015) in some Saharan regions such as the case of Biskra, new oasis farms, combining greenhouses and date palms, are the driving force behind the agricultural boom and territorial expansion of Saharan agriculture in Biskra. These evolving farms allow for territorial expansion and the cultivation of new lands. The oasis of Ouargla consists of traditional palm groves with small family farms (0.5 ha on average). The palm plantations are unevenly spaced, at a rate of 160–300 ft/ha. The landscape of the oasis of Ouargla is composed of the Ksar (the name of the ancient city) and the palm grove with three levels; the arborescent stratum is the most important and represented by the date palm. The arborescent stratum is composed of fruit trees (fig tree, pomegranate, etc.), and the herbaceous stratum is composed of crops, fodder, cereals, condiments… etc. (Zenkhri and Karabi, 2020). Like the region of Ouargla, the oasis system of the region of Touggourt, the valley of Oued Righ is characterized, for the most part, by a system of oasis type consisting of three plant strata, organized into three levels which are: the date palm, fruit trees and market gardening. This agro-ecosystem is located essentially, in the form of oases, along the Oued Righ in the vicinity of water drainage where the water table is shallow and not very salty (Berrakbia and Remini, 2020). The similarities between the oasis system of Ourgla and Touggourt are similar since the latter is also made of a group of small family farms, located near the ksours, where each palm grove bears the name of the neighboring ksar. These palm groves form an agricultural model of self-subsistence, to ensure primarily the survival of the farmer and his family, far from being intended to produce marketed yield excess. On the other hand, the Souf region is famous for its traditional hydro-agricultural system called Ghouts, which means the basin, the funnel. They symbolize the completion of man's genius in this region with a very hostile climate, they represent an ancestral agricultural technique, unique in the world, worthy of protection and encouragement. The Ghout system in the Souf is a concrete case of a traditional human landscape, where Sufis have been able to domesticate nature and turn the Erg into a dynamic environment. The Ghout system extends over some 9,500 Ghouts representing nearly 3,543,000 palm trees. However, nearly one-fifth of the Ghouts are threatened, representing nearly 708,600 palm trees. Peri-Oasian agriculture is represented by four production systems, namely crop cultures, integrated intensive systems, crops associated with semi-intensive livestock and fodder associated with cattle. Finally, an extra-oasis agriculture (Figures 10.3–10.5).

FIGURE 10.3 The integrated intensive monoculture system of date palms.

FIGURE 10.4 Crops associated with semi-intensive livestock and fodder associated with cattle.

FIGURE 10.5 Culture crops and extra-oasian agriculture.

10.2.1.1.2 Agricultural Land Development

The agricultural land development in the Algerian Sahara is even considered as the solution for the reduction of the country's food dependency. Indeed, its immense potential has permitted us to be able to extricate Saharan agriculture from its traditional oasis mode to move towards an intensive agriculture on newly developed lands. The installation of new farms and the development of greenhouse crop cultures have transformed the landscape of the Biskra region, moving from a multi-level oasis

Environmental Issues in Sustainable and Resilient Agriculture 283

to greenhouse cultures that are developing next to monoculture date palm plantations. The area and levels of production reveal a clear progression, with crop production ranking first in terms of agriculture in the Saharan region in 2019, with more than 2.6 million tons in addition to 12,000 tons of peanuts. The wilayas of El-Oued, Touggourt and Biskra remain by far the most important in terms of the production of vegetable crops. According to Bouder and Chella (2017), the Sahara was able to participate in national production in 2015 with 20.77% of the national production of crops and 4.4% of cereals, 4.89% of dairy production, 4.9% of fruit cultures, 2.7% production of olives and olive oil, 1% in industrial crops which are currently expanding, and pulses at about 0.07% (Figures 10.6 and 10.7).

FIGURE 10.6 Intensive culture system of potatoes in El Oued region.

FIGURE 10.7 Intensive culture system of olives in Touggourt region.

10.3 CLIMATES, SOIL, WATER RESOURCES

10.3.1 CLIMATE

Random climate fluctuations are constraints that can directly or indirectly affect economic performance. They limit the actions of decision-makers, whether at the macroeconomic or microeconomic level and even more so when it comes to elaborating development strategies (Bachir et al., 2021, 2022) The Saharan regions are characterized by a contrasting climate, with a very hot and dry summer season and a moderate winter season. It is also characterized by significant temperature differences between day and night temperatures, as well as by the frequency and intensity of winds. Rainfall remains very scarce and insufficient, not exceeding 120 mm, hence the need to rely on irrigation for the practice of any agricultural activity. The Saharan climate is characterized by an acute deficit in the water balance (Dorize, 1982). Indeed, rainfall is low, and temperatures are high, which leads to a very intense evaporative power of the atmosphere. Generally, the climate is arid. The lowest average monthly temperatures are observed in January (10°C) and the highest average monthly temperatures are in July (34.6°C) and August (34.2°C) with peaks that can reach in some periods 47°C. Rainfall is relatively low and poorly distributed, the annual total is about 133 mm for Biskra and does not exceed 100 mm for El Oued and Ouegla, and rainfall is very irregular, and can be torrential. A recent study conducted at the University of Biskra conducted by Boucetta (2018), showed a significant change in all climatic factors (including temperatures). Climatic indexes indicate that the studied region tends to become increasingly arid. Climate projections for the year 2060 predict an increase in temperature (about +1.7°C) and a decrease in precipitation (about −12%) in the Biskra region. They also show the existence of influences of the current climate change on the water parameters. Farmers in Biskra are increasingly aware of these climate changes, particularly with regard to water resources, but adaptation strategies remain limited and personal.

The wind is one of the most important climatic components in the Souf El Oued region. It is an important factor to consider in agriculture due to its essential role in the phenomenon of pollination, as it can provoke the wilting of some sensitive plant species.

About the prevailing winds in the region of El Oued, Touggourt and Ouargla, according to Kitous et al. (2012), the direction of the winds is East-North coming from the Libyan Mediterranean, loaded with moisture called "El-bahri" and blowing very strongly in spring. They are little appreciated despite their freshness because they cause sandy winds. While the Sirocco or "Chihili" winds appear during the summer period, with a South-North and South-West direction, they manifest themselves in hot and very dry weather.

10.3.2 SOIL

Saharan soils are known to be, in their overall nature, poor in organic matter and in mineral elements essential to plants. Indeed, it is described that arid regions are the preferred domain of the pedogenesis of salts with as consequences; soils in general not very fertile, not very deep, too calcareous or too gypsum, salty, poor in organic matter, with an unfavorable structure with a basic pH and a neutral salinization pathway and a sensitivity to degradation and desertification (Halitim and Robert, 1992; Hamdi-Aissa et al., 2004). According to Mostephaoui and Bensaid (2014), in the northern Algerian Sahara, the soils are essentially of eolian origin and consist mainly of sand; this is the case of the soils of the Ouargla region. These soils, given their characteristics, have poor physical and chemical characteristics. A possible solution to improve their properties is to add organic matter to these soils, in order to improve both their physical and chemical properties (Gueziz, and Hamouta, 2020). In the Touggourt region, the soils are of much more alluvial, colluvial and aeolian origin. They are derived from the alteration of Quaternary and Miocene geological outcrops. The successive phases of erosion and deportation have installed on the sedimentation bottom of the valley a textural heterogeneity of the soils (Conrad, 1969). The classes of soils existing in the regions are of the type: rough mineral, little evolved hydromorphic, solonstchaks and hyper solonstchaks. The chemical facies of the soils are of the sulfate-calcium to chloride-sodium type and carbonates and bicarbonates are low in the profile. The presence of gypsum or gypsum-limestone crusts is observed at a depth greater than 70 cm, a zone under the influence of the water table (Durand and Guyot, 1955).

The soils of the Ziban region (Biskra) are relatively homogeneous due to an initial gypsum and saline dominance that has standardized the different surfaces following the progressive and continuous leaching of different salts. At the pedological level, the soils are generally not very evolved, of sandy texture, very poor in clay and organic matter (Azzouzi et al., 2017). The soils of Biskra, differ mainly by their texture, their morphology, the mode of pedogenetic evolution and by the level and mode of salinization. Their spatial extension is very variable. But this diversity should not hide their main and quasi-general character. The coarse fraction is generally abundant. The sand content varies between 65.82% and 91.66% with an increase in the fine sand content. The silty-clay fraction varies between 8% and 30% (Boumadda, 2019). In the higher areas, the soils are sandy and characterized by good infiltration and low water retention capacity. In the lower parts, the soils are siltier, and their retention capacity tends to increase (Kadri and Van Ranst, 2002). These sandy materials, eolian by origin, are very rich in gypsum from the evaporites. These materials are likely to evolve in the form of calcareous-gypseous accumulations, sometimes indurated, superficial crusts or gypseous crusts of table (Halilat, and Tessier, 2002). The region of Oued Souf is characterized by light soils, predominantly sandy, with a particular structure. They are characterized by: Superficial land, of variable depth, ranging from 30 to 50 m, corresponding to dune sands; Land with variable depth, ranging from 50 to 80 m, corresponding to clayey sands and sandy clays; separated by an impermeable substratum of clay.

10.3.2.1 Soil Fertility Status

In these areas, organic matter contents are low (1% < OM ≤ 2%) to very low (OM ≤ 1%), and almost absent in some parts of the region where the soils are bare; this is due to the absence of sources of organic matter, the plant cover in spontaneous plants being very sparse (Daoud and Halitim, 1994; Halilat, 1998). Soil organic matter comes mainly from manure applied in the palm grove and agricultural extension areas, a situation that explains the variability of organic matter content of cultivated soils (Bakari et al., 2017).

Consequently, their adsorbent complex is limited because of their low levels of mineral and organic colloids. The soils in the north-eastern part of Algeria are characterized by alkaline pH, high salinity and low water and nutrient retention capacity, explaining the low biological activity in these soils (Drouet, 2010). The ionic garniture of saline soils and the resulting physical properties considerably influence the rate of biodegradation of organic matter (mineralization of nitrogen and carbon) and reduce the processes of their humification, thus producing poorly polymerized substances responsible for the relative instability of the structure of the soils of drylands (Hafhouf, 2022).

Organic matter generally decomposes faster in sandy soils than in clay soils where the amount of soil microbial biomass is greater, suggesting greater physical protection of SOM in clay soils (Bouguettaya, 2020). Nevertheless, the coarse texture of the soils, limits the formation of well-humified soil organic matter and stable organo-mineral complex and its potential in carbon sequestration. As a consequence, its carbon sequestration potential is limited and its water holding capacity is also low. This coarse texture makes soil organic carbon quantities subject to erosion and rapid mineralization of organic matter. Up to 4.7% of organic carbon may be lost annually in sandy soils in the Sahara region. This physical degradation consequently limits the density and activity of microorganisms in oasis soils (Karabi et al., 2016). To this must be combined the problems related to climates in different agrosystems (Lal, 2018). However, despite the poverty of the soils in organic matter, humification rates are high. This is quite normal, given the environmental conditions: alkaline pH, abundance of exchangeable Ca^{2+} with a largely saturated adsorbent complex. Indeed, Ca^{2+} ions act and participate in the insolubilization and polymerization of "humic precursors". However, although the sandy soil blocks the polymerization of fulvic acids into humic acids, some quantities of the latter are recorded in the Ouargla soils (Kahlaoui., 2021). This can be explained by the fact that the saturated and Ca^{2+} rich environment favored a degree of polymerization of humic compounds (Bouguettaya, 2020; Soltner, 2003).

10.3.2.2 Soil Microbial Biodiversity and Impact of Salinity

The organic matters are subjected, in contact with the soil, to a series of transformation by telluric microorganisms. One of the main processes concerns the degradation of organic matter and the major cycles that depend directly on it, such as that of carbon and nitrogen (Lal, 2004a). With respect to nutrients, microorganisms can act both as a source and as a reservoir. But microorganisms also participate in the realization of organo-mineral association of the soil (Lal, 2004b).

The richness of symbiotic nitrogen-fixing bacteria has been studied particularly in the region of Biskra in faba beans and alfalfa (Mouffok, 2010). This author indicates, however, that even if the varietal effect is notable on the diversity of these bacteria, the effect of the site of culture is much more important where an edaphic effect is much more important than the variety as much for the broad bean as the alfalfa. This conclusion is illustrated by the clear superiority of nodular dry biomass recorded with a variety of lucerne, while the number of specific rhizobia is lower than the threshold of induction of nodulation, or even an absence of such rhizobia, in some localities of Biskra.

In the same region, Hakkoum (2018) observes the same effect among the fungal microflora, he notes a variation in the presence of the genera of fungi where it is found that the density of the genera *Aspergilus* is the highest followed by the *Rhizopus* and finally the *Alternaria*. Ghadbane et al. (2013) and Bedjadj (2011) indicate the abundance of soils of Biskra and Ouargla, respectively, in actinomycetes, especially the genus *Streptomyces* sp. as well as the genus *Bacillus*. These microorganisms are widely known to be resistant to adverse effects especially salinity and high temperatures. Their densities are very important even in bare soil and at alkaline pH. This indicates the great capacity of these microorganisms to adapt in the most difficult climates as well as their aptitudes to degrade organic substances which are difficult to decompose; including some rare species that are interesting for the pharmaceutical industry because of their production of antibiotics.

The existence of halophilic species or strains capable of resisting high salt levels is possible thanks to certain adaptive strategies. The microbial exploration of these particular ecosystems has revealed the presence of archaeal and halophilic bacterial strains that have been isolated, purified and characterized. The archaeal strains belong to the family *Halobacteriaceae* the order of *Halobacteriales*. Extreme halophilic, aerobic, Gram-negative, these strains are also alkalithermotolerant. The phylogenetic study revealed their affiliation with the genera Natronorubrum and Halorubrum (Khallef et al., 2018). On the other hand, Bazzine and Hamdi-Aissa (2014), reported that the microbiological analyses of the biological crusts of gypsum soils show that all the soils studied are populated by diverse microflora adapted to the harsh conditions of the Saharan environment. From a qualitative point of view, all the soils studied contain a diversified microflora: bacteria, actinomycetes, fungi, especially the following fungal species: *Aspergillus* sp, *Penicillium* sp, *Alternaria* sp, and *Trichoderma* sp. and also unicellular and filamentous algae. Most of the microbial species of the studied soils remain poorly known from a taxonomic and functional point of view. The observations on the biodiversity of gypsum crusts in Ouargla and Biskra are similar to those in El Oued. According to Souadkia and Souadkia (2017), the extreme pedoclimatic conditions that characterize the soils of the three sites studied disadvantage the installation of an intense plant cover, which causes a rarefaction of the accumulation of organic matter whose action is fundamental on the properties, physical and physicochemical soil and consequently on the microflora telluric and its activity. Bedjadj (2011) in the region of Ouargla, reports that the ratio C/N, indicator of biological activity, varies from 6.14 to 8.18. This rate is less than 10.0 translating to a reduced mineralization of nitrogen and carbon existing in these soils. The climatic and pedological variations affect in a significant way the microbial biomass of the soil. The combined effect of low precipitation, high temperatures, high evaporation and strong winds strongly reduces the microbial density. In addition, the physicochemical conditions of the soil (moisture content, salinity, gypsum content… etc.), as well as significant variations in biochemical factors (nutritional and energetic) contribute to varying the microbial density from one site to another (Oustani, 2006). Furthermore, it is accepted that there is more interesting microbial biodiversity in cultivated soils than in bare soils. Karabi et al. (2016) explain this phenomenon by the fact that the cultivation of soils has led to a substantial increase in soil moisture compared to uncultivated soils, even though the EC is higher in cultivated soils. This salinity recorded in cultivated soils is justified, by poor management of irrigation facilities, poor internal drainage and poor quality of irrigation water something that can be corrected and improved. From another point of view, the most limiting factor of the microbial mass in the soil is the insufficiency of the energetic substrates for the microflora, whether carbon for the heterotrophs or reduced mineral substances for the chemolithrophs. This density is even lower in Saharan soils where energetic and nutrient substrates are reduced, combined with the effects of extreme pedoclimatic conditions, namely: too high temperature, low humidity and too alkaline soil pH (Karabi et al., 2015, 2016). The same authors report the important microbial density recorded in the soil under the culture of alfalfa which is certainly due to the rates of moisture, relatively high organic matter, and the salinity which is less important compared to the soils cultivated by a cereal and the bare soil. Nitrogen levels are relatively higher in the plots under legumes (alfalfa) and significantly lower in the uncultivated soil. Thus, it can be said that alfalfa crops promote nitrogen retention in this type of soil, while others remove it more quickly. In addition, the negative effect of tillage on the abundance of microorganisms can be explained by an alteration of microbial habitats (especially macroaggregates which represent habitats for fungi) and/or of the trophic status of the soil (content and quality of organic matter, electrical conductivity) even if the impact of tillage on the density of the telluric microflora in a loamy textured soil is weak compared to that in a clayey soil.

10.3.2.3 Effect of Exogenous Organic Matter Input

The ploughing of organic matter represents a possible solution to improve both the physical and chemical properties of soils (Lorenz and Lal, 2022). The soils of essentially eolian origin, containing mainly sand up to 90% (case of the soils of the regions of Ouargla, Biskra, Touggourt and El Oued) present poor physical and chemical qualities. To this, one of the best strategies is the exogenous contribution of organic matter to allow a re-texturization of the inert soils characterizing the agricultural regions of the Algerian Sahara but also they allow reviving the biological life of the soils through the microbial arsenal telluric. In the region of Ouargla, practices of adding raw clay, the natural bentonite of Mostaganem (located in the northwest of Algeria), improves the water retention capacity of the soil. Thus, the greater the addition of clay, the greater the retention capacity. However, given the quasi constitution of the soil of sandy origin, the effect of clay can be significant only from 12%. In the region of Biskra, the use of ovine manure, even in the presence of brackish irrigation water, allows to maintain the soil respiration. The rate of CO_2 released, following the mineralization of organic carbon, increases proportionally with the dose of organic matter (OM) brought to the soil until an EC of 8.64 dS/m beyond which, this accumulation decreased. Indeed, the contribution of sheep manure, the highest value of CO_2 released (89.47 mg of CO_2/100 g), is recorded for the soil amended with 2% OM and irrigated with water charged to 15.35 dS/m while the lowest value of CO_2 released (21.08 mg of CO_2/100 g) was recorded for the dose of 0% OM and with distilled water to 0.005 dS/m (Mancer et al., 2020). However, the addition of manures (sheep and cattle) significantly modified the physical and chemical properties of the soil. In fact, the organic contribution at increasing levels decreases the pH in a highly significant way. However, regarding electrical conductivity, organic matter increased soil salinity through the mineralization of these organic compounds. Koull and Halilat (2016) reported that soil water retention capacity showed very highly significant differences with the different organic matter treatments. These authors indicated that the organic matter increases the water retention capacity of the soil during the whole experimental period, the highest percentage being recorded with sheep manure with 39.45% on average. Thus, two distinct phases of the influence of organic matter on the properties of the Saharan soil: a first period, varying from 4 to 5 weeks, characterized by a rapid change in soil properties; a second phase characterized by a decline in soil characteristics. As for the increase of the EC, irrigation is essential for crops in arid regions, it can, if it is well managed, contribute to solving the problem of increasing salinity, but provided that the irrigated area is well drained. On the other hand, the use of date palm compost can improve the organic matter content of the region's soils.

Romani et al. (2007) reported that the C/N ratio of dried palm compost is very high, reaching a value of 12.53 after 7 weeks. It is also considered that a value below 25 characterizes a mature compost, while a ratio below 20 and even 15 is preferable (Biddlestone and Gray, 1988). In the same line, Aissaoui and Barkat (2020) reported that the use of compost, poultry manure and humic acid significantly improves the texture of the soil as well as its organic matter rate. Nevertheless, the comparison between the organic matter contents obtained after the experimental period and the initial contents shows a significant reduction in the organic matter values. It is the weakest with the poultry manures where the organic matter rate is more limited in comparison with the initial contents, and which does not exceed 1.5%. This is explained by the fact that a greater leaching of soils by irrigation after amendment of manures than of humic acid of plant origin and compost is observed. In addition, fertilizers are more rapidly degraded than plant amendments, resulting in an increase of EC in the soil. Furthermore, a study conducted on the impact of using bio-coal, based on residues of date palm, on a barley crop in the region of El Oued by Guesseire et al. (2022) and Chemsa (2019), reported an improvement of nitrogen stocks in the soil, among other things by a reduction of ammonium leaching due to its adsorption on the surface of bio-coal. They also evaluated its impact on soil microbial proliferation. Also, they noted a more rapid decomposition of organic matter and correlated with increased microbial, bacterial and fungal biomass. Changes in soil physicochemical properties through the introduction of metabolically available carbon compounds such as bio-carbon can influence microbial community structure and soil biogeochemical functions (Anderson et al., 2011). Among farmers, O.M. application is done on average 6 days before planting (6 days for poultry, 5 days for sheep and 7 days for cattle). The quantities brought by the farmers vary from one region to another, they use organic matter in each season for vegetable crops like potatoes, peppers, and others. The origin of the O.M. comes from the different breeding practiced in the bordering wilayates. Poultry manure is preferred, but more than two-thirds mix it with sheep and/or cattle manure. On average, each farmer contributes between 30 and 50 m^3 of organic matter per hectare, which corresponds to 20–35 equivalent tons of cattle manure, or 17–27 equivalent tons of sheep manure and 15–25 tons of poultry droppings. The direct financial costs of the organic matter are very important and are estimated between 1,000 and 1,250 $/ha.

10.3.2.4 Overview on Inorganic Fertilizer's Application

As for the use of inorganic fertilizers in intensive agriculture in these regions, in reality, there are no real studies on the impact of their use, on organic matter. Farmers producing market garden crops in El Oued and for plasticulture in Biskra, Touggourt and Ouargla, have not done real comparative tests with the different formulations that exist on the market. For each fertilizer, they justify their choices by a series of observations. They perceive an advantage with monoammonium phosphate (MAP 12.52.0) which is a granulated fertilizer; it contains 12% ammoniacal nitrogen and 52% water-soluble phosphoric anhydride. This fertilizer has an important acidifying power, which allows the release and assimilation of nutrients. It is ideal for improving the availability of phosphorus, particularly at the onset of cereal cultures in arid zones, especially for root development and in a mixture with M.O. For the sulfate-based $N_{15}P_{15}K_{15}$, it is above all the balanced development of the plant and tuberization that are perceived. Urea is appreciated for its effect on the vegetative growth of the aerial parts. The $N_8P_{10}K_{30}$ favors tuberization and the substantial size of the tubers.

10.3.3 Water Resources

The water resources are essentially underground due to the climate characterization, low rainfall and high temperatures that generate intense evaporation. The few rainfalls, certainly not negligible in this environment, are not enough, by far, to satisfy the water needs. On the other hand, the geological aquifer formations store very large quantities of water. These are located at variable depths and are characterized by two major constraints. The first one is linked to their weak renewal or even their non-renewal and the second one is their mediocre quality because of their mineralization.

10.3.3.1 Superficial Water Resources

The surface water resources are expressed by the many rivers, called "oueds", that drain the various watersheds. The oueds are directly linked to the dynamics of precipitation because in the absence of vegetation cover, the more intense the precipitation is, the more violent and often devastating the flooding is generated. These flash floods feed the aquifers as well as the closed depressions in which water areas accumulate. This is the case of Chotts and sebkhas. In dry periods, the watercourses are practically dry. The main courses in the region of Biskra are the Oued Djeddi, Oued El Arab and Oued Biskra which join the Chott Melrhir. Still in the district of Biskra, two reservoir barrages, Foum El Gherza and Fontaine des Gazelles, were built respectively on the Oued El Abiod and the Oued El Ham for agricultural purposes. The Foum El Gherza barrage has an initial storage capacity of 47 million m³ irrigating a small part of the oasis of Sidi Okba (irrigation of 850 ha) while Fontaine des Gazelles is 55.5 million m³ ensuring the irrigation of 1,600 ha of agricultural land situated in perimeter of M'Kimnet. These reservoirs are experiencing significant mud accumulation rates, which has considerably reduced the volumes of water mobilized. According to Dubost (1986), the water reserves of Foum El Gherza reach only 5–10 million m³ and irrigate only 250–300 ha.

This region of the Sub Sahara has many topographic depressions occupied by salinized waters, the most extensive and known is the chott Melghir which is at −33 m altitude compared to sea level. This place functions as an endoreic basin where many rivers converge.

10.3.3.2 Groundwater Resources

Most of the water resources are underground, found in the hydrogeological reservoirs of the northern Sahara, which extends from the Saharan Atlas in the north, to the Reggane - In Amenas line, south of the Saoura in the west to the Libyan border in the east. Various hydrogeological studies have been conducted (UNESCO, 1972; Besbes et al, 2003; Ould Baba Sy, 2005) on this transboundary basin, shared between Algeria, Libya and Tunisia (Lasmar et al., 2022). It is defined by two major aquifer units at the scale of the Sahara: the Terminal Complex (TC) and the Intercalar Continental (IC). The TC and IC (Table 10.1, Figure 10.8) constitute one of the largest hydraulic reservoirs in the world,

TABLE 10.1
Stratigraphic and Aquifer Units (OSS, 2003)

Stratigraphic unit		Aquiferes
Plio-quaternaire	Mio-	Second sand's groundwater
Miocene	Pliocene	First sand's groundwater
Oligocene		Semi - permeable
Middle eocene		Semi - permeable
Inferior eocene		Calcareous groundwater
Paleocene		
Superior senonian	Maestrichtian	
	Campanian	
	Santonian	
Inferior senonian		Non-pemeable
Turonian		Turonian groundwater
Cenomanian		Imperméable
Albian		Intercantinal groundwater
Aptian		
Barremian		
Neocomian		Salinized water
Malm	Kimmeridgaen	Jurassic groundwater
	Callovo- Oxfordian	
Dogger	Bathonien	
Lias		Non-permeale crust
Keuper		
Muschelkalk		
Buntsandstein		Trias salinized groundwater

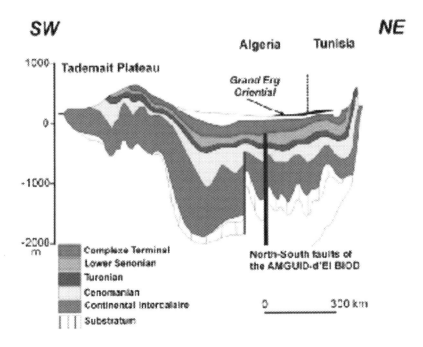

FIGURE 10.8 Longitudinal section of the Continental Intercalaire and Terminal Complexe.

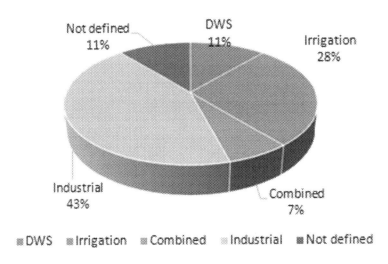

FIGURE 10.9 Groundwater utilization from the CI.

estimated at between 31,000 and 60,000 billion m³ (UNESCO, 1972; Lemarchand, 2008). Although of more modest dimensions compared to these two large reservoirs, we also have the water table, which in certain localities is of hydrogeological interest.

The temperature of the groundwater varies between 25°C and 70°C, with an estimated average level of salinity between 0.5 and 6 g/L. In Algerian Sahara, the industrial sector accounts for 41% of the boreholes in operation, which is still very important, followed by irrigation with 30% currently in operation against only 13% for drinking water supply (DWS) and 7% combined between irrigation and DWS (Figure 10.9).

According to Etsouri et al. (2022), irrigation is most often accompanied by an increase in the supply of drinking water due to the population's mobility to these areas. This is reflected in the drilling of boreholes for drinking water in the agglomerations. Dual-use boreholes were made to support small population with modest agricultural requirements. Most of the identified boreholes are distributed between the wilaya of Ouargla and the wilaya of El Oued with 79% and 11% respectively, followed by the wilaya of Touggourt with 5%. The rest of the boreholes (17% of the total number of identified boreholes) are distributed among several other wilayas such as In Sallah, Biskra and Ghardaia.

10.3.3.2.1 IC Groundwater

As reported by the Besbes (2003), the term Intercalary Continental (Interncontinal), is attributed to a localized continental episode between two marine sedimentary cycles:

- At the base, the Paleozoic cycle that completes the Hercynian orogeny,
- At the top, the cycle of the Upper Cretaceous

This complex is formed mainly by continental sandstone-clay formations of the inferior Cretaceous to which must be added marine or lagoon sediments, post-Paleozoic and ante-Cenomanian intercalated in the IC. The first hydraulic horizon, located at the base, is the Barremian, which is of sandstone nature and is an important aquifer of the large hydraulic reservoir of the IC with a thickness of about 100 m. The Aptian is located between the continental formations of the Barremian and the Albian and corresponds to a marine invasion materialized by a dolomitic bar of 20–30 m depth.

The Albian is a thick aquifer of up to 600 m depth, of sandy-stone nature, containing good quality water. The clayey and carbonated Cenomanian constitutes the crust of the IC. The water table is enclosed in the northern part of the Sahara, particularly in the localities of Biskra, Touggourt, El Oued and Ouargla, selected in this study. According to the Besbes (2003), the waters of the Intercalary Continental took place in the aquifer during the major wet phase of the Lower Pleistocene, which is recognized in several aquifer systems around the world. This period is the main period of constitution of the geological reserves of the Saharan aquifers whose ages vary between 20 and 40,000 years. The chemical exchanges of the waters of this great aquifer with the surrounding formations are determined by the high pressures and temperatures that are exerted. The water temperature reaches 60°C at the surface and the average salinity of the water is about 1.5 g/L (Guendouz et al., 2006). The direction of the flow seems to be from the North West to the South East. The waters converge towards the Tunisian outlet through the Mednine gap. Boussaada et al (2023), studied the physico-chemical properties of groundwater in the Ouargla CI, the mechanisms of mineralization of these waters and the evaluation of their qualities for human consumption, irrigation and industry. The results indicate that the waters of the intercalary continental aquifer (IC) present physicochemical parameters varying between; T (50°C–55°C), EC (2,165–3,240 µS/cm), TDS (1,502–2,658 mg/L), pH (7.66–8.4), Ca^{2+} (83–207 mg/L), Mg^{2+} (73–240 mg/L), Na^+ (175–320 mg/L), K^+ (18–54 mg/L), Cl^- (375–565 mg/L), SO^{2-} (463–1,050 mg/L), HCO^- (67–281), NO^- (0–15.3 mg/L) and TH (713–1,318 mg/L CaCO). The IC is artesian gushing, the pressure at the head of the drillings is about 15 bars (Etsouri et al., 2022).

10.3.3.2.2 TC Groundwater

The term "Terminal Complex" (Bel and Cuche, 1969) used in the ERESS project, with the term "Terminal Complex aquifer" (Bel and Cuche, 1969) is used in the ERESS project to refer to several aquifers located in different geological formations because these aquifers are part of the same hydraulic unit. The intercommunications between the Senonian, Eocene and Mio-Pliocene are evident throughout the Basin The intercommunications between the Senonian, Eocene and Mio-Pliocene are obvious on the whole basin, except for the region of the chotts where the impermeable middle and Superior Eocene comes to be intercalated. The Turonian nappe is more individualized due to the impermeable cover of the Senonian lagoon. However, its levels are consistent with those of the Senonian or Mio-Pliocene.

From a lithological point of view, the Terminal Complex is made up of carbonate formations from the Superior Cretaceous and detrital levels from the Tertiary, particularly from the Miocene. Two hydrogeological formations have been identified. The first is the Mio-Pliocene sands and the second is the underlying Senonian carbonate hydraulic reservoir. They are separated for the most part by an evaporitic intermediate layer attributed to the Eocene. The reserves of the TC covering the entire northern Sahara (Algeria, Tunisia, Libya) are very large (Ould Baba Sy, 2005). The water table of the Terminal Complex is quite deep (250 m) and captive in the Souf region, its supply areas are located in the south on the Grand Erg Oriental where it becomes shallower and free.

The reload is old and is made from rainwater having undergone changes in its isotopic composition over time, which would date from the inferior Pleistocene (Guendouz et al., 2003). They are therefore "fossil" water reserves and therefore in the medium term not renewable. Roughly speaking, the water flow converges towards the Souf (El Oued) before joining the Chott El Djerid (Tozeur, Tunisia). The Mio-Pliocene waters show a highly variable salinity (above 2 g/L), but the waters are generally more mineralized than those of the CI aquifer (Guendouz and Moulla, 1992).

10.3.3.2.3 The Water Table Groundwater

The water tables groundwater are negligible in size compared to the two large hydraulic reservoirs of the CI and CT. However, on a local scale, these surface water tables are of hydrogeological interest, particularly when the water is of good quality and shallow.

These types of groundwaters are generally located in the alluvial accumulations that are found at the level of palm groves and they are best known with a depth close to the surface of the ground and which can reach 100 m. They are also encountered in the beds of oueds where they are directly supplied by the violent floods that fall on the watershed. They are also replenished by direct rainfall that falls on the impluvium and by the return of irrigation water. This is the case in many areas (Ouled Djellal, Doucen, Sidi Khaled, Tolga) and oueds like Oued Djeddi or Oued Biskra. From a lithological point of view, they are made of sands, gravels, stones and sometimes clays. The water tables are shallow, being between 5 and 60 m and are widely exploited in the irrigation of palm groves.

The flows converge towards the Chott Melghir. The substratum of the water table rests, generally, on the impermeable clay formations of the Pliocene or in discordance on the middle Eocene lagoon, when the Mio-Pliocene is absent, as it is the case at Ouled Djellal and Sidi Khaled. In the region of El Oued (Souf), there is a "free" water table groundwater, contained in fine sands, interspersed locally by lenses of sandy and gypsum clays. It rests on a clayey substratum, impermeable, with a depth of 200 m of the pliocene or in discordance on the average lagoon eocene. The water table groundwater extends over almost the entire territory of the valley (Khechana et al., 2011). This water table is exploited by about 10,000 traditional wells, at an average depth of 40 m. It is included in the fine sandy deposits of eolian type, locally interspersed with lenses of sandy and gypsum clays.

10.4 IRRIGATION

Despite its low efficiency, the gravitational irrigation system is still frequent, particularly in old palm groves where most growers are reluctant to adopt localized drip irrigation. Water saving is only ensured in the water transport phase between the water withdrawal point and the palm lines. The farmers use the same dose of water with the same frequency of irrigation to irrigate the crops without taking into consideration the varietal aspect. The adoption of localized irrigation is more frequent in new palm groves. This practice is supported and disseminated to develop water saving. Water sources are drawn from various aquifers, and their number is increasing despite official restrictions, which has led to a pressure on the water resources and has aggravated the drawdown of the aquifers in the wilaya of Biskra. In this region of the Sahara, given the diversity of crops, there are no major

Environmental Issues in Sustainable and Resilient Agriculture

FIGURE 10.10 Development of multi-chapel greenhouses and organic agriculture in the Algerian Sahara.

differences between market vegetable crop operations. In fact, the cultivation system practiced by producers is identical for all farms. The extensive system is relatively more present in small farms, but in large and medium farms an intensification of the systems, using modern techniques, are practiced such as localized irrigation, fertigation, phytosanitary treatment, burying of the M.O. and heating of the greenhouses during the cold periods particularly in multi chapel greenhouse (Figure 10.10).

In the Oued Souf region, market vegetable crops (potatoes, peanuts, tomatoes, etc.) and open field forage crops are irrigated exclusively by sprinklers using a locally manufactured system. This system now offers farmers the opportunity to save water and energy, but it is still less efficient and lacks uniformity in its use.

10.5 ENVIRONMENTAL ISSUES

The development of Saharan cultivated lands has led to various environmental degradations that are often irreversible or very costly to rehabilitate and to remedy. Indeed, the uncontrolled irrigation of the perimeters with groundwater of lower quality has led to negative repercussions. These are manifested by the salinization of the soil, the rise of water from the surface and pollution. In addition, the intensive exploitation of the deep aquifers leads to a decrease in the pressure to which the hydrogeological reservoirs are subjected, resulting in the drying up of springs and the weakening of artesianism.

10.5.1 Soil Salinity Issues

The increased demand for freshwater resources is a global concern, but for Algeria and particularly in its Saharan regions, it becomes a serious challenge. Lands subjected to irrigation with salt-loaded water can aggravate the situation and lead to secondary salinization of soils, which can be accentuated by the neglected use of drainage techniques and the cleaning of the existing network (Figure 10.11).

FIGURE 10.11 Drain, poorly maintained, invaded by weeds.

10.5.1.1 Soil Salinization

In arid and semi-arid regions, the problem of soil salinization is a serious issue and constitutes both an agricultural and environmental constraint because it reduces large areas of cultivated land and affects crop productivity and quality (Yamaguchi and Blumwald, 2005). Soil salinization can occur naturally or be induced by man. In all cases, there are various geochemical processes that drive this phenomenon. Salts can be derived from the alteration of rocks, from the dissolution of minerals brought in by irrigation water and often by fertilizers. The process by which the soil becomes enriched with soluble salts is salinization. The other process is alkalinization, which occurs when the soil pH increases to values above 8.5. This process is often accompanied by the phenomenon of sodization when exchangeable sodium becomes the majority element adsorbed on the colloidal complex. The superficial water table groundwater can take part in the salinization of the soils by the phenomenon of capillary rise. When water inputs by precipitation or irrigation with fresh water are insufficient to leach ions from the soil profile, salts accumulate in the soil, resulting in soil salinity (Blaylock, 1994).

10.5.1.1.1 Primary Salinization

When salinization affects the natural environment without human involvement, it is called primary salinization. In Saharan soils, crystals of gypsum, limestone or halides are often found, transported by water or wind, and are due to the legacy of saliferous geological formations found in the geological layers. Moreover, geological studies (Busson, 1970; Fabre, 1976) have mentioned and described the saliferous levels as is the case of the Senonian lagoon or the saliferous Triassic. The Outaya region, located about 20 km from the city of Biskra, displays an imposing Triassic salt dome called Djebel El Melah, 4 km long. The lithology is marked by the presence of clay, marls of calcairo-dolomitic blocks, and especially by evaporites with the predominance of salt and gypsum (Guiraud, 1973).

Groundwater levels close to the surface, as is the case in many Saharan oases, are subject to capillary rise and once they reach the top of the profile, they are subject to a strong evaporative power of the atmosphere. This results in a deposit of salt by precipitation following the evaporation of the solution. According to Guendouz et al. (2006), the origin of salinity in El Oued comes from the dissolution of evaporites (halite). The SO_4/Cl ratio remains largely above 1 in the southern part and is close to unity in the north. This suggests that gypsum-halite dissolution reactions occur along the flow path.

10.5.1.1.2 Secondary Salinization

Irrigation with salt-charged water, in a Saharan environment where evaporation power is intense, leads to soil salinization. This phenomenon is not specific to the Algerian Sahara but occurs in several regions of the world (Negacz et al., 2022). In arid regions, salinity is essentially of geological origin. The causes of salinization of cultivated soils are the poor quality of irrigation water, insufficient water for the leaching of salts, poor soil drainage and the water table groundwater superficial localization (Ben Hassine, 2005; Hatira et al., 2007; Askri and Bouhlila, 2010). Some appropriately sized plots are equipped with drainage systems, but smaller plots are often not. Often, these drainage systems are poorly maintained. Secondary salinization is only the result of human activity, which has not been able to control the use of water in irrigation, in quantity and quality, in this fragile environment, characterized by a high evaporative capacity. Indeed, since the 1990s, the Saharan regions have experienced a strong agricultural change from a traditional agriculture or oasis based mainly on phoeniculture to a modern agriculture associating market gardening to date palm.

Without supervision and permanent monitoring of this agricultural dynamic, soil degradation by salinization quickly appeared and was accelerated by the severity of the climate. This phenomenon affects many regions and the extension of the salinity of irrigated soils increases every year. According to Ghassemi et al. (1995), 1 billion ha of land is affected by salinity, of which more than 20% is irrigated arable land. Recent work by Negacz et al. (2022), based on mapping, estimates that the total area of saline soils with EC greater than 4 dS/m was 16 646,000 km^2, which exceeds estimates made by previous studies. Several factors combine to cause salinization, in addition to evaporation, and mineralization of irrigation water, the lack of drainage or its poor design and maintenance are all parameters involved in this degradation process. The global area of soils affected by salinity is 424 Mha of topsoil (0–30 cm) and 833 Mha of subsoil (30–100 cm), based on 73% of mapped land. In Algeria, as early as the 1970s, the UNESCO report (1972) warned the public authorities about the evolution of the salinity of the exploited Saharan aquifers, which are contaminated by saline aquifers and consequently their use in the agricultural sector will be compromised. The report also emphasized that if intensive pumping is practiced, there is a great risk of displacement of salt water from the chotts to the groundwater. This is the case of the Chott Melghir which borders the oases of Biskra, Oued Rhir and El Oued. The effects of irrigation water quality on soil properties are widely studied in the literature (Boivin et al., 2002; Hachicha et al., 2000; Ayers and Westcot, 1985; Subyani, 2005; Al-Ghobari, 2011a, b). Saline soils are generally characterized by physical, chemical, and biological properties unfavorable to plant growth. Salts from irrigation water accumulate in the soil (Figure 10.12). These accumulations cause the osmotic pressure to increase and can lead to soil sterilization. In general, the relationship between relative crop yield and salinity is roughly linear based on comparisons between the yield of the same crop in saline and non-saline soils (Katerji et al., 2008). According to Ayers and Westcot (1985), the threshold for salinity in irrigation water is 3 dS/m but given the evaporative capacity on the one hand and fine soil texture on the other, this standard is quickly exceeded, and severe symptoms may appear on the plants.

Research conducted by Abdennour et al. (2020) in the irrigated perimeter of El Ghrous, located in the eastern part of the Biskra district, revealed soil pH values ranging from 8.0 to 9.0 and that 75% of the samples have EC values exceeding 1 dS/m. Calcium and sulfate seem to dominate the ionic pattern of the soil solution. In the authors' opinion, it appears that the salinity pathway is a sulfate-dominated neutral saline pathway, and that this pathway presents the major hazards for evolution to low calcium concentrations, and thus a greater likelihood of degradation of the soil structure. He adds that saturation of the soil solution with calcium decreases the assimilability of phosphorus by its insolubilization in the form of tricalcium phosphate and causes a deficiency of trace elements (Fe, Mn, Cu and Zn) in less assimilable forms due to the high pH.

In the Ouargla region, irrigation water and subsoil water solution salinity values were close (EC=3.07 and EC=3.16 dS/m, respectively) but the saturated horizon salinity was clearly higher (EC=9.11 dS/m) (Touahri et al., 2022). The use of these waters requires more attention to sustain agricultural development because they require processes never observed before in dry areas

FIGURE 10.12 Accumulation of salts on the topsoil surface after irrigation.

(Semar et al., 2019). Water quality for irrigation in Biskra is poor quality for irrigation because it presents a very high danger of salinization with a risk of alkalization very strong. The majority of groundwater in the region is in the poor-quality category for irrigation with a very low risk of soil sodization (Semar et al., 2021). Nitrate concentrations were found to be elevated almost to concentrations above 45 mg/L. The presence of high nitrate concentrations in groundwater is not only the result of massive use of artificial fertilizers. This fact is unfortunately favored by the absence of a sewage system in the whole region of El Oued where more than 1,000 septic tanks are in use.

10.5.2 Drawdown of the Groundwater Table

After the heavy exploitation of the Terminal Complex (TC) aquifer during the last 30 years, signs of degradation of the TC aquifer are noticed, presenting mainly in the low piezometric level and the increase of its salinity (Kharroubi et al., 2022). Forage flows are insufficient to meet the water demand, especially after the creation of new irrigation perimeters. Recent drilling is oriented towards the exploitation of the Continental Intercalary aquifer (CI) to meet the high demand for drinking water supply and agricultural use. In many localities, the water tables, not very deep, are affected by significant drawdowns following inadequate pumping. The recharge of these aquifers is low compared to their exploitation. This situation is observed along the Oued Djedi at Biskra and other small valleys of the Rhir or Ouargla. In the locality of Doucen (Biskra), we met a farmer who complained about the unacceptable lowering of the water level of his 200m-deep forage and no longer meets his irrigation water needs. This situation is due to the realization by his neighbor, far from a 100m of his plot, of a forage exceedingly largely the 200m (300m according to the farmer) and under continuous pumping. In the opinion of our interlocutor, the solution found is to carry out a much deeper forage to have water in quantity. We note here the interferences between the pumping forages lead to a race for the underground water resource by digging deeper and deeper.

10.5.3 Water Rising in Some Regions

The evolution over time of the domestic discharge flow and irrigation water backflow correspond to the beginning of the development of many lands in the region. The flow of forages in the deep aquifers increased in parallel, resulting in the discharge of wastewater and irrigation backflow that have

Environmental Issues in Sustainable and Resilient Agriculture 297

FIGURE 10.13 Rising water issue.

only one path, through the sand to reach the water table groundwater (Remini, 2001). Following the rise of water to the surface of the ground, the houses have been deteriorated, the Ghouts are flooded in urban and peri-urban areas. In the absence of a natural outflow, the evacuation of wastewater has become very difficult (Figure 10.13) and the oasis environment is covered by real puddles of waste water where weeds and reeds have taken the place of palm trees.

10.6 GHOUT SYSTEM

The oases based on the Ghout system are common on the eastern erg in the Souf region. The principle is based on the creation of a crater, about 10 m deep, in relation to the initial level of the ground. A Ghout is a technique for growing date palms specific to the Souf region. The palm trees are planted in groups of 20–100 palm trees in the center of an artificial basin, 10 m deep and 80–200 m in diameter, the bottom of which has been brought to less than 1 m above the water table. The oases dig progressively into the soil so that the palms have their roots constantly in the water and therefore do not need irrigation (Remini, 2004). These plantation systems are currently witnessing a know-how that unfortunately remains both poorly known in the world and in the regions that contain the same physical and natural conditions and characteristics (Figure 10.14). It is classified as Ingenious Systems of World Agricultural Heritage (ISWAH), it proves a system much more complex than it shows.

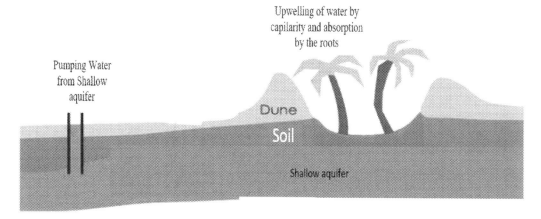

FIGURE 10.14 Ghout System scheme.

FIGURE 10.15 Ghout System at El Oued.

The Ghouts is a complex system that works in synergy with the environment and the very harsh conditions of the region (Figure 10.15).

The excavation of the sand out of the basin is done manually by men called "the Rammals." The digging stops when the roof of the water table groundwater is reached. At the bottom of the depression, the palm grove is installed. The palm tree roots absorb water from the soil horizons moistened by capillary action from the saturated layer. Thus, without having to resort to a conventional water mobilization, often budgetary, and to any conventional watering system, thousands of palm trees develop there and create an autonomous oasis life base (Figure 10.16). The advantage of this original

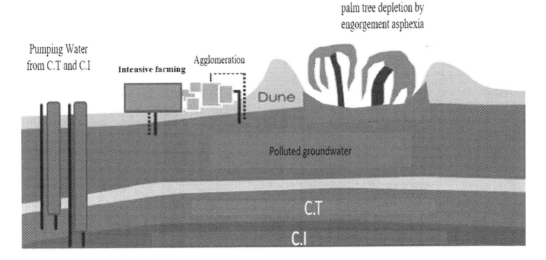

FIGURE 10.16 Impact of the rising water process whithin the ghouts (Remini, 2006).

technique is that it creates a microclimate protected from the siroccos where other crops can develop. Another danger spreads from below, that of the rise of the water table groundwater. The lack of a drainage network has favored the rise of the water table groundwater, as all the domestic and agricultural water discharges are added regularly. Several oases have already died by asphyxiation in this region where there are some 9,500 ghouts. A mega project is launched recently by the authorities to mitigate this phenomenon by evacuating the excess water to the adjacent chott. It is a real challenge for local populations and the survival of the oasis. These efforts of the public authorities have been able to turn the situation back to normal. Nevertheless, farmers must constantly monitor the level of the water table groundwater and pump water out of the ghout in case of overheating.

10.7 PERSPECTIVES FOR SUSTAINABLE MANAGEMENT OF SAHARAN AGRICULTURE

The impact of climate change is clearly perceived by local farmers, particularly in the region of Biskra, through rising temperatures, drought, disturbance of the crop culture cycle, lower yields, loss of soil fertility and increased pests and diseases of crops. The main effects on water resources are the depletion of wells, the drawdown of water levels in forages, the disappearance of artesianism and an increase in water salinity. To remedy this problem, farmers have implemented strategies to adapt to climate change, the most common of which are improved cultivation techniques, increased irrigation frequency with water-saving techniques (irrigation management), the use of early and drought-resistant varieties, the proper application of inputs and fertilizers, and the development of other activities parallel to agricultural production. The choice of development agricultural lands, the choice of irrigation methods, the control of water inputs and the monitoring of irrigated areas are the most effective measures for reducing soil salinization, controlling water loss, environmental degradation, and maintaining a sustainable and productive agriculture. This, of course, can only be done by promoting local expertise and research and development (Daoud and Halitim, 1994). In addition, encouraging the participation of farmers in the process of identifying crops that are best suited to the Saharan environment and climatic conditions could yield tangible results and extend the impact to other regions. There is a need to generalize this incentive to help decision-makers better redirect and frame their development strategies by adopting a participatory policy. From another point of view, it is interesting to learn, to correct and to innovate in order to propose new practices that are respectful of the environment and easy to implement. In addition, the establishment of the living-lab is also seen as an initiative to identify the water productivity of crops and thus tolerance of some crops with respect to water and soil salinity. Figure 10.17 shows a living-lab illustrated by a trial field in the wilaya of Touggourt with

FIGURE 10.17 Living-lab situated in Touggourt. Photo taken during the visit of Pr. Lal to Algeria in December (2022).

the establishment of some crops (sunflower, alfalfa, corn and clover, market crops and quinoa) to know their levels of adaptability to abiotic stress (Yaiche et al, 2016) and mastery of water-saving techniques with the technical supervision of the Direction des Services Agricoles (DSA) of Touggourt and the Algerian National Institute of Agronomic Research (INRAA).

Local awareness campaigns and impact and evaluation studies of the systems and policies put in place are urgently needed. Universities and research centers alone cannot carry out such work given the vastness of the territory and the complexity of the problems of the environment. The effective involvement of institutes through the creation of bridges between existing institutions in the region allows to mutualize the means and strengthen them as well as to share knowledge, experiences and know-how. Sustainable management of water resources requires, among other things, support for water saving to encourage collective irrigation and the adoption of water-saving systems, and the regularization of the legal status of certain farms. An institutional arrangement to create an interconnection between institutions would be a collaborative idea, built around a common charter with a vision and targeted objectives. Land development in these areas, increasing productivity per unit area, and conservation require detailed studies here more than elsewhere (Halitim, 1988). The expansion of vegetable crop plasticulture in the Saharan regions should be thought about by focusing on more ecological and sustainable solutions. The use of multi-hooded greenhouses can be a solution to control inputs and irrigation, including the use of O.M. in adequate doses and mixtures. Farmers frequently use mineral fertilizers such as NPK with standard sulfate-based formulations that are widely available in the Algerian market. The adoption of other fertilizers more appropriate to the nature of the region and with a composition specific to the crops grown will certainly alleviate the problem of groundwater pollution by nitrates. They bring O.M. to be buried in the soil from neighbouring and distant wilayas. Indeed, the rate of accumulation of soil M.O depends on rainfall, temperature and soil texture, which remains a crucial factor for the level of accumulation of O.M. in the soil (An et al., 2019). Furthermore, the control of groundwater pollution by nitrates could not be conceived without a program to raise awareness among farmers and to popularize good cultural practices that would allow for optimal yields. Soil fertility management through Organic Agriculture (OA) practices can affect soil inorganic carbon. Indeed, the formation of pedogenic carbonates can be enhanced by increasing biogenic processes through the application of manure, composts, crop residues on the soil, especially if these are rich in cations (Ca^{2+}, Mg^{2+}) (Lal, 2004a, b). In addition, O.M. can improve the conservation of the water column and be increased by the application of lime. According to Koull and Halilat (2016), increasing doses significantly decreases the pH to 7.73 by the dose of 113.4 T/ha of sheep manure after 2 months. Regarding EC, on the other hand, M.O increased soil salinity through mineralization of these organic compounds EC = 7.83 mS/cm obtained with the dose of 37.8 T/ha after 3 months. In addition, and to considerably reduce pollution from septic tanks and in the absence of a natural outflow that further complicates the rejection of excess water and wastewater; the solution to the problem of the rise and mixing with wastewater requires the removal of septic tanks and the generalization of the sewerage network in the entire region of El Oued with domestic wastewater treatment plants. Failing that, a very effective technique has already been tried on a real scale in the wilaya of Touggourt, in the Oued Rhir valley, designed by INRAA researchers (Hafouda et al., 2008). This technique has given very satisfactory results and could even be duplicated in the region of El Oued, which has the same natural characteristics, particularly topographical, where the slopes are low. This is the phytodepuration technique that uses plant species with very high purifying power, such as vetiver and other indigenous grasses (Figure 10.18). In addition, the planting of eucalyptus trees, as a bio-drainage tree gives good results due to its root suction force and the capacity of this tree to dry out the swamps (Khezzani et al., 2022).

Under salinity conditions, the extension of drainage networks, cleaning and maintenance of these canals are considered as a vital solution for the sustainability of the agricultural systems in place, such as the largest drainage canal in the Oued Rhir valley, which extends over three major cities (Touggourt, Djamaa and El Meghaïer) and crosses many wet and saline depressions in the North-East Sahara (Berrakbia and Remini, 2020). This maintenance and clearing considerably reduce

FIGURE 10.18 Wastewater plant treatment located in the wilaya of Touggourt.

the concentrations of salts in the palm groves and neighboring plots (Sayah and Remini, 2019). Currently, with the advancement of technology, the use of the reverse osmosis desalination technique for saline water is considered an affordable and less expensive technique (Arroyo and Shirazi, 2012). The average costs vary between 0.25 and 0.43 US$, depending on the technology used and the concentration of salts to be diluted (Liu et al., 2020). Powering these devices with clean energy will further reduce the costs of desalination of brackish water. Choosing a reliable energetic transition system by integrating green renewable energies in combination with fossil energies and electric energies from the distribution network will increase the time of use per day. It should be mentioned that the Algerian Sahara has one of the highest solar irradiation levels in the world, where the average annual sunshine duration can reach over 3,500 hours/year (Fathi et al., 2017). This recourse will also help farmers, located far from the electrical grid, to move to more favorable regions. Moving away from current unsustainable forms of Saharan agriculture will require a practical vision of agriculture in the desert and a revision of the ambitions and ideology of agricultural policies in the Sahara (Saidani et al., 2022). Indeed, circular agriculture is based on the idea of reducing resource consumption and emissions to the environment by closing the loops of a complex, multidimensional system, which is in line with the broader philosophy of the circular economy (De Boer and Van Ittersum, 2018). The concept of circularity is that reuse requires fewer resources and energy and is therefore more economical than recycling and therefore more energy efficient (Korhonen et al., 2018). This approach will reduce the carbon footprint of products and increase the book value of resources in terms of their diversification of use and the added value of by-products. Several examples can be cited in this sense. It is important to note that Saharan farmers practice burying O.M. in the soil and making compost. The latter is made from a mixture of cattle manure (ratio two-thirds) and chicken droppings (ratio one-third) mixed with date palm bunches and gives better results. Other formulations are being tested with research organizations such as INRAA and universities in the region. These formulation trials attempt to obtain compost products capable of offering better physico-chemical fertility of the soil and also capable of offering comfort to the plant with respect to the salinization brought by the water and the degradation of the OM. Also, it is necessary to speak about fertigation by water from fish farming containing fish excrements diluted in water and which proves to be a model to be followed in regions with similar conditions (Figure 10.19). This technique allows not only organic fertigation but also a diversification of production and also an offer of protein source of fish in the Saharan markets far from the coasts.

The valorization of the rejections of date palms consists of the use of the palms to protect the oases against the violent winds and against the silting. Also, they are used against the silting of roads and paths in oasis localities as shown in Figure 10.20.

FIGURE 10.19 System of fish farming and fertigation.

FIGURE 10.20 Use of palm waste as a windbreaker and sand barrier.

10.8 CONCLUSION

Saharan agriculture is a large-scale agricultural enterprise that defies climatic and edaphic conditions based on the availability of groundwater resources and human intelligence. How to take advantage of these billions of cubic meters of water buried in the subsoil without damaging the environment and ensuring sustainability in time and space?

At present, the agricultural production of various consumer products – mainly vegetables and fruits is encouraging given their availability throughout the country at reasonable prices. These Saharan regions are considered as the vegetable crop farm of Algeria. This fact is made possible by the efforts of the government in terms of facilitation and encouragement (finance, legislation…). Although the results are convincing, the objectives are not yet achieved.

Indeed, the objectives assigned are to produce more, which in the first instance is to ensure food security before moving on to a second stage of export of quality agricultural products. This trajectory must be accompanied by the sustainability of these agrosystems while preserving the environment. In this Saharan environment, the constraints are strong and numerous. The arid climate is characterized by strong evapotranspiration if we associate the plant with the physical phenomenon. The projections of climate change do not help the situation since the forecasts see an increase in temperatures. Therefore, in relation to this climatic aspect, irrigation techniques must be adapted to limit water losses. Still in relation to the climate, winds are a recurring factor and must be taken into consideration by their impacts. Their adverse effects on crops are not negligible, especially on the "ghouts" in the region of El Oued-Souf. After the climatic constraints, those of the grounds are very worrying. The soil which is the support of the plant is dominated by the sandy fraction accompanied by saliferous minerals of geological origin of which gypsum and limestone are widely spread. In its natural state, the O.M. is insignificant. Anthropic intervention through unsuitable irrigation techniques, without any calculation of irrigation doses, without any drainage system or lack of maintenance when it exists, results in a visible salinization of the soil. Layers of salts accumulate on the surface and in the rhizosphere and the consequences are reduced yields and even crop failure of sensitive crops. The groundwater resource of the Sahara is very important from a quantitative point of view, but two major constraints must be noted: The first is related to their low renewal or even non-renewal, and therefore exhaustible over time. Moreover, within the framework of this Saharan project, many forages are carried out in an illicit way, without authorization from the administration; the intensive pumping at the level of certain localities has led to a significant drawdown of the water table. The second constraint is their mineralization and consequently their use must be done with care, especially in the agricultural field. For the very deep aquifers, the waters are not only mineralized but also heated, which requires a refreshment before their use. In view of this, monitoring of piezometric levels and the chemical composition of the water in the various aquifers, in time and space, must be carried out regularly and carefully. Limiting depths or critical flows, which must not be exceeded, must be the basis for the exploitation of groundwater. Also, the combination of surface water, although not abundant, with groundwater is a significant advantage that must be encouraged. The possibilities of recharging the aquifers from floods or water from wastewater treatment plants (treatment and purification plants) must be favored. In relation to the salinity of the water, judicious choices of crops must be made to avoid crop failures or unacceptable yield reductions. The know-how of the oasis people who have lived in harmony in this hostile environment for thousands of years is of great interest that must be exploited and adapted to the requirements of modernity. Indeed, one of the examples of genius of the man of the Sahara is the idea of the creation of the "Ghout", of its development, implementation, monitoring, and maintenance until adulthood. This invention is justified by an essential balance and an almost permanent maintenance. The modernity, badly adapted, especially by the increasing demography, following the elevation of the water levels in the topographic depressions, these "Ghouts" and sometimes blocks of houses are flooded and sometimes drowned by the rise of the water table groundwater, which has led to the death of thousands of young palm trees.

This conclusion is intended to be a report on the state of Algerian Saharan agriculture, highlighting the advantages and constraints encountered, both natural and man-made. To achieve national food security in the southern regions, the assistance and involvement of all specialists, especially academics, is required to meet the challenge. The use and exploitation of local know-how will certainly be able to participate and contribute to a better adaptation to the environment. The use and exploitation of local know-how can certainly contribute to a better adaptation to the environment. The ancestral systems, implemented for centuries by the local population, summarize an extract of knowledge transmitted from one generation to another. Ultimately, the exchange and sharing of knowledge and preliminary knowledge about the success of the implementation of this agriculture with other countries with similar conditions can mitigate the risks of undernourishment and achieve food security in these countries.

ACKNOWLEDGMENTS

The authors thank:

- M. Ali Ferrah Director of INRAA for his support and encouragement.
- All researchers from INRAA's research station of Touggourt to have participated in the success of our investigation in the region especially: Lakhdari, W; Dahliz, A, Hafouda, L; Berrakbia, M.
- The directors of the agricultural services of Touggourt and El Oued respectively Mrs. Hanan Labyoudh and Mr. Matallah.
- The faculty of the Biology Department at the University of El Oued.
- The farmers of the region who collaborated positively and also the security services
- All the people who have contributed in any way to the realization of this work
- M. Bakor Habib for his help for the conception of some figure in this chapter

REFERENCES

Abdennour, M.A., Douaoui, A., Piccini, C., Pulido, M., Bennacer, A., Bradaï, A., ... & Yahiaoui, I. (2020). Predictive mapping of soil electrical conductivity as a proxy of soil salinity in south-east of Algeria. *Environmental and Sustainability Indicators*, 8 (100087), 1–13.

Adair, P., Lazreg, M., Bouzid, A., & Ferroukhi, S.A. (2022). L'agriculture Algérienne, l'héritage du passé et les défis contemporains. *Les cahiers du CREAD*, 38(3), 413–440. https//doi.org/10.4314/cread.v38i3.15.

Aissaoui, H., & Barkat, D. (2020). Physico-chemical characterizations and impact of organic matter on the dynamics of heavy metals (Cu, and Zn) in some soils of Biskra (Algeria). *Journal of King Saud University-Science*, 32(1), 307–311. https//doi.org/10.1016/j.jksus.2018.05.016.

Al-Ghobari, H.M. (2011a). The effect of irrigation water quality on soil properties under center pivot irrigation systems in central Saudi Arabia. *WIT Transactions on Ecology and the Environment*, 145, 507–516.

Al-Ghobari, H.M. (2011b). Effect of irrigation water quality on soil salinity and application uniformity under center pivot systems in arid region. *Australian Journal of Basic and Applied Sciences*, 5(7), 72–80.

Amichi, F., Bouarfa, S., Lejars, C., Kuper, M., Hartani, T., Daoudi, A., Amichi, H., & Belhamra, M. (2015). Des serres et des hommes, des exploitations motrices de l'expansion territoriale et de l'ascension socio-professionnelle sur un front pionnier de l'agriculture saharienne en Algérie. *Cahiers de l'Agriculture*, 24, 11–19. https//doi.org/10.1684/agr.2015.0736.

Amrani, K. (2021). Durabilité des agrosystèmes oasiens, évaluation et perspectives de développement, cas de la palmeraie de Ouargla (Algérie). [Doctoral Dissertation, Grenoble Alpes University, France] HAL open science Repository. https://theses.hal.science/tel-03329706/document.

An, H., Wu, X., Zhang, Y., & Tang, Z. (2019). Effect of land-use change on soil inorganic carbon, a meta-analysis. *Geoderma*, 353, 273–282. https//doi.org/10.1016/j.geoderma.2019.07.008.

Anderson, C.R., Condron, L.M., Clogh, T.J., Fiers, M., Stewart, A., Hill, R.A., & Scherlock, R.R. (2011). Biochar induced soil microbial community change, implications for biogeochemical cycling of carbon, nitrogen and phosphorus. *Pedobiologia*, 54(5–6), 309–320. https//doi.org/10.1016/j.pedobi.2011.07.005.

Aragüés, R., Urdanoz, V., Çetin, M., Kirda, C., Daghari, H., Ltifi, W., Lahlou, M., & Douaik, A. (2011). Soil salinity related to physical soil characteristics and irrigation management in four Mediterranean irrigation districts. *Agricultural Water Management*, 98(6), 959–966. https://doi.org/10.1016/j.agwat.2011.01.004.

Arroyo, J., & Shirazi, S. (2012). Cost of brackish groundwater desalination in Texas. *Texas Water Development Board*, 1–8.

Askri, B., & Bouhlila, R. (2010). Evolution de la salinité dans une oasis moderne de Tunisie. *Etude et Gestion des Sols*, 17(3–4), 197–212.

Ayers, R.S., & Westcot, D.W. (1985). *Water Quality for Agriculture: FAO Irrigation and Drainage Paper 29. Food and Agriculture Organization of the United Nations Rome*. 130 p.

Azzouzi, S.A., Vidal-Pantaleoni, A., & Bentounes, H.A. (2017). Desertification monitoring in Biskra, Algeria, with Landsat imagery by means of supervised classification and change detection methods. *IEEE Access*, 5, 9065–9072.

Bachir, H., Etsouri, S., Smadhi, D., & Semar, A. (2022). Representation of rainfall in regions with a low distribution of rain gauging stations. In Gökçekuş, H., & Kassem, Y. (eds), *Climate Change, Natural Resources and Sustainable Environmental Management. NRSEM 2021. Environmental Earth Sciences* (pp. 262–271). Springer, Cham. https://doi.org/10.1007/978-3-031-04375-8_30.

Bachir, H., Kezouh, S., Semar, A., Smadhi, D., & Ouamer-ali, K. (2021). Improvement of interpolation using information from rainfall stations and comparison of hydroclimate changes (1913–1938)/(1986–2016). *Al-Qadisiyah Journal for Agriculture Sciences*, 11(1), 54–67.

Bakari, N.E., Halis, Y., Benhaddya, M.L., & Saker, M.L. (2017). Étude de l'impact des activités agricoles sur l'environnement Oasien de la région de l'Oued Righ. *Journal Algérien des Régions Arides*, 14, 49–59.

Bazzine, M., & Hamdi-Aissa, B. (2014). Etude Des Croutes Biologiques De Quelques Sols Gypseux Et Salins Du Milieu Saharien, Cas De La Cuvette De Ouargla (Sahara Septentrional Est Algerien). *Algerian Journal of Arid Environment*, 4(1), 45–52.

Bedjadj, S. (2011). Contribution à l'étude des caractéristiques microbiologiques des sols dans la région de Ouargla (Cas de l'exploitation de l'Université de Ouargla). [Ingeneering Dissertation. Kasdi Merbah University, Ouargla]. DSpace Repository. https://dspace.univ-ouargla.dz/jspui/handle/123456789/4561.

Bel, F., & Cuche, D. (1969). *Mise au point des connaissances sur la nappe du Complexe Terminal*. ERESS, Ouargla. 20 p.

Ben Hassine, H. (2005). Effets de la nappe phréatique sur la salinisation des sols de cinq périmètres irrigués en Tunisie. *Etude et Gestion des Sols*, 12(4), 281–300.

Bencharif, S. (2018) Origines et transformations récentes de l'élevage pastoral de la steppe algérienne. *Revue internationale des études du développement*, 236, 55–79.

Berrakbia, M., & Remini, B (2020) Evaluation of amounts of salts brought by righ canal (Algeria), Analysis of correlation and multiple linear regression. *PONTE International Journal of Science and Research*, 76(2). 24–46.

Bessaoud, O., Pellissier, J.P., Rolland, J.P., & Khechimi, W. (2019). Rapport de synthèse sur l'agriculture en Algérie. Rapport de recherche CIHEAM-IAMM. https://www.iamm.ciheam.org/ress_doc/opac_css/doc_num.php?explnum_id=18246.

Biddlestone, A.J., & Gray, K.R. (1988). A review of aerobic biodegradation of solid wastes. *Biodeterioration*, 7, 825–839.

Blaylock, A.D. (1994). *Soil Salinity, Salt Tolerance and Growth Potential of Horticultural and Landscape Plants*. Co-operative Extension Service, University of Wyoming, Department of Plant, Soil and Insect Sciences. College of Agriculture, Laramie, Wyoming. 4p.

Boivin, P., Favre, F., Hammecker, C., Maeght, J.L., Delarivière, J., Poussin, J.C., & Wopereis, M.C.S. (2002). Processes driving soil solution chemistry in a flooded rice-cropped vertisol: Analysis of long-time monitoring data. *Geoderma*, 110(1–2), 87–107.

Bouammar, B., & Bakhti, B. (2008) Le développement de l'économie agricole oasienne, entre la réhabilitation des anciennes oasis et l'aménagement des nouvelles palmeraies. *El-Bahith Review*, 6(1), 19–24. https//www.elbahithreview.edu.dz/index.php/bahith/article/view/707.

Boucetta, D. (2018). Effets des changements climatiques sur les cultures pratiquées et les ressources en eau dans la région de Biskra. [Doctoral Dissertation, Mohamed Khider University, Biskra]. University of Biskra Theses Repository. https://thesis.univ-biskra.dz/4571/.

Bouder, A., & Chella, T. (2017). Contribution de l'agriculture saharienne a la sécurité alimentaire en Algérie: mythe ou réalité?. *Lucrările Seminarului Geografic" Dimitrie Cantemir"*, 44, 159–174.

Bouguettaya, H. (2020). Essai de caractérisation du statut humique dans les régions arides (Cas de la région d'Ouargla). [Master dissertation, Kasdi Merbah University, Ouargla]. DSpace Repository. https://dspace.univ-ouargla.dz/Repository/handle/123456789/28683.

Boumadda, A. (2019). Dynamique et durabilité des systèmes agricoles oasiens dans le Sahara Septentrional Algérien, Cas du Pays de Ouargla et du Souf. [Doctoral Dissertation. Kasdi Merah University, Ouargla]. DSpace Repository. https://dspace.univ-ouargla.dz/Repository/handle/123456789/22720.

Boussaada, N., Bouselsal, B., Benhamida, S.A., Hammad, N., & Kharroubi, M. (2023). Geochemistry and water quality assessment of continental intercalary aquifer in Ouargla Region (Sahara, Algeria). *Journal of Ecological Engineering*, 24(2), 279–294.

Busson, G. (1970). Le Mésozoïque saharien. 2ème partie: Essai de synthèse des données des sondages algéro-tunisiens. *Centre Rech. Zones Arides Géol.Eds. C.N.R.S. 11*, 811 p.

Chabour, N., Dib, H., Bouaicha, F., Bechkit, M.A., & Messaoud Nacer, N. (2021). A conceptual framework of groundwater flowpath and recharge in Ziban aquifer: South of Algeria. *Sustainable Water Resources Management*, 7, 1–15.

Chemsa, Y. (2019). Contribution à l'étude de l'évolution d'un sol sableux amélioré avec bio- charbon d'origine végétale dans la région d'El-oued. [Master dissertation. Echahid Hamma Lakhdar University, El-Oued]. DSpace Repository. https://dspace.univ-xeloued.dz/xmlui/handle/123456789/4315.

Conrad, G. (1969). L'évolution continentale post-hercynienne du Sahara Algérien. *CRZA-CNRS*. https://www.cnrseditions.fr/catalogue/ecologie-environnement-sciences-de-la-terre/evolution-continentale-post-hercynienne-du-sahara-algerien/.

Côte, M. (2002). Des oasis aux zones de mise en valeur, l'étonnant renouveau de l'agriculture saharienne. *Méditerranée*, 99(3), 5–14.

Daoud, Y., & Halitim, A. (1994). Irrigation et salinisation au Sahara algérien. *Science et changements planétaires/Sécheresse*, 5(3), 151–160.

de Boer, I.J.M., & van Ittersum, M.K. (2018). *Circularity in Agricultural Production*. Wageningen University & Research Eds. 74. https://edepot.wur.nl/470625

Dorize L. (1982). L'aridité saharienne (Saharan aridity). *Bulletin de l'Association de géographes français*, 483, 30–33. https//doi.org/10.3406/bagf.1982.529.

Drouet, T.H. (2010). Pédologie. *Book Pédologie. BING-F-302*. Ed.Lagev, Paris. 140 p.

Dubost, D. (1986). Nouvelles perspectives agricoles du Sahara algérien. *Revue des mondes musulmans et de la méditerranée*, 41(1), 339–356.

Dubost, D., & Moguedet, G. (2002). La révolution hydraulique dans les oasis impose une nouvelle gestion de l'eau dans les zones urbaines. *Méditerranée, tome 99.3-4. Le sahara, cette "autre Méditerranée" (Fernand Braudel)*. Paris. 15–20. https://doi.org/10.3406/medit.2002.3254.

Durand, J.H., & Guyot, J. (1955). *Irrigation des cultures dans l'Oued Righ. Travaux de l'IRS. Tome 13*. Université d'Alger.

Etsouri, S., Bachir, H., Bouaziz, M., Kaci, F., Malkia, R., & Etsouri, K.. (2022). The continental intercalary in Algeria, analysis, survey and perspectives for green agriculture development. *Al-Qadisiyah Journal for Agriculture Sciences*, 12(1), 20–29.

Fabre, J. (1976). Introduction à la géologie du Sahara algérien. Ed, Société Nationale d'Édition et de Diffusion, Algeirs, Algeria . 421 p.

FAO. 2022. World Food and Agriculture – Statistical Yearbook 2022. Rome. https://doi.org/10.4060/cc2211en

Fathi, M., Abderrezek, M., & Grana, P. (2017). Technical and economic assessment of cleaning protocol for photovoltaic power plants, case of Algerian Sahara sites. *Solar Energy*, 147, 358–367.

Ghadbane, M., Harzallah, D., Ibn Laribi, A., Jaouadi, B., & Elhadj, H. (2013). Purification and biochemical characterization of a highly thermostable bacteriocin isolated from *Brevibacillus brevis* strain GM100. *Bioscience Biotechnology and Biochemistry*, 77(1), 151–160.

Ghassemi, F., Jakeman, A.J., & Nix, H.A. (1995). *Salinisation of Land and Water Resources: Human Causes, Extent, Management and Case Studies*. Centre for Resource and Environmental Studies CAB International. 544p.

Guendouz, A., & Moulla, A.S. (1992). *Hydrochemical and Isotopic Investigation of Ouargla Depression Groundwaters, Algeria, 1993–1995*. 2nd *Tech Rep CDTN/DDHI, Algiers*. 107p.

Guendouz, A., Moulla, A.S., Edmunds, W.M., Zouari, K., Shand, P., & Mamou, A. (2003). Hydrogeochemical and isotopic evolution of water in the complexe terminal aquifer in the Algerian Sahara. *Hydrogeology Journal*, 11(4), 483–495.

Guendouz, A., Moulla, A.S., Remini, B., & Michelot, J.L. (2006). Hydrochemical and isotopic behaviour of a Saharan phreatic aquifer suffering severe natural and anthropic constraints (case of Oued-Souf region, Algeria). *Hydrogeology Journal*, 14(6), 955–968.

Guesseire, S., Fethiza, T.I., BenMoussa, I., & Bourasse, O. (2022). Contribution à l'étude du comportement chimique d'un sol sableux (ph et pouvoir tampon) amélioré avec bio-harbon d'origine végétale dans la région d'el oued. [Master dissertation. Echahid Hamma Lakhdar University, El-Oued]. https://dspace.univ-eloued.dz/xmlui/handle/123456789/12342?show=full.

Gueziz, W., & Hamouta, M. (2020). Contribution à l'évaluation de la teneur en matière organiques sols de la région de Ouargla. [Master dissertation. Kasdi Merbah University, Ouargla]. DSpace Repository. https://dspace.univ-ouargla.dz/jspui/handle/123456789/29155.

Guiraud, R. (1973). Evolution post-triasique de l'avant-pays de la chaîne alpine en Algérie. [Doctoral dissertation, Toulouse, France]. Toulouse University, Catalogue SUDOC. https://www.sudoc.abes.fr/cbs/DB=2.1/SRCH?IKT=12&TRM=009016031.

Hachicha, M., Cheverry, C., & Mhiri, A. (2000). The impact of long-term irrigation on changes of ground water level and soil salinity in northern Tunisia. *Arid Soil Research and Rehabilitation, 14*(2), 175–182.

Hafhouf, I. (2022). Étude de l'endommagement des sols salés soumis à des sollicitations d'humidification Séchage (température-humidité). [Doctoral Dissertation, Batna 2 University, Batna]. University of Batna Theses Repository. https://eprints.univ-batna2.dz/2066/.

Hafouda, L., Hadad, M., Arif, Y., Djafri, K., Balleche, O., Talab, B., & Cattin, F. (2008). L'épuration des eaux usées domestiques par les plantes, une alternative à encourager pour une préservation durable de l'environnement en zones arides, cas de la station pilote du vieux Ksar de Témacine, Touggourt. *International Symposium on Dryland Cultivation, Optimization of Agricultural Production and Sustainable Development (Vol. 13)*, Colloque International sur l'ARIDOCULTURE Optimisation des Productions Agricoles et Développement Durable. Centre for Scientific and Technical Research on Arid Regions (CRSTRA), Biskra, Algeria. 75–84.

Hakkoum, H. (2018). Contribution à l'étude de la microflore fongique du sol dans deux stations de la région de Biskra. [Master dissertation. Mohamed Khider University, Biskra]. Biskra University Theses and Dissertations Archive. https://archives.univ-biskra.dz/bitstream/123456789/13365/1/hamed_hakoum.pdf.

Halilat, M. (1998). Étude expérimentale de sable additionné d'argile. Comportement physique et organisation en conditions salines et sodiques. [Doctoral Dissertation, Institut National Agronomique Paris-Grignon, France]. BeL- INRAe online Library. 229p. https://belinra.inrae.fr/index.php?lvl=notice_display&id=179015.

Halilat, M.T., & Tessier, D. (2002). Amélioration Des Propriétés Physiques Des Sols Sableux Du Sahara Algerien Par Ajout D'argile. *Journal Algérien des Régions Arides, 1*(1), 50–61.

Halitim, A., & Robert, M. (1992). *Geneses of Gypseous and Calcareous Formations in Arid Zone (Algeria). Dynamics and Effects in Soil Properties. Workshop on Gypseous Soils*. ICARDA-FAO, Aleppo, Syria. 11p.

Hamdi-Aissa, B., Vallès, V., Aventurier, A., & Ribolzi, O. (2004). Soils and brine geochemistry and mineralogy of hyperarid desert playa, Ouargla Basin, Algerian Sahara. *Arid Land Research and Management, 18*(2), 103–126.

Hatira, A., Baccar, L., Grira, M., & Gallali, T. (2007). Analyse de sensibilité du système oasien et mesures de sauvegarde de l'oasis de Métouia (Tunisie). *Revue des sciences de l'eau/Journal of Water Science, 20*(1), 59–69.

Houichiti, R., Bouammar, B., & Bissati, S. (2020) Dynamique de L'agriculture Saharienne en Algérie (Cas De La Région De Ghardaïa). *Revue des bio ressources, 10*(1), 10.

Kadri, A., & Van Ranst, E. (2002). Contraintes de la production oasienne et stratégies pour un développement durable. Cas des oasis de Nefzaoua (Sud tunisien). *Science et changements planétaires/Sécheresse, 13*(1), 5–12.

Kahlaoui, B. (2021). Contribution to the evaluation of organic matter content of natural and cultivated soils in Ouargla region. [Master Dissertation, Kasdi Merbah University, Ouargla]. Ouargla University, DSpace Repository. https://dspace.univ-ouargla.dz/jspui/handle/123456789/29064.

Karabi, M., Aissa, B.H., & Zenkhri, S. (2016). Microbial diversity and organic matter fractions under two arid soils in Algerian Sahara. *AIP Conference Proceedings, 1758*, 030006. https://doi.org/10.1063/1.4959402.

Karabi, M., Aissa, B.H., Zenkhri, S., Kemassi, A., & Bouras, N. (2015). Seasonal variations affect microbiocenose arid soils in the Ouargla basin (Algerian Sahara). *Ciência e Técnica Vitivinícola, 30*(8), 176–187.

Katerji, N., Mastrorilli, M., & Rana, G. (2008). Water use efficiency of crops cultivated in the Mediterranean region: Review and analysis. *European Journal of Agronomy, 28*(4), 493–507.

Khallef, S., Lestini, R., Myllikallio, H., & Houali, K. (2018). Isolation and identification of two extremely halophilic archaea from sebkhas in the Algerian Sahara. *Cellular and Molecular Biology, 64*(4), 83–91. https://doi.org/10.14715/cmb/2018.64.4.14.

Kharroubi, M., Bouselsal, B., Ouarekh, M., Benaabidate, L., & Khadri, R. (2022). Water quality assessment and hydrochemical characterization of the ouargla complex terminal aquifer (Algerian Sahara). *Arabian Journal of Geosciences, 1*, 24.

Khechana, S., Derradji, F., & Mega, N. (2011). Caractéristiques Hydrochimiques Des Eaux De La Nappe Phréatique Du Vallée d'Oued-Souf (SE Algérien). *European Journal of Scientific Research, 62*(2), 207–215.

Khezzani, B., Khechekhouche, E.A., Zaater, A., Guezzoun, N., Tliba, B., Brahim, A.B., & Zeghdi, A. (2022). Eucalyptus sp. as biodrainage system in an arid region, a case study from the Souf Oasis (South Algeria). *Macedonian Journal of Ecology and Environment*, 24(1), 31–38.

Kitous, S., Bensalem, R., & Adolphe, L. (2012). Airflow patterns within a complex urban topography under hot and dry climate in the Algerian Sahara. *Building and Environment*, 56, 162–175. https//doi.org/10.1016/j.buildenv.2012.02.022.

Korhonen, J., Nuur, C., Feldmann, A., & Birkie, S.E. (2018). Circular economy as an essentially contested concept. *Journal of Cleaner Production*, 175, 544–552.

Koull, N., & Halilat, M.T. (2016). Effets de la matière organique sur les propriétés physiques et chimiques des sols sableux de la région d'Ouargla (Algérie). *Etude et Gestion des sols*, 23, 9–23.

Kouzmine, Y., Fontaine, J., Yousfi, B.E., & Otmane, T. (2009). Étapes de la structuration d'un désert, l'espace saharien algérien entre convoitises économiques, projets politiques et aménagement du territoire. *Annales de géographie*, 6, 659–685. https//doi.org/10.3917/ag.670.0659.

Lal, R. (2004a). Carbon sequestration in dryland ecosystems. *Environmental Management*, 33, 528–544. https//doi.org/10.1007/s00267-003-9110-9.

Lal, R. (2004b). Carbon sequestration in dryland agriculture. *Challenges and Strategies of Dryland Agriculture*, 32, 315–334. https//doi.org/10.2135/cssaspecpub32.c20.

Lal, R. (2018). Digging deeper, a holistic perspective of factors affecting soil organic carbon sequestration in agroecosystems. *Global Change Biology*, 24(8), 3285–3301. https//doi.org/10.1111/gcb.14054.

Lasmar, R.B., Guellala, R., & Inoubli, M.H. (2022). New elements on the "Continental Intercalaire," the most important aquifer in North Africa, joint use of seismic reflection and borehole data. *Arabian Journal of Geosciences*, 15(8), 1–22.

Lemarchand, F. (2008). Les nappes fossiles du Sahara, L'eau, Comment l'utiliser, comment la préserver. *La Recherche (Imprimé)*, 421, 60–61.

Li, P., Qian, H., & Wu, J., (2018). Conjunctive use of groundwater and surface water to reduce soil salinization in the Yinchuan Plain, North-West China. *International Journal Water Resources Development*, 34(3), 337–353. https//doi.org/10.1080/07900627.2018.1443059.

Liu, X., Shanbhag, S., Bartholomew, T.V., Whitacre, J.F., & Mauter, M.S. (2020). Cost comparison of capacitive deionization and reverse osmosis for brackish water desalination. *Acs Es&T Engineering*, 1(2), 261–273.

Lorenz, K., & Lal, R. (2022). Introduction to organic agriculture. *Organic Agriculture and Climate Change* (pp. 1–38). Springer International Publishing, Cham. https//doi.org/10.1007/978-3-031-17215-1_1.

Mancer, H., Bettiche, F., Chaib, W., Dekki, N., Benaoun, S., & Rechachi, M.Z. (2020). Influence de la salinité des eaux d'irrigation sur la minéralisation du carbone organique dans le sol. *Algerian Journal of Arid Regions*, 14(1), 48–55.

Margat, J., Foster, S., & Droubi, A. (2006). Concept and importance of non-renewable resources. *Non-Renewable Groundwater Resources: A Guidebook on Socially-Sustainable Management for Water-Policy Makers*, 10, 13–24.

M. Besbes, B. Abdous, B. Abidi, A. Ayed, M. Bachta, M. Babasy, B. Ben Baccar, D. El Batti, Y. Ben Salah, M. Biet Charreton, F. Biout, A. Douma, C. Fezzani, M. Gadhi, F. Horriche, S. Kadri, A. Khadraoui, R. Khanfir, V Kinzelbach, A. Larbes, D. Latrech, J. Margat, G. De Marsily, A. Mamou, M. El Mejerbi, A. Mekrazi, A. Mhiri, L. Moumni, M. Nanni, P. Pallas, G. Pizzi, A. Salem, O. M. Salem, R. Taibi & M. Zammouri (2003) Système Aquifère du Sahara septentrional Gestion commune d'un bassin transfrontière, *La Houille Blanche*, 89:5, 128–133, DOI: 10.1051/lhb/2003102

Mohamed, N.H., Benziada, R., Benlahbib, F.Z., Benkhedda, F., & Benlahbib, M. (2022). Evolution of the place of plants and water elements in the Saharan City. What results?. Rosso, F., Morea, D., & Pribadi, D.O. (eds). *Innovations in Green Urbanization and Alternative Renewable Energy. Advances in Science, Technology & Innovation*. Springer, Cham. https//doi.org/10.1007/978-3-031-07381-6_10.

Mostephaoui, T., & Bensaid, R. (2014). Caractérisation des sols gypseux dans les zones arides par télédétection. Cas du sous-bassin versant d'oued djedi-biskra. *Lebanese Science Journal*, 15(1), 99–115.

Mouffok, A. (2010). La symbiose à rhizobia chez la fève (*Vicia faba* L.) et la luzerne (*Medicago sativa* L.) dans la région de Biskra. [Master Dissertation. Mohamed Khider Uniersity, Biskra]. Biskra University Thesis Repository. https://thesis.univ-biskra.dz/894/.

Negacz, K., Malek, Ž., Vosb de, A., & Vellinga, P. (2022). Saline soils worldwide: Identifying the most promising areas for saline agriculture. *Journal of Arid Environments*, 203, 104775. https://doi.org/10.1016/j.jaridenv.2022.104775.

Omari, C., Moisseron, J.Y., & Arlène, A. (2012). L'agriculture algérienne face aux défis alimentaires. *Trajectoire historique et perspectives. Revue Tiers Monde*, 2(128), 123–141.

Ould Baba Sy, M. (2005). *Recharge et paléorecharge du système aquifère du Sahara septentrional*. Thèse Faculté des Sciences de Tunis, Tunisie, 277p.

Oustani, M. (2006). Contribution A L'étude De L'influence Des Amendements Organique (Fumier De Volailles Et Fumier De Bovins) Sur L'amélioration Des Propriétés Microbiologiques Des Sols Sableux Non Salés Et Salés Dans Les Régions Sahariennes (Cas d'Ouargla). Master Dissertation, Kasdi Merbah University, Ouargla, 187p.

Qureshi, A.S., Ahmad, W., & AlFalahi, A. (2013). Optimum groundwater table depth and irrigation schedules for controlling soil salinity in central Iraq. *Irrigation and Drainage*, 62(4), 414–424. https//doi.org/10.1002/ird.1746.

Remini B. (2001). Mega obstacles : leur influence sur la dynamique Eolienne et l'ensablement des espaces oasiens. Doctorat thesis. University of Reims Champagne-Ardenne. 188 p.

Remini B. 2004. The upwelling in the region of El Oued. Environment Vector Review (Canada)

Remini, B. (2006). La disparition des Ghouts dans la région d'El Oued (Algérie). *LARHYSS Journal*, 5, 49–62.

Romani, M., Bezzala, N., & Lakhdari, F. (2007). Valorisation des sous-produits du palmier dattier comme amendement des sols. *Journal Algérien des Régions Arides*, 6, 49–59.

Saidani, A., Kuper, M., Hamamouche, F.M., & Benmihoub, A. (2022). Reinventing the wheel, adapting a traditional circular irrigation system to 'modern' agricultural extensions in Algeria's Sahara. *New Medit*, 21(5), 35–53. https//doi.org/10.30682/nm2205c.

Sayah Lembarek, M., & Remini, B. (2019). Evolution of the flow of drainage waters in the Oued Righ canal, Algeria. *Journal of Water and Land Development*, 41, 133–138. https//doi.org/10.2478/jwld-2019-0036.

Semar, A., Bachir, H., & Bourafai, S. (2021). Hydrochemical characteristics of aquifers and their predicted impact on soil properties in Biskra region, Algeria. *Egyptian Journal of Agricultural Research*, 99(2), 205–220. https//doi.org/10.21608/ejar.2021.56750.1068.

Semar, A., Hartani, T., & Bachir, H. (2019). Soil and water salinity evaluation in new agriculture land under arid climate, the case of the Hassi Miloud area, Algeria. *Euro-Mediterranean Journal for Environmental Integration*, 4(1), 40. https//doi.org/10.1007/s41207-019-0130-0.

Singh, A. (2015). Soil salinization and waterlogging, a threat to environment and agricultural sustainability. *Ecological Indicators*, 57, 128–130. https//doi.org/10.1016/j.ecolind.2015.04.027.

Soltner, D. (2003). Les bases de la production végétale, Tome I, le sol et son amélioration. *Edit Collection Science Technique Agricole*, 23, 472p.

Souadkia, C., & Souadkia, H. (2017). Etude des croûtes biologiques des sols des écosystèmes arides (Cas de la Wilayad'El Oued). [Master dissertation. Echahid Hamma Lakhdar University, El-Oued]. El-Oued University, Sécheresse info, https://www.secheresse.info/spip.php?article103711.

Subyani, A.M. (2005). Hydrochemical identification and salinity problem of ground-water in Wadi Yalamlam basin, Western Saudi Arabia. *Journal of Arid Environments*, 60, 53–66.

Touahri, M., Belksier, M.S., Boualem, B., & Kebili, M. (2022). Groundwater quality assessment of hassi messaoud region (Algerian Sahara). *Ecological Engineering Journal*, 23(11), 165–178. https//doi.org/org/10.12911/22998993/153396.

UNESCO. (1972). Etude des ressources en eau du Sahara Septentrional. *Algérie - Tunisie, Rapport sur les résultats du projet, conclusions et recommandations, Paris, France*. 116p.

Wang, H., & Jia, G. (2012). Satellite-based monitoring of decadal soil salinization and climate effects in a semi-arid region of China. *Advances in Atmospherique Sciences*, 29, 1089–1099. https//doi.org/10.1007/s00376-012-1150-8.

Yaiche, M.R., Bouhanik, A., & Bekkouche, S.M.A. (2016). A new modeling approach intended to develop maps of annual solar irradiation and comparative study using satellite data of Algeria. *Journal of Renewable and Sustainable Energy*, 8(4), 043702. https://doi.org/10.1063/1.4958993.

Yamaguchi, T., & Blumwald, E. (2005). Developing salt-tolerant crop plants: Challenges and opportunities. *Trends in Plant Science*, 10, 615–620. https://doi.org/10.1016/j.tplants.

Zenkhri, S., & Karabi, M. (2020). L'oasis de Ouargla (sud est algérien) entre déclin et programmes d'Etat de réhabilitation. *Revue des Bioressources*, 10(1), 85–91.

LIST OF ABBREVIATIONS

CI	Continental intercalary
CT	Compexe terminal
M.O	Organic matter
O.A	Organic agriculture
OSS	Observatory of Sahara and Sahel
SASS	Saharan aquifer septentrional system

11 Soil and Water Management by Climate Smart and Precision Agriculture Systems under Arid Climate

Y.G.M. Galal and S.M. Soliman

11.1 INTRODUCTION

A few couples of decades ago, more attention was paid to improving soil fertility status and gaining the most proper management strategies for irrigation water to combat severe climate changes and water scarcity, especially in arid and semi-arid zones. Under Egyptian conditions, the agricultural sector was suffering from a decline in fertile soil area and a shortage in required water for irrigation due to population and incorrect use of water resources. In addition, the rise of temperature in correlation with severe climate changes leads to soil desertification and modification in crop pattern and rotation (The authors). Water and land management is an important scope to face the irregular insufficient rainfall as a key factor to agricultural production especially in drylands, where productivity levels are generally low. Great attention has been paid to improve water use efficiency and productivity in conjunction with enhancement of fertilizers potentiality. In arid and semi-arid regions, water scarcity in addition to meager irrigation facilities leads to low and unstable yields. These factors are the main constraints in agricultural production. Such constraints and means to overcome were discussed by Singh et al. (2015). In dealing with irrigation water management, nuclear and isotopic techniques are found to play an important and sometimes unique role in providing information essential to developing strategies aimed at improving agricultural water use efficiency, and hence in providing solutions to mitigate the increasing water scarcity. The soil moisture neutron probe (SMNP) was found to be ideal for the measurement of soil water in the immediate vicinity of the crop roots and for providing accurate data on the accessibility to the crop of available water to establish optimal irrigation schedules.

Research and development priorities in addition to site-specific research are essential to identify practices that can adapt soil, water and crop management to changing climate. Earlier, Lal (2012a), pointed out that a strategy could be considered to validate and fine-tune the practices for sustainable management of soil (i.e., conservation tillage, mulch farming, nutrient requirement, AWC, and SOM management), crop (i.e., choice of species and varieties; time, configuration, and methods of sowing; consumptive water use; CO_2 fertilization effect; hydraulic lift; and growing season duration), and microclimate (i.e., temperature, precipitation, extreme event, hydraulic balance, and radiation budget). He added that adaptation is possible when basic information on soil, crops, microclimate, and mesoclimate is known. This information is needed at soilscape, landscape, and watershed (agricultural watersheds and river basin) scales. From our experiences, we can confirm that the application of nuclear techniques either stable or radioactive sources could offer a chance to understand what happened in soil (nutrient cycling and dynamics), soil moisture content and related parameters and help in identifying the most proper management strategies (Soliman and Galal, 2020).

Water as a dynamic resource, its quantity, quality and factors affecting availability such as population growth, land-use change, environmental pollution, and climate change were reviewed in a special issue of *Environmental Management J.* (Albay et al., 2021). They noted that there is a strong and consistent association between climate change and water scarcity. In addition to climate change, human activities also have direct effects on water resources, threatening ecosystems and human health.

By 2050, the total water demand for agriculture is expected to increase to 8,500–11,000 km^3/year (Rockström et al., 2010), depending on the development and adoption of new water-saving technology. The competing demand for water for industry and urban uses and climate change are likely to reduce the availability of water for irrigation. Thus, enhancing WUE and WP by decreasing losses, especially of soil water, is more important now than ever before. The goal of water management is to produce more crop per drop of water by managing the climate, water, soil, and crop. Optimizing the use of irrigation water is essential to saving dwindling water resources via different strategies to enhance the WUE in irrigated agriculture (Figure 11.1).

In addition to their use in investigating the effectiveness of management on crop water requirements, neutron moisture gauges permit long-term studies of soil moisture such as an assessment of the need for, or testing the performance of, drainage systems. The information gained through such studies is invaluable in making an overall plan to increase crop yields with improved economy in water use. Management of irrigation with saline water (Kassab et al., 2018), reflected a positive effect of mulching whereas the wheat yield and water use efficiency of mulched treatments were higher than the non-mulched ones for both 6 and 8 dS/m. In non-mulched conditions, the plants irrigated with 6 dS/m recorded a higher production accounted for 2411 kg/ha than those irrigated with 8 dS/m recording 2231 kg/ha.

The use of root growth hormones and enhanced nodulation can also increase the agronomic yield in water-deficit environments (Diaz-Zorita and Fernández-Canigia, 2009; Belimov et al., 2009). Water and plant nutrients (fertigation) must be delivered directly to the plant roots by drip subirrigation to minimize losses. Condensation irrigation, delivering water to the plant roots as vapors (as is the case in desert plants from the subsoil to the roots in the surface layer at night), is another option that must be explored. Strategies to enhance the WUE in irrigated agriculture are outlined in Figure 11.2.

Increasing the water supply in Egypt is questionable. Policy to achieve water security and food security is to increase water use efficiency and water productivity, producing more with less water in all water sectorial uses particularly the agriculture sector receiving nearly 85% of the available water resources, but with poor on farm water efficiency not exceeding the 50%. Major efforts are directed toward the agriculture through increasing crop water productivity, reducing water losses and raising water use efficiency. Technically, several approaches are now implemented for better water saving in irrigated agriculture among them the introduction of new irrigation techniques such as surface and subsurface drip irrigation, sprinkler irrigation and pivot systems. Those are the dominant systems for irrigation totally covering all the sandy new reclaimed areas accounting for nearly 1 million ha. In those new areas, irrigation is relatively expensive, there is cost for energy, cost for water, besides the cost of the system itself.

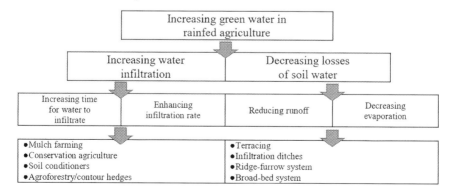

FIGURE 11.1 Principles and practices of enhancing green water in rainfed agriculture. Source: Cited from Lal (2012a).

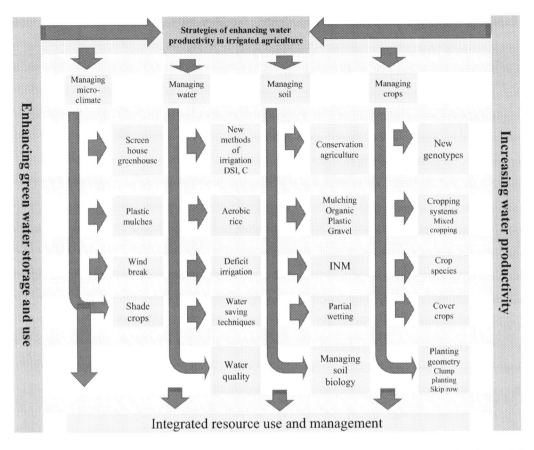

FIGURE 11.2 Strategies of enhancing water-use efficiency in irrigated agriculture. Source: Cited from Lal (2012a).

11.2 NUCLEAR EQUIPMENT FOR FIELD MEASUREMENT OF SOIL MOISTURE AND DENSITY

Nuclear equipment for field measurement of soil moisture and density has received wide acceptance from scientists and engineers. The neutron scattering method for measuring soil water content has been used extensively in soil water studies. Its wide acceptance results from the greater precision and rapidity of measurement than is possible with the gravimetric method. Additional major advantages of the neutron method over other methods include measurement in a relatively large volume of soil and over the full range of soil water.

An early and perhaps the most widely accepted application of the neutron technique for measuring moisture content is in agriculture. Because of threatened food shortages over the entire world, soil management, conservation of water and increased crop yield are of major importance. Frequently a water shortage, especially at critical periods of crop growth, is the main cause of low yield. Consequently, conservation practices in water control for all types of agriculture have become important.

In addition to their use in investigating the effectiveness of management on crop water requirements, neutron moisture gauges permit long-term studies of soil moisture such as an assessment of the need for, or testing the performance of, drainage systems. Information gained through such studies is invaluable in making an overall plan to increase crop yields with improved economy in water use. In agriculture, neutron moisture gauges are used mainly in the following types of study:

a. Plant-soil-water relations and plant growth; water use by plants; soil moisture deficiency and water use by native vegetation; soil moisture stream flow-tree growth relations; tree-soil water relations; and short-term effects of water on vegetation
b. Soil moisture storage and movement, including water content of soils, soil moisture accretion and depletion; soil moisture storage and unsaturated drainage; "moisture movement in stratified soils; water balance and movement; moisture retention capacity of soils; translocation of soil water in soil profiles; water movement through soil and rock; soil shrinkage studies and water losses due to unsaturated flow in alluvium
c. Irrigation and drainage, including measurement of moisture content to determine amounts of water used and water applied during irrigation, and evaluation of effectiveness of different drainage treatments
d. Evaporation and evapotranspiration, including consumptive use on watersheds; soil moisture depletion in relation to vegetation, weather and streamflow; evapotranspiration from dry farming plots under various plant cover densities and vegetation types; evapotranspiration from irrigated crops; effects of soil covers (plastic, straw, gravel, chemicals) in control of evaporation, and on plant growth and production; effects of land treatment on soil moisture use; and research to reduce evaporation
e. Recharge and yield, including natural recharge, of ground water in watershed areas; soil moisture recharge from snow; specific yield studies; and water drained from a water-table aquifer within the cone of depression of a pumped well (vertical drainage studies)
f. Soil moisture, including control of density and moisture content of embankments; studies of sediment density in situ and sediment concentration in streams; moisture and density in compaction experiments and crop rotation studies; water content and bulk density; and the relationship between soil moisture and density.and the quantity of runoff.

11.3 MANAGING WATER IN IRRIGATED AGRICULTURE

Soil moisture content information is needed for studies across a variety of disciplines, such as hydrology, soil science, ecology, meteorology and agronomy. Conventional volumetric water content measurement methods at these small scales include gravimetric, frequency and time domain reflectometry (FDR and TDR), neutron probe and capacitance probe techniques.

Under the title "Sustainable Water Management Practices for Intensified Agriculture" Yadav et al. (2022) reviewed that agriculture is the major consumer of water resources (Green et al., 2015). Irrigation accounts for over 70% of worldwide freshwater withdrawals, ensuring global food production (Rockström et al., 2017). Irrigated regions make for 18% of world farmland yet account for over 40% of world food production (Chartzoulakis and Bertaki, 2015; FAO, 2019). Simultaneously, 40% of worldwide irrigation techniques are non-sustainable due to the depletion of ecological and groundwater reserves (Wada and Berkens, 2014; Rosa et al., 2018). The strong link between economic and environmental demands for water reserves raises critical questions at the heart of the water sustainability debate: how can human appropriation of water reserves support economic deeds, such as agriculture, without depleting water reserves, habitats of aquatic ecosystems and other relevant ecosystem services? (Rosa et al., 2018; D'Odorico et al., 2018).

Sustainable irrigation strategies in agriculture must allow for increased crop production to satisfy growing food demands while also ensuring that natural resources (groundwater reserves, freshwater and water quality) are not irreparably depleted (Rosa et al., 2018; Borsato et al., 2020). The term "sustainability" is frequently used to refer to the management, use, and protection of natural assets in such a way that forthcoming generations will have access to them (FAO, 2013; Borsato et al., 2020). One of the most effective factors on crop growth is the water deficit and efficiency. In this respect, Bhattacharya (2021) reviewed that water deficit restricts the growth, development, and yield of crops worldwide, consequently making losses exceeding the total caused by all other adverse factors (Ray et al., 2002; Neumann, 2008; Zivcak et al., 2008; Farooq et al., 2009; Bajgain et al., 2016; Petrov et al., 2018). Also, Bhattacharya (2021) added that water deficit has a frequent effect on nutrients availability and dynamics in soil and consequently uptake by plants. Many investigators found that water deficit had enhanced the N use efficiency of wheat (Fan and Li, 2001) and rice (Pirmoradian et al., 2004), but the effect was only significant at high fertilizer application rates. Regarding soil moisture detection, Mujere and Muhoyi (2022) revealed that the use of remote sensing in measuring soil moisture is a promising technique, especially in data-scarce and water-limited regions. In this respect, a comparison between remote sensing and conventional methods in soil moisture estimation was pointed out by Moran et al. (2004) and listed in Table 11.1.

11.4 USE OF UNCONVENTIONAL WATER RESOURCES IN CROP IRRIGATION

Due to shortage in fresh water per capita and increasing demand for fresh water which becomes a serious challenge, especially in arid and semi-arid regions attributable to the fast rate in population, attention has been paid to reuse of the unconventional water resources like saline one. In Egypt, about 5,000 million m^3 of saline drainage water is used for irrigating about 405,000 ha of land. About 75% of the drainage water discharged into the sea has a salinity of less than 3,000 mg/L. The policy of the Government of Egypt is to use drainage water directly for irrigation if its salinity is less than 700 mg/L; to mix it 1:1 with Nile water (180–250 mg/L) if the concentration is 700–1,500 mg/L; or 1:2 or 1:3 with Nile water if its concentration is 1,500–3,000 mg/L; and to avoid reuse if the salinity of the drainage water exceeds 3,000 mg/L. The annual average volume of available drainage water is about 14,000 million m^3. The policy of the ministry of water resources and irrigation is to make full use of each drop of drainage water by the year 2017. Some large scales projects are currently executed depending on the drainage water as the main source for irrigation (El Gamal, 2007). He added that Egypt has a policy to use brackish and saline ground water with EC up to 4.5 dS/m and reuse of drainage ground water with EC 6 dS/m with SAR values of 10 to 15 in blending or cyclic mode with good quality water. The reuse of treated wastewater in Egypt

TABLE 11.1
A Comparison of Remote Sensing and Conventional Methods in Soil Moisture Estimation

Approach	Merits	Limitations
Optical remote sensing (visible, near infrared, short-wave infrared)	Fine spatial resolution; broad coverage; multiple sensors including hyperspectral sensors	Weak estimation of surface moisture, works only on clear weather; affected by vegetation cover, issue of temporal resolution
Thermal infrared	Fine spatial resolution with broad coverage; multiple sensors available	Minimal surface penetration, affected by high humid weather, vegetation and atmospheric gas density
Microwave remote sensing	Broad coverage; satellite sensors available; strongly related to surface moisture estimation; surface penetration up to 5 cm and it is insensitive to weather patterns and atmospheric conditions	Affected by vegetation cover and surface rough roughness; ideally has got a coarse spatial resolution of approximately 30 km
Field-based methods (combined)	Near accurate estimates, addresses one's needs at a time; depth depends on the need; insensitive to earth's roughness and atmospheric condition	Point-based; time consuming and laborious;

Source: Cited by Mujere and Muhoyi (2022), from Moran et al. (2004).

started in 1915 in the eastern desert northeast of Cairo. An area of 2,500 acres is still under irrigation with wastewater, which receives only primary treatment. With the scarcity of water resources, it is planned to irrigate 150,000 acres with treated wastewater. All urban wastewater projects include facilities for treatment up to the tertiary level and allow reuse for irrigation. It is estimated that the present amount of wastewater from major cities and urban areas is about 5 billion m³/year.

Effect of irrigation water salinity on the growth and production of some legumes, i.e., faba bean, chickpea, and lentil as well as tuber plants like sugar beet and potatoes was investigated in a lysimeter system under greenhouse-controlled conditions using a drip irrigation system with the application of ^{15}N tracer technique (Gadalla et al., 2007). It seems that, in general, increase in water salinity levels depressed the seed yield compared to freshwater treatment (Table 11.2). The response to salinity of irrigation water varied between different legume crops. Stable nitrogen (N-15) is used for distinguishing between the different sources of nitrogen derived to the legume plants and estimating exactly how much N could be compensated by the different sources. Seed-N derived from fertilizer tended to decrease with increasing water salinity levels. This was true with all tested crops. Under these adverse conditions of salinity, the tested crops tended to depend on the portion of nitrogen gained from air than those derived from mineral fertilizer. Another picture was drawn with a sugar beet crop in the field experiment, whereas the sugar % had been increased with increasing salinity levels of irrigation water up to 12 dS/m under both N_1 as 112.5 kg/ha in one full dose and N_2 as 56.25 kg/ha in split equal two doses treatments. It seems that the sugar % was higher in case of N_1 than N_2 under different salinity levels, when W_1 (100% WHC) was concerned. Opposite direction was noticed with W_2 (75% WHC) water treatment (Table 11.3).

11.5 ENHANCING WATER PRODUCTIVITY

The purpose of irrigation scheduling is to inform the farmer when to irrigate and how much water to apply to obtain a desired objective. The objective usually involves maximizing crop yield or profit, enhancing water use and productivity, but it may also include taking advantage of irrigation opportunities, minimizing deep percolation or leaching, and soil and water salinity management. Evapotranspiration (ET) is one of the most important components in field water cycles. The

TABLE 11.2
Changes in Seed Yield, Nitrogen Derived from Fertilizer and Air as Affected by Different Water Salinity Levels dS/m

Crop	Salinity Level dS/m	Seed Yield g/lysimeter		N Derived from			
		Total	100 Seed	Fertilizer - Ndff		Air - Ndfa	
				%	g/lysimeter	%	g/lysimeter
Chickpea	F.W.	182.0	24.2	8.62	0.14	71.28	1.18
	3 dS/m	59.8	21.4	8.04	0.16	71.00	1.46
	6 dS/m	35.2	9.8	7.72	0.08	67.10	0.70
	9 dS/m	19.2	5.8	6.20	0.04	72.87	0.48
Faba bean	F.W.	98.4	87.4	12.1	0.61	59.69	2.99
	3 dS/m	98.9	86.4	10.02	0.50	63.82	3.21
	6 dS/m	87.2	77.0	5.96	0.25	74.52	3.10
	9 dS/m	79.2	82.2	4.20	0.11	81.62	2.00
Lentil	F.W.	51.5	3.0	6.84	0.16	0.16	87.47
	3 dS/m	38.0	2.6	6.20	0.11	0.11	77.62
	6 dS/m	24.4	2.0	5.74	0.05	0.05	75.52
	9 dS/m	15.0	1.6	5.70	0.03	0.03	73.06

TABLE 11.3
Effect of Salinity Levels, N Doses and Water Percent on Sugar Yield (% and kg/ha) of Sugar Beet Plants

	W1				W2			
	(%)		kg/ha		(%)		kg/ha	
Salinity Levels	N_1	N_2	N_1	N_2	N_1	N_2	N_1	N_2
FW	16.0	14.0	5886.1	5192.2	13.7	15.7	4753.4	6366.0
4 dS/m	17.4	16.3	6375.3	5878.6	15.3	18.2	5928.1	6745.1
8 dS/m	19.0	17.8	5788.1	5787.9	18.8	20.5	7226.0	7604.4
12 dS/m	21.2	20.4	6340.2	6912.2	19.6	23.3	6607.5	7717.2

accurate measurement or estimation of ET in fields is of great importance for quantifying soil hydrological processes and, thus, making appropriate decisions regarding irrigation measures to improve water use efficiency levels in agricultural areas, particularly in arid or semi-arid regions. Therefore, information on evapotranspiration (ET), or consumptive water use, is significant for water resources planning and irrigation scheduling. Evapotranspiration is important as a term in the hydrological cycle, e.g., in soil water and groundwater balances, and in salinization. In irrigation and drainage engineering, we therefore need to devote proper attention to its determination, particularly in arid and semi-arid areas.

Salama et al. (2017) showed the superiority of the soil-based irrigation method (M2) over M1 (climatic-based method) when the growth and yield of spring wheat were concerned. Moreover, it could be used since it is more efficient with water saving (25.44%) in comparison to the M1 method. The highest value of deep percolation was detected with the M1 (77.83 mm) causing a decrease of irrigation efficiency (1.02 kg/m^3). Therefore, selecting suitable irrigation method is needed in order to improve irrigation efficiency. Another lysimeter experiment conducted by Gaber et al. (2017) reflected that scheduled irrigation regime (drip irrigation) equal to 80% ETc (440 mm/season) (W2) combined with either N2 (156 kg N/ha, 100% recommended rate) or N3 (124.8 kg N/ha, 80% recommended rate) fertilizer rates applied at S3 splitting doses (50%, 25%, 25%) resulted in a higher barley grain yield than other treatments. While the water regime W1 (100% ETc) interacted with either N1 (187.2 kg N/ha, 120% recommended rate) or N2 achieved the best values of N uptake by grains. They added that nitrogen and water management strategy composite of W1 (100%ETc) × N2 (100% N rate) × S3 (50, 25, 25) resulted in the remarkable N derived from labeled ammonium sulfate enriched with 2% atom excess (Ndff) gained by grains as indicated by overall means of the tested factors. Nitrogen fertilizer was efficiently used by barley grains treated with combined management of W1 × N1 × S3 achieving 67%. Fertilizer-N remained in the soil after harvest was not affected by water regime or splitting modes whereas it increases with the lowest N rate (N3). Full irrigation regime of pea crop assists the availability of nitrogen fertilizer to plants especially when added at a high rate of urea form compared to ammonium sulfate. On the other hand, N derived from ammonium sulfate was better than that of urea under full irrigation water regime. Additionally, the efficient use of both N forms was related to water regimes and application rates. For example, the efficient use of ammonium sulfate, to some extent, was affected by different water regimes, while urea added at the rate of 75% was more efficiently used by seeds than those added at the rate of 100% (Fahmy et al., 2016). In the same direction, Hekal et al. (2018) indicated that the optimum grain yield of maize in addition to the most efficient use of fertilizer-N and water productivity was induced by a moderate N rate equal to 248 kg N/ha combined with deficit irrigation water regime (4,612 m^3/season). Use of the unconventional water resources such as saline water was examined by Kassab et al. (2018), who revealed that the proportion and absolute values of nitrogen derived from fertilizer (Ndff) as well as nitrogen use efficiency (NUE) gained by wheat grains or straw were severely reduced by increasing water salinity level as compared to those irrigated with fresh water. Mulching has a positive effect on combating salinity stress. It seems that, in general, mulching made the plants more able to derive more nitrogen from fertilizer and improved NUE.

11.6 STRATEGIES OF WATER MANAGEMENT

A renewable freshwater supply is a finite resource. Similar to soil, water is also prone to misuse, contamination/pollution, and eutrophication. As much as 70% of total water withdrawal has been used for agriculture, mostly irrigation. However, there are numerous competing and essential uses (i.e., domestic, industrial, recreational, and aquaculture). While equipping some arable land in sub-Saharan Africa with irrigation facilities, the WP of existing irrigated land (i.e., China, India, Pakistan, Egypt, and Iran) must be improved, and new and innovative irrigation methods must be adopted. Because of its strong interaction, WP can be enhanced by improving soil quality and increasing the efficient use of N and other nutrients (Lal, 2012b).

In arid and semiarid regions, already faced with the severe problems of water stress, the loss of blue water must be minimized by storing it for future use within the watershed. Because of the high evaporative demands both now and in the future, storage in aboveground impoundments is prone to high losses by evaporation. Thus, recharging the aquifers and creating belowground storage are preferred strategies. Gray water must be recycled and used for enhancing the production of the third generation of biofuels (i.e., algae and cyanobacteria) and for promoting urban agriculture, both of which are a high priority to meet the growing demands of an increasing and urbanizing human population (Lal, 2012b).

In addition, agriculturists want to have information on the effects of a water supply on crop production. As there is often a direct relation between the ratio of actual to potential evapotranspiration and actual to potential crop yield, agriculturists want to know the specific water requirements of a crop, and whether these requirements are being met under the prevailing environmental conditions. Regular estimates of evapotranspiration may reveal water shortages and/or waterlogging, which can then lead to technical measures to improve irrigation and drainage, and, again ultimately, to an increase in crop yields.

Strategies for improving water use efficiency were compiled by Chaudhari (2021), who stated that the main components needed for improving water-use efficiency at the farm level and bridging the gap between irrigation potential created and utilization are:

1. Accelerated Irrigation Benefits Programme (AIBP): focuses on faster completion of ongoing major and medium irrigation, including National projects.
2. Facilitate and provide assured irrigation supplies to each farm. The schemes include (i) new minor irrigation schemes, (ii) repair, renovation and restoration of water bodies, (iii) Command Area Development (CAD), (iv) groundwater development in potential areas, (v) diversion schemes from plenty to scarce areas, (vi) creating and reviving water tanks, pond, etc.
3. PMKSY (Per Drop More Crop): This component emphasizes to promote micro-irrigation (sprinkle, drip, pivots, and rain guns), efficient water conveyance and application, precision irrigation systems, topping up of input cost beyond permissible limits, secondary storage including canal storages for tail-ends of canals, water lifting devices (like diesel/electric/solar pump sets), extension activities, coordination and management.
4. PMKSY (Watershed): It involves ridge area treatment, drainage line treatment, soil and moisture conservation, water harvesting structure, livelihood support activities and other watershed works.

Recently, the use of organic mulches, which contribute to soil water management and fertilization practices, has shown promise as an eco-friendly approach in enhancement of taro (*Colocasia esculenta* L. Schott) crops nutrition and growth (Juang et al., 2021). Under field experiment, they found that taro cultivation under the flooding regime had a higher level of soil fertility. The flooding regime promoted the taro plant growth and further enhanced the yields of the harvested corms;

also, sugarcane bagasse and rice husk would be the superior mulch to obtain better corm attributes. By contrast, cultivation under the upland regime enhanced the nutritive values of taro corm more than that under the flooding regime. Adversely, the nutritive value of taro corm was more enhanced by upland water regime than the flooding regime. Previously, the soil water regime was found to be one of the most important practices in taro corm production (Busari et al., 2019), and simultaneously the organic mulch breaks down into the soil contributed in improving fertility and increase the bulk of organic matter (McIntyre et al., 2000). Under Egyptian field conditions, Kassab et al. (2018) found that mulching of sandy soil under wheat crop irrigated with saline water contributed to the enhancement of nitrogen uptake in other turn, it made plants tolerant, to some extent, to the adverse effects of saline water.

Additionally, the main effects of organic mulching, besides water efficiency, limit soil erosion and water evaporation, maintain soil fertility, reduce the soil temperature, conserve soil moisture by slowing evaporation, and improve crop yield and quality (Adekalu et al., 2007; Azad et al., 2015; Cao et al., 2012; Kosterna, 2014; Lordan et al., 2015; Miyasaka et al., 2001; Nachimuthu et al., 2017; Ruíz-Machuca et al., 2014).

11.7 SOIL RESTORATION, A WAY TO ENHANCE WATER PRODUCTIVITY

Nutrient deficiency, shallow rooting depth, and excessive soil erosion are constraints that negatively affect water productivity in both rainfed and irrigated agriculture. Therefore, modification of nutrient status in the soil can improve water–plant–soil relationships. In this regard, N availability, at the critical stage, is crucial to improve WP and rational use of fertilizers enhanced the WP in the irrigated cropland (Fang et al., 2010). In addition to N, the application of P can also improve WP. Also, soils supplied with organic wastes (Oldare et al., 2011) can be improved and at the same time increase the WP. The goal is to reduce water consumption without reducing the yields through agronomic management. This is achievable by (i) reducing water delivery losses, (ii) improving soil water availability to crop roots, and (iii) increasing the WP. The agronomic practices that are effective in conserving soil water (green water), which also alter the soil surface energy balance, include mulch farming, plastic sheeting, and conservation tillage (Hatfield et al., 2001). The adoption of conservation-effective cropping systems, which reduce the soil erosion risks, is also relevant to improving the WP of soils prone to accelerated erosion.

11.8 SOIL MANAGEMENT

Soil and water are two vital resources for agricultural development and sustaining life on the earth. Additionally, intensive use of soil and water resources is inevitable to meet food and nutritional security of the nation, but post-green revolution, there are concerns about sustainability arising from the deterioration of soil chemical, physical and biological health due to imbalanced fertilizer application, low organic manureing and soil degradation. Therefore, efficient soil management is required to achieve success in site-specific nutrient management according to soil conditions and needs. This in turn improves sustainability and lowers the production costs in field (Mukhopadhyay and Thakur, 2021). Many programs were accepted in many countries to sustain soil health and quality including soil test kits with geo-referenced soil fertility maps, integrated plant nutrient system (IPNS) packages incorporating organics, micro and secondary nutrients documented for major cropping systems in different agro-climatic regions to promote balanced fertilization. In addition, standardized vermi/bio-enriched composting technology was considered to prepare various types of organic manures such as phosphor compost, vermicompost, bio-enriched compost, municipal solid waste compost, etc. from various organic wastes. Improvement of the efficient strains of biofertilizers specific to different crops and soil types was also included in developed programs. Biofertilizers may increase crop productivity by 10%, saving of 20%–25% of chemical fertilizers, improvement of nutrient use efficiency by 15%–25%, and produce quality and soil health.

Lately, FAO has prompted a sustainable agricultural production system, i.e., Climate-Smart Agriculture (CSA) an alternative to conventional agriculture and aimed to increase productivity, resilience to climate change, and reduce greenhouse gas emission. It improves the efficiency of natural resources, increased resilience and productivity of agriculture, and reduces greenhouse gas emissions (Totin et al., 2018). It is estimated that, by 2050, the population will reach 9.7 billion and will require about 70% more food to feed human than what is consumed today (Figure 11.3). According to Alexandratos and Bruinsma (2012), the world will need to respond to an increased demand of global food security by 2050 due to population and income growth.

The CSA has three focal areas, viz., (i) agronomic and economic productivity, (ii) adapting and building resilience to climate change, and (iii) Climate change mitigation (Palombi and Sessa, 2013). The key concept related to raising productivity is increasing food production sustainably from existing farmland while minimizing pressure on the environment (Totin et al. 2018). This first principle is strongly connected to the second one. Adapting and building resilience to climate change ensures food sufficiency despite unsuitable conditions. Dealing with climate-smart soil concept, the agriculture land is considered a major source of all three biogenic GHGs: carbon dioxide (CO_2), methane (CH_4), and nitrous oxide (N_2O). Soils contribute a major share (37%, mainly as N_2O and CH_4) of agricultural emissions (Tubiello et al., 2015), while land use contributes 25% of the total global anthropogenic GHG emission (Smith et al., 2014).

Plant nutrients are present in soil in the form of positively charged ions (i.e., cations). Apart from this, during decomposition of SOM, the multitude of organisms in the soil food web release nitrogen (in the form of ammonia ions), potassium, calcium, magnesium, and a range of other nutrients which is necessary for plant growth. The negative charges on the surface of the clay particles and organic matter attract cations and thus plants obtain many of their nutrients from the soil by cation exchange where the exchange of hydrogen ions with the cations adsorbed on the soil particles occurs by the root hairs of the plant. Mechanical soil disturbance such as plowing is detrimental for buildup of organic matter in the soil. It was observed that improved management practices can reduce GHG's emission and increase soil carbon stock to soil. Soil carbon sequestration is one of a few strategies that could be applied at large scales and at low cost (Ciais et al., 2013). It will not only stabilize climate but will also make agricultural production more sustainable and maintain the ecosystem services that are supported by soils.

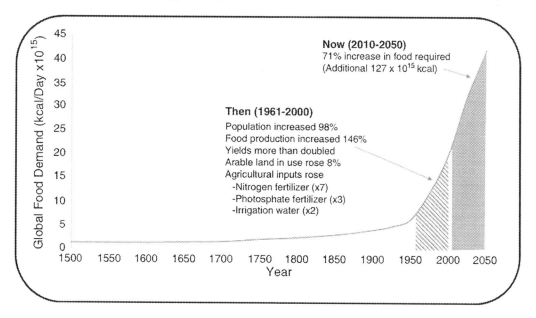

FIGURE 11.3 Framing the food security challenge. Source: Adopted by Mukherjee and Batabyal (2021), After Keating et al. (2014).

11.9 NUTRIENTS MANGEMENT

Ahmed et al. (2022) demonstrated the vital nutrients needed for biological functions. In this regard, Fahad et al. (2016); Ahmed et al. (2020), classified the nutrients into macronutrients that required at higher than 1–150 g/kg (>1,000 mg/kg dry weight) of plant dry matter, including N, P, K, Ca, Mg, and S. However, nutrients which are required at the concentration of 0.1–100 mg/kg (<100 mg/kg dry weight) of plant dry matter are called as micronutrients including Fe, Zn, Mn, Cu, B, Mo, and Cl. In addition, elements such as Al, Si, Co, Na, and Se are not essential according to the criteria but are widely taken up by the plants to perform different metabolic functions. Excessive use of applied nutrients may cause environmental pollution like global warming, greenhouse gas (GHG) emission (Ahmed, 2020; Hammad et al., 2018); and also cause above and underground water deterioration. Nutrient dynamics is the process by which nutrients are taken up by the plants. The role of different forms of nutrients taken up by plants on growth is illustrated in Figure 11.4.

This paragraph is quoted from Srivastava (2020), who explained that crops require 16 elements to grow properly. The elements like carbon, hydrogen and oxygen are derived from air and water. Other remaining nutrients used by the plants come from soil in the form of inorganic salts. Legumes can fix nitrogen from the air. Plant growth and development may be retarded if any of these elements is lacking in soil or not adequately balanced with other nutrients. Physical and chemical

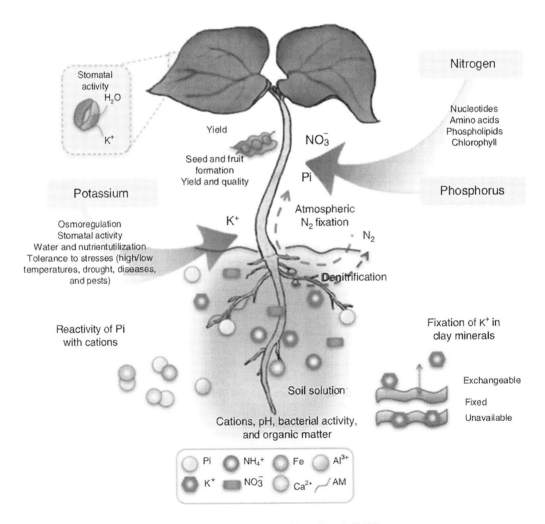

FIGURE 11.4 Plant growth and nutrients role. Source: Ahmed et al. (2022).

characteristics of soil affect the availability and uptake of micronutrients. Soil temperature and moisture are important factors. As soil pH increases, the availability of micronutrients decreases with the exception of molybdenum. The deficiency or increase of micronutrients/trace elements in soil has a direct bearing on the agricultural produce which in turn adversely affects the health of live stocks and human beings. Additionally, soils may be enriched or deficient in certain elements depending upon the source of the parent material. If the parent rock/alluvial soil or older soil is rich in certain elements, then it leads to enrichment of that element due to soil erosion. Physical laws of soil matter through erosion by water and wind action result in the removal of fertile element constituent of surface soil. The significant factors for the soil erosion are clay-ratio, clay-silt ratio, ratio of colloid to moisture equivalent of pH and organic matter, slope, and forest cover.

In a chapter entitled *"Deficiency of Essential Elements in Crop Plants"*, Tiwari et al. (2020), represented the role and deficiency symptoms of essential nutrients which are indispensable for the growth and development of plants. Indiscriminate use of NPK-based fertilizers in agriculture leads to diminution in the micronutrients bioavailability that exhausts the soil reserve of native micronutrients (Sidhu and Sharma, 2010; Shukla et al., 2016). Having considered the above backgrounds, there is an urgent need to regulate agro-techniques, which are the most suitable edaphic conditions that deteriorate the micronutrient deficiency in plants (Figure 11.5, Table 11.4).

The effect of N and P either individually or in combination on the enhancement of carbon sequestration in soil rhizosphere or bulk soil was traced by Zhran et al. (2021) using ^{13}C stable isotope (natural ^{13}C abundance). They indicated an enhancement of net assimilated ^{13}C due to N and NP fertilization on day 14 (D14), with maximum C assimilation occurring on day 22 (D22)

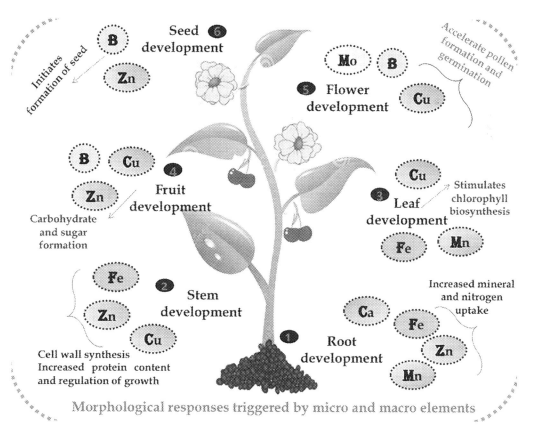

FIGURE 11.5 Schematic representation of morphological responses triggered by micro- and macronutrients in plants. Source: Tiwari et al. (2020).

TABLE 11.4
Potential Role, Toxicity Symptoms of Some Micro- and Macronutrients in Plants

Macro-micronutrients	Beneficial Role	Deficiency Symptom	Important Constituent	References
Nitrogen (N)	• Biosynthesis of macromolecules (proteins and nucleic acids) • Chlorophyll biosynthesis and amino acid synthesis • Nitrogen and ammonia assimilation	• Yellowing of leaves • Cell death • Reduced photosynthesis and thus reduced biomass • Inhibition of seed germination • Unregulated cell division • Reduced yield	• Protein • Amino acid • Chloroplast • Phospholipids • Enzymes	Nawaz et al. (2018) Parween et al. (2011) Xiong et al. (2018) Zhai and Li (2005)
Phosphorous (P)	• Regulation of protein synthesis • Stimulates the growth of new tissue • Regulation of cell division • Development of the roots and hastening of maturity • Participates in key structure of plant	• Affects early developmental stages • Stunted or weak growth • Dark to blue-green coloration appears in leaves • Purpling of stems • Reduced CO_2 assimilation	• Primary component of the energy molecule ATP (adenosine triphosphate) • Nucleotide • Component of DNA, RNA and phospholipids	Hammond and White (2008) MacDonald et al. (2011)
Potassium (K)	• Regulation of growth and development • Maintains membrane potential • Up-regulates enzymes NR, GDH, GS GOGAT	• Reduced photosynthesis • Inhibits the process of translation • Inhibits seed germination • Yellowing of leaves • Leaf curling • Poor root development • Poor yield • Plants are more prone to disease • Inhibits activity of aquaporins	• Responsible for many vital processes such as water and nutrient transportation • Protein and starch synthesis	Wang et al. (2013) Çokkizgin and Bölek (2015)
Magnesium (Mg)	• Acts as a co-factor • Needed for ribosomal stability • Biosynthesis of chlorophyll • Regulates photosynthesis • Activation of RuBisCo • Regulates cellular functions • Regulation of carbon assimilation	• Primary symptoms • Yellowing of leaves • Discoloration of green vein in to red tissues • Inter-veinal necrosis • Grain falling • Fruit expansion • Decreased carbon fixation • Reduced biomass	• Essential component of chlorophyll • Stackness of thylakoid membrane • Key component of cell membrane • Component of enzymes	Yang et al. (2012) Jezek et al. (2015)

(Continued)

TABLE 11.4 (*Continued*)
Potential Role, Toxicity Symptoms of Some Micro- and Macronutrients in Plants

Macro-micronutrients	Beneficial Role	Deficiency Symptom	Important Constituent	References
Sulphur (S)	• Regulated growth and development • under stress and un-stressed condition • Reduces ROS generation and oxidative stress • Regulation of phytohormone • Maintains Na/K ratio	• Lowers photosynthetic pigment content • Alters photosynthesis • Reduces activity of RuBisCo • Inhibits the protein synthesis • Poor yield and growth	• Integral part of different biologically active compounds such as some amino acids, phytohormones and coenzymes • Integrally involved in cysteine, methionine, sulphoxides and glucosinolates	Fatma et al. (2016) Takahashi et al. (2011) Khan et al. (2014)
Iron (Fe)	• Acts as a co-factor • Transports oxygen • Protein stability	• Reduction of growth • Poor yield • Dropping quality of crop plant • Swollen root tip • Lateral root development • Reduced pigment synthesis • Disturbance in thylakoid membrane • Affects PS II photochemistry • Inhibits activity of Cyt b6f • Reduced photosynthesis	• Important component of iron-sulphur (FeS)-rich proteins of PS I and ferredoxins	Couturier et al. (2013) Abadía et al. (2011) Laganowsky et al. (2009)
Manganese (Mn)	• Regulates growth and development • Regulates cellular function such as photosynthesis • Nitrogen fixation • DNA synthesis • Regulates biosynthetic processes • Acts as a catalyst • Responsible for the release of oxygen • Regulates biosynthesis of protein, carbohydrate, lipids and lignin	• Alteration in photosynthesis • Chlorosis of leaves • Stunted growth • Reduced biomass • Affects the PS II proteins D1, PsbP and PsbQ • Down-regulation of antioxidant • Excessive accumulation of reactive oxygen species • Less production of secondary metabolites • Increased transpiration	• Important cellular component • Raises enzyme activity • Important component of glutathione synthase and superoxide dismutase • Key player in oxygen-evolving complex (OEC) of photosystem II (PS II)	Hänsch and Mendel (2009) Nouet et al. (2011) Millaleo et al. (2010) Schmidt et al. (2016) Marschner (2012)

(*Continued*)

TABLE 11.4 (*Continued*)
Potential Role, Toxicity Symptoms of Some Micro- and Macronutrients in Plants

Macro-micronutrients	Beneficial Role	Deficiency Symptom	Important Constituent	References
Molybdenum (Mo)	• Acts as a co-factor • Activates several enzymes such as nitrogenase, nitrate reductase, xanthine dehydrogenase, aldehyde oxidase, and sulphate oxidase	• Stunted growth • Chlorosis • Yellowing of veins • Involuted leaf • Reduced growth • Reduced biomass • Disturbance in equilibrium N uptake and assimilation • Early senescence • Pale coloring of leaves	• Structural component of pterin	Yohe et al. (2016)
Zinc (Zn)	• Acts as a co-factor • Binds with oxidative radicle • Acts as co-factor for alcohol dehydrogenase (ADH) and carbonic anhydrase (CA)	• Morphological alteration in plants • Development of chlorosis • Cell death • Poor root development • Short intermodal growth • Down-regulation of enzymatic activity • Decreased biomass • Altered metabolic processes	• Essential component of plant cell membrane • Involves in cell repair system • Associated with post-translational changes • Involves in protein trafficking	Rafique et al. (2012) Nagajyoti et al. (2010)

Source: Adapted from Tiwari (2020).

under NP. More ^{13}C was accumulated in the aboveground biomass than belowground biomass and this holds true with all fertilization treatments. ^{13}C incorporation into the rhizosphere exceeded those found in bulk soil, with the maximum ranging from 6% to 10% found under N addition. Newly assimilated ^{13}C incorporated into particulate organic matter (POM) was increased in the rhizosphere under N and NP conditions, whereas mineral fraction (MIN) remained largely unaffected. ^{13}C-MBC proportion in the total microbial biomass C (MBC) pool revealed that N and NP stimulated microbial activity to a greater degree than P. The distribution of ^{13}C pool into different plant parts, microbial biomass C and incorporation into rhizosphere or bulk soil; mineral or particulate organic matter is illustrated by Fig. 11.6. Recently, Liu et al. (2022) referred to the necessity for obtaining soil nutrient information quickly and accurately under crop and soil sensing for precision crop production technology. Near-infrared spectroscopy (NIRS) with high-efficiency and nondestructive characteristics has great potential in soil nutrition detection. According to the NIR absorption of the hydrogen bonds, soil total nitrogen content and soil organic matter content can be estimated. Multiple linear regression, partial least square regression (PLSR), and principal component analysis (PCA) are commonly used to establish the estimation models of soil nutrient contents based on NIRS. Moreover, the modern algorithms of wavelet algorithm (WA), genetic algorit (GA), uninformative variable elimination (UVE), support vector machine (SVM), etc., are used to reduce the multicollinearity of the NIR spectra to improve estimation accuracy. Laser-induced breakdown spectroscopy (LIBS) is a promising spectral detection technology with high sensitivity, fast speed, and the ability to measure multiple elements simultaneously. It can also be used to detect both soil macronutrients and micronutrients.

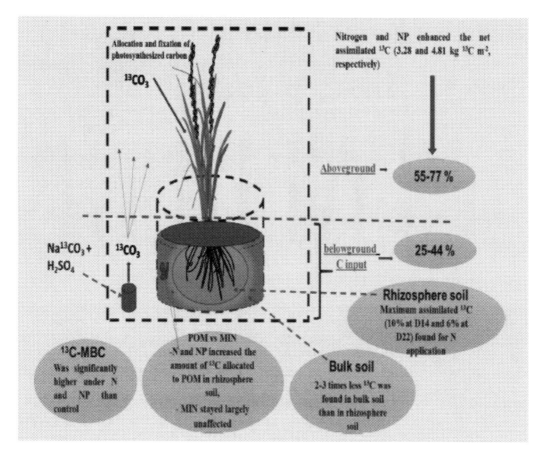

FIGURE 11.6 Conceptual ^{13}C distribution pattern in the soil without fertilization and with N and P fertilization on day 14 (D14) and 22 (D22) of ^{13}C-CO_2 continuous labeling. Source: Zhran et al., 2021

11.10 COMBINING INORGANIC AND ORGANIC NUTRIENT SOURCES

Although it is recognized that the use of fertilizers will continue to be important for agricultural production, there is a need to increase our knowledge of chemical and biological processes that render nutrients more available to plants from soil organic matter and other nutrient sources, such as manures, native vegetation, and crop residues. This will help optimize nutrient cycling and recovery, minimize fertilizer requirements, and maximize the efficiency of nutrient use. Such research is consistent with what farmers actually do with their available nutrient resources. In this respect, Moursy and Abdel Aziz (2013), examined the role of *Aspergillus flavus* and *Trichoderma* sp. in laboratory incubation experiment, to enhance the decomposition rate of maize stalks, wheat straw, chickpea straw, as plant residues beside cow manure. They indicated that the use of such fungus isolates is useful as biodegradable agents and helps enrich the poor sandy soil with available NH_4-N and NO_3-N forms. These available forms were increased with increasing incubation time intervals up to 90 days and simultaneously the organic-N tended to decrease. It means that the incorporation of such plant residues into the soil 3 months prior sowing could be accepted as the preferred strategy especially when inoculated with fungal or proper microbial inoculums.

Review article written by Soliman and Galal (2017) compiled a series of field experiments carried out by staff of Soil and Water Research Department, EAEA, Egypt for the evaluation of nitrogen derived to different strategic crops from either chemical or organic sources. They proved that wheat, maize and sesame crops were significantly varied in their response to organic compost and ammonium sulfate forms either individually or in combination (Figure 11.7). Cow manure and local

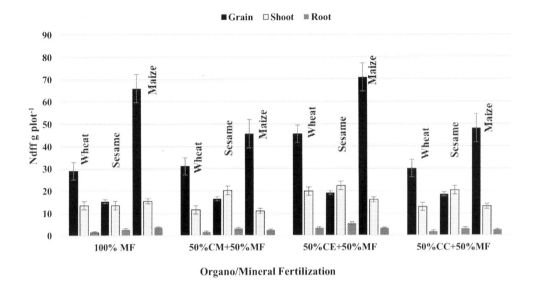

FIGURE 11.7 Nitrogen derived from fertilizer (g/plot) by wheat, sesame and maize crops as affected by organic manure and mineral fertilizer. Source: Adapted from Soliman and Galal (2017).

composted plant materials combined with 50% mineral fertilizer induced the highest %NUE by maize grain (58.9%) followed by wheat grain (46%) then seeds of sesame (42.3%) (Soliman et al., 2014; Ghabour et al., 2015). Application of 20 kgN plus inoculation resulted in higher nodule number and dry weight than those recorded with 50 kgN plus inoculation. On the other hand, application of 50 kgN plus rhizobial inoculation induced the highest Ndfa value where it doubled those of 20 kgN and/or organic N plus inoculation treatments. At the same time, chickpea dry weight was increased by increasing nitrogen fertilizer rates (Rizkalla et al., 2014).

Distribution of N proportion within the soil profile under different tested crops indicated that addition of a half rate of mineral fertilizer in combination with a half rate of organic compost significantly reduced mineral-N losses as compared to fully mineral fertilized ones. Also, it could be concluded that the portions of N that remained in the soil, the N uptake by plant and that lost from soil media were related to the rate of mineral fertilizer added. A little bit of difference was recorded with maize crop whereas N remained localized to some extent in 20–40 cm and 40–60 cm when a full dose of mineral fertilizer was applied. Also, the combined treatment of 50% CM+50% MF showed accumulation of remaining N in the top layer 0–20 cm while the quantities remained in 20–40 cm and 40–60 cm were nearly close to each other (Ghabour et al., 2015) (Table 11.5).

Soliman et al. (2012) declare that organic additives have an effective role in the enhancement of sorghum dry matter yield, increased N uptake by different plant parts and at the same time overcome the negative effects of irrigation water salinity levels. The response of sorghum plants to different water salinity levels was fluctuated according to organic additive type especially leuceana residues and chicken manure. In addition, the amelioration of sandy soil via the application of organic additives could be enhanced by spraying the proline acid on sorghum plants to combat the salinity stress of irrigation water (Galal et al., 2012). Generally, organic compost and biofertilizers have improved growth and seed quality of quinoa (Mousa et al., 2022). They attributed the highest physiological content of nitrogen (5.52%), crude protein (34.5%), phosphorus (0.81%) and potassium (1.65%) to the application of 200 kgN/ha combined with dual inoculum of *Azotobacter chroococcum* and *Bacillus polymyxa*).

Some of the recently published data dealt with the application of isotopic tracer techniques in different topics of soil sciences and plant nutrition are compiled and presented in Table 11.6. These trials were conducted on different experimental scales, field or green-house pot experiments, on different soil types and strategic or vegetable crops. Under Mo-deficient soils, the application of Mo significantly increased ^{15}N uptake efficiency in winter wheat with varied values according to N fertilizer

TABLE 11.5
Fertilizer-N Balance as Affected by Inorganic and Organic Fertilizers under Wheat, Sesame and Maize Crops

Treatments (T)	Crops (C)					
	Wheat		Sesame		Maize	
	N Remained in Soil					
	%	g/plot	%	g/plot	%	g/plot
100% MF	22.6	121.1	21.1	50.9	20.1	129.2
50%CM+50%MF	17.7	47.4	21.0	25.4	20.1	64.6
50%CE+50%MF	17.7	47.4	20.8	25.1	20.2	64.8
50%CC+50%MF	17.8	47.7	20.7	24.9	29.4	94.4
LSD (0.05)	T, 2.512	T, 3.512	C, 2.369	C, 2.369	TxC, 4.738	TxC, 4.738
	N Uptake and Derived from Fertilizer					
100% MF	46.8	250.8	45.7	110.1	47.7	306.7
50%CM+50%MF	44.5	119.3	41.9	50.7	43.6	139.9
50%CE+50%MF	43.4	116.3	43.2	52.3	42.3	135.8
50%CC+50%MF	42.6	113.9	41.9	50.7	44.6	143.2
LSD (0.05)	T, 2.809	T, 2.723	C, 2.43	C, 2.358	TxC, 4.86	TxC, 4.716
	N Losses					
100% MF	30.6	164.0	33.2	80.0	32.2	207.1
50%CM+50%MF	37.8	101.3	37.1	44.9	36.3	116.5
50%CE+50%MF	38.9	104.3	36.0	43.6	37.5	120.4
50%CC+50%MF	39.6	106.1	37.4	45.3	26.0	83.5
LSD (0.05)	T, 3.512	T, 5.250	C, 3.041	C, 4.54	TxC, 6.082	TxC, 9.094

forms and rates added as illustrated by Figure 11.8. At the same time, the total ^{15}N derived from fertilizer (Ndff) by different wheat organs (i.e., root, stem, flag leaf, leaf2, leaf3, leaf4, leaf5, glumes, and grains) were significantly affected by the supplied N form and rates (Moussa et al., 2021).

Additionally, Mo application increased the activities of the enzymes of nitrate reductase, nitrogenase, and microbial biomass of N, implying that Mo application increased N fixation in soil and N bioavailability to plant uptake (Moussa et al., 2022). Moreover, the measurement of ^{15}N natural abundance has provided a greater understanding of the role of Mo application on N bioavailability in natural ecosystems. Their findings demonstrated the vital role of Mo application on the N bioavailability in the soil-wheat system in Mo-insufficient cultivar and Mo-deficient soil. They concluded that Mo application may play an important role in soil N pools by improving the natural ^{15}N abundance and thus increasing N bioavailability to plant (Fig. 11.9).

Green pea grown on sandy soil under field conditions (Hashim and Hekal, 2022), was positively significantly affected by spraying Mn and Cu where its yield and growth attributes increased compared to the untreated plants. Interaction between different rates of N, Mn and Cu reflected an increase in pods fresh weight, pods dry weight, seeds dry yield, pods cover dry weight and shell out as well as protein content in dry seed, nitrogen, manganese and copper uptake. All measurements tended to increase with increasing fertilization rates.

Similarly, NUE, N recovery% and N derived from soil and fertilizer (Ndfs, Ndff) tended to increase with increasing N fertilizer rates. Another field trial conducted on virgin sandy soil (Hekal et al., 2021), using sunflower as a tested crop, showed that the highest N uptake was given by plants receiving a high N rate of 175 kgN/ha in the absence of microbial inoculation with either *Azotobacter chroococcum*, *Azospirillum brasilense* or *Bacillus megaterium* which relatively increased by about

TABLE 11.6
Some of Trials Conducted Using Tracer Techniques under Egyptian Soil Conditions by EAEA Soil Researcher's Staff

Crop	Experimental Conditions	Topic	Isotopic Technique	Aim of the Work	Treatments	Response	Recommendations	References
Winter wheat Mo-inefficient	Greenhouse with Mo-Deficient soil	Plant nutrition	^{15}N stable isotope	The Mo role on N uptake efficiency and recovery - N bioavailability (biological N_2 fixation) and N acquisition	N rates, and N forms, +Mo, –Mo,	Positive effects on N uptake and recovery in wheat particularly with Mo supply - Increase of enzymes activities, N fixation in soil and N bioavailability to plant	Mo-induced improvement in the N uptake efficiency and recovery; vital role of Mo application on the N bioavailability in the soil-wheat system	Moussa et al. (2021, 2022)
Wheat (*Triticum aestivum* L. Giza 171)	Field on virgin sand soil	Soil Microbiology	^{15}N stable isotope	Evaluate the role of different organic additives and inoculation with *B. megatherium* (PSB) in improving wheat growth, NPK uptake and P availability from natural rock-P	Inoculation with *B. megatherium* with different Rock-P rates & animal manure, peanut straw and quail feces	Organic sources reflected an effective role in improving wheat production & increased P fertilizer levels enhanced grain yield & Efficient use of N derived from organic sources was enhanced by increasing P rates and *B. megatherium* inoculation	Organic fertilization combined with P-solubilizing bacteria and recommended rate of rock-P could be accepted as a promising management practice taking environmental risks into consideration	Zaki et al. (2021b)

(Continued)

TABLE 11.6 (Continued)
Some of Trials Conducted Using Tracer Techniques under Egyptian Soil Conditions by EAEA Soil Researcher's Staff

Crop	Experimental Conditions	Topic	Isotopic Technique	Aim of the Work	Treatments	Response	Recommendations	References
Wheat (*Triticum aestivum* L. cv. Giza 168)	Field Experiments	Plant Nutrition & Water Management	^{15}N stable isotope	Find out the proper integral management of irrigation water and nitrogen fertilizer applied in different splitting modes and follow up their effects on nitrogen status and wheat yields under different soils	Clay loam and loamy sand soils & Three nitrogen fertilization rates & splitting Mode & Three water regimes	Cultivated wheat crop (*Triticum aestivum* L.) under 75% of crop water requirements (CWR) combined with 80% of the recommended nitrogen rate splitting into three doses (Mode A), recording higher nitrogen use efficiency (NUE) without significant reduction in yields in both clay loam and loamy sand soils	Further research considering the Nitrogen balance in soil-plant-environment using the ^{15}N tracer technique is encouraged, as it could be a promised tool to identify the nitrogen status in different environments	Hamed et al. (2019)
Wheat (*Triticum aestivum* L.) Masr 2	Field on virgin sand soil	Plant Nutrition & Water Management	^{15}N stable isotope	Evaluation of straw mulching effects on nitrogen status in wheat plants irrigated with saline water of different salinity levels using surface drip irrigation system	Fresh water F (0.5 dSm-1), S1 (6 dSm-1) and S2 (8 dSm-1) & Mulching with rice straw	Mulching has a positive effect on combating salinity stress. It seems that, in general, mulching made the plants more able to derive more nitrogen from fertilizer	Mulching strategy has the ability to conserve irrigation water and enhanced NDFF and its efficient use by wheat crop	Kassab et al. (2018)

(*Continued*)

TABLE 11.6 (Continued)
Some of Trials Conducted Using Tracer Techniques under Egyptian Soil Conditions by EAEA Soil Researcher's Staff

Crop	Experimental Conditions	Topic	Isotopic Technique	Aim of the Work	Treatments	Response	Recommendations	References
Maize (*Zea mays* cv. treble hybrid-329)	Field experiment	Plant Nutrition	^{15}N stable isotope	Response to water regimes and N-fertilizer rates	Three water regimes & three N rates	Irrigation water and N fertilizer were efficiently used by grains at moderate water regime and 248 kgN/ha	Moderate N fertilizer rate and irrigation water could be accepted for achieving considerable grain yield of maize crop	Hekal et al. (2018)
Pea (*Pisum sativum* L.)	Field on poor sand soil	Plant Nutrition	^{15}N stable isotope	Evaluation of ^{15}N, Mn and Cu rates on pea yield and chemical composition	Different rates of N and sprayed Mn and Cu	High yield, nutrients uptake and NUE, Ndff	Considering the application of microelements	Hashim and Hekal (2022)
Pea (*Pisum sativum*) cv. Kafr Elsheikh 1	Field on virgin sand soil	Plant Nutrition & Irrigation Management	^{15}N stable isotope & Neutron Scattering Probe	Explore the effects of different fertilizer regimes (forms and rates), and water regime (optimum and deficit water) on pea crop	Nitrogen fertilizer forms and rates & Irrigation water regimes	Fertilizer-N was more efficiently used under full irrigation and ammonium form surpass urea form	Further research on large scale is needed to recognize the most proper management of water and nitrogen fertilization practices	Fahmy et al. (2016)
Pea (*Pisum sativum* L.)	Field experiment	Plant nutrition & Soil Microbiology	^{15}N stable isotope	Trace the distribution and balance of fertilizer-N either added solely or in combination with organic residues and microbial inoculants	Organic manure; plant residues; *Rhizobium*, *Azospirillum* and Mycorrhizae fungi; ^{15}N-labeled ammonium sulfate	*Rhizobium*, Arbuscular mycorrhizae and *Azospirillum* compensate a considerable amount of fixed nitrogen;	Application of organic farming approach either alone or in combination with bio or mineral fertilizers contribute to improvement of nitrogen nutrition and fertilizer nitrogen balance in soil	El-Sherbiny et al. (2014)

(*Continued*)

TABLE 11.6 (Continued)
Some of Trials Conducted Using Tracer Techniques under Egyptian Soil Conditions by EAEA Soil Researcher's Staff

Crop	Experimental Conditions	Topic	Isotopic Technique	Aim of the Work	Treatments	Response	Recommendations	References
Sunflower (*Helianthus annuus*=L.)	Field on poor sand soil	Plant nutrition & Soil Microbiology	^{15}N stable isotope	Biofertilization against mineral-N fertilization	Free-living microorganisms and N rates	All measurements enhanced by increasing N rates but doesn't affect by bacterial inoculation	Mineral N fertilization more effective than microbial inoculums under given conditions	Hekal et al. (2021)
Peanut (*Arachis hypogaea*, var. GIZA 5)	Field on poor virgin sand soil	Soil Microbiology	^{15}N stable isotope	Achieving the best way for maximizing the groundnut yield without environmental risks and benefits from the low-cost agriculture approach	Bradyrhizobium inoculation, animal manure, leuceana residues and quail feces	Positive and significant effects on production and nutritional value of groundnut crop.	Microbial inoculation of legumes could be useful with assist of organic farming approach to substitute agrochemicals and prevent environmental hazardous effects	Zaki et al. (2021a)
Peanut (*Arachis hypogaea*) (GIZA 6)	Field on poor sand soil	Soil Microbiology	^{15}N stable isotope	Biofertilization and crop residues incorporation effects on peanut growth and yield	*Azotobacter* sp. and *Bradyrhizobium* sp. & different crop residues	Stimulation of peanut growth and yield by inoculation with symbiotic *Bradyrhizobium* sp. inoculum	Substitute organic residues instead of mineral form & bio-fertilization with either associative or symbiotic N_2 fixers	El-Sherbeny et al. (2022)

(*Continued*)

TABLE 11.6 (Continued)
Some of Trials Conducted Using Tracer Techniques under Egyptian Soil Conditions by EAEA Soil Researcher's Staff

Crop	Experimental Conditions	Topic	Isotopic Technique	Aim of the Work	Treatments	Response	Recommendations	References
Faba bean (*Vicia faba* L. var. Misr 1)	Field on poor sand soil	Soil Fertility & Plant Nutrition	Gamma irradiation	Investigates the implications of organic-N vs mineral-N in addition to seed irradiation on faba bean performance under poorly fertile sandy soils	Four irradiation doses in addition to un-irradiated control; five organic and mineral N treatments	Usefulness of gamma irradiation at low to median doses plus organo-mineral-N fertilization	Give highlight on the importance of amending poor sandy soils with combined Organic+Mineral-N-sources to increase the productivity of legume crop grown under given conditions.	Farid et al. (2021)
Broad beans, (*Vicia faba* L. (cv. Sakha 1)	Pot experiment	Plant Nutrition	^{15}N stable isotope	Trace the effect of urea fertilization and Ni levels on growth and N-uptake by broad bean plants	Three different rates of N; three levels Ni and two levels of acetic acid	Interaction between Ni and N-urea had a significant effect on broad bean growth	Addition of acetic acid can combat the adverse effect of high Ni rate and contribute to enhance the NUE% and Ndff % by broad bean plants.	Zhran et al. (2020)

(Continued)

TABLE 11.6 (Continued)
Some of Trials Conducted Using Tracer Techniques under Egyptian Soil Conditions by EAEA Soil Researcher's Staff

Crop	Experimental Conditions	Topic	Isotopic Technique	Aim of the Work	Treatments	Response	Recommendations	References
Common bean (*Phaseolus vulgaris* L.) cv. Nebraska	Field Experiment	Plant Nutrition & Biogation	^{15}N stable isotope & Neutron Scattering Techniques	Evaluation of water regime, nitrogen fertilizer (fertigation) and rhizobium inoculation (biogation) effects on common bean productivity and water, nitrogen use efficiency and water movement under sand soil condition	Two irrigation regime; three N rates; Rhizobium inoculants	Scheduling irrigation according to water consumptive use calculations under running drip irrigation system saved 25% from the used water in 100% ETc. for irrigating common bean plant	The best values of nitrogen derived from air by seeds of inoculated plants were detected with high rate (63.8 kg N/ha) of N fertilizer added in combination with low water regime (W2) 75% ETc.	Moussa et al. (2016)
Chickpea (*Cicer arietinum* L.)	Lab and Field	Soil Microbiology and Fertility	Gamma irradiation	Evaluation and selection of the most potent and active microbial inoculums	Different gamma radiation doses and microbial inoculums	Good performance of chickpea inoculated with irradiated Streptomyces combined with *Mesorhizobium ciceri*	Gamma irradiation at medium doses may improve microbial activities in soil and well implemented in chickpea-microbes relationship	Zaghloul et al. (2021a, b)

(*Continued*)

TABLE 11.6 (Continued)
Some of Trials Conducted Using Tracer Techniques under Egyptian Soil Conditions by EAEA Soil Researcher's Staff

Crop	Experimental Conditions	Topic	Isotopic Technique	Aim of the Work	Treatments	Response	Recommendations	References
Castor bean (*Ricinus communis* L.)	Field Experiment	Plant Nutrition	Gamma irradiation	Following up the effect of both gamma ray and irrigation with wastewater on macronutrients uptake and oil content in castor bean crop	Six doses of gamma rays & wastewater and fresh water	Castor bean Growth, nutrients uptake, oil content were enhanced due to synergetic effect of wastewater irrigation. 50 Gy resulted in the highest values of the estimates.	Gamma irradiation could help in activated plant growth at proper dose and sewage wastewater may be used, especially under water scarcity, with the non-edible crops	Abbas et al. (2015)
Chickpea (*Cicer arientinium* cv Giza 195)	Green-house pot experiment	Soil Microbiology	^{15}N stable isotope	Trace the release of nitrogen from different organic plant residues added to inoculated chickpea in different application methods in order to achieve the organic farming (low cost) concept.	^{15}N-labelled faba bean and wheat green residues in capsule or incorporation & Different Rhizobium inoculums	Fixed-N_2 (Ndfa) were fluctuated according to organic additives type and application method & *Mesorhizobium* inoculation treatments (individuals or dual inoculants).	Incorporation of green plant residues in combination with potent Rhizobium inoculum to minimize and substitute the inorganic fertilizers	Habib et al. (2017)

(*Continued*)

TABLE 11.6 (Continued)
Some of Trials Conducted Using Tracer Techniques under Egyptian Soil Conditions by EAEA Soil Researcher's Staff

Crop	Experimental Conditions	Topic	Isotopic Technique	Aim of the Work	Treatments	Response	Recommendations	References
Chickpea (*Cicer arietinum* L.) Giza 195	Pot experiment	Soil Microbiology	^{15}N stable isotope	Estimation of N2-fixation contribution to nitrogen nutrition of chickpea as influenced by *Mesorhizobium* inoculation and fertilizer nitrogen rates and forms	*Mesorhizobium ciceri* namely ICARDA 36 and ARC Nobaria; ^{15}N labelled ammonium sulfate and labeled barley residues applied in different rates	Addition of 20 kg N solely or in combination with rhizobial inoculum resulted in higher %NUE than those of 50 kg N either alone or in combination with inoculum	Low rate of fertilizer-N is useful for N2 fixation by Rhizobial inoculants & under such conditions, contribution of organic-N released from labeled residues was very low	Rizkalla et al. (2014)
Cowpea (*Vigna unguiculata*) var. Kareem 1	Field experiment on salt affected soil	Plant Nutrition & Soil Fertility	^{15}N stable isotope	Evaluate the impact of mineral fertilizer, fulvic acid, and seaweeds with or without bio-inoculation to combat soil salinity stress.	Urea^{15}N-fertilizer, seaweeds and fulvic acid & *Aspergillus terreus* and *Bradyrhizobium spp.*	Seaweeds were superior to fulvic acid. So, the use of inoculation leads to increase mineral availability to plant so improve crop productivity and increase grain yield	Necessity of inoculation with bio-effectors as well as fungi as eco-friendly agents to help in recognizing proper and low-cost management strategy with special emphasis on environmental impact	Abd El-Hakim et al. (2017)

(*Continued*)

TABLE 11.6 (Continued)
Some of Trials Conducted Using Tracer Techniques under Egyptian Soil Conditions by EAFA Soil Researcher's Staff

Crop	Experimental Conditions	Topic	Isotopic Technique	Aim of the Work	Treatments	Response	Recommendations	References
Sorghum (*Sorghum bicolor* L.)	Green-house pot	Soil Fertility	^{15}N stable isotope	Elucidate the effect of bio-organic fertilizers and mineral-N rates on growth and micronutrients uptake by sorghum crop	Different organic amendments; mineral-N at rates of 50 and 100 mgN/kg soil: *Azoobacter chrooccocum* inoculum	Positive effect of 400 mg N combined with inoculation and organic additives of dry matter yield and micronutrients uptake which contribute to plant biofortification	Bio-organic fertilization management strategy could help in improving plant nutritional value especially with needed micronutrients	Galal et al. (2017)
Sorghum (*Sorghum bicolor* L.) Dorado	Green-house pot	Plant Nutrition	^{15}N stable isotope	Evaluate the role of proline on combating water salinity stress in addition to role of organic additives in sandy soil amelioration	Rates of proline & Residue of leucaena (LU) and chicken manure (ChM)	Application of ^{15}N technique indicated that proline acid has a synergic effect on nitrogen derived from fertilizer (Ndff) and uptake by stalks and roots. This effect was more pronounced under salinity con	Spraying proline acid and organic additives achieved good impact on sorghum growth under different water salinity stress	Galal et al. (2012)

(*Continued*)

TABLE 11.6 (Continued)
Some of Trials Conducted Using Tracer Techniques under Egyptian Soil Conditions by EAEA Soil Researcher's Staff

Crop	Experimental Conditions	Topic	Isotopic Technique	Aim of the Work	Treatments	Response	Recommendations	References
Barley (*Hordeum vulgare* L., var. Giza 126)	Lysimetr Experiment	Plant Nutrition & Irrigation Management	^{15}N stable isotope & Neutron Scattering Probe	Identify the suitable and proper strategy that theoretically and technically provides a quantified basis for applicable irrigation, nitrogen fertilization practices that achieve the optimum yield	Nitrogen fertilizer rates and spelitting mode & Irrigation water regimes	Yield and N uptake were frequently affected by water regimes and N rates plus spelitting modes	Moderate amounts of N fertilizer (156 kgN/ha) spelitted into 50%, 25%, 25% and 80% Etc irrigation water regime could be accepted for reasonable barley production	Gaber et al. (2017), Samak et al. (2016)
Onion (*Allium cepa* L.) C.v. Giza 20	Field Experiment	Plant Nutrition	^{15}N stable isotope	Elucidating the impact of different rates of boron and N on onion growth, yield and N utilization	Six levels of boron (Control, 0.8, 1.6, 2.4, 3.2 and 4 ppm) & Three N levels	Foliar boron enhanced growth traits and yield and N utilization	2.4 ppm boron + 80 kg N/fed had a significant influence on growth traits and yield	El-Sherbeny and Hashim (2021)

(Continued)

TABLE 11.6 (Continued)
Some of Trials Conducted Using Tracer Techniques under Egyptian Soil Conditions by EAEA Soil Researcher's Staff

Crop	Experimental Conditions	Topic	Isotopic Technique	Aim of the Work	Treatments	Response	Recommendations	References
Summer squash (*Cucurbita moschata* cv.)	Field Experiment on virgin sand soil under drip irrigation	Plant nutrition & Soil Microbiology	^{15}N stable isotope	Assess the effect of different organic composts, bacterial inoculants, mineral fertilizer on growth and yield of squash crop	Mineral-N; plant-animal compost; commercial compost; combinations of both under inoculation with *Azospirillum brasilense* (Sp 245) or non-inoculated one	bacterial inoculation could compensate some N from air which let some of mineral N be kept in soil, especially under drip irrigation, for the successive crop; Nitrogen derived from air (Ndfa) was positively affected by addition of organic compost	Application of mineral-N plus Organic-N at 50:50 in combination with bacterial inoculation could be accepted as fertilization management strategy for achieving good crop nutrition and minimizing the environmental pollution risks and agriculture cost.	Habib et al. (2012)
Rice (*Oryza sativa* L. Fengliangyou 4)	Pot Experiment	Resource and Environment	isotope ratios tracing fallout $^{114/111}Cd$, $^{112/111}Cd$, $^{207/206}Pb$, $^{208/206}Pb$ and $^{207/206}Pb$	Clearing the mechanisms of foliar Cd and Pb uptake via the stomata of rice leaves exposed to atmospheric fallout	Four exposure treatments (T1, all day exposure without geotextile membranes; T2, all day exposure with geotextile membranes; T3, daytime exposure with geotextile membranes; and T4, night exposure with geotextile membranes	Foliar uptake atmospheric deposition had substantial effect on Cd and Pb accumulation in rice grains	Control of heavy metal foliar uptake should be paid more attention to maintain rice safety production	Zhu et al. (2022)

(Continued)

TABLE 11.6 (Continued)
Some of Trials Conducted Using Tracer Techniques under Egyptian Soil Conditions by EAEA Soil Researcher's Staff

Crop	Experimental Conditions	Topic	Isotopic Technique	Aim of the Work	Treatments	Response	Recommendations	References
Rice (*Oryza sativa* L. Zhongzao 39)	Pot Experiment	Plant Nutrition	$^{13}CO_2$ stable isotope	N and P will promote the allocation and fixation of newly assimilated C into both the POM and mineral fraction in the rhizosphere and bulk soil	Four treatments, Control, N, P and NP	N and NP fertilization enhanced the net photosynthesized rice 13C in the rice-soil system compared to that of the unfertilized treatment (or control)	NP treatments, indicating that this is a more promising fertilizer treatment for sustaining SOM levels under rice production	Zhran et al. (2021)
Prickly pears and olives	Field Experiment	Soil Erosion	Beryllium-7, ^{7}Be	Assess the potential for Be-7 to estimate soil erosion caused by intensive rain event	Three un-cultivated or cultivated successive fields along the direction of the different slope	The erosion was caused by intensive rain event & prickly pear helps to preserve the soil from erosion in the significant rainfall event compared to olive cultivation and it could be a useful agricultural practice to combat erosion in that area	The ^{7}Be technique can be used as a robust tool to estimate the soil erosion caused by rain, in an arid area due to relatively intensive rain event.	Kassab et al. (2022)

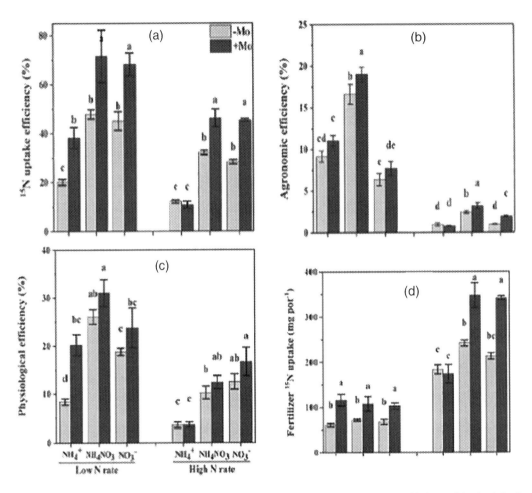

FIGURE 11.8 Impact of Mo application on ^{15}N uptake efficiency (a), agronomic N efficiency (b), physiological N efficiency (c) and ^{15}N derived from fertilizer (b) to winter wheat under different N form ($^{15}NH_4^+$, $^{15}NH_4\,^{15}NO_3$, and $^{15}NO_3^-$) and N rate (low: 0.05 or high: 0.25 g/kg of soil). −Mo and +Mo treatments represent the 0 and 0.15 mg Mo/kg soil, respectively. Source: Adapted from Moussa et al. (2021).

141.4% over the un-inoculated, un-fertilized control. Similarly, the highest fertilizer N recovery recorded 18.45% was detected with the same treatment. Incorporation of different cereal and legume crop residues compensated a considerable amount of nitrogen to peanut plant grown on poorly fertile sandy soil and inoculated with *Azotobacter* sp. and *Bradyrhizobium* sp. Bacterial inoculation boost more than 60% of peanut-N content derived from air (Ndfa%). The contribution of the N-fixed by inoculated peanut was significantly higher than those detected in the un-inoculated ones (El-Sherbeny et al., 2022). Following the ^{15}N Isotope Dilution Concept, Zaki et al. (2021a) recorded that nearly 70% of N content in seeds of peanut crop was derived from the air due to inoculation with *Bradyrhizobium* sp., and this percentage seems to be varied according to organic additives, i.e. animal manure, Leuceana residues and quail feces. In this respect, Ndfa% was higher in case of animal manure and quail feces followed by leuceana residues. They confirmed that the application of organic additives has positive and significant effects on the production and nutritional value of groundnut crop. Organic sources were differentiated among themselves in increasing crop production and nitrogen content in different organs. This may be attributed to their ability to release nutrients to plants and also their contribution in improving soil conditions. In general, leuceana residues were more effective in releasing organic-N to crop (Ndforg) and at the same time reducing the portion of N

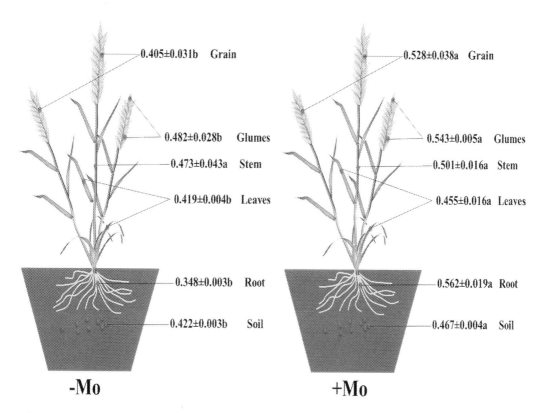

FIGURE 11.9 Effect of Mo application on the natural ^{15}N abundance (‰) in soil and wheat grown under insufficient and sufficient Mo supply (0(−Mo) and 0.15 (+Mo) mg/kg) respectively. All data show the means ± SD of three replicates. The small letters beside values indicate significant (P < 0.05) differences between treatments according to independent t-test analysis. Source: Adapted from Moussa et al. (2022).

derived from mineral fertilizer (Ndff), minimizing N remained in soil after harvest and consequently reducing N losses from soil media. In conclusion, their results reflected promising perspectives for achieving remarkable yield while minimizing environmental risks and production costs. At the same time, these prospects should be taken with precautions. In the same direction, organic fertilization of wheat inoculated with *B. megatherium* as PSB seems to be a proper management scenario in the presence of rock-P as a natural source applied in different rates (Zaki et al., 2021b). Also, they indicated that N derived from peanut straw and quail feces and urea were efficiently used by inoculated plants compared to the un-inoculated ones under all rock-P levels. In the same way, earlier Galal et al. (2012) indicated that a moderate rate of mineral fertilizer combined with an equal quantity of organic compost seems to be the most proper management fertilization practice that offers considerable cucumber yield. Microbial inoculum generally compensated a reasonable amount of nitrogen via N_2 fixation, enhancement of soil N absorption, growth-promoting substances that developed plant growth and nitrogen uptake, and promotion of organic N mineralization from the added compost that provided cucumber plants by remarkable amount of mineralized nitrogen.

Seed production of cowpea was significantly higher in case of seaweeds applied as bio-effector in combination with *Aspergillus terreus* inoculation and urea-^{15}N comparable to fulvic acid treatment. Inoculation with *Bradyrhizobium* spp., enhanced the portion of N gained from the air (Ndfa) especially when plants inoculated with fungi compared to the un-inoculated ones. Nitrogen derived from air (%Ndfa) ranged from 35% up to more than 60% depending on plant organ and fungal or non-fungal inoculation treatments (Abd El-Hakim et al., 2017).

Gamma irradiation technique especially when applied at low to medium doses has a synergistic effect on crop yield performance. This phenomenon was proved when seeds of faba bean were exposed to 20 and 40 Gy compared to 60 and 80 Gy γ doses (Farid et al., 2021). In addition to gamma ray, the incorporation of nitrogen at 25% mrl + 75% org and 75% mrl + 25% org, induced the best grain yield. Also, gamma irradiation of the most potent *Streptomyces alfalfae* strain XY25, *Streptomyces litmocidini* strain NRRL B-3635 and *Streptomyces hawaiiensis* strain ISP 5042, with increasable doses, i.e. 5, 10, 15, and 20 kGy revealed that 15 Gy dose was the proper one that enhanced zinc solubilization efficiency (ZSE), P-solubilizing rate, cytokinin and gibberellin production (Zaghloul et al., 2021a). Selection of the most potent strain *Streptomyces alfalfae* strain XY25, later exposed to gamma radiation and applied in combination with *Mesorhizobium ciceri* inoculum under field conditions for improving the nutritional value of chickpea plant (Biofortification approach), reflected a positive effect on microbial soil enzymes, i.e. dehydrogenase, phosphatase and nitrogenase. Also, they increased the plant's content of phytohormone, micro-elements (Zn, Fe and Mn), phosphorus, potassium, total carbohydrates, and crude protein (Zaghloul et al., 2021b). Previously, Abbas et al. (2015) revealed that seeds of castor bean exposed to different gamma radiation doses reflected variable responses when cultivated in poor fertile sandy soil and irrigated with wastewater. In this respect, they found that gamma irradiation resulted in higher values of both fresh and dry weight of castor bean. This effect seemed more obvious by increasing the dose of gamma ray up to 50 Gy beyond which a gradual decrease occurred. Also, they indicated an increased trend in nitrogen, phosphorus, potassium and oil contents in castor bean plants irrigated with wastewater compared to the corresponding values recorded in plants irrigated with fresh water. Concerning the bio-organic strategy, Galal et al. (2017) indicated that this strategy has an effective role in improving Fe, Zn and Cu uptake by sorghum crop and these microelements particularly increased with increasing mineral-N fertilizer rate. Similarly, the incorporation or capsule form of green wheat and faba bean residues (Habib et al., 2017), in addition to *Rhizobium* inoculums frequently affected the growth and nitrogen uptake by chickpea shoots. In this respect, the superiority of one incorporation method over other seems to be related to type of organic additives. *Mesorhizobium ciceri, 1148 ICARDA* strain either inoculated individually or in dual inoculants with *Sinorhizobium* sp. *36 ICARDA* achieved the highest values of nitrogen derived from air by shoots. The percentages and absolute values of nitrogen derived from organic additives (Ndforg) were gradually decreased with bacterial inoculation treatments but still higher in case of faba bean residues than wheat residues. To combat the adverse effect of Ni on NUE, Zhran et al. (2020) indicated the enhancement of NUE with application of 50 mg Ni but it tends to decrease with 100 mg Ni and up to 60 mg N with or without acetic acid. They concluded that acetic acid application could be accepted as a solution for combating the adverse effect of high levels of Ni on growth and efficient use of urea-N by broad bean crop.

11.11 IMPACT OF CLIMATE CHANGE ON SOIL PROPERTIES, FUNCTIONALITY, AND PRODUCTIVITY

Globally, great attention has been paid to avoid the adverse effects of extreme climate changes on agricultural soil productivity and functionality since the global warming becomes visible. Mondal (2021) summarized the adverse effects of climate changes in Table 11.7.

The physical properties, i.e., structure, water retention and hydraulic conductivity of soil have a great effect on its fertility and are deeply interrelated to the extreme changes in climate features like variations in seasonal temperatures or precipitation intensities, which affect the soil water regime (Horel et al., 2014, 2015). Consequently, the soil moisture regime seems to be strongly influenced by the regional climate changes and climate-induced changes in capillary water movement from groundwater to the root zone. These effects were more vigorous in silt-textured soils than the clayey ones (Bormann, 2012), reflecting that the soil texture is the dominant influencing factor determining the response of soil to regional climate change. Soil texture is defined as the mineral particle's size or the correlated proportions of several groups of mineral size present in a given soil sample. The soil texture is of three types, that is, clay, silt, and sand (Climate and Soil Considerations, 2020). Accordingly, soil conditions were frequently affected by both an excessive amount of rainfall (e.g., waterlogging) and its scarcity (drought) in natural and agro-ecosystems (Farkas et al., 2014).

TABLE 11.7
Impact of Climate Change on Soil Processes and Properties

Climatic Factors	Effects
Rise in temperature	• Salinization of soil • Soil organic matter decomposition increases • Loss of soil organic matter • Decreases soil porosity • Increases soil compactness • Reduction of soil CEC • Reduction of soil fertility • Deterioration of soil structure • Increases risk of soil erosion • Reduction of water retention capacity • Increases CO_2 release from soil • Reduction of soil organic C • Increases ammonia volatilization • Increases rhizospheric temperature • Stimulation of nutrient acquisition • Enhances soil microbial activity • Increases bioavailability of N and P from organic matter
Heavy and intensive rainfall	• Destruction of soil aggregate • Increases risk of soil erosion • Increases leaching of basic cations • Soil acidification • Reduces soil CEC • Loss of soil nutrients, especially N • Development of hypoxic condition in poorly drained soil • Toxicities of Fe, Mn, Al, and B • Loss of N through denitrification
Decreased rainfall	• Increases salt content • Soil moisture deficit • Decreases diffusion and mass flow of water-soluble nutrients • Possibility of occurring drought • Loss of nutrient from rooting zone through erosion • Reduces nutrient acquisition capacity of root system • Reduces N-fixation in legumes
Increase in atmospheric CO_2	• Increases soil C availability • Increases soil microbial activity • Increase soil fungal population

Source: Mondal (2021).

In this respect, infiltration is considered as one of the most important soil properties which help to increase soil water retention, soil erosion mitigation, and decrease the risk of flash floods and droughts (Singh et al., 2011). Advisable note commented by Pandey (2021), indicated that careful use of soil and proper land management can be helpful in mitigation of climate change and combating soil degradation. Interestingly, Lal (2004) showed that a large part of depleted soil organic carbon pool can be restored by using crops residue and mulching and promoting natural cycling of nutrients by using compost and manure.

11.12 CLIMATE SMART AGRICULTURE - CSA

Climate smart agriculture (CSA) was described as "agriculture that sustainably raises production and incomes, in addition to adapting and promoting resilience to CC and decreasing greenhouse gasses (GHG) emissions (Amin et al., 2015)". In other turn, Rashid et al. (2021) identified Climate Smart Agriculture (CSA) as an approach in which technological, strategic and investment conditions are developed to reach sustainable agricultural development for food security under climate change. The extent to which climate change is affecting agricultural systems necessitates ensuring comprehensive consolidation of these effects into national agricultural planning, investments and programs (Figure 11.10).

Recently, Hussain et al. (2022) pointed out three important trade-offs or objectives of CSA according to FAO (2015):

1. Increasing agricultural incomes and productivity on sustained basis
2. Building resilience to CC
3. Decreasing GHG emissions according to IPCC standards

Nowadays, the need for CSA approaches becomes urgent for the rapid transition of the current agriculture production system especially with increasing risks from CC and climate-associated disasters (Scherr et al., 2012). For improving the agriculture production system in CSA, the following five points should be put into consideration:

1. Food demand is increasing but food has to be produced with the same amount of resources like water, vegetation, and land.
2. Farmers are exposed to the effects of CC and there is urgency for sustainable adaptation to CC.
3. There is a total degradation and depletion of natural resources that sustain agriculture production.

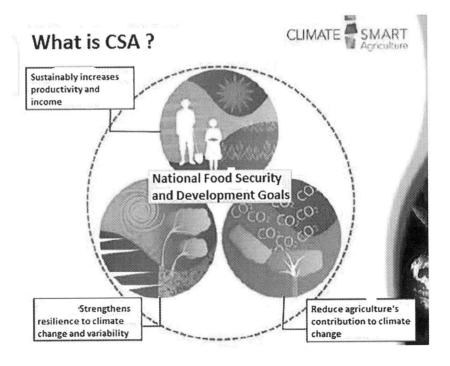

FIGURE 11.10 Climate smart agriculture – CSA. Source: Cited from Rashid et al. (2021).

4. The expression of CC in agricultural production systems has to be productive, strong to risks, efficient, and better constancy in their outputs in long-term CC.
5. There is an essential need for increasing food production along with mitigating CC as well as protecting the natural resource (IPCC, 2007).

One of the most important strategies for CSA is efficient resource management including soil, water and land use. Climate factors, agricultural practices, market prices, technological advances, and management practices cause intensive changes in agriculture. In this respect, the most efficient ways to use water, land, and soil nutrients should be achieved to confirm and attain the aforementioned three main objectives of CSA. Changes in soil genesis elements due to local climate changes lead to changes in its properties such as topography, geology, vegetation, and variable landscapes in depth. The abundance of nutrients that are available in the soil bring more diversity. A handful of soil that can hold various organisms can also play an important role in preserving soil health (Corsi et al., 2012).

Water management under severe climate changes becomes more urgent since the agriculture sector consumes about 80% of the freshwater resources on a global scale. In this regard, the CSA technology has been considered as a friendly method that applies a political aspect to attain sustainable development goals (Mundial, 2012). Water demand has been increasing in industries and cities as an outcome of fast economic development in developing countries (Mubeen et al., 2020). Pollution has been increased by industries, cities, and agricultural activities which have affected the ecosystems. Water demand is projected to increase in the whole world in the coming years as population will spread by 9 billion by 2050. Currently, the agricultural sector and rural communities are facing great challenges due to increase of temperature, water scarcity, rising sea level, flood and drought and losses of soil fertility which increase food insecurity (FAO, 2012), and governed us to search for high-level adaptation plan that decrease the risk of CC and protects the livelihood of the rural communities. Earlier information presented by FAO (2008) showed that agriculture crop production was impacted attributable to climate change in the components of weather/climate such as temperature, precipitation, cyclones, sea level etc. (Table 11.8).

As quoted by Hussain et al. (2022), sustainable land and water management (SLWM) contains a mixture of inorganic and organic soil fertility as well as water management more than compensates for the impact of CC on crop yields under existing management methods. In addition, SLWM is also profitable and increases family income to combat poverty (Sandhu et al., 2010). Due to the weak knowledge of advisory services for the management and development of irrigation systems (Mubeen et al., 2021), water loss in irrigation systems in Africa is over 50%. Short-term training with an operative focus on these significant topics will be more operative and practical than long-term training.

TABLE 11.8
Impacts of Climate Change on Crop Production (FAO 2008)

Event	Potential Impact
Day and night temperature increased over most of the land areas as cold periods become shorter and warmer (virtually certain)	High yields in low temperature areas; while in high temperature areas yield reduces; increased outbreaks of different new insect pests as well as pathogens causing notable effect on crop production
High frequency of precipitation over most areas (very likely)	Crop damages; soil erosion; waterlogged soils making land unable for cultivation
Increased drought affected areas (likely)	Soil erosion and degradation; reduced yields due to crop failure or damage; arable soil loss
Increased tropical cyclone frequency (likely)	Crop damage
Extremely increased level of sea water (excludes tsunami) (likely)	Saline irrigation water, fresh and estuaries water systems; arable land loss

Target for CSA improvement strategies, to some extent, depends on the role of institutions to create useful information and its involvement in programs that help people interpret, understand, and work on new technologies for combating extreme climate changes. In this respect, the programs and policies of the Egyptian Atomic Energy Authority are also concerned with peaceful uses of nuclear techniques in improving soil and water management strategies under the CSA approach and hope to be effective if their application is supported by reliable institutions. Therefore, it is important to increase the institutional capacity to apply and replicate CSA strategies. Institutions are also crucial for the development of agriculture and the creation of sustainable livelihoods. They are not only a tool for decision-makers and farmers but also the most important way to improve and maintain climate-friendly growing methods (FAO, 2013). The benefits of CSA technologies in improving food security, sustainable use of all products and natural resources and conservation of environment were excellently concluded by Hussain et al. (2022).

Recent argument about the use of crop models presented by Franke (2021), under Southern Africa conditions, showed that yield predictions by crop models used in climate change studies could become more accurate by improving the representation of processes simulating the impact of evolved CO_2 on growth and water use, and the impact of heat stress on reproductive processes. These models if applied in more multi-disciplinary settings and in interaction with farmers and their organizations, taking into account farmers' current capacities, may lead to useful insights into farmers' ability to cope with climate change now and in future. So, to increase crop production and, at the same time, reduce the negative effects of excessive fertilization, the precise use of fertilizers became a focal point in world agriculture through the application of a smart fertilizer management approach. This approach could be applied using imaging technologies and implemented plant biomarkers (Agrahari et al., 2021). For example, smart nitrogen fertilizer management utilizing GIS information in conjunction with plant-soil parameters was guaranteed for the development of soil management protocols (Say et al., 2018; Söderström et al., 2016). Several new VIs have been proposed by the data mining of HSI, which allow the prediction of the N content in field crops (Table 11.9).

The N status of crops (i.e., endogenous N level that supports physiological processes such as photosynthesis), which is associated with plant growth and yield, is more useful than tissue N content for efficient N management. To improve the accuracy of N status prediction, multispectral imaging (MSI) is combined with data obtained by other methods, such as HSI, visible red, green, and blue (RGB) images, and data obtained by sensors (Figure 11.11). Portable sensors that estimate the amount of chlorophyll and N status are currently used at farming sites (Cardim Ferreira Lima et al., 2020).

Transcriptional or RNA, in addition to biochemical biomarkers, are correlated with various nutrients status in plants. Although transcriptional biomarkers have some limitations in terms of cost and handling RNA samples, they have high specificity and robustness among a wide range of plant species (in terms of gene ontology; Ito et al., 2019; Agrahari et al., 2020a), as well as high reliability for diagnosing various plant conditions other than nutrient deficiencies. Nutrient deficiency in plants leads to changes that affect their metabolic pathways, resulting in major changes in particular metabolites (e.g., amino acids, organic acids, lipids, and sugars). Such metabolites act as biomarkers for nutrient deficiencies (Steinfath et al., 2010; Sung et al., 2015). Several signature metabolites have been identified as biochemical biomarkers for plant nutrient responses (Table 11.10).

The aforementioned information lead Agrahari et al. (2021) to conclude that the recent development of high-throughput technologies, particularly in transcriptomics and metabolomics, has facilitated the identification of plant biomarkers that can monitor plant responses to nutrients at the early stage of plant development in real time. The development of such approaches can be helpful for further development of smart fertilization under a precision agriculture system.

Currently, Varinderpal-Singh et al. (2021) revealed that precision N management using leaf color chart (LCC), chlorophyll meter (SPAD), and GreenSeeker optical sensor (GS) has a benefit role in sustaining wheat grain yield equivalent to the soil-test-based N fertilizer recommendation with the average savings of 20% N fertilizer. Precision N management strategies improved mean recovery efficiency (REN) and partial factor productivity (PFPN) of applied N fertilizer, respectively

TABLE 11.9
Evaluation of Nutrients Status and Content in Crops by Imaging Techniques

Imaging Technique	Detection Method/ Algorithm	Nutrients	Analyzed Plants	Experimental Setup	Combination of Additional Data	References
MSI multispectral imaging	Stepwise-MLR (prediction model)	N-content	Rice (canopy)	Field (using UAV)	Hyperspectral data (Ground sensor)	Zheng et al. (2018)
	LRM (prediction model)	N use efficiency for variety	Winter Wheat (canopy)	Field (using UAV)	No	Yang et al. (2020)
HSI Hyperspectral imaging	SAE-FNN (Predicted model)	N-content	Oilseed rape-leaf	Field & Lab	No	Yu et al. (2018)
	PLSR (prediction model)	N-status	Wheat (leaf)	Lab	No	Bruning et al. (2019)
	Normalized difference chlorophyll index	N-use efficiency for variety	Wheat (leaf)	Lab	RGB	Banerjee et al. (2020)
	MLR (prediction model)	N-content	Apple (leaf & canopy)	Lab and field	No	Ye et al. (2020)
RGB Red-green-blue	PLSR (Image feature evaluation)	N-status	Wheat (canopy)	Field	Reflectance sensor data	Elsayed et al. (2018)
	Linear regression (Yield model)	N-management	Maize (canopy)	Field	No	Zhang et al. (2020)
	BPNN (Damage recognition)	N, P, K, Mg, Fe, Zn deficiency	Rice (canopy)	Field	No	Anami et al. (2020)

HSI, hyperspectral imaging; MSI, multispectral imaging; RGB, red-green-blue; MLR, multiple linear regressions; RF, random frog; PLSR, partial least squares regression; SG, Savitzky–Golay; LNC, leaf nitrogen concentration; SAE, stacked auto-encoders; FNN, fully connected neural network; SAM, spectral angle mapper; LRM, linear regression model; VIs, vegetation indices; ELR, extreme learning machine based regression; PCA, principal component analysis; SVM, support vector machine; CNN, convolutional neural network; BPNN, back propagation neural network; UAV, unmanned aerial vehicle; SPAD, soil plant analysis development.
Source: Adapted from Agrahari et al. (2021).

by 26.0% and 26.4% over the soil-test-based N management. Spectral properties measured with LCC, SPAD and GS showed good correlation (R2 > 0.71) with grain yield, depicting the great potential of optical sensing tools in predicting grain yield and inferring need-based fertilizer N topdressings decisions in wheat. So, the aforementioned results proved that precision N management provides a potential solution to improve N nutrition in wheat while reducing nitrous oxide (N_2O) and total GHG emissions by 23.2% and 23.6%, respectively, in comparison to soil-test-based N application.

Concerning P fertilizer, the precise detection and accurate quantification of org-P compounds are among the major obstructions to soil org-P research (Turner et al., 2015). To meet these challenges, a much-improved analytical framework has been so far developed. This includes various wet-chemical techniques, chromatography, and recently nuclear magnetic resonance (NMR) spectroscopy (McLaren et al., 2015). Among these, solution ^{31}P NMR spectroscopy is predominantly adopted and extensively used (Cade-Menun and Liu, 2014; Turner et al., 2015). In addition, other methodological approaches, such as data mining (i.e., principal component analysis), meta-analysis, and modeling, can be of great value (Nash et al., 2014; Haygarth et al., 2018). Approaches could be adopted to enhance soil organic-P cycling including several promising and innovative techniques. Some of these techniques were reviewed by Sulieman and Mühling (2021) and listed in Table 11.11.

Soil-based factors involved in organic-P cycling and dynamics include a wide spectrum of soil microorganisms comprised numerous types of soil fungi (e.g., *Aspergillus* and *Penicillium*) and

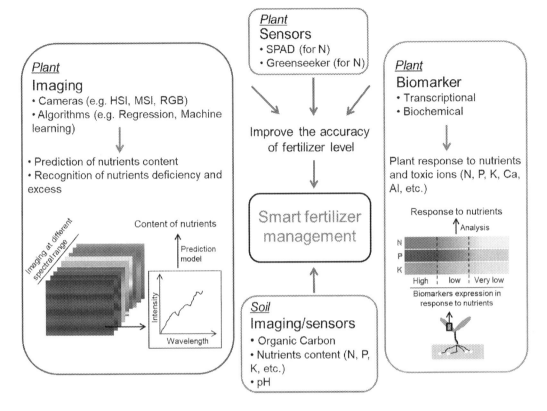

FIGURE 11.11 Diagrammatic representation of smart fertilizer management. Smart fertilizer management utilizes imaging, sensors, and biomarkers to phenotype the nutrient status in plants and soils. HSI, hyperspectral imaging; MSI, multispectral imaging; RGB, red-green-blue; SPAD, soil plant analysis development. Source: Agrahari et al. (2021).

TABLE 11.10
Biomarkers Used for Plant Nutrients Status Analysis

Approach	Biomarkers	Response	Plants	References
Transcriptional	112 genes set	N-status	Maize	Yang et al. (2011)
	miR399, miR827	P-deficiency	Barley	Hackenberg et al. (2013)
	14 genes set 19 genes set	P-status N-status	Rice	Takehisa and Sato (2019)
	DDF1, TSPO, RRTF1	Mg-deficiency	Arabidopsis	Hermans et al. (2010)
	ALS3	Al-toxicity level in acid soil	Arabidopsis, Soybean	Sawaki et al. (2016), Agrahari et al. (2020b)
Biochemical	NR, GDH	N-deficiency	Tomato	Urbanczyk-Wochniak and Fernie (2005)
	Anthocyanin	P-deficiency	Various plant species	Schachtman and Shin (2007)
	Putrescine	K-deficiency	Various plant species	Cui et al. (2020)

ALS3, aluminum sensitive 3; DDF1, dwarf and delayed flowering 1; TSPO, tryptophan-rich sensory protein related; RRTF1, redox responsive transcription factor 1; miR, microRNA; NR, nitrate reductase; GDH, glutamate dehydrogenase.
Source: Agrahari et al. (2021).

TABLE 11.11
Approaches and Technologies Used to Promote Soil Organic Phosphorus (Porg) Cycling

Approach/Technology	Concept	Porg Substrate	Test Plant	References
Placement (encapsulation) of phytase in the root vicinity using mesoporous silica nanoparticles materials	Phytases are stable and resistant to soil degradation	Phytate	*Medicago truncatula*	Trouillefou et al. (2015)
Cultivation of efficient agroforestry species	Enhancement of P org solubilization and mineralization	NAa	*Tithonia, tephrosia*	George et al. (2002)
Inoculation of plants with soil isolates/microorganisms that possess efficient phytase activity	Mineralization of complex organic substrates by phytases	Myo-inositol hexaphosphate	Pasture legume (subterranean clover, white clover, alfalfa, burr medic) and pasture grass (wallaby grass, phalaris) species	Richardson et al. (2001c)
Application of bacterial grazer (nematodes) together with mycorrhiza and P-solubilizing bacteria	Interaction of bacterial grazers with mycorrhiza and phosphobacteria promotes P org solubilization	Phytate	Maritime pine	Irshad et al. (2012)
Biochar addition to agricultural soils	Biochar enhances Pi-solubilizing bacteria	NAa	Ryegrass	Anderson et al. (2011)
	Biochar promotes soil enzymatic activities (phosphatases)	NAa	NAa	Bailey et al. (2011)
Genetic transformation of plants to overexpress extracellular phytases in root cells	Transgenic lines display better P i nutrition owing to the efficient release of extracellular root phytases	Phytate	*Arabidopsis*	Richardson et al. (2001a), Mudge et al. (2003)
			Subterranean clover	Richardson et al. (2001b), George et al. (2004)
			Potato	Zimmermann et al. (2003)
			Tobacco, *Arabidopsis*	Lung et al. (2005)

Source: Adapted after Sulieman and Mühling (2021).

bacteria (e.g., *Actinomycetes*, *Pseudomonas*, and *Bacillus*). Some rhizospheric microorganisms are potentially able to act as plant growth-promoting rhizobacteria (PGPR) (Hinsinger et al., 2015). In addition to soluble Pi, these PGPRs can provide plants with nitrogen, hormones, and iron through the strain-specific siderophores. A series of physicochemical and biochemical processes are responsible for modulating the stock of organic-P in agricultural soils (Figure 11.12). Additionally, several physiological and biochemical traits associated with plant roots can be exploited to facilitate the mobilization of soil organic-P at the field scale. Among them, the exudation of root substances (e.g., phosphatases and organic anions) has been well-documented and cited by Sulieman and Mühling (2021).

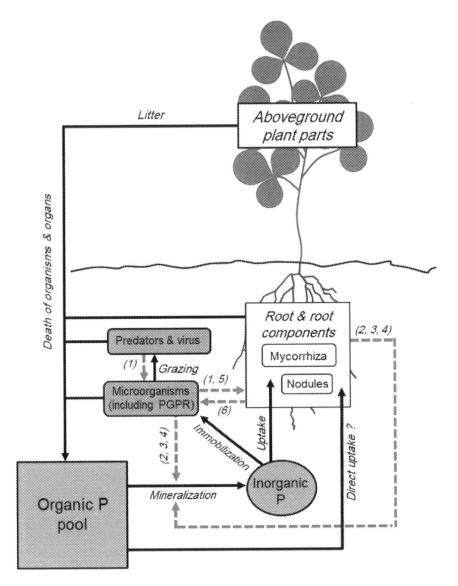

FIGURE 11.12 Plant and soil-related biological processes (*dotted* arrows) that can ultimately modify the pool size of soil organic P in the rhizosphere: (1) promotion and selection of soil microbes, (2) phosphatase release, (3) pH alteration, (4) organic acid exudation, (5) rhizodeposition, (6) hormonal and signaling effects. ? = unidentified uptake system. Source: Modified from Hinsinger et al. (2015) and cited from Sulieman and Mühling (2021).

They added that admittedly, the plants' potential ability to utilize sparingly available organic-P sources is greatly determined by their genetic background. At the same time, several plant species are presumed to be relatively efficient in organic-P utilization without assistance from soil microorganisms.

11.13 CONCLUSION

Attribute to wide gaps between population and food production, more attention has been paid toward suitable and proper management of soil and water resources especially in hot spots of the world that suffer from aridity and desertification. In this regard, more agronomists and plant breeders as well as soil scientists turned their efforts to utilize the problem soils and find ecofriendly and cost-effective ways to attenuate the impact of severe climate changes on soil degradation and irrigation water scarcity. Nuclear techniques may play an effective and unique role in recognition of soil problems and accurate determination of soil nutrients and water movement and its availability to crops. As concluded by Rajwar et al. (2021), the intensive usage of synthetic fertilizers, systematic degradation, soil erosion and the impact of rainfall or storms, the depletion of organic matter, and many other factors have resulted in growing desertification, the loss of millions of tons of productive surface soil and, consequently, the silting up of rivers and dams, the salinization of soils, climate change, and the loss of biodiversity. Decline in soil fertility due to the loss of organic matter, sudden increased erosion, and increase in soil salinity, soil compaction and reduced biodiversity are the major threats affecting soil health directly. Delayed implementation of sustainable approaches and strategies may make the situation worse by increasing the rate of land degradation due to climate change. Recently, many researchers reported that irrigated areas will increase in forthcoming years, while freshwater supplies will be diverted from agriculture to meet the increasing demand for domestic use and industry. Furthermore, the efficiency of irrigation is very low, since less than two-thirds of the applied water is actually used by the crops. The sustainable use of irrigation water is a priority for agriculture in arid areas. Nowadays, precision agriculture or smart agriculture is the cornerstone of smart agriculture applications depending on the acquisition of field information including the environment, crops, and soil, and the accuracy of sensing data. Soil and crop sensing technology involves the exploration of sensing mechanism, spectroscopy, biology, microelectronics, remote sensing, sensors, and information processing methods. It helps in solving soil and crop production problems. Intelligent, convenient, accurate, and energy-saving information acquisition technology will continue to be one of the research hotspots in the field of smart agriculture. In this regard, and to solve these problems, the scientific staff of Soil and Water Research Department, EAEA-Egypt spent more than 50 years in research for improving soil fertility and productivity parallel with increasing water use efficiency and productivity through the application of different nuclear techniques including stable and radioactive isotopes as well as moisture neutron probe gauge and ^{18}O. Also, ^{7}Be fallout was followed to estimate soil erosion. Advanced biological (bio-fertilizers, bio-fortification and BNF) and chemical technologies were applied to minimize dependence on agricultural chemicals and at the same time trials to improve its use efficiency through the application of nanomaterials. Nanotechnology has revolutionized all agricultural practices at research labs. The use of nanofertilizers, nanopesticides, nanoherbicides, and eventually nanosensors has led to a new form of agriculture which is precision agriculture. Experiences gained in the era of bio-agriculture let us confirm, with the aid of ^{15}N stable isotope, its positive role in achieving clean and safe agriculture via improving N nutrition through biological N_2 fixation and minimizing losses and harmful risks.

REFERENCES

Abadía, J., Vázquez, S., Rellán-Álvarez, R., El-Jendoubi, H., Abadía, A., and Álvarez Fernández, A. (2011). Towards a knowledge based correction of iron chlorosis. *Plant Physiol Biochem* 49, 471–482.

Abbas, H.H., Farid, I.M., Soliman, S.M., Galal, Y.G.M., Ismail, M.M., Kotb, E.A., and Moslhy, S.H. (2015). Growth and some macronutrients uptake by castor bean irradiated with gamma ray and irrigated with wastewater under sandy soil condition. *J. Soil Sci. and Agric. Eng., Mansoura Univ.*, 6 (4), 433 – 444.

Abd El-Hakim, A.M., El-Sherbiny, A.E., Dahdouh, S.M., and Galal, Y.G.M. (2017). Role of bioeffectors and soil ameliorates on cowpea yield grown on saline soil with aid of ^{15}N isotope dilution technique. *Zagazig J Agric Res* 44(6A), 2105–2015.

Adekalu, K.O., Olorunfemi, I.A., and Osunbitana, J.A. (2007). Grass mulching effect on infiltration, surface runoff and soil loss of three agricultural soils in Nigeria. *Bioresour Technol* 98(4), 912–917. https://doi.org/10.1016/j.biortech.2006.02.044.

Agrahari, R.K., Kobayashi, Y., Borgohain, P., Panda, S.K., and Koyama, H. (2020a). Aluminum-specific upregulation of GmALS3 in the shoots of soybeans: A potential biomarker for managing soybean production in acidic soil regions. *Agronomy* 10(9), 1228. https://doi.org/10.2134/precisionagbasics.2016.0093.

Agrahari, R.K., Kobayashi, Y., Tanaka, T.S.T., Panda, S.K. and Koyama, H. (2021). Smart fertilizer management: The progress of imaging technologies and possible implementation of plant biomarkers in agriculture. *Soil Sci Plant Nutr* 67(3), 248–258. https://doi.org/10.1080/00380768.2021.1897479.

Agrahari, R.K., Singh, P., Koyama, H., and Panda, S.K. (2020b). Plant-microbe Interactions for sustainable agriculture in the post-genomic era. *Curr Genom* 21(3), 168–178. https://doi.org/10.2174/1389202921999200505082116.

Ahmed, M. (2020). Introduction to modern climate change. Andrew E. Dessler: Cambridge University Press, 2011, 252 pp, ISBN-10: 0521173159. *Sci Total Environ* 734, 139397. https://doi.org/10.1016/j.scitotenv.2020.139397.

Ahmed, M., Aslam, M.A., Fayyaz-ul-Hassan, Hayat, R., Nasim, W., Akmal, M., Mubeen, M., Hussain, S., and Ahmad, S. (2022). Nutrient dynamics and the role of modeling. In: W. N. Jatoi et al. (eds.), *Building Climate Resilience in Agriculture*, Springer Nature Switzerland AG 2022, https://doi.org/10.1007/978-3-030-79408-8_20. pp. 297–316.

Ahmed, M., Hasanuzzaman, M., Raza, M.A., Malik, A. and Ahmad, S. (2020). Plant nutrients for crop growth, development and stress tolerance. In: R. Roychowdhury et al. (eds.), Sustainable Agriculture in the Era of Climate Change, p. 43-92. Springer International Publishing;Springer, Year: 2020 ISBN: 9783030456689,9783030456696 https://doi.org/10.1007/978-3-030-45669-6_3.

Albay, M., Ozbayram, E.G., Camur, D., and Topbaş, M. (2021). Recent trends in water and health studies on the focus of global changes. *Environ Manag* 67, 437–438.

Alexandratos, N., and Bruinsma, J. (2012). *World Agriculture towards 2030/2050: The 2012 Revision*. FAO, Rome.

Amin, A., Mubeen, M., Hammad, H.M., and Nasim, W. (2015). Climate smart agriculture: An approach for sustainable food security. *Agric Res Commun* 2(3), 13–21.

Anami, B.S., Malvade, N.N., and Palaiah, S. (2020). Classification of yield affecting biotic and abiotic paddy crop stresses using field images. *Inf Process Agric* 7, 272–285. https://doi.org/10.1016/j.inpa.2019.08.005.

Anderson, C.R., Condron, L.M., Clough, T.J., Fiers, M., Stewart, A., Hill, R.A., and Sherlock, R.R. (2011). Biochar induced soil microbial community change: Implications for biogeochemical cycling of carbon, nitrogen and phosphorus. *Pedobiologia* 54, 309–320.

Azad, B., Hassandokht, M.R., and Parvizi, K. (2015). Effect of mulch on some characteristics of potato in Asadabad, Hamedan. *International Journal of Agronomy and Agricultural Research* 6(3), 139–147.

Bailey, V.L., Fansler, S.J., Smith, J.L., and Bolton Jr., H. (2011). Reconciling apparent variability in effects of biochar amendment on soil enzyme activities by assay optimization. *Soil Biol Biochem* 43, 296–301.

Bajgain, R., Xiao, X., Basara, J., Wagle, P., Zhou, Y., Zhang, Y., and Mahan, H. (2016). Assessing agricultural drought in summer over Oklahoma Mesonet sites using the water-related vegetation index from MODIS. *Int J Biometeorol* 61, 1–14.

Banerjee, B., Joshi, S., Thoday-Kennedy, E., Pasam, R.K., Tibbits, J., Hayden, M., Spangenberg, G., and Kant, S. (2020). High-throughput phenotyping using digital and hyperspectral imaging-derived biomarkers for genotypic nitrogen response. *J Exp Bot* 71(15), 4604–4615. https://doi.org/10.1093/jxb/eraa143.

Belimov, A.A., Dodd, I.C., Hontzeas, N., Theobald, J.C., Safronova, V.I. and Davies, W.J. (2009). Rhizosphere bacteria containing l-aminocyclopropane-1-carboxylate deaminase increase yield of plants grown in drying soil via both local and systemic hormone signalling. *N Phytolog* 181, 413–423.

Bhattacharya, A. (ed.) (2021). *Soil Water Deficit and Physiological Issues in Plants*. Springer Nature Singapore Pte Ltd. ISBN 978-981-33-6275-8, ISBN 978-981-33-6276-5(eBook), p. 717. https://doi.org/10.1007/978-981-33-6276-5.

Bormann, H. (2012). Assessing the soil texture-specific sensitivity of simulated soil moisture to projected climate change by SVAT modelling. *Geoderma* 185, 186, 73–83.

Borsato, E., Rosa, L., Marinello, F., Tarolli, P., and D'Odorico, P. (2020). Weak and strong sustainability of irrigation: A framework for irrigation practices under limited water availability. *Front Sustain Food Syst* 4, 17.

Bruning, B., Liu, H., Brien, C., Berger, B., Lewis, M., and Garnett, T. (2019). The development of hyperspectral distribution maps to predict the content and distribution of nitrogen and water in wheat (*Triticum aestivum*). *Front Plant Sci* 10. 1-16.https://doi.org/10.3389/fpls.2019.01380.

Busari, T.I., Senzanje, A., Odindo, A.O., and Buckley, C.A. (2019). Evaluating the effect of irrigation water management techniques on (taro) madumbe (*Colocasia esculenta* (L.) Schott) grown with anaerobic filter (AF) effluent at Newlands, South Africa. *J Water Reuse Desalin* 9(2), 203–212. https://doi.org/10.2166/wrd.2019.058.

Cade-Menun, B., and Liu, C.W. (2014). Solution phosphorus-31 nuclear magnetic resonance spectroscopy of soils from 2005 to 2013: A review of sample preparation and experimental parameters. *Soil Sci Soc Am J* 78, 19–37.

Cao, J., Zhou, X., Zhang, W., Liu, C., and Liu, Z. (2012). Effects of mulching on soil temperature and moisture in the rain-fed farmland of summer corn in the Taihang Mountain of China. *J Food Agric Environ* 10(1), 519–523.

Cardim Ferreira Lima, M., Krus, A., Valero, C., Barrientos, A., Del Cerro, J., and J, R.-G. (2020). Monitoring plant status and fertilization strategy through multispectral images. *Sensors* 20(2), 435. https://doi.org/10.3390/s20020435.

Chartzoulakis, K. and Bertaki, M. (2015). Sustainable water management in agriculture under climate change. *Agricult Agricult Sci Procedia* 4, 88–98.

Chaudhari, S.K. (2021). Soil and water management in India: Challenges and opportunities. In: A. Rakshit, et al. (eds.), *Soil Science: Fundamentals to Recent Advances*, Springer Nature Singapore Pte Ltd, https://doi.org/10.1007/978-981-16-0917-6_36, pp. 751–764.

Ciais, P., Sabine, C., Bala, G., Bopp, L., Brovkin, V., Canadell, J., Chhabra, A., De Fries, R., Galloway, J., Heimann, M., Jones, C., Le Quéré, C., Myneni, R.B., Piao, S. and Thornton, P. (2013). Carbon and Other Biogeochemical Cycles. In: *Climate Change 2013: The Physical Science Basis. Contribution of Working Group I to the Fifth Assessment Report of the Intergovernmental Panel on Climate Change* [Stocker, T.F., D. Qin, G.K. Plattner, M. Tignor, S.K. Allen, J. Boschung, A. Nauels, Y. Xia, V. Bex and P.M. Midgley (eds.)]. Cambridge University Press, Cambridge, United Kingdom and New York, NY, USA, p. 465-570.

Climate and Soil Considerations (2020). https://smallfarms.cornell.edu/guide/2-climateandsoilconsiderations/. Accessed 25 Oct 2020.

Çokkizgin, H., and Bölek, Y. (2015). Priming treatments for improvement of germination and emergence of cotton seeds at low temperature. *Plant Breed Seed Sci* 71, 121–134.

Corsi, S., Friedrich, T., Kassam, A., Pisante, M., and de Moraes Sà, J.C. (2012). Soil organic carbon accumulation and greenhouse gas emission reductions from conservation agriculture: A literature review, integrated crop management (101 pp.), 16, AGP/FAO, Rome.

Couturier, J., Touraine, B., Briat, J.F., Gaymard, F., and Rouhier, N. (2013). The iron-sulfur cluster assembly machineries in plants: Current knowledge and open questions. *Front Plant Sci* 4, 259.

Cui, J., Pottosin, I., Lamade, E., and Tcherkez, G. (2020). What is the role of putrescine accumulated under potassium deficiency? *Plant Cell Environ* 43(6), 1331–1347. https://doi.org/10.1111/pce.13740.

D'Odorico, P., Davis, K.F., Rosa, L., CarrJoel, A., Chiarelli, D., Dell'Angelo, J. et al. (2018). The global food-energy-water nexus. *Rev Geophys* 56(3), 456–531.

Diaz-Zorita, M. and Fernández-Canigia, M.V. (2009). Field performance of a liquid formulation of *Azospirillum basilense* on dryland wheat productivity. *Euro J Soil Biol* 45, 3–11.

El Gamal, F. (2007). Use of nonconventional water resources in irrigated agriculture. In: N. Lamaddalena et al. (eds.), Water *Saving* in Mediterranean *Agriculture* and *Future Research Needs*, vol. 2, CIHEAM, Bari, pp. 33–43 (Options Méditerranéennes: Série B. Etudes et Recherches; Vol. II, no. 56).

Elsayed, S., Barmeier, G., and Schmidhalter, U. (2018). Passive reflectance sensing and digital image analysis allows for assessing the biomass and nitrogen status of wheat in early and late tillering stages. *Front Plant Sci* 9, 1478. https://doi.org/10.3389/fpls.2018.01478.

El-Sherbeny, T.M.S. and Hashim, M. E. (2021). Effect of boron and nitrogen on growth and yield of onion (*Allium cepa L.*) plant using N-15 technique. *Arab J Nucl Sci Appl* 54(1), 105–112.

El-Sherbeny, T.M.S., Mousa, A.M., and Zhran, M.A. (2022). Response of peanut (*Arachis hypogaea* L.) plant to bio-fertilizer and plant residues in sandy soil. *Environ Geochem Health* 13, 1–13.

El-Sherbiny, A.E., Galal, Y.G.M, Soliman, S.M., Dahdouh, S.M. Ismail, M.M., and Fathy, A. (2014). Fertilizer nitrogen balance in soil cultivated with pea (*Pisum sativum* L.) under bio and organic fertilization system using ^{15}N stable isotope. *4th Int Con Rad Res Appl Sci Taba Egypt* 2014, 85–96.

Fahad, S., Hussain, S., Saud, S., Hassan, S., Tanveer, M., Ihsan, M.Z., Shah, A.N., Ullah, A., Nasrullah, Khan, F., Ullah, S., Alharby, H., Nasim, W., Wu, C., and Huang, J. (2016). A combined application of biochar and phosphorus alleviates heat-induced adversities on physiological, agronomical and quality attributes of rice. *Plant Physiol Biochem* 103, 191–198.

Fahmy, A.E., Al-Gindy, A.M., Arafa, Y.E., and Abdel Aziz, H.A. (2016). Field trial on pea (*Pisum Sativum* L.) grown on sand soil and subjected to water regimes and nitrogen forms with aid of ^{15}N stable isotope. *J Nucl Tech Appl Sci* 4(2), 65–74.

Fan, X.L. and Li, Y.K. (2001). Effect of water deficit and water deficit tolerancec heredity on nitrogen efficiency of winter wheat. In: Horst WWJ, Schenk MK, Burkert A, Claasen N, Flessa H, Frommer WB, Goldbach HE, Olfs H-W, Romheld W, Sattelmacher B, Schmidhalter U, Chubert S, von Wiren N, Wittenmayer L (eds.), *Plant Nutrition: Food Security and Sustainability of Agro-Ecosystems*. American Society of Agronomy, Madison, WI, pp. 62–63.

Fang, Q.X., Ma, L., Green, T.R., Yu, Q., Wang, T.D., and Ahuja, L.R. (2010). Water resources and water use efficiency in the North China Plain: Current status and agronomic management options. *Agric Water Manag* 97, 1102–1116.

FAO (2008). Climate change adaptation and mitigation in the food and agriculture sector. High-level conference on food security-the challenges of climate change and bioenergy (Available https://www.preventionweb.net/files/8314_HLC08bak1E.pdf.

FAO (2012). Mainstreaming climate-smart agriculture into a broader landscape approach. In: *Second Global Conference on Agriculture, Food Security and Climate Change*, Hanoi, Vietnam, 3-7 September 2012. pp.34. FAO, Viale delle Terme di Caracalla 00153 Rome, Italy, www.fao.org/climatechange, climate-change@fao.org

FAO (2013). *Sustainability Assessment of Food and Agriculture Systems. Guidelines Version 3.0.* FAO, Rome.

FAO (2015). *Regional Overview of Food Insecurity: African Food Insecurity Prospects Brighter than Ever.* Accra, Ghana, www.fao.org/contact-us/licence-request, , ISBN 978-92-5-108781-7, pp.35.

FAO (2019). FAOSTAT. Food and Agriculture data [WWW Document].

Farid, I.M., El-Nabarawy, A.A.A., Abbas, M.H.H., Moursy, A.A., Afify, M.H.E., Abbas, H.H., and Hekal, M.A. (2021). Implications of seed irradiation with γ-rays on the growth parameters and grainyield of faba bean. *Egypt J Soil Sci* 61(2), 175–186.

Farkas, C., Gelybó, G., Bakacsi, Z., Horel, A., Hagyó, A., Dobor, L., Kása, I., and Tóth, E. (2014). Impact of expected climate change on soil water regime under different vegetation conditions. *Biologia* 69, 1510–1519.

Farooq, M., Wahid, A., Kobayashi, N., Fujita, D., and Basra, S.M.A. (2009). Plant drought stress: Effects, mechanisms and management. *Agron Sustain Dev* 29, 185–212.

Fatma, M., Masood, A., Per, T.S., and Khan, N.A. (2016). Nitric oxide alleviates salt stress inhibited photosynthetic performance by interacting with sulfur assimilation in mustard. *Front Plant Sci* 7, 521.

Franke, A.C. (2021). Assessing the impact of climate change on crop production in southern Africa: A review. *South African Journal of Plant and Soil* 38(1), 01–12.

Gaber, E.I., Samak, M.R., Galal, Y.G.M., and Mohamed, M.A. (2017). Grain yield and nitrogen uptake by barley subjected to fertilization and water regimes using ^{15}N stable isotope. *Arab J Nucl Sci Appl* 50(2), 55–66.

Gadalla, A.M., Hamdy, A., and Galal, Y.G.M. (2007). Use of saline irrigation water for production of some legumes and tuber plants. In: N. Lamaddalena et al. (eds.), *Water Saving in Mediterranean Agriculture and Future Research Needs*, Vol. 2. CIHEAM, Bari, pp. 85–97 (Options Méditerranéennes: Série B. Etudes et Recherches; n. 56 Vol. II)

Galal, Y.G.M., AbdelAziz, H.A., and Degwy, S.M. (2017). Effect of organic/mineral-N sources on Fe, Zn and Cu uptake by sorghum grown on clay soil. *Arab J Nucl Sci Appl* 50(2), 30–36.

Galal, Y.G.M., El-Sherbiny, A.E., Soliman, S.M., Dahdouh, S.M., and Fathy, A. (2012). Benefits from bio and organic fertilization by cucumber (*Cucumis sativus*) with application of ^{15}N stable isotope. In: Proceedings of Eleventh Arab Conference on the Peaceful Uses of Atomic Energy, *Khartoum, Sudan, 23–27 December 2012*, pp. 1–17., Arab Atomic Energy Authority, Tunis

Galal, Y.G.M., Soliman, S.M., Abou El-Khair, R.A., El-Mohtasem Bella, M.O., Abdel Aziz, H.A., Kotb, E.A., and Abd El-Latteef, E.M. (2012). Effect of sprayed proline acid and organic amendments on improvement of sorghum growth against water salinity with application of ^{15}N isotope dilution Technique. In: *Proceedings of Minia International Conference for* Agriculture and Irrigation in the Nile Basin Countries,*Minia University, 26th–29th March 2012, El-Minia, Egypt*, pp. 1207–1221.

George, T.S., Gregory, P.J., Robinson, J.S., and Buresh, R.J. (2002). Changes in phosphorus concentrations and pH in the rhizosphere of some agroforestry and crop species. *Plant Soil* 246, 65–73.

George, T.S., Richardson, A.E., Hadobas, P.A., and Simpson, R.J. (2004). Characterization of transgenic *Trifolium subterraneum* L. which expresses phyA and releases extracellular phytase: Growth and P nutrition in laboratory media and soil. *Plant Cell Environ* 27, 1351–1361.

Ghabour, S., Galal, Y.G.M, Soliman, S.M., El-Sofi, D.M., Morsy, A.A., and El-Sofi, M.M. (2015). Nitrogen distribution in soil profile under sesame, maize and wheat crops as affected by organic and inorganic nitrogen fertilizers using ^{15}N technique. *Int J Plant Soil Sci* 6(5), 294–302.

Green, P.A., Vörösmarty, C.J., Harrison, I., Farrell, T., Sáenz, L., and Fekete, B.M. (2015). Freshwater ecosystem services supporting humans: Pivoting from water crisis to water solutions. *Glob Environ Chang* 34, 108–118.

Habib, A.A.M., El-Sherbiny, A.E., Dahdouh, S.M., and Galal, Y.G.M. (2012). Organic and mineral fertilization of squash plant with application of ^{15}N stable isotope. *Zagazig J Agric Res* 39(5), 909–919.

Habib, F.M., Abd El-Hameed, A.H., Galal, Y.G.M., Abdel-Aziz, O.A., Mousa, A.A., and Zaki, Z.A. (2017). Contribution of ^{15}N-labeled organic manures in chickpea-rhizobium symbiosis performance. *J Nucl Tech Appl Sci* 5(3), 131–141.

Hackenberg, M., Shi, B.J., Gustafson, P., and Langridge, P. (2013). Characterization of phosphorus-regulated miR399 and miR827 and their isomirs in barley under phosphorus-sufficient and phosphorus-deficient conditions. BMC Plant Boil 13(1), 214. https://doi.org/10.1186/1471-2229-13-214.

Hamed, L.M.M., Galal, Y.G.M., Soliman, M.A.E., and Emara, E.I.R. (2019). Optimum applications of nitrogen fertilizer and water regime for wheat (*Triticum aestivum* L.) using ^{15}N tracer technique under mediterranean environment. *Egyptian J Soil Sci* 59(1), 41–52.

Hammad, H.M., Abbas, F., Saeed, S., Fahad, S., Cerdà, A., Farhad, W., Bernardo, C.C., Nasim, W., Mubeen, M., and Bakhat, H.F. (2018). Offsetting land degradation through nitrogen and water management during maize cultivation under arid conditions. *Land Degrad Dev* 29(5), 1366–1375.

Hammond, J.P., and White, P.J. (2008). Sucrose transport in the phloem: Integrating root responses to phosphorus starvation. *J Exp Bot* 59, 93–109.

Hänsch, R., and Mendel, R.R. (2009). Physiological functions of mineral micronutrients (Cu, Zn, Mn, Fe, Ni, Mo, B, Cl). *Curr Opin Plant Biol* 12, 259–266.

Hashim, M.E. and Hekal, M.A. (2022). Nitrogen fertilization and foliar application with Mn and Cu in green pea (*Pisum sativum* L.) using ^{15}N stable isotope. *J Soil Sci Agric Eng Mansoura Univ* 13(9), 311–316.

Hatfield, J.L., Sauer, J.T., and Prueger, J.H. (2001). Managing soils to achieve greater water use efficiency: A review. *Agronomy J* 93, 271–280.

Haygarth, P.M., Hinsinger, P., and Blackburn, D. (2018). Organic phosphorus: Potential solutions for phosphorus security. *Plant Soil* 427, 1–3.

Hekal, M.A., Abdel-Salam, A.A., Soliman, S.M., Galal, Y.G.M., Abd-El-Moniem, M., Zahra, W.R., and Moursy, A.A. (2018). Efficient use of N and water for maize (*Zea Mays* L.) crop under drip irrigation system using ^{15}N stable isotope. In: *Proceedings of 4th International Conference on Biotechnology Applications In Agriculture* (ICBAA) 2018, Organized by Faculty of Agriculture, Benha University, 4-7 April 2018, Hurghada, Egypt. *Bio-Fertilizers Section*, pp. 615–622.

Hekal, M.A., Abdel-Salam, A.A., Soliman, S.M., Galal, Y.G.M., Moursy, A.A., and Zahra, W.R. (2021). Bio and mineral-N fertilization of sunflower (*Helianthus annuus* L.) grown on sandy soil using ^{15}N technique. *Ann Agric Sci* 59(4), 1077–1082.

Hermans, C., Vuylsteke, M., Coppens, F., Craciun, A., Inzé, D., and Verbruggen, N. (2010). Early transcriptomic changes induced by magnesium deficiency in arabidopsis thaliana reveal the alteration of circadian clock gene expression in roots and the triggering of abscisic acid-responsive genes. *N Phytol* 187(1), 119–131. https://doi.org/10.1111/j.14698137.2010.03258.x.

Hinsinger, P., Herrmann, L., Lesueur, D., Robin, A., Trap, J., Waithaisong, K., and Plassard, C. (2015). Impact of roots, microorganisms and microfauna on the fate of soil phosphorus in the rhizosphere. *Annu Plant Rev* 48, 377–407.

Horel, A., Lichner, L., Alaoui, A., Czachor, H., Nagy, V., and Tóth, E. (2014). Transport of iodide in structured clay-loam soil under maize during irrigation experiments analyzed using HYDRUS model. *Biologia* 69, 1531–1538.

Horel, Á., Tóth, E., Gelybó, G.Y., Kása, I., Bakacsi, Z.S., and Farkas, C.S. (2015). Effect of land use and management on soil hydraulic properties. *Open Geosci* 1, 742–754. https://doi.org/10.3390/horticulturae3030042.

Hussain, S., Amin, A., Mubeen, M., Khaliq, T., Shahid, M., Hammad, H.M., Sultana, S.R., Awais, M., Murtaza, B., Amjad, M., Fahad, Sh., Amanet, K., Ali, A., Ali, M., Ahmad, N., and Nasim, W. (2022). Climate smart agriculture (CSA) technologies. In: W. N. Jatoi et al. (eds.), *Building Climate Resilience in Agriculture*, Springer Nature Switzerland AG 2022, pp. 319–338, https://doi.org/10.1007/978-3-030-79408-8_20.

Intergovernmental Panel on Climate Change (IPCC) (2007). Technical summary. In: *Climate Change 2007: Mitigation. Contribution of Working Group III to the Fourth Assessment Report of the IPCC*. Cambridge University Press, Cambridge and New York, NY.

Irshad, U., Alain Brauman, A., Villenave, C., and Plassard, C. (2012). Phosphorus acquisition from phytate depends on efficient bacterial grazing, irrespective of the mycorrhizal status of *Pinus pinaster*. *Plant Soil* 358, 155–168.

Ito, H., Kobayashi, Y., Yamamoto, Y.Y. and Koyama, H. (2019). Characterization of NtSTOP1-regulating genes in tobacco under aluminum stress. *Soil Sci Plant Nutr* 65(3), 251–258. https://doi.org/10.1080/00380768.2019.1603064.

Jezek, M., Geilfus, C.M., Bayer, A., Mühling, K.H., and Struik, P.C. (2015). Photosynthetic capacity, nutrient status, and growth of maize (*Zea mays* L.) upon MgSO4 leaf-application. *Front Plant Sci* 5, 1–10.

Juang, K-W., Lin, M-C., and Hou, C-J. (2021). Influences of water management combined with organic mulching on taro plant growth and corm nutrition. *Plant Prod Sci* 24(2), 152–169, https://doi.org/10.1080/1343943X.2020.1820877.

Kassab, M.F., Galal, Y.G. M., El-Gindy, A.M., El-Bagoury, K.F., and El-Tohory, Sh.K. (2018). Improvement of nitrogen fertilization practices for efficient use by wheat irrigated with saline water under mulching using ^{15}N technique. *Arab J Nucl Sci Appl* 51(1), 99–103.

Keating, B.A., Herrero, M., Carberry, P.S., Gardner, J., and Cole, M.B. (2014). Food wedges: Framing the global food demand and supply challenge towards 2050. *Glob Food Sec* 3(3–4), 125–132.

Khan, N.A., Khan, M.I.R., Asgher, M., Fatma, M., Masood, A., and Syeed, S. (2014). Salinity tolerance in plants: Revisiting the role of sulfur metabolites. *J Plant Biochem Physiol* 2, 2.

Kosterna, E. (2014). The effect of covering and mulching on the temperature and moisture of soil and broccoli yield. *Acta Agrophys* 21(2), 165–178.

Laganowsky, A., Gómez, S.M., Whitelegge, J.P., and Nishio, J.N. (2009). Hydroponics on a chip: Analysis of the Fe deficient Arabidopsis thylakoid membrane proteome. *J Proteom* 72, 397–415.

Lal, R. (2004). Soil carbon sequestration to mitigate climate change. *Geoderma* 123(1–2), 1–22.

Lal, R. (2012a). Toward enhancing storage of soil water and agronomic productivity. In: R. Lal and B.A. Stewart (eds.), *Soil Water and Agronomic Productivity*, CRC Press Taylor & Francis Group, Boca Raton, FL. pp. 559–568.

Lal, R. (2012b). Soil water and agronomic production. In: R. Lal and B.A. Stewart (eds.), *Soil Water and Agronomic Productivity*, CRC Press Taylor & Francis Group, Boca Raton, FL. pp. 43–60.

Liu, F., He, X., and He, Y. (2022). Theories and methods for soil nutrient sensing. In: M. Li et al. (eds.), *Soil and Crop Sensing for Precision Crop Production*, Springer Nature Switzerland AG 2022, pp. 49–73. https://doi.org/10.1007/978-3-030-70432-2_3.

Lordan, J., Pascual, M., Villar, J.M., Fonseca, F., Papió, J., Montilla, V., and Rufat, J. (2015). Use of organic mulch to enhance water-use efficiency and peach production under limiting soil conditions in a three-year-old orchard. *Spanish J Agric Res* 13(4), e0904. https://doi.org/10.5424/sjar/2015134-6694.

Lung, S.-C., Chan, W.-L., Yip, W., Wang, L., Yeung, E.C., and Lim, B.L. (2005). Secretion of beta-propeller phytase from tobacco and *Arabidopsis* roots enhances phosphorus utilization. *Plant Sci* 169, 341–349.

MacDonald, G.K., Bennett, E.M., Potter, P.A., and Ramankutty, N. (2011). Agronomic phosphorus imbalances across the world's croplands. *Proc Natl Acad Sci USA* 108, 3086–3091.

Marschner, H. (2012). *Marschner's Mineral Nutrition of Higher Plants*, 3rd edition. Academic Press, London.

McIntyre, B.D., Speijer, P.R., Riha, S.J., and Kizito, F. (2000). Effects of mulching on biomass, nutrients, and soil water in banana inoculated with nematodes. *Agrono J* 92(6), 1081–1085. https://doi.org/10.2134/agronj2000.9261081x.

McLaren, T.I., Smernik, R.J., McLaughlin, M.J., McBeath, T., Kirby, J.K., Simpson, R.J., Guppy, C.N., Doolette, A., and Richardson, A. (2015). Complex forms of soil organic phosphorus – A major component of soil phosphorus. *Environ Sci Technol* 49, 13238–13245.

Millaleo, R., Reyes-Díaz, M., Ivanov, A.G., Mora, M.L., and Alberdi, M. (2010). Manganese as essential and toxic element for plants: Transport, accumulation and resistance mechanisms. *J Soil Sci Plant Nutr* 10, 476–494.

Miyasaka, S.C., Hollyer, J.R., and Kodani, L.S. (2001). Mulch and compost effects on yield and corm rots of taro. *Field Crops Res* 71(2), 101–112. https://doi.org/10.1016/S0378-4290(01)00154-X.

Mondal, S. (2021). Impact of climate change on soil fertility. In: D.K. Choudhary et al. (eds.), *Climate Change and the Microbiome, Soil Biology 63*, Springer Nature Switzerland AG 2021, pp. 551–569, https://doi.org/10.1007/978-3-030-76863-8_28.

Moran, M.S., Peters-Lidard, C.D., Watts, J.M., and McElroy, S. (2004). Estimating soil moisture at the watershed scale with satellite-based radar and land surface models. *Canad J Remote Sens* 30, 23.

Moursy, A.A. and Abdel Aziz, H.A. (2013). Contribution of fungal inoculation in degradation of organic residues and nitrogen-mineralization using ^{15}N technique. *Int J Curr Microbiol App Sci* 2(10), 452–466.

Moussa, M.G., El-Gindy, A.M., Abdel Aziz, H.A., and Arafa, Y.E. (2016). Effect of fertigation and biogation on common bean yield grown on sand soil with application of ^{15}N and neutron scattering techniques. *J Nucl Tech Appl Sci* 4(1), 51–64.

Moussa, M.G., Hu, C., Elyamine, A.M., Ismael, M.A., Rana, M.S., Imran, M., Syaifudin, M., Tan, Q., Marty, C., and Sun, X. (2021). Molybdenum-induced effects on nitrogen uptake efficiency and recovery in wheat (*Triticum aestivum* L.) using ^{15}N-labeled nitrogen with different N forms and rates. *J Plant Nutr Soil Sci* 184, 613–621. https://doi.org/10.1002/jpln.202100040.

Mousa, A.M., Ibrahim, R.M., and Galal, Y.G. (2022). Contribution of organic compost and bacterial inoculation in improving quality of quinoa seeds. *Al-Azhar Bull Sci C* 33(1), 133–142. https://doi.org/10.21608/absb.2022.119792.1170.

Moussa, M.G., Sun, X., El-Tohory, S., Mohamed, A., Saleem, M.H., Riaz, M., Dong, Z., He, L., Hu, C., and Ismael, M.A. (2022). Molybdenum role in nitrogen bioavailability of wheat-soil system using the natural ^{15}N abundance technique. *J Soil Sci Plant Nutr* 22, 3611–3624. https://doi.org/10.1007/s42729-022-00913-w.

Mubeen, M., Ahmad, A., Hammad, H.M., Awais, M., Farid, H.U., Saleem, M., and Nasim, W. (2020). Evaluating the climate change impact on water use efficiency of cotton-wheat in semi-arid conditions using DSSAT model. *J Water Climate Change* 11(4), 1661–1675.

Mubeen, M., Bano, A., Ali, B., Islam, Z.U., Ahmad, A., Hussain, S., Fahad, S., Nasim, W. (2021) Effect of plant growth promoting bacteria and drought on spring maize (*Zea mays* L.). *Pakistan J Bot* 53(2), 731–739. https://doi.org/10.30848/PJB2021-2(38).

Mudge, S.R., Smith, F.W., and Richardson, A.E. (2003). Root-specific and phosphate-regulated expression of phytase under the control of a phosphate transporter promoter enables *Arabidopsis* to grow on phytate as a sole P source. *Plant Sci* 165, 871–878.

Mujere, N. and Muhoyi, H. (2022). Estimating soil moisture using remote sensing in Zimbabwe: A review. In: S. K. Dubey et al. (eds.), Soil-Water, Agriculture, and Climate Change, *Water Science and Technology Library*, Vol. 113, pp. 79–91. https://doi.org/10.1007/978-3-031-12059-6_5.

Mukhopadhyay, D. and Thakur, P. (2021). Natural resource management and conservation for smallholder farming in India: Strategies and challenges. In: A. Rakshit et al. (eds.), *Soil Science: Fundamentals to Recent Advances*, Springer Nature Singapore Pte Ltd, pp. 731–749. https://doi.org/10.1007/978-981-16-0917-6_36.

Mukherjee, A.K. and Batabyal, K. (2021). Climate-Smart Soil Management: Prospect and Challenges in Indian Scenario. In: A. Rakshit et al. (eds.), Soil Science: Fundamentals to Recent Advances, p. 875-902, Springer Nature Singapore Pte Ltd. 2021. https://doi.org/10.1007/978-981-16-0917-6_42

Mundial, B. (2012). *Agricultural Innovation Systems: An Investment Sourcebook*. World Bank, Washington, DC.

Nachimuthu, G., Halpin, N.V., and Bell, M.J. (2017). Productivity benefits from plastic mulch in vegetable production likely to limit adoption of alternate practices that deliver water quality benefits: An on-farm case study. *Horticulturae* 3(3), 42.

Nagajyoti, P.C., Lee, K.D., and Sreekanth, T.V.M. (2010). Heavy metals, occurrence and toxicity for plants: A review. *Environ Chem Lett* 8, 199–216.

Nash, D.M., Haygarth, P.M., Turner, B.L., Condron, L.M., McDowell, R.W., Richardson, A.E., Watkins, M., and Heaven, M.W. (2014). Using organic phosphorus to sustain pasture productivity: A perspective. *Geoderma* 221, 11–19.

Nawaz, M.A., Chen, C., Shireen, F., Zheng, Z., Sohail, H., Afzal, M., Ali, M.A., Bie, Z., and Huang, Y. (2018). Genome-wide expression profiling of leaves and roots of watermelon in response to low nitrogen. *BMC Genomics* 19:456.

Neumann, P.M. (2008). Coping mechanisms for crop plants in drought-prone environments. *Ann Bot* 101, 901–907.

Nouet, C., Motte, P., and Hanikenne, M. (2011). Chloroplastic and mitochondrial metal homeostasis. *Trends Plant Sci* 16, 395–404.

Oldare, M., Arthurson, V., Pell, M., Svensson, K., Nehrenheim, E., and Abubaker, J. (2011). Land application of organic waste-effects on the soil ecosystem. *Appl Energy* 88, 2210–2218.

Palombi, L. and Sessa, R. (2013). *Climate-Smart Agriculture: Sourcebook*. FAO, Rome.

Pandey, D. (2021). Impact of climate change on soil functionality. In: D.K. Choudhary et al. (eds.), *Climate Change and the Microbiome, Soil Biology*, Vol. 63, Springer Nature Switzerland AG, pp. 597–604. https://doi.org/10.1007/978-3-030-76863-8_28.

Parween, T., Jan, S., Mahmooduzzafar, M., and Fatma, T. (2011). Alteration in nitrogen metabolism and plant growth during different developmental stages of green gram *Vigna radiata* L. in response to chlorpyrifos. *Acta Physiol Plant* 33, 2321–2328.

Petrov, P., Petrova, A., Dimitrov, I., Tashev, T., Olsovska, K., Brestic, M., and Misheva, S. (2018). Relationships between leaf morpho-anatomy, water status and cell membrane stability in leaves of wheat seedlings subjected to severe soil drought. *J Agron Crop Sci* 204(3), 219–227.

Pirmoradian, N., Sepaskhah, A.R., and Maftoun, M. (2004). Deficit irrigation and nitrogen effects on nitrogen-use efficiency and grain protein of rice. *Agronomie* 24, 143–153.

Rafique, E., Rashid, A., and Mahmood-Ul-Hassan, M. (2012). Value of soil zinc balances in predicting fertilizer zinc requirement for cotton-wheat cropping system in irrigated aridisols. *Plant Soil* 361, 43–55.

Rajwar, J., Joshi, D., Suyal, D.C., and Soni, R. (2021). Factors affecting soil ecosystem and productivity. In: R. Soni et al. (eds.), *Microbiological Activity for Soil and Plant Health Management*. Springer Nature Singapore Pte Ltd, pp. 437–457. https://doi.org/10.1007/978-981-16-2922-8_18.

Rashid, S., BinMushtaq, M., Farooq, I., and Khan, Z. (2021). Climate smart crops for food security. In: Harris A.A. (ed.), *The Nature, Causes, Effects and Mitigation of Climate Change on the Environment*, IntechOpen, IntechOpen Limited 5 Princes Court,Gate London, SW7, UNITED KINGDOM p. 20. https://doi.org/10.5772/intechopen.99164.

Rashid, S., BinMushtaq, M., Farooq, I., and Khan, Z. (2021). Climate smart crops for food security. In: *The Nature, Causes, Effects and Mitigation of Climate Change on the Environment*, IntechOpen, p. 20. https://doi.org/10.5772/intechopen.99164.

Ray, J.D., Gesch, R.W., Sinclair, T.R. and Allen, L.H. (2002). The effect of vapor pressure deficit on maize transpiration response to a drying soil. *Plant Soil* 239, 113–121.

Richardson, A.E., Hadobas, P.A., and Hayes, J.E. (2001a). Extracellular secretion of *Aspergillus phytase* from Arabidopsis roots enables plants to obtain phosphorus from phytate. *Plant J* 25, 641–649.

Richardson, A.E., Hadobas, P.A., Hayes, J.E., O'Hara, C.P., and Simpson, R.J. (2001c). Utilization of phosphorus by pasture plants supplied with myo-inositol hexaphosphate is enhanced by the presence of soil micro-organisms. *Plant Soil* 229, 47–56.

Richardson, A.E., Hadobas, P.A., and Simpson, R.J. (2001b). Phytate as a Source of Phosphorus for the Growth of Transgenic *Trifolium subterraneum*. In: W.J. Horst et al. (eds.), *Progress in Plant Nutrition: Food Security and Sustainability of Agro-Ecosystems through Basic and Applied Research*, Kluwer Academic, Dordrecht, pp. 560–561.

Rizkalla, M.G., Ali, M.A., Galal, Y.G.M., Abo Taleb, H.H., and ALhudaiji, M.A. (2014). Biological nitrogen fixation by chickpea as affected by nitrogen fertilizer using ^{15}N technique. *J Nucl Tech Appl Sci* 2(5), 539–548.

Rockström, J., Karlberg, L., Wani, S.P., Barron, J., Hatibu, N., Oweis, Th., Bruggeman, A., Farahani, J., and Qiang, Z. (2010). Managing water in rainfed agriculture-The need for a paradigm shift. *Agric Water Manag* 97, 543–550.

Rockström, J., Williams, J., Daily, G., Noble, A., Matthews, N., Gordon, L., and Smith, J. (2017). Sustainable intensification of agriculture for human prosperity and global sustainability. *Ambio* 46(1), 4–17.

Rosa, L., Rulli, M.C., Davis, K.F., Chiarelli, D.D., Passera, C., and D'Odorico, P. (2018). Closing the yield gap while ensuring water sustainability. *Environ Res Lett* 13(10), 104002.

Ruíz-Machuca, L.M., Ibarra-Jiménez, L., Valdez-Aguilar, L.A., Robledo-Torres, V., Benavides-Mendoza, A., and Cabrera-De La Fuente, M. (2014). Cultivation of potato - use of plastic mulch and row covers on soil temperature, growth, nutrient status, and yield. *Acta Agric Scand B Soil Plant Sci* 65(1), 30–35. https://doi.org/10.1080/09064710.2014.960888.

Salama, M.A., Mostafa, A.Z., and Yousef, K.H.M. (2017). Water use efficiency of wheat crop under two water application methods. *Arab J Nucl Sci Appl* 50(3), 77–84.

Samak, R.R. Magdy, E., Gaber, I., Galal, Y.G.M., and Mohamed, M.A. (2016). Barley nitrogen acquisition as affected by water regime, fertilizer rates and application mode using ^{15}N stable isotope. *Int J Curr Microbiol App Sci* 5(1), 116–135.

Sandhu, H.S., Wratten, S.D., and Cullen, R. (2010). Organic agriculture and ecosystem services. *Environ Sci Policy* 13, 1–7.

Sawaki, K., Sawaki, Y., Zhao, C.R., Kobayashi, Y., and Koyama, H. (2016). Specific transcriptomic response in the shoots of arabidopsis thaliana after exposure to Al rhizotoxicity: Potential gene expression biomarkers for evaluating Al toxicity in soils. *Plant Soil* 409(1–2), 131–142. https://doi.org/10.1007/s11104-016-2960-8.

Say, S.M., Keskin, M., Sehri, M. and Sekerli, Y.E. (2018). Adoption of precision agriculture technologies in developed and developing countries. *Online J Sci Technol* 8, 7–15.

Schachtman, D.P. and Shin, R. (2007). Nutrient sensing and signaling: NPKS. *Annu Rev Plant Biol* 58(1), 47–69. https://doi.org/10.1146/annurev.arplant.58.032806.103750.

Scherr, S.J., Shames, S., and Friedman, R. (2012). From climate-smart agriculture to climate-smart landscapes. *Agric Food Secur* 1, 12. https://www.agricultureandfoodsecurity.com/content/1/1/12.

Schmidt, S.B., Jensen, P.E., and Husted, S. (2016). Manganese deficiency in plants: The impact on photosystem II. *Trends Plant Sci* 21, 622–632.

Shukla, A.K., Tiwari, P.K., Pakhare, A., and Prakash, C. (2016). Zinc and iron in soil, plant, animal and human health. *Ind J Fertil* 12, 133–149.

Sidhu, G.S., and Sharma, B.D. (2010). Diethylene triamine penta acetic acid-extractable micronutrients status in soil under a rice-wheat system and their relationship with soil properties in different agroclimatic zones of Indo-Gangetic plains of India. *Commun Soil Sci Plant Anal* 41, 29–51.

Singh, B.P., Cowie, A.L., and Chan, K.Y. (eds.) (2011). *Soil Health and Climate Change, Soil Biology*. Springer, Heidelberg, p. 414.

Singh, R.J., Mandal, D., Ghosh, B.N., Chand, L., Alam, N.M., and Sharma, N.K. (2015). Efficient soil and water management under limited water supply condition. In: D.J., Rajkhowa, Das, Anup, S.V., Ngachan, A. K. Sikka, and M. Lyngdoh (eds.), *Integrated Soil and Water Resource Management for Livelihood and Environmental Security. 2015*, ICAR Research Complex for NEH Region, Meghalaya, pp. 1–19.

Smith, P., Bustamante, M., Ahammad, H., Clark, H., Dong, H., Elsiddig, E.A., Haberl, H., Harper, R., House, J., Jafari, M., Masera, O., Mbow, C., Ravindranath, N.H., Rice, C.W., Robledo Abad, C., Romanovskaya, A., Sperling, F. and Tubiello, F. (2014). Agriculture, forestry and other land use (AFOLU). In: O. Edenhofer, R. Pichs-Madruga, Y. Sokona, E. Farahani, S. Kadner, K. Seyboth, A. Adler, I. Baum, S. Brunner, P. Eickemeier, B. Kriemann, J. Savolainen, S. Schlömer, C. von Stechow, T. Zwickel, and J.C. Minx (eds.), *Climate Change 2014: Mitigation of Climate Change. Contribution of Working Group III to the Fifth Assessment Report of the Intergovernmental Panel on Climate Change*, Cambridge University Press, Cambridge and New York, NY. pp. 811–922.

Söderström, M., Sohlenius, G., Rodhe, L., and Piikki, K. (2016). Adaptation of regional digital soil mapping for precision agriculture. *Precis Agric* 17(5), 588–607. https://doi.org/10.1007/s11119-016-9439-8.

Soliman, S.M. and Galal, Y.G.M. (2017). Contribution of nuclear science in agriculture sustainability. *Arab J Nucl Sci Appl* 50(2), 37–48.

Soliman, S.M. and Galal, Y.G.M. (2020). Applications of isotopes in fertilizer research. In: R. Lal, (ed.), *Soil and Fertilizers: Managing the Environmental Footprint*, First edition, CRC Press, Boca Raton, FL, CRC Press is an imprint of Taylor & Francis Group, LLC, pp. 209–.

Soliman, S.M., Galal, Y.G.M., Abou El-Khair, R.A., El-Mohtasem Bella, M.O., Abdel Aziz, H.A., Kotb, E.A., and Abd El-Latteef, E.M. (2012). Role of organic additives and inorganic fertilizer in combating irrigation water salinity stress. In: Anaç D. et al., (Eds.), Proceedings of 8th International Soil Science Congress, Vol V, Nutrient Management for Soil Sustainability, Food Security and Human Health, May 15-17, 2012 Çeşme – Izmir, Turkey, pp. 191–198.

Soliman, S.M., Galal, Y.G.M., Abou El-Khair, R.A., El-Mohtasem Bella, M.O., Abdel Aziz, H.A., Kotb, E.A., and Abd El-Latteef, E.M. (2012). Role of organic additives and inorganic fertilizer in combating irrigation water salinity stress. In: Proceedings of 8th International Soil Science Congress, Vol V, Nutrient Management for Soil Sustainability, Food Security and Human Health, pp. 191–198.

Soliman, S.M., Ghabour, S., Galal, Y.G.M., El-Sofi, D.M., Moursy, A.A. and El-Sofi, M.M. (2014). Alternative strategies for improving nitrogen nutrition of some economical crops using ^{15}N stable isotope. *Int J Curr Microbiol Appl Sci* 3(7), 970–983.

Srivastava, V.C. (2020). Elemental concentrations in soil, water and air. In: K. Mishra et al. (eds.), *Sustainable Solutions for Elemental Deficiency and Excess in Crop Plants*, Springer Nature Singapore Pte Ltd, pp. 3–18, https://doi.org/10.1007/978-981-15-8636-1_1.

Steinfath, M., Strehmel, N., Peters, R., Schauer, N., Groth, D., Hummel, J., Steup, M., Selbig, J., Kopka, J., Geigenberger, P., and van Dongen, J.T. (2010). Discovering plant metabolic biomarkers for phenotype prediction using an untargeted approach. *Plant Biotechnol J* 8(8), 900–911. https://doi.org/10.1111/j.1467-7652.2010.00516.x.

Sulieman, S. and Mühling, K.H. (2021). Game changer in plant nutrition utilization of soil organic phosphorus as a strategic approach for sustainable agriculture. *J Plant Nutr Soil Sci* 184, 311–319.

Sung, J., Lee, S., Lee, Y., Ha, S., Song, B., Kim, T., and Krishnan, H.B. (2015). Metabolomic profiling from leaves and roots of tomato (*Solanum lycopersicum* L.) plants grown under nitrogen, phosphorus or potassium-deficient condition. *Plant Sci Int J Exp Plant Biol* 241, 55–64. https://doi.org/10.1016/j.plantsci.2015.09.027.

Takahashi, H., Kopriva, S., Giordano, M., Saito, K., and Hell, R. (2011). Sulfur assimilation in photosynthetic organisms: Molecular functions and regulations of transporters and assimilatory enzymes. *Annu Rev Plant Biol* 62, 157–184.

Takehisa, H., and Sato, Y. (2019). Transcriptome monitoring visualizes growth stage-dependent nutrient status dynamics in rice under field conditions. *Plant J* 97(6), 1048–1060. https://doi.org/10.1111/tpj.14176.

Tiwari, S., Patel, A., Pandey, N., Raju, A., Singh, M., and Prasad, S.M. (2020). Deficiency of essential elements in crop plants. In: K. Mishra et al. (eds.), *Sustainable Solutions for Elemental Deficiency and Excess in Crop Plants*, Springer Nature Singapore Pte Ltd, pp. 19–52. https://doi.org/10.1007/978-981-15-8636-1_1.

Totin, E., Segnon, A.C., Schut, M., Aognon, H., Zougmoré, R., Rosenstock, T., and Thornton, P. (2018). Institutional perspectives of climate-smart agriculture: A systematic literature review. *Sustainability* 10, 1990.

Trouillefou, C.M., Le Cadre, E., Cacciaguerra, T., Cunin, F., Plassard, C., and Belamie, E. (2015). Protected activity of a phytase immobilized in mesoporous silica with benefits to plant phosphorus nutrition. *J Sol-Gel Sci Technol* 74, 55–65.

Tubiello, F.N., Salvatore, M., Ferrara, A.F., House, J., Federici, S., Rossi, S., Biancalani, R., Golec, R.D.C., Jacobs, H., Flammini, A., Prosperi, P., Cardenas-Galindo, P., Schmidhuber, J., Sanchez, M.J.S., Srivastava, N., and Smith, P. (2015). The contribution of agriculture, forestry and other land use activities to global warming, 1990–2012. *Glob Chang Biol* 21(7), 2655–2660.

Turner, B.L., Cheesman, A.W., Condron, L.M., Reitzel, K., and Richardson, A.E. (2015). Introduction to the special issue: Developments in soil organic phosphorus cycling in natural and agricultural ecosystems. *Geoderma* 257, 1–3.

Urbanczyk-Wochniak, E., and Fernie, A.R. (2005). Metabolic profiling reveals altered nitrogen nutrient regimes have diverse effects on the metabolism of hydroponically-grown tomato (*Solanum lycopersicum*) plants. *J Exp Bot* 56(410), 309–321. https://doi.org/10.1093/jxb/eri059.

Varinderpal-Singh, K., Gosal, S.K., Choudhary, R., Singh, R., and Adholeya, A. (2021). Improving nitrogen use efficiency using precision nitrogen management in wheat (*Triticum aestivum* L.). *J Plant Nutr Soil Sci* 184, 371–377.

Wada, Y. and Berkens, M.F. (2014). Sustainability of global water use: Past reconstruction and future projections. *Environ Res Lett* 9(10), 104003.

Wang, M., Zheng, Q., Shen, Q., and Guo, S. (2013). The critical role of potassium in plant stress response. *Int J Mol Sci* 14, 7370–7390.

Xiong, Q., Tang, G., Zhong, L., He, H., and Chen, X. (2018). Response to nitrogen deficiency and compensation on physiological characteristics, yield formation, and nitrogen utilization of rice. *Front Plant Sci* 9, 1075.

Yadav, M., Vashisht, B.B., Jalota, S.K., Kumar, A., and Kumar, D. (2022). Sustainable water management practices for intensified agriculture. In: S.K. Dubey et al. (eds.), *Soil-Water, Agriculture, and Climate Change, Water Science and Technology Library 113*, Springer Nature Switzerland AG, pp. 131–161. https://doi.org/10.1007/978-3-031-12059-6_8.

Yang, G.H., Yang, L.T., Jiang, H.X., Li, Y., Wang, P., and Chen, L.S. (2012). Physiological impacts of magnesium-deficiency in Citrus seedlings: Photosynthesis, antioxidant system and carbohydrates. *Trees Struct Funct* 26, 1237–1250.

Yang, M., Hassan, M.A., Xu, K., Zheng, C., Rasheed, C.A., Zhang, Y., Jin, X., Xia, X., Xiao, Y., and He, Z. (2020). Assessment of water and nitrogen use efficiencies through UAV-based multispectral phenotyping in winter wheat. *Front Plant Sci* 11, 927. https://doi.org/10.3389/fpls.2020.00927.

Yang, X.S., Wu, J., Ziegler, T.E., Yang, X., Zayed, A., Rajani, M.S., Zhou, D., Basra, A.S., Schachtman, D.P., Peng, M., Armstrong, C.L., Caldo, R.A., Morrell, J.A., Lacy, M., and Staub, J.M. (2011). Gene expression biomarkers provide sensitive indicators of in planta nitrogen status in maize. Plant Physiol 157(4), 1841–1852. https://doi.org/10.1104/pp.111.187898.

Ye, X., Abe, S., and Zhang, S. (2020). Estimation and mapping of nitrogen content in apple trees at leaf and canopy levels using hyperspectral imaging. *Precis Agric* 21(1), 198–225. https://doi.org/10.1007/s11119-019-09661-x.

Yohe, S.L., Choudhari, H.J., Mehta, D.D., Yohe, S.L., Choudhari, H.J., Mehta, D.D., Dietrich, P.J., Detwiler, M.D., Akatay, C.M., Stach, E.A., Miller, J.T., Delgass, W.N., Agrawal, R., and Ribeiro, F.H. (2016). High pressure vapor-phase hydrodeoxygenation of lignin-derived oxygenates to hydrocarbons by a PtMo bimetallic catalyst: Product selectivity, reaction pathway, and structural characterization. *J Catal* 344, 535–552.

Yu, X., Lu, H., and Liu, Q. (2018). Deep-learning-based regression model and hyperspectral imaging for rapid detection of nitrogen concentration in oilseed rape (*Brassica Napus* L.) leaf. *Chemom Intell Lab Syst* 172, 188–193. https://doi.org/10.1016/j.chemolab.2017.12.010.

Zaghloul, R.M., Galal, Y.G.M., Abdel Aziz, H.A., Abdel Rahman, H.M., Salem, A.A., Mousa, A.M., and Weesa, S.E. (2021a). Activities profile of irradiated *Streptomyces Alfalfae* strain XY25 in vitro. *Env Biodiv Soil Security* 5, 289–303.

Zaghloul, R.M., Galal, Y.G.M., Abdel Aziz, H.A., Abdel Rahman, H.M., Salem, A.A., Mousa, A.M., and Weesa, S.E. (2021b). Effects of dual inoculation with irradiated or non irradiated *Streptomyces alfalfa* strain XY25 and *Mesorhizobium ciceri* on yield and plant performance of chickpea. *Egypt J Soil Sci* 61(4), 413–432.

Zaki, Z.A., Habib, F.M., Galal, Y.G.M., and Abd El-Hameed, A.H. (2021a). Importance of bio-organic fertilizers on peanut (*Arachis hypogaea* L.) nutrition following organic farming approach with application of ^{15}N isotope dilution concept. *Env Biodiv Soil Security* 5, 15–29.

Zaki, Z.A., Habib, F.M., Galal, Y.G.M., and Abd El-Hameed, A.H. (2021b). Effect of P rates combined with PDB on nutrients uptake by wheat grown under organic farming of sandy soil using ^{15}N tracer technique. *Arab J Nucl Sci Appl* 54(4), 67–82.

Zhai, B.N., and Li, S. (2005). Response to nitrogen deficiency and compensation on growth and yield of winter wheat. *J Plant Nutr Fertil Sci* 11, 308–313.

Zhang, J., Xie, T., Yang, C., Song, H., Jiang, Z., Zhou, G., Zhang, D., Feng, H., and Xie, J. (2020). Segmenting purple rapeseed leaves in the field from UAV RGB imagery using deep learning as an auxiliary means for nitrogen stress detection. *Remote Sens* 12(9), 1403. https://doi.org/10.3390/rs12091403.

Zheng, H., Cheng, T., Li, D., Yao, X., Tian, Y., Cao, W., and Zhu, Y. (2018). Combining unmanned aerial vehicle (UAV)-based multispectral imagery and ground-based hyperspectral data for plant nitrogen concentration estimation in rice. *Front Plant Sci* 9, 936. https://doi.org/10.3389/fpls.2018.00936.

Zhran, M., Ge, T., Tong, Y., Zhu, Z., Deng, Y., Fahmy, A., Chen, M., Lynn, T.M., Wu, J., and Gunina, A. (2021). Effect of N and P fertilization on the allocation and fixation of photosynthesized carbon in paddy soil. *Ecosyst Health Sustain* 7(1), 1941271. https://doi.org/10.1080/20964129.2021.1941271.

Zhran, M., Moursy, A., Lynn, T.M., and Fahmy, A. (2020). Effect of urea fertilization on growth of broad bean (Vicia faba L.) under various nickel (Ni) levels with or without acetic acid addition, using ^{15}N-labeled fertilizer. Environ Geochem Health 43, 2423–2431.

Zhu, Z., Xu, Z., Peng, J., Fei, J., Yu, P., Wang, M., Tan, Y., Huang, Y., Zhran, M., and Fahmy, A. (2022). The contribution of atmospheric deposition of cadmium and lead to their accumulation in rice grains. *Plant Soil* 477, 373–387. https://doi.org/10.1007/s11104-022-05429-x.

Zimmermann, P., Zardi, G., Lehmann, M., Zeder, C., Amrhein, N., Frossard, E., and Bucher, M. (2003). Engineering the root-soil interface via targeted expression of a synthetic phytase gene in trichoblasts. *Plant Biotechnol J* 1, 353–360.

Zivcak, M., Brestic, M., Olsovska, K., and Slamka, P. (2008). Performance index as a sensitive indicator of water stress in *Triticum aestivum* L. *Plant Soil Environ* 54(4), 133–139.

12 Managing Soil Drought in Agro-Ecosystems of North China Plains

*Zheng-Rong Kan, Xing Wang, Jian-Ying Qi, Ling-Tao Zhong, Rattan Lal, and Hai-Lin Zhang**

12.1 INTRODUCTION

Widespread droughts over many land areas result from a prolonged deficiency in precipitation (known as meteorological drought) or increased evaporation demand under global warming (Dai, 2013; Mishra and Cherkauer, 2010; Zhang et al., 2023). Under climate change, soil drought is the main constraint limiting crop production and has a devastating impact on agricultural production as compared to other environmental stresses (Ali et al., 2017; Dang, 2023; Lal, 2023). The risk of yield reduction induced by drought for major crops will increase significantly, up to 50% in 2050 and almost 90% in 2100 globally (Li et al., 2009). Thus, supplementary irrigation could alleviate soil drought effectively. However, the shortage of water resources has become a limiting factor for agricultural development around the world, causing a severe threat to global food security and sustainability. Furthermore, croplands are crucial in tackling both great challenges: ensuring food security and mitigating greenhouse gas (GHG) emissions (Shang et al., 2021). Droughts not only severely affect crop growth and food production (Li et al., 2009), but drought-induced salinization may also increase GHG emissions (Chamberlain et al., 2020; O'Connell et al., 2018). Thus, the risk of drought is the top natural disaster in China and causes a wide variety of sectors in agricultural, ecological, and environmental processes (Geng et al., 2015; Liu et al., 2018).

The North China Plain (NCP) is classified as a semi-arid monsoon area of the warm temperate zone, mainly including Beijing, Tianjin, Heibei, Shandong, and Henan provinces. The NCP is not only the political, economic, and cultural center but also an important commercial grain production area in China. The NCP, one of the largest agricultural production areas in China, covers ~18.3% of the national total farmlands and provides ~25% of the total production in the country (Guan et al., 2015). Winter wheat (*Triticum aestivum* L.) -summer maize (*Zea mays* L.) rotation is the main crop production system in the NCP, which has been extensively produced, and it has played a vital role in China's food security. The rapid expansion of wheat and maize in recent decades has greatly increased irrigation water consumption (Zhang et al., 2020). However, high productivity with irrigation is obtained at the expense of depleting groundwater. However, as much as 70% of the water demand for agricultural production in China depends on groundwater (Guan et al., 2015). Excessive exploitation of groundwater resources has caused a rapid fall in the groundwater table at the rate of ~1 m/year (Wang et al., 2018a,b,c), threatening agricultural sustainable development.

Large-scale, intense, and frequent droughts have led to a decrease in wheat and maize production in this region (Liu et al., 2018). In addition, the climate warming and drying phenomenon in the NCP are particularly prominent. The drought-affected area of major provinces in the NCP accounts for about 50% of the total natural disaster area (Figure 12.1). The drought in the NCP has continued to intensify, and it has occurred for consecutive years since the late 1990s (Zhang

---

* Corresponding author. Email address: hailin@cau.edu.cn Tel: +86 1062733376. Fax: +86 1062733316.

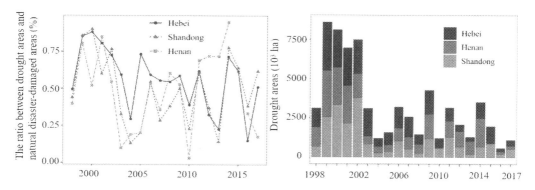

FIGURE 12.1 The ratio between drought areas and natural disaster-damaged areas (%) and drought areas (10^3 ha) in three major provinces (Hebei, Shandong, and Henan) from 1998 to 2017 in the NCP. Source: Data were sourced from China Statistical Yearbook.

et al., 2021; Liu et al., 2022). Although the agricultural drought area has decreased significantly since 2000 due to the strengthening of social disaster reduction capabilities, the drought area in the three main provinces of the NCP still affects as much as 2 M ha (Figure 12.1). Numerous studies have shown that northern China was frequently affected by drought in the past 50 years, with 40–50 drought events observed in the NCP (Piao et al., 2010; Xiao and Tao, 2014; Yu et al., 2018). In addition, the current drought pattern has been dominated by winter and spring droughts (Liu et al., 2018; Wang et al., 2018a,b,c), and drought events in the NCP will still occur frequently in the future (Sang et al., 2018).

Under the conditions of soil drought in the NCP, it is imperative to adopt water-saving or rain-fed agriculture to obtain the possible decrease in water consumption and increase in agricultural water use efficiency (WUE) while maintaining the crop yield in the NCP. In this context, with a decreasing trend in groundwater consumption, wheat production may be transformed from conventional irrigated cultivation to cultivation with reduced irrigation during the growth stage. For example, limited irrigation, reducing irrigation frequency and amount, and irrigation methods (e.g., micro-irrigation) could be considered as promising strategies to save water and improve WUE in the NCP (Xu et al., 2018a,b). In addition, the use of recent cultivars of wheat that can extract more water from the deep soil profile to increase grain yield and WUE is also a key strategy (Thapa et al., 2018). Input of biochar and adoption of and no-till (NT) practice could improve total porosity, proportion of continuous pores and available water capacity (AWC) to mitigate drought (Gao et al., 2019; Li et al., 2020a,b; Obia et al., 2016). Thus, many management practices were implemented to reduce water consumption and improve WUE, but controversial results were reported in the NCP. The objective of this chapter is to review and synthesize the frequency of drought and its disadvantages on crop production and the environment and to summarize the effective approaches that could be applied to improve WUE and alleviate soil drought in the NCP.

12.2 THE FREQUENCY OF DROUGHT AND ITS DISADVANTAGES IN THE NCP

The spatial and temporal distribution of precipitation in the NCP is uneven, with large inter-annual variability, and the frequency of droughts ranks first (Figure 12.2). Overall, the water demand for winter wheat and effective precipitation decreased first and then increased. Thus, water deficit varies over different growth stages. The first three stages (sowing-heading) required less water, and thus the water deficit was lower than that at the heading–mature stage. During the total growth stages, the effective precipitation cannot meet the water demand that wheat required, of which the water deficit was ~350 mm.

Liu et al. (2018) investigated the spatial distribution of drought using the Standardized Precipitation Evapotranspiration Index (SPEI) in the NCP (Figure 12.3). During the winter wheat

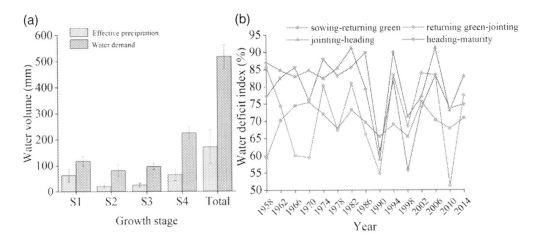

FIGURE 12.2 Water demand and effective precipitation of winter wheat (a) and its water deficit index (%) (b) in the NCP. S1, sowing-returning green; S2, returning green-jointing; S3, jointing-heading; S4, heading-maturity. Source: Data were sourced from Zhang et al. (2020).

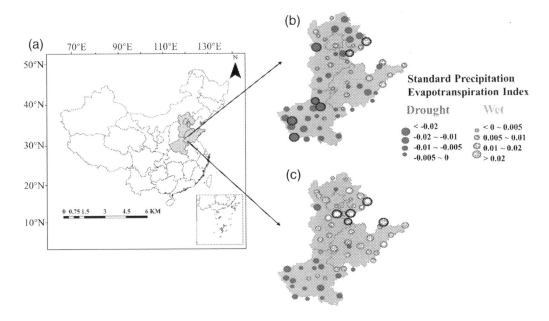

FIGURE 12.3 Geographic locations of the NCP (a) and spatial distribution of the SPEI trend during the winter wheat (b) and summer maize (c) growing seasons. Source: Data were sourced from Liu et al. (2018).

season, a drought trend was observed in the northern and western parts of the Henan Province, while a wetting trend occurred in the Shandong Province. During the summer maize season, a wetting trend was dominant, indicating a lower drought risk during the maize season. However, the drought risk has remained in the southern, western, and northern parts of Henan Province. Remarkably, the flood risk in the Shandong Province and the increasing drought effects under global warming should be further investigated (Liu et al., 2018).

The spatial-temporal distribution of drought in wheat and maize growing seasons in the NCP from 1961 to 2017 is shown in Figure 12.4 (Ma et al., 2020). The average frequency of light drought in the growing season of wheat was 34.3%, which was characterized by high frequency in the northern part of NCP and low frequency in the central and eastern parts of NCP. The average frequency

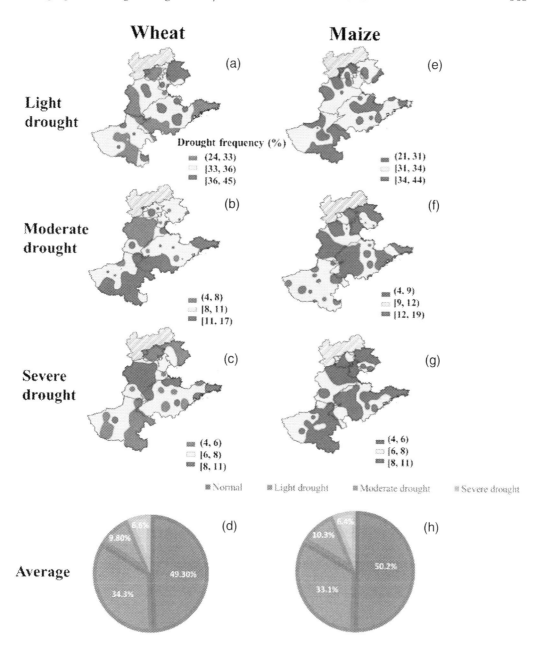

FIGURE 12.4 Distribution of drought frequency of the wheat (a, b, and c) and maize (d, e, and f) growing season and its average frequency (d and h) in the NCP during 1961–2017. Source: The figure and data were sourced from Ma et al. (2020).

of moderate drought is 9.8%, and it occurs mostly in eastern Shandong and southern Henan. The average frequency of severe drought was 6.6%, mostly in central Hebei province. The frequency of severe drought in Henan Province was lower at 6.1%.

In the growing season of maize in NCP, the frequency of light drought was 21.1%–43.9%, moderate drought was 3.5%–19.3%, and severe drought was 3.5%–10.5%. The average frequency of light drought was 33.1%, mostly occurring in central and eastern NCP. The average frequency of moderate drought was 10.3%, which occurred in the eastern and central parts of Hebei province and

Shandong province. The average frequency of severe drought was 6.4%, and the average frequency of severe drought was 7.1% in the middle of Shandong province.

Soil drought directly affects the morphological, physiological, and biochemical processes of crops (Hanaka et al., 2021). In addition, drought can reduce soil enzyme activity, weaken nutrient cycling (such as C, N, P), decrease soil fertility, and exacerbate the decline of crop productivity (Abdul Rahman et al., 2021; Nguyen et al., 2018). The effect of drought stress on crops is the result of many factors, including potential evapotranspiration, root water absorption, root distribution, canopy structure, and other plant and environmental factors. Soil conditions under drought stress are strictly related to plant growth. Soil drought can cause the acceleration or deceleration of crop development, cause stomatal closure to reduce the total assimilation rate, alter the distribution of dry matter among different components of crops (Hoogenboom, 2000), and reduce seed germination and seedling growth (Khan et al., 2019).

12.3 EFFECTS OF SOIL DROUGHT ON SOIL FUNCTIONS AND GHGS

The most important consequence of climate change is aggravation of drought stress. Soil drought is one of the important, worrying, and prevalent abiotic stresses (Siebielec et al., 2020). It alters soil microbial activity and functional composition and affects soil nutrient cycling and crop productivity in the NCP (Abdul Rahman et al., 2021; Kränzlein et al., 2021; Zhang et al., 2021).

12.3.1 Soil Physicochemical Properties

Soil drought has a significant negative impact on nutrient cycling (such as C, N, and P), resulting in loss of soil fertility, thereby affecting crop productivity, especially that of drought-susceptible crops (Chandra et al., 2021). As the most active soil process between plant and soil, nutrient cycling in the rhizosphere exhibits high sensitivity to environmental changes. Water is one of the most important abiotic factors affecting nutrient transformation and availability in the soil rhizosphere. Drought has an important impact on soil carbon and nitrogen mineralization (Geng et al., 2015; Chandra et al., 2021). Many studies have reported that drought stress not only affects the rate of soil carbon and nitrogen mineralization but also affects the mineralization process and efficiency, especially for rhizosphere soils in the NCP (Liu et al., 2018). Soil available phosphorus is an indicator of soil nutrient supply, which is of great significance to plant growth and the biological cycle of phosphorus in the ecosystem. However, soil drought would significantly reduce the availability of soil phosphorus (Al-Kaisi et al., 2012). Although soil drought has a negative impact on soil nutrient cycling, there are some differences in specific research results, which may be due to differences in crop species, duration, and intensity of drought stress.

12.3.2 Soil Microbial Properties and Enzyme Activities

Soil drought can reduce soil microbial abundance, interfere with the microbial community structure, and reduce microbial activity, including enzyme generation (e.g., β-glucosidase, oxidoreductase, catalase, urease, and phosphatase) and nutrient cycling, directly leading to the decrease of soil fertility, and may even result in reduced crop productivity in the NCP (Zhang et al., 2021). Severe and prolonged drought stress can significantly disturb soil microbial communities, affect crop root structure, root secretion release, and availability of nutrients (Figure 12.5). The composition of the plant rhizosphere microbial community changes during drought. Soil drought favors actinomycetes and many gram-positive species, which displaces the predominantly gram-negative groups present (Breitkreuz et al., 2021). In general, soil microorganisms would improve crop drought resistance through different underlying mechanisms, including the generation of polysaccharides to improve soil structure and water retention capacity, and the synthesis of deaminase, aminoacylacetic acid, and proline (Pro) to induce plant drought tolerance (Milosevic et al., 2012). Studies have shown that

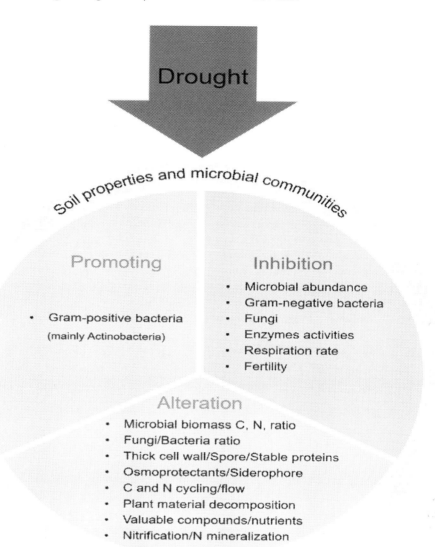

FIGURE 12.5 Schematic representation on the effects of soil drought on microbial communities and activities. Source: Adapted from Bogati and Walczak (2022).

the negative effects of drought on microorganisms can be mitigated by adding organic matter (e.g., biochar and manure) to the soil (Pascual et al., 1998). In addition, microbial activity (respiration rate) and soil microbial community would be altered by drought-related phytohormones such as abscisic acid, jasmonic acid, and 1-aminocyclopropane-1-carboxylic acid. However, persistent soil drought also changes the response degree of soil microbial communities to these phytohormones (Sayer et al., 2021).

12.3.3 GHG Emissions

CO_2, CH_4, and N_2O are the main GHGs that contribute to climate warming, which has exacerbated the risk of soil drought in the NCP (Zhang et al., 2021). How do soil respiration, CH_4 production and consumption, and N_2O emissions respond to soil drought? These have become very important scientific issues. Some studies have reported that warming would reduce the cumulative uptake of CH_4 in soil. Soil drought due to soil moisture loss would reduce the activity of methane-oxidizing bacteria, thereby reducing soil CH_4 uptake. At the same time, soil drought would negatively affect soil respiration. Moreover, soil drought would slow down soil microbial operation and reduce the denitrification rate, resulting in a significant reduction in cumulative N_2O emission fluxes throughout the year (Liu et al., 2015, 2019). The effects of soil drought on GHG emissions are regulated by many factors, such as soil texture, drought duration, and soil organic matter (SOM) content.

12.4 EFFECTS OF SOIL DROUGHT ON CROP GROWTH

Drought has become an important limiting factor for the stable increase of grain yield in the NCP. It was estimated that the area prone to drought stress will expand with an increase in temperature from 15.4% to 44.0% in the future (Li et al., 2009). Consequently, the rates of major crop yield loss induced by drought will increase by >50% in 2050 and almost 90% in 2100 (Li et al., 2009). Drought stress could lead to some negative effects on crop growth and development (Chandra et al., 2021). Drought-induced inhibition of cell enlargement and division could alter plant basic physiological functions, e.g., ion uptake leaves photosynthesis process and nutrient uptakes (Li et al., 2009; Chandra et al., 2021). Due to decrease in cell elongation in the beginning stages, drought stress can also restrain seed germination. Another drought-induced plant response is low turgor pressure, which restrains plant growth. Insufficient water restrains leaf growth and the photosynthetic process, which generates a negative impact on the photosynthetic pigments (Krieger-Liszkay and Roach, 2014).

Severe drought has a significant negative effect on crop yield. Some crops are vulnerable to soil drought, while some crops have more resistance to soil drought. Water uptake is the main factor influencing yield under drought conditions (Blum, 2009). Wheat and maize are the main cereal crops and contribute to world food security (Kan et al., 2020a). Wheat and maize cropping areas in the NCP occupy 45% and 33% of the whole nation (Guan et al., 2018; Kan et al., 2020a). However, the NCP is prone to drought, and the increased occurrence and severity of drought may be caused by higher evapotranspiration as a result of increasing temperature (Liu et al., 2018). Maize and wheat are also produced in North and northwest China, where drought is also a serious problem for agricultural production (Li et al., 2019a,b). Therefore, strengthening knowledge about the response of wheat and maize production to drought is critical for food production and security.

12.4.1 Wheat

Drought stress could decrease wheat growth by restraining plant height, spike length, and root and shoot dry biomasses, which have negative effects on wheat yields (Abbas et al., 2017). Abbas and colleagues also found that drought stress stimulated oxidative reactions and reduced antioxidant enzyme activities while improving Cd absorption by plants. Therefore, irrigation management of wheat production should be planned for contaminated soils. In addition, soil drought intensity and occurrence period could impact wheat growth. Zhang et al. (2004) demonstrated that severe soil drought decreases wheat and maize yield, while slight soil water deficit at a growth stage from spring green up to grain-filling may not decrease wheat yield and water use efficiency. Therefore, based on the water requirement period, the irrigation schedules were recommended for practical winter wheat production in the NCP (Li et al., 2005). The optimal irrigation management must include pre-sowing irrigation, pre-sowing to the jointing (or booting); and pre-sowing, jointing, and flowering period irrigation (Sza et al., 2011). If the irrigation facility fails to provide sufficient water for wheat production, regulating plant density could improve wheat productivity under insufficient water conditions in the NCP. Li et al. (2020a,b) reported that, under insufficient irrigation conditions, higher numbers of spikes and yield of wheat were achieved at high plant density. Under insufficient water supply, higher plant density could offset yield loss by translocating the pre-anthesis accumulated photosynthate to grains. Higher wheat plant densities could also improve nitrogen accumulation, which can contribute to wheat yield. Therefore, increasing wheat plant density could decrease the loss of wheat yield in dry years.

12.4.2 Maize

Crop yield loss induced by drought would be offset less for wheat but much for maize in the NCP (Hu et al., 2014). A good case for investigation of drought conditions was reported for Iowa, USA, by Al-Kaisi et al. (2012) based on regional scales, which showed that 43% of Iowa experienced moderate drought, and nearly 10% experienced a serious drought in 2011. Unfortunately, a more severe drought situation was recorded in 2012: 100% of Iowa experienced serious drought and 65% suffered extreme drought. The predicted Iowa maize yields in 2012 were 8.8 mg/ha, which was 22% below the recorded average yield. Based on the severe drought situation, Al-Kaisi et al. (2012) found several impacts of drought on soil and crop cultivation and demonstrated the development of appropriate drought mitigation strategies for crop management, mainly by adoption of no-till for maize soybean rotation. Crop residue mulch and no-till could improve water infiltration and increase soil water content by decreasing soil cracking and improving soil structure. Continuous maize cropping system suffered more yield losses than did the maize-soybean rotation and thus had a lower yield than the maize-soybean rotation system. Strategies for mitigating drought stress were also identified for the Loess Plateau of China. Zhang et al. (2011) showed that higher soil water recorded in no-till with crop residue mulch plots stimulated maize growth, and thus the highest grain yield and WUE. Similarly, Kashif et al. (2018) reported that straw mulching at the rate of 5,000 kg/ha improved soil water storage during a drought period during the maize season, with an increase in plant growth and evapotranspiration. In addition, subsoiling could improve water infiltration rate (Kuang et al., 2020), and increase soil water storage in drought conditions. Subsoiling could increase water consumption, by increasing the crop utilization of soil water. Compared with rotary tillage at a depth of 15 cm, maize yield and yield composition were significantly increased in subsoiling management at a depth of 35 cm. Therefore, both grain yield and crop water productivity were significantly improved.

Bai et al. (2006) reported that severe drought significantly decreased the relative water content and increased leaf membrane permeability. Furthermore, under extreme drought conditions, antioxidant enzyme activities decreased, e.g., superoxide dismutase and peroxidase catalase. Buljovcic and Engels (2001) observed that drought could significantly influence the root growth of maize. However, the moderate levels of 10d soil drought (10% w/w water content) neither altered maize roots nor uptake of N. When soil water decreased to 5%, the N uptake of maize root was severely reduced (Buljovcic and Engels, 2001).

12.5 AGRICULTURAL MANAGEMENT PRACTICES TO ALLEVIATE DROUGHT AT A REGIONAL SCALE

Droughts occur naturally, and climate change has generally accelerated the hydrological processes to intensify drought. Globally, it is estimated that drought-affected areas will globally increase from 15.4% to 44.0% with the projected increase in temperature by 2,100 (Li et al., 2009). It is important to improve the ability of forecasting of occurrence of drought under climate change, but this is difficult. For example, downscaling techniques, available climate data and potential evapotranspiration, baseline period, and anthropogenic forcing are challenges for a reliable drought assessment under climate change (Mukherjee et al., 2018). Thus, management practices that could alleviate aggravated drought have been widely investigated and developed.

By using meta-analysis, our team summarized several strategies that could restrain soil drought in the NCP. Liu et al. (2022) listed the factors such as nitrogen input, irrigation, tillage practices, and residue management (Table 12.1). Liu and colleagues observed that wheat

TABLE 12.1
Effects of Agricultural Management on Winter Wheat Yield and Water Use Efficiency

Management Practices	n	Total Water Consumption Effect Size (%)	95% CI (%)	WUE Effect Size (%)	95% CI (%)
Straw return vs. removal	46	−3.83	−6.63 to −0.95	9.11	2.38–18.91
Subsoiling with rotary tillage vs. rotary tillage	22	12.08	8.10–16.16	8.53	6.01–10.86
Subsoiling vs. rotary tillage	11	−5.31	−9.38 to −1.71	13.38	8.17–18.05
No-till vs. rotary tillage	6	−15.68	−26.55 to −6.46	2.63	−13.11–18.35
Plow tillage vs. rotary tillage	31	2.84	0.82–5.11	3.96	0.67–6.8
Irrigation vs. No irrigation	657	31.34	29.15–33.44	−0.97	−2.8–1.21
Nitrogen input vs. No nitrogen	141	5.56	3.85–7.28	41.62	34.82–49.38

Data were collated and adapted from Liu et al. (2022).

production, by adoption of promising strategies in the NCP, can conserve irrigation water and reduce the negative environmental feedback. Indeed, judicious management of N could significantly increase wheat yield by 70%, evapotranspiration by 6%, and WUE by 42%. These results indicate that science-based management of nitrogen application is critical to improving the WUE of wheat in the NCP. Compared to rain-fed cultivation of wheat, irrigation significantly improved its yield by 30% but also increased the evapotranspiration by 31%, and thus no significant effect was observed on WUE (Liu et al., 2022). These results are similar to the findings of Ai et al. (2020). Yu et al. (2021) also observed that improvement in soil fertility is one of the most effective strategies to alleviate pedological drought. Tillage can change soil physicochemical properties (Kan et al., 2020a), but the WUE of wheat responded differently to changes in tillage practices. Compared with rotary tillage, no-till significantly decreased the wheat yield and evapotranspiration, whereas it had no significant effect on WUE. However, subsoiling significantly improved the wheat yield by 12% and WUE by 13% due to reduced evapotranspiration. Straw retention significantly improved wheat yield by 4% and WUE by 9% due to reduced evapotranspiration. Similar results were also reported in previous studies (Xiao et al., 2021; Meyer et al., 2019). Consequently, subsoiling and straw retention are among promising tillage methods, providing a win-win strategy for wheat yield and consumptive water use.

The aforementioned sustainable management option of reducing soil water loss is also useful in many regions of the world with similar climate conditions, such as north-western India and eastern Pakistan (Hammad et al., 2012; Mahajan et al., 2012). In addition, Deng et al. (2006) reviewed water-conserving agricultural strategies to increase agricultural WUE in the water-limited areas of China, including furrow irrigation, mulching with drip irrigation, precipitation conservation, and terracing. Deng et al. (2006) addressed the compensatory effect of limited soil water and fertilizer management supplementation on WUE and emphasized the need to investigate new varieties with high WUE. There exists a large potential for further improvements in agricultural WUE and crop growth on effective storage of moisture and optimal utilization of the limited water (Figure 12.6).

In addition, several other water-saving strategies can be summarized from the literature review. In general, increasing SOM content can conserve soil water, and decrease risks of agricultural yield reduction induced by drought. Using data from 12,376 county years in the USA, Kane et al. (2021) showed that soils with higher SOM content are associated with less yield reduction under drought stress. Moreover, under extreme drought conditions, a 1% SOM increment was associated with a yield improvement of maize by 2.2±0.33 mg/ha. These results suggest that SOM is related to yield resiliency at regional scales.

Straw and film mulch could increase soil water storage. Hu et al. (2019) evaluated four treatments: without mulch (CK) and with straw mulch (SM), with plastic film mulch (PM), and with both film and

FIGURE 12.6 The photos of wheat growth using recommended management practices (A), and improper management practices (B) under limited irrigation in the North China Plain (Liu et al., 2022).

straw mulch (PSM). The results showed that all mulch managements improved yields and WUE of both wheat and maize compared with CK. The PSM treatment had the highest grain yield and WUE over two wheat-maize cycles. The soil water storage in 0–200 cm soil depth was 1.31%–3.82% higher under mulch treatments compared with that under CK. Similar results were also reported by Li et al. (2020a,b), who recommended that the biodegradable film and straw mulching with furrow tillage are positive for soil conditions and maize growth in areas frequently subject to drought.

Lignin hydrogel, which was usually used for drug transportation and seed coating formulations (Li and Pan, 2010), may also decrease drought-induced yield loss of crops. Mazloom et al. (2019) showed that lignin hydrogel absorbed 34 g water/g, without inducing phytotoxicity in wheat seeds. The lengthy biodegradation of lignin may prolong its working life. Thus, lignin hydrogel would also increase the accumulation of SOM content (Passauer et al., 2011). Mazloom et al. (2020) showed that only 0.6% lignin hydrogel added into soil can produce more maize biomass and increase relative leaf water content, with decreased leaf proline and electrolyte leakage under extreme drought stress.

Biochar is currently attracting much attention as an amendment, which has been widely evaluated due to its ability to adsorb heavy metals, and carbon and mitigate GHG emissions (Abbas et al., 2017; Rizwan et al., 2016). Application of biochar could improve soil water holding capacity (Ali et al., 2017), and improve maize yield and available soil water reserves (Cornelissen et al., 2013). Abbas et al. (2017) reported that the input of biochar altered the morphological and physiological conditions of wheat plants subjected to drought and Cd stress. In addition, biochar reduces oxidative stress and stimulates antioxidant enzymes. The decreased Cd by biochar in drought-stressed plants may result from the improved crop biomass and mitigation of oxidative stress (Deng et al., 2006).

12.6 CASE STUDIES TO ALLEVIATE DROUGHT AND IMPROVE WATER USE EFFICIENCY IN WUQIAO COUNTY

Multiple field experiments were conducted to improve WUE at Wuqiao Experimental Station of China Agricultural University (37°36′–37°41′ N, 116°21′–116°36′ E), Hebei Province, China. This region had mean annual precipitation of 500–600 mm, temperature of ~13°C, and daylight of ~2,340 h over the past 30 years (temperate continental climate). Winter wheat and summer maize is the dominant crop rotation system. The growing season of wheat is from early October to mid-June, and that of maize from mid-June to later September. Because over 75% of the precipitation is received from June to September and precipitation does not occur in synchronization with wheat growth stages. Because the precipitation received in the region is not enough for adequate crop growth, supplemental irrigation is necessary (Xu et al., 2018a,b).

12.6.1 No-Till

Kan et al. (2020a) conducted a field experiment over 10-year period for different tillage practices in Wuqiao and reported that the adoption of no-till (NT) decreased the total water consumption of wheat by 8.5%–17.2% and 8.6%–10.5% compared with conventional tillage and rotary tillage, respectively (Table 12.2). The WUE of wheat was the highest under NT and rotary tillage compared with that under conventional tillage. However, Guan et al. (2015) argued that no significant difference in total water consumption of wheat was observed among different tillage practices. Lower WUE was observed under NT and rotary tillage compared with that under conventional tillage (Guan et al., 2015). These contradictory results may be due to the different durations of NT adoption. Experiment duration was 10 years for the study reported by Kan et al. (2020a) but was only 4 years for that reported by Guan et al. (2015). A meta-analysis conducted in China based on a wheat-maize cropping system also observed that WUE in both wheat and maize were improved with the adoption of NT for more than 6 years compared with that under conventional tillage (Wang et al., 2018a,b,c), consistent with the experimental results of a similar study conducted in Wuqiao. These results indicated that long-term NT has the potential to enhance water retention and reduce water consumption and further increase WUE in the NCP. Wang et al. (2018a,b,c) reported that NT effects on total water consumption and WUE for wheat and maize depend on environmental conditions (e.g., precipitation and air temperature) and management practices (e.g., residue retention and crop rotation). Thus, climate-smart agriculture, such as conservation agriculture including NT, residue retention, crop rotation, and balanced nutrient management, should be adopted to reduce the adverse effects of drought under climate change (Jat et al., 2020).

12.6.2 Crop Variety

The regulation of shoot transpiration and favorable root architecture is of importance to maintaining optimal water relationships (Dietz et al., 2021). Breeding resilient varieties are underway across the world. Important target traits of improved varieties for drought tolerance include reduced plant height (dwarf), which is related to a high harvest index; shortened time of anthesis and maturity, which enable the crop to evade terminal drought stress; and root architectural traits, which promote effective water uptake (Mwadzingeni et al., 2016). Eight varieties representing different years of

TABLE 12.2
Effects of Tillage Practices on Total Water Consumption and Water Use Efficiency

Year	Treatment	Total Water Consumption (mm)	Water Use Efficiency (kg/ha mm)	References
2017–18	No-till	387.4c	16.1a	Kan et al. (2020a)
	Conventional tillage	468.1a	12.5b	
	Rotary tillage	423.8b	16.1a	
2018–19	No-till	383.1b	13.9a	
	Conventional tillage	418.5a	12.0b	
	Rotary tillage	427.9a	15.1a	
2011–12	No-till	366.1a	17.7b	Guan et al. (2018)
	Conventional tillage	353.7a	20.3a	
	Rotary tillage	373.8a	18.6b	
2012–13	No-till	435.2a	17.9b	
	Conventional tillage	427.4a	19.2a	
	Rotary tillage	429.2a	18.7ab	

national approval were selected to investigate the effects of wheat varieties on water consumption and WUE under NT and rotary tillage in Wuqiao (Table 12.3). Under different wheat varieties, higher total water consumption was observed in rotary tillage by 4.4%–6.5%, while higher WUE was obtained in rotary tillage by 0.6%–6.4% compared with those under NT, indicating that rotary tillage had higher grain yield. Under rotary tillage, Yannong 19 and Shannong 29 had lower water consumption and higher WUE. Under NT, Yannong 19 also had higher WUE compared with that for the other varieties. The year of national approval for Yannong 19 was 2001, earlier than for the other varieties, indicating that development in breeding technology cannot satisfy the requirement of mitigating drought through improved WUE.

12.6.3 CROP ROTATION

The responses of different crops to drought also differ widely (Table 12.4). When maize plants were under drought conditions, the higher membrane stability index and water retention capacity were presented (Moussa et al., 2008; Zhao et al., 2016). Peanut had no significant effects on chlorophyll fluorescence values, indicating photosynthetic capacity was maintained under drought.

New rotation patterns with a 2-year cycle of "one-year traditional winter-maize+one-year other crops" were designed in Wuqiao to identify a water-saving cropping system (Table 12.5). This study involves nine different crops, including food crops, oil crops, and feed crops, thus it is not advisable to use the traditional WUE index (the ratio of economic yield to total water consumption) to measure their water-saving potential. Economic WUE is calculated according to the total water consumption and economic benefit of each crop, that is, the income that can be obtained from the consumption of water per unit volume of farmland (Van Duivenbooden et al., 2000). It can realize the comparison of WUE between different crops on the same research level. During the 2018–19, spring maize → wheat-maize, spring sweet potato (*Solanum tuberosum* L.) →wheat-maize, and spring peanut (*Arachis hypogaea* L.) →wheat-maize had the lower total water consumption compared with the other cropping systems. The highest economic WUE was observed in spring peanut →

TABLE 12.3
Effects of Wheat Varieties on Total Water Consumption and Water Use Efficiency

Year	Tillage	Variety	Year of National Approval	Total Water Consumption (mm)	Water Use Efficiency (kg/ha mm)
2015–16	Rotary tillage	Yannong 19	2001	462.9b	10.5a
		Jimai 22	2006	469.5ab	7.9bc
		Lumai 21	2006	499.1a	8.4bc
		Qingmai 6	2007	488.3ab	6.7c
		Wennong 14	2010	488.9a	8.2bc
		Luyuan 502	2011	483.3ab	9.2ab
		Nongda 399	2012	492.3a	9.7ab
		Shannong 29	2016	448.5b	10.1a
		Mean values		479.1	8.8
	No-till	Yannong 19	2001	466.4a	9.5a
		Jimai 22	2006	427.1b	7.3bc
		Lumai 21	2006	441.5ab	8.9a
		Qingmai 6	2007	467.4a	6.7c
		Wennong 14	2010	440.5ab	7.5bc
		Luyuan 502	2011	483.6a	8.2b
		Nongda 399	2012	453.0a	8.4ab
		Shannong 29	2016	418.4b	9.9a
		Mean values		449.7	8.3

(*Continued*)

TABLE 12.3 (*Continued*)
Effects of Wheat Varieties on Total Water Consumption and Water Use Efficiency

Year	Tillage	Variety	Year of National Approval	Total Water Consumption (mm)	Water Use Efficiency (kg/ha mm)
2016–17	Rotary tillage	Yannong 19	2001	463.0b	14.3a
		Jimai 22	2006	408.4b	13.2a
		Lumai 21	2006	471.6b	13.4a
		Qingmai 6	2007	479.3ab	10.9b
		Wennong 14	2010	504.0a	13.0ab
		Luyuan 502	2011	480.4ab	12.8ab
		Nongda 399	2012	492.2ab	12.7ab
		Shannong 29	2016	463.6b	14.1a
		Mean values		470.3	13
	No-till	Yannong 19	2001	454.1a	14.3a
		Jimai 22	2006	387.5b	11.6b
		Lumai 21	2006	441.5ab	14.3a
		Qingmai 6	2007	464.3a	13.8a
		Wennong 14	2010	449.6a	13.2ab
		Luyuan 502	2011	471.7a	12.3b
		Nongda 399	2012	474.3a	12.1b
		Shannong 29	2016	461.0a	12.2b
		Mean values		450.5	13

Source: Data were collected from Liu et al. (2019).

TABLE 12.4
Responses of Different Crops to Drought

Crops	Responses to Drought	References
Wheat	Decrease in chlorophyll and carotenoid content, membrane stability, and nitrate reductase activity; increase in proline and abscisic acid (ABA)	Chandrasekar et al. (2008)
Maize	increase in gylycinebetain (GB), free proline (PRO), H_2O_2, and malondialdehyde (MDA) and higher membrane stability index and high water retention capacity	Moussa and Abdel-Aziz (2008), Zhao et al. (2016)
Peanut	increase in proline accumulation and activities of ascorbate peroxidase (APX), no significant variation in chlorophyll fluorescence values	Celikkol Akcay et al. (2010)
Potato	Decrease in plant height and leaf number, and root depth is considered the only significant cause of potato drought susceptibility	Nasir and Toth (2021, 2022)
Soybean	Decrease in plant height and leaf area, reduced the chlorophyll content and relative water content in the soybean leaves and increased the osmolyte contents, antioxidant potential, and peroxidation of the membrane lipids	Dong et al. (2019)

wheat-maize, followed by winter wheat-summer peanut → wheat-maize. During 2019–20, all rotation patterns were winter wheat-summer maize cropping system, and thus the difference between treatments was small. wheat-maize → wheat-maize and spring sweet potato → wheat-maize had the lower total water consumption compared with that for the other cropping systems. The highest

TABLE 12.5

Effects of Rotation Pattern on Total Water Consumption and Economic Water Use Efficiency

Rotation Pattern	2018–19 Total Water Consumption (mm)	2018–19 Economic Water Use Efficiency (Yuan/ha mm)	2019–20 Total Water Consumption (mm)	2019–20 Economic Water Use Efficiency (Yuan/ha mm)	2018–20 Total Water Consumption (mm)	2018–20 Economic Water Use Efficiency (Yuan/ha mm)
Wheat-maize → wheat-maize	747.3ab	28.7c	687.4b	33.7a	1434.8b	31.1c
Spring maize → wheat-maize	613.7d	25.2c	790.1a	28.7bc	1403.8b	27.2c
Winter wheat → wheat-maize	707.8bc	9.6d	741.1ab	28.7bc	1449.0b	19.4d
Spring sweet potato → wheat-maize	592.4d	27.8c	675.6b	33.4ab	1268.0c	30.8c
Spring peanut → wheat-maize	612.9d	63.8a	744.5ab	31.6ab	1357.4bc	46.1a
Winter wheat-summer peanut → wheat-maize	809.7a	45.6b	742.1ab	31.1ab	1551.8a	38.7b
Potato-silage maize → wheat-maize	642.3cd	32.4c	744.4ab	25.5c	1386.8b	28.7c

Source: Data were collated and adapted from Zhao et al. (2022).

economic WUE was observed in wheat-maize → wheat-maize, followed by spring sweet potato → wheat-maize, spring peanut → wheat-maize, and winter wheat-summer peanut → wheat-maize. During the 2018–20, winter wheat-summer peanut → wheat-maize had the highest water consumption while spring sweet potato → wheat-maize and spring peanut → wheat-maize had the lowest water consumption. Economic WUE of spring peanut → wheat-maize and winter wheat-summer peanut → wheat-maize were 1.5 times and 1.2 times significantly higher than that of wheat-maize → wheat-maize, respectively. Thus, the 2-year rotation patterns of spring peanut → wheat-maize could reduce the farmland water consumption, meanwhile, improve the economic WUE, and could be implemented to partially replace the winter wheat and summer maize double-cropping system in the NCP. The possible reason is that peanut and maize had the higher water retention capacity and more crop production under limited water supply (Table 12.4).

12.6.4 Biochar Application

Biochar is a predominantly stable, refractory organic carbon compound produced through the anoxic thermochemical transformation of biomass between 300°C and 1,000°C (Jeffery et al., 2011; Lehmann et al., 2021). Biochar is highly porous, thus its addition to soil is regarded as an improvement of a range of soil physical properties, including total porosity, soil density, water holding capacity, and hydraulic conductivity (Basso et al., 2013). Biochar as a soil amendment has been documented to mitigate climate change through sustainable sequestration of carbon while improving soil functions simultaneously (Hailegnaw et al., 2019; Hardie et al., 2014; Kan et al., 2020b; Virk et al., 2021). A field experiment was established to investigate the effects of biochar addition on soil water consumption and WUE in Wuqiao (Table 12.6). The results showed that, compared to CK (no biochar addition), biochar addition decreased total water consumption during the wheat season by 3.7%–6.7%, 3.4%–9.7%, and 1.0%–6.1% under BH, BM, and BL, respectively; during the maize season by 22.8%–28.3%, 21.1%–21.4%, and 11.8%–12.5% under BH, BM, and BL, respectively. Higher WUE was observed under biochar addition, especially for BM and BL treatments. Thus, suitable application of biochar could reduce water consumption and improve WUE and grain yield in the NCP.

12.6.5 Irrigation Strategy

Only about 20% of the cropland is irrigated around the world, however, this irrigated cropland contributes about 40% of total food production (Basso et al., 2013). Thus, the WUE in agriculture should be substantially increased to meet the growing demand for food through an improved irrigation strategy. In normal precipitation (550 mm) years of NCP, irrigation of the maize is not required, and soil water contents are abundant after the summer maize season. However, precipitation cannot meet the demands of winter wheat and additional water should be provided to obtain a high grain yield (Figure 12.1). In this context, we should ensure the water supply at critical growth stages of wheat. A previous study reported that enough water supply at the critical growth stage and a moderate water deficit at the non-critical stage of wheat can effectively improve WUE by regulating the population structure and achieving higher utilization efficiency of irrigation water (Li et al., 2018). In Wuqiao, field experiments about different irrigation timing and frequency and methods were conducted to investigate the effects of saving-water irrigation on WUE and to provide a practical approach for wheat production in the NCP.

12.6.5.1 Irrigation Timing and Frequency

Generally, local conventional irrigation management practice of winter wheat is 75 mm of water applied at sowing, joint stage, and anthesis stage, respectively. Summer maize is only 75 mm of water applied at sowing. Xu et al. (2018a,b) optimized the timing of two irrigations after sowing to improve wheat yield and WUE (Table 12.7). Compared with irrigation at the other stages, irrigation at joint and anthesis stages optimized crop characteristics with reasonable leaf area index and prolonged grain-filling duration by 1–3 days, then increased biomass of post-anthesis and harvest index.

TABLE 12.6
Effects of Biochar Addition on Total Water Consumption and Economic Water Use Efficiency

		Winter Wheat		Summer Maize	
Year	Treatment	Total Water Consumption (mm)	Water Use Efficiency (kg/ha mm)	Total Water Consumption (mm)	Water Use Efficiency (kg/ha mm)
2015–16	BH	457.3bc	17.4ab	268.4c	38.6a
	BM	442.7c	19.5a	274.2c	39.6a
	BL	460.4b	18.6a	304.2b	38.0a
	CK	490.2a	16.4b	347.6a	33.4b
2016–17	BH	369.6b	20.8b	271.4d	39.4a
	BM	370.6b	21.5a	297.4c	37.2a
	BL	379.7ab	20.8b	334.0b	36.8a
	CK	383.7a	19.6b	378.5a	31.7b

BH, BM, BL, and CK refer to 7,200, 3,600, 2,800, and 0 kg/ha year, respectively.
Source: Data were collected from Kan et al. (2019a,b).

TABLE 12.7
Effects of Irrigation Timing and Frequency on Total Water Consumption and Economic Water Use Efficiency

Irrigation Times	Treatment	2013–14		2014–15		2015–16		References
		Total Water Consumption (mm)	WUE (kg/ha mm)	Total Water Consumption (mm)	WUE (kg/ha mm)	Total Water Consumption (mm)	WUE (kg/ha mm)	
Single irrigation	No-irrigation after sowing	371.4d	20.6bc	389.4d	18.3bc	396.1c	17.4c	Xu et al. (2018a)
	Upstanding	430.8a	20.5c	442.3a	17.7c	452.0a	17.8bc	
	Jointing	421.3ab	21.8a	427.7ab	19.2a	431.5b	19.8a	
	Booting	412.8bc	21.6a	424.8bc	19.3a	428.7b	19.7a	
	Anthesis	404.3c	21.3ab	418.9bc	18.5ab	423.9b	18.4b	
	Medium milk	399.8c	21.0abc	410.4c	18.3bc	417.8b	17.7c	
Double irrigations	No-irrigation after sowing	371.4d	20.6ab	389.4d	18.3abc	396.1d	17.4b	Xu et al. (2018b)
	Late tillering and booting	475.7a	19.9b	482.1a	18.1bc	490.1a	18.3b	
	Late tillering and anthesis	467.6ab	19.7b	473.3ab	18.0bc	484.6a	18.2b	
	Late tillering and medium milk	457.2b	20.0ab	469.9ab	17.3c	478.6ab	18.0b	
	Jointing and anthesis	451.8b	21.6a	462.4abc	19.3a	466.8bc	19.6a	
	Jointing and medium milk	443.4bc	21.3ab	452.6bc	19.0ab	458.4c	19.6a	
	Booting and medium milk	434.2c	21.3ab	444.3c	18.9ab	453.0c	19.4a	

Irrigation at jointing and anthesis stages could improve grain yield and WUE by increasing biomass post-anthesis and obtained the highest grain yield (9267.6 kg/ha) and WUE (20.2 kg/ha mm), which is consistent with the findings of Zhang et al. (2011).

Xu et al. (2018a,b) optimized a single irrigation scheme after wheat sowing to improve WUE by manipulating winter wheat sink-source relationships in the NCP. Double irrigation resulted in more water consumption and grain yield during wheat production compared with single irrigation. Thus, there was no difference in WUE between double irrigation and single irrigation. This trend indicated a linear relationship between irrigation amount and wheat grain yield. Irrigation at jointing or at booting stage could coordinate pre- and post-anthesis water use and reduce total water consumption. Irrigation at jointing or at booting stage can optimize population quantity, prolonged grain-filling period, and improve post-anthesis biomass. These results demonstrated that single irrigation at jointing or booting is a promising strategy to improve grain yield and WUE effectively compared with irrigation at other stages. Thus, it is concluded that the promising irrigation scheme in the NCP is single irrigation at jointing or booting, and double irrigation at jointing and anthesis stages. In addition, controlling soil water before jointing not only can stimulate wheat roots to grow into deeper soil layers, but also regulate the leaf size and tiller number and decrease unnecessary water consumption (Gao et al., 2021).

In the NCP, young farmers leave farming for higher incomes; thus, the remaining labor sources for fieldwork are elderly men and children. It is hard to apply timely irrigation to wheat when droughts occur, and it is meaningful for promising WUE under lower irrigation frequency. Zhang et al. (2011) reported that lower irrigation frequency causes reduced total water consumption and increased WUE. As shown in Table 12.7, the water consumption and WUE of double irrigation at jointing and anthesis stages on average were 460.3 mm and 20.2 kg/ha mm, respectively. The water consumption and WUE of single irrigation at the jointing stage on average were 426.8 mm and 20.3 kg/ha mm, respectively. These results indicate that reduced total water consumption under lower irrigation frequency cannot contribute to the obvious improvement of WUE.

Wang et al. (2018a,b,c) investigated the effects of reduced irrigation (no-irrigation after wheat sowing) on WUE of wheat and maize. Wang and colleagues reported that the growth period of wheat was shortened and the maize sowing date advanced, and the flowering date subsequently advanced by 2–8 days, effectively prolonging the duration of maize grain-filling stage. Thus, the yield and WUE of maize were higher under reduced irrigation than conventional irrigation, which compensated for the wheat yield loss under reduced amounts of irrigation. Although precipitation could satisfy the requirement of water demand for summer maize, reducing water consumption during the maize season could retain more water resources in the soil for wheat.

12.6.5.2 Irrigation Methods

Flood irrigation is the primary irrigation method in the NCP. It is difficult to increase the WUE under the traditional flood irrigation strategy. Recently, novel irrigation methods (e.g., drip and sprinkler irrigation), have been developed and applied to field crops (Man et al., 2014). These irrigation methods significantly resist drought and improve WUE, particularly in regions of groundwater deficits. Compared with traditional irrigation, drip irrigation could decrease the irrigation water consumption of wheat and maize, and increase grain yield and WUE (Mon et al., 2016; Sui et al., 2018). Micro-sprinkling irrigation is a novel irrigation method developed through a combination of drip and sprinkler irrigation (Man et al., 2014). A previous study reported that micro-sprinkling irrigation could decrease pre-anthesis water consumption while increasing post-anthesis water consumption and ensuring the water supply in the upper soil depths during the critical stage (Li et al., 2018). On average, drip irrigation and micro-sprinkling irrigation had lower total water consumption by 2.2% and 3.1%, higher grain yield by 9.79% and 14.1%, and more WUE by 12.3% and 17.7% compared with traditional flood irrigation (Li et al., 2018). Root length density (RLD) of traditional flood irrigation in the 0–80-cm soil depth was significantly higher than that of micro-sprinkling irrigation, whereas micro-sprinkling irrigation had higher RLD than traditional flood irrigation below the 80-cm soil layer, which effectively promoted the water absorption from the su- soil (Li et al., 2018). Furthermore, with the flooding irrigation and one-time topdressing before irrigation event, nitrogen losses increased with the increase in fertilizer application rates (Li et al., 2019a,b). Nitrogen as topdressing was completely dissolved in a fertilization device and used together with the irrigation water under micro-irrigation, which helped nitrogen use and improved wheat yield and WUE (Li et al., 2018, 2019a,b).

Furthermore, a field experiment was conducted using micro-sprinkling irrigation with different irrigation amounts in the NCP (Li et al., 2019a,b). Irrigation amount of 60 mm using micro-sprinkling had the lowest total water consumption (408.7 mm), but the lowest WUE (1.71 kg/ha mm), which cannot satisfy the requirement of food demand. A suitable irrigation amount (i.e. ≥90 mm) under micro-sprinkling irrigation of 60 mm can ensure the grain yield of winter wheat and efficient utilization of irrigation water in the NCP.

12.6.6 Seeding Density

Seeding density and irrigation are limiting factors affecting the population architecture and wheat yield (Gao et al., 2021). The crop canopy architecture changes the leaf area index and radiation use efficiency, related to the photosynthetic efficiency. A reasonable seeding density can mitigate the competition between populations and individuals, and in return, it reduces water consumption and improves WUE (Ma et al., 2018). A field experiment was conducted at Wuqiao to identify the effects of seeding density on population architecture under limited irrigation schemes (Gao et al., 2021). Compared with conventional high-yield cultivation (plant densities of 225 plants/m^2 and irrigation of 75 mm at returning green, jointing, and anthesis stages, respectively), the water-saving cultivation (plant densities of 525 plants/m^2 and irrigation of 75 mm at the jointing stage) improved the spike number and grains per unit area because water-saving cultivation created a canopy architecture with higher leaf area index and no change in leaf area index. Overall, the water-saving cultivation obtained a comparable yield and meanwhile improved the WUE by 14.2%–39.2%.

12.7 FUTURE PROSPECTS

The NCP is a water-scarce area which is prone to pedological drought but is a crucial grain production region in China. Soil drought is caused by lower-than-average precipitation (known as meteorological drought) or higher evapotranspiration (known as crop/agronomic drought). However, the contribution of meteorological and crop drought to soil drought is not thoroughly understood and needs further investigation. Previous studies have mainly concentrated on the effects of single management practices (such as irrigation or tillage) on crop water use. Therefore, it is important to study the interaction between several management practices to improve crop production and WUE. In addition, soil drought could induce negative effects on nutrient cycling, soil fertility, and GHG emissions. Management practices to mitigate soil drought have mainly focused on lower water consumption and higher crop production, while the environmental effects have been mostly ignored. For example, a meta-analysis showed that WUE was highly improved when the nitrogen application rate was 220–250 kg/ha (Liu et al., 2022). However, increased nitrogen application could induce a linear increase in N2O emissions and soil acidification (Zhang et al., 2020). Thus, comprehensive management practices should be considered to mitigate soil drought and obtain positive environmental effects. Apart from management practices, taken together this review highlights the importance of breeding efforts to improving crop responses to drought depending on limited water supply stage (e.g., wheat jointing).

12.8 CONCLUSIONS

Soil drought has become one of the most influential, and widely concerned natural disasters in the NCP due to climate change and severe water resources shortage. During the total growth stages of winter wheat, the effective precipitation cannot meet the water demand, of which the total water deficit was ~350 mm and the drought area is still as large as 2 M ha due to lower precipitation and more crop planting in the NCP. Soil drought due to soil moisture loss would reduce the activity of methane-oxidizing bacteria and soil CH_4 uptake. Meanwhile, drought negatively affects soil respiration, while drought could reduce the denitrification rate and cumulative N_2O emission fluxes. The effects of soil drought on GHG emissions are regulated by many factors, such as soil texture, drought duration, and SOM content. Optimizing management practices to improve crop WUE is a possible strategy to manage soil drought. Subsoiling, residue retention, crop rotation, single irrigation at the jointing or booting stage, and double irrigation at jointing and anthesis stages can mitigate the adverse effects of drought on crop yield by reducing evapotranspiration and improving WUE under climate change. Synergistic effects are ignored when management practices are investigated separately. Thus, comprehensive management should be built to alleviate the negative effects induced by soil drought.

LIST OF ABBREVIATIONS

AWC   Available water capacity
GHG   Greenhouse gas
NCP   North China plain
NT   No-till
RLD   Root length density
SOM   Soil organic matter
SPEI   Standardized precipitation evapotranspiration index
WUE   Water use efficiency

REFERENCES

Abbas, T., M. Rizwan, S. Ali, M. Adrees, and M.F. Qayyum. 2017. Biochar application increased the growth and yield and reduced cadmium in drought stressed wheat grown in an aged contaminated soil. *Ecotoxicology & Environmental Safety*, 148: 825.

Abdul Rahman, N.S.N., N.W. Abdul Hamid, and K. Nadarajah. 2021. Effects of abiotic stress on soil microbiome. *International Journal of Molecular Sciences*, 22(16): 9036.

Ai, Z., Q. Wang, Y. Yang, K. Manevski, and X. Zhao. 2020. Variation of gross primary production, evapotranspiration and water use efficiency for global croplands. *Agricultural and Forest Meteorology*, 287: 107935.

Ali, S., M., Rizwan, M.F. Qayyum, Y.S. Ok, M. Ibrahim, M. Riaz, M.S. Arif, F. Hafeez, M.I. Al-Wabel, and A.N. Shahzad, 2017. Biochar soil amendment on alleviation of drought and salt stress in plants: A critical review. *Environmental Science and Pollution Research*, 24(14): 12700–12712.

Al-Kaisi, M.M., R. Elmore, J. Guzman, M. Hanna, C.E. Hart, M.J. Helmers, E. Hodgson, A. Lenssen, A. Mallarino, A. Robertson, and J. Sawyer, 2012. Drought impact on crop production and the soil environment: 2012 experiences from Iowa. *Journal of Soil and Water Conservation*, 68(1): 19–24.

Bai, L.P., F.G. Sui, T.D. Ge, Z.H. Sun, Y.H. Lu, and G.S. Zhou, 2006. Effect of soil drought stress on leaf water status, membrane permeability and enzymatic antioxidant system of maize. *Pedosphere*, 16(3): 7.

Basso, A.S., F.E. Miguez, D.A. Laird, R. Horton, and M. Westgate, 2013. Assessing potential of biochar for increasing water-holding capacity of sandy soils. *GCB Bioenergy*, 5(2): 132–143.

Blum, A., 2009. Effective use of water (EUW) and not water-use efficiency (WUE) is the target of crop yield improvement under drought stress. *Field Crops Research*, 112(2–3): 119–123.

Bogati, K. and M. Walczak, 2022. The impact of drought stress on soil microbial community, enzyme activities and plants. *Agronomy*, 12(1): 189.

Breitkreuz, C., L. Herzig, F. Buscot, T. Reitz, and M. Tarkka, 2021. Interactions between soil properties, agricultural management and cultivar type drive structural and functional adaptations of the wheat rhizosphere microbiome to drought. *Environ Microbiol*, 23(10): 5866–5882.

Buljovcic, Z., and C. Engels, 2001. Nitrate uptake by maize roots during and after drought stress. *Plant and Soil*, 229(1): 125–135.

Celikkol Akcay, U., O. Ercan, M. Kavas, L. Yildiz, C. Yilmaz, H.A. Oktem, and M. Yucel, 2010. Drought-induced oxidative damage and antioxidant responses in peanut (*Arachis hypogaea* L.) seedlings. *Plant Growth Regulation*, 61: 21–28.

Chamberlain, S.D., K.S. Hemes, E. Eichelmann, D.J. Szutu, J.G. Verfailie, and D.D. Baldocchi, 2020. Effect of drought-induced salinization on wetland methane emissions, gross ecosystem productivity, and their interactions. *Ecosystems*, 23(3): 675–688.

Chandra, P., A. Wunnava, P. Verma, A. Chandra, and R.K. Sharma, 2021. Strategies to mitigate the adverse effect of drought stress on crop plants-influences of soil bacteria: A review. *Pedosphere*, 31(3): 496–509.

Chandrasekar, V., R.K. Sairam, and G.C. Srivastava, 2008. Physiological and biochemical responses of hexaploid and tetraploid wheat to drought stress. *Journal of Agronomy and Crop Science*, 185(4): 219–227.

Cornelissen, G.V.M., V. Shitumbanuma, V. Alling, G.D. Breedveld, D.W. Rutherford, M. Sparrevirk, S.E. Hale, A. Obia, and J. Mulder, 2013. Biochar effect on maize yield and soil characteristics in five conservation farming sites in Zambia. *Agronomy*, 3(2): 256–274.

Dai, A., 2013. Increasing drought under global warming in observations and models. *Nature Climate Change*, 3(1): 52–58.

Dang, Y.P., 2023. Preserving soil health for generations. Farming System, 1: 100044.

Deng, X.P., L. Shan, H. Zhang, and N.C. Turner, 2006. Improving agricultural water use efficiency in arid and semiarid areas of China. *Agricultural Water Management*, 80, 23–40.

Dietz, K.J., C. Zrb, and C.M. Geilfus, 2021. Drought and crop yield. *Plant Biology*, 23(6):881–893.

Dong, S., Y. Jiang, Y. Dong, L. Wang, W. Wang, Z. Ma, C. Yan, C. Ma, and L. Liu, 2019. A study on soybean responses to drought stress and rehydration. *Saudi Journal of Biological Sciences*, 26: 2006–2017.

Gao, L., B. Wang, S. Li, H. Wu, X. Wu, G. Liang, D. Gong, X. Zhang, D. Cai, and A. Degre, 2019. Soil wet aggregate distribution and pore size distribution under different tillage systems after 16 years in the Loess Plateau of China. *Catena*, 173: 38–47.

Gao, Y., M. Zhang, C. Yao, Y. Liu, Z. Wang, and Y. Zhang, 2021. Increasing seeding density under limited irrigation improves crop yield and water productivity of winter wheat by constructing a reasonable population architecture. *Agricultural Water Management*, 253: 106951.

Geng, S.M., D.H. Yan, T.X. Zhang, B.S. Weng, Z.B. Zhang, and T.L. Qin, 2015. Effects of drought stress on agriculture soil. *Natural Hazards*, 75(2): 1997–2011.

Guan, D., Y. Zhang, M.M. Al-Kaisi, Q. Wang, M. Zhang, and Z. Li, 2015. Tillage practices effect on root distribution and water use efficiency of winter wheat under rain-fed condition in the North China Plain. *Soil and Tillage Research*, 146: 286–295.

Hailegnaw, N.S., F. Mercl, K. Pračke, J. Száková, and P. Tlustoš, 2019. High temperature- produced biochar can be efficient in nitrate loss prevention and carbon sequestration. *Geoderma*, 338: 48–55.

Hammad, H.M., A. Ahmad, F. Abbas, and W. Farhad, 2012. Optimizing water and nitrogen use for maize production under semiarid conditions. *Turkish Journal of Agriculture and Forestry*, 36(5): 519–532.

Hanaka, A., E. Ozimek, E. Reszczyńska, J. Jaroszuk-Ściseł, and M. Stolarz, 2021. Plant tolerance to drought stress in the presence of supporting bacteria and fungi: An efficient strategy in Horticulture. *Horticulturae*, 7(10): 390.

Hardie, M., B. Clothier, S. Bound, G. Oliver, and D. Close, 2014. Does biochar influence soil physical properties and soil water availability? *Plant and Soil*, 376(1–2): 347–361.

Hoogenboom, G., 2000. Contribution of agrometeorology to the simulation of crop production and its applications. *Agricultural and Forest Meteorology*, 103(1): 137–157.

Hu, Y., P. Ma, B. Zhang, R.L. Hill, S. Wu, Q. Dong, and G. Chen, 2019. Exploring optimal soil mulching for the wheat-maize cropping system in sub-humid drought-prone regions in China. *Agricultural Water Management*, 219: 59–71.

Hu, Y.N., Y.J. Liu, H.J. Tang, Y.L. Xu, and J. Pan, 2014. Contribution of drought to potential crop yield reduction in a wheat maize rotation region in the North China Plain. *Journal of Integrative Agriculture*, 13(7): 1509–1519.

Jat, M.L., D. Chakraborty, J.K. Ladha, D.S. Rana, and B. Gerard, 2020. Conservation agriculture for sustainable intensification in South Asia. *Nature Sustainability*, 3(4): 336–343.

Jeffery, S., F.G.A. Verheijen, M.V.D. Velde, and A.C. Bastos, 2011. A quantitative review of the effects of biochar application to soils on crop productivity using meta-analysis. *Agriculture, Ecosystems & Environment*, 144(1): 175–187.

Kan, Z.R., Q.Y. Liu, C. He, Z.H. Jing, A.L. Virk, J.Y. Qi, X. Zhao, and H.L. Zhang, 2020a. Responses of grain yield and water use efficiency of winter wheat to tillage in the North China Plain. *Field Crops Research*, 249: 107760.

Kan, Z.R., Q.Y. Liu, G. Wu, S.T. Ma, A.L. Virk, J.Y. Qi, X. Zhao, and H.L. Zhang, 2020b. Temperature and moisture driven changes in soil carbon sequestration and mineralization under biochar addition. *Journal of Cleaner Production*, 265: 121921.

Kan, Z.R., C. Pu, J.Y. Qi, S.T. Ma, P. Liu, X. Zhao, and H.L. Zhang, 2019a. Effects of biochar on soil water and grain yield of winter wheat in the North China Plain. *Journal of China Agricultural University*, 24(4): 01–10 (in Chinese with English Abstract).

Kan, Z.R., S.T. Ma, J.Y. Qi, C. Pu, X. Wang, X. Zhao, H.L. Zhang, 2019b. Effects of biochar addition on photosynthetic potential and grain yield of winter wheat. *Journal of Triticeae Crops*, 39(9): 719–727 (in Chinese with English Abstract).

Kane, D.A., M.A. Bradford, E. Fuller, E.E. Oldfield, and S.A. Wood, 2021. Soil organic matter protects US maize yields and lowers crop insurance payouts under drought. *Environmental Research Letters*, 16: 044018.

Kashif, A., W. Wang, K. Ahmad, G. Ren, and G. Yang, 2018. Wheat straw mulching with fertilizer nitrogen: An approach for improving soil water storage and maize crop productivity. *Plant, Soil and Environment*, 64(7): 330–337.

Khan, M.N., J. Zhang, T. Luo, J. Liu, F. Ni, M. Rizwan, S. Fahad, and L. Hu, 2019. Morpho-physiological and biochemical responses of tolerant and sensitive rapeseed cultivars to drought stress during early seedling growth stage. *Acta Physiologiae Plantarum*, 41(2): 1–13.

Kränzlein, M., C.M. Geilfus, B.L. Franzisky, X. Zhang, M.A. Wimmer, and C. Zörb, 2022. Physiological responses of contrasting maize (*Zea mays* L.) hybrids to repeated drought. *Journal of Plant Growth Regulation*, 41: 2708-2718 https://doi.org/10.1007/s00344-021-10468-2.

Krieger-Liszkay, A. and T. Roach, 2014. Regulation of photosynthetic electron transport and photo inhibition. *Current Protein & Peptide Science*, 15(4): 351–362.

Kuang, N., D. Tan, H. Li, Q. Gou, Q. Li, and H. Han, 2020. Effects of subsoiling before winter wheat on water consumption characteristics and yield of summer maize on the North China Plain. *Agricultural Water Management*, 227: 105786.

Lal, R., Farming systems to return land for nature: It's all about soil health and re-carbonization of the terrestrial biosphere. Farming System, 1: 100002.

Lehmann, J., A. Cowie, C.A. Masiello, C. Kammann, D. Woolf, J.E. Amonette, M.L. Cayuela, M. Camps-Arbestain, and T. Whitman, 2021. Biochar in climate change mitigation. *Nature Geoscience*, 14(12): 883–892.

Li, D., D. Zhang, H. Wang, H. Li, Q. Fang, H. Li, and R. Li, 2020a. Optimized planting density maintains high wheat yield under limiting irrigation in North China Plain. *International Journal of Plant Production*, 14(1): 107–117.

Li, J., S. Inanaga, Z. Li, and A.E. Eneji, 2005. Optimizing irrigation scheduling for winter wheat in the North China Plain. *Agricultural Water Management*, 76(1): 8–23.

Li, J., X. Xu, G. Lin, Y. Wang, Y. Liu, M. Zhang, J. Zhou, Z. Wang, and Y. Zhang, 2018. Micro-irrigation improves grain yield and resource use efficiency by co-locating the roots and N-fertilizer distribution of winter wheat in the North China Plain. *Science of the Total Environment*, 643: 367–377.

Li, J., Z. Zhang, Y. Liu, C. Yao, W. Song, X. Xu, M. Zhang, X. Zhou, Y. Gao, Z. Wang, Z. Sun, and Y. Zhang, 2019a. Effects of micro-sprinkling with different irrigation amount on grain yield and water use efficiency of winter wheat in the North China Plain. *Agricultural Water Management*, 224: 105736.

Li, J., Y. Wang, M. Zhang, Y. Liu, X. Xu, G. Lin, Z. Wang, Y. Yang, and Y. Zhang, 2019b. Optimized micro-sprinkling irrigation scheduling improves grain yield by increasing the uptake and utilization of water and nitrogen during grain filling in winter wheat. *Agricultural Water Management*, 211: 59–69.

Li, R., X. Hou, Z. Jia, and Q. Han, 2020b. Soil environment and maize productivity in semi-humid regions prone to drought of Weibei Highland are improved by ridge-and-furrow tillage with mulching. *Soil and Tillage Research*, 196: 104476.

Li, S., J. Liu, G. Liang, X. Wu, M. Zhang, E. Plougonven, Y. Wang, L. Gao, A.A. Abdelrhman, X. Song, X. Liu, and A. Degré, 2021. Factors governing soil water repellency under tillage management: The role of pore structure and hydrophobic substances. *Land Degradation & Development*, 32: 1046–1059

Li, X. and X. Pan, 2010. Hydrogels based on hemicellulose and lignin from lignocellulose biorefinery: A mini-review. *Journal of Biobased Materials and Bioenergy*, 4(4): 289–297.

Li, Y., Y. Wei, W. Meng, and X. Yan, 2009. Climate change and drought: A risk assessment of crop-yield impacts. *Climate Research*, 39(1): 31–46.

Liu, B.Y., W.S. Liu, B.J Lin, W.X. Liu, S.W. Han, X. Zhao, and H.L. Zhang, 2022. Sustainable management practices to improve the water use efficiency of winter wheat in the North China Plain: A meta-analysis. *Agronomy for Sustainable Development*, 42: 33.

Liu, C., Z. Yao, K. Wang, X. Zheng, and B. Li, 2019. Net ecosystem carbon and greenhouse gas budgets in fiber and cereal cropping systems. *Science of the Total Environment*, 647: 895–904.

Liu, L., C. Hu, P. Yang, Z. Ju, J.E. Olesen, and J. Tang, 2015. Effects of experimental warming and nitrogen addition on soil respiration and CH4 fluxes from crop rotations of winter wheat-soybean/fallow. *Agricultural and Forest Meteorology*, 207: 38–47.

Liu, X., Y. Pan, X. Zhu, T. Yang, and Z. Sun, 2018. Drought evolution and its impact on the crop yield in the North China Plain. *Journal of Hydrology*, 564: 984–996.

Ma, S., T. Wang, X. Guan, and X. Zhang, 2018. Effect of sowing time and seeding rate on yield components and water use efficiency of winter wheat by regulating the growth redundancy and physiological traits of root and shoot. *Field Crops Research*, 221: 166–174.

Ma, X., Q. Hu, J. Wang, X. Pan, J. Zhang, X. Wang, L. Hu, H. He, M. Xing, and L. Li, 2020. Spatiotemporal variation characteristics of drought trend at annual of wheat-maize in the North China Plain based on SPEI_KC index. *Transactions of the Chinese Society of Agricultural Engineering*, 36(21): 164–174 (in Chinese with English Abstract).

Mahajan, G., B.S. Chauhan, J. Timsina, P.P. Singh, and K. Singh, 2012. Crop performance and water- and nitrogen-use efficiencies in dry-seeded rice in response to irrigation and fertilizer amounts in northwest India. *Field Crops Research*, 134: 59–70.

Man, J., J. Yu, P.J. White, S. Gu, Y. Zhang, Q. Guo, Y. Shi, and D. Wang, 2014. Effects of supplemental irrigation with micro-sprinkling hoses on water distribution in soil and grain yield of winter wheat. *Field Crops Research*, 161: 26–37.

Mazloom, N., R. Khorassani, G.H. Zohuri, H. Emami, and J. Whalen, 2019. Development and characterization of lignin-based hydrogel for use in agricultural soils: preliminary evidence. *CLEAN - Soil Air Water*, 47(11): 1900101.

Mazloom, N., R. Khorassani, G.H. Zohury, H. Emami, and J. Whalen, 2020. Lignin-based hydrogel alleviates drought stress in maize. *Environmental and Experimental Botany*, 175: 104055.

Meyer, N., J.E. Bergez, J. Constantin, and E. Justes, 2019. Cover crops reduce water drainage in temperate climates: A meta-analysis. *Agronomy for Sustainable Development*, 39: 3.

Milosevic, N., J. Marinkovic, and B. Tintor, 2012. Mitigating abiotic stress in crop plants by microorganisms. *Zbornik Matice Srpske za Prirodne Nauke*, 123: 17–26.

Mishra, V. and K.A. Cherkauer, 2010. Retrospective droughts in the crop growing season: Implications to corn and soybean yield in the Midwestern United States. *Agricultural and Forest Meteorology*, 150(7–8): 1030–1045.

Mon, J., K.F. Bronson, D.J. Hunsaker, K.R. Thorp, J.W. White, and A.N. French, 2016. Interactive effects of nitrogen fertilization and irrigation on grain yield, canopy temperature, and nitrogen use efficiency in overhead sprinkler-irrigated durum wheat. *Field Crops Research*, 191: 54–65.

Moussa, H.R., and S.M. Abdel-Aziz, 2008. Comparative response of drought tolerant and drought sensitive maize genotypes to water stress. *Australian Journal of Crop Science*, 1(1): 31–36.

Mukherjee, S., A. Mishra, and K.E. Trenberth, 2018. Climate change and drought: a perspective on drought indices. *Current Climate Change Reports*, 4(2): 145–163.

Mwadzingeni, L., H. Shimelis, E. Dube, M.D. Laing, and T.J. Tsilo, 2016. Breeding wheat for drought tolerance: Progress and technologies. *Journal of Integrative Agriculture*, 15(5): 935–943.

Nasir, M.W., and Z. Toth, 2021. Response of different potato genotypes to drought stress. *Agriculture*, 11, 763.

Nasir, M.W., and Z. Toth, 2022. Effect of drought stress on potato production: A review. *Agronomy*, 12, 635.

Nguyen, L.T.T., Y. Osanai, I.C. Anderson, M.P. Bange, D.T. Tissue, and B.K. Singh, 2018. Flooding and prolonged drought have differential legacy impacts on soil nitrogen cycling, microbial communities and plant productivity. *Plant and Soil*, 431(1): 371–387.

O'Connell, C.S., L. Ruan, and W.L. Silver, 2018. Drought drives rapid shifts in tropical rainforest soil biogeochemistry and greenhouse gas emissions. *Nature Communications*, 9: 1348.

Obia, A., J. Mulder, V. Martinsen, G. Cornelissen, and T. Børresen, 2016. In situ effects of biochar on aggregation, water retention and porosity in light-textured tropical soils. *Soil and Tillage Research*, 155: 35–44.

Pascual, J.A., T. Hernandez, C. Garcia, and M. Ayuso, 1998. Enzymatic activities in an arid soil amended with urban organic wastes: Laboratory experiment. *Bioresource technology*, 64(2): 131–138.

Passauer, L., K. Fischer, and F. Liebner, 2011. Preparation and physical characterization of strongly swellable oligo(oxyethylene) lignin hydrogels. *Holzforschung*, 65(3): 309–317.

Piao, S., P. Ciais, Y. Huang, Z. Shen, S. Peng, J. Li, L. Zhou, H. Liu, Y. Ma, Y. Ding, P. Friedlingstein, C. Liu, K. Tan, Y. Yu, T. Zhang, and J. Fang, 2010. The impacts of climate change on water resources and agriculture in China. *Nature*, 467(7311): 43–51.

Rizwan, M., S. Ali, M.F. Qayyum, M. Ibrahim, and S.O. Yong, 2016. Mechanisms of biochar- mediated alleviation of toxicity of trace elements in plants: a critical review. *Environmental Science and Pollution Research*, 23(3): 2230–2248.

Sang, Y.F., V.P. Singh, Z. Hu, P. Xie, and X. Li, 2018. Entropy-aided evaluation of meteorological droughts over China. *Journal of Geophysical Research. Atmospheres*, 123(2): 740–749.

Sayer, E.J., J.A. Crawford, J. Edgerley, A.P. Askew, I.C. Dodd, 2021. Adaptation to chronic drought modifies soil microbial community responses to phytohormones. *Communications Biology*, 4: 516.

Shang, Z., M. Abdalla, L. Xia, F. Zhou, and P. Smith, 2021. Can cropland management practices lower net greenhouse emissions without compromising yield? *Global Change Biology*, 27(19): 4657–4670.

Siebielec, S., G. Siebielec, A. Klimkowicz-Pawlas, A. Galazka, J. Grzadziel, and T. Syuczynski, 2020. Impact of water stress on microbial community and activity in sandy and loamy soils. *Agronomy*, 10(1429): 1429.

Sui, J., J. Wang, S. Gong, D. Xu, Y. Zhang, and Q. Qin, 2018. Assessment of maize yield- increasing potential and optimum N level under mulched drip irrigation in the Northeast of China. *Field Crops Research*, 215: 132–139.

Sza, B., B. Pl, B. Xy, B. Zw, and A. Xc, 2011. Effects of tillage and plastic mulch on soil water, growth and yield of spring-sown maize. *Soil and Tillage Research*, 112(1): 92–97.

Thapa, S., S.K. Reddy, M.P. Fuentealba, Q. Xue, and S. Liu, 2018. Physiological responses to water stress and yield of winter wheat cultivars differing in drought tolerance. *Journal of Agronomy and Crop Science*, 204(4): 347–358.

Van Duivenbooden, N., M. Pala, C. Studer, C.L. Bielders, and D.J. Beukes, 2000. Cropping systems and crop complementarity in dryland agriculture to increase soil water use efficiency: a review. *NJAS - Wageningen Journal of Life Sciences*, 48(3): 213–236.

Virk, A.L., Z.R. Kan, B.Y. Liu, J.Y. Qi, and H.L. Zhang, 2021. Impact of biochar water extract addition on soil organic carbon mineralization and C fractions in different tillage systems. *Environmental Technology & Innovation*, 21: 101193.

Wang, S., X. Mo, S. Hu, S. Liu, and Z. Liu, 2018a. Assessment of droughts and wheat yield loss on the North China Plain with an aggregate drought index (ADI) approach. *Ecological Indicators*, 87: 107–116.

Wang, Y., Y. Zhang, Z. Rui, J. Li, and Z. Wang, 2018b. Reduced irrigation increases the water use efficiency and productivity of winter wheat-summer maize rotation on the North China Plain. *Science of the Total Environment*, 618: 112–120.

Wang, Y., Y. Zhang, S. Zhou, and Z. Wang, 2018c. Meta-analysis of no-tillage effect on wheat and maize water use efficiency in China. *Science of the Total Environment*, 635: 1372–1382.

Xiao, D. and F. Tao, 2014. Contributions of cultivars, management and climate change to winter wheat yield in the North China Plain in the past three decades. *European Journal of Agronomy*, 52: 112–122.

Xiao, D., D.L. Liu, P. Feng, B. Wang, and J. Tang, 2021. Future climate change impacts on grain yield and groundwater use under different cropping systems in the North China Plain. *Agricultural Water Management*, 246(106685): 14.

Xu, X., M. Zhang, J. Li, Z. Liu, Z. Zhao, Y. Zhang, S. Zhou, and Z. Wang, 2018a. Improving water use efficiency and grain yield of winter wheat by optimizing irrigations in the North China Plain. *Field Crops Research*, 221: 219–227.

Xu, X., Y. Zhang, J. Li, Z. Meng, and Z. Wang, 2018b. Optimizing single irrigation scheme to improve water use efficiency by manipulating winter wheat sink-source relationships in Northern China Plain. *PLoS One*, 13(3): e0193895.

Yu, H., Q. Zhang, P. Sun, and C. Song, 2018. Impact of droughts on winter wheat Yield in different growth stages during 2001-2016 in Eastern China. *International Journal of Disaster Risk Science*, 9(3): 376–391.

Yu, L., X. Zhao, X. Gao, R. Jia, and K. Siddique, 2021. Effect of natural factors and management practices on agricultural water use efficiency under drought: A meta-analysis of global drylands. *Journal of Hydrology*, 594(3): 125977.

Zhang, H.L., Y.P. Dang, and L. Li, 2023. Farming system: A systemic solution to sustainable agricultural development. Farming System, 1: 100007.

Zhang, L., F. Chen, and Y. Lei, 2020. Climate change and shifts in cropping systems together exacerbate China's water scarcity. *Environmental Research Letters*, 15(10): 104060.

Zhang, L., Q. Chu, Y. Jiang, F. Chen, and Y. Lei, 2021. Impacts of climate change on drought risk of winter wheat in the North China Plain. *Journal of Integrative Agriculture*, 20(10): 2601–2612.

Zhang, X., J. Guo, R.D. Vogt, J. Mulder, and X. Zhang, 2020. Soil acidification as an additional driver to organic carbon accumulation in major Chinese croplands. *Geoderma*, 366: 114234.

Zhang, Y., E. Kendy, Q. Yu, C.M. Liu, Y.J. Shen, and H.Y. Sun, 2004. Effect of soil water deficit on evapotranspiration, crop yield, and water use efficiency in the North China Plain. *Agricultural Water Management*, 64: 107–122.

Zhang, Y., Y. Zhang, Z. Wang, and Z. Wang, 2011. Characteristics of canopy structure and contributions of non-leaf organs to yield in winter wheat under different irrigated conditions. *Field Crops Research*, 123(3): 187–195.

Zhao, F., D. Zhang, Y. Zhao, W. Wang, H. Yang, F. Tai, C. Li, and X. Hu, 2016. The difference of physiological and proteomic changes in maize leaves adaptation to drought, heat, and combined both stresses. *Frontiers in Plant Science*, 7: 1471.

Zhao, Y.X., B. Wang, Q. Liu, T. Song, X.P. Zhang, Y.Q. Chen, and P. Sui, 2022. Characteristics of farmland water consumption under two-year wheat-maize interannual rotation patterns in heilonggang plain. *Acta Agronomica Sinica*, 48(7): 1787–1799 (in Chinese with English Abstract).

13 Soil Drought and Human Health

Marium Husain

13.1 INTRODUCTION

It is estimated that by 2050, 75% of the planet's population will be affected by drought (U. N. C. t. C. Desertification, 2022). Drought is a complicated phenomenon that is intersectional, not only in its causes but also in its effects. The United Nations Intergovernmental Negotiating Committee for the Elaboration of an International Convention to Combat desertification in countries that were experiencing serious drought and/or desertification produced a report in 1994 where they defined drought as "the naturally occurring phenomenon that exist when precipitation has been significantly below normal recorded levels, causing serious hydrological imbalances that adversely affect land resource production systems" (U. N. G. A. I. N. Committee, 1994). Another term that is commonly associated with drought is desertification: "land degradation in arid, semi-arid and dry sub-humid areas resulting from various factors, including climatic variations and human activities" (U. N. G. A. I. N. Committee, 1994). They are not technically the same phenomenon and should be used distinctly and separately.

The "hydrological balance" in the definition of drought refers to the water cycle, one of the main complex and intersectional systems on the planet. A system is a "regularly interacting or interdependent group of items forming a unified whole" (Merriam-Webster, 2022). The National Oceanic and Atmospheric Administration (NOAA) describes the water cycle as the "continuous movement of water within the Earth and atmosphere" but also that is a complex phenomenon that includes air, clouds, oceans, lakes, vegetation, and glaciers (NOAA, 2019) (Figure 13.1).

The two main components to this definition of drought that exist in balance are precipitation and land. Precipitation is "any liquid or frozen water that forms in the atmosphere and falls back to Earth"

FIGURE 13.1 The water cycle (NOAA).

Soil Drought and Human Health

(Geographic, 2022) like rain or snow. Land, as defined by the Intergovernmental Negotiating Committee is the "terrestrial bio-productive system that comprises soil, vegetation, other biota, and the ecological and hydrological processes that operate within the system" (U. N. G. A. I. N. Committee, 1994).

Both definitions of precipitation and land also reference resource production and ecology. The Committee also defined land degradation as "reduction or loss, in arid, semi-arid and dry sub-humid areas, of the biological or economic productivity and complexity of rainfed cropland, irrigated cropland, or range, pasture, forest and woodlands resulting from land uses or from a process or combination of processes, including processes arising from human activities and habitation patterns, such as (i) soil erosion caused by wind and/or water; (ii) deterioration of the physical, chemical and biological or economic properties of soil; and (iii) long-term loss of natural vegetation" (U. N. G. A. I. N. Committee, 1994).

Although precipitation and land are modeled as distinct entities, they are inextricably linked and can be considered to exist on the same continuum that is the water cycle/system. When there are imbalances in this system, there is not just one effect but multiple as everything is, truly, connected.

Therefore, the objectives of this chapter will be to understand how drought leads to worsened human health. Our hypotheses are that drought is associated with worsened malnutrition, it leads to increased risk of toxicity from soil, water, and air contamination, and drought can exacerbate underlying social dynamics and indirectly worsen human health. We reviewed the literature to provide a holistic introduction to the relationships between soil drought and human health.

13.2 SOIL–HUMAN INTERFACE

Humans interact with the soil every day, directly or indirectly. One primary interaction is through water, and it is impacted by the water cycle balance (Figure 13.1). Soil water content is the measure of the amount of water (in mm) present in the soil (at a specific depth) (FAO) (Figure 13.2).

This can also be measured as percentage (depth of water over surface area of soil). Soil water content not only depends on the amount of soil and other biological contents, but also at a certain point in time, as soil is a dynamic entity that does not remain static (FAO). The amount and availability of water in the soil, along with humidity in the air, impacts how much water is biologically used within the soil, by plants and their growth, and ultimately available for humans (Scherer et al., 2022).

The importance of water for humans may seem intuitive, but it is important to highlight how intersectional water is in our lives, individually and as a society. The typical, intuitive purpose of water is for ingestion; we all drink it, and we are made of it. In general, the human body is about 55%–65% water, with about 20% of it inside blood vessels (Taylor & Jones, 2022). Water is

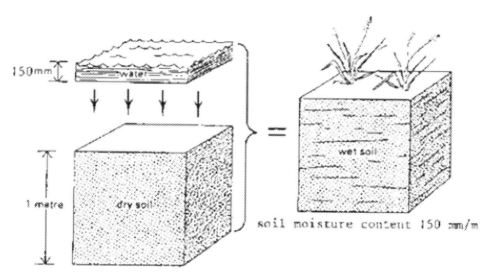

FIGURE 13.2 Soil moisture content of 150 mm/m assuming no loss by evaporation, runoff or deep seepage.

absorbed in the gastrointestinal tract and is lost via the kidneys (urine), skin (sweat), gastrointestinal tract (e.g., diarrhea) and the lungs. Due to a number of illnesses and environmental conditions, humans are at risk for dehydration from excess water losses, despite a complex balance system within the body regulating water balance. This can lead to significant morbidity and mortality if there isn't adequate replacement in a timely manner (Taylor & Jones, 2022). Water is also used for agriculture and irrigation to grow the food we eat or process it to create other foods (see next section); hydropower, as an alternate energy source with dams and reservoirs; industrial applications, like manufacturing, fashion/clothing, chemical production; healthcare, like dialysis machines and hospital hygiene; and also, recreation, such as a water parks or swimming (NOAA, 2019).

Another major interaction is with food. Our foods are ultimately derived from agriculture, such as corn grown from the soil and processed into a derivative food, like cereal. Although we buy many foods from grocery stores, separated from farmland and farmers, our food system is dependent upon the soil to grow basic crops to prevent famine and improve food security (Lal, 2021). The U.S. grows particular crops that are staples of our economy: corn, rice, soybeans, sugar/sweeteners, wheat, and fruit and tree nuts (USDA, 2022). The soil not only provides food for us humans but also for animals, plants, and even microorganisms. These microorganisms also serve as models from which we can generate new antibiotics (Francis, 2018).

However, when the natural balance with the soil is disrupted by deforestation, loss of biodiversity, natural resources exploitation and climate change leading to drought, desertification, and land degradation, we see negative impacts build up geographically and over time that affect different aspects of society (FAO, 2019; WMO, 2012; Stanke et al., 2013). There are four major categories of impacts in which drought affects humans: health, environmental, cultural and social, and economic (NIDIS, 2022; FAO). Inevitably, these negative consequences disproportionately affect vulnerable populations around the world (e.g., women/children, rural communities, racial minorities/indigenous communities, and comorbid conditions) (Barbier and Hochard, 2016; Ebi and Bowen, 2016). For this chapter, we will discuss health impacts.

13.3 THE NEGATIVE IMPACTS OF DROUGHT ON HUMAN HEALTH

In reality, these negative impacts are intersectional and have indirect and direct consequences on human health (Figure 13.3; Sena et al., 2014; Sena and Ebi, 2020).

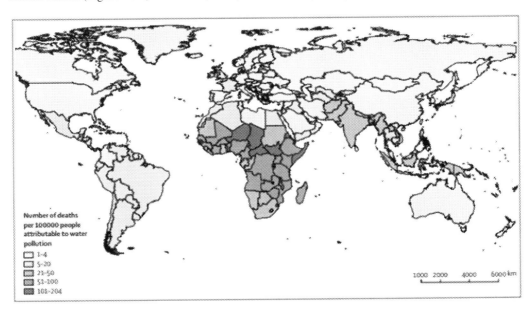

FIGURE 13.3 Number of deaths per 100,000 people due to water pollution.

13.3.1 Malnutrition

Approximately 144 million children (younger than 5) worldwide were affected by malnutrition in 2020 (WB, 2019, 2020). Malnutrition leads to a variety of health disorders: increased risk of infectious diseases due to compromised immune systems, micronutrient deficiencies leading to blindness (Vitamin A), anemia (iron), muscle wasting (protein), prematurity, intrauterine growth retardation (IUGR) and maternal/infant mortality (Black et al., 2013; Ebi and Bowen, 2016; Stanke et al., 2013). Before detailing the impacts of drought on malnutrition, we will define malnutrition and common associated terms.

13.3.1.1 Types of Malnutrition

Malnutrition can be described in various forms and typical categories are acute and chronic. Chronic malnutrition is manifested by stunting, which is decreased height, with a 22% prevalence worldwide (WHO, 2020). Moderate stunting is defined as a height or length Z-score of −2 to −3 and severe stunting is Z-score<−3 (WHO, 2009). A Z-score represents weight-for-height and height-for-age relative to the population. Acute malnutrition is further categorized into moderate acute malnutrition and severe acute malnutrition. Moderate acute malnutrition is characterized by a mid-upper arm circumference (MUAC) of 115–124 mm (typically >220 mm) and a weight-for-length Z-score −2 to −3. Severe acute malnutrition is characterized by an MUAC <115 mm, weight-for-length Z-score<−3. Severe acute malnutrition has specific syndromes that are well-characterized: marasmus, kwashiorkor and sometimes the combination of the two (marasmic kwashiorkor). Marasmus is a type of protein-energy malnutrition where people suffer from wasting due to decreased protein and calorie intake. It is characterized by low weight-for-height and decreased MUAC. Kwashiorkor is also a type of protein-energy malnutrition like marasmus, but people also display fluid retention (peripheral pitting edema). Children suffering from this typically manifest with a large, distended belly (abdomen). There are various clinical manifestations as well, including compromised immune systems and cardiovascular dysfunction.

In western Rajasthan in the 2003 drought, studies revealed a 28% prevalence of wasting (low weight for height) in children younger than 5, compared to the WHO's 15% cutoff that indicates severe malnutrition in a region (Singh et al., 2006). Interestingly, rates of Vitamin A/B deficiency and anemia were declining, due to the state government intervening with supplement replacement. During the 2011 Ethiopian drought, 46% of children younger than 5 were underweight, with 51.1% being short for their age (more than 2 standard deviations from the average) (Kaluski et al., 2002). These same patterns were seen in the Tharparker region in Pakistan during the 2013–15 drought. Not only were children under 7 years of age dying of marasmus (severe malnutrition) and wasting, but the malnutrition was also leading to deaths from infectious diseases like pneumonia and gastroenteritis (Khan et al., 2015).

The opposite end of the malnutrition spectrum is consuming processed foods or high calorie foods that do not provide adequate nutrients. This can lead to obesity/being overweight and toxicity. This is typical in situations of imported foods due to limited food production or drastic changes in diets due to crop yield losses (Black et al., 2013). During the 1976–77 Ethiopian drought in the northern Wello region, there was an outbreak of neurolathyrism, spastic paralysis of the lower extremities (legs) that is caused by degeneration of upper motor neurons in the nervous system. Researchers investigated 228 patients (mainly under 40 years old) and found that ingestion of beta-ODAP, an AMPA-receptor-specific neuro-excitatory amino acid, via unripe or roasted rip grasspea seeds was correlated with incidence of neurolathyrism (Getahun et al., 1999). Grasspea was the main crop at this time due to the other crops being wiped out during the 1995–96 drought. This is another example of the temporal impacts of drought. During the 1970–72 Afghanistan drought, there was a 22.6% incidence rate of liver disease (along with weight loss and fluid accumulation in the abdomen or ascites) in a community of 7,200 people in the northwestern region of the country (Mohabbat et al., 1976). Hepatic veno-occlusive disease was found to be correlated with the ingestion of pyrrolizidine alkaloids, which are liver toxic in large amounts, found in the seeds of *Heliotropium* plants. These seeds had contaminated the wheat that was used to make bread and were not properly removed from the wheat before processing. In the 1974 western Indian drought, there was an increased mortality rate of villagers with liver disease, particularly ascites, portal hypertension (increased pressure in the blood system of the liver) and jaundice (Krishnamachari et al., 1977). This was found to be caused by ingestion of maize contaminated with the fungus, *Aspergillus flavus*, which creates aflatoxins that are harmful to humans. It was reported that people had been ingesting 2–6 mg/g of aflatoxin daily for weeks, where typical concentrations in the area are less than 0.1 mg/g. The study attributed the increased concentration of *Aspergillus* to "unseasonal rains" leading to improper storage facilities where increased moisture accumulated. The researchers further note: "In view of the isolated situation of the villages in these areas, there is little possibility of changing over to some other staple or for purchase of good maize from the market in the event of grain spoilage." Similar to India and Ethiopia, the eastern, southern, western and lake zones of Tanzania have been affected by multiple outbreaks of Konzo since 1985, another upper motor neuron degenerative disease that also leads to spastic paralysis of the extremities, but in variable patterns that can be permanent (Mlingi et al., 2011). This is also true for the Democratic Republic of Congo, Mozambique, Cameroon and the Central African Republic. Konzo has been attributed to the ingestion of increased cyanide levels in the bitter cassava root due to improper cutting methods because of food shortages. And what is more representative of how interconnected food systems truly are, the increased effects of cyanide were also more prevalent due to food shortages of protein-rich foods in the area that provide sulfur amino acids that detoxify the cyanide. Not only is drought contributory, but these outbreaks are also seen as a result of war and, such as in the DRC and Mozambique (Cliff et al., 2011; Mlingi et al., 2011).

Of note, malnutrition is not always equally prevalent across a drought-stricken region. For example, the 1973–74 Ethiopian drought hit the North Ogaden region the most with the highest rates of malnutrition, compared to the Issa desert which had the lowest rates (Seaman et al., 1978). These differences are not only based on geography but also on socio/economic/political considerations that must be incorporated into any interventions.

13.3.1.2 Crop Health

As discussed previously, our food systems are reliant on main crops that are integrated within the U.S. and global economies. These crops, which are plants, are also affected by drought.

Drought affects the plant's ability to absorb nutrients as these are typically carried to the roots by water (Selvakumar et al., 2012). These nutrients impact the growth of the plants and their ability to fight off disease. It also impacts photosynthesis by decreasing chlorophyll amounts as identified in certain species (Astorga and Alcaraz, 2010; Beinsan et al., 2003). These changes in the soil can also impact the microorganisms, which have a direct relationship with plants, providing environments to help plants grow as well as be resilient against shifts in the ecosystem.

However, with time, the composition and balance of microorganisms change with drought and the new microorganisms do not display the same beneficial characteristics as plants (Cherif et al., 2015; Schmidt et al., 2014). We are reliant on the health of plants for our food systems and food security, both now and in the future (Vurukonda et al., 2016).

13.3.1.3 Food Insecurity

The world population is projected to be 8.5 billion by 2030 (United Nations, 2019). Drought is also linked to decreased crop production and yield, which limits the food supply in local and global markets, placing millions of people at risk of hunger, malnutrition and starvation (FAO, 2017; Ebi and Bowen, 2016). Livestock is also at risk for malnutrition and death, which is a major economic source for lower-income farmers (FAO, 2015; Centers for Disease Control and Prevention, 2020). Not only does a decreased food supply lead to food insecurity, but it also causes the prices of whatever food is produced to increase, posing a threat to those populations that are of lower socioeconomic status (Green et al., 2013). This not only becomes a health crisis but also one of health equity.

With decreasing crop production and rising food prices, this also leads to migration episodes we are seeing in different parts of the world, as people migrate to find food (FAO, 2018b). This is a reality regardless of political party or affiliation.

13.3.2 TOXICITIES

13.3.2.1 Soil Quality

Land degradation, either through drought or poor agricultural practices, can lead to poor soil quality. As previously discussed, the soil has a balance between water, the atmosphere, microorganisms, animals and humans. When that is disrupted by decreased soil water content due to climate change, or contamination from industrial practices (e.g., heavy metals) or excess fertilizers and pesticides, there is dysfunction (FAO, 2018a; Rodríguez-Eugenio et al., 2018): Crop yield is impacted; Heavy metals are not filtered out in the soil and can potentially end up in the actual crop due to dysfunction in nutrient absorption. In addition, restricting our crop diversity to just a few crops as part of our economy ("monocropping") leads to decreased soil quality, decreased microorganism diversity in the soil (SARE, 2019; UCD, 2016). Since the soil is not as fertile, it necessitates the use of synthetic fertilizers that also negatively impact the balance of nutrients in the soil (Paungfoo-Lonhienne et al., 2015; Zhou et al., 2017).

Humans are exposed to contaminants in the soil through various mechanisms. Typically, this is through ingestion, inhalation or skin (dermal) exposure (Rodríguez-Eugenio et al., 2018). These contaminants can also end up in the water and/or air. These contaminants, like heavy metals, dioxin and dioxin-like substances, and highly hazardous pesticides, are listed under the World Health Organization's (WHO) chemicals of major public health concern (Rodríguez-Eugenio et al., 2018). Heavy metals, including cadmium, arsenic, and lead, can cause significant diseases like cancer, neurological deficits and hematologic disorders (FAO, 2018a; Toth et al., 2016). Cadmium is mainly ingested via food and can cause liver and kidney disease, as well as disrupting endocrine systems. It can also impact the developing fetus as it can cross the placenta. Arsenic can be ingested or inhaled and is carcinogenic, as well as can cause muscle and nerve disorders as it accumulates in these organs. It can also cause anemia, like lead. Lead poisoning can cause neurological problems in children as well as liver, kidney and spleen problems. Other heavy metals include mercury, which is very neurotoxic at high levels (Brevik, 2013).

Dioxins are chemical compounds also known as persistent organic pollutants (POPs) by the WHO (2016). They are typically produced from the manufacturing industry, like chlorine bleaching of paper palp and the production of pesticides. However, they are also naturally occurring through volcanic eruptions or wildfires/forest fires. Long-term exposure to dioxins can lead to compromises in the immune system as well as impact the nervous system. They can also affect the developing fetus and are also present in breast milk. It also carries a cancer risk at high doses, deemed a "known human carcinogen" by the International Agency for Research on Cancer (IARC).

Pesticides/herbicides are used on crops to decrease damage by insects and microorganisms like fungi. Their residue can remain on crops and in the soil. Different countries have various regulatory procedures in place to ensure that there are acceptable levels of pesticide residue on food; however, this is not always the case for every country. In addition, contaminant levels can proportionally increase during a drought. Pesticide exposure has been associated with mild conditions like skin irritation to more severe illnesses, like hypersensitivity reactions, endocrine problems and even some reports of cancer (Burgess, 2013; Rodríguez-Eugenio et al., 2018).

The result of chronic exposure to these compounds is manifested in agricultural workers. There are more studies noting the impact of chemicals used in farming and heavy metal exposure on the incidence of chronic kidney disease (Chapman et al., 2019; Hoy et al., 2017; Obrador et al., 2017). Whereas this is not specific to drought, the increasing temperatures and decreased water supply in drought are likely to exacerbate these conditions.

13.3.2.2 Water Quality

Due to the imbalance of the water cycle in drought, there is decreased supply of water, not only in the soil but also in the groundwater, which is a source of water for many communities (McLeman, 2017; Centers for Disease Control and Prevention, 2020). As migration due to drought increases and the global population continues to grow requiring more land use for housing and agriculture, the demand for water will only increase and strain nonrenewable sources (WWAP, 2016; McLeman, 2017; Centers for Disease Control and Prevention, 2020; Smoyer-Tomic et al., 2004). In areas like southeast Asia and Brazil that are undergoing high deforestation rates for a variety of reasons, there are increasing concerns for the high use of water that is being used in non-agriculture or monocrop uses (McLeman, 2017; Oliveira, 2013; Rulli et al., 2013). The poor soil practices described earlier can also lead to poor water quality due to runoff of fertilizer and heavy metals from contaminated soil. Heavy rain events can also lead to contamination of water sources, and then during drought conditions, the concentrations of the contaminants increase in a smaller water supply (WHO, 2017). This has led to an increase in heavy metals like arsenic, manganese and iron in developing countries like India, Bangladesh and Latin American countries (Jiménez Cisneros et al., 2014). This phenomenon also impacts developed countries, like the Netherlands and France. For example, the Meuse river basin was investigated for water quality effects from prolonged drought between 1976 and 2003 (van Vliet and Zwolsman, 2008). Due to low river flows and high temperatures, there was an increase in some heavy metals and decreased oxygen concentration that led to harmful algal blooms. In another drought period, they found an increase in pharmaceutical contaminants (e.g. ibuprofen, sulfamethoxazole) in the drinking water (Wolff & van Vliet, 2021). They did note that there was a decrease in the concentration of other heavy metals (lead, cadmium, mercury, chromium) due to their high affinity to adsorb onto other solid matter. The researchers do note that each river will have its own characteristics. This impacts humans through ingestion of contaminants and aquatic life that have been exposed to these contaminants; it also negatively impacts animal and microorganism life in water systems that are dependent on clean water (Centers for Disease Control and Prevention, 2020). For example, algal blooms are made up of microscopic algae, also known as phytoplankton, which are already present in rivers and oceans. Certain species produce neurotoxins that are harmful to humans and can cause death in extreme cases (DiLiberto, 2022). Shellfish and other plankton-eating fish can also ingest these toxic algae and transfer the neurotoxin up the food stream, accumulating with each transfer and causing more risk to human health. In north-east Brazil during the 1996 drought, 60 people died out of 126 patients receiving hemodialysis due to water contaminated by toxic algae (Pouria et al., 1998).

During prolonged drought, communities are forced to obtain water from these contaminated sources, like lakes and reservoirs, due to no other options (Sena et al., 2014). Concurrent practices regardless of drought, like mining, can pollute existing water supplies with heavy metals like arsenic and lead, contributing to health risks and reasons for migration (Dooyema et al., 2012; Epstein et al., 2011; McLeman, 2017). There are increased rates of deaths from water pollution (GBD Risk Factors Collaborators, 2016; Landrigan et al., 2018).

13.3.2.3 Air Quality

In the same way that soil contamination impacts water quality, these contaminants can become airborne in dusty and dry conditions during drought (Landrigan et al., 2018). In the Canadian Prairie Provinces, the 1930s were termed the "dirty thirties" due to dust storms from soil erosion and drought conditions (NMDC, 2002; Smoyer-Tomic et al., 2004). This air contamination can lead to respiratory illnesses, like asthma, chronic bronchitis as well as cancer and heart disease (Landrigan et al., 2018). This is typically through particulate matter (PM) that can include minerals and organic material. Organic dust has been studied and found to have several components: dust from machinery/facilities, mold, bacteria, fungi, and animal/plant-derived materials, polyaromatic hydrocarbons (PAHs) (Government of Canada, 2004; Lang, 1996). Drought conditions also lead to increased risk of wildfires which also lead to increased contaminant and PM inhalational exposure, leading to similar respiratory illnesses (Xu et al., 2020). For example, in California and reviewing the impact of the drought in the early 1990s, the fungus, *Coccidioides immitis* was found to have an increased prevalence specifically correlated with drought conditions. It causes infection in humans specifically once its spores are inhaled (SMA, 1996; Fisher et al., 2000; Pappagianis, 1994). The United Nations Environmental Programme (UNEP) released a report on achieving a pollution-free world in 2017, outlining the source and impact of land/soil pollution (Figure 13.4) (UNEP, 2017).

13.3.2.4 Socioeconomic Impacts

The intersectional impacts of drought not only harm human health but also livelihoods which can all lead to poor mental health (Berry et al., 2010; Vins et al., 2015). There is increased risk of anxiety, depression and suicide due to income loss, unemployment, food/water insecurity and forced migration (Austin et al., 2018; Vins et al., 2015). Studies out of Australia reveal an increased incidence of mental health disorders and stress in farmers and farm workers who are suffering from drought in the region (Austin et al., 2018; Edwards et al., 2015). There is also increased psychological distress in communities where they are eating more processed/high-calorie foods in response to drought and food shortages (Friel et al., 2014).

Climate change is projected to increase interpersonal violence by 4% and intergrous conflict by 14% for 1 standard deviation change in global temperatures or extreme rainfall (Hsiang et al., 2013). As stated earlier, rainfall/precipitation has a direct impact on the hydrological balance and imbalances lead to drought. Since drought is linked to climate change, there are concerns for increasing violence with increasing drought. A study evaluated risk models for intimate-partner violence (IPV) in sub-Saharan Africa, South and Southeast Asia, Latin America, and the Caribbean due

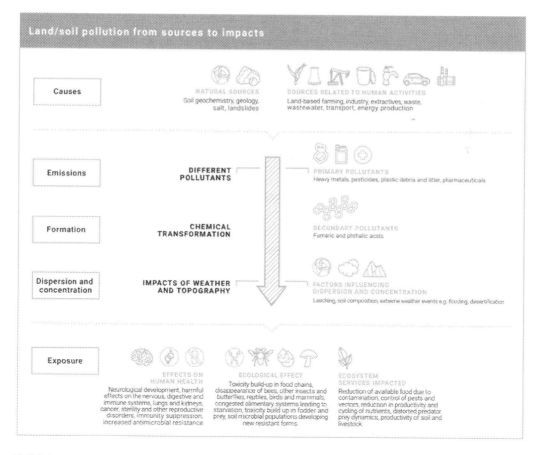

FIGURE 13.4 Land and soil pollution – from sources to impacts on human health and ecosystems.

to previously published conflicting reports on IPV and its relationship to drought (Cooper et al., 2021). They identified a relationship between controlling behaviors (a specific behavior of IPV) and drought, but not with other behaviors associated with IPV (physical, sexual and emotional violence). This may appear counterintuitive compared with the previously stated data. The authors conclude that when evaluating the impacts of drought, there needs to be a specific focus on spatial factors as well as taking into the temporal impacts of drought: "Future work should explore different ways to measure drought, accounting for different timescales and hydrological processes" (Cooper et al., 2021).

In addition, health systems that are already vulnerable or ineffective can become more stressed due to drought (Figure 13.5) (Ebi and Bowen, 2016).

Health centers, like hospitals, need clean water for operations and dialysis, as previously discussed in the 1996 drought in Brazil where patients were exposed to toxic algae and died from contaminated water (Ebi and Bowen, 2016; Oliveira, 2013). This can also impact triage of limited resources and patients that normally could have received care are now restricted from it due to poor access to resources in a disaster situation like drought. Vulnerable groups in current healthcare systems are at risk for increased vulnerability during disasters, like drought. Women, children, the elderly and those with physical disabilities face decreased access for their particular healthcare needs and may face the limited ability to migrate to areas with more resources, like water and food during a drought (WHO, 2014; Ebi and Bowen, 2016; Scandlyn et al., 2010).

13.4 RESEARCH PRIORITIES

There is a wealth of knowledge on soil drought and analyses on human health conditions during the drought. There is also research on the temporality of drought and the impacts before and after such conditions that can impact humans and population health. The United Nations, World Health Organization and other convening bodies generate guidance documents on soil health, soil pollution and preparedness for drought. This field can benefit from more prospective studies analyzing solutions to the already-researched problems related to drought. Drought-affected communities and countries are on the ground and can influence best practices and strategies for ameliorating the negative effects of drought. One research priority, as well as funding priority, is on regenerative agriculture to decrease poor soil practices, such as reduced fertilizer use, and evaluate the impact on crop yield and nutritional content. Another priority area is on behavior change recommendations

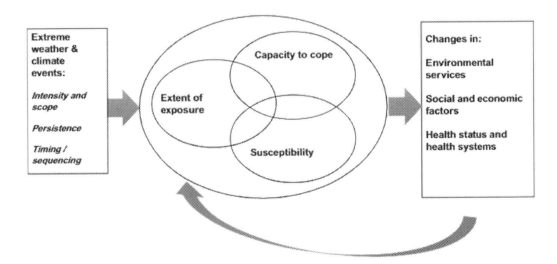

FIGURE 13.5 Key drivers of health vulnerability to extreme weather and climate events.

and guidelines, such as changes in food preparation in drought to reduce exposure to toxic contaminants, like in Tanzania and Ethiopia. A third research priority is on health systems strengthening and capacity building, particularly in larger urban areas that will absorb many drought/climate refugees. Due to the progression of climate change, droughts will become more common and prolonged, and urban areas/cities will need to prepare around the world to ensure health systems are not overrun and stressed, leading to emergency situations that could have been prevented.

13.5 CONCLUSION

Drought is an increasing threat for millions in semi-arid and arid regions but has been demonstrated to lead to several intersectional consequences that ultimately impact human health around the world. These include decreased food production/food insecurity, poor soil quality, decreased freshwater access, exacerbation of chronic illness, poor air quality, and increasing mental health disorders. These consequences will only worsen as the global population increases and land and water supplies become limited, leading to a worsening imbalance in the water cycle and making drought even worse. Efforts to improve soil quality and reduce global temperatures can help humanity decrease the incidence of drought and the resulting negative consequences.

REFERENCES

Astorga, G.I., Alcaraz, L. (2010). Salinity effects on protein content, lipid peroxidation, pigments and proline in *Paulownia imperialis* and *Paulowina fortune* grown in vitro. *Electron J Biotechnol*, 115.

Austin, E.K., Handley, T., Kiem, A.S., Rich, J.L., Lewin, T.J., Askland, H.H., Askarimarnani, S.S., Perkins, D.A., Kelly, B.J. (2018). Drought-related stress among farmers: Findings from the Australian Rural Mental Health Study. *Med J Aust*, 209, 159–165. https://www.mja.com.au/journal/2018/209/4/drought-related-stress-among-farmers-findings-australian-rural-mental-health.

Barbier, E.B., Hochard, J.P. (2016). Does land degradation increase poverty in developing countries? *PLoS One*. https://journals.plos.org/plosone/article/file?id=10.1371/journal.pone.0152973&type=printable.

Beinsan, C., Camen, D., Sumalan, R, Babau, M. (2003). Study concerning salt stress effect on leaf area dynamics and chlorophyll content in four bean local landraces from Banat areas. *Fac Hortic*, 119, 416–419.

Berry, H.L., Bowen, K., Kjellstrom, T. (2010). Climate change and mental health: a causal pathways framework. *Int J Public Health*, 55(2), 123–132. https://doi.org/10.1007/s00038-009-0112-0.

Black, R.E., Victora, C.G., Walker, S.P., Bhutta, Z.A., Christian, P., de Onis, M., Ezzati, M., Grantham-McGregor, S., Katz, J., Martorell, R., Uauy, R., Maternal, & Child Nutrition Study, G. (2013). Maternal and child undernutrition and overweight in low-income and middle-income countries. *Lancet*, 382(9890), 427–451. https://doi.org/10.1016/S0140-6736(13)60937-X

Brevik, E.C. (2013). Soils and human health: An overview. In *Soils and Human Health* (pp. 29–58).

Burgess, L.C. (2013). Organic pollutants in soil. In *Soils and human health* (pp. 83–106). CRC Press.

Centers for Disease Control and Prevention (2020). Health implications of drought. https://www.cdc.gov/nceh/drought/implications.htm.

Chapman, E., Haby, M.M., Illanes, E., Sanchez-Viamonte, J., Elias, V., Reveiz, L. (2019). Risk factors for chronic kidney disease of non-traditional causes: A systematic review. *Rev Panam Salud Publica*, 43, e35. https://doi.org/10.26633/RPSP.2019.35.

Cherif, H., M. R., Rolli, E., Ferjani, R., Fusi, M., Souss, A., et al. (2015). Oasis desert farming selects environment-specific date palm root endophytic communities and cultivable bacteria that promote resistance to drought. *Environ Microbiol*, 7, 668–678.

Cliff, J., Muquingue, H., Nhassico, D., Nzwalo, H., Bradbury, J.H. (2011). Konzo and continuing cyanide intoxication from cassava in Mozambique. *Food Chem Toxicol*, 49(3), 631–635. https://doi.org/10.1016/j.fct.2010.06.056.

Cooper, M., Sandler, A., Vitellozzi, S., Lee, Y., Seymour, G., Haile, B., Azzari, C. (2021). Re-examining the effects of drought on intimate-partner violence. *PLoS One*, 16(7), e0254346. https://doi.org/10.1371/journal.pone.0254346.

DiLiberto, T.N. (2022). Record-breaking algal bloom expands across the North Pacific. https://www.climate.gov/news-features/event-tracker/record-breaking-algal-bloom-expands-across-north-pacific.

Dooyema, C.A., Neri, A., Lo, Y.C., Durant, J., Dargan, P.I., Swarthout, T., Biya, O., Gidado, S.O., Haladu, S., Sani-Gwarzo, N., Nguku, P.M., Akpan, H., Idris, S., Bashir, A.M., & Brown, M.J. (2012). Outbreak of fatal childhood lead poisoning related to artisanal gold mining in northwestern Nigeria, 2010. *Environ Health Perspect*, 120(4), 601–607. https://doi.org/10.1289/ehp.1103965.

Ebi, K., Bowen, K. (2016). Extreme events as sources of health vulnerability: Drought as an example. *Weather Clim Extrem*. https://www.sciencedirect.com/science/article/pii/S221209471530030X.

Edwards, B., Gray, M., Hunter, B. (2015). The impact of drought on mental health in rural and regional Australia. *Soc Indic Res*, 121, 177–194. https://link.springer.com/article/10.1007%2Fs11205-014-0638-2.

Epstein, P.R., Buonocore, J.J., Eckerle, K., Hendryx, M., Stout Iii, B.M., Heinberg, R., Clapp, R.W., May, B., Reinhart, N.L., Ahern, M.M., Doshi, S.K., Glustrom, L. (2011). Full cost accounting for the life cycle of coal. *Ann N Y Acad Sci*, 1219, 73–98. https://doi.org/10.1111/j.1749-6632.2010.05890.x.

FAO (2015). *The Impact of Natural Hazards and Disasters on Agriculture and Food Security and Nutrition: A Call for Action to Build Resilient Livelihoods*, Food and Agriculture Organization of the United Nations. https://www.fao.org/3/a-i4434e.pdf.

FAO (2017). *The Future of Food and Agriculture-Trends and Challenges*, Food and Agriculture Organization of the United Nations. https://www.fao.org/3/a-i6583e.pdf.

FAO (2018a). Global soil partnership, 2018, polluting our soils is polluting our future. https://www.fao.org/global-soil-partnership/resources/highlights/detail/en/c/1127952/.

FAO (2018b). The state of food and agriculture 2018. Migration, agriculture and rural development. https://www.fao.org/3/I9549EN/i9549en.pdf.

FAO (2019). Climate change and land. An IPCC special report on climate change, desertification, land degradation, sustainable land management, food security, and greenhouse gas fluxes in terrestrial ecosystems, summary for policymakers. https://www.ipcc.ch/site/assets/uploads/2019/08/4.-SPM_Approved_Microsite_FINAL.pdf.

FAO. Chapter 2 – Soil and water. https://www.fao.org/3/r4082e/r4082e03.htm.

Fisher, M.C., Koenig, G.L., White, T.J., Taylor, J.W. (2000). Pathogenic clones versus environmentally driven population increase: analysis of an epidemic of the human fungal pathogen Coccidioides immitis. *J Clin Microbiol*, 38(2), 807–813. https://doi.org/10.1128/JCM.38.2.807-813.2000.

Francis, C. (2018). NIH director's blog: Powerful antibiotics found in dirt. https://directorsblog.nih.gov/2018/02/20/powerful-antibiotics-found-in-dirt/.

Friel, S., B. H., Dinh, H., O'Brien, L, Walls, H.L. (2014). The impact of drought on the association between food security and mental health in a nationally representative Australian sample. *BMC Public Health*, 14.

GBD Risk Factors Collaborators (2016). Global, regional, and national comparative risk assessment of 79 behavioural, environmental and occupational, and metabolic risks or clusters of risks, 1990–2015: A systematic analysis for the Global Burden of Disease Study 2015. *Lancet*, 388(10053), 1659–1724. https://doi.org/10.1016/S0140-6736(16)31679-8.

Getahun, H., Mekonnen, A., TekleHaimanot, R., Lambein, F. (1999). Epidemic of neurolathyrism in Ethiopia. *Lancet*, 354(9175), 306–307. https://doi.org/10.1016/S0140-6736(99)02532-5.

Government of Canada. (2004). Canada country study: Summary, Ottawa. https://www.climatechange.gc.ca/english/publications/ccs/.

Green, R., Cornelsen, L., Dangour, A.D., Turner, R., Shankar, B., Mazzocchi, M., Smith, R.D. (2013). The effect of rising food prices on food consumption: systematic review with meta-regression. *BMJ*, 346, f3703. https://doi.org/10.1136/bmj.f3703.

Hoy, W.G., Martinez, R., Reveiz, L., Ordunez, P. (2017). *Setting the Context. Epidemic of Chronic Kidney Disease in Agricultural Communities in Central America. Case Definitions, Methodological Basis and Approaches for Public Health Surveillance*. Pan American Health Organization (PAHO). https://iris.paho.org/xmlui/handle/123456789/34132.

Hsiang, S.M., Burke, M., Miguel, E. (2013). Quantifying the influence of climate on human conflict. *Science*, 341(6151), 1235367. https://doi.org/10.1126/science.1235367

Jiménez Cisneros, B.E., Oki, T., Arnell, N.W., Benito, G., Cogley, J.G., Döll, P., et al. (2014). Freshwater resources. In: *Climate Change 2014: Impacts, Adaptation, and Vulnerability. Part A: Global and Sectoral Aspects. Contribution of Working Group II to the Fifth Assessment Report of the Intergovernmental Panel on Climate Change*. CU Press.

Kaluski, D.N., Ophir, E., Amede, T. (2002). Food security and nutrition – The Ethiopian case for action. *Public Health Nutr*, 5(3), 373–381. https://doi.org/10.1079/phn2001313.

Khan, N., Yousuf ul Islam, M., Siddiqui, M., Iqbal Mufti, B.A. (2015). Thar drought: A complete public health failure. *J Infect Public Health*, 8(5), 506–507. https://doi.org/10.1016/j.jiph.2015.01.004.

Krishnamachari, K.A., Bhat, V.R., Nagarajan, V., Tilak, T.B., Tulpule, P.G. (1977). The problem of aflatoxic human disease in parts of India-epidemiological and ecological aspects. *Ann Nutr Aliment*, 31(4–6), 991–996. https://www.ncbi.nlm.nih.gov/pubmed/566071.

Lal, R. (2021). Opinion: How soil can save us all. https://www.devex.com/news/sponsored/opinion-how-soil-can-save-us-all-101619.

Landrigan, P.J., Fuller, R., Acosta, N.J.R., Adeyi, O., Arnold, R., Basu, N.N., Balde, A.B., Bertollini, R., Bose-O'Reilly, S., Boufford, J.I., Breysse, P.N., Chiles, T., Mahidol, C., Coll-Seck, A.M., Cropper, M.L., Fobil, J., Fuster, V., Greenstone, M., Haines, A., ... Zhong, M. (2018). The Lancet Commission on pollution and health. *Lancet*, 391(10119), 462–512. https://doi.org/10.1016/S0140-6736(17)32345-0

Lang, L. (1996). Danger in the dust. *Environ Health Perspect*, 104(1), 26–30. https://doi.org/10.1289/ehp.9610426.

McLeman, R. (2017). Migration and land degradation: Recent experiences in future trends. In *The Global Land Outlook Working Paper*, United Nations Convention to Combat Desertification (UNCCD). https://knowledge.unccd.int/sites/default/files/2018-06/8.%20Migration%2Band%2BLand%2BDegradation__R_McLeman.pdf.

Merriam-Webster. (2022). System. https://www.merriam-webster.com/dictionary/system.

Mlingi, N.L., Nkya, S., Tatala, S.R., Rashid, S., Bradbury, J.H. (2011). Recurrence of konzo in southern Tanzania: Rehabilitation and prevention using the wetting method. *Food Chem Toxicol*, 49(3), 673–677. https://doi.org/10.1016/j.fct.2010.09.017.

Mohabbat, O., Younos, M.S., Merzad, A.A., Srivastava, R.N., Sediq, G.G., Aram, G.N. (1976). An outbreak of hepatic veno-occlusive disease in north-western Afghanistan. *Lancet*, 2(7980), 269–271. https://doi.org/10.1016/s0140-6736(76)90726-1.

National Geographic (2022). Precipitation. https://education.nationalgeographic.org/resource/precipitation.

NIDIS (2022). Advancing drought science and preparedness across the nation. https://www.drought.gov/

NMDC (2002). What is drought? Drought in the dust bowl years. https://www.drought.unl.edu/whatis/dustbowl.htm.

NOAA (2019). Water cycle. https://www.noaa.gov/education/resource-collections/freshwater/water-cycle.

Obrador, G.T., Schultheiss, U.T., Kretzler, M., Langham, R.G., Nangaku, M., Pecoits-Filho, R., Pollock, C., Rossert, J., Correa-Rotter, R., Stenvinkel, P., Walker, R., Yang, C.W., Fox, C.S., Kottgen, A. (2017). Genetic and environmental risk factors for chronic kidney disease. *Kidney Int Suppl* (2011), 7(2), 88–106. https://doi.org/10.1016/j.kisu.2017.07.004.

Oliveira, G. de L.T. (2013). Land Regularization in Brazil and the Global Land Grab. *Development and Change Development and Change*, 44(2), 261–283.

Pappagianis, D. (1994). Marked increase in cases of coccidioidomycosis in California: 1991, 1992, and 1993. *Clin Infect Dis*, 19(Suppl 1), S14–18. https://doi.org/10.1093/clinids/19.supplement_1.14.

Paungfoo-Lonhienne, C., Yeoh, Y.K., Kasinadhuni, N.R., Lonhienne, T.G., Robinson, N., Hugenholtz, P., Ragan, M.A., Schmidt, S. (2015). Nitrogen fertilizer dose alters fungal communities in sugarcane soil and rhizosphere. *Sci Rep*, 5, 8678. https://doi.org/10.1038/srep08678.

Pouria, S., de Andrade, A., Barbosa, J., Cavalcanti, R.L., Barreto, V.T., Ward, C.J., Preiser, W., Poon, G.K., Neild, G.H., Codd, G.A. (1998). Fatal microcystin intoxication in haemodialysis unit in Caruaru, Brazil. *Lancet*, 352(9121), 21–26. https://doi.org/10.1016/s0140-6736(97)12285-1.

Rodríguez-Eugenio, N., McLaughlin, M., Pennock, D. (2018). Soil pollution: A hidden reality. https://www.fao.org/3/I9183EN/i9183en.pdf.

Rulli, M.C., Saviori, A., D'Odorico, P. (2013). Global land and water grabbing. *Proc Natl Acad Sci U S A*, 110(3), 892–897. https://doi.org/10.1073/pnas.1213163110.

SMA (1996). Life cycle and epidemiology of Coccidioides immitis.

SARE (2019). Rotations and soil organic matter levels. https://www.sare.org/Learning-Center/Books/Building-Soils-for-Better-Crops-3rd-Edition/Text-Version/Crop-Rotations/Rotations-and-Soil-Organic-Matter-Levels.

Scandlyn, J.S., C.N., Thomas, D.S.K., Brett, J. (2010). Theorical framing of worldviews, values, and structural dimensions of disasters.

Scherer, T.F., F.D., Cihacek, L. (2022). Soil, water and plant characteristics important to irrigation. https://www.ndsu.edu/agriculture/sites/default/files/2022-03/ae1675.pdf

Schmidt, R., K.M., Mostafa, A., Ramadan, E.M., Monschein, M., Jensen, K.B., Bauer, R., Berg, G. (2014). Effects of bacterial inoculants on the indigenous microbiome and secondary metabolites of chamomile plants. *Front Microbiol*, 5, 64.

Seaman, J., Holt, J., Rivers, J. (1978). The effects of drought on human nutrition in an Ethiopian province. *Int J Epidemiol*, 7(1), 31–40. https://doi.org/10.1093/ije/7.1.31.

Selvakumar, G., P.P., Ganeshamurthy, A.N. (2012). Bacterial mediated alleviation of abiotic stress in crops. In D. K. Maheshwari (ed), *Bacteria in Agrobiology: Stress Management* (pp. 205–224). Springer-Verlag.

Sena, A, B.C., Freitas, C., Corvalan, C. (2014). Managing the health impacts of drought in Brazil. *Int J Environ Res Public Health*, 11, 10737–10751. https://doi.org/10.3390/ijerph111010737.

Sena, A., Ebi, K. (2020). When land is under pressure health is under stress. *Int J Environ Res Public Health*, 18. https://doi.org/10.3390/ijerph18010136.

Singh, M.B., Fotedar, R., Lakshminarayana, J., Anand, P.K. (2006). Studies on the nutritional status of children aged 0-5 years in a drought-affected desert area of western Rajasthan, India. *Public Health Nutr*, 9(8), 961–967. https://doi.org/10.1017/s1368980006009931

Smoyer-Tomic, K.E., Klaver, J.D., Soskolne, C.L., et al. (2004). Health consequences of drought on the Canadian prairies. *EcoHealth*, 1. https://doi.org/10.1007/s10393-004-0055-0.

Stanke, C., Kerac, M., Prudhomme, C., Medlock, J., Murray, V. (2013). Health effects of drought: A systematic review of the evidence. *PLoS Curr*, 5. https://doi.org/10.1371/currents.dis.7a2cee9e980f91ad7697b570bcc4b004.

Taylor, K., Jones, E.B. (2022). Adult dehydration. *StatPearls*. https://www.ncbi.nlm.nih.gov/pubmed/32310416.

Toth, G., Hermann, T., Da Silva, M.R., Montanarella, L. (2016). Heavy metals in agricultural soils of the European Union with implications for food safety. *Environ Int*, 88, 299–309. https://doi.org/10.1016/j.envint.2015.12.017

UCD (2016). Why insect pests love monocultures, and how plant diversity could change that. https://www.sciencedaily.com/releases/2016/10/161012134054.htm.

U. N. C. t. C. Desertification (2022). Drought. https://www.unccd.int/land-and-life/drought/overview.

U. N. G. A. I. N. Committee (1994). Elaboration of an international convention to combat desertification in countries experiencing serious drought and/or desertification, particularly in Africa https://www.unccd.int/sites/default/files/relevant-links/2017-01/English_0.pdf.

UNDESAPD (United Nations, Department of Economic and Social Affairs, Population Division) (2019). World population prospects 2019: Highlights (ST/ESA/SER.A/423). https://population.un.org/wpp/Publications/Files/WPP2019_Highlights.pdf.

UNEP (2017). Towards a pollution-free planet background report. https://wedocs.unep.org/bitstream/handle/20.500.11822/21800/UNEA_towardspollution_long%20version_Web.pdf?sequence=1&isAllowed=y.

USDA. (2022). Crops. https://www.ers.usda.gov/topics/crops/.

van Vliet, M.T.H., Zwolsman, J.J.G. (2008). Impact of summer droughts on the water quality of the Meuse river. *J Hydrol*, 353(1–2), 1–17. https://doi.org/10.1016/j.jhydrol.2008.01.001.

Vins, H., Bell, J., Saha, S., Hess, J.J. (2015). The mental health outcomes of drought: A systematic review and causal process diagram. *Int J Environ Res Public Health*, 12(10), 13251–13275. https://doi.org/10.3390/ijerph121013251

Vurukonda, S.S., Vardharajula, S., Shrivastava, M., Sk, Z.A. (2016). Enhancement of drought stress tolerance in crops by plant growth promoting rhizobacteria. *Microbiol Res*, 184, 13–24. https://doi.org/10.1016/j.micres.2015.12.003.

WB (2019). Levels and trends in child malnutrition: Key findings of the 2019 edition of the joint child malnutrition estimates. https://www.who.int/nutgrowthdb/jme-2019-key-findings.pdf?ua=1.

WB (2020). Levels and trends in child malnutrition: Key findings of the 2020 edition of the joint child malnutrition estimates. https://www.who.int/publications-detail/jme-2020-edition.

WHO (2009). WHO child growth standards and the identification of severe acute malnutrition in infants and children. https://apps.who.int/iris/bitstream/10665/44129/1/9789241598163_eng.pdf?ua=1.

WHO (2014). Gender, climate change and health. https://www.who.int/globalchange/GenderClimateChangeHealthfinal.pdf.

WHO (2016). Dioxins and their effects on human health. https://www.who.int/news-room/fact-sheets/detail/dioxins-and-their-effects-on-human-health.

WHO (2017). Climate-resilient water safety plans: Managing health risks associated with climate variability and change. https://apps.who.int/iris/bitstream/handle/10665/258722/9789241512794-eng.pdf:jsessionid=F4B5E121316C78D4BAB1C92FB4971F75?sequence=1.

WHO (2020). UNICEF/WHO/The World Bank Group joint child malnutrition estimates: Levels and trends in child malnutrition: Key findings of the 2020 edition. https://www.who.int/publications/i/item/jme-2020-edition.

WMO (2012). Atlas of Health and Climate, World Health Organization and World Meteorological Organizational. https://www.who.int/globalchange/publications/atlas/report/en/.

Wolff, E., van Vliet, M.T.H. (2021). Impact of the 2018 drought on pharmaceutical concentrations and general water quality of the Rhine and Meuse rivers. *Sci Total Environ*, 778, 146182. https://doi.org/10.1016/j.scitotenv.2021.146182.

WWAP (2016). The United Nations World Water Development Report 2016: Water and Jobs, United Nations World Water Assessment Programme. https://www.womenforwater.org/uploads/7/7/5/1/77516286/wwdr_2016_report_-_water_and_jobs.pdf.

Xu, R., Yu, P., Abramson, M.J., Johnston, F.H., Samet, J.M., Bell, M.L., Haines, A., Ebi, K.L., Li, S., Guo, Y. (2020). Wildfires, global climate change, and human health. *N Engl J Med*, 383(22), 2173–2181. https://doi.org/10.1056/NEJMsr2028985.

Zhou, J., Jiang, X., Wei, D., Zhao, B., Ma, M., Chen, S., Cao, F., Shen, D., Guan, D., Li, J. (2017). Consistent effects of nitrogen fertilization on soil bacterial communities in black soils for two crop seasons in China. *Sci Rep*, 7(1), 3267. https://doi.org/10.1038/s41598-017-03539-6.

14 Drought and Soil Structure

Naba R. Amgain and Rattan Lal

14.1 DROUGHT

Drought is an extended period of unusually dry weather due to low precipitation such as rain, snow, or sleet, resulting in a water deficit. Drought is a natural process, but human activities including the usage and management of water have increased the frequency and severity of drought making the situation worse. Drought is one of the major problems affecting agriculture and the food supply system, which is only getting worse because of the changing climate. Droughts are more common in semiarid areas such as Northern and Southwestern Africa, Central Asia, Australia, the western United States, and the Iberian Peninsula (Mishra et al., 2021). Extreme drought can have a significant impact on the function (Liu et al., 2010), structure (Zak et al., 2003), and productivity (Lal et al., 2013) of soil ecosystems. Drought losses are often quantified using social and economic indicators such as economic loss and crop yield loss (Geng et al., 2015). Drought also affects the soil stability, soil nutrient cycle, and water cycle. However, there have not been many studies done on how drought affects the resilience of the soil ecosystem.

This chapter aims to discuss drought, its causes and effects, and soil structure and its dynamics in relation to drought. The specific focus of this chapter is on a wide range of environmental and anthropogenic factors that impact drought, and outline management strategies to overcome the effects of drought. Impacts of drought on soil structure and moisture retention are also discussed in relation to crop growth and agronomic productivity.

Drought can be of different forms, each with its own set of consequences as depicted in Figure 14.1. Wilhite and Glantz (1985) classified droughts into four kinds based on their effects: meteorological drought, is a period of a lack of precipitation for long time which produce water scarcity;

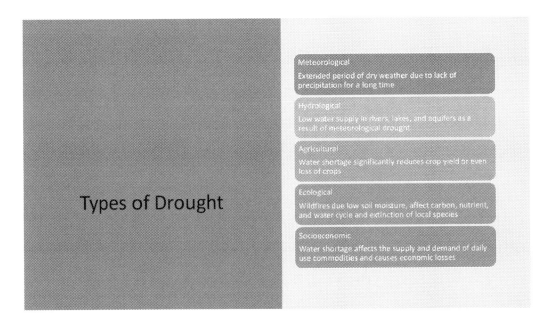

FIGURE 14.1 Types of droughts.

Drought and Soil Structure

hydrological drought is related to low water flow in stream, lakes, and groundwater levels; agricultural drought result in insufficient soil moisture in rootzone to support a crop; and socioeconomic drought interferes with the normal functioning of society. In addition, there is also a pedological drought, which specifically refers to soil water depletion during the critical stages of crop growth (Lal, 2020).

14.2 CAUSES OF DROUGHT

Drought is caused by natural factors such as climate change, changes in temperature, and local topography modifications. However, human activity including deforestation, and high-water demand have increased the frequency and severity of drought. Previous studies suggest global warming due to human activities and increased water use for irrigation and human consumption are responsible for metrological and hydrological drought (Dai, 2013; Wisser et al., 2010). Figure 14.2 illustrates how drought, climate change, and land degradation are interconnected. Land degradation is often an outcome of the combined effects of climate, physical processes, and land use practices (Kiage, 2013). Physical factors that contribute to land degradation include droughts, heat waves, and changes in seasonal precipitation patterns. Climate change is expected to make these factors more frequent and severe (Hermans & Mcleman, 2021; Reed & Stringer, 2016). Land degradation reduces the carbon sequestration in soil and vegetation which contributes to

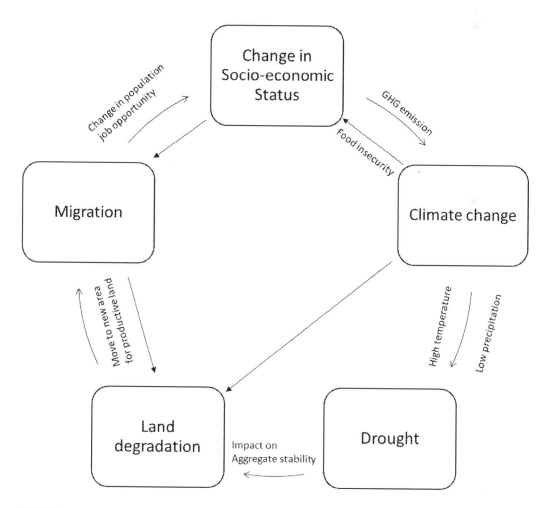

FIGURE 14.2 Linkage between climate change, drought, land degradation and migration.

climate change (Hermans & Mcleman, 2021). Drought and land degradation has a negative impact on people due to increasing competition for the few natural resources which results in migration (Hermans & Mcleman, 2021). Migration changes the socio-economic status of the society. Low crop yield due to land degradation and scarcity of water resources are push factors for migration which changes the socio-economic and political status of the society (Hermans & Mcleman, 2021). Out-migration reduces the population density and creates labor shortage whereas in-migration increases the population density and decreases the employment opportunities. High population density results in deforestation, extensive use of natural resources, and increased emission of greenhouse gases (GHGs) which results in climate change.

14.2.1 Climate Change

Climate change is a natural phenomenon but sudden change in climate has increased the likelihood of extreme weather (Ashraf et al., 2021). Climate change has a significant impact on precipitation, temperature, biodiversity, water resources, and agriculture production (Ashraf et al., 2021; Murthy et al., 2015). Climate change has already increased the severity of drought intensity and is expected to bring about more frequent extreme weather events such as relatively shorter but intense rainfall events (Cook et al., 2018). Changes in regional climate and variability in sea surface temperature cause global droughts (Mukherjee et al., 2018). Global warming due to anthropogenic factors such as increased CO_2 and other GHGs increases the evapotranspiration rate which is directly related to increase in surface heating (Dai, 2011). Climate change and the attendant global warming is making dry regions dryer and wet regions wetter (Trenberth et al., 2014). Warm air absorbs more water in wetter places, resulting in greater rain showers. In contrast, in arid regions, high temperature increases evapotranspiration which makes the surface dry. When a region becomes extremely dry, an impermeable crust develops, which causes water to flow off the surface when it rains, leading to low soil moisture. Abrupt climate change significantly increases the vulnerability of individual farming risking the production quality and quantity (Arshad et al., 2020; Ashraf et al., 2021).

14.2.2 Deforestation

Tropical forest is a key source of global hydrological fluxes. Change in forest cover significantly alters the climate and hydrological cycle (Avissar & Werth, 2005; Kanae et al., 2001; Nobre et al., 1991). Between 2000 and 2012, 7.5% of the world's forests disappeared which resulted in species loss and climate change (Hansen et al., 2013). Forest and vegetation play an important role in water cycle by providing moisture through evapotranspiration to the atmosphere which forms clouds and returns to the ground as rain. Deforestation reduces the amount of water available for the water cycle which reduces the frequency of rainfall and makes the region vulnerable to drought (Bennett, 2017). In addition, deforestation exposes soil, soil dries out faster, decreases infiltration and water holding capacity, and increases runoff rate (Bennett, 2017). The main cause of deforestation is crop land expansion. Land and rainfall are two critical inputs for agricultural production. A lack of rainfall can have serious consequences for yields, resulting in migration to new areas with more rainfall, which increases the deforestation rates (Desbureaux & Damania, 2018). Farmers convert forests to farmland as a strategy to combat the negative impact of drought. Desbureaux and Damania (2018) reported deforestation in Madagascar increased by 7.6% in drought year compared to normal weather year, and the rate of deforestation was higher in dry and semi-arid area.

14.2.3 High Water Demand

Over the past 50 years, the human water use rate has doubled, which has affected streamflow in many regions of the world (Wada et al., 2013). Over the period of 1960–2010, human water consumption significantly decreased streamflow over Europe, North America, and Asia, which in turn

Drought and Soil Structure

increased the severity of hydrological droughts by 10%–500%. The frequency of droughts increased by 27(±6)% due to human water use alone. Drought frequency is increasing most severely in Asia (35%) followed by North America (25%) and Europe (20%) (Wada et al., 2013). While industrial and household consumption have a much larger impact on the intensification of hydrological droughts over the eastern US and western and central Europe, irrigation is responsible over the western and central US, southern Europe, and Asia. When surface water runs out, people start pumping more from groundwater, which depletes valuable water resources and permanently impacts future water availability. Drought in lower downstream locations may also result from increased water demand upstream in rivers for irrigation or constructions of dams for electricity.

14.3 HOW ARE PEOPLE AFFECTED BY DROUGHT?

Droughts can occur anywhere in the world. The impacts of drought, however, differ by location. What is regarded as a drought in one region may not be considered a drought in another region. Drought is most likely to affect areas that are already in danger of desertification. Drought has a greater socioeconomic impact as shown in Figure 14.3 in developing countries where agriculture is the main occupation. Droughts have a direct impact on the environment and ecosystem. The environmental effects of drought include lack of accessibility to water and water resources, soil drying, and erosion resulting in negative effects on vegetation, and increasing the risk of forest fire which increases the dust and declines the air quality. Drought may also cause economic loss to farmers. Access to water and water resources are directly impacted by droughts which reduces crop yield and livestock production. Farmers may be required to spend more money on irrigation or drill more wells if their water supply is inadequate. Reduced agriculture activities including livestock and farming reduce the job opportunities in agriculture-related industries which reduces the household income and purchasing power of the family. Drought also has a negative impact on agro-tourism. The social consequence of drought includes a negative impact on the welfare and standard of living, increasing cost of living and poverty rate, negative impact on family and social relationship, increased violence and conflict in the society, migration, and change in population structure. Mdemu (2021) reported that drought occurred in Tanzania's central and north-eastern semi-arid areas between 2007 and 2011 increased water scarcity and food insecurity. Residents in vulnerable communities spend more time and money getting water for themselves and their animals. Food insecurity was found to be influenced by household social circumstances, low resource ownership, and income sources. Families in excellent social standing who earned their livings from non-agriculture, crop-livestock integration, and resourcefulness were less vulnerable and hence better equipped to deal with the effects of droughts. Drought can also have an impact

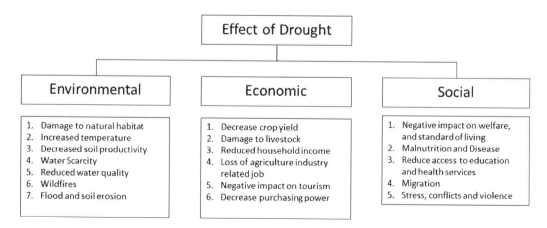

FIGURE 14.3 Effect of drought.

on human and animal health due to a shortage of healthy food and safe drinking and sanitary water. In addition to having a detrimental influence on agriculture and animals, drought also has a significant impact on increasing food insecurity, and malnutrition. It has a significant impact on mental health conditions including stress, tension, depression, social disputes, and violence (Mehdipour et al., 2022).

14.3.1 Reduce Crop Yield

Drought has been and will continue to be a major cause of yield loss (Lobell et al., 2014). The relative impact of droughts on crop yield is projected to worsen if the trend toward wetter springs and dryer summers continues (Fu et al., 2022). Extreme droughts have historically resulted in significant agricultural losses and a drop in crop yield, both in rain-fed and irrigated soils (Ding et al., 2013). Drought can affect photosynthesis, nutrient uptake by crop, and also increases soil temperature which affects the plant growth and development and reduces crop yield. Drought-induced water stress decreases crop root growth, delays maturation, and diminishes agricultural yield (Ge et al., 2012; Piscitelli et al., 2021)

14.3.2 Wildfires

Wildfires can have serious consequences for society and the environment harming ecosystems, agricultural fields, and urban settlements. In addition, fires can degrade air quality and cause chronic respiratory disorders. Weather and environment have a big impact on wildfire activity. Extreme dry conditions can cause wildfires that burn vegetation, wildlife, and homes. Each year large hectares of forest fires occur in the USA, Australia, and other parts of the world. The western United States, Australia, the southern and central regions of Africa, the southern-northern part of South America, and central Asia are more prone to fires (Mishra et al., 2021). There have been severe droughts throughout most of the western United States, which have led to an increase in big wildfires (Murphy et al., 2020; Rust et al., 2019). The occurrence and severity of wildfires have considerably increased as a result of changes in the climate, forest area, and land-use patterns (Mishra et al., 2021). Wildfires that destroy forest cover may result in less evapotranspiration, interceptions, and soil degradation (Ebel & Moody, 2017). Mishra et al. (2021) reported higher concentrations of nitrogen (N), phosphorus, dissolved organic carbon, manganese, and turbidity in rivers and streams after wildfires. High concentrations of nutrient and organic matter (OM) may result in excessive algae growth and increased turbidity (Rust et al., 2019). Spencer et al. (2003) reported a huge algae bloom after a wildfire.

14.3.3 Migration

Droughts are major causes of soil degradation, which has a negative impact on resource-dependent rural people and may result in livelihood losses and eventual migration out of affected areas (Hermans & McLeman, 2021). As a result of drought, individuals have to travel further in search of safe water. Women and children are often responsible for this and even miss school and other employment opportunities just to get water for drinking. Many people are forced to leave their houses permanently because they lack access to clean water or food. This is the main issue affecting rural communities in most of the developing countries. Climate change is having a global impact and is causing push factors for inter and intra-continent migration that cannot be reversed. It is anticipated that by 2050, new climate-related migrants from South Asia, Latin America, and Sub-Saharan Africa will reach up to 143 million (Rigaud et al., 2018). Droughts and periods of excessive heat harm grazing areas, reduce feed crop yields, increase livestock mortality rates, and decrease revenue from livestock sales, all of which reduce household incomes and increase food insecurity (Hermans & McLeman, 2021). People will be forced to move because of a lack of water and low crop yield. Threats of climate are already visible in most of the Sub-Saharan African countries and have an impact on the local livelihoods of those who depend on agriculture and fisheries (Azumah & Ahmed, 2023). Kaczan and Orgill-Meyer (2020) reported that slow-onset climate changes such as droughts are more likely to induce increased migration than rapid-onset changes such as floods.

14.3.4 MALNUTRITION AND DISEASE

Crop losses due to drought may cause malnutrition. Children and women are at risk of suffering from malnutrition and often die. Those that survive will be sick for the rest of their lives. Also, access to safe drinking water is severely impacted by drought. As a result, people may consume contaminated water, causing epidemics of diseases such as cholera and typhoid. Another result of not having access to clean water is that these diseases can spread in areas with inadequate sanitation. Depending on whether the exposure was 6-month mild or severe drought, it increased the risk of diarrhea by 5% or 8% respectively (Wang et al., 2022). Children who did not have access to either water or soap/detergent for handwashing were more likely to get disease (Wang et al., 2022). In 2016, the Global Burden of Disease Research projected that diarrhea was the fifth most common cause of mortality in children under the age of 5, accounting for 446 thousand deaths globally. Among these, sub-Saharan Africa and South Asia have the highest contribution (90%) due to inadequate water, sanitation, and hygiene standards (Wang et al., 2022). The World Health Organization predicted that diarrheal deaths among children aged 0–15 would be approximately 32,954 by 2050 (Wang et al., 2022). Lack of precipitation increases the risk of diarrhea by concentrating pathogens in water sources, whereas extreme precipitation and floods spread the sewages and animal waste which increases the disease.

14.3.5 WATER QUANTITY AND QUALITY

A major contributing factor to the water crisis in many areas of the world is drought (Masroor et al., 2020). It is expected that when temperatures rise, the frequency of droughts will increase due to increased evaporation and transpiration water loss (Zambreski et al., 2018). Droughts can have a serious impact on drinking and irrigation water quality and quantity, waste load allocation, and aquatic ecosystems (Mishra et al., 2021). Low flows and a lack of water availability on surface and groundwater resources could lead to a decline in water quality. Drought reduces stream flow, which causes N concentrations to rise due to a lack of dilution, resulting in an algal bloom. By accelerating the decomposition of OM and sediments and discharging them into receiving streams, the drought-waterlogging cycles may also have an impact on the water quality.

14.3.6 FLOODING AND SOIL EROSION

We may believe that rain after a prolonged drought is beneficial, but it can cause dangerous flash flooding and soil erosion. This is because prolonged drought hardens the ground and forms a crust, preventing rain from soaking and infiltrating. Runoff and wind accelerate soil erosion in drought-prone locations. High temperature and low rainfall under drought circumstances impact the cohesiveness of soil OM which reduces aggregation and hence increases the soil erodibility (Sardans & Peñuelas, 2013). Extreme drought conditions impact the soil texture, which reduces the soil's ability to hold water (Lin et al., 2018). Less crop residue as a result of decreased crop growth due to drought increases the chance of soil erosion from wind and heavy rain. (Bodner et al., 2015).

14.4 SOIL STRUCTURE AND ITS MANAGEMENT FOR REDUCING RISKS OF DROUGHT

Soil structures are the arrangement of soil aggregates of different shapes and sizes with pores space between them. Soil aggregates are formed by sand, silt, and clay glued by organic matter and roots exude. Well-structured soil with stable aggregates creates a channel for air and water movement, can hold more water, crop roots can penetrate to a deeper depth, and crops are less prone to drought as water can be drawn to the surface from subsoils through channels and deep roots system. In contrast, poorly structured soils with less stable soil aggregates collapse filling pore spaces with small soil aggregates which causes surface sealing, crusting, reduced permeability, and infiltration contributing to runoff and erosion. Aggregate stability and soil structure are affected by soil texture, climate, and quantity and diversity of microorganisms living in the soil. Soil high in silt and clay can form more stable aggregates than sand as silt and clay can bind particles better than sand. Climatic conditions including rainfall patterns and temperature also affect the soil structure. Irregular and intense rainfall can lead to the breakdown of aggregate stability. Rainfall after a long dry period can also increase the risk of surface runoff and erosion due to surface crusting. Organic matter decomposition rate will be high in warmer temperature, which can lead to loss of aggregate stability as OM binds soil aggregates together.Soil structure is also influenced by plant roots and microorganisms living in the soil. Sticky substances excreted by plant roots and soil microbes bind soil particles together which contributes to stable aggregate and well-structured soil.

Soil structure effects on agriculture can be felt at the local, regional and global levels (Lal, 1991). At the local or farm level, the decline in soil structure results in a reduction in soil productivity, increased cost of production, lack of sustainability, and reduction of profit margin affecting the economic well-being of the family. At the regional level, soil structure affects the water quality of surface and groundwater, disturb landscape due to gully formation and damage other infrastructures. Global impact of soil structure includes changes in water and energy balance, carbon and nutrient cycles, greenhouse effect, and global economy. Therefore, the maintenance of well-structured soil is necessary for sustainable agriculture at local, regional, and global scale. A decline in soil structure results in soil drying, soil compaction, surface crusting, and soil erosion (Lal, 1998). Agricultural operations including the use of heavy machinery and industrial activity are major causes of soil compaction (Alaoui et al., 2018). Soil management practices, with an emphasis on managing the dynamic biotic and abiotic processes of soils and OM, have been shown to increase an ecosystem's resilience to local or regional drought impacts (Bodner et al., 2015; Saco et al., 2021). In agriculture, various soil management strategies can be adopted to reduce the impact of drought effects. For example, cover cropping and using mulch cover the soil surface which reduces runoff and minimizes evaporation water loss (Bodner et al., 2015). Use of mulches can increase soil water storage by 10% (Saco et al., 2021). The increase in soil moisture is due to decreased evapotranspiration and water retention by mulches. Verhulst et al. (2011) reported that non-tillage fields had higher soil moisture compared to conventional tillage.

Root-associated microbes have the potential to improve plant growth and increase crop resilience to future droughts (de Vries et al., 2020). Plant-soil organism interactions are critical for the functioning of terrestrial ecosystems and their ability to adapt to changing climate (de Vries et al., 2020). Soil microbes feed on leaves, roots litters, and roots exudates and release nutrients for plant growth and development. In the rhizosphere, plants and microorganisms interact directly, forming symbiotic relationships such as mycorrhizae or by promoting plant growth through phytohormone production or reducing plant stress signaling. Arbuscular mycorrhizal (AM) fungi and other microbes improve the drought tolerance of plants increasing the activity of antioxidant enzymes and lowering oxidative stress which enhances water use efficiency (de Vries et al., 2020). Wu et al. (2008) reported the beneficial effect of AM symbiosis on soil structure. AM colonization improves the moisture retention of soil which enhances plant development under drought stress.

Conservation agriculture may also help to reduce drought. Approaches such as integrated water management, reduction of soil erosion and soil compaction, agricultural diversification, and biodiversity conservation have demonstrated positive effects on natural hazard mitigation, including droughts (McElwee et al., 2020). Furthermore, alternative cropping systems, such as large-scale conservation tillage, may be beneficial for drought mitigation (Saco et al., 2021). Outside of agricultural systems, improved forest management and afforestation are landscape management measures that have shown potential to improve drought resiliency (McElwee et al., 2020).

Improved forest management, regeneration, and afforestation are landscape management practices that have the potential to increase drought resistance outside of agricultural systems (McElwee et al., 2020).

14.4.1 Cover Crops

Cover crops, when planted between or among cash crops, provide a number of water and soil conservation benefits. Live cover crops limit surface exposure which reduces soil erosion and soil moisture losses due to evaporation. Cover crops improve cash crop access to water by opening root channels through compacted soil (Chen & Weil, 2011). Cover crops can also affect the crop yield by affecting the dynamics of soil nutrients (Sainju et al., 2006). Leguminous cover crops acquire atmospheric N through biological N fixation and add it to the soil which is available for cash crops (Hunter et al., 2021). When incorporated in the soil, cover crop residue reduces rain drop impact, reduces runoff, holds water, and promotes infiltration. Organic matter from cover crop residue decomposition contributes to soil aggregate stability which increases infiltration rate and water holding capacity (Basche & Delonge, 2017).

14.4.2 Crop Residue

Crop residue can have significant advantages in reducing drought by enhancing soil moisture through holding water, increasing soil water infiltration, and increasing groundwater recharge. Another benefit of residue is that it acts as an insulating layer to lower soil temperature by increasing the soil's surface reflectivity to solar radiation. Changes in microclimatic conditions such as increase in soil moisture and lower soil temperature impact soil biological and chemical properties. Crop residue enhances the OM in the soil, which improves the soil's ability to retain water. Previous studies have reported that every 1% increase in OM results in increases in maximum water holding capacity of 2.9% (Xu et al., 2022), and 5% depending on soil texture (Emerson et al., 1994). Increased water holding capacity makes more water available to crops during the dry season, which also improves fertilizer uptake.

14.4.3 Crop Rotation

Crop rotation is one of the most important techniques for improving soil health and reducing drought conditions. Crop diversity alters the microclimatic conditions of soil, impact soil biological and physicochemical process, enhance soil OM and structure, and improve soil infiltration and moisture retention. Healthy soil with improved soil structure impacts root growth and distribution at different depths, which may make it easier for roots to spread out and find water (Renwick et al., 2021). In contrast to monocropping, the inclusion of multiple crops in the cropping system increases the diversity of root systems increases the OM pool, improves nutrient uptake, and encourages a diverse microbial community. During a drought, rotating shallow-rooted crops with deep-rooted crops helps to absorb water from the lower soil profile and avoid drought.

14.4.4 Minimum Tillage

One of the most significant effects of conservation tillage is improved soil structure. When compared to conventional tillage, minimum tillage keeps crop residue on the soil surface, which acts as mulch that protects the soil from wind and water erosion, and runoff. The decomposition of crop residue increases the OM in the soil and strengthens soil aggregates, both of which improve infiltration rates. Minimal tillage saves water because it exposes less soil, resulting in less evaporation and a slower rate of soil drying at the surface. However, weeds can grow in the absence of tillage, so this approach must be supplemented with appropriate weed management measures.

14.4.5 Sustainable Farming Practices and Resistant Variety

Adoption of improved agronomic practices such as water management, adjusted planting density and time of showing, and nutrient management (Raza et al., 2022), as well as the use of more robust crops that require less water can be an effective management strategy for increasing water use efficiency. Early sowing can avoid the potential drought and high temperature increasing the yield of the crops. Increasing plant density can preserve water by shading the soil surface (Raza et al., 2022) and this water can be used by plants during later growth stages. A maize crop with a high plant density had a lower leaf area index, which reduced evapotranspiration and improved water use efficiency (Guo et al., 2021). Management strategies, such as drip irrigation, mulching, and reduced tillage result in greater water retention in the field and crop water availability. Drip irrigation methods deliver water directly to the roots of plants, reducing evaporation. To withstand drought crop production must be more stable and sustainable (Varshney et al., 2021). Drought reduces crop yield by affecting physiological, biochemical, and reproduction process (Cui et al., 2020; Rai et al., 2021). Plant ability to tolerate droughts depends on drought duration, intensity, plant development stages, and genetical potential (Varshney et al., 2021). Deep rooted plant can absorb water from lower profile during drought which increases the yield and production (Soriano & Alvaro, 2019). Selection of drought resistance variety can be another option to withstand short-term drought. A variety of new technology advances may aid in the reduction of drought risk. One logical and promising alternative is to combine and modify current technologies and people experience, such as various modeling and satellite-based systems to predict drought and assist people in making wiser judgments about the risks associated with drought.

14.4.6 Rainwater Harvesting

Rainwater harvesting is the technique of collecting rainwater when it rains and storing it for later use. Rainwater harvesting practices are common in countries having drought and water problems. Collecting rainwater can be an effective and sustainable method to mitigate the effect of drought. Harvested rainwater can be used to irrigate plants, feed animals, as well as wash clothes and take showers. Harvest rainwater can also be used to recharge the well, pond, as well as groundwater. Rainwater can be harvested in three different ways: in situ water conservation, runoff farming or flood water harvesting, and collection and storage of runoff water into different structures such as ponds, dams, and lakes (Prinz, 2002; Rockström, 2002). In situ soil water conservation aims to promote infiltration and water is stored in the rootzone for crop. In runoff farming, water is directed to cultivated land from runoff gully or catchment scale.

The goal of water harvesting includes restoring the productivity of land, increasing the yield minimizing the risk of crop failure in drought-prone areas and supplying drinking water to humans and animals. Water harvesting is particularly advantageous in dry regions with poorly distributed rainfall, in areas where water supply to humans and animals is not sufficient, and in arid land suffering from desertification.

14.5 SOIL STRUCTURE AND MOISTURE RETENTION

Soil structure refers to the arrangement of soil particles such as sand, silt and clay in aggregate form that differ in shape, size, stability, and degree of adhesion to one another. Organic compounds released by roots and hyphae, enzymes released by microorganisms and abundance of OM in the soil provide glues which form stable soil aggregates (Denef & Six, 2005). The interaction of many factors, including the environment, soil management factors, plant species and roots, soil properties such as mineral composition, texture, soil organic carbon (SOC) concentration, microbial activities, exchangeable ions, nutrient reserves, and moisture availability, results in the complex dynamics of aggregation (Kay, 1998). Aggregate stability is used to assess soil structure (Six et al., 2000).

Aggregates can be found in a variety of shapes and sizes which are classified as macroaggregates (>250 m) and microaggregates (<250 m) (Tisdall & Oades, 1982).

Soil function is governed by its ability to retain water, which has a significant impact on soil management. Soil water retention is influenced by many factors, such as soil physical properties, soil chemical properties, plant roots, and freeze-thaw cycles. Soil with smaller particles such as clay and slit has a larger surface which allows soil to hold more water compared to larger sand particles. The structural property of soil can influence the severity of seasonal drought. According to D'Angelo et al. (2014), clayey red soils have a high susceptibility to drought due to a significant fraction of water that is rarely accessible to plants. Micro-aggregation properties of clay soil facilitated downward water movement, resulting in low available water storage (Assi et al., 2019; He et al., 2019). The ability of the soil to hold and lose water may also vary with seasonal climate changes. During the rainy season, aggregate disperses and reduces soil porosity, whereas aggregate formation with large pores occurs during the dry season. (He et al., 2018; Kursar et al., 1995). Changes in soil aggregation and water retention or depletion during seasonal climate changes may be caused by variations in aggregate-binding substances such as soil OM and Fe oxides (Xue et al., 2019; Zhu et al., 2010). Precipitation after a drought stimulates the breakdown of organic fertilizers (De Neve & Hofman 2002), lowers soil OM content, and speeds up the transformation of Fe oxides (He et al., 2022), which impact aggregate stability and water retention.

Well-structured and aggregated soil has a higher infiltration rate, which allows more water to be transmitted to a deeper soil profile, increasing the soil's total ability to hold water. Infiltrated water can be held between aggregate spaces or as a water film around each individual soil aggregates. Water is held tightly in the microscopic pores to avoid soil percolation loss, but it is loose enough for plant roots to absorb. Soil must allow excess water to pass through while also holding water for later use. Soil compaction and surface crusting reduce the soil's ability to infiltrate water, and water is lost as runoff. Compacted soil will have lower porosity and lower saturated water content compared to uncompacted soil. Ponding, runoff, and soil loss result from impermeable soil, which also reduces plant water availability. Plants experience drought if soil is unable to retain water. Soil high on OM has high moisture retention properties compared to soil with low OM. Soil organic matter attached to soil aggregates absorbs water, which reduces surface runoff and enhances infiltration resulting in increased groundwater recharge. During a drought, OM in the soil dries out and turns water-repellent or hydrophobic, decreasing the soil's ability to absorb moisture.

Cation abundance and their level of hydration play the main roles in controlling the adsorption of water molecules around soil particles in dry soil (Khorshidi et al., 2016). Soil cations can have a direct impact on soil aggregation and structure, which are important factors in soil capillary water retention (Guber et al., 2003). Positively charged cations interact with negatively charged clay surfaces, holding clay particles together and facilitating aggregate formation (Kallenbach et al., 2019). Ca^{2+} is well known for its ability to flocculate clays and aid in the formation of soil aggregates, thereby improving water infiltration and movement (Kögel-Knabner et al., 2008; Rowley et al., 2018).

14.6 EFFECT OF DROUGHT ON AGRICULTURE SOIL

Drought often affects the soil ecosystem. The effect of drought on soil is illustrated in Figure 14.4. Drought results in soil particle contraction, which leads to aggregate breakdown. Drought significantly decreases soil macroaggregates while increasing microaggregates (Su et al., 2020). During the period of drought, medium soil aggregates increase in the expense of larger aggregates whereas post-drought small aggregate fraction increases (Quintana et al., 2023). The size and stability of the pores are also impacted by soil shrinkage, which is particularly noticeable in soils with a high percentage of macropores. Macroaggregate disintegration may lead to changes in SOC storage (Zhang et al., 2019). Carbon (C) in the form of OM bonded in macroaggregates regulates C storage and the stability of the C pool in the soil (Six et al., 2000). Although the

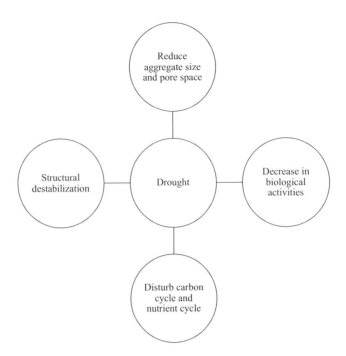

FIGURE 14.4 Effects of drought on soil.

majority of the C is typically associated with the larger aggregate fraction, the drought and the year after it resulted in a large decrease in the C attached to macroaggregates. This was associated not only with the decrease in weight of the large aggregate fraction but also with the decrease in the amount of C linked to macroaggregates (Quintana et al., 2023). The decrease in total C following drought could be attributed to microbial degradation of previously physically and chemically protected organic materials.

The majority of enzyme activity and soil OM processes take place within macroaggregates (Bach & Hofmockel, 2014). Quintana et al. (2023) found positive and significant correlations between the weight of the macroaggregates, and the enzyme activity related to the C and N cycles before the drought period. In the post-drought year, significant structural destabilization of the soil resulted in the breakdown of the link between macroaggregates and enzyme activity. Breakdown of the soil aggregates has an impact on soil microbial activity and greenhouse gas emissions (Smith et al., 2017), as well as a disruption in the carbon cycle. The fragmented soil aggregates shift from sink to source of C, which is indicated by the drop in soil total organic C concentration (Quintana et al., 2021).

With the severity of drought soil aggregate stability and structure become weak, biological activities reduced, and soil organic matter and organic carbon also decline. In addition to abiotic factors like clay content, mineralogy, and organic matter content, the formation and stabilization of soil aggregates are also influenced by biotic factors like roots, fungal hyphae, and products exuded by the roots and soil microbiota that promote aggregate formation (Denef & Six, 2005). During drought, activities of biotic factors are reduced which impact the aggregation formation and soil stability. Since soil aggregation provides physical protection against organic matter breakdown, it is one of the most important abiotic factors regulating C storage in the soil and the stability of the C pool. When the soil shrinks during a drought, the aggregates disintegrate, exposing previously protected organic materials which have a significant impact on soil organic C and carbon sequestration potential (Quintana et al., 2023).

Climate change including global warming, change in precipitation, and rainfall patterns all have an effect on soil microbial growth and reproduction (Geng et al., 2015). Different microorganisms react differently to these stresses. Soil microbial activity and community structure could be damaged or even disappear under adverse drought conditions. Drought alters the structure, size, and activity of soil microbial communities. Impact of drought on microbial community is more noticeable in soil with low OM content (Bogati & Walczak, 2022). The unavailability of nutrient due to reduce diffusion of soluble nutrients and soil acidity might be the contributing factors to decrease the microbial activity (Van Meeteren et al., 2008; Rietz & Haynes, 2003). In addition to plant OM such as plant residue and roots, microbial activities have a significant impact on soil C and N. SOM decomposition by microbes results in CO_2 efflux and gaseous N emissions. Drought changes the soil's microbial population and enzyme activity, which has an impact on the C and N balance. Field investigations, modeling, and theoretical studies have reported drought can speed up C loss from soil in comparison to what would be lost under constant conditions (Deng et al., 2021). Reducing microbial activity slows down the decomposition of nutrients and mineralization of other nutrients which create nutrient deficiency in the soil.

Drought can affect the pool, fluxes and process of the C and N cycle. In addition to altering above-ground C-producing processes like leaf photosynthesis, net primary productivity, and plant variety, drought also has an impact on soil C stocks via below-ground processes including biomass and rhizodeposition as well as C outputs from soil OM (Deng et al., 2021). The greatest C pool in terrestrial ecosystems is found in soil, hence even little reductions in the soil C pool can result in a sizable rise in atmospheric CO_2 levels. Increased soil C losses have a significant positive impact on global warming, which may eventually lead to an increase in the number of drought episodes worldwide. Drought also affects the N dynamics directly and indirectly. The drought impacts the process of N_2 fixation, nitrification, denitrification, N mineralization and immobilization as well as N uptake.

14.7 WAY FORWARD AND POLICY RECOMMENDATIONS

Climate change and drought will continue to increase pressure on the environment and livelihoods of people. Significant international efforts are being made to reduce GHG emissions, and sustainable development by adapting climate resilience and disaster risk reduction strategies. The 2030 Agenda for Sustainable Development and the 17 Sustainable Development Goals are an illustration of this joint approach and emphasize the significance of adaptable and resilient methods to climate-related hazards and disasters (Hervás-Gámez & Delgado-Ramos, 2019).

Although policymakers recognize the importance of water resource protection and make plans to protect them, this may not be adapted at all local, regional, and national scales. Therefore, policy should be set by those who are at the forefront of adaptation, such as local governments, cities, or municipalities. Each nation should develop a drought risk management policy and work on a drought reduction plan. Encourage the public to reduce activities that produce carbon and greenhouse gases that have a negative impact on climate change. Government can offer incentives to those that use green energy and operate organic farms. Carbon credit is another way to reduce the emission of carbon. Farmers can make money by reducing their carbon emission and trading their credit to others that produce more carbon. To prevent deforestation in developing countries, strict laws must be implemented. During the dry season, based on data on local water flow and demand, local government and water boards may develop standard rules for water use rates. Another way to reduce water consumption is to raise the price of water during times of scarcity. The use of recycled water and stormwater capture could be another option for minimizing the overuse of groundwater. Groundwater should be used as drought reserve.

14.8 CONCLUSIONS

Drought is a natural process that occurs in different parts of the world. However, climate change and human activities have increased the frequency and severity of drought. Human activities including deforestation and greenhouse gas emission contributes to climate change which increases the frequency and severity of drought. Drought has a greater socioeconomic impact on local, regional, and global levels. Drought reduces the aggregate stability and impacts carbon and nutrient cycles which results in climate change. Irregular rainfall patterns and high temperatures result in drought and land degradation. Despite a natural process and beyond our reach to stop drought, we could manage the effect of drought by successful pre-drought planning, and reduction in water use. We could adopt short-term and long-term management strategies for effective water use rate and drought stress resistance. Short-term measures include the use of mulch, minimum tillage, and early seeding. The long-term strategies include activities reducing the production of greenhouse gases, which contribute to climate change, and increasing the organic matter in the soil, which promotes aggregate stability and enhances water-holding capacity.

REFERENCES

Alaoui, A., Rogger, M., Peth, S., & Blöschl, G. (2018). Does soil compaction increase floods? A review. *Journal of Hydrology*, 557, 631–642.

Arshad, A., Ashraf, M., Sundari, R. S., Qamar, H., Wajid, M., & Hasan, M. U. (2020). Vulnerability assessment of urban expansion and modelling green spaces to build heat waves risk resiliency in Karachi. *International Journal of Disaster Risk Reduction*, 46, 101468.

Ashraf, M., Arshad, A., Patel, P. M., Khan, A., Qamar, H., Siti-Sundari, R., ... & Babar, J. R. (2021). Quantifying climate-induced drought risk to livelihood and mitigation actions in Balochistan. *Natural Hazards*, 109, 2127–2151.

Assi, A. T., Blake, J., Mohtar, R. H., & Braudeau, E. (2019). Soil aggregates structure-based approach for quantifying the field capacity, permanent wilting point and available water capacity. *Irrigation Science*, 37, 511–522.

Avissar, R., & Werth, D. (2005). Global hydroclimatological teleconnections resulting from tropical deforestation. *Journal of Hydrometeorology*, 6(2), 134–145.

Azumah, S. B., & Ahmed, A. (2023). Climate-induced migration among maize farmers in Ghana: A reality or an illusion?. *Environmental Development*, 45, 100808.

Bach, E. M., & Hofmockel, K. S. (2014). Soil aggregate isolation method affects measures of intra-aggregate extracellular enzyme activity. *Soil Biology and Biochemistry*, 69, 54–62.

Basche, A., & DeLonge, M. (2017). The impact of continuous living cover on soil hydrologic properties: A meta-analysis. *Soil Science Society of America Journal*, 81(5), 1179–1190.

Bennett, L. (2017). Deforestation and climate change. *A Publication of Climate Institute*, 1-16.

Bodner, G., Nakhforoosh, A., & Kaul, H. P. (2015). Management of crop water under drought: A review. *Agronomy for Sustainable Development*, 35, 401–442.

Bogati, K., & Walczak, M. (2022). The impact of drought stress on soil microbial community, enzyme activities and plants. *Agronomy*, 12(1), 189.

Chen, G., & Weil, R. R. (2011). Root growth and yield of maize as affected by soil compaction and cover crops. *Soil and Tillage Research*, 117, 17–27.

Cook, B. I., Mankin, J. S., & Anchukaitis, K. J. (2018). Climate change and drought: From past to future. *Current Climate Change Reports*, 4(2), 164–179. https://doi.org/10.1007/s40641-018-0093-2.

Cui, J., Shao, G., Lu, J., Keabetswe, L., & Hoogenboom, G. (2020). Yield, quality and drought sensitivity of tomato to water deficit during different growth stages. *Scientia Agricola*, 77, e20180390. https://doi.org/10.1590/1678-992x-2018-0390.

Dai, A. (2011). Drought under global warming: A review. *Wiley Interdisciplinary Reviews: Climate Change*, 2(1), 45–65.

Dai, A. (2013). Increasing drought under global warming in observations and models. *Nature Climate Change*, 3(1), 52–58.

D'Angelo, B., Bruand, A., Qin, J., Peng, X., Hartmann, C., Sun, B., ... & Muller, F. (2014). Origin of the high sensitivity of Chinese red clay soils to drought: Significance of the clay characteristics. *Geoderma*, 223, 46–53.

De Neve, S., & Hofman, G. (2002). Quantifying soil water effects on nitrogen mineralization from soil organic matter and from fresh crop residues. *Biology and Fertility of Soils*, 35, 379–386.

de Vries, F. T., Griffiths, R. I., Knight, C. G., Nicolitch, O., & Williams, A. (2020). Harnessing rhizosphere microbiomes for drought-resilient crop production. *Science*, 368(6488), 270–274.

Denef, K., & Six, J. (2005). Clay mineralogy determines the importance of biological versus abiotic processes for macroaggregate formation and stabilization. *European Journal of Soil Science*, 56(4), 469–479.

Deng, L., Peng, C., Kim, D. G., Li, J., Liu, Y., Hai, X., ... & Kuzyakov, Y. (2021). Drought effects on soil carbon and nitrogen dynamics in global natural ecosystems. *Earth-Science Reviews, 214*, 103501.

Desbureaux, S., & Damania, R. (2018). Rain, forests and farmers: Evidence of drought induced deforestation in Madagascar and its consequences for biodiversity conservation. *Biological Conservation, 221*, 357–364.

Ding, R., Kang, S., Li, F., Zhang, Y., & Tong, L. (2013). Evapotranspiration measurement and estimation using modified Priestley-Taylor model in an irrigated maize field with mulching. *Agricultural and Forest Meteorology, 168*, 140–148.

Ebel, B. A., & Moody, J. A. (2017). Synthesis of soil-hydraulic properties and infiltration timescales in wildfire-affected soils. *Hydrological Processes, 31*(2), 324–340.

Emerson, W. W., Foster, R. C., Tisdall, J. M., & Weissmann, D. (1994). Carbon content and bulk-density of an irrigated natrixeralf in relation to tree root growth and orchard management. *Soil Research, 32*(5), 939–951.

Fu, P., Jaiswal, D., McGrath, J. M., Wang, S., Long, S. P., & Bernacchi, C. J. (2022). Drought imprints on crops can reduce yield loss: Nature's insights for food security. *Food and Energy Security, 11*(1), e332.

Ge, T., Sui, F., Bai, L., Tong, C., & Sun, N. (2012). Effects of water stress on growth, biomass partitioning, and water-use efficiency in summer maize (*Zea mays* L.) throughout the growth cycle. *Acta Physiologiae Plantarum, 34*, 1043–1053. https://doi.org/10.1007/s11738-011-0901-y.

Geng, S. M., Yan, D. H., Zhang, T. X., Weng, B. S., Zhang, Z. B., & Qin, T. L. (2015). Effects of drought stress on agriculture soil. *Natural Hazards, 75*, 1997–2011.

Guber, A. K., Rawls, W. J., Shein, E. V., & Pachepsky, Y. A. (2003). Effect of soil aggregate size distribution on water retention. *Soil Science, 168*(4), 223–233.

Guo, Q., Huang, G., Guo, Y., Zhang, M., Zhou, Y., & Duan, L. (2021). Optimizing irrigation and planting density of spring maize under mulch drip irrigation system in the arid region of northwest China. *Field Crops Research, 266*, 108141. https://doi.org/10.1016/j.fcr.2021.108141.

Hansen, M. C., Potapov, P. V., Moore, R., Hancher, M., Turubanova, S. A., Tyukavina, A., ... & Townshend, J. (2013). High-resolution global maps of 21st-century forest cover change. *Science, 342*(6160), 850–853.

He, Y., Gu, F., Xu, C., & Chen, J. (2019). Influence of iron/aluminum oxides and aggregates on plant available water with different amendments in red soils. *Journal of Soil and Water Conservation, 74*(2), 145–159.

He, Y., Xu, C., Gu, F., Wang, Y., & Chen, J. (2018). Soil aggregate stability improves greatly in response to soil water dynamics under natural rains in long-term organic fertilization. *Soil and Tillage Research, 184*, 281–290.

He, Y., Yang, M., Chen, B., Zhao, J., & Ali, W. (2022). Soil aggregation influences soil drought degree after long-term organic fertilization in red soil. *Arabian Journal of Geosciences, 15*(5), 410.

Hermans, K., & McLeman, R. (2021). Climate change, drought, land degradation and migration: Exploring the linkages. *Current Opinion in Environmental Sustainability, 50*, 236–244.

Hervás-Gámez, C., & Delgado-Ramos, F. (2019). Drought management planning policy: From Europe to Spain. *Sustainability, 11*(7), 1862.

Hunter, M. C., Kemanian, A. R., & Mortensen, D. A. (2021). Cover crop effects on maize drought stress and yield. *Agriculture, Ecosystems & Environment, 311*, 107294.

Kaczan, D. J., & Orgill-Meyer, J. (2020). The impact of climate change on migration: A synthesis of recent empirical insights. *Climatic Change, 158*(3–4), 281–300.

Kallenbach, C. M., Conant, R. T., Calderón, F., & Wallenstein, M. D. (2019). A novel soil amendment for enhancing soil moisture retention and soil carbon in drought-prone soils. *Geoderma, 337*, 256–265.

Kanae, S., Oki, T., & Musiake, K. (2001). Impact of deforestation on regional precipitation over the Indochina Peninsula. *Journal of Hydrometeorology, 2*(1), 51–70.

Kay, B. D., 1998. Soil structure and organic carbon: A review. In: Lal, R., Kimble, J. M., Follett, R. F., Stewart, B. A. (eds.), *Soil Processes and the Carbon Cycle*. CRC Press, Boca Raton, FL, pp. 169–197.

Khorshidi, M., Lu, N., & Khorshidi, A. (2016). Intrinsic relationship between matric potential and cation hydration. *Vadose Zone Journal, 15*(11), 1–12.

Kiage, L. M. (2013). Perspectives on the assumed causes of land degradation in the rangelands of Sub-Saharan Africa. *Progress in Physical Geography, 37*(5), 664–684.

Kögel-Knabner, I., Guggenberger, G., Kleber, M., Kandeler, E., Kalbitz, K., Scheu, S., ... & Leinweber, P. (2008). Organo-mineral associations in temperate soils: Integrating biology, mineralogy, and organic matter chemistry. *Journal of Plant Nutrition and Soil Science, 171*(1), 61–82.

Kursar, T. A., Wright, S. J., & Radulovich, R. (1995). The effects of the rainy season and irrigation on soil water and oxygen in a seasonal forest in Panama. *Journal of Tropical Ecology, 11*(4), 497–515.

Lal, R. (1991). Soil structure and sustainability. *Journal of Sustainable Agriculture, 1*(4), 67–92.

Lal, R. (1998). Soil erosion impact on agronomic productivity and environment quality. *Critical Reviews in Plant Sciences*, *17*(4), 319–464.

Lal, R. (2020). Soil organic matter and water retention. *Agronomy Journal*, *112*(5), 3265–3277.

Lal, S., Bagdi, D. L., Kakralya, B. L., Jat, M. L., & Sharma, P. C. (2013). Role of brassinolide in alleviating the adverse effect of drought stress on physiology, growth and yield of green gram (*Vigna radiata* L.) genotypes. *Legume Research-An International Journal*, *36*(4), 359–363.

Lin, B. B., Egerer, M. H., Liere, H., Jha, S., & Philpott, S. M. (2018). Soil management is key to maintaining soil moisture in urban gardens facing changing climatic conditions. *Scientific Reports*, *8*(1), 17565.

Liu, Z., Fu, B., Zheng, X., & Liu, G. (2010). Plant biomass, soil water content and soil N: P ratio regulating soil microbial functional diversity in a temperate steppe: A regional scale study. *Soil Biology and Biochemistry*, *42*(3), 445–450.

Lobell, D. B., Roberts, M. J., Schlenker, W., Braun, N., Little, B. B., Rejesus, R. M., & Hammer, G. L. (2014). Greater sensitivity to drought accompanies maize yield increase in the U.S. Midwest. *Science*, *344*(6183), 516. https://doi.org/10.1126/science.1251423.

Masroor, M., Rehman, S., Avtar, R., Sahana, M., Ahmed, R., & Sajjad, H. (2020). Exploring climate variability and its impact on drought occurrence: evidence from Godavari Middle sub-basin, India. *Weather and Climate Extremes*, *30*, 100277.

McElwee, P., Calvin, K., Campbell, D., Cherubini, F., Grassi, G., Korotkov, V., … & Smith, P. (2020). The impact of interventions in the global land and agri-food sectors on Nature's Contributions to People and the UN Sustainable Development Goals. *Global Change Biology*, *26*(9), 4691–4721.

Mdemu, M. V. (2021). Community's vulnerability to drought-driven water scarcity and food insecurity in central and northern semi-arid areas of Tanzania. *Frontiers in Climate*, *3*, 737655..

Mehdipour, S., Nakhaee, N., Khankeh, H., & Haghdoost, A. A. (2022). Impacts of drought on health: A qualitative case study from Iran. *International Journal of Disaster Risk Reduction*, *76*, 103007.

Mishra, A., Alnahit, A., & Campbell, B. (2021). Impact of land uses, drought, flood, wildfire, and cascading events on water quality and microbial communities: A review and analysis. *Journal of Hydrology*, *596*, 125707.

Mukherjee, S., Mishra, A., & Trenberth, K. E. (2018). Climate change and drought: a perspective on drought indices. Current climate change reports, 4, 145–163.

Murphy, S. F., McCleskey, R. B., Martin, D. A., Holloway, J. M., & Writer, J. H. (2020). Wildfire-driven changes in hydrology mobilize arsenic and metals from legacy mine waste. *Science of the Total Environment*, *743*, 140635.

Murthy, C. S., Laxman, B., & Sai, M. S. (2015). Geospatial analysis of agricultural drought vulnerability using a composite index based on exposure, sensitivity and adaptive capacity. *International Journal of Disaster Risk Reduction*, *12*, 163–171.

Nobre, C. A., Sellers, P. J., & Shukla, J. (1991). Amazonian deforestation and regional climate change. *Journal of Climate*, *4*(10), 957–988.

Piscitelli, L., Colovic, M., Aly, A., Hamze, M., Todorovic, M., Cantore, V., & Albrizio, R. (2021). Adaptive agricultural strategies for facing water deficit in sweet maize production: A case study of a semi-arid mediterranean region. *Water*, *13*(22), 3285.

Prinz, D. (2002). The role of water harvesting in alleviating water scarcity in arid areas. In: Keynote *Lecture, Proceedings, International Conference on Water Resources Management in Arid Regions, 23–27 March, 2002*, Kuwait Institute for Scientific Research, Kuwait, vol III, pp. 107–122.

Quintana, J. R., Martín-Sanz, J. P., Valverde-Asenjo, I., & Molina, J. A. (2023). Drought differently destabilizes soil structure in a chronosequence of abandoned agricultural lands. *Catena*, *222*, 106871.

Quintana, J. R., Molina, J. A., Diéguez-Antón, A., & Valverde-Asenjo, I. (2021). Interannual climate variability determines the efficiency of functional recovery in dry Mediterranean abandoned vineyards. *Land Degradation & Development*, *32*(4), 1883–1900.

Rai, G. K., Parveen, A., Jamwal, G., Basu, U., Kumar, R. R., Rai, P. K., Sharma, J. P., Alalawy, A. I., Al-Duais, M. A., Hossain, M. A., & Habib ur Rahman, M. (2021). Leaf proteome response to drought stress and antioxidant potential in tomato (*Solanum lycopersicum* L.). *Atmosphere*, *12*, 1021. https://doi.org/10.3390/atmos12081021.

Raza, A., Mubarik, M. S., Sharif, R., Habib, M., Jabeen, W., Zhang, C., … & Varshney, R. K. (2022). Developing drought-smart, ready-to-grow future crops. *The Plant Genome*, *16*(1), e20279.

Reed, M. S., & Stringer, L. C. (2016). *Land Degradation, Desertification and Climate Change: Anticipating, Assessing and Adapting to Future Change*. Routledge, New York, NY.

Renwick, L. L., Deen, W., Silva, L., Gilbert, M. E., Maxwell, T., Bowles, T. M., & Gaudin, A. C. (2021). Long-term crop rotation diversification enhances maize drought resistance through soil organic matter. *Environmental Research Letters*, *16*(8), 084067.

Rietz, D. N., & Haynes, R. J. (2003). Effects of irrigation-induced salinity and sodicity on soil microbial activity. *Soil Biology and Biochemistry*, 35(6), 845–854.

Rigaud, K. K., de Sherbinin, A., Jones, B., Bergmann, J., Clement, V., Ober, K., ... & Midgley, A. (2018). *Groundswell: Preparing for Internal Climate Migration*. World Bank, Washington, DC.

Rockström, J. (2002). *Potential of Rainwater Harvesting to Reduce Pressure on Freshwater Resources International Water Conference, Dialogue on Water, Food and Environment Hanoi, Vietnam, October 14–16,*

Rowley, M. C., Grand, S., & Verrecchia, É. P. (2018). Calcium-mediated stabilisation of soil organic carbon. *Biogeochemistry*, 137(1–2), 27–49.

Rust, A. J., Saxe, S., McCray, J., Rhoades, C. C., & Hogue, T. S. (2019). Evaluating the factors responsible for post-fire water quality response in forests of the western USA. *International Journal of Wildland Fire*, 28(10), 769–784.

Saco, P. M., McDonough, K. R., Rodriguez, J. F., Rivera-Zayas, J., & Sandi, S. G. (2021). The role of soils in the regulation of hazards and extreme events. *Philosophical Transactions of the Royal Society B*, 376(1834), 20200178.

Sainju, U. M., Whitehead, W. F., Singh, B. P., & Wang, S. (2006). Tillage, cover crops, and nitrogen fertilization effects on soil nitrogen and cotton and sorghum yields. *European Journal of Agronomy*, 25(4), 372–382.

Sardans, J., & Peñuelas, J. (2013). Plant-soil interactions in Mediterranean forest and shrublands: impacts of climatic change. *Plant and Soil*, 365, 1–33.

Six, J., Elliott, E. T., & Paustian, K. (2000). Soil structure and soil organic matter II. A normalized stability index and the effect of mineralogy. *Soil Science Society of America Journal*, 64(3), 1042–1049.

Smith, A. P., Bond-Lamberty, B., Benscoter, B. W., Tfaily, M. M., Hinkle, C. R., Liu, C., & Bailey, V. L. (2017). Shifts in pore connectivity from precipitation versus groundwater rewetting increases soil carbon loss after drought. *Nature Communications*, 8(1), 1335.

Soriano, J. M., & Alvaro, F. (2019). Discovering consensus genomic regions in wheat for root-related traits by QTL meta-analysis. *Scientific Reports*, 9, 10537.

Spencer, C. N., Gabel, K. O., & Hauer, F. R. (2003). Wildfire effects on stream food webs and nutrient dynamics in Glacier National Park, USA. *Forest Ecology and Management*, 178(1–2), 141–153.

Su, X., Su, X., Zhou, G., Du, Z., Yang, S., Ni, M., ... & Deng, J. (2020). Drought accelerated recalcitrant carbon loss by changing soil aggregation and microbial communities in a subtropical forest. *Soil Biology and Biochemistry*, 148, 107898.

Tisdall, J. M., & Oades, J. M. (1982). Organic matter and water-stable aggregates in soils. *Journal of Soil Science*, 33(2), 141–163.

Trenberth, K. E., Dai, A., Van Der Schrier, G., Jones, P. D., Barichivich, J., Briffa, K. R., & Sheffield, J. (2014). Global warming and changes in drought. *Nature Climate Change*, 4(1), 17–22.

Van Meeteren, M. J. M., Tietema, A., Van Loon, E. E., & Verstraten, J. M. (2008). Microbial dynamics and litter decomposition under a changed climate in a Dutch heathland. *Applied Soil Ecology*, 38(2), 119–127.

Varshney, R. K., Barmukh, R., Roorkiwal, M., Qi, Y., Kholova, J., Tuberosa, R., ... & Siddique, K. H. (2021). Breeding custom-designed crops for improved drought adaptation. *Advanced Genetics*, 2(3), e202100017.

Verhulst, N., Nelissen, V., Jespers, N., Haven, H., Sayre, K. D., Raes, D., ... & Govaerts, B. (2011). Soil water content, maize yield and its stability as affected by tillage and crop residue management in rainfed semi-arid highlands. *Plant and Soil*, 344, 73–85.

Wada, Y., Van Beek, L. P., Wanders, N., & Bierkens, M. F. (2013). Human water consumption intensifies hydrological drought worldwide. *Environmental Research Letters*, 8(3), 034036.

Wang, P., Asare, E., Pitzer, V. E., Dubrow, R., & Chen, K. (2022). Associations between long-term drought and diarrhea among children under five in low-and middle-income countries. *Nature Communications*, 13(1), 3661.

Wilhite, D. A., & Glantz, M. H. (1985). Understanding: the drought phenomenon: The role of definitions. *Water International*, 10(3), 111–120.

Wisser, D., Fekete, B. M., Vörösmarty, C. J., & Schumann, A. H. (2010). Reconstructing 20th century global hydrography: a contribution to the Global Terrestrial Network-Hydrology (GTN-H). *Hydrology and Earth System Sciences*, 14(1), 1–24.

Wu, Q. S., Xia, R. X., & Zou, Y. N. (2008). Improved soil structure and citrus growth after inoculation with three arbuscular mycorrhizal fungi under drought stress. *European Journal of Soil Biology*, 44(1), 122–128.

Xu, N., Amgain, N. R., Rabbany, A., Capasso, J., Korus, K., Swanson, S., & Bhadha, J. H. (2022). Interaction of soil health indicators to different regenerative farming practices on mineral soils. *Agrosystems, Geosciences & Environment*, 5(1), e20243.

Xue, B., Huang, L., Huang, Y., Yin, Z., Li, X., & Lu, J. (2019). Effects of organic carbon and iron oxides on soil aggregate stability under different tillage systems in a rice-rape cropping system. *Catena, 177*, 1–12.

Zak, D. R., Holmes, W. E., White, D. C., Peacock, A. D., & Tilman, D. (2003). Plant diversity, soil microbial communities, and ecosystem function: Are there any links?. *Ecology, 84*(8), 2042–2050.

Zambreski, Z. T., Lin, X., Aiken, R. M., Kluitenberg, G. J., & Pielke Sr, R. A. (2018). Identification of hydroclimate subregions for seasonal drought monitoring in the US Great Plains. *Journal of Hydrology, 567*, 370–381.

Zhang, Q., Shao, M., Jia, X., & Wei, X. (2019). Changes in soil physical and chemical properties after short drought stress in semi-humid forests. *Geoderma, 338*, 170–177.

Zhu, H., Wu, J., Huang, D., Zhu, Q., Liu, S., Su, Y., … & Li, Y. (2010). Improving fertility and productivity of a highly-weathered upland soil in subtropical China by incorporating rice straw. *Plant and Soil, 331*, 427–437.

Index

Note: **Bold** page numbers refer to tables and *italic* page numbers refer to figures.

abiotic stress 235
    intricate nature of 236
    tolerance in crops 238–239, 247
abiotic tolerance in crops 236
abscisic acid (ABA) 235, 245
Accelerated Irrigation Benefits Programme (AIBP) 317
aerial drones 30
aerosols 105
aggregated soil 409
agricultural development, plans and programs promoting 278–279
agricultural drought 42, 89
    in semi-arid ecosystems 173, 175, **176**
agricultural field
    nutrient management in 123
    wind erosion control in 105–106, 108
agricultural GDP 45
agricultural land development, Saharan agriculture 281, *283*
agricultural management practices, to alleviate drought 369–371, **370**
agricultural production, and cropping systems 278–283
agriculture
    conservation 406
    Oasian 280–281, *281*
    sustainable irrigation strategies in 313
    sustainable water management in 271–272, *272*
    and water quality 272–273
agriculture soil, effect of drought on 409–411, *410*
agrivoltaic system 124, *124*
agro-advisory services, based on meteorological forecasts 121
agrobacterium-mediated plant transformation 239
agronomic approaches 14
agro-pastoral system, flood farming in 187, *188, 189*
AIBP *see* Accelerated Irrigation Benefits Programme (AIBP)
air quality 392–393
alcohols, sugar 244–245
algal blooms 392
alleviate drought, agricultural management practices to 369–371
alleviate drought and improve water use efficiency 371
    biochar application 375–376, **376**
    crop rotation 373–375, **374, 375**
    crop variety 372–373, **373–374**
    irrigation methods 378–379
    irrigation strategy 376
    irrigation timing and frequency 376, **377**
    no-till 372, **372**
    seeding density 379
alluvial aquifer 255, 257, 259, 262, 263, 266, 269, 270, 273
Ambitious public policies 277
amelioration of sandy soil 326
ANN *see* artificial neural network (ANN)
anthesis-silking interval (ASI) 17

anti-oxidant enzymes 245–246
arbuscular mycorrhizal (AM) fungi 406
aridity index (AI) 212, *212*
aridity in meteorology 212
arid western India (AWI) 87, *87*
artificial intelligence and IoT applications 122
artificial neural network (ANN) 122
ASI *see* anthesis-silking interval (ASI)
Aspergillus flavus 390
Astian 257
Australian Grain industry 8
available water capacity (AWC) 363
available water holding capacity (AWHC) 133
AWI *see* arid western India (AWI)

biochar 371
    application 375–376, **376**
    Indian rainfed drylands 149
biodegradation of lignin 371
biological nitrogen fixation (BNF) 15
biomarkers, for plant nutrients status analysis 348, **348**
biomolecules 243
BNF *see* biological nitrogen fixation (BNF)
breeding, for enhancing drought tolerance 22–30
    characterizing target environment 22–23
    genomic selection 30
    marker-assisted selection 29–30
    molecular techniques 26–30
    selection environment 23–25
    trait-QTL association 26–29
breeding program, genetic gains of 31
broad bed and furrow (BBF) system 191, **191**
Bundelkhand Region, Central India 183–186, **184,** *184, 185*

CA *see* conservation agriculture (CA)
Cabinet Committee on Security (CCS) 48
cadmium 391
Calamity Relief Fund 59
canal water resources 99–100
carbon-energy footprint 180–182
carbon footprint of solar PV pumping systems 118
carbon storage and water use efficiency 145–147, *146, 147*
cellular homeostasis 235
Central Drought Relief Commissioner (CDRC) 49
cereal-legume crop rotation 110
cereals 14–15, 215–216
    area, production and productivity of major dryland 14, **15**
chickpea (*Cicer arietinum* L.) 15, **15,** 217
    'QTL hotspot' region in 30
China, dryland of Northwest 9
chloro-alkaline indices 260–261
chromatin remodeling 243
chronic malnutrition 389
Cicer species 20

417

CIMMYT Asia maize program 25
CIMMYT Maize Lines (CMLs) 25
clayey red soils 409
climate
    models 213
    resilient technologies, field scale 197
    Saharan agriculture 284
climate change 14, 85, 161, 402, 410
    on crop production **345**
    impact of 299
    on semi-arid agroecosystem 168–170
    on soil properties, functionality, and productivity 342–343, **343**
"climate-resilient villages" concept 9
climate smart agriculture (CSA) 319, *344,* 344–350
climatic constraints 88
    high evaporation 89
    high radiation 88
    high wind speed 89
    low humidity 88–89
    low rainfall 88
    wide-ranging diurnal temperatures 88
Clustered Regularly Interspaced Short Palindromic Repeats (CRISPR) 238
    for enhancing drought tolerance 241
    for enhancing heat/cold tolerance 241
    for enhancing salt tolerance 240–241
conservation agriculture (CA) 3–4, 111, 190, 406
    for carbon storage and water use efficiency 145–147, *146, 147*
conservation agriculture system intensification (CASI) technologies 111
conservation tillage methods
    conservation agriculture 3–4
    traditional tillage 4, *4*
contingency crop planning 57
contour farming, for erosion control 145
conventional drought 90
cover crops 219, 406, 407
cowpea (*Vigna unguiculata* (L.) Walp.) 217
    seed production of 341
Crisis Management Group (CMG)
    for drought management 57
    functions 49
Crisis Management Plan (CMP) 49, 56
CRISPR/Cas9 31
    enhancing abiotic stress tolerance using 240–241
CRISPR/Cas system 239
crop diversification, using alternate and drought-stress-tolerant crops 214–215, *215*
crop growth, effects of soil drought on 368
    maize 369
    wheat 368–369
crop health 390
crop improvement
    drought stress tolerance through 219–224
    programs 23
crop intensification, impact of landscape interventions on 186
crop irrigation, use of unconventional water resources in 314–315
croplands 362
crop management 5–6, *6*, 7
    for sustainable dry farming 7, *8*
    technologies 66
crop modelling approach 23
cropping systems 121–122
    agricultural production and 278–283
    diversification of 177
crop planning, contingency 57
crop production
    climate change on **345**
    in drylands 14
    rainwater harvesting for 119
crop residue 407
    as mulch 1–2
crops
    abiotic stress tolerance in 247
    abiotic tolerance in 236
    dryland 32
    gene editing in **240**
    genome editing approach to generate abiotic stress tolerance in 238–239
    response to drought 16
    rotation 6, 373–375, **374, 375**, 407
    variety 372–373, **373–374**
    water productivity 272
Crop Weather Watch Group (CWWG) 48
Crop Weather Watch Group for Drought Management (CWWGDM) 49
crop wild relatives (CWRs) 20
cryptobiotic crusts, at Thar desert 97
CSA *see* climate smart agriculture (CSA)
CSM-CERES-Pearl millet model 23
culturable command area (CCA) 99
CWRs *see* crop wild relatives (CWRs)

DACPs *see* District Agriculture Contingency Plans (DACPs)
decomposition rate, organic matter 406
Deficiency of Essential Elements in Crop Plants (Tiwari) 321
deficit irrigation (DI) 117
deforestation 402
degradation, land 391
dehydration-responsive element binding transcription factors 224
Department of Agriculture and Farmers Welfare (DAFW) 49, 57, 69
Desert Development Programme (DDP) 48, **54, 55**
DI *see* deficit irrigation (DI)
diethylenetriamine pentaacetate (DTPA)-extractable iron (Fe) 94
differentially methylated regions (DMRs) 242
diffused reflectance spectroscopy (DRS) 197
dioxins 391
direct seeded rice (DSR) *192,* 192–193, *193*
disaster management 60
    institutional arrangements for 49–50
    institutional framework for 48–49
disaster relief, financial mechanism and provision for 67–68
disaster risk reduction, risk prevention and mitigation 63
    development and spread of drought mitigation technologies 66
    financial products 66
    government programs 65
    need for soil health act in India 64–65

Index

soil, water and land resources management 63–64
dissection of quantitative trait loci pertaining to drought tolerance (DT-QTLs) 28
District Agriculture Contingency Plans (DACPs) 56, 57
District Disaster Management Authority (DDMA) 49
diversification of cropping/farming systems 177
DM Act (2005) 48, 49, 73
DMRs *see* differentially methylated regions (DMRs)
DNA
  markers 27
  methylation 242
downscaling techniques 369
downy mildews 25
drainage structures, field bunding with field 183–184
drip irrigation
  methods 408
  with organic mulch 143
drought 14, 32, 400–401, 411–412
  across drylands, spatial and temporal variation 90, 92
  agricultural 42, 89
  avoidance 17
  conventional 90
  crop response to 16
  in dryland environments 16
  in drylands of India 90
  escape 17
  future strategies of crop planning in case of occurrence of 121
  hydrological 89
  impacts 44–46
  indexes in South Asia **225**
  intersectional impacts of 393
  and land degradation 402
  and magnitude in drylands 89–90
  meteorological 42, 89
  proofing 121–122
  relief 60–63, *61*
  resilient rainfed technologies 7–8
  scenarios for semi-arid regions 213–214
  socio-economic 42, 89
  soil structure and management for reducing risks of 405–408
  spatial-temporal distribution of 364
  types 41–42, 400, *400*
drought, causes of *401*, 401–402
  climate change 402
  deforestation 402
  high water demand 402–403
drought-coping mechanism, of dryland crops 17–18
drought declaration 58–60
  rainfall and impact indicators for **60**
drought ecology, field crop species for 14–18
drought hazard, risk and vulnerability assessment
  drought prone areas in India 53–54
  droughts in past 50–52, **51**, *51, 52*
  drought to famine **52**, 52–53
drought, in semi-arid ecosystems 170–171
  agricultural drought 173, 175, **176**
  classification of drought 171–172, *172*
  hydrological drought 172–173
  meteorological drought 172
  socio-economical drought 175–176
drought management fund utilization 69–72

drought management policies, evolution of 46
  post-independence era 47–48
  pre-independence era 46–47
drought mitigation
  financial provision for 68–69
  funds for 69, *72*
  impact of landscape interventions on 186
drought monitoring and early warning 57–58
Drought Monitoring Centre (DMC) 57
drought, people affected by *403*, 403–404
  crop yield reduction 404
  flooding and soil erosion 405
  malnutrition and disease 405
  migration 404–405
  water quantity and quality 405
  wildfires 404
Drought Prone Areas Programme (DPAP) 48, **54, 55**
drought-sensitive regions 214
drought stress 16, 22, 224
  enhancement of genetic gain under 247
drought stress tolerance, through crop improvement 219
  breeding and molecular approaches 219–223
  transgenic approaches 223–224
drought-stress-tolerant crops, crop diversification using alternate and 214–215, *215*
drought tolerance 17–18, 28
  breeding for enhancing 22–30
  CRISPR for enhancing 241
  genetic dissection of 28
  noncoding RNAs for enhancing 243
  phenotypic traits associated with 20–22
  QTL in different field crops for **221–222**
  in transgenic crops 237
drought tolerance, genetic improvement for 18–30
  breeding for enhancing drought tolerance 22–30
  genetic resources of dryland crops 18–20
  phenotypic traits associated with drought tolerance 20–22
drought-tolerant field crops and distribution 14–15, **15**
Drought Tolerant Population (DTP) 25
drought-tolerant species 6–7, *8*
dryland 85
  crop production in 14
  drought and magnitude in 89–90
  inhabitants of 85
  of Northwest China 9
  relative humidity in 88
  of semi-arid tropics 161
  soils of 161
  spatial and temporal variation in occurrence of drought across 90, *92*
dryland crops 32
  drought-coping mechanism of 17–18
  genetic resources of 18–20, 32
dryland environments 102
  drought in 16
dryland farming 1, 9
  impact of drought in 90
  in India 86, *86*
  innovative crop management for sustainable 7, *8*
  sustainable, strategies of 2, *3*
dry land soils, nutrient deficiencies in 103, **104**
DSR *see* direct seeded rice (DSR)

DTPA-extractable copper 96
DTPA-extractable zinc 94
dust 105

economic policy of drought management 66–67
    financial mechanism and provision for disaster relief 67–68
    financial provision for mitigation 68–69
    framework for drought management fund utilization 69–72
    funds for drought mitigation 69
edaphic constraints, of drylands
    canal water resources 99–100
    groundwater resources 97–99
    limited surface water resources 97
    low concentration of biota 96–97
    non-judicious use of water resources 100–101
    nutrient-depleted soils 94, 96
    poor response of rainfall to runoff 101
    scarce water resources 97–101
    soil physical conditions 96, **96**
    soil types and spread 93–94, *94*
EDI *see* Effective Drought Index (EDI)
EDVs *see* essentially derived varieties (EDVs)
Effective Drought Index (EDI) 171
Egyptian soil conditions, tracer techniques under **328–339**
El Harrach formation 257
El Nino 42
El-Nino Southern Oscillation (ENSO) 42
emissions
    GHG 367–368
    greenhouse gas 362
entisols 93
environmental issues, Saharan agriculture 293
Environmental Quality Incentives Program 64
enzymes
    activities, soil microbial properties and 366–367, *367*
    anti-oxidant 245–246
    for membrane lipid biosynthesis 245
    for production of protective metabolites 243–244
enzyme phosphatidylglycerol (PG) synthase 245
epigenetic modifications 241–243, **242**
    of histone proteins 243
eroded soil mass 105
erosion control, contour farming for 145
essentially derived varieties (EDVs) 30
Ethiopia, water spreading weirs in drought-affected areas in 186–189
evapotranspiration (ET) 315, 316
ex-situ rainwater harvesting, through farm ponds 138

faba bean (*Vicia faba* L.) 217
Famine Code (FC) (1883) 46
Famine Enquiry Commissions 53
farming systems 121–122
fertigation systems 115
fertilizer control order (FCO) **115**
fertilizer-N balance **327**
fertilizers
    mineral 300
    water soluble 115
field crops, genetic improvement of 14
field scale climate resilient technologies 197
financial products 66

flood farming in agro-pastoral systems 187, *188, 189*
flooding and soil erosion 405
flood irrigation 378
flood-spreading weirs 188
    impacts of applying 188–189
'Food for Work' program 47
food insecurity 390–391
food security 72
    challenge 319, *319*
footprints, reducing water-energy-carbon 198
forest
    tropical 402
    and vegetation 402
fragile natural resources in drylands 93–101
functional proteins, overproduction of 243
fungi, arbuscular mycorrhizal (AM) 406

gamma irradiation technique 341–342
GEBV 30
GenBank 31
gene editing
    in crops **240**
    graft-mobile 240
genetically modified organisms (GMOs) 238
genetic improvement
    for drought tolerance 18–30
    of field crops 14
genetic resources of dryland crops 18–20
genome editing, to generate abiotic stress tolerance 238–239
genome-wide association studies (GWAS) 27, 28
genomics-assisted breeding techniques 220
genomic selection (GS), breeding for enhancing drought tolerance 30
genotypes 17
genotyping-by-sequencing (GBS) strategies 29
geochemical evolution, interaction and 268–270, *268–271*
geochemical indicators 255
geographic information system (GIS) 23
Ghout system, Saharan agriculture *297*, 297–299, *298*
Global Burden of Disease Research 405
global warming 402
glycine betaine (GB) 244
graft-mobile gene editing 240
grain yield 17
grass cover, in rangelands 105
grasslands, semi-arid 212
Grasspea 389
gravel mulch 2, *3*
greenhouse gas (GHG) emission 118, 362, 367–368
green manuring, for SOC and moisture storage 143–144
Green Revolution 47
    of 1960 in India 9
green water 5
    in rainfed agriculture *311*
    storage 5
groundwater
    chemistry, mechanism controlling 267
    constrains, poor-quality 109
    irrigation 120
    recharge, rainwater harvesting 120
    sustainable management of 278
groundwater resources 97–99, 117
    Indian rainfed drylands 138

Index

Saharan agriculture 289–291
groundwater's geochemical status in agricultural and sustainable use 255
   chemical composition 261–262
   chemical types 260
   chloro-alkaline indices 260–261, 266, *266*
   climate 256
   geographical framework 256, *258*
   geology and hydrogeology 256–259, *259*
   Gibbs diagram 261
   hydrochemical properties 261–267
   hydrochemical types 263, *264*
   marine influence 265–266
   saturation index 260, 266–267, *267*
   water analysis 259–260
   water hardness 260
   water sampling 259
'Grow-more-food' campaign 66
GWAS *see* genome-wide association studies (GWAS)

HaHB4 transcription factor 237
harvesting, rainwater 118–120, 408
health vulnerability, weather and climate events 394, *394*
heat/cold tolerance, CRISPR for enhancing 241
heat shock protein (HSP) 238
heat stress 235
heat tolerance, in transgenic crops 238
hepatic veno-occlusive disease 389–390
herbicides 391
high water demand 402–403
histone proteins, epigenetic modifications of 243
homeostasis, cellular 235
homology-directed repair (HDR) 238
HSP *see* heat shock protein (HSP)
human health, negative impacts of drought on 388, *388*
   air quality 392–393
   crop health 390
   food insecurity 390–391
   malnutrition 389–390
   socioeconomic impacts 393–394
   soil quality 391
   toxicities 391–394
   water quality 392
hydrogel, lignin 371
hydrological balance 386
hydrological drought 89
   in semi-arid ecosystems 172–173, *173*
hyperspectra-based soil assessment 123

IGNP 99, 100, **100**
India
   agriculture 41
   deficiency of soil nutrients in 165
   drought in drylands of 90
   dryland agricultural region in 86, *86*
   Green Revolution of 1960 in 9
   integrated nutrient management crops in rainfed regions of **112–114**
   rainfall seasons and distribution in **42**, 42–43, *43*, **44**
   rainfed production system, climate, soil type, growing period and mean annual rainfall in **91**
Indian Council of Agricultural Research (ICAR) 121
Indian Meteorological Department (IMD) 121

Indian Ocean Dipole (IOD) 42
Indian rainfed drylands
   biochar 149
   conservation agriculture for carbon storage and water use efficiency 145–147, *146*, *147*
   contour farming for erosion control 145
   cover crops and market wastes 149–150
   drip irrigation with organic mulch 143
   ex-situ rainwater harvesting through farm ponds 138
   green manuring for SOC and moisture storage 143–144
   ground water resources 138
   in-situ moisture conservation 138, **140**
   integrated nutrient management 147–148
   intercropping as efficiency production system under droughts 144–145
   mulch cum manuring of soils 143
   national and state government initiations 150–151
   organic amendments 148–149
   rainfall trends in rainfed drylands 138, **139**
   rainwater management 138–140
   soil moisture storage through SOC buildup 140–141, *141*
   soil organic matter and key contributing factors 141–142
   soil water storage through building SOM 142–143
   surface water resources 138
   tank silt 149
   water resources availability in 136–137, *137*
Indian summer monsoon (ISM) 42
India's National Action Plan on Climate Change 64
Indo-Gangetic Plain (IGP) 168
Ingenious Systems of World Agricultural Heritage (ISWAH) 297
INM *see* integrated nutrient management (INM)
inorganic fertilizer's application, Saharan agriculture 288
insecurity, food 390–391
in-situ moisture conservation, Indian rainfed drylands 138, **140**
Integrated Child Development Service 62
integrated nutrient management (INM) 110–111, **112**, 147–148
   crops in rainfed regions of India **112–114**
integrated plant nutrient system (IPNS) 318
Integrated Watershed Management Programme (IWMP) 151
Intercalary Continental (IC) groundwater, Saharan agriculture 291
internal nutrient circulation 110
International Agency for Research on Cancer (IARC) 391
International Crops Research Institute for the Semi-Arid Tropics (ICRISAT) 142, 172, *173*, *175*, **176**
International Service for the Acquisition of Agri-biotech Applications (ISAAA) 236
intimate-partner violence (IPV) 393–394
IPCC's Sixth Assessment Report 168
IPV *see* intimate-partner violence (IPV)
iron **323**
irrigation
   managing water in 313–314
   Saharan agriculture 292–293
   water-use efficiency in *312*
ISWAH *see* Ingenious Systems of World Agricultural Heritage (ISWAH)

Kachch 87
K fertilizer 8

land
   abuse 45
   and rainfall 402
   and soil pollution *393*
   urbanization 272–273
land degradation 64, 161–165, *163,* 387, 391
   drought and 402
landscape management 186
Land Utilization Policy 64
laser-induced breakdown spectroscopy (LIBS) 324
laser land levelling technology **194,** 194–195
leaf rolling 17
LeasyScan 24, 31
legumes 216–218
   area, production and productivity of major dryland 14, **15**
leguminous crops (pulses) 16
leguminous trees 111
lignin hydrogel 371
Liguleless1 (LG1) gene to edit 31
lipid biosynthesis, enzymes for membrane 245
living labs establishment 198

MAAS *see* micro-level agromet advisories (MAAS)
machine-harvestable plant type 31
machine learning (ML) tools 122
magnesium **322**
magnetofection 239
Mahatma Gandhi National Rural Employment Guarantee Act (MGNREGA) 62, 150
Mahatma Gandhi National Rural Employment Guarantee Scheme (MGNREGS) 47, 65
Maison Carree 257
maize crop 15, **15,** 369, 408
Mali (case study) 177–178
   carbon-energy footprint 180–182
   landscape rejuvenation approaches 182–183
   nutrient management from biomass and manure 178–179
   prospects for SBISs development in 180–181
   sustainability of the current agricultural production system 179–180
malnutrition
   chronic 389
   and disease 405
manganese **323**
Manual for Drought Management 59
marginal lands 9
marker-assisted backcrossing (MABC) methods 30
market-driven water budget-based cropping/farming system 198
Marusthali 87
mechanical soil disturbance 319
membrane lipid biosynthesis, enzymes for 245
meteorological drought 42, 89, 362
   in semi-arid ecosystems 172
meteorological forecasts, agro-advisory services based on 121
meteorology, aridity in 212
methylation, DNA 242
Meuse river basin 392

MGNREGA *see* Mahatma Gandhi National Rural Employment Guarantee Act (MGNREGA)
MGNREGS *see* Mahatma Gandhi National Rural Employment Guarantee Scheme (MGNREGS)
micro- and macronutrients, in plants **322–324**
microbes, root-associated 406
microbial biomass C (MBC) pool 324
micro-level agromet advisories (MAAS) 121
microRNAs (miRNAs) 246–247
mid-upper arm circumference (MUAC) 389
migration 404–405
millets 15, **15,** 216
mineral fertilizers 300
minimum tillage 407
Ministry of Agriculture and Farmers Welfare (MoAFW) 49, 63, 69
Ministry of Home Affairs (MHA) 54
Mitidja Basin 271
Moisture Index (MI) 54
moisture retention, soil structure and 408–409
moisture storage, green manuring for SOC and 143–144
molecular marker 223
molybdenum **324**
moth bean (*Vigna aconitifolia* (Jacq.) Marechal) 217
multi-crop planter, zero tillage 193–194
multi-nutrient deficiency 102–103, **104**

NADP *see* National Agricultural Development Program (NADP)
nano-fertilizers 123
nanoparticle 239
Narmada Canal Project (NCP) 99, 100
National Action Plan on Climate Change (NAPCC) 150
National Agricultural Development Program (NADP) 278
National Agricultural Drought Assessment and Monitoring System (NADAMS) 58
National Commission on Agriculture 42, 64
National Crisis Management Committee (NCMC) 48
National Disaster Management Authority (NDMA) 48, 50
National Disaster Response Force 48
National Disaster Response Fund (NDRF) 62, 69–71
national drought management 49
National Executive Committee (NEC) 48
National Food for Work Programme 47
National Food Security Act (2013) 62
National Mission for Sustainable Agriculture (NMSA) 65, 150
National Oceanic and Atmospheric Administration (NOAA) 386, *386*
National Policy on Disaster Management (NPDM) 48
National Rainfed Area Authority (NRAA) 48
National Water Resources Agency (NWRA) 255
National Watershed Development Programme for Rainfed Areas 48
natural resource management (NRM) 66, 166
natural resources, despite limitations in 93
NCED genes, upregulation of 237
NCMC *see* National Crisis Management Committee (NCMC)
NCP *see* Narmada Canal Project (NCP); North China Plain (NCP)
NDMA *see* National Disaster Management Authority (NDMA)

Index

NDVI *see* Normalized Difference Vegetation Index (NDVI)
near-infrared spectroscopy (NIRS) 324
need-based fertilizer application 114
neutron moisture gauges 313
nitrogen **322**
nitrogen derived from fertilizer (Ndff) 316
nitrogen-fixing trees 111
nitrogen use efficiency (NUE) 316
noncoding RNAs, for enhancing drought tolerance 243
non-GM genome editing approaches 239
    agrobacterium-mediated plant transformation 239
    graft-mobile gene editing 240
    nanoparticle 239
    virus-like particles 239
non-homologous end joining (NHEJ) 238
non-judicious use of water resources 100–101, **100**
Normalized Difference Vegetation Index (NDVI) 57, 166–167
North China Plain (NCP) 362, 368, 379
    frequency of drought and disadvantages in 363–366, *364, 365*
northeast monsoon (NEM) 42
nuclear equipment, for field measurement of soil moisture and density 312–313
NUE *see* nitrogen use efficiency (NUE)
nutrient deficiencies, in dry land soils 103, **104**
nutrient-depleted soils 94, 96
nutrient management *320,* 320–324
    in agricultural field 123
    soil test-based 112, 114
nutrient sources, combining inorganic and organic 325–327
nutrients status, evaluation of **347**
nutrient-use-efficient (NUE) crop cultivars 31
nutrition 162–165

Oasian agriculture 280–281, *281, 282*
oilseeds 218
optimum quantity, application of 110
Organic Agriculture (OA) 300
organic amendments, Indian rainfed drylands 148–149
organic amendments/live mulching, role in semi-arid tropics soils 195, *195*
organic matter decomposition rate 406
organic mulch, drip irrigation with 143
ornithine (Orn) 244
osmotic potential 116
osmotic regulation 216

Palmer Drought Severity Index (PDSI) 90
particulate matter (PM) 392
PAWC *see* plant-available water capacity (PAWC)
PDSI *see* Palmer Drought Severity Index (PDSI)
peanut 218
perennial cereals 7
perennial legumes 7
perennial wheat 7
Peri-Oasian agriculture 280
pesticides 391
phenomics platforms 30
phenotypic traits 22
    associated with drought tolerance 20–22
phosphorous **322**

photovoltaic (PV) pumps 117
pigeon pea 15, **15**
plant-available water capacity (PAWC) 1, 5
plant growth-promoting rhizobacteria (PGPR) 349
plant nutrients 319
    biomarkers used for **348**
plant pathogens 166
plants, micro- and macronutrients in **322–324**
plant-soil organism interactions 406
plant transformation, agrobacterium-mediated 239
plastic mulch 2
PM *see* particulate matter (PM)
policy and institutional support 8–9
policy framework for drought management 46–66
policy interventions 8–9
policymakers 411
polyacrylamide (PAM) 5
polyamines (PAs) 244
polyethylene glycol (PEG) 24
potassium **322**
potential evaporation ratio 212
Pradhan Mantri Krishi Sinchai Yojana (PMKSY) 62, 65
precipitation 386, 387
pressurized irrigation 117
    integrating solar PV pumping system with 117–118
primary salinization, Saharan agriculture 294
proline 244
protective metabolites, enzymes for production of 243–244
protein-encoding transcripts, transportation of 240
proteins 243
psamments-orthids 93
pyrroline-5-carboxylate reductase (P5CR) 244
pyrroline-5-carboxylate synthase (P5CS) 244

quantitative trait loci (QTL) 27–29
    in different field crops for drought tolerance **221–222**
Quaternary Mitidja formations 257

radiation use efficiency (RUE) 23
rainfall
    deficiency/uncertainty in 136
    land and 402
    reduction 135
    seasons and distribution in India **42,** 42–43, *43,* **44**
    trends in rainfed drylands 138, **139**
    variations 213
rainfall-runoff relationships 101
rainfall-use efficiency (RUE) 192
rainfed agriculture 1, 8
    green water in *311*
Rainfed Area Development Programme 65
rainfed drylands, rainfall trends in 138, **139**
rain-fed farming 1
rainfed soils of India 135–136
rainfed technologies, drought resilient 7–8
rainwater harvesting 118–119, 408
    for crop production 119
    drainage network for decentralised 185, *185*
    for groundwater recharge 120
    rejuvenation of traditional 184–185
rainwater management, Indian rainfed drylands 138–140
random amplified polymorphic DNA (RAPD) 217
rangelands, grass cover in 104

RapidGen 31
reactive oxygen species (ROS) 16, 245
recombinant inbred lines (RILs) 28
red soils, clayey 409
relative humidity in drylands 88
relative water content (RWC) 18
remote sensing and conventional methods **314**
residual sodium carbonate (RSC) 108
resource use efficiency 191–192
rhizosphere 97, 406
ridge-furrow plastic film mulching (RFM) 2
RILs see recombinant inbred lines (RILs)
roadmap, for drought management 197–198
root architecture 21
root-associated microbes 406
root growth hormones 311
root length density (RLD) of traditional flood irrigation 378
root system architecture (RSA) 18
ROS see reactive oxygen species (ROS)
RUE see radiation use efficiency (RUE)
Rural Infrastructure Development Fund (RIDF) 151

safflower (*Carthamus tinctorius* L.) 218
Saharan agriculture 277–278, 303
    agricultural land development 282–283, *283*
    agricultural production and cropping systems 278–283
    climate 284
    drawdown of groundwater table 296
    effect of exogenous organic matter input 287–288
    environmental issues 293
    farming activities 279–280
    Ghout system *297*, 297–299, *298*
    groundwater resources 289–291
    IC groundwater 291
    inorganic fertilizer's application 288
    irrigation 292
    Oasian agriculture 280–281, *281*, *282*
    perspectives for sustainable management of *299*, 299–302, *301*, *302*
    plans and programs promoting agricultural development 278–279
    primary salinization 294
    secondary salinization 295–296
    soil 284–285
    soil fertility status 285
    soil microbial biodiversity and impact of salinity 286–287
    soil salinity issues 293, *294*
    soil salinization 294
    superficial water resources 288–289
    TC groundwater 291–292
    water resources 288
    water rising 296–297, *297*
    water table groundwater 292
Saharan climate 277
saline water, management of irrigation with 311
salinity levels, effect of **316**
salinization, soil 277
salt-affected water, for specific production system 120
salt overly sensitive (SOS) 237
salt stress 235
salt tolerance
    CRISPR for enhancing 240–241
    in transgenic crops 236–237
sand dune stabilization 105
sandy soil, amelioration of 326
saturation index (SI) 260, 266, *267*
scarce water resources 97–101
SDMA see State Disaster Management Authority (SDMA)
SDRF see State Disaster Response Fund (SDRF)
secondary salinization, Saharan agriculture 295–296
Second Irrigation Commission 48
seed production of cowpea 341
semi-arid agroecosystem, impact of climate change on 168–170
semi-arid grasslands 212
semi-arid regions, drought scenarios for 213–214
semi-arid tropics (SAT)
    characteristic features of 166
    drought management strategies/options in soils of 176–196
    drylands of 161
    soil functions and soil ecosystem services 176–177
    soils of 166–168, **167**, *167*
    soils, organic amendments/live mulching role 195, *195*
    vegetation index 166–168
semi-tropics, roadmap for drought management in soils of 197–198
sensor-based smart farming 122–123
sesame 218
shelterbelt plantations, along canals and roads 105, **107**
SHM see Soil Health Management (SHM)
single nucleotide polymorphisms (SNPs) 27
site-specific nutrient management (SSNM) 114
smart farming, sensor-based 122–123
SMNP see soil moisture neutron probe (SMNP)
SNPs see single nucleotide polymorphisms (SNPs)
SOC see soil organic carbon (SOC)
socio-economic drought 42, 89
    in semi-arid ecosystems 175–176
soil
    aggregation 410
    assessment, hyperspectra-based 123
    carbon management 102
    cations 409
    disturbance, mechanical 319
    of drylands 161
    fertility, Saharan agriculture 285
    macro and micronutrient percentage deficiencies in **103**
    nutrient-depleted 94, 96
    and nutrient management, interventions 102–115
    restoration 318
    Saharan agriculture 284–285
soil ecosystem services, soil functions and 176–177
soil erosion 64
    control mechanisms to reduce 103–106, 108
    flooding and 405
soil function 409
    and GHGS, effects of soil drought on 366–368
    and soil ecosystem services 176–177
Soil Health Card (SHC) Mission 150–151
Soil Health Law 64
Soil Health Management (SHM) 64
soil-human interface *387*, 387–388
soil management 318–319
    crop residues as mulch 1–2

Index

gravel mulch 2, *3*
managing pedological drought through innovative 4, *4*
plastic mulch 2
soil mass, eroded 105
soil microbial biodiversity and impact of salinity, Saharan agriculture 286–287
soil microbial properties, and enzyme activities 366–367, *367*
soil moisture 313
   and density, nuclear equipment for field measurement 312–313
   estimation **314**
   storage through SOC buildup 140–141, *141*
soil moisture neutron probe (SMNP) 310
soil nutrients in India, deficiency 165
soil organic carbon (SOC) 6, 93, 133
   win-win for 147–148
soil organic matter (SOM) 1, 6, 64, 133, 370
   and key contributing factors 141–142
   soil water storage through building 142–143
soil organic phosphorus (porg) cycling 349
soil physicochemical properties 366
soil pollution, land and *393*
soil profile, improving biological activities in 109–110
soil salinity
   issues, Saharan agriculture 293, *294*
   management 108–109
soil salinization 277
   Saharan agriculture 294
soil structure
   and management for reducing risks of drought 405–408
   and moisture retention 408–409
soil test-based nutrient management 111
soil water
   consumption 6
   content 387
   retention 115–116, 409
soil water storage (SWS) 147
   through building SOM 142–143
solar-based irrigation systems (SBIS) 181–182
   in sub-Sahara Africa 180
solar PV pumping systems 117
   carbon footprint of 118
SOM *see* soil organic matter (SOM)
sorghum 15, **15**, 22
   landraces 20
   production 16
South Asia, drought indexes in **225**
southwest monsoon rainfall, deficiency of 42
SPEI *see* Standardized Precipitation Evapotranspiration Index (SPEI)
SSNM *see* site-specific nutrient management (SSNM)
stabilization, by vegetation cover sand dune 105
Standardized Precipitation Evapotranspiration Index (SPEI) 90, 171, 363
State Disaster Management Authority (SDMA) 49, 50
State Disaster Response Fund (SDRF) 62, 70
State Executive Committee (SEC) 49
stay-green 18, 21
stomatal closure 17
stomatal conductance 17
sub-Sahara Africa (SSA) 15
   as game changer for smallholder agriculture in 181–182
   solar-based irrigation systems in 180
sub-surface drip irrigation 143
sugars and sugar alcohols 244–245
sulphur **323**
superficial water resources, Saharan agriculture 288–289
superoxide dismutase (SOD) 245
surface cover management, for controlling wind erosion 103–105
surface water resources
   Indian rainfed drylands 138
   limited 97
sustainability 313
   in drylands 101–120
Sustainable Agriculture Mission 64
sustainable dry farming, innovative crop management for 7, *8*
sustainable farming practices, and resistant variety 408
sustainable intensification 177
sustainable irrigation 313
sustainable management
   of groundwater 278
   of water resources 300
sustainable water management, in agriculture *271*, 271
SWS *see* soil water storage (SWS)
synoptic-scale weather system 42

tank silt (TS) 149
target population of environment (TPE) 22, 23
Terminal Complex (TC) groundwater, Saharan agriculture 291–292
Thar desert, cryptobiotic crusts at 97
torripsamments 93–94
toxicities 391–394
TPE *see* target population of environment (TPE)
traditional tillage 4, *4*
transcription factor 246
   HaHB4 237
transgenic crops
   drought tolerance in 237
   heat tolerance in 238
   salt tolerance in 236–237
transportation of protein-encoding transcripts 240
transporters 246
triticale 216
tRNA-like sequence (TLS) motifs 240
tropical forest 402
TS *see* tank silt (TS)
Turcicum leaf blight (TLB). 25

Union Territories (UTs) 48
United Nations Environmental Programme (UNEP) 393
United Nations Intergovernmental Negotiating Committee 386
urbanization, land 272–273
US Drought Monitor (USDM) 90

vegetation cover, sand dune stabilization by 105
Vegetation Drought Response Index 57
vegetation, forest and 402
vegetation index, semi-arid tropics (SAT) 166–168
Vegetation Temperature Condition Index 57–58
veno-occlusive disease, hepatic 389–390

vertisols, swelling and shrinking in 93
virus-like particles 239

water
    balance in plants 17
    cycle balance *386*
    in irrigated agriculture 313–314
    and land management 310
    potential gradients 235
    productivity 271, 315–318
    quantity and quality 405
    rising, Saharan agriculture 296–297, *297*
    salinity levels **315**
    shortages 161
    soluble fertilizers 115
    spreading weirs in drought-affected areas in Ethiopia 186–189
    storage in soil profile 147–148
    stress 16
water-efficient crops, adoption of 121–122
water-energy-carbon footprints, reduction 198
water management, interventions on
    in condition of scarcity 271
    harvesting rainwater 118–120
    improving water use efficiency/water productivity 117
    integrating solar PV pumping system with pressurized irrigation 117–118
    strategies of 317–318
    using salt-affected water for specific production system 120
    water retention, improving 115–116, *116*
water quality 392
    agriculture and 272–273
water resources
    in crop irrigation 314–315
    in Indian rainfed drylands 136–137, *137*
    management 62
    non-judicious use of 100–101, **100**
    Saharan agriculture 288
    sustainable management of 300
water scarcity 136, 162–165, *164*
    management irrigation 271–272
Watershed Development Project in Shifting Cultivation Areas (WDPSCA) 151
water spreading weirs (WSWs) 187, 188
water table groundwater, Saharan agriculture 292
water use efficiency (WUE) 1, 5, 18, 196, *196*, 311, 363, 370, 372, 378
    in irrigated agriculture *312*
Water Use Efficiency Initiative 8
well-structured soil 406, 409
wheat 368–369
whole genome re-sequencing (WGRS) 28
wildfires 404
wind erosion
    control in agricultural field 105–106, 108
    surface cover management for controlling 103–105
wind strip cropping 108
World Meteorological Organization (WMO) 44
WSWs *see* water spreading weirs (WSWs)
WUE *see* water use efficiency (WUE)

xylem 235

yield gap, systems approach for decomposing 197

zeolite 8
zero-tillage (ZT) 145
    multi-crop planter 193–194
zero-till multi-planter 192, *192*
zinc **324**
zinc solubilization efficiency (ZSE) 342
zwitterionic amino lipids (ZALs) 239